BIOINORGANIC CHEMISTRY

BIOINORGANIC CHEMISTRY

IVANO BERTINI
University of Florence

HARRY B. GRAY
California Institute of Technology

STEPHEN J. LIPPARD
Massachusetts Institute of Technology

JOAN SELVERSTONE VALENTINE
University of California, Los Angeles

University Science Books
Mill Valley, California

University Science Books
20 Edgehill Road
Mill Valley, CA 94941
Fax (415) 383-3167

Library of Congress Cataloging-in-Publication Data
Bioinorganic chemistry / authors/editors Ivano Bertini, Harry B. Gray,
   Stephen Lippard, Joan Valentine.
     p.   cm.
   Includes bibliographical references.
   ISBN 0-935702-57-1 : $58.00
   1. Bioinorganic chemistry.   I. Bertini, Ivano.
QP531.B543   1994
574.19'214—dc20   91–67870
                                              CIP

Printed in the United States of America
10 9 8 7 6 5 4 3 2 1

Contents

List of Contributors

*Numbers in parentheses indicate the pages on which
the authors' contributions begin.*

JACQUELINE K. BARTON (455), Division of Chemistry and Chemical Engineering, California Institute of Technology, Pasadena, California 91125.

IVANO BERTINI (37), Department of Chemistry, University of Florence, Via Gino Capponi 7, 50121 Florence, Italy.

WALTHER R. ELLIS, JR. (315), Department of Chemistry, University of Utah, Salt Lake City, Utah 84112.

STURE FORSÉN (107), Physical Chemistry 2, Chemical Centre, University of Lund, P. O. Box 124, S-221 00, Lund, Sweden.

GRAHAM N. GEORGE (365), Stanford Synchrotron Radiation Laboratory, P. O. Box 4349, Bin 69, Stanford, California 94309.

HARRY B. GRAY (315), Beckman Institute, California Institute of Technology, Pasadena, California 91125.

JAMES A. IBERS (167), Department of Chemistry, Northwestern University, Evanston, Illinois 60208.

GEOFFREY B. JAMESON (167), Department of Chemistry, Georgetown University, Washington, D. C. 20057.

JOHAN KÖRDEL (107), Physical Chemistry 2, Chemical Centre, University of Lund, P. O. Box 124, S-221 00 Lund, Sweden.

STEPHEN J. LIPPARD (505), Department of Chemistry, Massachusetts Institute of Technology, Cambridge, Massachusetts 02139.

CLAUDIO LUCHINAT (37), Institute of Agricultural Chemistry, University of Bologna, Viale Berti Pichat 10, 40127 Bologna, Italy.

KENNETH N. RAYMOND (1), Department of Chemistry, University of California, Berkeley, California 94720.

EDWARD I. STIEFEL (365), Exxon Research and Engineering Company, Clinton Township, Rt. 22 East, Annandale, New Jersey 08801.

ELIZABETH C. THEIL (1), Department of Biochemistry, North Carolina State University, Raleigh, North Carolina 27695-7622.

JOAN SELVERSTONE VALENTINE (253), Department of Chemistry and Biochemistry, University of California, Los Angeles, California 90024.

Preface

This book covers material that could be included in a one-quarter or one-semester course in bioinorganic chemistry for graduate students and advanced undergraduate students in chemistry or biochemistry. We believe that such a course should provide students with the background required to follow the research literature in the field. The topics were chosen to represent those areas of bioinorganic chemistry that are mature enough for textbook presentation. Although each chapter presents material at a more advanced level than that of bioinorganic textbooks published previously, the chapters are not specialized review articles. What we have attempted to do in each chapter is to teach the underlying principles of bioinorganic chemistry as well as outlining the state of knowledge in selected areas.

We have chosen not to include abbreviated summaries of the inorganic chemistry, biochemistry, and spectroscopy that students may need as background in order to master the material presented. We instead assume that the instructor using this book will assign reading from relevant sources that is appropriate to the background of the students taking the course.

For the convenience of the instructors, students, and other readers of this book, we have included an appendix that lists references to reviews of the research literature that we have found to be particularly useful in our courses on bioinorganic chemistry.

Acknowledgments

The idea of preparing a bioinorganic chemistry textbook was conceived by one of us (IB) at a "Metals in Biology" Gordon Conference in January, 1986. The contributing authors were recruited to the project shortly thereafter. The project evolved as a group effort, with substantial communication among the authors at all stages of planning and execution. Both first and revised drafts of the book were class-tested at UCLA, Caltech, and the University of Wisconsin and modified in response to the reviews of students and teachers. Particularly valuable suggestions were made by Professor Judith N. Burstyn (University of Wisconsin); Ken Addess, Raymond Ho, Kathy Kinnear, Clinton Nishida, Roger Pak, Marlene Sisemore (UCLA); and Deborah Wuttke (Caltech) during the review process. We thank them for their contributions.

Even with all this help, the book would never have seen the light of day had it not been for the dedication and hard work of Debbie Wuttke. With HBG, Debbie checked every line through four rounds of galleys and pages. Grazie mille, Debbie!

Ivano Bertini
Harry B. Gray
Stephen J. Lippard
Joan Selverstone Valentine

BIOINORGANIC CHEMISTRY

1

Transition-Metal Storage, Transport, and Biomineralization

ELIZABETH C. THEIL
Department of Biochemistry
North Carolina State University

KENNETH N. RAYMOND
Department of Chemistry
University of California at Berkeley

I. GENERAL PRINCIPLES

A. Biological Significance of Iron, Zinc, Copper, Molybdenum, Cobalt, Chromium, Vanadium, and Nickel

Living organisms store and transport transition metals both to provide appropriate concentrations of them for use in metalloproteins or cofactors and to protect themselves against the toxic effects of metal excesses; metalloproteins and metal cofactors are found in plants, animals, and microorganisms. The normal concentration range for each metal in biological systems is narrow, with both deficiencies and excesses causing pathological changes. In multicellular organisms, composed of a variety of specialized cell types, the storage of transition metals and the synthesis of the transporter molecules are not carried out by all types of cells, but rather by specific cells that specialize in these tasks. The form of the metals is always ionic, but the oxidation state can vary, depending on biological needs. Transition metals for which biological storage and transport are significant are, in order of decreasing abundance in living organisms: iron, zinc, copper, molybdenum, cobalt, chromium, vanadium, and nickel. Although zinc is not strictly a transition metal, it shares many bioinorganic properties with transition metals and is considered with them in this chapter. Knowledge of iron storage and transport is more complete than for any other metal in the group.

The transition metals and zinc are among the least abundant metal ions in the sea water from which contemporary organisms are thought to have evolved (Table 1.1).[1–5] For many of the metals, the concentration in human blood plasma

Table 1.1
Concentrations of transition metals
and zinc in sea water and human
plasma.[a]

Element	Sea water $(M) \times 10^8$	Human plasma $(M) \times 10^8$
Fe	0.005–2	2230
Zn	8.0	1720
Cu	1.0	1650
Mo	10.0	1000
Co	0.7	0.0025
Cr	0.4	5.5
V	4.0	17.7
Mn	0.7	10.9
Ni	0.5	4.4

[a] Data from References 1–5 and 12.

greatly exceeds that in sea water. Such data indicate the importance of mechanisms for accumulation, storage, and transport of transition metals and zinc in living organisms.

The metals are generally found either bound directly to proteins or in cofactors such as porphyrins or cobalamins, or in clusters that are in turn bound by the protein; the ligands are usually O, N, S, or C. Proteins with which transition metals and zinc are most commonly associated catalyze the intramolecular or intermolecular rearrangement of electrons. Although the redox properties of the metals are important in many of the reactions, in others the metal appears to contribute to the structure of the active state, e.g., zinc in the Cu-Zn dismutases and some of the iron in the photosynthetic reaction center. Sometimes equivalent reactions are catalyzed by proteins with different metal centers; the metal binding sites and proteins have evolved separately for each type of metal center.

Iron is the most common transition metal in biology.[6,7] Its use has created a dependence that has survived the appearance of dioxygen in the atmosphere ca. 2.5 billion years ago, and the concomitant conversion of ferrous ion to ferric ion and insoluble rust (Figure 1.1 *See color plate section, page C-1.*). All plants, animals, and bacteria use iron, except for a lactobacillus that appears to maintain high concentrations of manganese instead of iron. The processes and reactions in which iron participates are crucial to the survival of terrestrial organisms, and include ribonucleotide reduction (DNA synthesis), energy production (respiration), energy conversion (photosynthesis), nitrogen reduction, oxygen transport (respiration, muscle contraction), and oxygenation (e.g., steroid synthesis, solubilization and detoxification of aromatic compounds). Among the transition metals used in living organisms, iron is the most abundant in the environment. Whether this fact alone explains the biological predominance of iron or whether specific features of iron chemistry contribute is not clear.

Many of the other transition metals participate in reactions equivalent to those involving iron, and can sometimes substitute for iron, albeit less effectively, in natural Fe-proteins. Additional biological reactions are unique to nonferrous transition metals.

Zinc is relatively abundant in biological materials.[8,9] The major location of zinc in the body is metallothionein, which also binds copper, chromium, mercury, and other metals. Among the other well-characterized zinc proteins are the Cu-Zn superoxide dismutases (other forms have Fe or Mn), carbonic anhydrase (an abundant protein in red blood cells responsible for maintaining the pH of the blood), alcohol dehydrogenase, and a variety of hydrolases involved in the metabolism of sugars, proteins, and nucleic acids. Zinc is a common element in nucleic-acid polymerases and transcription factors, where its role is considered to be structural rather than catalytic. Interestingly, zinc enhances the stereoselectivity of the polymerization of nucleotides under reaction conditions designed to simulate the environment for prebiotic reactions. Recently a group of nucleic-acid binding proteins, with a repeated sequence containing the amino acids cysteine and histidine, were shown to bind as many as eleven zinc atoms necessary for protein function (transcribing DNA to RNA).[10] Zinc plays a structural role, forming the peptide into multiple domains or "zinc fingers" by means of coordination to cysteine and histidine (Figure 1.2A *See color plate section, page C-1.*). A survey of the sequences of many nucleic-acid binding proteins shows that many of them have the common motif required to form zinc fingers. Other zinc-finger proteins called steroid receptors bind both steroids such as progesterone and the progesterone gene DNA (Chapter 8). Much of the zinc in animals and plants has no known function, but it may be maintaining the structures of proteins that activate and deactivate genes.[11]

Copper and iron proteins participate in many of the same biological reactions:

(1) reversible binding of dioxygen, e.g., hemocyanin (Cu), hemerythrin (Fe), and hemoglobin (Fe);

(2) activation of dioxygen, e.g., dopamine hydroxylase (Cu) (important in the synthesis of the hormone epinephrine), tyrosinases (Cu), and catechol dioxygenases (Fe);

(3) electron transfer, e.g., plastocyanins (Cu), ferredoxins, and *c*-type cytochromes (Fe);

(4) dismutation of superoxide by Cu or Fe as the redox-active metal (superoxide dismutases).

The two metal ions also function in concert in proteins such as cytochrome oxidase, which catalyzes the transfer of four electrons to dioxygen to form water during respiration. Whether any types of biological reactions are unique to copper proteins is not clear. However, use of stored iron is reduced by copper deficiency, which suggests that iron metabolism may depend on copper proteins, such as the serum protein ceruloplasmin, which can function as a ferroxidase, and the cellular protein ascorbic acid oxidase, which also is a ferrireductase.

Cobalt is found in vitamin B_{12}, its only apparent biological site.[12] The vitamin is a cyano complex, but a methyl or methylene group replaces CN in native enzymes. Vitamin-B_{12} deficiency causes the severe disease of pernicious anemia in humans, which indicates the critical role of cobalt. The most common type of reaction in which cobalamin enzymes participate results in the reciprocal exchange of hydrogen atoms if they are on adjacent carbon atoms, yet not with hydrogen in solvent water:

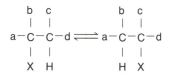

(An important exception is the ribonucleotide reductase from some bacteria and lower plants, which converts ribonucleotides to the DNA precursors, deoxyribonucleotides, a reaction in which a sugar —OH is replaced by —H. Note that ribonucleotide reductases catalyzing the same reaction in higher organisms and viruses are proteins with an oxo-bridged dimeric iron center.) The cobalt in vitamin B_{12} is coordinated to five N atoms, four contributed by a tetrapyrrole (corrin); the sixth ligand is C, provided either by C5 of deoxyadenosine in enzymes such as methylmalonyl-CoA mutase (fatty acid metabolism) or by a methyl group in the enzyme that synthesizes the amino acid methionine in bacteria.

Nickel is a component of a hydrolase (urease), of hydrogenase, of CO dehydrogenase, and of S-methyl CoM reductase, which catalyzes the terminal step in methane production by methanogenic bacteria. All the Ni-proteins known to date are from plants or bacteria.[13,14] However, about 50 years elapsed between the crystallization of jack-bean urease in 1925 and the identification of the nickel component in the plant protein. Thus it is premature to exclude the possibility of Ni-proteins in animals. Despite the small number of characterized Ni-proteins, it is clear that many different environments exist, from apparently direct coordination to protein ligands (urease) to the tetrapyrrole F430 in methylreductase and the multiple metal sites of Ni and Fe-S in a hydrogenase from the bacterium *Desulfovibrio gigas*. Specific environments for nickel are also indicated for nucleic acids (or nucleic acid-binding proteins), since nickel activates the gene for hydrogenase.[15]

Manganese plays a critical role in oxygen evolution catalyzed by the proteins of the photosynthetic reaction center. The superoxide dismutase of bacteria and mitochondria, as well as pyruvate carboxylase in mammals, are also manganese proteins.[16,17] How the multiple manganese atoms of the photosynthetic reaction center participate in the removal of four electrons and protons from water is the subject of intense investigation by spectroscopists, synthetic inorganic chemists, and molecular biologists.[17]

Vanadium and chromium have several features in common, from a bioinorganic viewpoint.[18a] First, both metals are present in only small amounts in most organisms. Second, the biological roles of each remain largely unknown.[18] Finally, each has served as a probe to characterize the sites of other metals, such as iron and zinc. Vanadium is required for normal health, and could act *in vivo* either as a metal cation or as a phosphate analogue, depending on the oxidation state, V(IV) or V(V), respectively. Vanadium in a sea squirt (tunicate), a primitive vertebrate (Figure 1.2B), is concentrated in blood cells, apparently as the major cellular transition metal, but whether it participates in the transport of dioxygen (as iron and copper do) is not known. In proteins, vanadium is a cofactor in an algal bromoperoxidase and in certain prokaryotic nitrogenases. Chromium imbalance affects sugar metabolism and has been associated with the glucose tolerance factor in animals. But little is known about the structure of the factor or of any other specific chromium complexes from plants, animals, or bacteria.

Molybdenum proteins catalyze the reduction of nitrogen and nitrate, as well as the oxidation of aldehydes, purines, and sulfite.[19] Few Mo-proteins are known compared to those involving other transition metals. Nitrogenases, which also contain iron, have been the focus of intense investigations by bioinorganic chemists and biologists; the iron is found in a cluster with molybdenum (the iron-molybdenum cofactor, or FeMoCo) and in an iron-sulfur center (Chapter 7). Interestingly, certain bacteria *(Azotobacter)* have alternative nitrogenases, which are produced when molybdenum is deficient and which contain vanadium and iron or only iron. All other known Mo-proteins are also Fe-proteins with iron centers, such as tetrapyrroles (heme and chlorins), Fe-sulfur clusters, and, apparently, non-heme/non-sulfur iron. Some Mo-proteins contain additional cofactors such as the flavins, e.g., in xanthine oxidase and aldehyde oxidase. The number of redox centers in some Mo-proteins exceeds the number of electrons transferred; reasons for this are unknown currently.

B. Chemical Properties Relative to Storage and Transport

1. Iron

Iron is the most abundant transition element in the Earth's crust and, in general, in all life forms. An outline of the distribution of iron in the Earth's crust[20,21] is shown in Table 1.2. As can be seen, approximately one-third of the Earth's mass is estimated to be iron. Of course, only the Earth's crust is relevant for life forms, but even there it is the most abundant transition element. Its concentration is relatively high in most crustal rocks (lowest in limestone, which is more or less pure calcium carbonate). In the oceans, which constitute 70 percent of the Earth's surface, the concentration of iron is low but increases with depth, since this iron exists as suspended particulate matter rather than as a soluble species. Iron is a limiting factor in plankton growth, and the rich

Table 1.2
Iron: Its terrestrial distribution.[a]

One third of Earth's mass, most abundant element by weight
  Distribution in crustal rocks (weight %):
    igneous   5.6
    shale   4.7
    sandstone 1.0
    limestone 0.4
  Ocean (70% of Earth's surface):
    0.003–0.1 ppb, increasing with depth; limiting factor in plankton
    growth
  Rivers:
    0.07–7 ppm
  K_{sp} for $Fe(OH)_3$ is approximately 10^{-39}, hence at pH 7 $[Fe^{3+}]$ 10^{-18} M

[a] Data from References 1a and 20.

fisheries associated with strong up-welling of ocean depths result at least in part from the biological growth allowed by these iron supplies. Properties that dominate the transport behavior of most transition metal ions are: (1) redox chemistry, (2) hydrolysis, and (3) the solubility of the metal ions in various complexes, particularly the hydroxides.

As an example of the effects of solubility, consider the enormous variation in the concentration of iron in rivers, depending on whether the water is from a clear mountain stream running over rock or a muddy river carrying large amounts of sediment. However, the amount of dissolved iron in the form of free ferric ion or its hydrolysis products, whatever the source of water, is extremely low. As can be seen from the solubility of hydrated Fe(III) ($K_s \sim 10^{-18}$ M) (Table 1.2), the concentration of free ferric ion is extraordinarily low at neutral pH; so significant concentrations of soluble iron species can be attained only by strong complex formation.

One example of the versatility of iron as a function of its environment is how the ligand field can strongly alter the structural and ligand exchange properties of the metal ion (Figure 1.3). The ligand field can also alter the redox properties. For high-spin ferric ion, as found in the aquo complex or in many other complexes (including the class of microbial iron-transport agents called siderophores, to be discussed later), the coordination geometry is octahedral or pseudo-octahedral. In the relatively weak ligand field (high-spin ground state), the complex is highly labile. In a strong ligand field, such as an axially ligated porphyrin complex of ferric ion, or the simple example of the ferrocyanide anion, the low-spin complex is exchange-inert. Similarly, the high-spin octahedral ferrous complexes are exchange-labile, but the corresponding axially ligated porphyrin complexes, or the ferrocyanide complexes, are spin-paired (diamagnetic) and ligand exchange-inert. Large, bulky ligands or constrained ligands, such as those provided by metalloprotein and enzyme sites, can cause a tetrahedral environment, in which both ferrous ion and ferric ion form high-spin complexes.

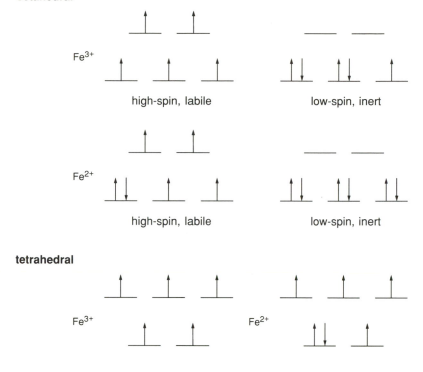

Figure 1.3
Versatility of Fe coordination complexes.

The distribution of specific iron complexes in living organisms depends strongly on function. For example, although there are many different iron complexes in the average human, the relative amounts of each type differ more than 650-fold (Table 1.3). The total amount of iron in humans is quite large, averaging more than three and up to five grams for a healthy adult. Most of the iron is present as hemoglobin, the plasma oxygen-transport protein, where the func-

Table 1.3
Average human Fe distribution.

Protein	Function	Oxidation state of Fe	Amount of Fe (g)	Percent of total
Hemoglobin	Plasma O_2 transport	2	2.6	65
Myoglobin	Muscle O_2 storage	2	0.13	6
Transferrin	Plasma Fe transport	3	0.007	0.2
Ferritin	Cell Fe storage	3	0.52	13
Hemosiderin	Cell Fe storage	3	0.48	12
Catalase	H_2O_2 metabolism	2	0.004	0.1
Cytochrome c	Electron transport	$\frac{2}{3}$	0.004	0.1
Other	Oxidases, other enzymes, etc.		0.14	3.6

tion of the iron is to deliver oxygen for respiration. A much smaller amount of iron is present in myoglobin, a muscle oxygen-storage protein. For transport, the most important of these iron-containing proteins is transferrin, the plasma iron-transport protein that transfers iron from storage sites in the body to locations where cells synthesizing iron proteins reside; the major consumers of iron in vertebrates are the red blood cells. However, at any given time relatively little of the iron in the body is present in transferrin, in much the same way that at any given time in a large city only a small fraction of the population will be found in buses or taxis. Other examples of iron-containing proteins and their functions are included in Table 1.3 for comparison.

An example of different iron-coordination environments, which alter the chemical properties of iron, is the difference in the redox potentials of hydrated Fe^{3+} and the electron-transport protein cytochrome c (Table 1.4). The coordina-

Table 1.4
Fe redox potentials.

Complex	Coord. no., type	Fe^{3+}/Fe^{2+} $E°$ (mV)
$Fe(OH_2)_6^{3+}$	6, aquo complex	770
Cytochrome a_3	6, heme	390
HIPIP	4, $Fe_4S_4(SR)_4^-$	350
Cytochrome c	6, heme	250
Rubredoxin	4, $Fe(SR)_4$	-60
Ferredoxins	4, $Fe_4S_4(SR)_4^{2-}$	-400

tion environment of iron in cytochrome c is illustrated in Figure 1.4. For example, the standard reduction potential for ferric ion in acid solution is 0.77 volts; so here ferric ion is quite a good oxidant. In contrast, cytochrome c has a redox potential of 0.25 volts. A wide range of redox potentials for iron is achieved in biology by subtle differences in protein structure, as listed in Table 1.4. Notice the large difference in the potential of cytochrome c and rubredoxin (Figure 1.5), 0.25 volts vs. -0.06 volts, respectively. In polynuclear ferredoxins, in which each iron is tetrahedrally coordinated by sulfur, reduction potentials are near -0.4 volts. Thus, the entire range of redox potentials, as illustrated in Table 1.4, is more than one volt.

2. Chemical properties of zinc, copper, vanadium, chromium, molybdenum, and cobalt

The chemical properties of the other essential transition elements simplify their transport properties. For zinc there is only the $+2$ oxidation state, and the hydrolysis of this ion is not a limiting feature of its solubility or transport. Zinc is an essential element for both animals and plants.[8,9,20,21] In general, metal ion uptake into the roots of plants is an extremely complex phenomenon. A cross-sectional diagram of a root is shown in Figure 1.6. It is said that both diffusion

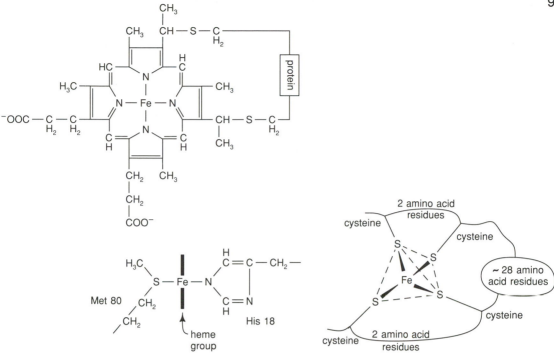

Figure 1.4
Heme group and iron coordination in cytochrome *c*.

Figure 1.5
$Fe^{3+/2+}$ coordination in rubredoxin.

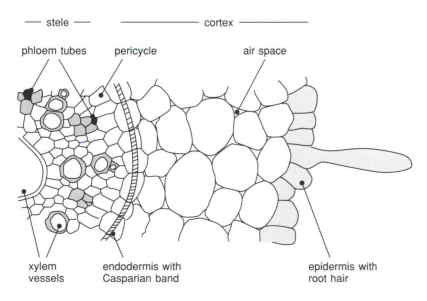

Figure 1.6
Transverse section of a typical root.[20] The complex features of the root hair surface that regulate reductase and other activities in metal uptake are only beginning to be understood.

and mass flow of the soil solution are of significance in the movement of metal ions to roots. Chelation and surface adsorption, which are pH dependent, also affect the availability of nutrient metal ions. Acid soil conditions in general retard uptake of essential divalent metal ions but increase the availability (sometimes with toxic results) of manganese, iron, and aluminum, all of which are normally of very limited availability because of hydrolysis of the trivalent ions.

Vanadium is often taken up as vanadate, in a pathway parallel to phosphate.[18] However, its oxidation state within organisms seems to be highly variable. Unusually high concentrations of vanadium occur in certain ascidians (the specific transport behavior of which will be dealt with later). The workers who first characterized the vanadium-containing compound of the tunicate, *Ascidia nigra*, coined the name tunichrome.[22] The characterization of the compound as a dicatecholate has been reported.[23]

Quite a different chemical environment is found in the vanadium-containing material isolated from the mushroom *Amanita muscaria*. Bayer and Kneifel, who named and first described amavadine,[24] also suggested the structure shown in Figure 1.7.[25] Recently the preparation, proof of ligand structure, and (by implication) proof of the complex structure shown in Figure 1.7 have been established.[26] Although the exact role of the vanadium complex in the mushroom

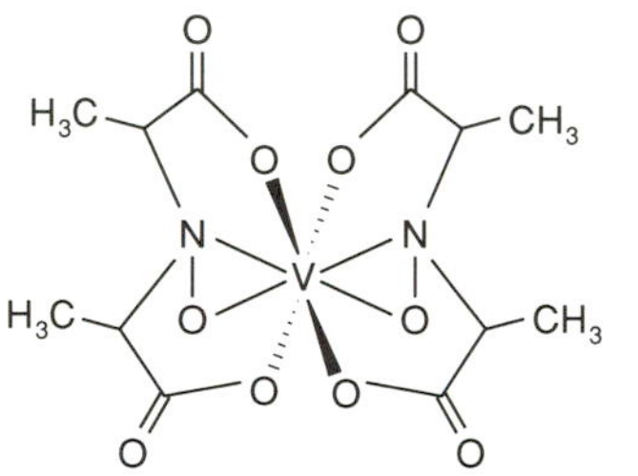

Figure 1.7
A structure proposed for amavadine.[27]

remains unclear, the fact that it is a vanadyl complex is now certain, although it may take a different oxidation state *in vivo*.

The role of chromium in biology remains even more mysterious. In human beings the isolation of "glucose tolerance factor" and the discovery that it contains chromium goes back some time. This has been well reviewed by Mertz, who has played a major role in discovering what is known about this elusive and apparently quite labile compound.[27] It is well established that chromium is taken up as chromic ion, predominantly via foodstuffs, such as unrefined sugar, which presumably contain complexes of chromium, perhaps involving sugar hydroxyl groups. Although generally little chromium is taken up when it is administered as inorganic salts, such as chromic chloride, glucose tolerance in many adults and elderly people has been reported to be improved after supplementation with 150–250 mg of chromium per day in the form of chromic chloride. Similar results have been found in malnourished children in some studies in Third World countries. Studies using radioactively labeled chromium have shown that, although inorganic salts of chromium are relatively unavailable to mam-

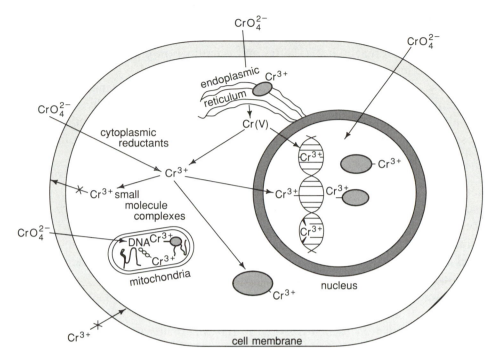

Figure 1.8
The uptake-reduction model for chromate carcinogenicity. Possible sites for reduction of chromate include the cytoplasm, endoplasmic reticulum, mitochondria, and the nucleus.[27]

mals, brewer's yeast can convert the chromium into a usable form; so brewer's yeast is today the principal source in the isolation of glucose tolerance factor and has been used as a diet supplement.

Although chromium is essential in milligram amounts for human beings as the trivalent ion, as chromate it is quite toxic and a recognized carcinogen.[30] The uptake-reduction model for chromate carcinogenicity as suggested by Connett and Wetterhahn is shown in Figure 1.8. Chromate is mutagenic in bacterial and mammalian cell systems, and it has been hypothesized that the difference between chromium in the $+6$ and $+3$ oxidation states is explained by the "uptake-reduction" model. Chromium(III), like the ferric ion discussed above, is readily hydrolyzed at neutral pH and extremely insoluble. Unlike Fe^{3+}, it undergoes extremely slow ligand exchange. For both reasons, transport of chromium(III) into cells can be expected to be extremely slow unless it is present as specific complexes; for example, chromium(III) transport into bacterial cells has been reported to be rapid when iron is replaced by chromium in the siderophore iron-uptake mediators. However, chromate readily crosses cell membranes and enters cells, much as sulfate does. Because of its high oxidizing power, chromate can undergo reduction inside organelles to give chromium(III), which binds to small molecules, protein, and DNA, damaging these cellular components.

In marked contrast to its congener, molybdenum is very different from chromium in both its role in biology and its transport behavior, again because of fundamental differences in oxidation and coordination chemistry properties. In contrast to chromium, the higher oxidation states of molybdenum dominate its chemistry, and molybdate is a relatively poor oxidant. Molybdenum is an essential element in many enzymes, including xanthine oxidase, aldehyde reductase, and nitrate reductase.[19] The range of oxidation states and coordination geometries of molybdenum makes its bioinorganic chemistry particularly interesting and challenging.

The chemistry of iron storage and transport is dominated by high concentrations, redox chemistry (and production of toxic-acting oxygen species), hydrolysis (pK_a is about 3, far below physiological pH), and insolubility. High-affinity chelators or proteins are required for transport of iron and high-capacity sequestering protein for storage. By comparison to iron, storage and transport of the other metals are simple. Zinc, copper, vanadium, chromium, manganese, and molybdenum appear to be transported as simple salts or loosely bound protein complexes. In vanadium or molybdenum, the stable anion, vanadate or molybdate, appears to dominate transport. Little is known about biological storage of any metal except iron, which is stored in ferritin. However, zinc and copper are bound to metallothionein in a form that may participate in storage.

II. BIOLOGICAL SYSTEMS OF METAL STORAGE, TRANSPORT, AND MINERALIZATION

A. Storage

1. The storage of iron

Three properties of iron can account for its extensive use in terrestrial biological reactions:

(a) facile redox reactions of iron ions;

(b) an extensive repertoire of redox potentials available by ligand substitution or modification (Table 1.4);

(c) abundance and availability (Table 1.1) under conditions apparently extant when terrestrial life began (see Section I.B.).

Ferrous ion appears to have been the environmentally stable form during prebiotic times. The combination of the reactivity of ferrous ion and the relatively large amounts of iron used by cells may have necessitated the storage of ferrous ion; recent results suggest that ferrous ion may be stabilized inside ferritin long enough to be used in some types of cells. As primitive organisms began to proliferate, the successful photosynthetic cells, which trapped solar energy by reducing CO_2 to make carbohydrates $(CH_2O)_n$ and produce O_2, exhausted from the environment the reductants from H_2 or H_2S or NH_3. The ability of primitive organisms to switch to the use of H_2O as a reductant, with the concomitant production of dioxygen, probably produced the worst case of environmental pollution in terrestrial history. As a result, the composition of the atmosphere, the course of biological evolution, and the oxidation state of environmental iron all changed profoundly. Paleogeologists and meteorologists estimate that there was a lag of about 200–300 million years between the first dioxygen production and the appearance of significant dioxygen concentrations in the atmosphere, because the dioxygen produced at first was consumed by the oxidation of ferrous ions in the oceans. The transition in the atmosphere, which occurred about 2.5 billion years ago, caused the bioavailability of iron to plummet and the need for iron storage to increase. Comparison of the solubility of Fe^{3+} at physiological conditions (about 10^{-18} M) to the iron content of cells (equivalent to 10^{-5} to 10^{-8} M) emphasizes the difficulty of acquiring sufficient iron.

Iron is stored mainly in the ferritins, a family* of proteins composed of a protein coat and an iron core of hydrous ferric oxide [$Fe_2O_3(H_2O)_n$] with various amounts of phosphate.[6,7] As many as 4,500 iron atoms can be reversibly stored inside the protein coat in a complex that is soluble; iron concentrations equivalent to 0.25 M [about 10^{16}-fold more concentrated than Fe(III) ions] can be easily achieved *in vitro* (Figure 1.1). Ferritin is found in animals, plants, and even in bacteria; the role of the stored iron varies, and includes intracellular use for Fe-proteins or mineralization, long-term iron storage for other cells, and detoxification of excess iron. Iron regulates the synthesis of ferritin, with large amounts of ferritin associated with iron excess, small or undetectable amounts associated with iron deficiency. [Interestingly, the template (mRNA) for ferritin synthesis is itself stored in cells and is recruited by intracellular iron or a derivative for efficient translation into protein.[31] Iron does not appear to interact directly with ferritin mRNA nor with a ferritin mRNA-specific regulatory (binding) protein; however, the specific, mRNA regulatory (binding) protein has sequence homology to aconitase, and formation of an iron-sulfate cluster prevents RNA binding.] Because iron itself determines in part the amount of ferritin in an organism, the environmental concentration of iron needs to be considered before one can conclude that an organism or cell does not have ferritin.

Ferritin is thought to be the precursor of several forms of iron in living organisms, including hemosiderin, a form of storage iron found mainly in animals. The iron in hemosiderin is in a form very similar to that in ferritin, but the complex with protein is insoluble, and is usually located within an intracellular membrane (lysosomes). Magnetite (Fe_3O_4) is another form of biological iron derived, apparently, from the iron in ferritin. Magnetite plays a role in the behavior of magnetic bacteria, bees, and homing pigeons (see Section II.C).

The structure of ferritin is the most complete paradigm for bioinorganic chemistry because of three features: the protein coat, the iron-protein interface, and the iron core.[6,7]

---

* A family of proteins is a group of related but distinct proteins produced in a single organism and usually encoded by multiple, related genes.

14

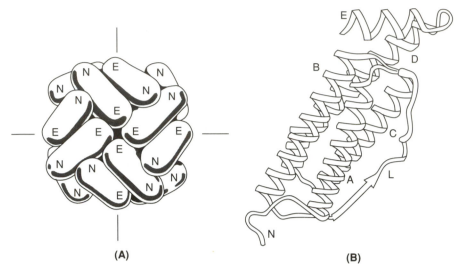

(A)   **(B)**

Figure 1.9
(A) The protein coat of horse spleen apoferritin deduced from x-ray diffraction of crystals of the protein.[32] The outer surface of the protein coat shows the arrangement of the 24 ellipsoidal poly-peptide subunits. N refers to the N-terminus of each polypeptide and E to the E-helix (see B). Note the channels that form at the four-fold axes where the E-helices interact, and at the three-fold axes near the N-termini of the subunits. (B) A ribbon model of a subunit showing the packing of the four main alpha-helices (A, B, C, and D), the connecting L-loop and the E-helix.

Protein Coat   Twenty-four peptide chains (with about 175 amino acids each), folded into ellipsoids, pack to form the protein coat,* which is a hollow sphere about 100 Å in diameter; the organic surface is about 10 Å thick (Figure 1.9). Channels which occur in the protein coat at the trimer interfaces may be involved in the movement of iron in and out of the protein.[62,63,65] Since the protein coat is stable with or without iron, the center of the hollow sphere may be filled with solvent, with $Fe_2O_3 \cdot H_2O$, or, more commonly, with both small aggregates of iron and solvent. Very similar amino-acid sequences are found in ferritin from animals and plants. Sorting out which amino acids are needed to form the shape of the protein coat and the ligands for iron core formation requires the continued dedication of bioinorganic chemists; identification of tyrosine as an Fe(III)-ligand adds a new perspective.[64]

Iron-Protein Interface   Formation of the iron core appears to be initiated at an Fe-protein interface where Fe(II)-O-Fe(III) dimers and small clusters of Fe(III) atoms have been detected attached to the protein and bridged to each other by oxo/hydroxo bridges. Evidence for multiple nucleation sites has been obtained

---

* Some ferritin subunits, notably in ferritin from bacteria, bind heme in a ratio of less than one heme per two subunits. A possible role of such heme in the oxidation and reduction of iron in the core is being investigated.

from electron microscopy of individual ferritin molecules (multiple core crystal-lites were observed) and by measuring the stoichiometry of binding of metal ions, which compete with binding of monoatomic iron, e.g., VO(IV) and Tb(III) (about eight sites per molecule). EXAFS (*Extended X-ray Absorption Fine Structure*) and Mössbauer spectroscopies suggest coordination of Fe to the pro-tein by carboxyl groups from glutamic (Glu) and aspartic (Asp) acids. Although groups of Glu or Asp are conserved in all animal and plant ferritins, the ones that bind iron are not known. Tyrosine is an Fe(III)-ligand conserved in rapid mineralizing ferritins identified by Uv-vis and resonance Raman spectro-scopy.[64]

Iron Core Only a small fraction of the iron atoms in ferritin bind directly to the protein. The core contains the bulk of the iron in a polynuclear aggregate with properties similar to ferrihydrite, a mineral found in nature and formed experimentally by heating neutral aqueous solutions of $Fe(III)(NO_3)_3$. X-ray dif-fraction data from ferritin cores are best fit by a model with hexagonal close-packed layers of oxygen that are interrupted by irregularly incomplete layers of octahedrally coordinated Fe(III) atoms. The octahedral coordination is con-firmed by Mössbauer spectroscopy and by EXAFS, which also shows that the average Fe(III) atom is surrounded by six oxygen atoms at a distance of 1.95 Å and six iron atoms at distances of 3.0 to 3.3 Å.

Until recently, all ferritin cores were thought to be microcrystalline and to be the same. However, x-ray absorption spectroscopy, Mössbauer spectroscopy, and high-resolution electron microscopy of ferritin from different sources have revealed variations in the degree of structural and magnetic ordering and/or the level of hydration. Structural differences in the iron core have been associated with variations in the anions present, e.g., phosphate[29] or sulfate, and with the electrochemical properties of iron. Anion concentrations in turn could reflect both the solvent composition and the properties of the protein coat. To under-stand iron storage, we need to define in more detail the relationship of the ferritin protein coat and the environment to the redox properties of iron in the ferritin core.

Experimental studies of ferritin formation show that Fe(II) and dioxygen are needed, at least in the early stages of core formation. Oxidation to Fe(III) and hydrolysis produce one electron and an average of 2.5 protons for iron atoms incorporated into the ferritin iron core. Thus, formation of a full iron core of 4,500 iron atoms would produce a total of 4,500 electrons and 11,250 protons. After core formation by such a mechanism inside the protein coat, the pH would drop to 0.4 if all the protons were retained. It is known that protons are released and electrons are transferred to dioxygen. However, the relative rates of proton release, oxo-bridge formation, and electron transfer have not been studied in detail. Moreover, recent data indicate migration of iron atoms during the early stages of core formation and the possible persistence of Fe^{2+} for periods of time up to 24 hours. When large numbers of Fe(II) atoms are added, the protein coat appears to stabilize the encapsulated Fe(II).[34a,b] Formation of the iron core of ferritin has analogies to surface corrosion, in which electrochemical gradients are known to occur. Whether such gradients occur during ferritin formation and how different protein coats might influence proton release or alter the structure of the core are subjects only beginning to be examined.

2. The storage of zinc, copper, vanadium, chromium, molybdenum, cobalt, nickel, and manganese

Ions of nonferrous transition metals require a much less complex biological storage system, because the solubilities are much higher ($\geq 10^{-8}$ M) than those for Fe^{3+}. As a result, the storage of nonferrous transition metals is less obvious, and information is more limited. In addition, investigations are more difficult than for iron, because the amounts in biological systems are so small. Essentially nothing is known yet about the storage of vanadium, chromium, molybdenum, cobalt, nickel, and manganese, with the possible exception of accumulations of vanadium in the blood cells of tunicates.

Zinc and copper, which are used in the highest concentrations of any of the non-ferrous transition metals, are specifically bound by the protein metallothionein[35,36] (see Figure 1.10). Like the ferritins, the metallothioneins are a family of proteins, widespread in nature and regulated by the metals they bind. In contrast to ferritin, the amounts of metal stored in metallothioneins are smaller (up to twelve atoms per molecule), the amount of protein in cells is less, and the template (mRNA) is not stored. Because the cellular concentrations of the metallothioneins are relatively low and the amount of metal needed is relatively small, it has been difficult to study the biological fate of copper and zinc in living organisms, and to discover the natural role of metallothioneins. However, the regulation of metallothionein synthesis by metals, hormones, and growth factors attests to the biological importance of the proteins. The unusual metal environments of metallothioneins have attracted the attention of bioinorganic chemists.

Metallothioneins, especially in higher animals, are small proteins[35,36] rich in cysteine (20 per molecule) and devoid of the aromatic amino acids phenylalanine and tyrosine. The cysteine residues are distributed throughout the peptide chain. However, in the native form of the protein (Figure 1.10), the peptide chains fold to produce two clusters of —SH, which bind either three or four atoms of zinc, cadmium, cobalt, mercury, lead, or nickel. Copper binding is distinct from zinc, with 12 sites per molecule.

In summary, iron is stored in iron cores of a complicated protein. Ferritin, composed of a hollow protein coat, iron-protein interface, and an inorganic core, overcomes the problems of redox and hydrolysis by directing the formation of the quasi-stable mineral hydrous ferric oxide inside the protein coat. The outer surface of the protein is generally hydrophilic, making the complex highly soluble; equivalent concentrations of iron are ≤ 0.25 M. By contrast to iron, storage of zinc, copper, chromium, manganese, vanadium, and molybdenum is relatively simple, because solubility is high and abundance is lower. Little is known

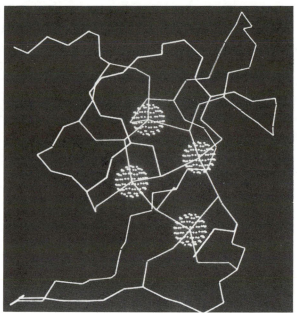

Figure 1.10
The three-dimensional structure of the α domain from rat cd_7 metallothionein-2, determined by NMR in solution (Reference 36a), based on data in Reference 36b. The four metal atoms, bonded to the sulfur of cysteine side chains, are indicated as spherical collections of small dots. A recent description of the structure of the cd_5Zn_2 protein, determined from x-ray diffraction of crystals, agrees with the structure determined by NMR (Reference 36c).

about the molecules that store these metals, with the possible exception of metallothionein, which binds small clusters of zinc or copper.

B. Transport

1. Iron

The storage of iron in humans and other mammals has been dealt with in the previous section. Only a small fraction of the body's inventory of iron is in transit at any moment. The transport of iron from storage sites in cellular ferritin or hemosiderin occurs via the serum-transport protein transferrin. The transferrins are a class of proteins that are bilobal, with each lobe reversibly (and essentially independently) binding ferric ion.[37–39] This complexation of the metal cation occurs via prior complexation of a synergistic anion that *in vivo* is bicarbonate (or carbonate). Serum transferrin is a monomeric glycoprotein of molecular weight 80 kDa. The crystal structure of the related protein, lactoferrin,[39] has been reported, and recently the structure of a mammalian transferrin[40] has been deduced.

Ferritin is apparently a very ancient protein and is found in higher animals, plants, and even microbes; in plants and animals a common ferritin progenitor is indicated by sequence conservation.[41] In contrast, transferrin has been in existence only relatively recently, since it is only found in the phylum Chordata. Although the two iron-binding sites of transferrin are sufficiently different to be distinguishable by kinetic and a few other studies, their coordination environments have been known for some time to be quite similar. This was first discovered by various spectroscopies, and most recently was confirmed by crystal-structure analysis, which shows that the environment involves two phenolate oxygens from tyrosine, two oxygens from the synergistic, bidentate bicarbonate anion, nitrogen from histidine, and (a surprise at the time of crystal-structure analysis) an oxygen from a carboxylate group of an aspartate.[39]

The transferrins are all glycoproteins, and human serum transferrin contains about 6 percent carbohydrate. These carbohydrate groups are linked to the protein, and apparently strongly affect the recognition and conformation of the native protein.

Although transferrins have a high molecular weight and bind only two iron atoms, transferrin is relatively efficient, because it is used in many cycles of iron transport in its interaction with the tissues to which it delivers iron. Transferrin releases iron *in vivo* by binding to the cell surface and forming a vesicle inside the cell (endosome) containing a piece of the membrane with transferrin and iron still complexed. The release of the iron from transferrin occurs in the relatively low pH of the endosome, and apoprotein is returned to the outside of the cell for delivery of another pair of iron atoms. This process in active reticulocytes (immature red blood cells active in iron uptake) can turn over roughly a million atoms of iron per cell per minute.[38] A schematic structure of the protein, deduced from crystal-structure analysis, is shown in Figure 1.11. Transferrin is an ellipsoidal protein with two subdomains or lobes, each of which binds iron. The two halves of each subunit are more or less identical, and are connected by a relatively small hinge. In human lactoferrin, the coordination site of the iron is the same as the closely related serotransferrin site. A major question that remains about the mechanism of iron binding and release is how the protein structure changes in the intracellular compartment of low pH to release the iron when it forms a specific complex with cell receptors (transferrin binding proteins) and whether the receptor protein is active or passive in the process. Recent studies suggest that the cell binding site for transferrin (a membrane, glycoprotein called the transferrin receptor) itself influences the stability of the iron-transferrin complex. The path of iron from the endosome to Fe-proteins has not been established; and the form of transported intracellular iron is not known.

Another major type of biological iron transport occurs at the biological opposite of the higher organisms. Although almost all microorganisms have iron as an essential element, bacteria, fungi, and other microorganisms (unlike humans and other higher organisms) cannot afford to make high-molecular-weight protein-complexing agents for this essential element when those complexing agents would be operating extracellularly and hence most of the time would be lost to the organism. As described earlier, the first life forms on the surface of the

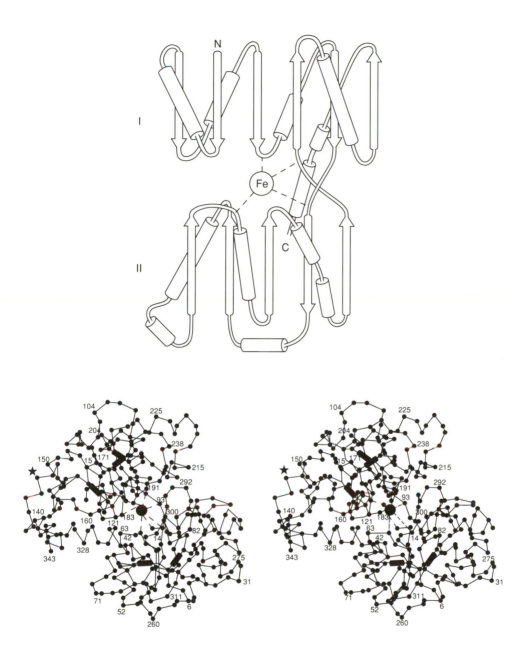

Figure 1.11

Three-dimensional structure of lactotransferrin. *Top*: schematic representation of the folding pattern of each lactoferrin lobe; Domain I is based on a beta-sheet of four parallel and two antiparallel domains; Domain II is formed from four parallel and one antiparallel strand. *Bottom*: stereo Cα diagram of the N lobe of lactoferrin; (●) iron atom between domain I (residues 6–90 +) and domain II (residues 91–251); (■■■) disulfide bridges; (★) carbohydrate attachment site. See Reference 39.

Earth grew in a reducing atmosphere, in which the iron was substantially more available because it was present as ferrous-containing compounds. In contrast to the profoundly insoluble ferric hydroxide, ferrous hydroxide is relatively soluble at near neutral pH. It has been proposed that this availability of iron in the ferrous state was one of the factors that led to its early incorporation in so many metabolic processes of the earliest chemistry of life.[6,38] In an oxidizing environment, microorganisms were forced to deal with the insolubility of ferric hydroxide and hence when facing iron deficiency secrete high-affinity iron-binding compounds called siderophores (from the Greek for iron carrier). More than 200 naturally occurring siderophores have been isolated and characterized to date.[42]

Most siderophore-mediated iron-uptake studies in microorganisms have been performed by using cells obtained under iron-deficient aerobic growth conditions. However, uptake studies in *E. coli* grown under anaerobic conditions have also established the presence of siderophore-specific mechanisms. In both cases, uptake of the siderophore-iron complex is both a receptor- and an energy-dependent process. In some studies the dependence of siderophore uptake rates on the concentration of the iron-siderophore complex has been found to conform to kinetics characteristic of protein catalysts, i.e., Michaelis-Menten kinetics. For example, saturable processes with very low apparent dissociation constants of under one micromolar (1 μM) have been observed for ferric-enterobactin transport in *E. coli* (a bacterium), as shown in Figure 1.12. Similarly, in a very

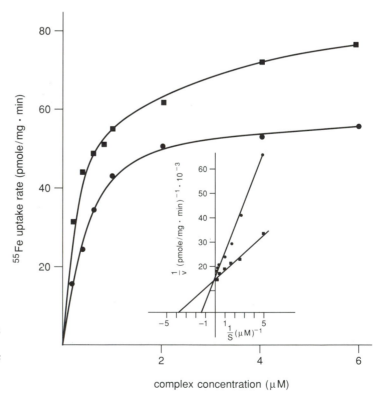

Figure 1.12
Effect of MECAM analogues on iron uptake from *E. coli*. Iron transport by 2 μM ferric enterobactin is inhibited by ferric MECAM.

different microorganism, the yeast *Rhodoturala pilimanae*, Michaelis-Menten kinetics were seen again with a dissociation constant of approximately 6 μM for the ferric complex of rhodotoroulic acid; diagrams of some representative siderophores are shown in Figure 1.13. The siderophore used by the fungus *Neurospora crassa* was found to have a dissociation constant of about 5 μM and, again, saturable uptake kinetics.

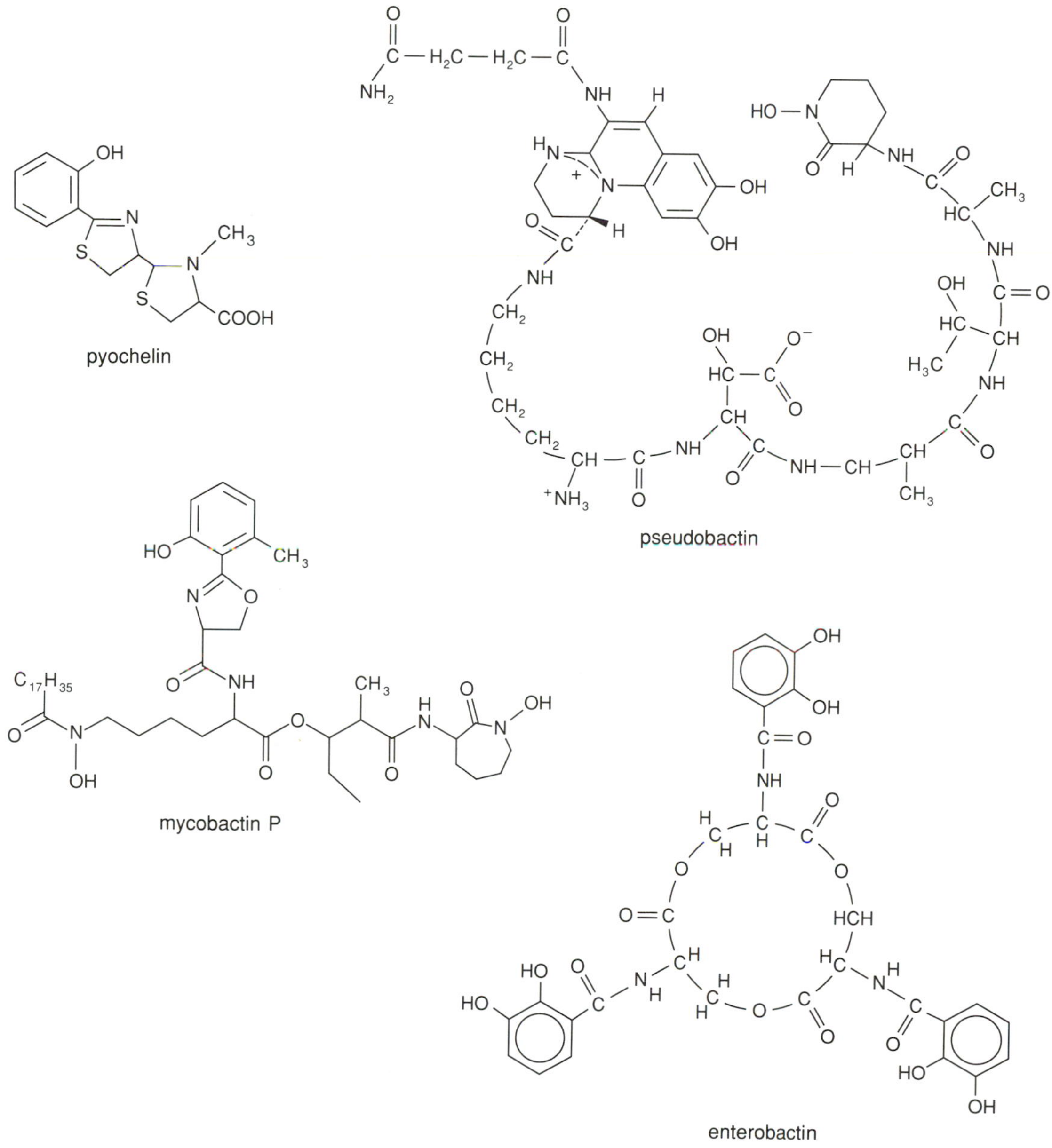

Figure 1.13
Examples of bacterial siderophores. See Reference 42.

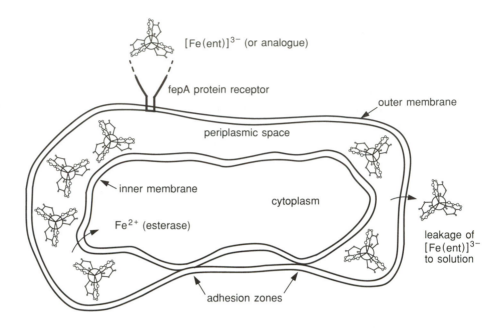

Figure 1.14
Model for enterobactin-mediated Fe uptake in *E. coli*.

Although the behavior just described seems relatively simple, transport mechanisms in living cells probably have several more kinetically distinct steps than those assumed for the simple enzyme-substrate reactions underlying the Michaelis-Menten mechanism. For example, as ferric enterobactin is accumulated in *E. coli*, it has to pass through the outer membrane, the periplasm, and the cytoplasm membrane, and is probably subjected to reduction of the metal in a low-pH compartment or to ligand destruction.

A sketch of a cell of *E. coli* and some aspects of its transport behavior are shown in Figure 1.14. Enterobactin-mediated iron uptake in *E. coli* is one of the best-characterized of the siderophore-mediated iron-uptake processes in microorganisms, and can be studied as a model. After this very potent iron-sequestering agent complexes iron, the ferric-enterobactin complex interacts with a specific receptor in the outer cell membrane (Figure 1.14), and the complex is taken into the cell by active transport. The ferric complexes of some synthetic analogs of enterobactin can act as growth agents in supplying iron to *E. coli*. Such a feature could be used to discover which parts of the molecule are involved in the sites of structural recognition of the ferric-enterobactin complex. Earlier results suggested that the metal-binding part of the molecule is recognized by the receptor, whereas the ligand platform (the triserine lactone ring; see Figure 1.13) is not specifically recognized.

To find out which domains of enterobactin are required for iron uptake and recognition, rhodium complexes were prepared with various domains of enterobactin (Figure 1.15) as ligands to use as competitors for ferric enterobactin.[44] The goal was to find out if the amide groups (labeled Domain II in Figure 1.15),

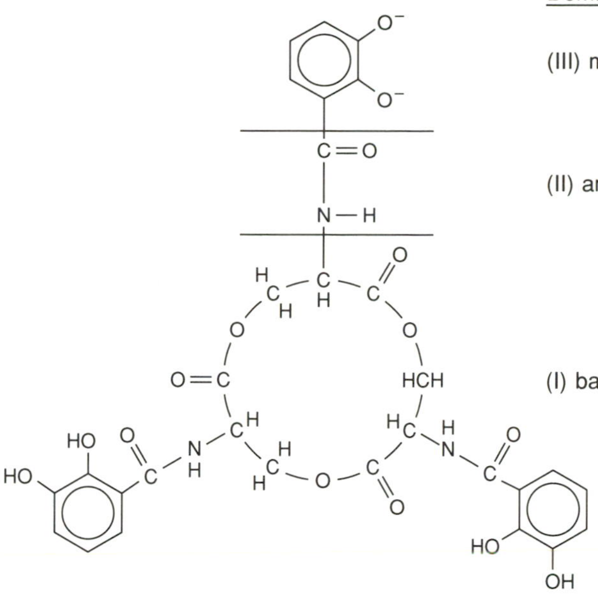

Figure 1.15
Definition of recognition domains in enterobactin.

which linked the metal-binding catechol groups (Domain III, Figure 1.15) to the central ligand backbone (Domain I, Figure 1.15), are necessary for recognition by the receptor protein. In addition, synthetic ligands were prepared that differed from enterobactin by small changes at or near the catecholate ring. Finally, various labile trivalent metal cations, analogous to iron, were studied to see how varying the central metal ion would affect the ability of metal enterobactin complexes to inhibit competitively the uptake of ferric enterobactin by the organism. For example, if rhodium MECAM (Figure 1.16) is recognized by the receptor for ferric enterobactin on living microbial cells, a large excess of rhodium MECAM will block the uptake of radioactive iron added as ferric enterobactin. In fact, the rhodium complex completely inhibited ferric-enterobactin uptake, proving that Domain I is not required for recognition of ferric enterobactin.

However, if only Domain III is important in recognition, it would be expected that the simple tris(catecholato)-rhodium(III) complex would be an equally good inhibitor. In fact, even at concentrations in which the rhodium-catechol complex was in very large excess, no inhibition of iron uptake was observed, suggesting that Domain II is important in the recognition process.

The role of Domain II in the recognition process was probed by using a rhodium dimethyl amide of 2,3-dihydroxybenzene (DMB) as a catechol ligand, with one more carbonyl ligand than in the tris(catecholato)-rhodium(III) complex. Remarkably, this molecule shows substantially the same inhibition of enterobactin-mediated iron uptake in E. coli as does rhodium MECAM itself. Thus, in addition to the iron-catechol portion of the molecule, the carbonyl groups

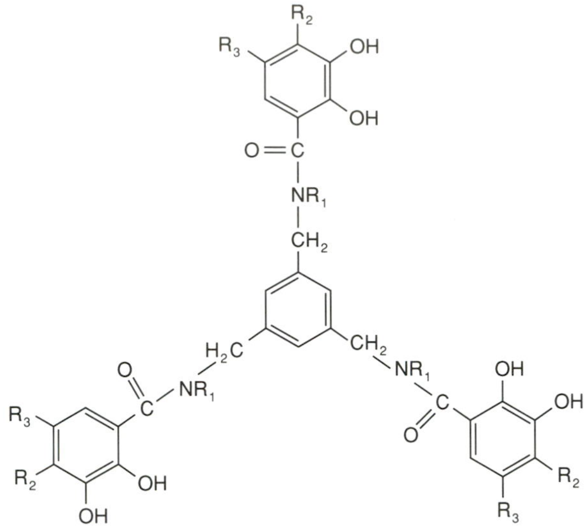

MECAM $(R_1 = R_2 = R_3 = H)$
MECAM-Me $(R_1 = R_3 = H, R_2 = CH_3)$
MECAMS $(R_1 = R_2 = H, R_3 = SO_3^-)$
Me$_3$MECAM $(R_2 = R_3 = H, R_1 = CH_3)$

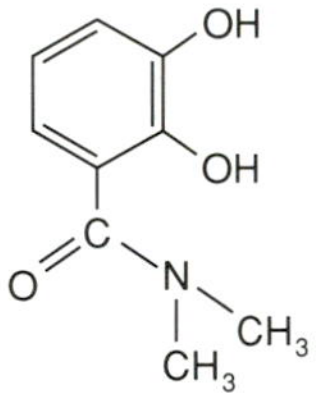

DMB

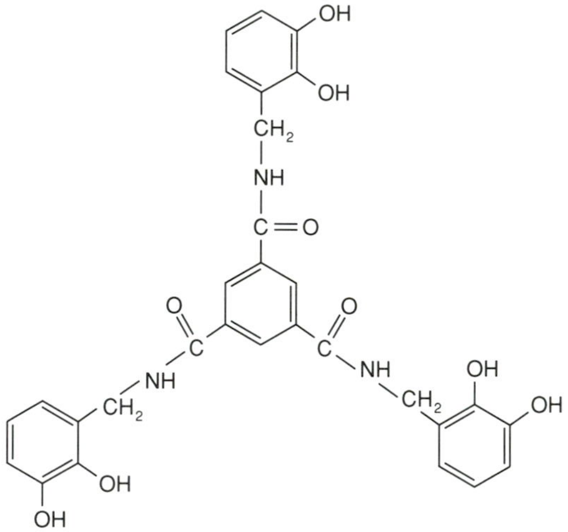

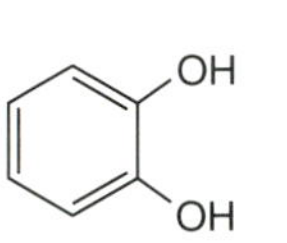

catechol

TRIMCAM

Figure 1.16
MECAM and related enterobactin analogues.

(Domain II) adjacent to the catechol-binding subunits of enterobactin and synthetic analogs are required for recognition by the ferric-enterobactin receptor. In contrast, when a methyl group was attached to the "top" of the rhodium MECAM complex, essentially no recognition occurred.

In summary, although the structure of the outer-membrane protein receptor of *E. coli* is not yet known, the composite of the results just described gives a sketch of what the ferric-enterobactin binding site must look like: a relatively rigid pocket for receiving the ferric-catecholate portion of the complex, and proton donor groups around this pocket positioned to hydrogen bond to the carbonyl oxygens of the ferric amide groups. The mechanisms of iron release from enterobactin, though followed phenomenologically, are still not known in detail.

2. Zinc, copper, vanadium, chromium, molybdenum, and cobalt

As described in an earlier section, transport problems posed by the six elements listed in the heading are somewhat simpler (with the exception of chromium) than those for iron. One very interesting recent development has been the characterization of sequestering agents produced by plants which complex a number of metal ions, not just ferric ions. A key compound, now well-characterized, is mugeneic acid (Figure 1.17).[45] The structural and chemical similari-

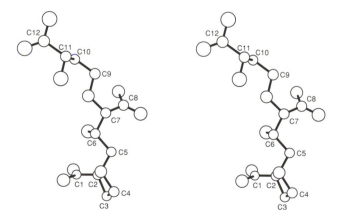

Figure 1.17
Structure and a stereo view of mugeneic acid. See Reference 42.

ties of mugeneic acid to ethylenediaminetetraacetic acid (EDTA) have been noted. Like EDTA, mugeneic acid forms an extremely strong complex with ferric ion, but also forms quite strong complexes with copper, zinc, and other transition-metal ions. The structure of the cobalt complex (almost certainly essentially identical with that of the iron complex) is shown in Figure 1.18. Like the siderophores produced by microorganisms, the coordination environment accommodated by mugeneic acid is essentially octahedral. Although the coordination properties of this ligand are well laid out, and it has been shown that divalent metal cations, such as copper, competitively inhibit iron uptake by this ligand, the detailed process of metal-ion delivery by mugeneic acid and related compounds has not been elucidated.

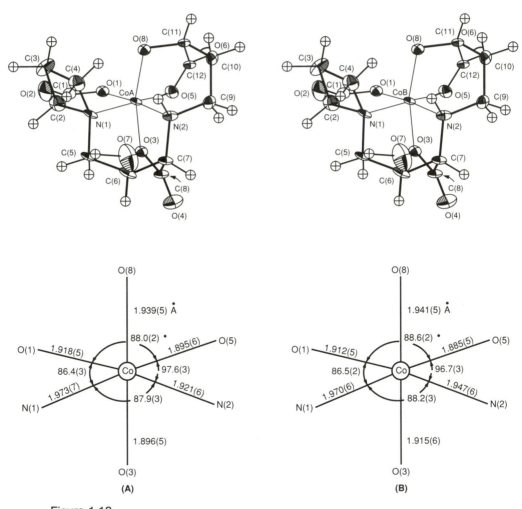

Figure 1.18
Molecular structures of the complexes (molecules A and B) and coordination about the cobalt ion in molecules A and B of the mugeneic acid-Co(III) complex. Bond lengths in Å; angles in degrees. See Reference 42.

As noted in an earlier section, the biochemistry of vanadium potentially involves four oxidation states that are relatively stable in aqueous solution. These are V^{2+}, V^{3+}, VO^{2+}, and VO_2^+ (the oxidation states 2, 3, 4, and 5, respectively). Since even without added sequestering agents, V^{2+} slowly reduces water to hydrogen gas, it presumably has no biological significance. Examples of the remaining three oxidation states of vanadium have all been reported in various living systems. One of the most extensively investigated examples of transition-metal-ion accumulation in living organisms is the concentration of vanadium in sea squirts (tunicates), which is reported to be variable; many species have vanadium levels that are not exceptionally high. Others such as *Ascidia nigra* show exceptionally high vanadium concentrations.[46]

In addition to showing a remarkable concentration of a relatively exotic transition-metal ion, tunicates are a good laboratory model for uptake experiments, since they are relatively simple organisms. They possess a circulation system with a one-chambered heart, and a digestive system that is essentially a pump and an inlet and outlet valve connected by a digestive tract. The organism can absorb dissolved vanadium directly from sea water as it passes through the animal. The influx of vanadate into the blood cells of *A. nigra* has been studied by means of radioisotopes. The corresponding influx of phosphate, sulfate, and chromate (and the inhibition of vanadate uptake by these structurally similar oxoanions) has been measured. In the absence of inhibitors, the influx of vanadate is relatively rapid (a half-life on the order of a minute near 0°C) and the uptake process shows saturation behavior as the vanadate concentration is increased. The uptake process (in contrast to iron delivery in microorganisms, for example, and to many other uptake processes in microorganisms or higher animals) is *not* energy-dependent. Neither inhibitors of glycolysis nor decouplers of respiration-dependent energy processes show any significant effect on the rate of vanadate influx.

Phosphate, which is also readily taken up by the cells, is an inhibitor of vanadate influx. Neither sulfate nor chromate is taken up significantly, nor do they act as significant inhibitors for the vanadate uptake. Agents that inhibit transport of anions, in contrast, were found to inhibit uptake of vanadate into the organism. These results have led to the model proposed in Figure 1.19:

(1) vanadate enters the cell through anionic channels; this process eliminates positively charged metal ion or metal-ion complexes present in sea water;

(2) vanadate is reduced to vanadium(III); since the product is a cation, and so cannot be transported through the anionic channels by which vanadate entered the cell, the vanadium(III) is trapped inside the cell—the net result is an accumulation of vanadium. [It has been proposed that the tunichrome could act either as a reducing agent (as the complex) or (as the ligand) to stabilize the general vanadium(III); however, this seems inconsistent with its electrochemical properties (see below).]

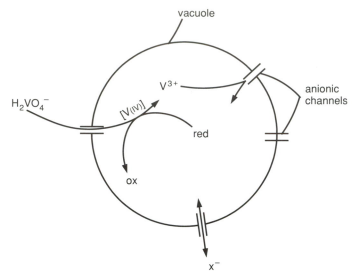

Figure 1.19
Diagram of a vanadium accumulation mechanism. Vanadium enters the vacuole within the vana-docyte as mononegative $H_2VO_4^-$, although it may be possible for the dinegative anion, HVO_4^{2-}, to enter this channel as well (X^- stands for any negative ion such as Cl^-, $H_2PO_4^-$, etc., that may exchange across the membrane through the anionic channel). Reduction to V^{3+} takes place in two steps, via a V(IV) intermediate. The resulting cations may be trapped as tightly bound complexes, or as free ions that the anionic channel will not accept for transport. The nature of the reducing species is unknown.

Synthetic models of tunichrome b-1 (Figure 1.20) have been prepared. Tun-ichrome is a derivative of pyrogallol whose structure precludes the formation of an octahedral complex of vanadium as a simple 1:1 metal:ligand complex. The close analogue, described as 3,4,5-TRENPAMH$_9$, also cannot form a simple octahedral 1:1 complex. In contrast, the synthetic ligands TRENCAM and 2,3,4-TRENPAM can form pseudo-octahedral complexes. The structure of the vana-dium TRENCAM complex shows that it is indeed a simple pseudo-octahedral tris-catechol complex.[47] The electrochemical behavior of these complexes is similar, with vanadium(IV/III) potentials of about -0.5 to -0.6 volts versus NHE. These results indicate that tunichrome b-1 complexes of vanadium(IV/III) would show similar differences in their redox couples at high pH. At neutral pH, in the presence of excess pyrogallol groups, vanadium(IV) can be expected to form the intensely colored tris-catechol species. However, comparison of the EPR properties reported for vanadium-tunichrome preparations with model van-adium(IV)-complexes would indicate predominantly bis(catechol) vanadyl co-ordination. In any case, the vanadium(III) complexes must remain very highly reducing. It has been pointed out that the standard potential of pyrogallol is 0.79 V and decreases 60 mV per pH unit (up to about pH 9), so that at pH 7 the potential is about 0.4 V. The potentials of the vanadium couples for the tunichrome analogs are about -0.4 V. It has been concluded, therefore, that tunichrome or similar ligands cannot reduce the vanadium(IV) complex; so the

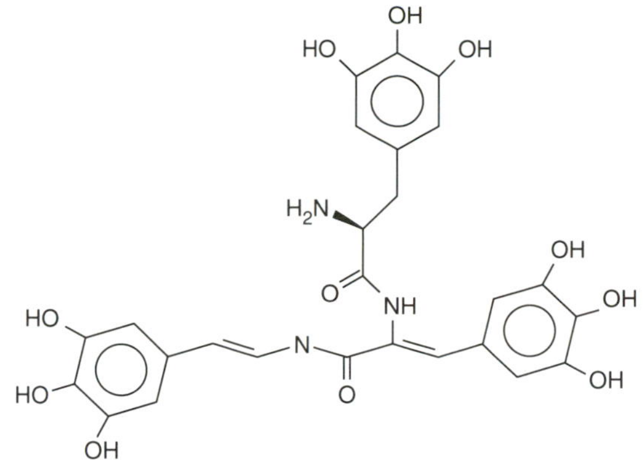

Figure 1.20
Structures of tunichrome b-1 and synthetic analogues.[43]

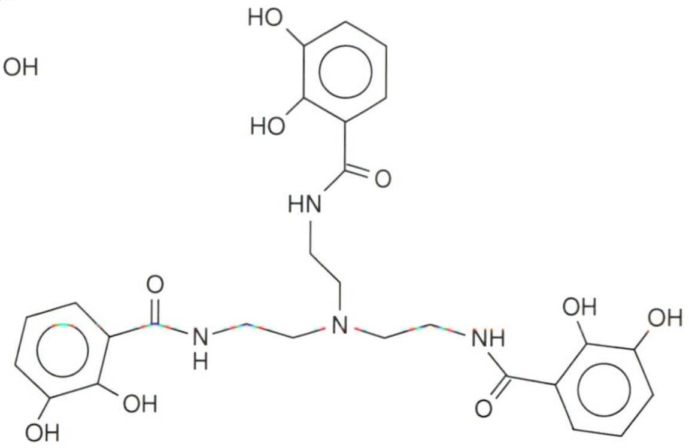

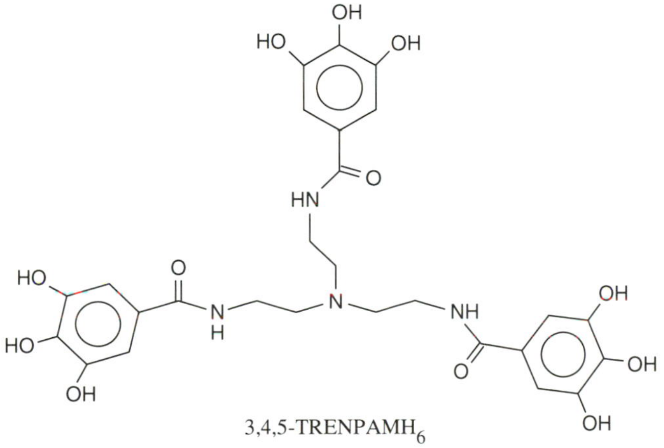

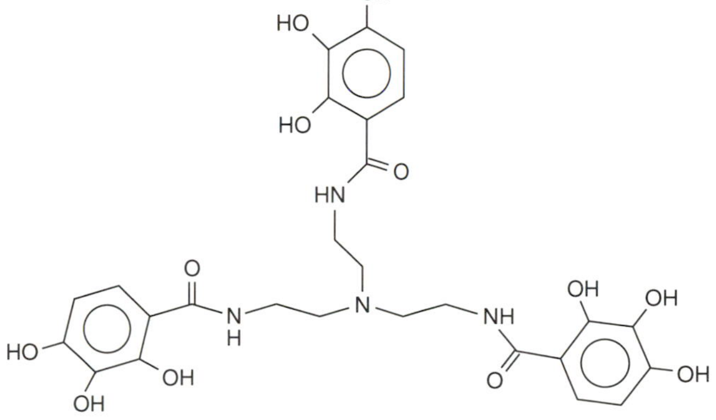

highly reducing vanadium(III) complex of tunichrome must be generated in some other way.[47]

Although a detailed presentation of examples of the known transport properties of essential transition-metal ions into various biological systems could be the subject of a large book, the examples that we have given show how the underlying inorganic chemistry of the elements is used in the biological transport systems that are specific for them. The regulation of metal-ion concentrations, including their specific concentration when necessary from relatively low concentrations of surrounding solution, is probably one of the first biochemical problems that was solved in the course of the evolution of life.

Iron is transported in forms in which it is tightly complexed to small chelators called siderophores (microorganisms) or to proteins called transferrins (animals) or to citrate or mugeneic acid (plants). The problem of how the iron is released in a controlled fashion is largely unresolved. The process of mineral formation, called biomineralization, is a subject of active investigation. Vanadium and molybdenum are transported as stable anions. Zinc and copper appear to be transported loosely associated with peptides or proteins (plants) and possibly mugeneic acid in plants. Much remains to be learned about the biological transport of nonferrous metal ions.

C. Iron Biomineralization

Many structures formed by living organisms are minerals. Examples include apatite [$Ca_2(OH)PO_4$] in bone and teeth, calcite or aragonite ($CaCO_3$) in the shells of marine organisms and in the otoconia (gravity device) of the mammalian ear, silica (SiO_2) in grasses and in the shells of small invertebrates such as radiolara, and iron oxides, such as magnetite (Fe_3O_4) in birds and bacteria (navigational devices) and ferrihydrite FeO(OH) in ferritin of mammals, plants, and bacteria. Biomineralization is the formation of such minerals by the influence of organic macromolecules, e.g., proteins, carbohydrates, and lipids, on the precipitation of amorphous phases, on the initiation of nucleation, on the growth of crystalline phases, and on the volume of the inorganic material.

Iron oxides, as one of the best-studied classes of biominerals containing transition metals, provide good examples for discussion. One of the most remarkable recent characterizations of such processes is the continual deposition of single-crystal ferric oxide in the teeth of chiton.[48] Teeth of chiton form on what is essentially a continually moving belt, in which new teeth are being grown and moved forward to replace mature teeth that have been abraded. However, the study of the mechanisms of biomineralization in general is relatively recent; a great deal of the information currently available, whether about iron in ferritin or about calcium in bone, is somewhat descriptive.

Three different forms of biological iron oxides appear to have distinct relationships to the proteins, lipids, or carbohydrates associated with their formation and with the degree of crystallinity.[49] Magnetite, on the one hand, often forms almost perfect crystals inside lipid vesicles of magneto-bacteria.[50] Ferrihydrite, on the other hand, exists as large single crystals, or collections of small crystals, inside the protein coat of ferritin; however, iron oxides in some ferritins that have large amounts of phosphate are very disordered. Finally, goethite [α-FeO(OH)] and lepidocrocite [γ-FeO(OH)] form as small single crystals in a complex matrix of carbohydrate and protein in the teeth of some shellfish (limpets and chitons); magnetite is also found in the lepidocrocite-containing teeth. The differences in the iron-oxide structures reflect differences in some or all of the following conditions during formation of the mineral: nature of co-precipitating ions, organic substrates or organic boundaries, surface defects, inhibitors, pH, and temperature. Magnetite can form in both lipid and protein/carbohydrate environments, and can sometimes be derived from amorphous or semicrystalline ferrihydrite-like material (ferritin). However, the precise relationship between the structure of the organic phase and that of the inorganic phase has yet to be discovered. When the goal of understanding how the shape and structure of biominerals is achieved, both intellectual satisfaction and practical commercial and medical information will be provided.

Synthetic iron complexes have provided models for two stages of ferritin iron storage and biomineralization:[51-59] (1) the early stages, when small numbers of clustered iron atoms are bound to the ferritin protein coat, and (2) the final stages, where the bulk iron is a mineral with relatively few contacts to the protein coat. In addition, models have begun to be examined for the microenvironment inside the protein coat.[54]

Among the models for the early or nucleation stage of iron-core formation are the binuclear Fe(III) complexes with [$Fe_2O(O_2CR_2)$]$^{2+}$ cores;[55,56] the three other Fe(III) ligands are N. The μ-oxo complexes, which are particularly accurate models for the binuclear iron centers in hemerythrin, purple acid phosphatases, and, possibly, ribonucleotide reductases, may also serve as models for ferritin, since an apparently transient Fe(II)-O-Fe(III) complex was detected during the reconstitution of ferritin from protein coats and Fe(II). The facile exchange of (O_2CR) for (O_2PR) in the binuclear complex is particularly significant as a model for ferritin, because the structure of ferritin cores varies with the phosphate content. An asymmetric trinuclear (Fe$_3$O)$^{7+}$ complex[57] and an (FeO)$_{11}$ complex (Figure 1.21) have been prepared; these appear to serve as models for later stages of core nucleation (or growth).[59]

Models for the full iron core of ferritin include ferrihydrite, which matches the ordered regions of ferritin cores that have little phosphate; however, the site vacancies in the lattice structure of ferrihydrite [FeO(OH)] appear to be more regular than in crystalline regions of ferritin cores. A polynuclear complex of iron and microbial dextran (α-1,4-D-glucose)$_n$ has spectroscopic (Mössbauer, EXAFS) properties very similar to those of mammalian ferritin, presumably because the organic ligands are similar to those of the protein (—OH, —COOH). In contrast, a polynuclear complex of iron and mammalian chondroitin sulfate (α-1,4-[α-1,3-D-glucuronic acid-N-acetyl-D-galactosamine-4-sulfate]$_n$) contains two types of domains: one like mammalian ferritin [FeO(OH)] and one like hematite (α-Fe$_2$O$_3$), which was apparently nucleated by the sulfate, emphasizing

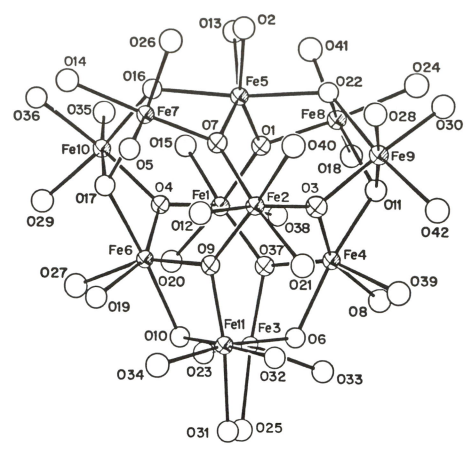

Figure 1.21
The structure of a model for a possible intermediate in the formation of the ferritin iron core.
The complex consists of 11 Fe(III) atoms with internal oxo-bridges and a coat of benzoate li-
gands; the Fe atoms define a twisted, pentacapped trigonal prism. See Reference 53.

the importance of anions in the structure of iron cores.[60] Finally, a model for
iron cores high in phosphate, such as those from bacteria, is Fe-ATP (4:1), in
which the phosphate is distributed throughout the polynuclear iron complex,
providing an average of 1 or 2 of the 6 oxygen ligands for iron.[61]

The microenvironment inside the protein coat of ferritin has recently been
modeled by encapsulating ferrous ion inside phosphatidylcholine vesicles and
studying the oxidation of iron as the pH is raised. The efficacy of such a model
is indicated by the observation of relatively stable mixtures of Fe(II)/Fe(III)
inside the vesicles, as have also been observed in ferritin reconstituted experimentally from protein coats and ferrous ion.[43,54]

Models for iron in ferritin must address both the features of traditional metal-
protein interactions and the bulk properties of materials. Although such modeling may be more difficult than other types of bioinorganic modeling, the difficulties are balanced by the availability of vast amounts of information on Fe-

protein interactions, corrosion, and mineralization. Furthermore, powerful tools such as x-ray absorption, Mössbauer and solid state NMR spectroscopy, scanning electron and proton microscopy, and transmission electron microscopy reduce the number of problems encountered in modeling the ferritin ion core.

Construction of models for biomineralization is clearly an extension of modeling for the bulk phase of iron in ferritin, since the major differences between the iron core of ferritin and that of other iron-biominerals are the size of the final structure, the generally higher degree of crystallinity, and, at this time, the more poorly defined organic phases. A model for magnetite formation has been provided by studying the coulometric reduction of half the Fe^{3+} atoms in the iron core of ferritin itself. Although the conditions for producing magnetite have yet to be discovered, the unexpected observation of retention of the Fe^{2+} by the protein coat has provided lessons for understanding the iron core of ferritin. Phosphatidyl choline vesicles encapsulating Fe^{2+} appear to serve as models for both ferritin and magnetite; only further investigation will allow us to understand the unique features that convert Fe^{2+} to $[FeO(OH)]$, on the one hand, and Fe_3O_4, on the other.

III. SUMMARY

Transition metals (Fe, Cu, Mo, Cr, Co, Mn, V) play key roles in such biological processes as cell division (Fe, Co), respiration (Fe, Cu), nitrogen fixation (Fe, Mo, V), and photosynthesis (Mn, Fe). Zn participates in many hydrolytic reactions and in the control of gene activity by proteins with "zinc fingers." Among transition metals, Fe predominates in terrestial abundance; since Fe is involved in a vast number of biologically important reactions, its storage and transport have been studied extensively. Two types of Fe carriers are known: specific proteins and low-molecular-weight complexes. In higher animals, the transport protein transferrin binds two Fe atoms with high affinity; in microorganisms, iron is transported into cells complexed with catecholates or hydroxamates called siderophores; and in plants, small molecules such as citrate, and possibly plant siderophores, carry Fe. Iron complexes enter cells through complicated paths involving specific membrane sites (receptor proteins). A problem yet to be solved is the form of iron transported in the cell after release from transferrin or siderophores but before incorporation into Fe-proteins.

Iron is stored in the protein ferritin. The protein coat of ferritin is a hollow sphere of 24 polypeptide chains through which Fe^{2+} passes, is oxidized, and mineralizes inside in various forms of hydrated Fe_2O_3. Control of the formation and dissolution of the mineral core by the protein and control of protein synthesis by Fe are subjects of current study.

Biomineralization occurs in the ocean (e.g., Ca in shells, Si in coral reefs) and on land in both plants (e.g., Si in grasses) and animals (e.g., Ca in bone, Fe in ferritin, Fe in magnetic particles). Specific organic surfaces or matrices of protein and/or lipid allow living organisms to produce minerals of defined shape and composition, often in thermodynamically unstable states.

IV. REFERENCES

1. (a) J. H. Martin and R. M. Gordon, *Deep Sea Research* **35** (1988), 177; (b) F. Egami, *J. Mol. Evol.* **4** (1974), 113.
2. J. F. Sullivan *et al., J. Nutr.* **109** (1979), 1432.
3. M. D. McNeely *et al., Clin. Chem.* **17** (1971), 1123.
4. A. R. Byrne and L. Kosta, *Sci. Total Env.* **10** (1978), 17.
5. A. S. Prasad, *Trace Elements and Iron in Human Metabolism*, Plenum Medical Book Company, 1978.
6. E. C. Theil, *Adv. Inorg. Biochem.* **5** (1983), 1.
7. E. C. Theil and P. Aisen, in D. van der Helm, J. Neilands, and G. Winkelmann, eds., *Iron Transport in Microbes, Plants, and Animals*, VCH, 1987, p. 421.
8. C. F. Mills, ed., *Zinc in Human Biology*, Springer-Verlag, 1989.
9. B. L. Vallee and D. S. Auld, *Biochemistry* **29** (1990), 5647.
10. J. Miller, A. D. McLachlan, and A. Klug, *EMBO J.* **4** (1985), 1609.
11. J. M. Berg, *J. Biol. Chem.* **265** (1990), 6513.
12. C. Sennett, L. E. G. Rosenberg, and I. S. Millman, *Annu. Rev. Biochem.* **50** (1981), 1053.
13. J. J. G. Moura *et al.,* in A. V. Xavier, ed., *Frontiers in Biochemistry*, VCH, 1986, p. 1.
14. C. T.Walsh and W. H. Orme-Johnson, *Biochemistry* **26** (1987), 4901.
15. H. Kim and R. J. Maier, *J. Biol. Chem.* **265** (1990), 18729.
16. Reference 5, p. 5.
17. V. L. Schramm and F. C. Wedler, eds., *Manganese in Metabolism and Enzyme Function,* Academic Press, 1986.
18. (a) D. W. Boyd and K. Kustin, *Adv. Inorg. Biochem.* **6** (1984), 312; (b) R. C. Bruening *et al., J. Nat. Products* **49** (1986), 193.
19. T. G. Spiro, ed., *Molybdenum Biochemistry,* Wiley, 1985.
20. L. L. Fox, *Geochim. Cosmochim. Acta* **52** (1988), 771.
21. M. E. Farago, in A. V. Xavier, ed., *Frontiers in Bioinorganic Chemistry*, VCH, 1986, p. 106.
22. I. G. Macara, G. C. McCloud, and K. Kustin, *Biochem. J.* **181** (1979), 457.
23. R. C. Bruening *et al., J. Am. Chem. Soc.* **107** (1985), 5298.
24. E. Bayer and H. H. Kneifel, *Z. Naturforsch.* **27B** (1972), 207.
25. E. Bayer and H. H. Kneifel, in A. V. Xavier, ed., *Frontiers in Bioinorganic Chemistry*, VCH, 1986, p. 98.
26. J. Felcman, J. J. R. Frausto da Silva, and M. M. Candida Vaz, *Inorg. Chim. Acta* **93** (1984), 101.
27. W. Mertz, *Nutr. Rev.* **3** (1975), 129.
28. E. C. Theil, *J. Biol. Chem.* **265** (1990), 4771; *Biofactors* **4** (1993), 87.
29. J. S. Rohrer *et al., Biochemistry* **29** (1990), 259.
30. P. H. Connett and K. Wetterhahn, *Struct. Bonding* **54** (1983), 94.
31. E. C. Theil, *Ann. Rev. Biochem.* **56** (1987), 289; *Adv. Enzymol.* **63** (1990), 421.
32. G. C. Ford *et al., Philos. Trans. Roy. Soc. Lond. B*, **304** (1984), 551.
33. G. D. Watt, R. B. Frankel, and G. C. Papaefthymiou, *Proc. Natl. Acad. Sci. USA* **82** (1985), 3640.
34. (a) J. S. Rohrer *et al., J. Biol. Chem.* **262** (1987), 13385; (b) J. S. Rohrer *et al., Inorg. Chem.* **28** (1989), 3393.
35. D. H. Hamer, *Ann. Rev. Biochem.* **55** (1986), 913.
36. A. H. Robbins, D. E. McRee, M. Williamson, S. A. Collett, N. H. Xuong, W. F. Furey, B. C. Want, and C. D. Stout, *J. Mol. Biol.* **221** (1991), 1269.
37. N. D. Chasteen, *Adv. Inorg. Biochem.* **5** (1983), 201.
38. P. Aisen and I. Listowsky, *Annu. Rev. Biochem.* **49** (1980), 357.
39. B. T. Anderson *et al., Proc. Natl. Acad. Sci. USA* **84** (1987), 1768; E. N. Baker, B. F. Anderson, and H. M. Baker, *Int. J. Biol. Macromol.* **13** (1991), 122.
40. S. Bailey *et al., Biochemistry* **27** (1988), 5804.
41. M. Ragland *et al., J. Biol. Chem.* **263** (1990), 18339.
42. B. F. Matzanke, G. Muller, and K. N. Raymond, in T. M. Loehr, ed., *Iron Carriers and Iron Proteins*, VCH, 1989, pp. 1–121.
43. L. Stryer, *Biochemistry*, Freeman, 1981, pp. 110–116.
44. D. J. Ecker *et al., J. Am. Chem. Soc.* **110** (1988), 2457.
45. Y. Sugiura and K. Nomoto, *Struct. Bonding* **58** (1984), 107.
46. S. Mann, *Struct. Bonding* **54** (1986), 125.

47. A. R. Bulls *et al., J. Am. Chem. Soc.* **112** (1990), 2627.
48. J. Webb, in P. Westbroek and E. W. de Jong, eds., *Biomineralization and Biological Metal Accumulation*, Reidel, 1983, pp. 413–422.
49. K. Kustin *et al., Struct. Bonding* **53** (1983), 139.
50. R. B. Frankel and R. P. Blakemore, *Philos. Trans. Roy. Soc. Lond. B* **304** (1984), 567.
51. S. J. Lippard, *Angew. Chemie* **22** (1988), 344.
52. K. E. Wieghardt, *Angew. Chemie* **28** (1989), 1153.
53. E. C. Theil, in R. B. Frankel, ed., *Iron Biomineralization*, Plenum Press, 1990.
54. S. Mann, J. P. Harrington, and R. J. P. Williams, *Nature* **234** (1986), 565.
55. W. H. Armstrong and S. J. Lippard, *J. Am. Chem. Soc.* **107** (1985), 3730.
56. L. Que, Jr. and R. C. Scarrow, in L. Que, ed., *Metal Clusters in Proteins*, ACS Symposium Series 372, American Chemical Society, Washington, DC, 1988, p. 152 and references therein.
57. S. M. Gorun and S. J. Lippard, *J. Am. Chem. Soc.* **107** (1985), 4570.
58. S. M. Gorun *et al., J. Am. Chem. Soc.* **109** (1987), 3337.
59. Q. Islam *et al., J. Inorg. Biochem.* **36** (1989), 51.
60. C.-Y. Yang *et al., J. Inorg. Biochem.* **28** (1986), 393.
61. A. N. Mansour *et al., J. Biol. Chem.* **260** (1985), 7975.
62. D. C. Harris and P. Aisen, in T. M. Loehr, ed., *Iron Carriers and Iron Proteins*, VCH, 1989, pp. 239–352.
63. P. M. Harrison and T. M. Lilley, in Reference 62, pp. 123–238.
64. G. S. Waldo *et al. Science* **259** (1993), 796.
65. J. Trikha, G. S. Waldo, F. A. Lewandowski, Y. Ha, E. C. Theil, P. C. Weber, and N. M. Allewell, *Protein* **18** (1994), issue #2, in press.

These references contain general reviews of the subjects indicated:

Chromium: 27, 30
Cobalt: 12
Copper: 8
Iron
   Biochemistry; 7, 31, 37, 42
   Biomineralization polynuclear models: 6, 42, 56, 57, 58
   Siderophores: 42
   Structure of storage and transport proteins: 32, 62, 63
Manganese: 17
Molybdenum: 19
Nickel: 13, 14
Vanadium: 18
Zinc: 8, 9, 11, 35

2

The Reaction Pathways of Zinc Enzymes and Related Biological Catalysts

IVANO BERTINI
Department of Chemistry
University of Florence

CLAUDIO LUCHINAT
Institute of Agricultural Chemistry
University of Bologna

I. INTRODUCTION

This chapter deals with metalloenzymes wherein the metal acts mainly as a Lewis acid; i.e., the metal does not change its oxidation state nor, generally, its protein ligands. Changes in the coordination sphere may occur on the side exposed to solvent.

The substrate interacts with protein residues inside the active cavity and/or with the metal ion in order to be activated, so that the reaction can occur. Under these circumstances the catalyzed reactions involve, as central steps with often complex reaction pathways, the following bond-breaking and/or formation processes:

peptide hydrolysis

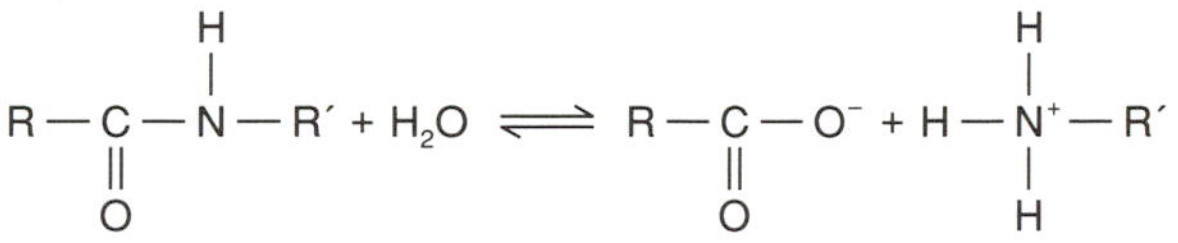

$$R - \underset{\underset{O}{\parallel}}{\underset{|}{C}} - \underset{\overset{H}{|}}{N} - R' + H_2O \rightleftharpoons R - \underset{\underset{O}{\parallel}}{C} - O^- + H - \underset{\overset{H}{|}}{\underset{\underset{H}{|}}{N^+}} - R' \qquad (2.1)$$

carboxylic ester hydrolysis

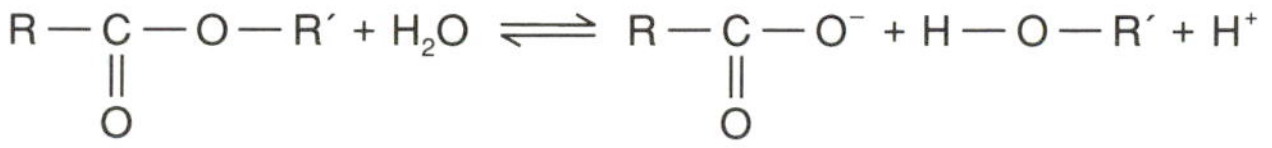

$$R - \underset{\underset{O}{\parallel}}{C} - O - R' + H_2O \rightleftharpoons R - \underset{\underset{O}{\parallel}}{C} - O^- + H - O - R' + H^+ \qquad (2.2)$$

37

phosphoric ester hydrolysis

$$H-O-\overset{\overset{O^-}{|}}{\underset{\underset{O}{\|}}{P}}-O-R' + H_2O \rightleftharpoons H-O-\overset{\overset{O^-}{|}}{\underset{\underset{O}{\|}}{P}}-O^- + H-O-R' + H^+ \qquad (2.3)$$

nucleophilic addition of OH^- and H^-

$$H-O^- \longrightarrow \overset{\overset{O}{\|}}{\underset{\underset{O}{\|}}{C}} \rightleftharpoons \overset{H}{\underset{O}{\diagdown}}O-C\overset{O^-}{\diagdown O} \quad ; O=\overset{\overset{R}{|}}{\underset{\underset{R'}{|}}{C}} \longleftarrow H^- \rightleftharpoons {}^-O-\overset{\overset{R}{|}}{\underset{\underset{R'}{|}}{C}}-H \qquad (2.4)$$

Scheme (2.3) also pertains to the reactions which need ATP hydrolysis to promote endoenergetic reactions.

We will also briefly deal with coenzyme B_{12}; this is a cobalt(III) complex that, by interacting with a number of proteins, produces an $R-CH_2$ radical by homolytic breaking of the $Co-C$ bond as follows:

$$Co(III)-\overset{\overset{H}{|}}{\underset{\underset{H}{|}}{C}}-R \longrightarrow Co(II) + {}^{\cdot}\overset{\overset{H}{|}}{\underset{\underset{H}{|}}{C}}-R \qquad (2.5)$$

After an $R-CH_2$ radical is formed, it initiates a radical reaction. This is the only system we treat in which the oxidation state changes.

II. THE NATURAL CATALYSTS

Table 2.1 lists metalloenzymes that catalyze hydrolytic and related reactions. According to the above guidelines the hydrolysis of peptide bonds is catalyzed by enzymes called peptidases that belong to the class of hydrolases (according to the official enzyme classification). Two peptidases (carboxypeptidase and thermolysin) are known in great detail, because their structures have been elucidated by high-resolution x-ray crystallography. They share many features; e.g., their metal ions coordinate to the same kind of protein residues. A discussion of the possible mechanism of carboxypeptidase A will be given in Section V.A. Metallopeptidases are zinc enzymes: generally they are single polypeptide chains with molecular weights in the range 30 to 40 kDa. Metallohydrolases of carboxylic and phosphoric esters are also often zinc enzymes. Alkaline phosphatase will be described in Section V.B as a representative of this class. Magnesium is sometimes involved in hydrolytic reactions. This is common when phosphate groups are involved, probably because the affinity of Mg^{2+} for phosphate groups is high.[1] However, hydrolytic reactions can be performed by other systems (not treated here) like urease, which contains nickel(II),[2] or acid phosphatase, which contains two iron ions,[3] or aconitase, which contains an Fe_4S_4 cluster.[4]

Table 2.1
Representative metalloenzymes catalyzing hydrolytic and related reactions.

Enzyme	Metal(s)	Function
Carboxypeptidase	Zn^{2+}	Hydrolysis of C-terminal peptide residues
Leucine aminopeptidases	Zn^{2+}	Hydrolysis of leucine N-terminal peptide residues
Dipeptidase	Zn^{2+}	Hydrolysis of dipeptides
Neutral protease	Zn^{2+}, Ca^{2+}	Hydrolysis of peptides
Collagenase	Zn^{2+}	Hydrolysis of collagen
Phospholipase C	Zn^{2+}	Hydrolysis of phospholipids
β-Lactamase II	Zn^{2+}	Hydrolysis of β-lactam ring
Thermolysin	Zn^{2+}, Ca^{2+}	Hydrolysis of peptides
Alkaline phosphatase	Zn^{2+}, Mg^{2+}	Hydrolysis of phosphate esters
Carbonic anhydrase	Zn^{2+}	Hydration of CO_2
α-Amylase	Ca^{2+}, Zn^{2+}	Hydrolysis of glucosides
Phospholipase A_2	Ca^{2+}	Hydrolysis of phospholipids
Inorganic pyrophosphatase	Mg^{2+}	Hydrolysis of pyrophosphate
ATPase	Mg^{2+}	Hydrolysis of ATP
Na^+-K^+-ATPase	Na^+, K^+	Hydrolysis of ATP with transport of cations
Mg^{2+}-Ca^{2+}-ATPase	Mg^{2+}, Ca^{2+}	
Phosphatases	Mg^{2+}, Zn^{2+}	Hydrolysis of phosphate esters
Creatine kinase	M^{2+}	Phosphorylation of creatine
Pyruvate kinase	M^+, M^{2+}	Dephosphorylation of phosphoenolpyruvate
Phosphoglucomutase	Mg^{2+}	Phosphate transfer converting glucose-1-phosphate to glucose-6-phosphate
DNA polymerase	Mg^{2+} (Mn^{2+})	Polymerization of DNA with formation of phosphate esters
Alcohol dehydrogenase	Zn^{2+}	Hydride transfer from alcohols to NAD$^+$

Examples of enzymes catalyzing nucleophilic addition of OH$^-$ (other than hydrolysis) and H$^-$ are carbonic anhydrase and alcohol dehydrogenase. Both are zinc enzymes. In the official biochemical classification of enzymes, carbonic anhydrase belongs to the class of lyases. Lyases are enzymes that cleave C—C, C—O, C—N, or other bonds by elimination, leaving double bonds, or conversely add groups to double bonds. Carbonic anhydrase has a molecular weight around 30 kDa, and is among the most-studied metalloenzymes. It catalyzes the deceivingly simple CO_2 hydration reaction. The subtleties of its biological function, unraveled by a combination of techniques, make it an ideal example for bioinorganic chemistry. Section IV is fully dedicated to this enzyme. Alcohol dehydrogenase is a 90-kDa enzyme that catalyzes the reversible transfer of a hydride ion from alcohols to NAD$^+$. Although it is a redox enzyme (in fact, classified as an oxidoreductase) and not a hydrolytic one, it will illustrate a different use that Nature makes of zinc to catalyze nucleophilic attack at carbon (Section V.C).

Finally, the enzymatic transfer of organic radicals by enzymes involving coenzyme B_{12} will be briefly considered.

III. STRATEGIES FOR THE INVESTIGATION OF ZINC ENZYMES

A. Why Zinc?

Zinc has a specific role in bioinorganic processes because of the peculiar properties of the coordination compounds of the zinc(II) ion.

(1) Zinc(II) can easily be four-, five-, or six-coordinate, without a marked preference for six coordination. The electronic configuration of zinc(II) is $3d^{10}$ with two electrons per orbital. In coordination compounds, there is no ligand-field stabilization energy, and the coordination number is determined by a balance between bonding energies and repulsions among the ligands. Tetrahedral four-coordinate complexes have shorter metal-donor distances than five-coordinate complexes, and the latter have shorter ones than six-coordinate complexes (Table 2.2), whereas the ligand repulsion increases in the same order. The re-

Table 2.2
Average zinc(II)-donor atom distances (Å) for some common zinc(II) ligands in four-, five-, and six-coordinate complexes.[5]

	Coordination number		
Ligand	4	5	6
H_2O	2.00	2.08	2.10
$R-COO^-$	1.95	2.02	2.07
Imidazole	2.02	—	2.08
Pyridine	2.06	2.12	2.11
$R-NH_2$	—	2.06	2.15
$R, R'NH$	2.19	2.27	—

pulsion can be both steric and electronic. In enzymes, zinc(II) usually has coordination numbers smaller than six, so that they have available binding sites in their coordination spheres. Substrate can in principle bind to zinc by substituting for a coordinated water or by increasing the coordination number. This behavior would be typical of Lewis acids, and, indeed, zinc is the most common Lewis acid in bioinorganic chemistry. Zinc could thus substitute for protons in the task of polarizing a substrate bond, e.g., the carbonyl C—O bond of peptides and esters, by accepting a substrate atom (oxygen) as a ligand. This has been shown to be possible in model systems. Relative to the proton, a metal ion with an available coordination position has the advantage of being a "superacid,"[6] in the sense that it can exist at pH values where the H_3O^+ concentration is extremely low. Also, relative to the proton, the double positive charge partly compensates for the smaller electrophilicity due to the smaller charge density.

(2) As a catalyst, zinc in zinc enzymes is exposed to solvent, which for enzymes is almost always water. A coordinated water molecule exchanges rap-

idly, because ligands in zinc complexes are kinetically labile. This, again, can be accounted for by zinc's lack of preference for a given coordination number. A six-coordinate complex can experience ligand dissociation, giving rise to a five-coordinate complex with little energy loss and then little energetic barrier. On the other side, four-coordinate complexes can add a fifth ligand with little energetic barrier and then another ligand dissociates.[7] The coordinated water has a pK_a sizably lower than free water. Suitable models have been synthesized and characterized in which a solvent water molecule coordinated to various dipositive metal ions has pK_a values as low as 7 (Table 2.3). This is the result of the formation of the coordination bond. The oxygen atom donates two electrons to

Table 2.3
The pK_a values of coordinated water in some metal complexes.

Complex	Note	Donor set	pK_a	Reference
$Ca(NO_3)_2(OH_2)_4{}^{2+}$		O_6	10.3	8
$Cr(OH_2)_6{}^{3+}$		O_6	4.2	8
$Cr(NH_3)_5OH_2{}^{3+}$		N_5O	5.1	8
$Mn(OH_2)_6{}^{2+}$		O_6	10.5	8
$Fe(OH_2)_6{}^{3+}$		O_6	1.4	8
$Co(OH_2)_6{}^{2+}$		O_6	9.8	8
$Co(dacoda)OH_2{}^{2+}$	(a)	N_2O_3	9.4	9
$Co(TPyMA)OH_2{}^{2+}$	(b)	N_4O	9.0	10
$Co(TMC)OH_2{}^{2+}$	(c)	N_4O	8.4	11
$Co(CR)OH_2{}^{2+}$	(d)	N_4O	8.0	12
$Co(NH_3)_5OH_2{}^{3+}$		N_5O	6.2	8
$Ni(OH_2)_6{}^{2+}$		O_6	10.0	8
$Cu(OH_2)_6{}^{2+}$		O_6	7.3	8
$Cu(DMAM\text{-}PMHD)OH_2^{2+}$	(e)	N_3O_2	7.1	13
$Cu(C\text{-}PMHD)OH_2^{+}$	(f)	N_2O_3	6.6	14
$Zn(OH_2)_6{}^{2+}$		O_6	9.0	8
$Zn(DMAM\text{-}PMHD)OH_2^{2+}$	(e)	N_3O	9.2	13
$Zn(C\text{-}PMHD)OH_2^{+}$	(f)	N_3O_2	7.1	14
$Zn(CR)OH_2{}^{2+}$	(d)	N_4O	8.7	12
$Zn([12]aneN_3)OH_2^{2+}$	(g)	N_3O	7.3	15,16
$Zn(HP[12]aneN_3)OH_2^{+}$	(h)	N_3O_2	10.7	16
$Zn(TImMP)OH_2^{2+}$	(i)	N_3O	<7	17
$Co(TImMP)OH_2^{2+}$	(i)	N_3O	7.8	17

(a) dacoda = 1,4-diaza-cyclooctane-1,4-diacetate. (b) TPyMA = tris(3,5-dimethyl-1-pyrazolylmethyl)amine. (c) TMC = 1,4,8,11-tetramethyl-1,4,8,11-tetraaza-cyclotetradecane. (d) CR = Schiff base between 2,6-diacetylpyridine and bis(3-aminopropyl)amine. (e) DMAM-PMHD = 1-[(6(dimethylamino)methyl)-2-pyridyl)methyl]hexahydro-1,4-diazepin-5-one. (f) C-PMHD = 1[(6-carboxy)-2-pyridyl)methyl]hexa-hydro-1,4-diazepin-5-one. (g) [12]aneN$_3$ = 1,5,9-triaza-cyclododecane. (h) HP[12]aneN$_3$ = 2-(2-hydroxyphenylate)-1,5,9-triaza-cyclododecane. (i)TImMP = tris(4,5-dimethyl-2-imidazolylmethyl)phosphinoxide.

the metal ion and formally becomes positively charged:

$$\begin{array}{c} H \\ \diagdown \\ \overline{O}| + M^{2+} \\ \diagup \\ H \end{array} \rightleftharpoons \begin{array}{c} N \\ \diagdown \\ O^+ \diagdown {}^{M^+} \\ \diagup \\ N \end{array}$$

Under these conditions a proton is easily released. The nucleophilicity of coordinated water is, of course, decreased with respect to free water, owing to the decreased electronic charge on the oxygen atom, but a significant concentration of M—OH species may exist in neutral solution. In turn, the coordinated hydroxide is a slightly poorer nucleophile than the free OH^- ion, but better than water. On the basis of recent MO calculations,[18] the order of nucleophilicity for solvent-derived species can be summarized as follows:

$$H_3O^+ < H_2O—M^{2+} \simeq H_2O < HO—M^+ \simeq HO^-$$

Therefore, at neutral or slightly alkaline pH, the small decrease in efficiency of coordinated vs. free hydroxide ions is more than compensated for by the higher concentration of reactive species available (i.e., $HO—M^+$ vs. HO^-). Another common role for zinc enzymes is thus to provide a binding site at which the substrate can be attacked by the metal-coordinated hydroxide:

$$\begin{array}{c} Zn \diagdown \overline{O}| \longrightarrow Substrate \\ \diagdown \\ H \end{array}$$

The pK_a of coordinated water in zinc complexes is controlled by the coordination number and by the total charge of the complex, in the sense that it decreases with decreasing coordination number and with increasing positive charge, because a zinc ion, bearing in effect a more positive charge, will have greater attraction for the oxygen lone pair, thus lowering the pK_a. Charged ligands affect water pK_a's more than does the number of ligands.[18] The pK_a in metalloproteins is further controlled by the presence of charged groups from protein side chains inside the cavity or by the binding of charged cofactors. The coordinated water may have a pK_a as low as 6, as in carbonic anhydrase (see later). On the other hand, the pK_a of the coordinated water is 7.6 in liver alcohol dehydrogenase (LADH) when NAD^+ is bound, 9.2 in the coenzyme-free enzyme, and 11.2 in the presence of NADH (see Section V.C).

(3) As mentioned before, Zn complexes show facile four- to five-coordinate interconversion. The low barrier between these coordination geometries is quite important, because the substrate may add to the coordination sphere in order to replace the solvent or to be coordinated together with the solvent. If the interconversion between four- and five-coordination is fast, catalysis is also fast.

Thus, to summarize, zinc is a good Lewis acid, especially in complexes with lower coordination numbers; it lowers the pK_a of coordinated water and is kinetically labile, and the interconversion among its four-, five-, and six-coordinate states is fast. All of these properties make zinc quite suitable for biological catalysis.[19]

1. The groups to which zinc(II) is bound

Zinc(II) is an ion of borderline hardness and displays high affinity for nitrogen and oxygen donor atoms as well as for sulfur. It is therefore found to be bound to histidines, glutamates or aspartates, and cysteines. When zinc has a catalytic role, it is exposed to solvent, and generally one water molecule completes the coordination, in which case the dominating ligands are histidines. It has been noted[20] recently that coordinated histidines are often hydrogen-bonded to carboxylates:

It is possible that the increase in free energy for the situation in which the hydrogen is covalently bound to the carboxylate oxygen and H-bonded to the histidine nitrogen is not large compared to $k_B T$. Under these circumstances the protein could determine the degree of imidazolate character of the ligand and therefore affect the charge on the metal.

The binding of zinc(II) (like that of other metal ions) is often determined by entropic factors. Water molecules are released when zinc(II) enters its binding position, thus providing a large entropy increase. Most commonly zinc is bound to three or four protein ligands. Large entropy increases are not observed, however, when zinc(II) binds to small polypeptides like the recently discovered zinc fingers, for here the binding site is not preformed (see Section III.B), and zinc(II) must be present for the protein to fold properly into the biologically active conformation.

2. The reactivity of zinc(II) in cavities

In the preceding section we discussed the properties of zinc(II) as an ion. These properties are, of course, important in understanding its role in biological catalysis, but it would be too simplistic to believe that reactivity can be understood solely on this basis. Catalysis occurs in cavities whose surfaces are constituted by protein residues. Catalytic zinc is bound to a water molecule, which often is H-bonded to other residues in the cavity and/or to other water molecules. The structure of the water molecules in the cavity cannot be the same as the structure of bulk water. Furthermore, the substrate interacts with the cavity residues through either hydrophilic (H-bonds or electric charges) or hydrophobic (London dispersive forces) interactions. As a result, the overall thermodynamics of the reaction pathway is quite different from that expected in bulk solutions. Examples of the importance of the above interactions will be given in this chapter.

3. The investigation of zinc enzymes

Direct spectroscopic investigation of zinc enzymes is difficult, because zinc(II) is colorless and diamagnetic; so it cannot be studied by means of electronic or

EPR spectroscopy. Its NMR-active isotope, ^{67}Zn, the natural abundance of which is 4.11 percent, has a small magnetic moment, and cannot (with present techniques) be examined by means of NMR spectroscopy at concentrations as low as 10^{-3} M. The enzymes could be reconstituted with ^{67}Zn. However, ^{67}Zn has a nuclear quadrupolar moment, which provides efficient relaxation times, especially in slow-rotating proteins and low-symmetry chromophores, making the line very broad.[210] Of course, ^{1}H NMR can be useful for the investigation of the native enzymes. However, often the molecular weight is such that the proteins are too large for full signal assignment given the current state of the art. At the moment the major source of information comes from x-ray data. Once the structure is resolved, it is possible to obtain reliable structural information on various derivatives by the so-called Fourier difference map. The new structure is obtained by comparing the Fourier maps of the native and of the derivative under investigation. Many x-ray data are now available on carboxypeptidase (Section V.A) and alcohol dehydrogenase (Section V.C).

The zinc ion can be replaced by other ions, and sometimes the enzymatic activity is retained fully or partially (Table 2.4). These new systems have attracted the interest of researchers who want to learn about the role of the metal and of the residues in the cavity, and to characterize the new systems per se.

Table 2.4
Representative metal-substituted zinc enzymes. Percent activities with respect to the native zinc enzyme in parentheses.[a]

Enzyme	Substituted metals
Alcohol dehydrogenase	Co(II)(70), Cu(II)(1), Cu(I)(8), Cd(II)(30), Ni(II)(12)
Superoxide dismutase	Co(II)(90), Hg(II)(90), Cd(II)(70), Cu(II)(100)
Aspartate transcarbamylase	Cd(II)(100), Mn(II)(100), Ni(II)(100)
Transcarboxylase	Co(II)(100), Cu(II)(0)
RNA polymerase	Co(II)(100)
Carboxypeptidase A	Mn(II)(30), Fe(II)(30), Co(II)(200), Ni(II)(50), Cu(II)(0),[b] Cd(II)(5), Hg(II)(0), Co(III)(0), Rh(II)(0), Pb(II)(0)
Thermolysin	Co(II)(200), Mn(II)(10), Fe(II)(60), Mg(II)(2), Cr(II)(2), Ni(II)(2), Cu(II)(2), Mo(II)(2), Pb(II)(2), Cd(II)(2), Nd(III)(2), Pr(III)(2)
Alkaline phosphatase	Co(II)(30), Cd(II)(1), Mn(II)(1), Ni(II)(0), Cu(II)(0), Hg(II)(0)
β-Lactamase II	Mn(II)(3), Co(II)(11), Ni(II)(0), Cu(II)(0), Cd(II)(11), Hg(II)(4)
Carbonic anhydrase[c]	Cd(II)(2), Hg(II)(0), Cu(II)(0), Ni(II)(2), Co(II)(50), Co(III)(0), Mn(II)(18), V(IV)O^{2+}(0)
Aldolase	Mn(II)(15), Fe(II)(67), Co(II)(85), Ni(II)(11), Cu(II)(0), Cd(II)(0), Hg(II)(0)
Pyruvate carboxylase	Co(II)(100)
Glyoxalase	Mg(II)(50), Mn(II)(50), Co(II)(50)

[a] Taken from Reference 21.
[b] Recent data indicate nonnegligible catalytic activity.[22]
[c] BCA II, except the value for Cd(II) obtained with HCA II.

Spectroscopic techniques can be appropriate for the new metal ions; so it is possible to quickly monitor properties of the new derivative that may be relevant for the investigation of the zinc enzyme.

B. Metal Substitution

With zinc enzymes, metallosubstitution is a convenient tool for monitoring the protein and its function by means of spectroscopic techniques. Furthermore, it is interesting to learn how reactivity depends on the nature of the metal ion and its coordination properties, because much of it depends on the protein structure, which seemingly remains constant. As discussed, zinc enzymes can be studied by replacing zinc with other spectroscopically useful metal ions, whose activities have been checked, and by transferring the information obtained to the native enzyme. The strategy of metal substitution is not limited to zinc enzymes, since it has been used for magnesium-activated enzymes and, occasionally, other metalloenzymes as well.

By dialyzing a protein solution against chelating agents, such as EDTA, 1,10-phenanthroline, or 2,6-dipicolinic acid at moderately acidic pH, or by reversibly unfolding the protein with denaturing agents (as has been done with alkaline phosphatase), one can cause zinc proteins to release their metal ions, giving rise to the corresponding but inactive apoprotein. Sometimes (e.g., by using alcohol dehydrogenase) dialysis against chelating agents can be applied to a suspension of protein microcrystals. In this way the chelating agent is still able to reach and remove the active site metal by slowly diffusing in the crystals through the hydration water, while the apoprotein is maintained in the native conformation by the crystal packing forces and denaturation is avoided. After the chelating agent is dialyzed out, often against a high-salt (e.g., ClO_4^-) buffer to reduce nonspecific binding, a new metalloprotein can obtained by addition of the appropriate metal salt.[23]

Cobalt(II)-substituted zinc proteins often show about as much activity as the native zinc enzymes (Table 2.4). This is a general characteristic of the cobalt-substituted zinc enzymes,[24] since the coordination chemistry of cobalt(II) is very similar to that of zinc(II). The two ions also show virtually identical ionic radii. Cobalt(II) derivatives generally display useful electronic spectra. High-spin cobalt(II) ions are paramagnetic, containing three unpaired electrons ($S = \frac{3}{2}$); thus they can also give rise to EPR spectra. The electronic relaxation times, i.e., the average lifetimes of the unpaired electrons in a given spin state of the S manifold ($-\frac{3}{2}, -\frac{1}{2}, \frac{1}{2}, \frac{3}{2}$), are very short ($10^{-11}$ to 10^{-12} s) at room temperature. In order to detect EPR spectra, the sample temperature is usually decreased, often down to liquid helium temperature, to increase the electronic relaxation times and sharpen the EPR linewidths. On the other hand, as the paramagnetic broadening of the NMR lines in such systems is inversely proportional to the electronic relaxation times (see Section IV.C.3), room-temperature ^1H NMR spectra of cobalt(II) complexes can be easily detected, even in the absence of chemical exchange. Therefore cobalt(II) is an exceptional probe to monitor the structure and reactivity of zinc enzymes. Of course, the transfer of information from the artificial to the native enzyme must be done with caution. However, if we can understand the functioning of the cobalt enzyme, we then have a reference frame by which to understand the kinetic properties of the native enzyme. The spectroscopic properties of cobalt(II) in cobalt-substituted proteins have been reviewed.[25]

Copper(II)-substituted zinc proteins are generally inactive with respect to the natural and most artificial substrates (Table 2.4). In model compounds copper(II) is often principally four-coordinate, with at most two more ligands present at metal-ligand distances that are longer than normal coordination bonds. As a consequence, the ability of zinc to switch between four- and five-coordinate species without any appreciable barrier and with usual metal-donor distances is not mimicked by copper. Furthermore, binding at the four principal coordination positions is generally stronger for copper than for zinc. It follows that substrates may have slow detachment kinetics. These properties are unfavorable for catalysis.

Copper(II) can be easily and meaningfully studied by means of electronic spectroscopy. Moreover, the EPR spectra can be recorded even at room temperature because of the long electronic relaxation times, which are of the order of 10^{-9} s. Because a protein is a macromolecule, it rotates slowly, and the EPR spectra in solution at room temperature look like those of crystalline powders or frozen solutions (powder-like spectra). ENDOR spectra are also easily obtained for copper proteins at low temperatures, because at low temperature the electronic relaxation times are even longer, and saturation of the EPR lines (which is a requirement to obtain ENDOR spectra) is easy to accomplish. The long electronic relaxation times make the broadening effects on the NMR lines of nuclei sensing the metal ion too severe; so these lines, unlike those of cobalt(II) complexes, generally escape detection. However, if the nucleus under investigation is in fast exchange between a free species in large excess and a bound species, the line may be observed, because the broadening effects are scaled down by a factor equal to the molar fraction of bound species. The nuclear relaxation parameters contain precious structural and/or dynamic information (see Section IV.C.3). The spectroscopic properties of copper(II) in proteins have been extensively reviewed.[26,27]

Cadmium-substituted zinc proteins may also be active (Table 2.4), although usually at higher pH. This observation is readily explained in terms of the pK_a of a coordinated water, which is expected to be higher than that of analogous zinc complexes because the cadmium ion is larger and polarizes the $Cd—OH_2$ bond less.

^{113}Cd and ^{111}Cd are nuclei with relatively high sensitivity for NMR spectroscopic study. The ^{113}Cd chemical shift spans from -200 to 800 ppm relative to $CdSO_4$ in H_2O, depending on the number and nature of donor atoms.[24,28] Sulfur donor atoms cause larger downfield shifts than oxygens or nitrogens, and the downfield shift increases with decreasing number of donor atoms. Therefore, ^{113}Cd NMR probes have been used extensively to study zinc enzymes, metal-storage proteins like thioneins, and other proteins with cysteine ligands, and chemical shifts in various cadmium proteins, together with the proposed ligand donor set, have been obtained (Figure 2.1).

Manganese(II)-containing proteins give rise to detectable EPR signals; however, their interpretation in terms of structure and dynamics is not always informative. The electronic relaxation times of Mn^{2+} are the longest among metal ions, of the order of 10^{-8} s at room temperature and at the magnetic fields of interest. This property and the large $S = \frac{5}{2}$ value account for a large NMR linewidth, even larger than in copper(II) systems. Manganese(II)- and nickel(II)-substituted zinc proteins have often been reported to have fractional activity (Table 2.4).[24] Several efforts have been devoted to Mn(II) derivatives, especially by studying the NMR signals of nuclei in molecules that exchange rapidly with the metalloprotein.

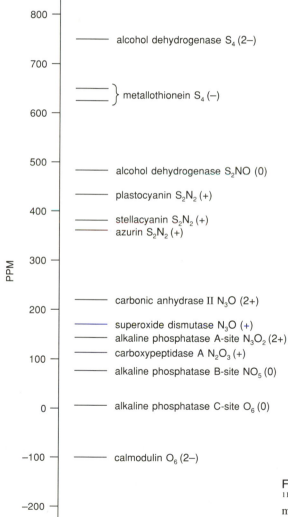

Figure 2.1
^{113}Cd chemical shifts in cadmium(II)-substituted metalloproteins.[24,28] Donor sets and overall charges (in parentheses) of the complexes are indicated.

Finally, several other metal-substituted zinc metalloprotein derivatives have been prepared, including those of VO^{2+}, Fe(II), Co(III), Pt(II), and $HgCl_2$. Although these systems add little directly to our understanding of the relationship between structure and function of the enzymes, nonetheless they represent new bioinorganic compounds and are of interest in themselves, or can add information on the coordinating capabilities, and reactivity in general, of the residues present in the active cavity.

Under the heading *zinc enzymes* there are several enzymes in which zinc is essential for the biological function, but is not present in the catalytic site. Among the most-studied enzymes, zinc has a structural role in superoxide dismutase, where the ligands are three histidines and one aspartate. In alcohol dehydrogenase there is a zinc ion that has a structural role, besides the catalytically active one. The former zinc has four cysteine ligands. Cysteine ligands are also present in zinc thioneins, which are zinc-storage proteins. The recently discovered class of genetic factors containing ''zinc fingers'' are zinc proteins in which the metal has an essentially structural role.[29] Such a role may consist of lowering the folding enthalpy of a protein to induce an active conformation or to stabilize a particular quaternary structure.

Zinc may also have a regulatory role; i.e., it does not participate in the various catalytic steps, but its presence increases the catalytic rate. This is a rather loose but common definition. Typically, zinc in the B site of alkaline phosphatase (Section V.B) has such a role, and the ligands are histidines, aspartates, and water molecules.

The enzymes in which zinc plays a structural or regulatory role will not be further discussed here, because they do not participate in the catalytic mechanisms; see the broader review articles.[23,29,30] Rather, we will describe in some detail the enzyme carbonic anhydrase, in order to show how researchers have investigated such complicated systems as enzymes. We will discover as we look at the details of the structures and mechanisms of enzymes that there are large differences between reactivities in solution and in enzymatic cavities. The fundamental properties underlying these differences are still not fully understood.

IV. ELUCIDATION OF STRUCTURE–FUNCTION RELATIONSHIPS: CARBONIC ANHYDRASE AS AN EXAMPLE

A. About the Enzyme

Carbon-dioxide hydration and its mechanism in living systems are of fundamental importance for bioinorganic chemistry. In 1932 the existence of an enzyme catalyzing CO_2 hydration in red blood cells was established.[31] The enzyme was named carbonic anhydrase (abbreviated CA). In 1939 the enzyme was recognized to contain zinc.[32] Because CO_2 is either the starting point for photosynthesis or the endpoint of substrate oxidation, carbonic anhydrases are now known to be ubiquitous, occurring in animals, plants, and several bacteria. Different enzymes from different sources, catalyzing the same reaction and usually having homologous structures, are termed isoenzymes. Sometimes the same organism has more than one isoenzyme for a particular function, as is true for human carbonic anhydrase.

CO_2 gas is relatively soluble in water (3×10^{-2} M at room temperature under $p_{CO_2} = 1$ atm), equilibrating with hydrogen carbonate at pK_a 6.1:

$$CO_2 + H_2O \rightleftharpoons HCO_3^- + H^+ \qquad (2.6)$$

The uncatalyzed reaction is kinetically slow around physiological pH ($k \simeq 10^{-1}$ s^{-1}), whereas, in the presence of the most efficient isoenzyme of CA, the maximal CO_2 turnover number (i.e., the number of substrate molecules transformed per unit time by each molecule of enzyme)[33] is $\simeq 10^6$ s^{-1}. The uncatalyzed attack by water on CO_2 may be facilitated by two hydrogen-bonded water molecules, one of which activates the carbon by means of a hydrogen bond to a terminal CO_2 oxygen, the other of which binds the carbon atom via oxygen:[34,35]

$$(2.7)$$

Only above pH 9 does the uncatalyzed reaction become fast, owing to direct attack of OH$^-$, which is a much better nucleophile than H_2O ($k \simeq 10^4$ M^{-1}s^{-1}, where M^{-1} refers to the OH$^-$ concentration):

$$CO_2 + OH^- \rightleftharpoons HCO_3^- \qquad (2.8)$$

On the other hand, the rate constant in the presence of the enzyme, called k_{cat}, is pH-independent above pH 8 in every CA isoenzyme (Figure 2.2).[33,36]

In vitro, carbonic anhydrase is quite versatile, catalyzing several reactions that involve both OH$^-$ and H$^+$, such as the hydrolysis of esters and the hydration of aldehydes. The various isoenzymes have been characterized to different degrees of sophistication. High-activity forms are labeled II ($k_{cat} \simeq 10^6$ s^{-1} at 25°C); low-activity forms I ($k_{cat} \simeq 10^5$ s^{-1}), and the very-low-activity forms III ($k_{cat} \simeq 10^3$ s^{-1}).[37] X-ray structural information at nominal 2 Å resolution is available for HCA I[38] and HCA II,[39] where H indicates human. The structure of HCA II has been refined recently.[40] High-resolution structures of mutants and of their substrate and inhibitor derivatives are being reported.[211] All isoenzymes are single-chain polypeptides, with M.W. about 30 kDa and one zinc ion per molecule. They have the shape of a rugby ball with a crevice 16 Å deep running through the south pole (Figure 2.3 *See color plate section, page C-2.*). At the bottom of the crevice, the zinc ion is anchored to the protein by three histidine nitrogen atoms and is exposed to solvent. Two histidines (His-94 and His-96, HCA I numbering) are bound to zinc via their Nε2 atoms, whereas one (His-

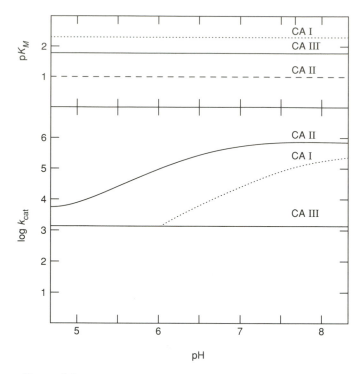

Figure 2.2
pH dependence of k_{cat} and K_m values for CO_2 hydration catalyzed
by carbonic anhydrase I, II, and III isoenzymes.[33,36]

119) is bound via its Nδ1 atom (Figure 2.4). It is quite general that histidines
bind zinc equally well by either of the two histidine nitrogens, the preference
being probably dictated by the steric constraints imposed by the protein folding.
The three histidine NH protons are all engaged in H-bonding (Figure 2.4). His-
tidine-119 is involved in H-bonding with a glutamate residue. As mentioned,
this could be a way of controlling the basicity of the metal ligands. A solvent
molecule bound to zinc is involved in an H-bond with Thr-199, which in turn
is H-bonded to Glu-106. This H-bonding network is important for understanding
the subtle structural changes that occur with pH changes; these could, in principle, account for the pH-dependent properties. Although the structure of crystals grown at pH 8 in sulfate-containing buffer gives some indication of a single
solvent molecule bound to zinc (Figures 2.3 and 2.5 *See color plate section,
pages C2, C3.*), theoretical studies indicate that two water molecules can be at
bonding distances.[42] Such a finding is consistent with spectroscopic studies on
other derivatives and with the concept that attachment and detachment of substrates occur through five coordination.

Just as is true for every zinc enzyme in which zinc is at the catalytic site,
activity is lost if the metal is removed, and is restored by zinc uptake. The
tertiary structure of carbonic anhydrase is maintained in the absence of zinc;
even the denatured apoprotein can refold spontaneously from a random coil to
a native-like conformation. Although such a process is accelerated by zinc,[43,44]

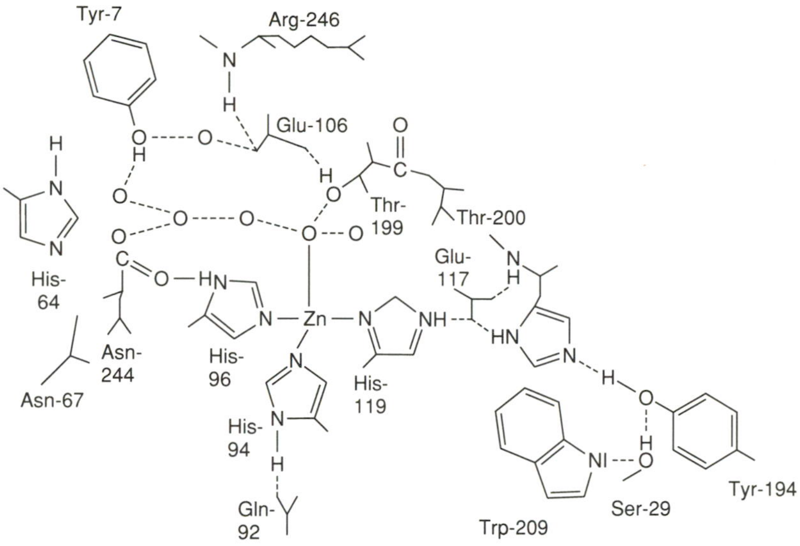

Figure 2.4
Schematic representation of the active site of human carbonic anhydrase II.
Hydrogen bonds (–––) and ordered water molecules (o) are indicated.[41]

the presence of the metal does not seem to be an absolute requirement for the correct folding of CA, whereas it is an absolute requirement for several other metalloproteins.[23,29,30]

Anions are attracted in the metal cavity by the positive $Zn(N_3OH_2)^{2+}$ moiety, and are believed to bind to zinc in carbonic anhydrase very effectively; so their use should be avoided as much as possible if the goal is to study the enzyme as it is. When the protein is dialyzed against freshly doubly distilled or carefully deionized water under an inert atmosphere, the pH of the sample approaches the isoelectric point, which is below 6 for HCA I and bovine (BCA II) enzymes. The pH can then be adjusted by appropriate additions of NaOH. All the measurements reported in the literature performed in acetate, phosphate, imidazole, or tris sulfate buffers are affected by the interference of the anion with the metal ion. However, buffer species containing large anions like Hepes (4[(2-hydroxyethyl)-1-piperazinyl]ethanesulfonic acid) can be used,[45] since these anions do not enter the cavity.

There are many indications that zinc in the high-pH form of CA is four-coordinate with an OH group in the fourth coordination site. At low pH the enzyme exists in a form that contains coordinated water; the coordination number can be four (one water molecule) or five (two water molecules). Of course, the occurrence of the low-pH species depends on the pK_a's of the complex acid-base equilibria.

B. Steady-State and Equilibrium Kinetics of Carbonic Anhydrase-Catalyzed CO_2/HCO_3^- Interconversion

The $CO_2 \rightleftharpoons HCO_3^-$ interconversion catalyzed by CA is extremely fast. The usual kinetic parameters describing an enzymatic reaction are the turnover number or kinetic constant for the reaction, k_{cat}, and the Michaelis constant K_m. In the simple catalytic scheme

$$E + S \underset{k_{-1}}{\overset{k_1}{\rightleftharpoons}} ES \overset{k_2}{\rightarrow} E + P,$$

where E stands for enzyme, S for substrate, and P for product, K_m^{-1} is given by $k_1/(k_{-1} + k_2)$. If k_2 is small, $k_{cat} = k_2$ and $K_m^{-1} = k_1/k_{-1}$, the latter corresponding to the thermodynamic affinity constant of the substrate for the enzyme. The pH dependences[46] of k_{cat} and K_m for CO_2 hydration for the high- and low-activity isoenzymes have been determined (Figure 2.2).[33,36] It appears that K_m is pH-independent, whereas k_{cat} increases with pH, reaching a plateau above pH 8. For bicarbonate dehydration (the reverse of Equation 2.6), H^+ is a cosubstrate of the enzyme. The pH dependence of k_{cat}/K_m for HCO_3^- dehydration is also mainly due to k_{cat}, which shows the same pH profile as that for CO_2 if the experimental kinetic data are divided by the available concentration of the H^+ cosubstrate.[47,48] Further measurements have shown that the pH dependence of k_{cat} reflects at least two ionizations if the measurements are performed in the absence of anions.[49] The value of k_{cat} reaches its maximum at alkaline pH only when buffer concentrations exceed 10^{-2} M.[50] In other words, the exchange of the proton with the solvent is the rate-limiting step along the catalytic pathway if relatively high concentrations of proton acceptors and proton donors are not provided by a buffer system. This limit results from the high turnover of the enzyme, which functions at the limit imposed by the diffusion rate of the H^+ cosubstrate. At high buffer concentration, k_{cat} shows an isotope effect consistent with the occurrence of an internal proton transfer as the new rate-limiting step.[51]

Measurements of the catalyzed reaction performed at chemical equilibrium starting from mixtures of ^{12}C-^{18}O-labeled HCO_3^- and ^{13}C-^{16}O-labeled CO_2 have shown the transient formation of ^{13}C-^{18}O-labeled species (both CO_2 and HCO_3^-) before ^{18}O-labeled water appears in solution.[52] These experiments provided evidence that, at chemical equilibrium, an oxygen atom can pass from HCO_3^- to CO_2 and vice versa several times before being released to water. Furthermore, maximal exchange rates are observed even in the absence of buffers.

Under chemical equilibrium conditions, ^{13}C NMR spectroscopy is particularly useful in investigating substrate interconversion rates, since the rates pass from a slow-exchange regime in the absence of enzyme to fast exchange at sufficient enzyme concentration. In the absence of enzyme two ^{13}C signals are observed, one for CO_2 and the other for HCO_3^-. In the presence of enzyme only one averaged signal is observed (Figure 2.6). Starting from the slow exchange situation, in the absence of enzyme, the increase in linewidth ($\Delta\nu$) of the substrate (A) and product (B) signals (caused by exchange broadening that is caused

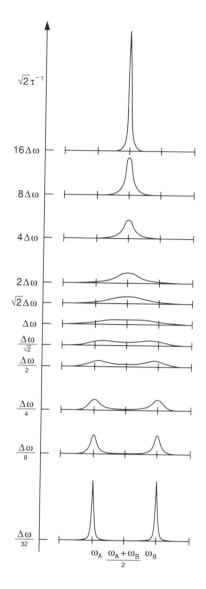

Figure 2.6

Calculated lineshape for the NMR signals of nuclei equally distributed between two sites ($[A] = [B]$), as a function of the exchange rate τ^{-1}. $\Delta\omega$ is the peak separation in rad s^{-1}.

in turn by the presence of a small amount of catalyst) depends on the exchange rate and on the concentration of each species, according to the following relation:

$$\Delta\nu_A[A] = \Delta\nu_B[B] = \tau_{exch}^{-1}. \qquad (2.9)$$

Therefore, the exchange rate τ_{exch}^{-1} can be calculated.[53] The appearance of the NMR spectrum for different τ_{exch} values is illustrated in Figure 2.6 under the condition $[A] = [B]$. For the high-activity enzyme it was found that the maximal exchange rates are larger than the maximal turnover rates under steady-state conditions; the ratio between k_{exch} of the high-activity (type II) and low-activity (type I) forms is 50, i.e., larger than the ratio in k_{cat}.[49,54] This result is consistent with the idea that the rate-limiting step in the steady-state process is an intramolecular proton transfer in the presence of buffer for type II enzymes, whereas it may not be so for the type I enzymes. The exchange is pH-independent in the pH range 5.7–8, and does not show a proton-deuteron isotope effect. The apparent substrate binding (HCO_3^-) is weaker than steady-state K_m values, indicating that these values are not true dissociation constants. Chloride is a competitive inhibitor of the exchange.[49]

A similar investigation was conducted for type I CoHCA at pH 6.3, where the concentrations of CO_2 and HCO_3^- are equal.[55] The two lines for the two substrates were found to have different linewidths but equal T_1 values. Measurements at two magnetic fields indicate that the line broadening of the HCO_3^- resonance is caused by substrate exchange and by a paramagnetic contribution due to bonding. The temperature dependence of the linewidth shows that the latter is determined by the dissociation rate. Such a value is only about 2.5 times larger than the overall $CO_2 \rightleftharpoons HCO_3^-$ exchange-rate constant. Therefore the exchange rate between bound and free HCO_3^- is close to the threshold for the rate-limiting step. Such an exchange rate is related to the higher affinity of the substrate and anions in general for type I isoenzymes than for type II isoenzymes. This behavior can be accounted for in terms of the pK_a of coordinated water (see below).

C. What Do We Learn from Cobalt Substitution?

1. Acid-base equilibria

It is convenient to discuss the cobalt-substituted carbonic anhydrase enzyme, since its electronic spectra are markedly pH-dependent and easy to measure (Figures 2.7 and 2.8).[56,57] The spectra are well-shaped, and a sharp absorption at 640 nm is present at high pH and absent at low pH. Whereas CoHCA I is almost entirely in the low-pH form at pH 5.7, this is not true for the CoBCA II isoenzyme. The acid-base equilibrium for Co-substituted carbonic anhydrase (deprotonation of the metal-coordinated water) involves three species:

$$
\begin{array}{c}
\text{N} \diagdown \text{OH}_2 \\
\text{N}-\overset{2+}{\text{Co}} \\
\text{N} \diagup \text{OH}_2
\end{array}
\underset{+H_2O}{\overset{-H_2O}{\rightleftharpoons}}
\begin{array}{c}
\text{N} \\
\text{N}-\overset{2+}{\text{Co}}-\text{OH}_2 \\
\text{N}
\end{array}
\underset{+H^+}{\overset{-H^+}{\rightleftharpoons}}
\begin{array}{c}
\text{N} \\
\text{N}-\overset{+}{\text{Co}}-\text{OH} \\
\text{N}
\end{array}
\qquad (2.10)
$$

The first equilibrium has never been directly monitored, and the conditions that determine it are quite vague. However, the five-coordinate species has been proposed in HCA I at low pH values.[48] Figure 2.7 at first seems to show isosbestic points* between 16,000 and 18,000 cm^{-1}, so that a single acidic group could

---

* An isosbestic point is a value of frequency where the two species in an $A \rightleftharpoons B$ equilibrium have the same absorption. As a consequence, all mixtures of A and B also show the same absorption at that frequency, and all the spectra along, e.g., a pH titration from A to B, plotted one on top of the other, cross at the isosbestic point. The presence of isosbestic points thus indicates the presence of only two species in equilibrium.

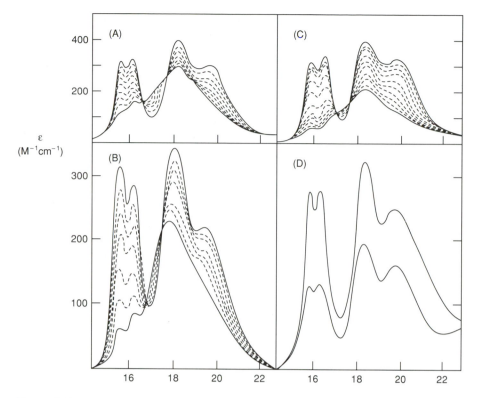

Figure 2.7
pH-variation of the electronic spectra of cobalt(II)-substituted BCA II (A), HCA II (B), HCA I (C), and BCA III (D). The pH values, in order of increasing $\epsilon_{15\cdot6}$, are (A) 5.8, 6.0, 6.3, 6.7, 7.3, 7.7, 7.9, 8.2, 8.8; (B) 6.1, 6.6, 7.1, 7.8, 8.3, 8.6, 9.5; (C) 5.3, 6.1, 6.6, 7.0, 7.3, 7.5, 7.9, 8.4, 8.6, 9.1, 9.6.[56]

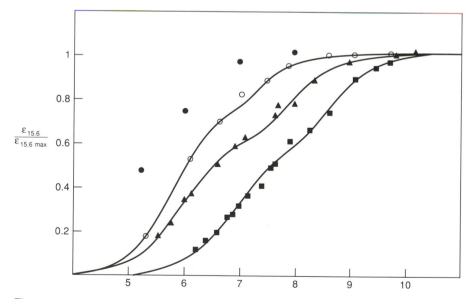

Figure 2.8
pH dependence of $\epsilon_{15\cdot6}$ for cobalt(II)-substituted BCA III (●), HCA II (o), BCA II (▲), and HCA I (■) isoenzymes. The high pH limit value of $\epsilon_{15\cdot6}$ is normalized to 1 for each isoenzyme.[57]

account for the experimental data. Therefore, CoHCA I would have a pK_a of about 8, CoBCA II and CoHCA II a pK_a of about 6.5, and CoBCA III a pK_a around 5.5. The analysis of the dependence of the absorbance on pH, however, clearly shows that two apparent pK_a's can be extracted from the electronic spectra of at least CoCA I and II (Figure 2.8). These kinds of isoenzymes contain at least another histidine in the cavity, which represents another acidic group, with a pK_a of about 6.5 in its free state. The interaction between such an acidic group and metal-coordinated water, for example, via a network of hydrogen bonds, provides a physical picture that can account for the observed experimental data.[49] Two apparent acid dissociation constants K_a can be obtained from the fitting of the curves of Figure 2.8. They are called apparent, because they do not represent actual acid dissociations at the microscopic level. When there are two acidic groups interacting with each other, the system must be described in terms of four constants, also called microconstants, because the dissociation of each of the two groups is described by two different pK_a's, depending on the ionization state of the other group (Figure 2.9); so the two apparent constants

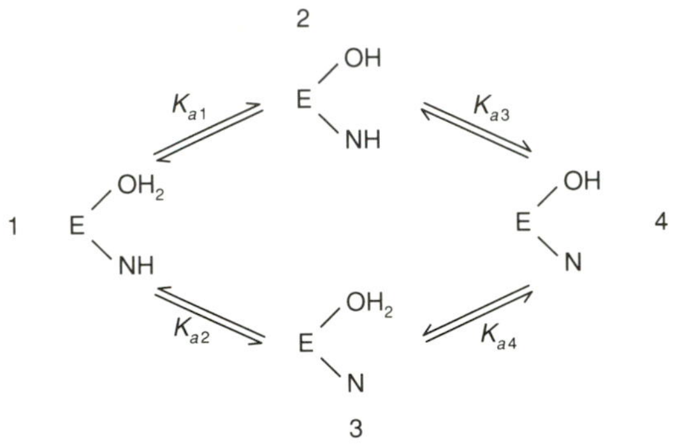

Figure 2.9
General scheme for two coupled acid-base equilibria applied to carbonic anhydrase. The two acid-base groups are the metal-coordinated water molecule and a histidine residue present in the active-site cavity.[58]

can be expressed in terms of four microconstants describing two interacting acidic groups. It is again a general feature of these systems that the four microconstants can be obtained only by making some assumptions. In one analysis the molar absorbances of species (1) and (3), and of species (2) and (4), were assumed to be equal.[58] In other words, it is assumed that the changes in the electronic spectra of cobalt(II) (Figure 2.7) are due entirely to the ionization of the coordinated water, not at all to the ionization state of the other group. This assumption accounts for the observation of approximate isosbestic points, even though there is an equilibrium between more than two species. With this assumption the four microconstants could be obtained (Table 2.5). Recall that the activity and spectroscopic profiles follow one another (see Figure 2.2 and Section IV.B). Furthermore, similar microconstant values had been obtained on

Table 2.5
Values of microconstants associated with
acid-base equilibria[a] in cobalt(II)-substituted
carbonic anhydrases.[58]

	pK_{a1}	pK_{a2}	pK_{a3}	pK_{a4}
CoHCA I	7.14	7.21	8.45	8.38
CoHCA II	5.95	5.62	6.62	6.95
CoBCA II	6.12	6.28	7.75	7.59

[a] As defined in Figure 2.9.

ZnHCA II by analyzing the pH dependence of the maximum velocity of the hydration reaction, V_{max}, assuming that the two hydroxo-containing species had the same activity.[49] The present analysis implies that species (2) and (3) of Figure 2.9 are distinguishable, although their interconversion may be fast.

Metal coordination lowers the pK_a of coordinated water. Factors affecting the acidity of the coordinated water are many, and their effects are probably overlapping, making the analysis quite complex (see also Section III.A). Nonetheless, the following factors probably contribute to the lowering of the pK_a:

(1) the charge of the chromophore, which in this case is $2+$, although it may be somewhat lowered by the H-bonding between a coordinated histidine and a negative glutamate residue;

(2) the coordination number (which is four), since a higher value leads to a larger electron density on the metal ion ligands;

(3) the presence of other acidic groups with which the coordinated water interacts;

(4) the presence of positively charged residues inside the metal binding cavity that favors the removal of a proton from the cavity.

This last factor is presumably operating for CA III, which contains several arginine residues in the cavity; the same factor may also induce changes in the microscopic properties of the solvent inside the active cavity. These considerations account for the observation that most model complexes have a significantly higher pK_a value than the protein itself.

2. Coordination Geometries

The binding of inhibitors is also pH-dependent. It is possible, however, to obtain fully inhibited systems by adjusting the inhibitor concentration and pH. In this manner the so-called limit spectra of CoCA derivatives are obtained. Many systems have been characterized, providing a variety of spectral characteristics[59] (Figure 2.10). The differences in molar absorbance are larger than expected for changing only one coordinated atom. A rationalization of the

58

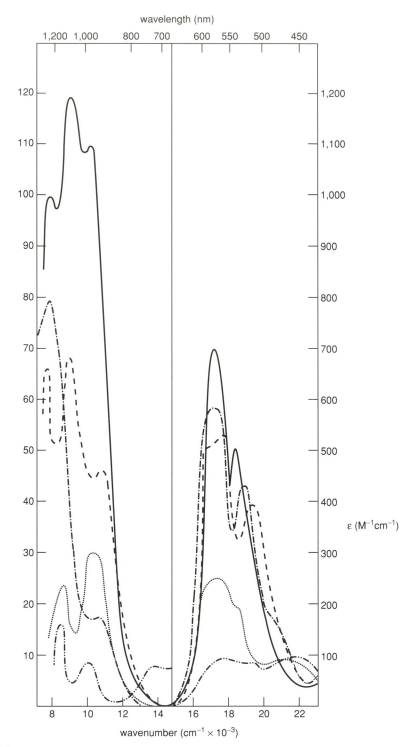

Figure 2.10
Electronic spectra of cyanide (———), cyanate (–·–·–), acetazolamide (– – –), azide (· · · · ·), and thiocyanate (–··–) adducts of cobalt(II)-substituted bovine carbonic anhydrase II.[59]

experimental data came by applying a criterion, first suggested by Gray,[60] according to which four-coordinate species have larger maximal absorption than five-coordinate species. This property theoretically arises from greater mixing of p and d metal orbitals in the four-coordinate case, which makes the d-d transitions partially allowed, neglecting other factors such as the covalency of the coordination bond, nephelauxetic effects,* or vicinity of charge transfer bands. Subsequent extension of the measurements to the near-infrared region was instructive:[59] the low-intensity spectra exhibited a weak absorption between 13,000 and 15,000 cm^{-1}. The latter band was assigned to the highest in energy of the F $\rightarrow$ F transitions, which increases in energy with the coordination number.† Therefore both the low intensity of the bands ($\epsilon_{max} < 200$ M^{-1} cm^{-1}) and the presence of the F $\rightarrow$ F transition at high energy were taken as evidence for five coordination. Spectra showing high maximal absorption ($\epsilon_{max} > 300$ M^{-1} cm^{-1}) were assigned as arising from four-coordinate species. The corresponding chromophores are $CoN_3In(OH_2)$ and CoN_3In, where In denotes inhibitor. Intermediate maximal absorptions may indicate an equilibrium between four- and five-coordinate species. In Table 2.6 some inhibitors are classified according to their behavior. Bicarbonate, which is a substrate of the enzyme, gives rise to an equilibrium between four- and five-coordinate species.[48,59]

The differences in the electronic spectra outlined above also have been detected in both CD and MCD spectra. In the latter, pseudotetrahedral species

Table 2.6
Classification of inhibitors of bovine carbonic anhydrase II according to the electronic spectral properties of the adducts with cobalt(II) derivatives.[a] [48,59]

Four-coordinate	Equilibria between four- and five-coordinate species	Five-coordinate
Sulphonamides (N_4)	Bicarbonate (N_3O—N_3O_2)	Carboxylates (N_3O_2)
Cyanide (N_3C)	Chloride (N_3Cl—N_3OCl)	Thiocyanate (N_4O)
Cyanate (N_4)	Bromide (N_3Br—N_3OBr)	Nitrate (N_3O_2)
Aniline (N_4)	Azide (N_4—N_4O)	Iodide (N_3OI)
Phenol (N_3O)		
Chlorate (N_3O)		

a Donor sets in parentheses.

* Nephelauxetic (literally, cloud-expanding) effects are due to partial donation of electrons by the ligand to the metal, and are stronger for less electronegative and more reducing ligands.

† By F $\rightarrow$ F transition we mean here a transition between two electronic states originating from the same F term (the ground term) in the free ion and split by the ligand field; the stronger the ligand field, the larger the splitting. For high-spin cobalt(II), the free-ion ground state ^4F (quartet F) is split in octahedral symmetry into $^4T_{2g}$, $^4T_{1g}$, and $^4A_{2g}$ states, the $^4T_{2g}$ lying lowest; in lower symmetries the T states are further split. The highest F $\rightarrow$ F transition is, therefore, that from the ground state $^4T_{2g}$, or the lowest of its substates in low symmetry, to the $^4A_{2g}$ state. For the same type of ligands, e.g., nitrogens or oxygens, the ligand field strength, and therefore the energy of the F $\rightarrow$ F transition, increases with the number of ligands.

give a sizably positive band in the high-energy region, whereas five-coordinate species show a much weaker positive band and six-coordinate complexes have only weak negative bands (Figure 2.11).[21,61] This additional empirical criterion may be helpful in assigning the coordination number. A further criterion is based on how much of the splitting of the $S = \frac{3}{2}$ ground state is caused by spin-orbit coupling (zero-field splitting). This splitting can be indirectly measured from the temperature dependence of the electronic relaxation times of the cobalt complexes, in turn estimated from their ability to saturate the EPR lines of the complexes at low temperatures.[62] There are theoretical reasons to predict that the above splitting increases in the order four coordination < five coordination < six coordination.[63]

Three binding sites have been identified in the cavity of CA[40,64-66] (Figure 2.12). The OH$^-$ binding site, which provides a tetrahedral structure around the metal ion, is called the A site. The hydrogen interacts via hydrogen bonding with the oxygen of Thr-199. Thr-199 and Thr-200, together with their protein backbone, identify a hydrophilic region that probably plays a fundamental role in the energetic balance of ligand binding. On the back of the cavity there is a hydrophobic region formed by Val-143, Leu-198, and Trp-209. Although this cavity is hydrophobic, the x-ray structure shows evidence of a water molecule, H-bonded to the coordinated water. Ligands with a hydrophobic end could easily be located in this binding position, which is called B. The coordinated water

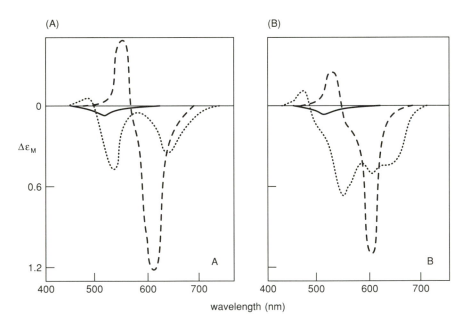

Figure 2.11
(A) MCD spectra of model six-coordinate (Co(Gly-Gly)$_2$, ———), five-coordinate ((Co-Me$_6$tren)Br$_2$, ... , and four-coordinate (Co(py)$_2$Br$_2$, – – –) cobalt(II) complexes and (B) MCD spectra of the cobalt(II) derivatives of pyruvate kinase (———), alkaline phosphatase (· · · · ·), and carbonic anhydrase in the presence of acetazolamide (– – –).[21,61]

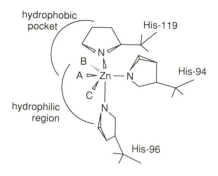

Figure 2.12
Schematic drawing of the active cavity of HCA II showing the three possible ligand binding sites.[64–66] Site A is the site of the OH$^-$ ligand in the active form; site B is the binding site of NCS$^-$, which gives rise to a five-coordinate adduct with a water molecule in the C site.[64–66]

molecule would change its position in order to make reasonable angles between coordinated groups. The new position is labeled C. The x-ray structure of the thiocyanate derivative of HCA II[40,64] illustrates the latter case (see Figure 2.5). The NCS$^-$ ion is in van der Waals contact with Val-143, Leu-198, and Trp-209. The water interacts with the hydroxyl group of Thr-199. The geometry of the five-coordinate derivative can be roughly described as a distorted square pyramid with His-94 in the apical position (Figure 2.13A). This could be a typical structure for those derivatives that have spectra typical of five-coordinate adducts, like the carboxylate derivatives.

In aromatic sulfonamide (Ar—SO$_2$—NH$_2$) derivatives, which probably bind as anions (see Section IV.C.4), the NH$^-$ group binds zinc in the A position,[64–66] giving rise to an H-bond with Thr-199. The oxygens do not interact with the metal; one of them sits in the hydrophobic pocket. The chromophore around zinc is pseudotetrahedral (Figure 2.13B). The energy involved in the coordination includes the coordination bond, the hydrophobic interactions of the aromatic sulfonamide ring, and the maintainance of the Zn-X-H-Thr-199 hydrogen bonding (X=N,O). It is interesting to note that cyanate, according to spectroscopic studies,[48,59] gives rise to tetrahedral derivatives, probably because the terminal oxygen can enter into H-bonds with the hydrophilic region of the cavity. ^{13}C NMR data on N^{13}CO$^-$ interacting with CoBCA indicate that the anion interacts directly with the metal ion.[67] We do not have direct information on where it binds.[212]

The fine balance between hydrophobic and hydrophilic interactions, as well as major steric requirements, play important roles in the binding of inhibitors. Cyanide is the only ligand that may bind in a 2:1 ratio.[68] It is likely that the bis-cyanide adduct has the same arrangement as the NCS$^-$—H$_2$O derivative. The spin state of the bis-cyanide adduct is $S = \frac{1}{2}$.[68]

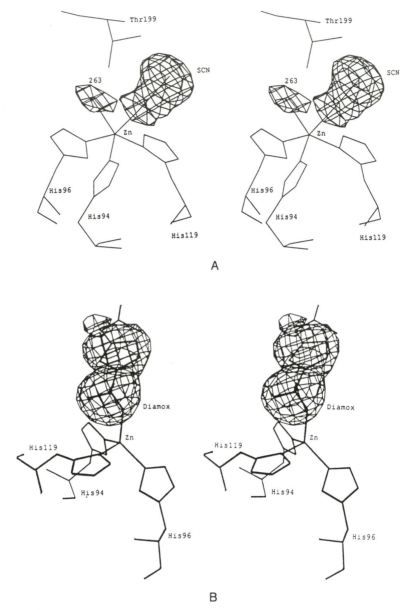

Figure 2.13
Stereo views of the NCS⁻ (A) and acetazolamide (B) adducts of HCA II.[40]

3. Coordinated water and NMR

It is quite relevant to know whether a water molecule is coordinated to the metal ion in a metalloenzyme, and whether it is still coordinated in the presence of substrates and inhibitors. The presence or absence of H_2O coordinated to a paramagnetic center can in principle be monitored by solvent water ^1H NMR,[69] by exploiting the occurrence of a magnetic interaction between the magnetic moments of the unpaired electrons and the nuclear magnetic moments of the water protons. When this interaction fluctuates with time, it causes a shortening of the water-proton relaxation times.*

The longitudinal relaxation rate values, T_1^{-1}, of all the solvent water protons increase when even a single water molecule interacts with a paramagnetic center, provided that this bound water exchanges rapidly with free water molecules. To obtain the necessary experimental data, a methodology has been developed based on the measurement of water ^1H T_1^{-1} values at various magnetic fields (Nuclear Magnetic Relaxation Dispersion, NMRD).[69–71] The experimental data contain information on the correlation time, i.e., the time constant for the dynamic process that causes the proton-unpaired electron interaction to fluctuate with time; furthermore, under certain conditions, they may provide quantitative information on the number of interacting protons and their distance to the metal. The enhancement of T_1^{-1}, called T_{1p}^{-1}, is caused by the paramagnetic effect on bound water molecules and by the exchange time τ_m, according to the relationship

$$(T_{1p})^{-1} = f_M (T_{1M} + \tau_M)^{-1}, \tag{2.11}$$

where f_M is the molar fraction of bound water and T_{1M} is the relaxation time of a bound water proton. Therefore we measure the water ^1H T_1^{-1}, subtract the diamagnetic effect (i.e., the water-proton relaxation rate measured in a solution of a diamagnetic analogue), obtain T_{1p}^{-1}, then check that τ_m is negligible with respect to T_{1M}. For high-spin cobalt(II), T_{1M} is of the order of 10^{-3} s, whereas τ_m is about 10^{-5} s. Then the experimental T_{1p} can be safely related to T_{1M}. It is now important, in order to proceed with the analysis, to define the correlation time for the interaction between proton nuclei and unpaired electrons, τ_c. Its definition is important in order to obtain a physical picture of the system, and to quantitatively analyze the obtained T_{1M} values.[69] τ_c is defined by

$$\tau_c^{-1} = \tau_r^{-1} + \tau_s^{-1} + \tau_m^{-1}, \tag{2.12}$$

where τ_r is the rotational correlation time, τ_s is the electronic relaxation time, and τ_m has been previously defined. τ_r depends on the size of the molecule, which can be calculated rigorously if the molecule is spherical, or approximately if it is not. The appropriate expression is

$$\tau_r = \frac{4\pi\eta a^3}{3k_B T}, \tag{2.13}$$

---

* The nuclear longitudinal relaxation time, T_1, can be defined as the rate constant by which the populations of the $M_I = \frac{1}{2}$ and $M_I = -\frac{1}{2}$ (for protons) levels reach their equilibrium value after an external perturbation (e.g., a radiofrequency pulse in an NMR experiment). The transverse relaxation time, T_2, can be defined as the average lifetime of a hydrogen nucleus in a given spin state. The NMR linewidth is inversely proportional to T_2. The relation $T_2 \leq T_1$ always holds.

where η is the microviscosity of the solution, a is the radius (or approximate radius) of the molecule, k_B is the Boltzmann constant, and T is the absolute temperature. For CA, τ_r can be safely calculated to be $\simeq 10^{-8}$ s at room temperature. Since the correlation time τ_c in high-spin cobalt proteins varies between 10^{-11} and 10^{-12} s, it must therefore be determined by the electronic relaxation time.

Water ^{1}H NMRD profiles are often analyzed by using the classical dipolar interaction approach, as first described by Solomon:[72]

$$T_{1M}^{-1} = \frac{2}{15}\left(\frac{\mu_0}{4\pi}\right)^2 \frac{\gamma_I^2 g_e^2 \mu_B^2 S(S+1)}{r^6}\left(\frac{7\tau_c}{1+\omega_S^2\tau_c^2}+\frac{3\tau_c}{1+\omega_I^2\tau_c^2}\right), \quad (2.14)$$

where μ_0 is the permeability of vacuum, γ_I is the nuclear magnetogyric ratio, g_e is the electron g-factor, S is the electron spin quantum number, r is the electron-nucleus distance, and ω_S and ω_I are the electron and nuclear Larmor frequencies, respectively. This equation describes the dipolar interaction between the magnetic moment of nucleus I ($\hbar\gamma_I\sqrt{I(I+1)}$) and the magnetic moment of the electrons S ($g_e\mu_B\sqrt{S(S+1)}$) as a function of the correlation time (τ_c) and of the magnetic field (expressed as ω_I and ω_S). Neglect of the zero-field splitting of the $S = \frac{3}{2}$ manifold may introduce an error in the quantitative estimates within a factor of two.[73]

Fitting of the data for pseudotetrahedral complexes shows that they have τ_s of 10^{-11} s, whereas five-coordinate complexes have a shorter τ_s, on the order of 10^{-12} s. The latter derivatives also have exchangeable protons that could correspond to a water molecule in the coordination sphere, whereas the former do not.[25] The τ_s values are thus proposed as indicators of the coordination number in low-symmetry, four- and five-coordinate cobalt complexes. The shorter electronic relaxation times are related to low-lying excited states, which, independently of the particular mechanism, favor electron relaxation.[74]

Short electronic relaxation times in paramagnetic compounds cause only minor broadening of ^{1}H NMR lines, whereas the isotropic shifts (i.e., the shifts due to the presence of unpaired electron(s), usually very large) are independent of the value of the electronic relaxation times. For cobalt-substituted carbonic anhydrase, the ^{1}H NMR spectra have been recorded for several derivatives, and the proton signals of histidines coordinated to the metal were found to be shifted well outside the diamagnetic region (Figure 2.14).[75] Five-coordinate species give sharper signals than four-coordinate ones. The spectra in D_2O for both kinds of derivatives show three fewer isotropically shifted signals than in H_2O. These signals are assigned to histidine NH protons, which are replaced by deuterons in D_2O. Five-coordinate species provide ^{1}H NMR spectra with many signals slightly shifted from the diamagnetic position. It is believed that such complexes have relatively large magnetic anisotropy, which, summed up to the external magnetic field, provides further differentiation in shifts of the protons. Such shift contributions are called pseudocontact shifts. These shifts depend on the third power of the distance from the metal and on the position of the proton

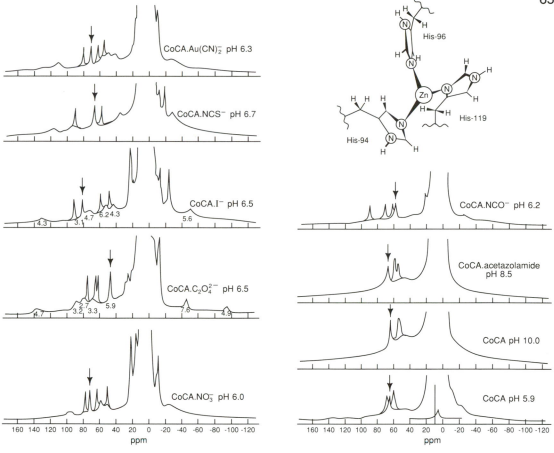

Figure 2.14
^{1}H NMR spectra of cobalt(II)-substituted bovine carbonic anhydrase II and some inhibitor derivatives. The three sharp downfield signals in each spectrum disappear in D_2O and are assigned to the exchangeable ring NH protons of the three coordinated histidines. The sharp signal labeled with an arrow is assigned to the Hδ2 proton of His-119, which is the only non-exchangeable ring proton in a meta-like rather than in an ortho-like position with respect to the coordinating nitrogen. The T_1 values (ms) of the signals for the I$^-$ and $C_2O_4^{2-}$ derivatives are also shown.[25,75]

with respect to the molecular axes. These signals belong to protons of noncoordinated residues from 5 to 10 Å from the metal. Their assignment in principle provides further information on the structure in the vicinity of the metal ion. The ^{1}H NMR spectra of cobalt(II) enzymes thus afford a powerful method for monitoring structure and reactivity of the metal-bound residues. This is one task for future investigations of the enzyme.

4. pH dependence of inhibitor binding

The ease with which electronic spectra can be obtained provides a simple way of determining the affinity constants of inhibitors for the cobalt-substituted enzymes. An aliquot of enzyme is diluted in a spectrophotometric cell up to a

66

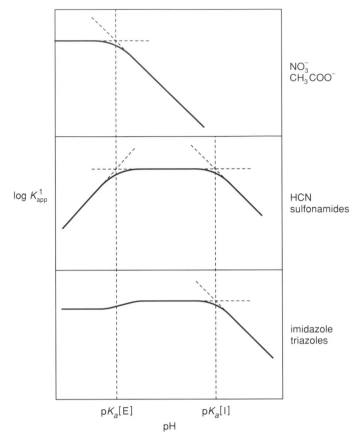

Figure 2.15
Types of pH dependences observed for the affinity constants of inhibitors for cobalt(II)-substituted carbonic anhydrases. $pK_a[E]$ represents the main pK_a value of the enzyme, $pK_a[I]$ that of the inhibitor, if present.[48]

fixed volume, and the spectrum is measured. Then the spectra are remeasured on samples containing the same amount of enzyme plus increasing amounts of inhibitor in the same cell volume. The pH is rigorously controlled. If solutions of enzyme and inhibitor have the same pH, the pH should be verified after the spectral measurements, in order to avoid contamination from the electrode salt medium. Both absolute values and pH dependences of affinity constants obtained from electronic spectra are the same as those obtained from inhibition measurements, where known, and are comparable to those obtained on the native enzyme.

Although affinity constant values reported in the literature were measured under different experimental conditions of, e.g., pH, buffer type, and buffer concentration, several pH-dependent trends are apparent. According to such dependences, three classes of inhibitors can be identified[48] (Figure 2.15). In the

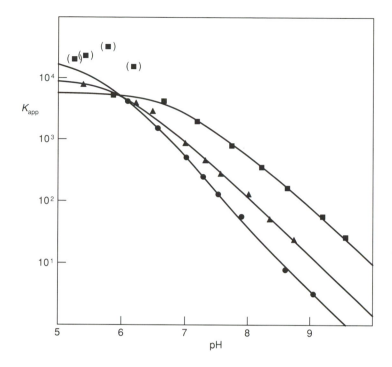

Figure 2.16
pH dependence of the apparent affinity constants of nitrate for human I (■),
bovine II (▲), and human II (●) carbonic anhydrases. The curves are best-fit
curves obtained assuming non-zero affinity of the anion for species 1 and 3
of Figure 2.9. The best-fit parameters are reported in Table 2.6. Points in
parentheses for HCA I reflect possible binding of a second nitrate ion and
have been excluded from the fit.[57]

first class, the affinity constant, expressed as log K, decreases linearly with
increasing pH. Anions that are weak Lewis bases (Cl^-, N_3^-, CH_3COO^-, NO_3^-,
etc.) behave in this manner, as do neutral ligands like CH_3OH and aniline. An
example is shown in Figure 2.16. A qualitative fit to such curves can be obtained using a single pK_a. This behavior could be accounted for by assuming
that the ligand binds only the low-pH form of the enzyme, in a simplified scheme
in which only one pK_a value determines the species distribution in CA. We
know, however, that the picture is more complex. If the species distribution
calculated according to the scheme of Figure 2.9 is assumed to hold, and if it
is assumed that only the two water-containing species (1) and (3) can be bound
by the ligand, then actual affinity constants can be evaluated for both species
(1) and (3)[57] (see Table 2.7). Such constants are similar for the three isoenzymes, whereas the apparent affinity constants at pH 7, for example, mainly
depend on the pK_a's of the coordinated water according to the values of Table
2.5. Therefore, the low-activity species CA I has larger affinity for anions like
nitrate (and bicarbonate) than do the high-activity forms at pH 7.

Table 2.7
Affinity constants of nitrate for species 1 and 3[a] of
cobalt(II)-substituted carbonic anhydrases.[57]

	HCA I	BCA II	HCA II
$\log K_1$	3.74 ± 0.04	4.01 ± 0.02	4.34 ± 0.04
$\log K_3$	2.62 ± 0.06	2.56 ± 0.04	2.61 ± 0.05

[a] As defined in Figure 2.9.

A second type of behavior occurs for weak acids like HCN, H_2S, and aromatic sulfonamides ($ArSO_2NH_2$).[76,77] Assuming that the anions (conjugated bases) bind the low-pH species of the enzyme, the bell-shaped plot of log K versus pH (Figure 2.15) can be accounted for. In fact, at low pH, the inhibitors are in the protonated form, which is not suitable for metal binding. At high pH the concentration of the low-pH species of the enzyme decreases. The maximal apparent affinity is experimentally halfway between the pK_a of the inhibitor and the "pK_a" of the enzyme, treated as if it were only one. The same type of curve is also expected if the high-pH species of the enzyme binds the weak acid. Indeed, kinetic measurements seem to favor this hypothesis for sulfonamides.[78]

A third type of behavior obtains for inhibitors like imidazole and triazoles,

imidazole   1,2,3-triazole   1,2,4-triazole

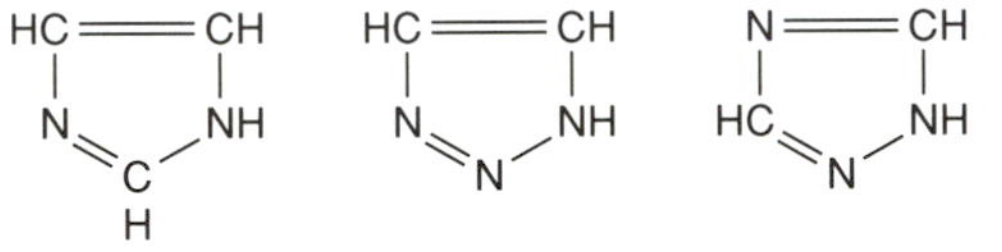

which bind the enzyme with similar affinities over a large range of pH (Figure 2.15),[79,80] because both the imidazolate anion and the neutral imidazole can bind to the aquo forms of the enzyme with essentially the same affinity,[48,80,81] and the reaction of imidazole with the Zn—OH species cannot be distinguished thermodynamically from the reaction of imidazolate with the aquo forms:

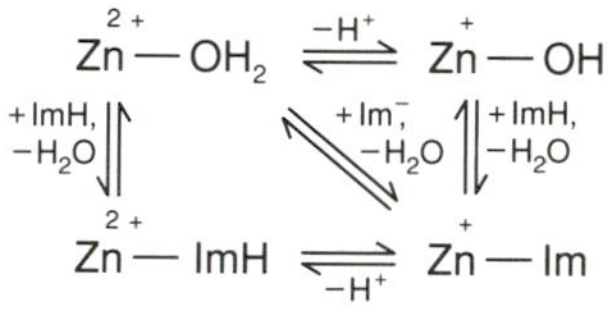

It is possible that the noncoordinated nitrogen can interact with a group in the protein via a hydrogen bond. A candidate could be the NH group of His-200 in HCA I or the hydroxyl group of Thr-200 in HCA II. Indeed, only imidazole and triazoles, which have two nitrogens in 1,3-positions, seem to have this ability.[213]

In summary, from cobalt substitution we have learned:

(1) the coordination geometry of the high- and low-pH forms by means of electronic spectroscopy;

(2) the values of the pK_a's from the pH dependence of the electronic spectra;

(3) the four and five coordination of the various derivatives with exogenous ligands;

(4) the affinity constants of exogenous ligands and their pH dependence;

(5) a fingerprint in the ^1H NMR spectra that can be used to monitor structural variations.

Most of these conclusions can be safely transferred to the native zinc enzyme, although minor differences can occur, for example, in the position of the equilibrium between four- and five-coordinate species.

D. What Do We Learn from Copper Substitution?

The coordination chemistry of CuCA is not yet fully understood, since the electronic spectra are not very pH-sensitive. Nevertheless, the affinity of anions is pH-dependent, as it is for CoCA.[82] As could be anticipated from Section III.B, the affinity of anions, including HCO_3^-, is higher than that of CoCA. Water is usually present in the coordination sphere, along with the anion, as checked by water ^1H NMRD.[83,84] The steric requirements of the three histidines and of the cavity allow the anion and the water molecule to arrange in an essentially square pyramidal geometry (Figure 2.17). This is consistent with the electronic and

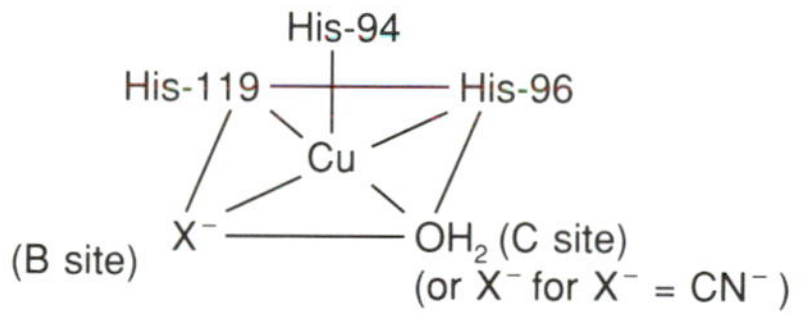

Figure 2.17
Schematic representation of the suggested coordination geometry for the anion adducts of CuCA.

EPR spectra. In particular, the EPR spectra are all axial, with g-values decreasing from 2.31 in the nonligated enzyme to 2.24 in the various anion adducts.[84] The water molecule would be in the C site or hydrophilic binding site, and the anion would be in the B site or hydrophobic pocket. His-94 would be in the apical position of the square pyramid. It has been shown by EPR spectroscopy that at low temperature two cyanide anions bind to copper. The donor atoms are two cyanide carbon and two histidine nitrogen atoms in the basal plane, and the third histidine nitrogen in the axial position.[85] The hyperfine splitting is observed only with nuclei in the basal plane. It is observed both with ^{13}C nuclei of ^{13}C-enriched CN^- and with the two ^{14}N of two histidines. The second cyanide may thus displace the coordinated water (Figure 2.17). Oxalate and sulfonamides displace water from the coordination sphere.[85,86] For the oxalate ion this may occur through bidentate behavior. Coordination to an oxygen of the sulfonamide cannot be ruled out, although the electronic and EPR spectra of the sulfonamide complex are more consistent with a pseudotetrahedral chromophore. The SO_2 moiety would in any case point toward the B binding site. It is likely that sulfonamides bind as in ZnCA. Bicarbonate also shows less water relaxivity than other monodentate anions.[83,84,86]

^{13}C NMR spectroscopy has been used to investigate the location of the two substrates, CO_2 and HCO_3^-, with respect to the metal ion in CuCA.[86–88] As was pointed out in Section IV.B, the interconversion between the two species is slow on the NMR timescale in the absence of catalysts. Therefore, two signals are observed (Figures 2.6 and 2.18). In the presence of the catalytically active CoCA, only one signal is observed at suitable enzyme concentrations, and individual information on CO_2 binding cannot be obtained.[89,90] In the presence of inactive CuCA, two signals are again observed, which are broadened to different extents.

For the HCO_3^- signal the T_2^{-1} values as estimated from the linewidth are much larger than T_1^{-1}. Since the equation for T_2^{-1}, analogous to Equation (2.14), would predict similar T_1 and T_2 values,[69,72] a sizeable broadening due to chemical exchange must be present. Indeed, unlike T_{1p}^{-1} (Equation 2.11), T_{2p}^{-1} may be a complicated function of the exchange time τ_M and of the isotropic shift, $\Delta\omega_M$,

$$T_{2p}^{-1} = \frac{f_M}{\tau_M} \frac{T_{2M}^{-2} + T_{2M}^{-1}\tau_M^{-1} + (\Delta\omega_M)^2}{(T_{2M}^{-1} + \tau_M^{-1})^2 + (\Delta\omega_M)^2}. \tag{2.15}$$

In the slow-exchange region, i.e., when two separate signals are observed and the broadening is due to exchange, $T_{2p}^{-1} = f_M\tau_M^{-1}$. This region is characterized

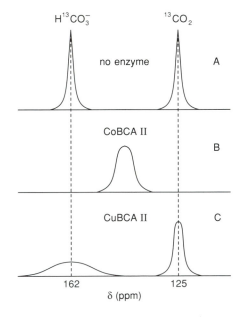

Figure 2.18
Schematic representation of the ^{13}C NMR spectra of the CO_2/HCO_3^- system (A) in pure water, (B) in the presence of CoCA, and (C) in the presence of CuCA.[56]

by a marked increase in linewidth with increasing temperature, as confirmed by measurements at 4 and 25°C. Therefore, T_{2p} gives a direct measure of τ_M.[56] The ^{13}C T_1^{-1} values of HCO_3^- are consistent with bicarbonate bound to the metal. The Cu—C distance would be 2.5 Å if the unpaired electron were completely on the copper ion, as estimated by using Equation (2.1) and a value of $\tau_c = 2.1 \times 10^{-9}$ s independently obtained from water ^1H NMRD.[83] This distance is much too short for a coordinated bicarbonate; however, electron delocalization on the bicarbonate ligand may account for such a short calculated distance; the possibility of a bidentate type of ligation cannot be discarded. The dissociation rate, which is very low, by itself accounts for the lack of activity of the derivative.

For CO_2, a carbon-copper distance could be calculated if the affinity constants of the substrate for the protein were known. When the binding site, if any, starts being saturated, fast exchange with excess ligand (in this case, CO_2) decreases the observed paramagnetic effect. From this behavior, the affinity constant may be estimated. For CO_2 the paramagnetic effect remained constant up to 1 M CO_2; i.e., the affinity constant is smaller than 1 M^{-1}. This means that practically there is no affinity for copper; yet the paramagnetic effect is paradoxically high.[88]

Another picture comes by analyzing the NMR data in terms of a pure diffusive model.[88] Here Hubbard's equation[91] has been used:

$$T_{2p}^{-1} = N_M \left(\frac{\mu_0}{4\pi}\right)^2 \frac{8\pi}{225} \frac{\gamma_I^2 n_e^2 \mu_B^2 S(S + 1)}{d(D_N + D_M)} \left(13f(\omega_S, \tau_D) + 3f(\omega_S, \tau_D)\right), \quad (2.16)$$

where

$$f(\omega, \tau_D) = \frac{15}{2} I(u);$$

$$u = [\omega\tau_D]^{1/2};$$
$$I(u) = u^{-5}\{u^2 - 2 + e^{-u}[(u^2 - 2)\sin u + (u^2 + 4u + 2)\cos u]\};$$

d is the distance of closest approach, D_N and D_M are the diffusion coefficients of the molecules containing the nucleus under investigation, and $\tau_D = 2d^2/(D_N + D_M)$. The experimental paramagnetic effect can be reproduced with a CO_2 concentration inside the cavity much larger than the one in the bulk solution. This result indicates that substrate does not bind to a specific site, but probably binds in the hydrophobic region. Note that CO_2 is more soluble in organic solvents than in water.

The effect of the cavity is to attract CO_2 by interaction either with the metal ion or with a hydrophobic part of the cavity itself. But the affinity constant is in any case lower than expected from the Michaelis constant (see Section IV.B) measured under steady-state conditions, indicating that the latter does not represent the dissociation constant of the enzyme-CO_2 system.

In summary, the main information concerning the catalytic cycle obtained from the copper derivative is the structural and kinetic characterization of both

CO_2 and HCO_3^- species when they are not interconverting but present within the cavity. In this way we have further proof that HCO_3^- is bound to the metal and that CO_2 is attracted inside the cavity either by hydrophobic interactions or by the metal ion or both. The data obtained on the geometry around copper are consistent with those obtained on cobalt.

E. What Do We Learn from Manganese and Cadmium Substitution?

Several studies have been performed on MnCA. Although CA is not the protein for which Mn(II) has been most extensively used as a paramagnetic probe to map substrates and inhibitors within the metal cavity, by measuring the T_{2M}^{-1} values of protons of the inhibitor N-acetyl-sulfanilamide, and by assuming that dipolar contributions are dominant, researchers have mapped the orientation of the inhibitor inside the active cavity (Figure 2.19).[92] This orientation is consistent with x-ray data on stronger binding sulfonamides.[64-66] MnCA is not completely inactive. ^{13}C NMR studies of the $CO_2 \rightleftharpoons HCO_3^-$ interconversion at pH 8.5 showed that the interconversion rate is about 4 percent that of the native enzyme.[93] The T_1^{-1} and T_2^{-1} values of $H^{13}CO_3^-$ suggest that bicarbonate might be bidentate in the central step of the catalytic cycle.[93]

Data from [113]Cd studies that have been performed on CdBCA II and CdHCA I are consistent with the general picture presented here.[94] The [113]Cd chemical shifts are indeed consistent with a donor set of three nitrogens and two oxygens. The cadmium(II) derivative could thus be five-coordinate with two water mole-

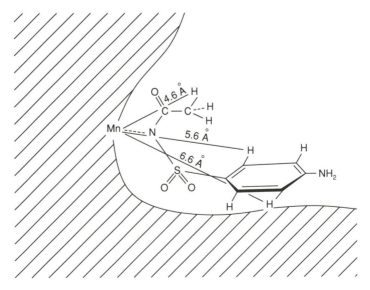

Figure 2.19
Schematic drawing of the geometric arrangement of the inhibitor N-acetyl-sulfanilamide in the active cavity of manganese(II)-substituted CA, as revealed by ^{1}H NMR spectroscopy.[92]

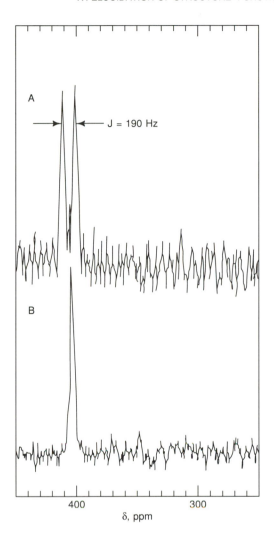

Figure 2.20
[113]Cd NMR spectra of Cd-substituted bovine carbonic anhydrase II in the presence of ^{15}N-enriched (A) or ^{14}N-enriched (B) benzenesulfonamide inhibitor.[95]

cules, in agreement with the expectation based on its ionic radius being larger than that of zinc(II). The [113]Cd signal of CdBCA II in the presence of benzene-sulfonamide enriched in ^{15}N is split into a doublet because of the nitrogen-cadmium coupling (Figure 2.20).[95] This result provides direct evidence for metal-nitrogen bonding in sulfonamides, which has been confirmed by x-ray data.[65]

F. Catalytic Mechanism

All the above structural and kinetic information obtained under a variety of conditions with different metal ions can be used to propose a catalytic cycle for carbonic anhydrase (Figure 2.21). As shown by studies on the pH-dependent properties of native and metal-substituted CAs, both type-I and type-II proteins have two acidic groups, the zinc-coordinated water and a free histidine. At

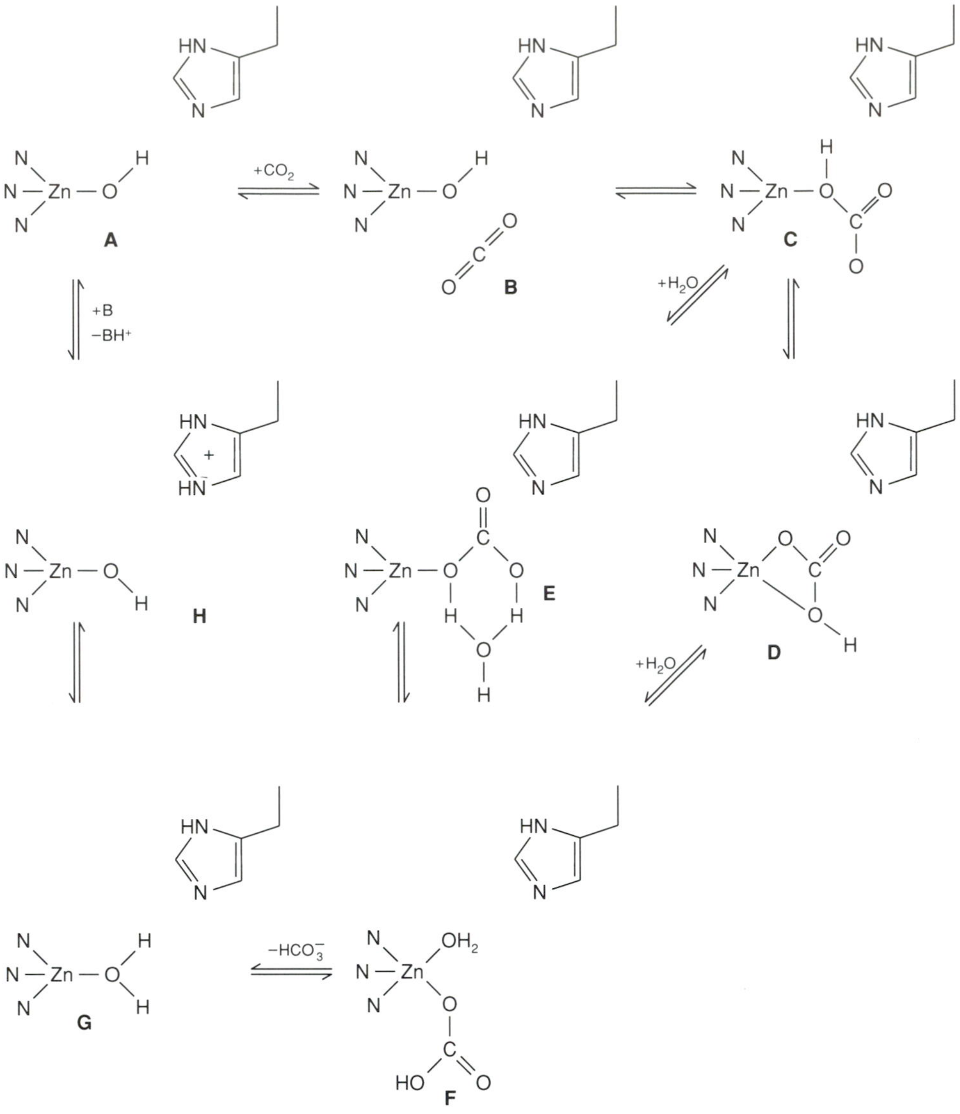

Figure 2.21
Proposed catalytic cycle of CA.

physiological pH the enzyme is essentially in the Zn—OH form (step A in Figure 2.21). A Zn—OH moiety is a relatively good nucleophile, poised for nucleophilic attack on carbon dioxide. It is possible that the hydrogen bond with Thr-199, which seems to be consistent with an sp^3 oxygen, orients the OH for attack at the substrate CO_2. Studies of the copper derivative indicated that the concentration of CO_2 in the cavity is higher than in bulk solution (step B).

Molecular dynamics calculations have shown that there are either three[96] or two[97] potential wells for CO_2 in the hydrophobic pocket. It was shown[98] that when Val-143 is replaced by the much larger Phe, the activity decreases by a factor of 10^3. Apparently the large Phe residue does not leave space within the cavity to accomodate CO_2.

It would also be nice if the enzyme were able to activate CO_2. There is no evidence that it does, even though the positive charge around zinc and the NH of Thr-199 would represent two electrostatic attraction points that could activate CO_2. It is well-known that CO_2's interactions with positive charges activate the carbon for nucleophilic attack.[99,100] The positioning of CO_2 between zinc and the peptide NH of Thr-199 would be ideal for the OH attack. Merz[97] locates it as shown in Figure 2.22.

It was believed that, once bicarbonate is formed (C), the proton has to transfer to a terminal oxygen atom, either via an intermediate in which bicarbonate is bidentate (D) or via a hydrogen-bond network (E). Indeed, in model compounds one would expect HCO_3^- to bind through a nonprotonated oxygen. However, the possibility of restoring the hydrogen bond with Thr-199 as in sulfonamide adducts could justify the presence of the hydrogen on the coordinating oxygen.[214] The bicarbonate derivative is presumably in equilibrium between four- and five-coordinate species (F), as shown by the electronic spectra of the cobalt derivative.[59] The five-coordinate species provides a low barrier for the substrate detachment step via an associative mechanism involving coordination of a water molecule (G). A possible five-coordinate species would contain bicarbonate in the B site and water in the C site (Figure 2.12). It is reasonable that the measured K_m for the reaction of bicarbonate dehydration is the thermodynamic dissociation constant of the $M—HCO_3^-$ species. Anionic or neutral

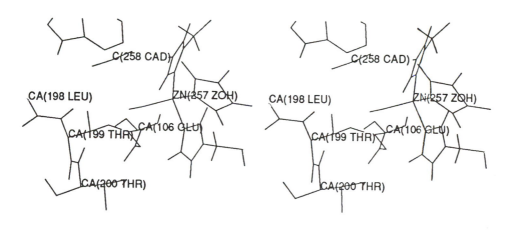

Figure 2.22
Stereo view of the site of activation of CO_2 in the cavity of CA as proposed by Merz.[97]

inhibitors are competitive with bicarbonate because they tend to bind at the same site. At this stage the second substrate, which is H^+, has to be released (H). It is reasonable that the water proton transfers to a group inside the cavity, e.g., the free histidine mentioned above, and subsequently to the solvent. In the absence of buffers the latter step is rate-limiting for the high-activity isoenzymes, since the diffusion rate cannot exceed the product of the concentration times the diffusion coefficient, i.e., 10^{-7} M $\times$ 10^{11} M^{-1} s^{-1}. Such a limit is then 10^4 s^{-1}, whereas the turnover rate is 10^6 s^{-1}. The presence of buffer can assist in proton transfer at this stage, in such a way that the rate-limiting step becomes the internal proton transfer. The release of H^+ from the Zn—OH_2 moiety is also the rate-limiting step for the low-activity CA III, as nicely shown by the electronic spectra of CoCA III. These spectra change from the basic form at the beginning of the reaction to the acidic form upon CO_2 addition (Figure 2.23).[101] After the interconversion of CO_2 into bicarbonate, there is an accumulation of the $CoOH_2$ species, the deprotonation of which is slower than the release of HCO_3^-.

G. Model Chemistry

Some efforts have been reported in the literature to simulate the activity of CA and therefore to obtain further information on the mechanism. The pK_a of Zn—OH_2 moieties in various complexes has been studied as discussed in Sec-

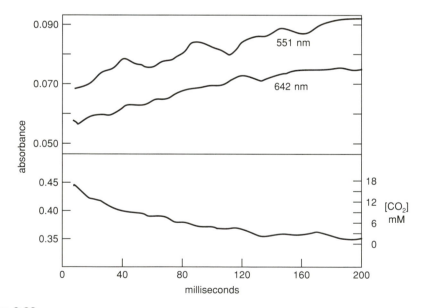

Figure 2.23
Time dependence of $\epsilon_{15.6}$ and $\epsilon_{18\cdot1}$ of cobalt(II)-substituted CA III after addition of CO_2 to a buffered enzyme solution at pH 8. The initial drop of absorbance reflects the accumulation of a $CoOH_2$ intermediate.[101]

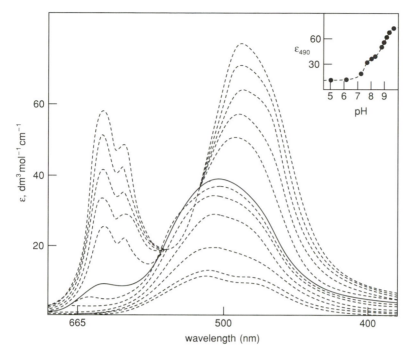

Figure 2.24
Electronic spectra of Co(TPyMA)OH$_2$$^{2+}$ (Table 2.3) at various pH values.[10] Note the similarity to the electronic spectra of cobalt-substituted carbonic anhydrase at various pH values as reported in Figure 2.7.[56]

tion III.A. The electronic spectra of some cobalt analogues have been found to be similar. One such example is shown in Figure 2.24; the complex Co(TPyMA)OH$_2$$^{2+}$ (Table 2.3)[10] provides a five-coordinate adduct with a weakly bound axial nitrogen (Figure 2.25A).

The interconversion between Co(TPyMA)OH$_2$$^{2+}$ and Co(TPyMA)OH$^+$ was studied by electronic spectroscopy (Figure 2.24). Despite the difference in the number of coordinated nitrogens, the difference between the high- and low-pH forms resembles that of the cobalt enzyme (cf. Figure 2.7).[10]

Table 2.3 shows that only one compound, with zinc(II) as the metal ion, seems to have three nitrogens and a water, whereas all the other models have a higher coordination number.[15–17] The simple [Co^{III}(NH$_3$)$_5$OH]$^{2+}$ complex has been shown to accelerate the formation of bicarbonate ($k = 2 \times 10^2 \, M^{-1} \, s^{-1}$), but, of course, bicarbonate remains coordinated to the metal because of the kinetic inertness of cobalt(III).[102,103] Some relatively ill-defined systems have been reported to have some kind of activity. The ligand shown in Figure 2.25B, with zinc(II) as the metal ion in H$_2$O, accelerates the attainment of the equilibrium[104]

$$CO_2 + H_2O \underset{k_{-1}}{\overset{k_1}{\rightleftharpoons}} HCO_3^- + H^+ \tag{2.17}$$

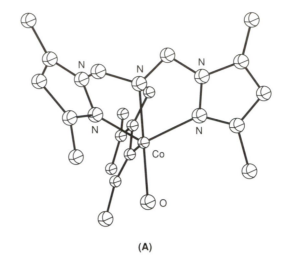

(A)

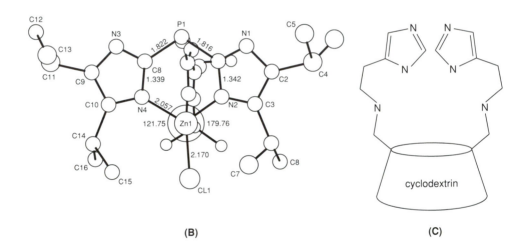

(B)   **(C)**

Figure 2.25
Some multidentate ligands as models of CA: (A) tris-(3,5-dimethyl-l-
pyrazolylmethyl) amine[4] (cobalt[II] complex); (B) tris (4,5-
diisopropylimidozal-2yl)phosphine[104] (zinc[II] complex); (C)
bis(histamino) β-cyclodextrin.

with $k_{obs} = k_1 + k_{-1} \simeq 10^3 \ M^{-1} \ s^{-1}$. The system in Figure 2.25C, with Zn^{2+}
and excess imidazole, promotes CO_2 hydration, though not the back reaction.[105]
The cobalt(II) analogue shows no activity.[106]

It can be concluded that the M—OH group can indeed be involved in one
step of the enzymatic pathway. The sophistication of the whole enzymatic function has not yet been fully achieved with the present generation of models, even
though the functionalization of both hydrophilic and hydrophobic molecules like
cyclodextrins (Figure 2.25C) has also been used.[105]

V. OTHER ENZYMATIC MECHANISMS AND MODEL CHEMISTRY

A. Peptide Hydrolysis

At neutral pH the uncatalyzed hydrolysis of amides or peptides

$$R—CO—NH—R' + H_2O \rightleftharpoons R—COO^- + R'—NH_3^+ \qquad (2.18)$$

is a slow process, with rate constants as low as 10^{-11} s^{-1}. Peptide hydrolysis catalyzed by carboxypeptidase or thermolysin can attain k_{cat} values of 10^4 s^{-1}. Organic chemistry teaches us that amide hydrolysis is relatively efficiently catalyzed by acids and bases. The general mechanisms involve protonation of the carbonyl oxygen (or amide nitrogen), and addition of OH$^-$ (or of a general nucleophile) to the carbonyl carbon atom. Several organic and inorganic bases have been found to be reasonably efficient catalysts. On the other hand, transition metal aquo-ions or small metal-ion complexes also display catalytic efficiency (Table 2.8).[107–114] A metal ion is a Lewis acid, capable of effectively polarizing the carbonyl bond by metal-oxygen coordination. Furthermore, the metal ion can coordinate a hydroxide group in such a way that there is a high OH$^-$ concentration at neutral or slightly alkaline pH. It is thus conceivable that a metalloenzyme may combine some or all of these features and provide a very efficient catalyst.

Much experimental work has been done on mimicking ester and especially peptide hydrolysis with model coordination compounds. Most of the work car-

Table 2.8
Rate constants for amide and ester hydrolysis catalyzed by acids, bases, or metal ions.

Compound	Catalyst and conditions	Rate constant	Reference
Glycine amide	pH 9.35	1.9×10^{-5} s^{-1}	107
	Cu^{2+}, pH 9.35	2.6×10^{-3} M^{-1} s^{-1}	107
[Co(en)$_2$(glycine amide)]$^{3+}$	pH 9.0^a	2.6×10^{-4} s^{-1}	108
D,L-phenylalanine ethylester	pH 7.3	5.8×10^{-9} s^{-1}	109
	Cu^{2+}, pH 7.3	3.4×10^{-2} M^{-1} s^{-1}	109
(tn)$_2$O$_3$PO—C$_6$H$_4$NO$_2$	OH$^-$	5.1 M^{-1} s^{-1}	110
Co^{III}—(tn)$_2$O$_3$PO—C$_6$H$_4$NO$_2$	—^{a}	7×10^{-5} s^{-1}	110
Ethyl-β-phenylpropionate	H$^+$	5×10^{-3} M^{-1} s^{-1}	111
	OH$^-$	1.3×10^{-6} M^{-1} s^{-1}	111
Adenosine triphosphate	pH 5.3	5.6×10^{-6} s^{-1}	112
	Cu^{2+}, pH 5.3	1.1×10^{-2} M^{-1} s^{-1}	112
Glycine methylester	Co^{2+}(1:1), pH 7.9	1.6×10^{-2} s^{-1}	113
	Cu^{2+}(1:1), pH 7.3	4.2×10^{-2} s^{-1}	113
Glycine propylester	Co^{3+}(1:1), pH 0	1.1×10^{-3} s^{-1}	114
	Co^{3+}(1:1), pH 8.5	$>1 \times 10^{-2}$ s^{-1}	114

^{a} autohydrolysis.

ried out has involved[108,110,115,116] cobalt(III). Although such an ion may not be the best conceivable model for zinc-promoted hydrolytic reactions (see Section IV.G), it has the great advantage of being substitutionally inert, thus removing mechanistic ambiguities due to equilibration among isomeric structures in the course of the reaction. Interesting amide hydrolysis reactions also have been described using complexes with other metal ions, such as copper(II)[117] and zinc(II)[118] itself. In recent years efforts have focused on the construction of bifunctional catalysts to better mimic or test the enzymatic function. For instance, phenolic and carboxylic groups can be placed within reach of Co(III)-chelated amides in peptidase models.[116] The presence of the phenolic group clearly accelerates amide hydrolysis, but carboxyl groups are ineffective. This model chemistry is too simple to provide insights into the actual enzymatic mechanism, which must start with recognizing the substrate through several steps, orienting it, activating it, performing the reaction, and finally releasing the products. See the more specialized reviews dealing with nonenzymatic reactivity.[119–121]

From basic knowledge of the chemistry of hydrolytic reactions, the x-ray structures of carboxypeptidase A and a variety of its derivatives with inhibitors as substrate analogues, product analogues, and transition-state analogues have revealed several features of the active site that are potentially relevant for the catalytic mechanism (Figures 2.26–2.28 *See color plate section, pages C4, C5.*).[122] The metal ion is coordinated to two histidine residues (His-69 and His-196), to a glutamate residue that acts as a bidentate ligand (Glu-72), and to a water molecule. The metal is thus solvent-accessible and, as such, can activate the deprotonation of a water molecule to form a hydroxide ion, or polarize the carbonyl oxygen of the substrate by coordinating it in the place of the solvent

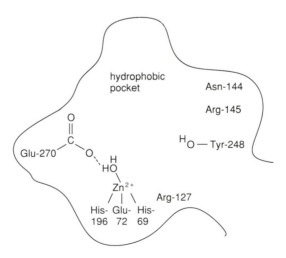

Figure 2.26
Schematic drawing of the active-site cavity of carboxypeptidase A.[122] Only the residues believed to play a role in the catalytic mechanism are shown.

molecule, or both, if some flexibility of the coordination sphere is allowed. Another glutamic-acid residue (Glu-270) is in close proximity to the metal center. If the role of the metal were mainly to polarize the carbonyl carbon, Glu-270 in its deprotonated form could be positioned to perform a nucleophilic attack on the carbonyl carbon, yielding an anhydride intermediate. Alternatively, the metal could mainly serve to provide a coordinated hydroxide ion that, in turn, could attack the carbonyl carbon; here Glu-270 would help form ZnOH by transferring the proton to the carboxylate group:

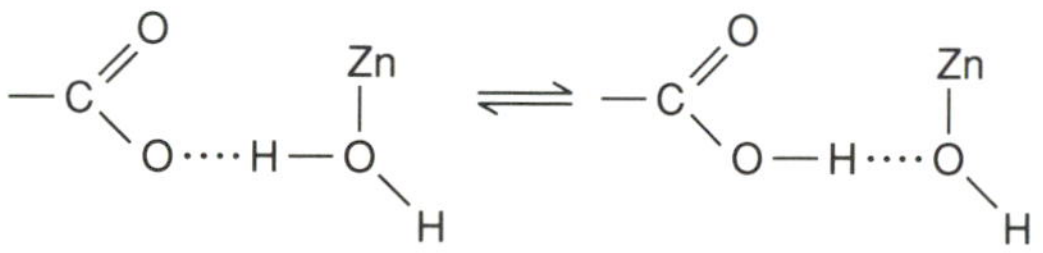

On the opposite side of the cavity is a tyrosine residue that has been shown to be quite mobile and therefore able to approach the site where the catalytic events occur. The cavity has a hydrophobic pocket that can accommodate the residue, R, of nonpolar C-terminal amino acids of the peptide undergoing hydrolysis (Figures 2.26 and 2.28), thereby accounting for the higher efficiency with which hydrophobic C-terminal peptides are cleaved. Finally, an Asn and three Arg residues are distributed in the peptide-binding domain; Asn-144 and Arg-145 can interact via hydrogen bonds with the terminal carboxyl group. Arg-127 can hydrogen-bond the carbonyl oxygen of the substrate.

All these features have enabled detailed interpretation of many chemical and physico-chemical data at the molecular level. The essential data are as follows:

(1) *Metal substitution*. Table 2.9 lists the divalent metals that have been substituted for zinc(II) in CPA, together with their relative peptidase (and esterase) activities.[22] For some of them, the available x-ray data

Table 2.9
Catalytic activities of metal-substituted carboxypeptidases.[a] [22]

	Peptidase	Esterase
Apo	0	0
Cobalt	200	110
Nickel	50	40
Manganese	30	160
Cadmium	5	140
Mercury	0	90
Rhodium	0	70
Lead	0	60
Copper	—[b]	—[b]

[a] Activities are relative to the native enzyme, taken as 100%.
[b] Some activity toward both peptides and esters has recently been observed.[22]

show[123] that the active-site structure is essentially maintained. Even the copper derivative is slightly active. The apoenzyme is completely inactive, however.

(2) *Active-site modifications.* Chemical modification and site-directed mutagenesis experiments suggest that Glu-270 is essential for catalysis.[124,125] Tyr-248,[126] Tyr-198,[127] and one or more of the arginines[124] are involved but not essential.

(3) *Kinetics.* k_{cat}/K_m pH profiles are bell-shaped, characterized by an acid pK_a limb around 6 and an alkaline pK_a limb around 9: k_{cat} increases with the pK_a of 6 and then levels off, and K_m increases with a pK_a of 9. Several lines of evidence suggest that the $pK_a \approx 6$ corresponds to the ionization of the Glu-270-coordinated H_2O moiety:

$$\begin{array}{ccc} A & \rightleftharpoons & B \end{array} \qquad (2.19)$$

Site-directed mutagenesis has ruled out Tyr-248 as the group with the pK_a of 9 in the rat enzyme.[125,126] Unfortunately, in this enzyme the pK_a of 9 is observed in k_{cat} rather than K_m; so the situation for the most-studied bovine enzyme is still unclear. Tyr-248 favors substrate binding three to five times more than the mutagenized Phe-248 derivative.[126] The three possible candidates for this pK_a are the coordinated water, Tyr-248, and the metal-coordinated His-196, whose ring NH is not hydrogen-bonded to any protein residue.[128] The x-ray data at different pH values show a shortening of the Zn—O bond upon increasing pH.[129] This favors the ZnOH hypothesis.

(4) *Anion binding.* The metal binds anionic ligands only below pH 6, i.e., when Glu-270 is protonated, when Glu-270 is chemically[130] or genetically[125] modified, or when aromatic amino acids or related molecules are bound in the C-terminal binding domain (Arg-145 + hydrophobic pocket).[131–134]

(5) *Intermediates.* An anhydride intermediate involving Glu-270 for a slowly hydrolyzed substrate may have been identified.[135] Some other intermediates have been observed spectroscopically at subzero temperatures with the cobalt(II) derivative.[22,136] Peptides bind in a fast step without altering the spectroscopic properties of cobalt(II), following which a metal adduct forms and accumulates.[22] Thus, if an anhydride intermediate is formed, it is further along the catalytic path.

On the basis of these data, and many related experiments, a detailed mechanism can be formulated (Figure 2.29). The incoming peptide interacts with

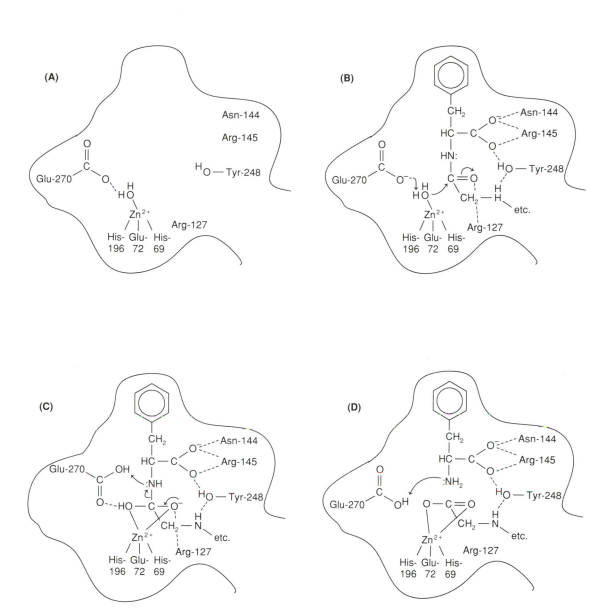

Figure 2.29
Possible catalytic cycle of CPA.

arginine residues through its terminal carboxylate group. The interaction could initially involve Arg-71 (not shown); then the peptide would smoothly slide to its final docking position at Arg-145, while the R residue, if hydrophobic, moves to the hydrophobic pocket (Figure 2.29B). The carbonyl oxygen forms a strong hydrogen bond with Arg-127. Additional stabilization could come from hydrogen bonding of Tyr-248 to the penultimate peptide NH. This adduct might be the first intermediate suggested by cryospectroscopy[22,136] (Figure 2.24).

At this point the metal-bound hydroxide, whose formation is assisted by Glu-270, could perform a nucleophilic attack on the carbonyl carbon activated by Arg-127 and possibly, but not necessarily, by a further electrostatic interaction of the carbonyl oxygen with the metal ion. The structure of the substrate analogue α-R-β-phenylpropionate shows that the carbonyl binds in a bidentate fashion:

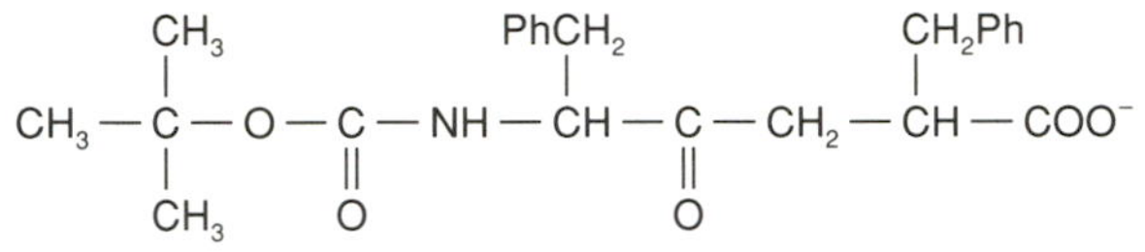

$$
\begin{array}{ccc}
CH_3 & PhCH_2 & CH_2Ph \\
| & | & | \\
CH_3-C-O-C-NH-CH-C-CH_2-CH-COO^- \\
| \;\; \| & \| & \\
CH_3 \;\; O & O &
\end{array}
$$

(Figure 2.30).[137] Five coordination is maintained by switching the Glu-72 metal

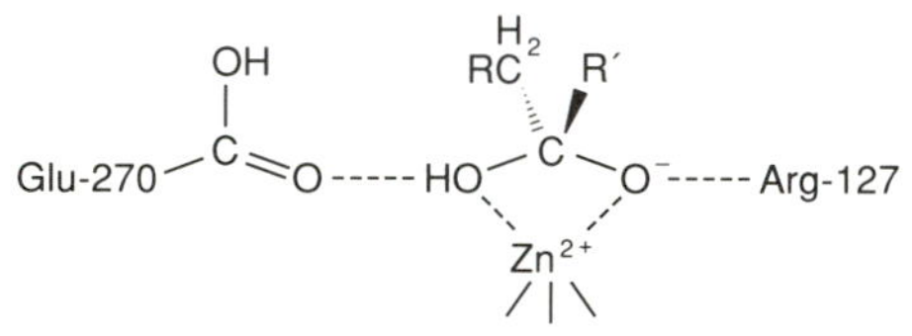

Figure 2.30
Binding mode of α-R-β-phenylpropionate to the zinc(II) ion in CPA.[137]

ligand from bidentate to monodentate, because the metal moves toward Arg-127. It is likely that this situation mimics an intermediate of the catalytic cycle. The resulting adduct might be the second spectroscopic intermediate (Figure 2.29C).

The system then evolves toward breaking of the C—N bond, caused by addition of a proton to the amino nitrogen. This proton could come from Glu-270, which thereby returns to the ionized state. The breaking of the peptide bond could be the rate-limiting step.[22] The second proton required to transform the amino nitrogen into an NH_3^+ group could come from the coordinated carboxylic group of the substrate, which now bears one excess proton, again through Glu-270 (Figure 2.29D). The system shown in Figure 2.29D can, in fact, be seen as a ternary complex with a carboxylate ligand and an amino-acid zwitterion, bound synergistically.[131-134] Finally, the metal moves back to regain a bidentate Glu-72 ligand, and the cleaved peptide leaves, while a further water molecule adds to the metal ion and shares its proton with the free carboxylate group of Glu-270.

Figure 1.1
The stabilization of Fe in aqueous solution by the protein coat of ferritin. In the absence of protein, at neutral pH, in air, flocculent precipitates of ferric hydrous oxide form. The equivalent concentration of Fe(III) in the solution of ferritin is about 10^{14} times greater than in the inorganic solution. *Left*: a solution of Fe(II)SO$_4$, pH 7, in air after 15 min. *Right*: the same solution in the presence of apoferritin, the protein coat of ferritin (reprinted from Reference 6).

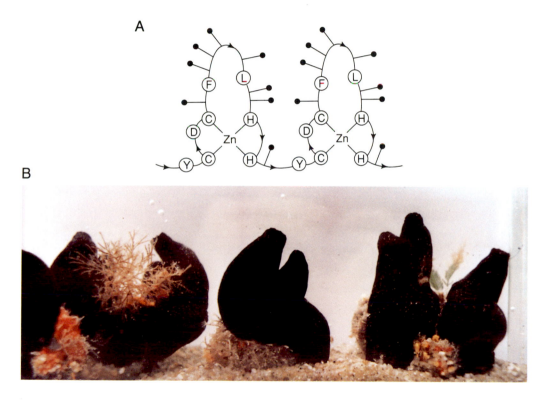

Figure 1.2
(A) Zinc-binding domains of a DNA-binding protein from frog eggs. Two of the nine repeating units of polypeptide with Zn-binding ligands are displayed, with two of the 7–11 zinc atoms per molecule in the configuration originally proposed by Klug and coworkers for "zinc fingers." The protein studied regulates the transcription of DNA to RNA and also binds to RNA, forming a storage particle.[9] Recently, the putative zinc-binding sequence has been shown to occur in many nucleic acid-binding proteins.[10] Each finger binds to a site on the double helix. However, other zinc-finger proteins function with fewer fingers, related apparently to the structure of the nucleic acid site. (B) Examples of the sea squirt (tunicate) *Ascidia nigra* (reproduced with permission from Reference 18a).

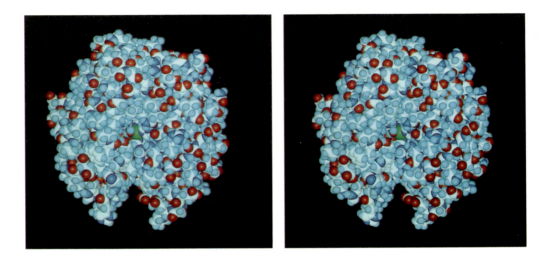

Figure 2.3
Human carbonic anhydrase II viewed as a CPK model. The zinc ion is the green sphere at the
bottom of the active site. The color codes for the other atoms are: C = white, H = cyan, N
= blue, O = red, S = yellow. Among the zinc ligands, the His-94 ring and the water molecule are clearly visible at the right- and left-hand side of the zinc ion, respectively. The presence
of a hydrophobic region above the zinc ion can also be discerned. The crevice that runs longitu-
dinally below the active site is obstructed by the histidine ring of His-64, one of the invariant
active-site residues.

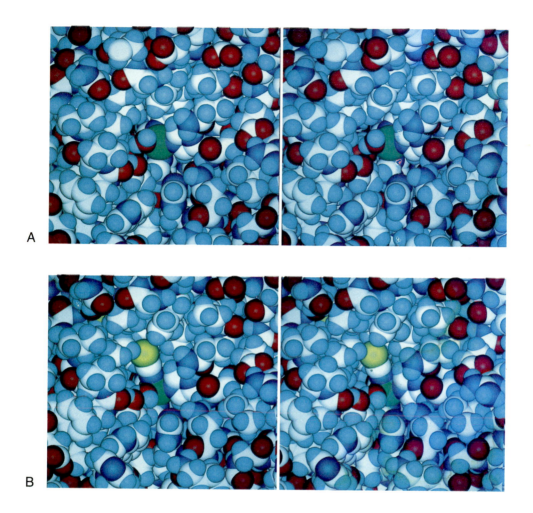

Figure 2.5
Active site of human carbonic anhydrase II (A) and its NCS[-] adduct (B) viewed as CPK models. It is apparent from the comparison that the NCS[-] ion occupies a binding site (B site) that is more buried than the binding site of the OH[-] ion in the active form (A site). The water molecule in the NCS[-] adduct occupies the C site, which is pointing more toward the entrance of the cavity.[64]

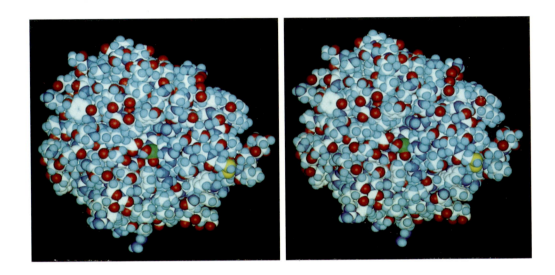

Figure 2.27
Carboxypeptidase A viewed as a CPK model. The color codes are as in carbonic anhydrase.
Note the shallower active-site cavity with respect to carbonic anhydrase.

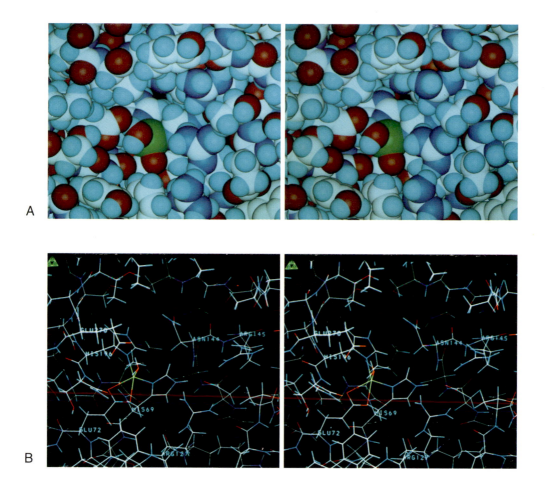

Figure 2.28
Active site of carboxypeptidase A viewed as CPK (A) and stick (B) models. The two views are taken from about the same perspective, i.e., from the entrance of the cavity. View B is self-explanatory. In view A, the zinc ion is the green sphere; its ligands are His-69 (on the right), His-196 (in the back), and Glu-72 (one of the two coordinated oxygens is clearly visible below the zinc ion). The coordinated water molecule (pointing outward) is hydrogen-bonded to Glu-270. On the opposite side of Glu-270 the three arginines (Arg-145, Arg-127, and Arg-71) can be seen, more or less on a vertical line, running from top to bottom of the figure. In the upper part of the figure, pointing outward, is the aromatic ring of Tyr-248. Behind it is the hydrophobic pocket.

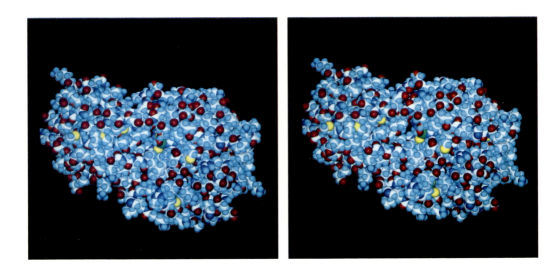

Figure 2.34
Liver alcohol dehydrogenase subunit viewed as a CPK model. The left-hand side of the molecule is the coenzyme binding domain and the right-hand side is the catalytic domain. The catalytic zinc ion is accessible from two channels located above (not visible) and below the coenzyme binding domain. The upper channel permits approach of the nicotinamide ring of the
coenzyme. The lower channel permits approach of the substrate. The substrate channel closes
up, trapping the substrate inside the molecule, when both coenzyme and substrate are present.

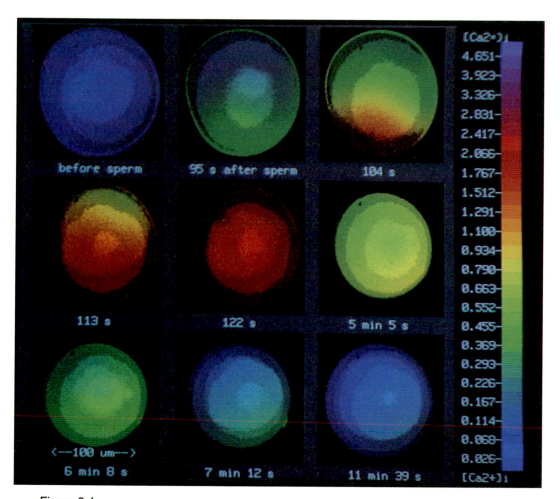

Figure 3.4
A Fura-2 study of the transient Ca^{2+} fluxes in an egg of the sea urchin (*Lutechinus pictus*). The diameter of this egg is about 120 μM. The fluorescent dye was injected into the egg, and the fluorescence intensity with excitation at 350 and 385 nm was measured with a lower-light-level television camera feeding a digital image processor (512 $\times$ 486 pixels). The image finally displayed in pseudocolor is the ratio of intensities at the two excitation wavelengths. The series of images shows a wave of high Ca^{2+} concentration that traverses the egg after it is fertilized by a sperm. Resting Ca^{2+} concentration is typically 100 nM and uniform through the cell. The fertilizing sperm sets off a transient wave of high Ca^{2+} that begins as a local elevation and thereafter spreads rapidly. After 20–30 seconds, the Ca^{2+} concentration of the entire egg is uniformly high ($\sim$2 μM). The figure is from an experiment by M. Poenie, J. Alderton, R. Steinhardt, and R. Tsien; see also Reference 26.

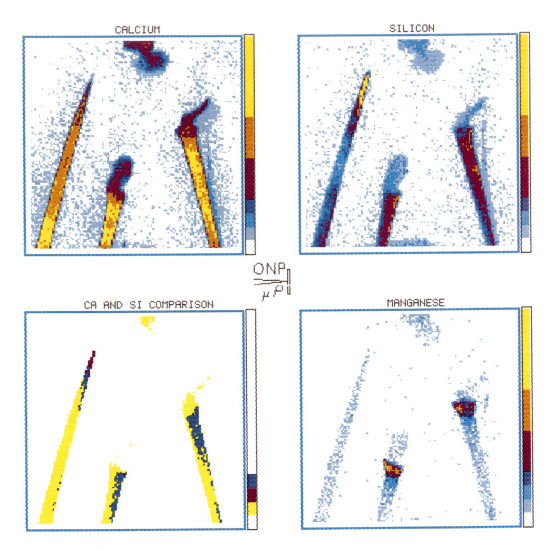

Figure 3.7

The elemental distribution of Ca, Si, and Mn in the hair of the common stinging nettle (*Urtica dioica*) obtained using the Oxford University PIXE microprobe. The color code for Ca is: yellow, >3.4 M; orange, 2.0–3.4 M; red, 1.5–2.0 M; dark blue, 1–1.5 M; blue, 0.5–1.0 M; light blue, 0.1–0.5 M; white, <0.1 M. The color codes of Si and Mn are similar. The PIXE data show that the tip mainly is made up of Si (presumably amorphous silica), but the region behind is largely made up of Ca (calcium oxalate crystals). The base of the hair contains substantial amounts of Mn. The pictures were kindly provided by R. J. P. Williams.

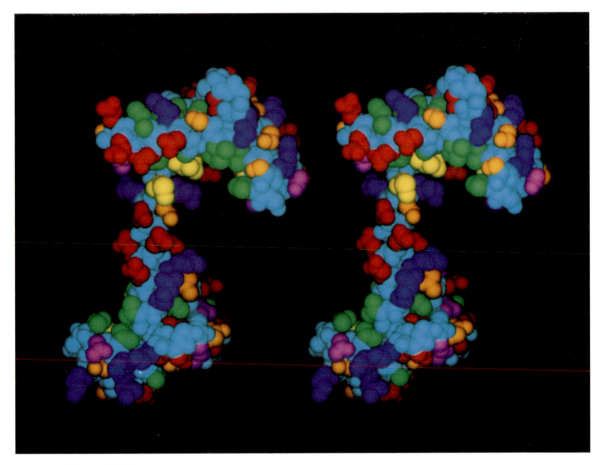

Figure 3.17
Space-filling stereo model of bovine brain calmodulin. Residues 5 to 147 are included, N-termi-nal half at the top. Positively charged side chains (Arg, Lys, His) are dark blue, negatively charged (Asp, Glu) red, hydrophobic (Ala, Val, Leu, Ile, Phe) green, Met yellow, Asn and Gln purple, Ser, Thr, and Tyr orange, and Pro, Gly, and main-chain atoms light blue. The figure was kindly provided by Y. S. Babu et al.[85]

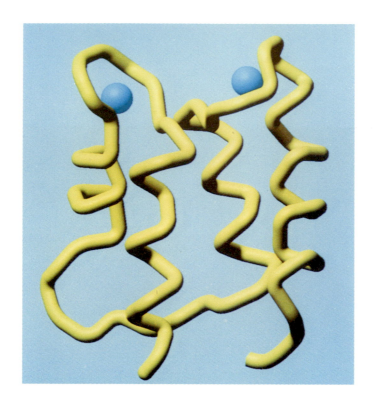

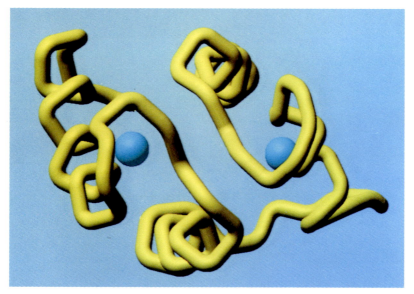

Figure 3.25
The backbone trace of the solution structure of porcine calbindin D_{9k} calculated from NMR data[124] shown in two different views. The position of the calcium ions (blue spheres) is modeled after the crystal structure[123] of bovine calbindin D_{9k}. Figure kindly provided by M. Pique, M. Akke, and W. J. Chazin.

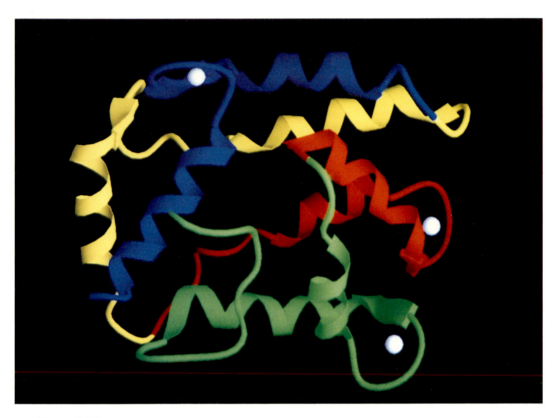

Figure 3.26
Drawing of the crystal structure of the sarcoplasmic calcium-binding protein from *Nereis diversicolor*.[130] The drawing is based on the α-carbon positions; helices are represented by thin ribbons, β-strands by thick ribbons, and Ca^{2+} ions by white spheres. Domain I (residues 1–40) is colored blue, domain II (41–86) yellow, domain III (87–126) red, and domain IV (127–174) green. Figure kindly provided by W. J. Cook.

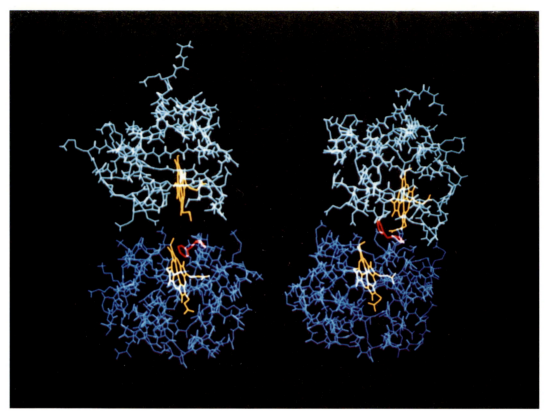

Figure 6.29
Computer-graphics models of the cytochrome b_5/cytochrome c complex: (left) static model produced by docking the x-ray structures of the individual proteins; (right) after extension by molecular dynamics simulations.[112] Reproduced with permission from Reference 112.

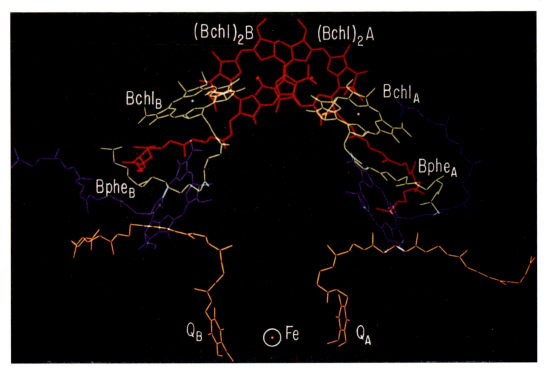

Figure 6.36
Rps. sphaeroides RC cofactors. Electron transfer proceeds preferentially along the A branch.[177]
Reproduced with permission from Reference 177.

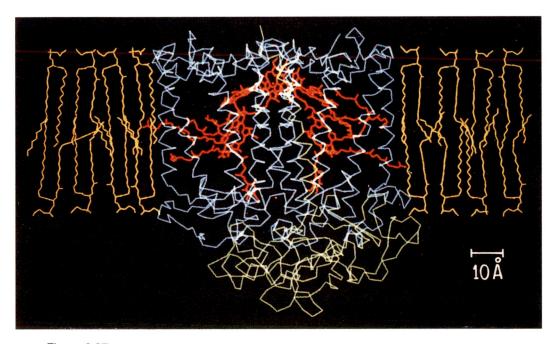

Figure 6.37
Position of the *Rps. sphaeroides* RC in the membrane bilayer. Cofactors are displayed in red, lipids in yellow, the H subunit in green, and L and M subunits in blue.[178] Reproduced with permission from Reference 178.

A

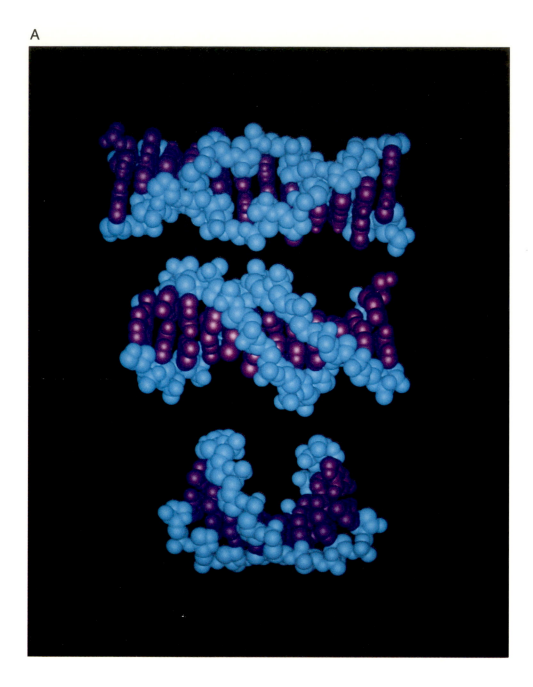

Figure 8.2

(A) Space-filling models depicting A- (left), B- (center), and Z- (right) DNA based on crystallographic data.[2-4] The sugar-phosphate backbone is shown in aqua and the base pairs in purple. This representation as well as subsequent graphics representations were obtained using the program *Macromodel*. (B) Schematic illustration of unusual conformations of DNA. (C) Space-filling models of tRNA[Phe] based on crystallographic data.[4] The sugar-phosphate backbone is shown in aqua and the bases in purple.

B

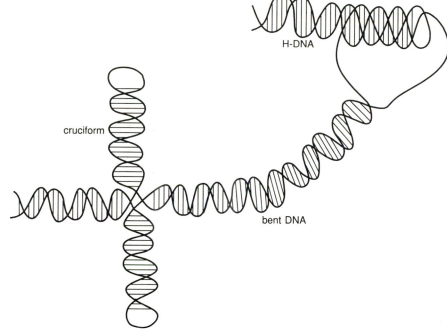

H-DNA

cruciform

bent DNA

C

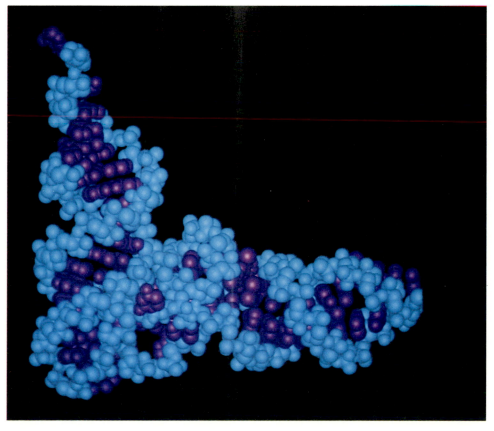

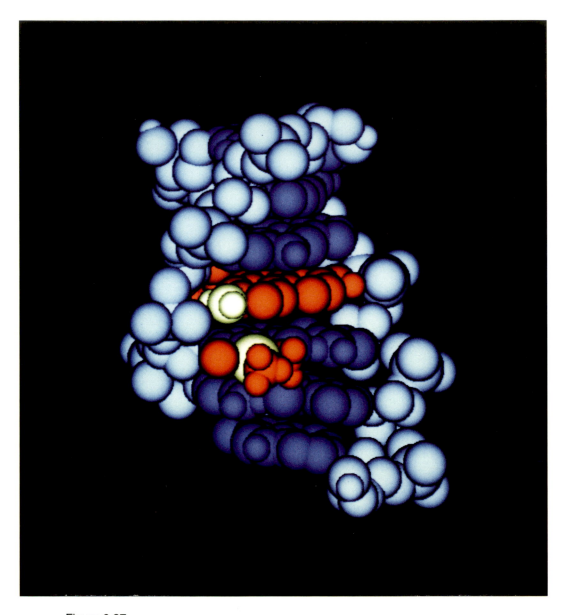

Figure 9.27
Proposed intermediate in the DNA-promoted reaction of *cis*-DDP with ethidium. The Etd$^+$ exocyclic amino group and the Pt atom are highlighted in the major groove (reproduced by permission from Reference 160).

Once the hydrolysis has been performed, the cleaved amino acid still interacts with Arg-145 and with the hydrophobic pocket, whereas the amino group interacts with Glu-270. The carboxylate group of the new terminal amino acid interacts with zinc. This picture, which is a reasonable subsequent step in the catalytic mechanism, finds support from the interaction of L- and D-phenylalanine with carboxypeptidase. [131–134,138]

This mechanism, essentially based on the recent proposal by Christianson and Lipscomb,[137] underlines the role of the Zn—OH moiety in performing the nucleophilic attack much as carbonic anhydrase does. This mechanism can apply with slight changes to thermolysin[139] and other proteases. Thermolysin cleaves peptidic bonds somewhere in the peptidic chain. The mechanism could be very similar, involving zinc bound to two histidines and Glu-166 (Figure 2.31).[139] Glu-166 is monodentate. The role of Glu-270 in CPA is played by Glu-143 and the role of Arg-127 is played by His-231.

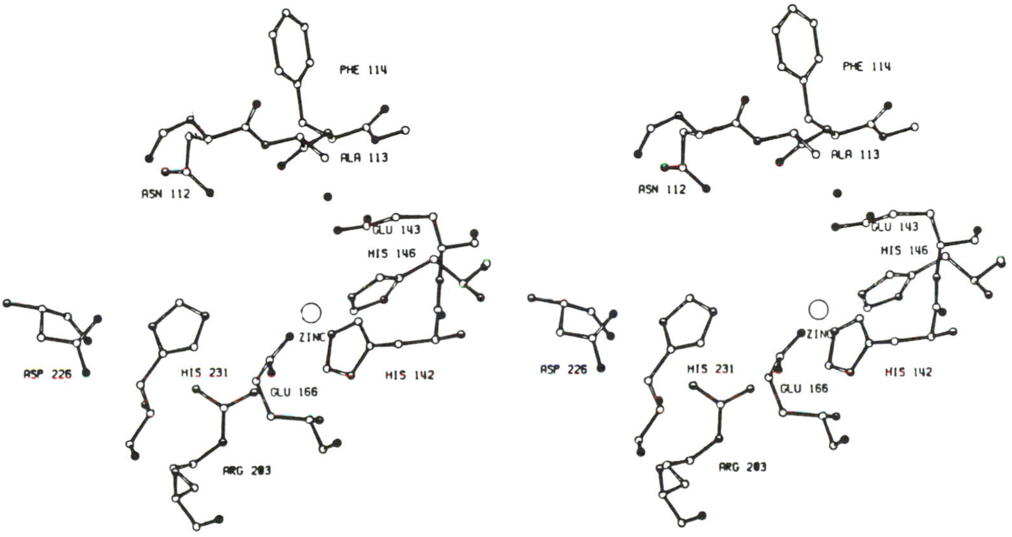

Figure 2.31
Stereo view of the active site of thermolysin.[139]

B. Ester Hydrolysis and Phosphoryl Transfer

Hydrolysis of carboxylic and phosphoric esters is also a slow process at neutral pH, and is catalyzed by acids and bases by mechanisms similar to those involved in amide and peptide hydrolysis. Metal ions are also good catalysts of both carboxylic and phosphoric ester hydrolysis, typically with rate increases much higher than those observed for hydrolysis of amides or peptides (Table

2.8). The ability of metal ions to coordinate to the carbonyl oxygen—which is higher in amides than in esters—is inversely correlated with their catalytic properties, perhaps because the main role of the metal ion is not in polarizing the carbonyl group, but in providing a metal-coordinated hydroxide as the attacking nucleophile.[108] For the hydrolysis of phosphate esters, it is difficult to draw conclusions based on experience with carboxylic esters, because, although the coordinating ability of the phosphoric oxygen may be higher, thus favoring the polarizing role of the metal, the nucleophilic attack is also likely to be easier, because the energy of the trigonal bipyramidal intermediate is probably rather low. Base-catalyzed hydrolysis of phosphate esters occurs with inversion of configuration, and this supports the existence of a trigonal bipyramidal intermediate.[140] The metal acts both as activator of substrate through binding and as Lewis acid to provide the OH moiety for the nucleophilic attack:

$$
H-\overline{O}| \longrightarrow \ \underset{M \underline{\hspace{1.5cm}} O}{\overset{O}{\underset{|}{\overset{\backslash}{O-P-OR}}}} \ \rightleftharpoons \ \underset{M-O\ O}{\overset{OH}{\underset{|}{\overset{|}{O-P-OR}}}} \ \rightleftharpoons \ \underset{M \quad O}{\overset{OH}{\underset{|}{\overset{/}{O-P}\diagdown_{O}}}} + RO^{-} \quad (2.20)
$$

As with peptide hydrolysis, several enzyme systems exist that catalyze carboxylic and phosphoric ester hydrolysis without the need for a metal ion. They generally involve a serine residue as the nucleophile; in turn, serine may be activated by hydrogen-bond formation—or even proton abstraction—by other acid-base groups in the active site. The reaction proceeds to form an acyl- or phosphoryl-enzyme intermediate, which is then hydrolyzed with readdition of a proton to the serine oxygen. Mechanisms of this type have been proposed for chymotrypsin.[141] In glucose-6-phosphatase the nucleophile has been proposed to be a histidine residue.[142]

Again by analogy with peptide hydrolysis, metalloenzymes catalyzing ester hydrolysis may take advantage of additional chemical features provided by amino-acid residues present in the active-site cavity. This situation occurs with carboxypeptidase,[143] which shows esterase activity *in vitro*. Although the rate-limiting steps for carboxylic esters and peptides may differ, several features, such as the pH dependences of k_{cat} and K_m and the presence of two spectroscopically observable intermediates, point to substantially similar mechanisms. On the other hand, carboxylic ester hydrolysis catalyzed by carbonic anhydrase seems to rely on fewer additional features of the active-site cavity, perhaps only on the presence of a metal-coordinated hydroxide that can perform the nucleophilic attack on the carbonyl carbon atom.[47]

Metalloenzyme-catalyzed phosphoric ester hydrolysis can be illustrated by alkaline phosphatase, by far the most-investigated enzyme of this class. The protein is a dimer of 94 kDa containing two zinc(II) and one magnesium(II) ions per monomer, and catalyzes, rather unspecifically, the hydrolysis of a variety of phosphate monoesters as well as transphosphorylation reactions. The x-ray structure at 2.8 Å resolution obtained on a derivative in which all the native metal ions were replaced by cadmium(II) reveals three metals in each subunit,

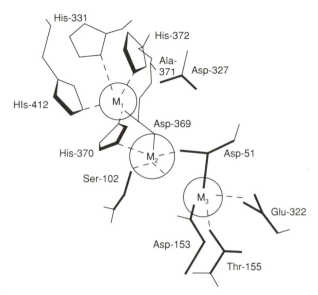

Figure 2.32
Schematic drawing of the active-site cavity of a subunit of alkaline phosphatase.[28,144,145]
The catalytic metal is labeled M_1. The M_1-M_2 distance is $\simeq 4$ Å, the M_2-M_3 distance
is $\simeq 5$ Å, and the M_1-M_3 distance is $\simeq 7$ Å.[144,145]

all located in a single binding region (Figure 2.32). In the native enzyme M_1
and M_2 sites are occupied by zinc and M_3 by magnesium.[144] M_1 was first reported to be coordinated to three histidine residues (His-331, 372, and 412 in
Figure 2.32). Further refinement indicated that Asp-327 could be a ligand to
M_1, in the place of His-372.[145] ^{1}H NMR spectroscopy of the enzyme with cobalt
substituted in the M_1 site shows that there are three exchangeable protons sensing the paramagnetic metal ion.[146] They could come from three histidine NHs,
or from two histidine NHs and another group containing the exchangeable proton very close to the metal ion, like an arginine. Protein ligands to M_2 are Asp-
369, His-370, and Asp-51, the latter probably bridging M_2 to M_3 with the other
carboxyl oxygen. M_3 is coordinated, in addition, to Asp-51, to Asp-153, to Thr-
155, and to Glu-322. Several spectroscopic pieces of evidence on the native and
metal-substituted derivatives indicate that M_1 is five-coordinate, but M_2 and M_3
are six-coordinate, probably with water molecules completing the coordination
spheres.[28]

M_1 is essential for activity, but full catalytic efficiency is reached only when
all metal ions are present. These data suggest that maximum activity is the result
of fine-tuning several chemical properties of the active site as a whole, including
the nature of the M_1 metal, which can be only zinc or cobalt (Table 2.4).

A further key feature of the active site is the presence of a serine residue
(Ser-102), the oxygen atom of which is close to the $M_1 - M_2$ pair (especially
to M_2), although not at direct binding distance according to the crystal structure.
There is ample and direct evidence that Ser-102 is reversibly phosphorylated

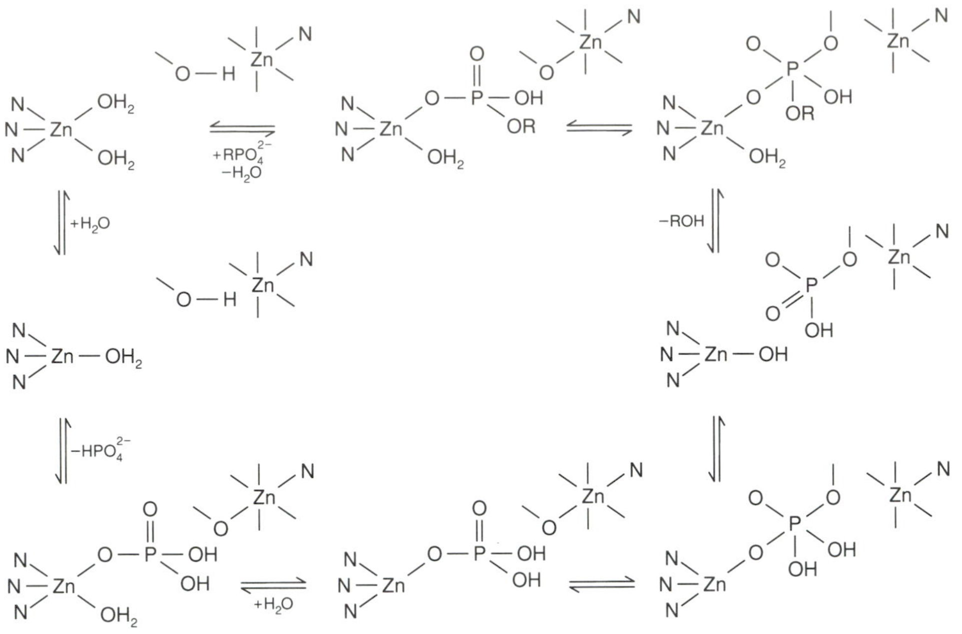

Figure 2.33
Possible catalytic cycle of AP.

during the course of the catalytic reaction, and that M_1 is able to coordinate a phosphate ion.[28]

Another crucial piece of information obtained by physico-chemical techniques is that the lability of the phosphoseryl intermediate and the catalytic activity increase with pH, depending on the state of ionization of an active-site group, which is most likely a water molecule coordinated to M_1.[147] Thus the active form of the enzyme is again a metal-hydroxide species. Furthermore, an inactive derivative with copper ions in the M_1 and M_2 sites shows evidence of magnetic coupling between the metal ions, of the magnitude expected if the two metals shared a common donor atom.[148] Likely candidates are a bridging hydroxide ion or Ser-102, which thus might be somewhat mobile relative to the position occupied in the x-ray structure, and demonstrate its potential ability to be activated for the nucleophilic attack by coordination to a metal ion. Such a mechanism would be an "inorganic" version of the type of activation postulated for chymotrypsin and other hydrolases.

A possible mechanism for alkaline phosphatase-catalyzed phosphoric ester hydrolysis could involve the following steps (Figure 2.33):

(1) Binding of the phosphate group to M_1—in the place of a water molecule—by one of the nonprotonated oxygens, and subsequent activation of the phosphorus atom for nucleophilic attack. The binding of the substrate may be strengthened by interaction with the positively charged

Arg-166 residue[149] (not shown). The steric alteration in the active site could cause movement of Ser-102 toward M_2, with deprotonation upon binding.

(2) Nucleophilic attack on phosphorus by the coordinated serine alkoxide, cleaving the ester bond and liberating the alcohol product.

(3) Formation of the phosphoseryl intermediate with cleavage of the M_1-phosphate bond, decreasing the pK_a of the second coordinated water molecule, the proton of which could be taken up by the leaving alcohol.

(4) Attack by the metal-coordinated hydroxide on the phosphoryl derivative, possibly with M_2 again polarizing the seryl oxygen, yielding a free phosphate ion coordinated to M_1. A further water molecule could aid in the liberation of phosphate via an associative mechanism.

In the presence of alcohols, alkaline phosphatase displays transphosphorylation activity, i.e., hydrolysis of the starting ester and esterification of the phosphate group with a different alcohol. This ability is easily understood if one keeps in mind that the reaction depicted above is reversible, and that a different alcohol may be involved in the formation of the ester bond. Most group-transfer reactions catalyzed by metalloenzymes are likely to proceed through the same elementary steps proposed for hydrolytic reactions.

C. Nucleophilic Addition of OH^- and H^-

Nucleophilic addition of OH^- ions as a step in enzymatic pathways is not restricted to hydrolytic processes; it often occurs in lyases, the class of enzymes catalyzing removal (or incorporation in the reverse reaction) of neutral molecules such as H_2O—but also NH_3, CO_2, etc.—from a substrate. It is outside the scope of this section to review all other mechanisms involved in lyase reactions, especially because they are not reducible to common steps and because several of them do not require the presence of a metal ion. We restrict ourselves to H_2O removal (or incorporation), a widespread feature of which seems to be the splitting of water into the constituents H^+ and OH^- ions at some step of the mechanism. As an example, the dehydration of 2-phospho-D-glycerate to phosphoenolpyruvate catalyzed by enolase, a Mg-activated enzyme,

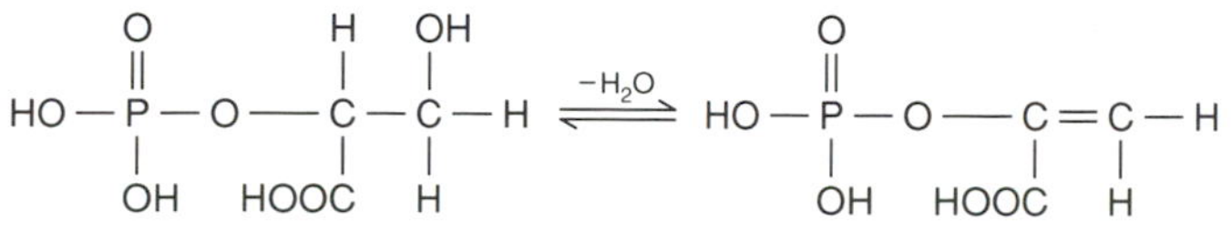

$$\text{HO} - \overset{\overset{\displaystyle O}{\|}}{\underset{\underset{\displaystyle OH}{|}}{P}} - O - \overset{\overset{\displaystyle H}{|}}{\underset{\underset{\displaystyle HOOC}{|}}{C}} - \overset{\overset{\displaystyle OH}{|}}{\underset{\underset{\displaystyle H}{|}}{C}} - H \underset{}{\overset{-H_2O}{\rightleftharpoons}} \text{HO} - \overset{\overset{\displaystyle O}{\|}}{\underset{\underset{\displaystyle OH}{|}}{P}} - O - \overset{}{\underset{\underset{\displaystyle HOOC}{|}}{C}} = \overset{}{\underset{\underset{\displaystyle H}{|}}{C}} - H \qquad (2.21)$$

has been shown by kinetic isotope-effect studies[150] to proceed via fast H^+ removal from substrate followed by slow release of the product, and finally by release of OH^-. The role of a metal ion like magnesium might be to activate the substrate by coordinating the phosphate group, rather than by providing a coordinated hydroxide for nucleophilic attack.

Other lyases, however, contain transition metal ions [often iron(II)], and their main role might well be that of lowering the pK_a of water. None of them, however, is yet known well enough to allow a detailed discussion of the molecular mechanism. A striking exception is carbonic anhydrase, which has been so extensively and successfully studied that it is ideal as a case study (Section IV).

Hydride transfer is another elementary process encountered in many enzymatic reactions. Although hydride transfer implies a redox reaction, it also involves nucleophilic attack on substrate as in the foregoing examples. Unlike OH^-, hydride ions do not exist in aqueous solutions as free ions. In biological systems hydride is always directly transferred from one organic moiety to another by simultaneous breakage and formation of covalent bonds. The activation energy for this process is much higher than, for example, that of H^+ transfer via the formation of hydrogen bonds. Moreover, unlike hydrogen-bonded species, there is no intermediate in the process that can be stabilized by the catalyst. Instead, reacting species can be destabilized in order to lower the activation energy barrier. The role of the enzyme, and of the metal ion when present, is to provide binding sites for both substrates. The enzyme achieves this both geometrically, by allowing for proper orientation of the groups, and electronically, by providing energy to overcome the activation barrier.

These general concepts can be exemplified by liver alcohol dehydrogenases (LADH), dimeric zinc enzymes of 80 kDa that catalyze the following class of reactions using the NADH/NAD$^+$ system as coenzyme (or, really, as cosubstrate):

$$(2.22)$$

In particular, LADHs catalyze the reversible dehydrogenation of primary and secondary alcohols to aldehydes and ketones, respectively. Other enzymatic activities of LADHs are aldehyde dismutation and aldehyde oxidation.[151] The physiological role, although surely related to the metabolism of the above species, is not definitely settled. Much effort is being devoted to understanding the mechanism of action of this class of enzymes, which have obvious implications for the social problem of alcoholism.

Each monomer unit of LADH contains two zinc ions: one coordinated to four cysteine sulfurs, the other coordinated to two cysteine sulfurs, one histidine nitrogen, and a water molecule. The former has no apparent role in catalysis; the latter is essential for catalytic activity. The x-ray structure of the metal-depleted enzyme from horse liver has been solved at 2.4 Å resolution, and that

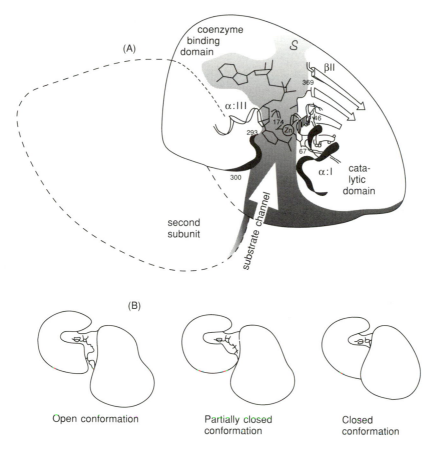

Figure 2.35
Schematic drawing of (A) the LADH dimer and (B) the domains constituting
the active site of a subunit.[154]

of the holoenzyme at 2.9 Å resolution (Figure 2.34 *See color plate section,
page C-6.*). Many crystal structures are also available for binary complexes with
substrates, pseudosubstrates, or coenzymes, as well as for ternary complexes
with coenzyme and substrates.[152] The very detailed picture emerging from such
structural information has helped us understand how LADH functions. As will
be evident from the following discussion, elucidation of this mechanism also
reveals some important fundamental chemistry.

A key property of the enzyme, established by x-ray data, is the existence of
two protein domains in each monomer that are relatively free to rotate relative
to each other. The apo- and holo-enzymes exist in the so-called open form,
whereas binding of NADH coenzyme induces rotation of one domain, resulting
in the so-called closed form[153,154] (Figures 2.34 and 2.35). Closure brings the
catalytic zinc ion into an ideal position to bind the aldehyde substrate in such a
way that the reactive CH_2 group of the nicotinamide ring of NADH points toward

the carbonyl carbon (Figure 2.35). The main functions of the metal are thus to orient the substrate geometrically and to polarize the carbon-oxygen bond. Although the latter makes obvious chemical sense for the aldehyde reduction reaction, since polarization of the C=O bond facilitates nucleophilic attack of hydride at the carbonyl carbon, coordination of an alcohol to a metal is expected to decrease the alcohol's tendency to transfer hydride to NAD^+, unless the hydroxyl proton is released upon coordination.[155]

$$S\diagdown \atop S{-}Zn-OH_2 \atop N\diagup \quad + \quad {R \atop |} \atop {C \atop HO{\diagup} {|\diagdown} H \atop R'} \quad \rightleftharpoons \quad S\diagdown \atop S{-}Zn-{}^-O-C{\diagup R \atop \diagdown H} \atop N\diagup \quad R' \quad + \quad H_3O^+ \qquad (2.23)$$

Formation of an alkoxide ion as an intermediate has often been questioned, because the pK_a of the alcohol would have to be reduced by about 10 units upon coordination.[156] The possibility that hydride transfer from alcohol to NAD^+ and hydroxyl proton release could occur simultaneously is attractive, but careful experiments have shown that the two steps must be kinetically separate.[157] We summarize here the key information that leads to a full, although circumstantial, rationalization of the chemical behavior of the enzyme.

(1) The activity versus pH profiles[156,158] are bell-shaped, with k_{cat} increasing with a pK_a below 7, reaching a plateau, and decreasing with a pK_a above 11, and K_m increasing with a pK_a of about 9.

(2) X-ray data show that the zinc ion is accessible to solvent in the open conformation, much less so in the closed conformation when the reduced coenzyme is bound, and inaccessible when the substrate is coordinated to the metal in the ternary complex, extruding all the water molecules from the active site.[152] None of the complexes has a coordinated water molecule as a fifth ligand when substrates or inhibitors are bound to the metal. The metal ion is always four coordinate and pseudotetrahedral. Computer graphics reveal beyond any doubt that there is no room for a fifth ligand in the active site, at least in the closed form.

(3) Many (although not all) spectroscopic data on metal-substituted derivatives and their binary and ternary complexes have also been interpreted as indicative of a four-coordinate metal.[159] Even nickel(II) and copper(II), which have little tendency to adapt to a pseudotetrahedral ligand environment, do so in LADH, the electronic structure of the latter resembling that of blue proteins (Figure 2.36).[160]

(4) The substrate binding site is actually "created" by the closure of the protein (Figure 2.34). The reactive species are thus trapped in an absolutely anhydrous environment.

The chromophoric aldehyde DACA

$$\qquad (2.24)$$

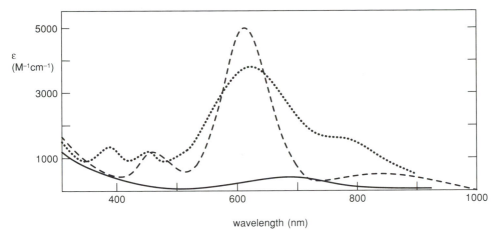

Figure 2.36
Electronic spectra of liver alcohol dehydrogenase substituted with copper at the catalytic site
($\cdots$),[160] together with the spectra of blue (stellacyanin, $---$)[161] and non-blue (superoxide dismutase, $\longrightarrow$)[162] copper proteins.

has been extensively used as an "indicator" of the polarity of the binding site. Large red shifts of the ligand π-π^* transition upon binding indicate the polarity of the site to be much higher than in water; there is a further sizeable increase in polarity when NAD^+ instead of NADH is bound in the ternary complex.[163]

(5) The electronic spectra of the cobalt-substituted derivative are characteristically different when different anions are bound to the metal (Figure 2.37).[164] A catalytically competent ternary complex intermediate displays the electronic absorption pattern typical of anion adducts.[166]

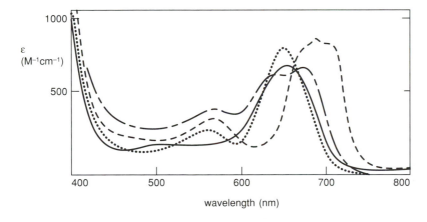

Figure 2.37
Electronic spectra of liver alcohol dehydrogenase substituted with cobalt at the catalytic site. Binary complex with NAD^+ ($\longrightarrow$);[165] ternary complex with NAD^+ and Cl^- ($---$);[165] binary complex with acetate ($\cdots$);[164] intermediate in the oxidation of benzyl alcohol with NAD^+ ($-----$).[166]

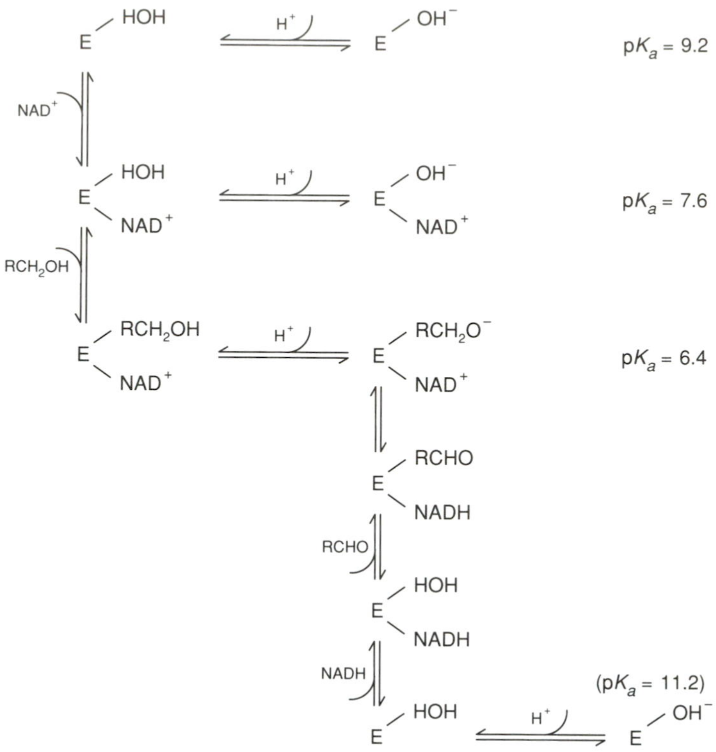

HOH   H⁺   OH⁻

E ... E $pK_a = 9.2$

NAD⁺

HOH   H⁺   OH⁻

E (NAD⁺) ... E (NAD⁺) $pK_a = 7.6$

RCH₂OH

RCH_2OH H⁺ RCH_2O^-

E (NAD⁺) ... E (NAD⁺) $pK_a = 6.4$

$RCHO$

E (NADH)

RCHO

HOH

E (NADH)

NADH

HOH H⁺ OH^- $(pK_a = 11.2)$

E ... E

Figure 2.38

Protonation scheme for LADH and its adducts with coenzymes and substrates.[157,167]

(6) From extended kinetics measurements a protonation scheme (Figure 2.38) has been proposed that accounts for the many pK_a values observed under different conditions.[167] This scheme again requires formation of a coordinated alkoxide intermediate, but has the advantage of rationalizing in a simple way a complex pattern. In essence, the only relevant acid-base group supplied by the enzyme is the metal-coordinated water, which has a pK_a of 9.2 in the free enzyme (open form). Upon binding of NADH the pK_a increases to 11.2. Since NADH dissociation is the last and rate-limiting step of the alcohol oxidation reaction, the decrease in k_{cat} with this pK_a is accounted for by a decrease in dissociation rate of NADH from the hydroxo form. On the other hand, the pK_a of water is decreased to 7.6 upon binding of NAD⁺. These rather large changes in both directions are best explained by a marked sensitivity of the coordinated water molecule to the polarity of the environment, which, with the possible exception of the unligated form that has a more or less "regular" pK_a value of 9.2, can be almost completely anhydrous and much different from that of bulk water. The nonpolar nicotinamide ring of NADH decreases the overall electrostatic interactions of the water molecule, whereas the positive charge of NAD⁺ drastically increases them. In this scheme, the association rates of both coenzymes are predicted to (and, in fact, do) decrease with a pK_a of 9.2, the dissociation

rate of NAD$^+$ is predicted to (and does) decrease with a pK_a of 7.6, and the dissociation rate of NADH is predicted to decrease with a pK_a of 11.2 (and, indeed, it is pH-independent up to and above pH 10).

The decrease of k_{cat} at low pH depends on an ionization that in turn depends on the substrate. This pK_a must be that of the coordinated alcohol; at too low a pH, deprotonation of the coordinated alcohol becomes the rate-limiting step. The pK_a values observed for this process range from 6.4 for ethanol to 4.3 for trifluoroethanol. What is surprising for aqueous-solution chemistry—that the pK_a of a coordinated alcohol is lower than the pK_a of a coordinated water molecule—can now be explained in terms of the different polarity of the two adducts in LADH. In the binary complex with NAD$^+$ (pK_a = 7.6), the water molecule is still free to interact through H-bonding with the solvent and partially dissipate the electrostatic charge. In the ternary complex with any alcohol, the R group may prevent access of the solvent to the cavity, decreasing the dielectric constant of the medium. As a consequence, the polarity of the environment is increased. It is interesting to speculate that Nature could have chosen a stronger Lewis acid than a zinc ion coordinated to two negatively charged residues to decrease the pK_a of a coordinated alkoxide, but then the pK_a of the coordinated water would have simultaneously undergone a parallel and possibly even stronger decrease. Instead, LADH provides a self-regulating environment that is tailored to decrease the pK_a of a coordinated alcohol, once properly positioned, more than that of a coordinated water. The full catalytic cycle for the dehydrogenation reaction at pH around 7 can be summarized as follows (Figure 2.39):

(1) NAD$^+$ binds to the open, water-containing form of the enzyme with a maximal on-rate. The pK_a of water is decreased to 7.6, but water is still mostly unionized.

(2) A neutral alcohol molecule enters the crevice between the two domains, and coordinates the zinc ion by displacing the water molecule. The protein is still in the open form.

(3) Domain rotation brings the protein into the closed form, excluding all the residual water molecules from the active site; the combined effect of the metal positive charge and of the unshielded positive charge of the nicotinamide ring lowers the pK_a of the coordinated alcohol below 7. A proton is expelled from the cavity, possibly via a hydrogen-bond network of protein residues.

(4) Direct hydride transfer takes place from the alcohol CH to the 4-position of the properly oriented nicotinamide ring. The resulting ternary complex is an NADH-aldehyde adduct. The polarity of the active site dramatically drops.

(5) The aldehyde product leaves and is replaced by a neutral water molecule (its pK_a now being 11.2). Additional water molecules can now enter the crevice, favoring the partial opening of the structure.

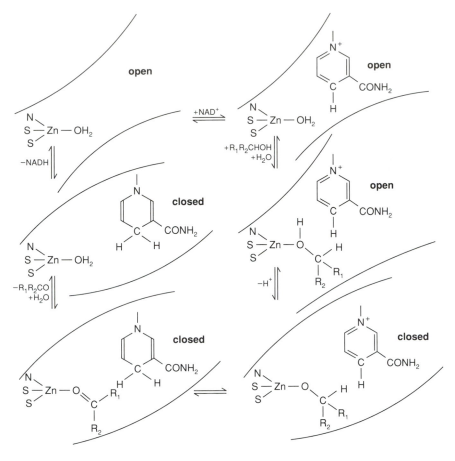

Figure 2.39
Possible catalytic cycle of LADH.

(6) The loss of contacts between the two halves of the channel favors a complete opening and then the release of NADH, whose dissociation rate is maximal and pH-independent.

D. Group Transfer and Vitamin B_{12}

1. Group transfer enzymes

The phosphodiester bond in ATP and in related molecules is a high-energy bond whose hydrolysis liberates a large quantity of energy:

$$\text{ATP} + H_2O \rightleftharpoons \text{ADP} + P_i + 30\text{-}50 \text{ kJ mol}^{-1} \tag{2.25}$$

In many systems, typically the ATPases, the terminal phosphoryl group is transferred to another acidic group of the enzyme, e.g., a carboxylate group, to form another high-energy bond whose energy of hydrolysis is needed later for some endoenergetic transformation. Therefore the first step of the reaction is the phosphoryl transfer to a group of the enzyme.

Kinases, a subset of the class of transferases, constitute a large group of enzymes that phosphorylate organic substrates:

$$ATP^{4-} + HX \rightleftharpoons ADP^{3-} + PX^{2-} + H^+ \qquad (2.26)$$

In some kinases, such as nucleoside diphosphate kinase,[168,169] an intermediate step is the phosphoryl transfer to a group belonging to the enzyme, as happens in ATPase and as was discussed in detail for alkaline phosphatase (Section V.B). In other kinases the phosphoryl transfer occurs directly from the donor to the acceptor in a ternary complex of the enzyme with the two substrates.[170] Often metal ions like magnesium or manganese are needed. These ions interact with the terminal oxygen of the ATP molecule, thus facilitating the nucleophilic attack by the acceptor. The metal ion is often associated with the enzyme. For mechanistic schemes, see the proposed mechanism of action of alkaline phosphatase, especially when a phosphoryl enzyme intermediate is involved.

2. The B_{12}-dependent enzymes

There are many enzymes that need a cobalt complex as cofactor in order to carry out vicinal 1,2 interchange:

$$(2.27)$$

or

$$(2.28)$$

For the former type of reactions, X can be a group containing either C or N. Typical reactions[171] include insertion of a secondary methyl group into a main chain

$$(2.29)$$

isomerization of an amino group from a primary to a secondary carbon

$$(2.30)$$

and deamination reactions

$$H_2N{\diagup}{\diagup}OH \longrightarrow \text{(acetaldehyde)} H + NH_3 \qquad (2.31)$$

A list of coenzyme-B_{12}-dependent enzymes is given in Table 2.10.

Table 2.10
Some coenzyme-B_{12}-dependent
enzymes.

MethylmalonylCoA mutase
Glutamate mutase
α-Methylene-glutarate mutase
Dioldehydrase
Glyceroldehydrase
Ethanoldeaminase
L-β-lysine mutase
D-α-lysine mutase
Ribonucleotide reductase
Methionine synthetase
Methane synthetase
Methyl transferase
Acetate synthetase

In coenzyme-B_{12}, cobalt is bound to a tetraazamacrocyclic ligand[172] (Figure 2.40). The cobalt atom lies approximately in the plane of the corrin ligand (shown in bold). Note that rings A and D are directly linked. The conjugation therefore extends over only 13 atoms, excluding the cobalt, and involves 14 π electrons. Complexes that possess the α-D-ribofuranose-3-phosphate and the terminal 5,6-dimethylbenzimidazole as an axial ligand are called cobalamins. The name cobamides applies to complexes that lack or have different heterocyclic groups. Finally, the upper or β position is occupied by another ligand, which may be water, OH^-, CN^-, an alkyl group, etc. The cyano derivative (ii) is vitamin B_{12}. 5'-deoxyadenosylcobalamin (i) is called coenzyme B_{12}. The cobalt atom in these complexes is a diamagnetic cobalt(III) system (d^6).

The aquo complex has a pK_a of 7.8, which compares with that of 5.7 for the aquopentaamminecobalt(III) complex at 298 K.[173] The difference has been mainly ascribed to the difference in solvation of the two complexes, although the corrin ligand bears a negative charge, which reduces the positive charge and therefore the Lewis acidity of the metal ion. The standard reduction potential between pH 2.9 and 7.8 is -0.04 V vs. SCE, featuring the conversion from aquocobalamin with bound benzimidazole (base on) to base-on cob(II)alamin.[174] The potential decreases with pH above pH 7.8 down to -0.3 V. The reduced form is five-coordinate, without the water molecule above pH 2.9, and low-spin.[175,176] The system can be further reduced at a potential of -0.85 V to

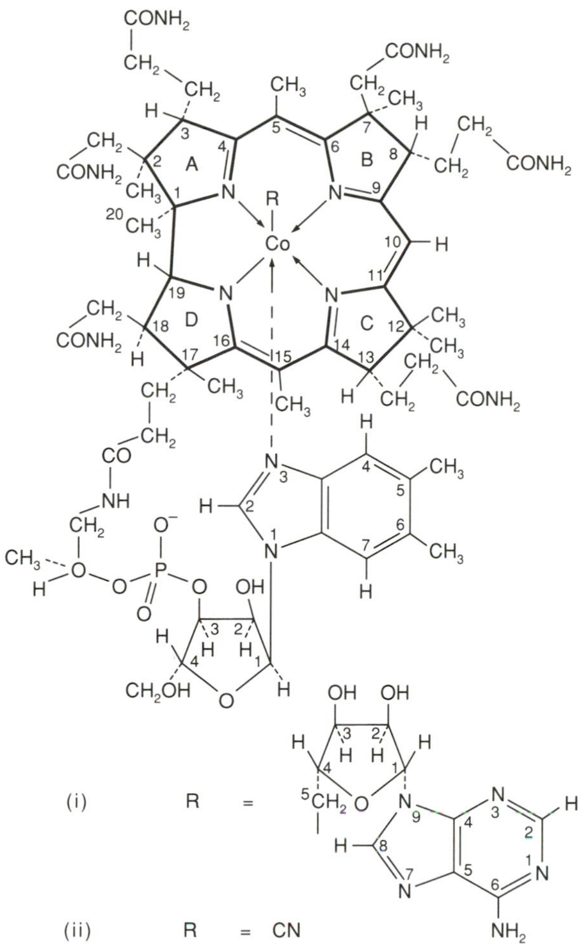

Figure 2.40
Structure of (i) coenzyme B_{12}, 5'-deoxyadenosylcobalamin, and (ii) vitamin B_{12}, cyanocobalamin.[172]

obtain cob(I)alamin, in which the metal ion is four-coordinate and low-spin (d^8). The standard reduction potential for the hexaaquocobalt(III) complex is 1.95 V, which is lowered to 0.10 for the hexaammine complex, to -0.13 for the tris-ethylenediamine complex, and to -0.80 for the hexacyanocobaltate(III) ion.[173] After reduction to cobalt(II), the model complexes are reduced to the metal.

The electronic spectrum of the metal-free corrin resembles that of metal derivatives; it seems therefore that the bands are essentially π-π* transitions modified by the central atom and by the axial ligands.[177,178] The cob(III)alamins are red, whereas the cob(II)alamins, which are brown, show an additional band at 600 nm.[179] The latter have an EPR spectrum typical of an unpaired electron in the d_z^2 orbital with some 4s mixing: the cob(II)alamin at pH 7 has $g_\parallel = 2.004$,

$g_\perp$ = 2.32, $A_\|(Co)$ = 0.0100, $A_\perp$ (Co) = 0.0027 cm^{-1}, and $A_\|$ (N) = 0.00173 cm^{-1}.

Both cobalt(I) and cobalt(II)-containing cobalamins readily react with alkyl derivatives to give alkylcob(III)alamins:

$$Cob(II)alamin + CH_3I \rightarrow CH_3\text{-}Cob(III)alamin + 1/2I_2 \qquad (2.32)$$

$$Cob(I)alamin + CH_3I \rightarrow CH_3\text{-}Cob(III)alamin + I^- \qquad (2.33)$$

These can formally be regarded as complexes of cobalt(III) with a carbanion. These are rare examples of naturally occurring organometallic compounds. The Co—C bond in alkylcobalamins is relatively weak (bond dissociation energy $\simeq$ 100 kJ mol^{-1}, though higher values are reported in the literature[182,183]) and can be broken thermally (by heating the complex above 100°C)[182-184] or photochemically, even in daylight exposure.[180,181] The energy of the Co—C bond is about 17 kJ mol^{-1} greater when the transaxial base is absent.[184]

The cobalamin coenzyme is bound by the apoenzyme with no significant change in the absorption spectrum.[185] This suggests that no major change occurs in the coordination of cobalt(III). The first step of the reaction involves homolytic fission of the Co—C bond:[182-184, 186-188]

$$B\text{—}Co^{III}\text{—}R \rightarrow B\text{—}Co^{II}\cdot + R\cdot \qquad (2.34)$$

where B and R are the ligands at the α and β apical positions. The 5′-deoxyadenosyl radical probably reacts with the substrate, generically indicated as SubH, to give the Sub· radical and RH. Then the rearrangement reaction proceeds along a not-well-established pathway. It is the protein-substrate binding that controls the subsequent chemistry. In the absence of protein the Co—C bond is kinetically stable; in the presence of protein and substrate the rate of labilization of the Co—C bond increases by a factor of 10^{11}–10^{12}.[182-185] By generating the radical in the coenzyme without the protein by means of photolysis or thermolysis, we enable the coenzyme to catalyze some rearrangement reactions without the protein. It may therefore be that the protein plays a major role in inducing the homolytic fission, but a relatively minor role in the subsequent steps, perhaps confined to preventing the various species from diffusing away from each other.

Studies on protein-free corrinoids and model complexes have shown that increasing the steric bulkiness around the coordinated $C\alpha$ atom can cause a dramatic labilization of the Co—C bond.[189] The protein-coenzyme adduct might contain the coenzyme in a resting state and the protein in a strained state; the substrate would then switch the system into a strained coenzyme and a relaxed enzyme with little thermodynamic barrier. The strained form of the coenzyme is then in labile equilibrium with base-on cobalt(II) and the free radical.[190] This hypothesis, that conformational changes in cobalamin can switch chemical reactions on and off, is closely analogous with the known aspects of hemoglobin function.

It has been suggested that the radical formation in the coenzyme is triggered by a steric perturbation involving an enzyme-induced conformational distortion of the corrin ring toward the deoxyadenosyl group, thereby weakening the cobalt-carbon bond.[187,190–194] Structural studies of different corrinoid complexes reveal highly puckered and variable conformations of the corrin ring, attesting to its flexibility.[195] For the dimethylglyoxime models, it has been shown that increasing the size of the axial ligand B does induce Co—C bond lengthening and weakening because of conformational distortion of the equatorial ligand away from B and toward the R group.[196] It has been proposed that the flexibility of the corrin ligand is the reason why Nature does not use the porphyrin ligand in vitamin B_{12}.[197] In an alternative explanation, the weakening of the Co—C bond would be an electronic effect associated with the labilization of the Co—N bond.[198]

VI. PERSPECTIVES

Although a great deal is known about the biophysical characteristics of the various enzyme derivatives mentioned in this chapter, we are still far from a clear understanding of their mechanisms of action, especially if we take into consideration the role of each amino-acid residue inside the active-site cavity. Although we can successfully discuss why certain metal ions are used in certain biological reactions, we still do not know why nickel(II), for example, is involved in the enzymatic hydrolysis of urea.[199,200] If we are content with the explanations given in Sections III.A or V.D, we would need model compounds that are good catalysts and perform the job in several steps. This latter requirement would make the various models much more interesting, and would represent a new objective in the investigation of the structure-function relationship of catalytically active molecules. Indeed, the synthesis of large polypeptides may in principle provide such models. In this respect we need to know more about protein folding, for which emerging techniques like protein computer graphics and molecular dynamics are very promising.

Chemical modifications of proteins like the alkylation of carboxylate[124,201] or histidine[202] residues have been performed for a long time. A newer approach toward modeling the function of a protein, and understanding the role of the active site, involves cleaving part of a naturally occurring protein through enzymatic or chemical procedures, and then replacing it with a synthetic polypeptide. The use of modern techniques of molecular genetics has allowed site-directed mutagenesis to become in principle a very powerful technique for changing a single residue in a cavity. Site-directed mutagenesis is a very popular approach, and its principal limitation with respect to the synthetic polypeptide route is that only natural amino acids can be used (aside from the technical difficulties in both approaches). Small quantities of site-directed mutants have been obtained for CPA[125–127] and AP,[203] whereas the expression of CA[204,205] is now satisfactory.

Predictions of the changes in structure needed to affect the reaction pathway can nowadays be made with the aid of computers. The occurrence of the predicted change can be checked through x-ray analysis and NMR. The latter spectroscopy is today well-recognized as being able to provide structural information on small ($\leq$20 kDa) proteins through 2- or 3-dimensional techniques.[206-208] These techniques are increasingly being applied to paramagnetic metalloproteins such as many of those discussed here.[208,209] The advantage of handling a paramagnetic metalloprotein is that we can analyze signals shifted far away from their diamagnetic positions, which correspond to protons close to the metal ion,[69] even for larger proteins. It is possible to monitor the distances between two or more protons under various conditions, such as after the addition of inhibitors or pseudosubstrates, chemical modification, or substitution of a specific amino acid.

VII. REFERENCES

1. H. Sigel and A. Sigel, eds., *Metal Ions in Biological Systems*, Dekker, **26** (1990).
2. R. K. Andrews, R. L. Blakeley, and B. Zerner, in Reference 1, **23** (1988).
3. K. Doi, B. C. Antanaitis, and P. Aisen, *Struct. Bonding* **70** (1988), 1.
4. M. M. Werst, M. C. Kennedy, H. Beinert, and B. M. Hoffman, *Biochemistry* **29** (1990), 10526, and references therein.
5. A. G. Orpen *et al.*, *J. Chem. Soc., Dalton Trans.* **S1** (1989).
6. F. H. Westheimer, *Spec. Publ. Chem. Soc.* **8** (1957), 1.
7. F. Basolo and R. G. Pearson, *Mechanisms of Inorganic Reactions*, Wiley, 2d ed., 1967.
8. L. G. Sillen and A. E. Martell, *Stability Constants of Metal-Ion Complexes*, Spec. Publ. Chem. Soc., London, **25**, 1971.
9. E. J. Billo, *Inorg. Nucl. Chem. Lett.* **11** (1975), 491.
10. I. Bertini, G. Canti, C. Luchinat, and F. Mani, *Inorg. Chem.* **20** (1981), 1670.
11. S. Burki, Ph.D. Thesis, Univ. Basel, 1977.
12. P. Wolley, *Nature* **258** (1975), 677.
13. J. T. Groves and R. R. Chambers, *J. Am. Chem. Soc.* **106** (1984), 630.
14. J. T. Groves and J. R. Olson, *Inorg. Chem.* **24** (1985), 2717.
15. L. J. Zompa, *Inorg. Chem.* **17** (1978), 2531.
16. E. Kimura, T. Koike, and K. Toriumi, *Inorg. Chem.* **27** (1988), 3687; E. Kimura *et al.*, *J. Am. Chem. Soc.* **112** (1990), 5805.
17. R. S. Brown *et al.*, *J. Am. Chem. Soc.* **104** (1982), 3188.
18. I. Bertini *et al.*, *Inorg. Chem* **29** (1990), 1460.
19. R. J. P. Williams, *Coord. Chem. Rev.* **100** (1990), 573.
20. D. W. Christianson, *Adv. Prot. Chem.* **42** (1991), 281.
21. I. Bertini, C. Luchinat, and M. S. Viezzoli, in I. Bertini *et al.*, eds., *Zinc Enzymes*, Birkhaüser, 1986, p.27.
22. D. S. Auld and B. L. Vallee, in Reference 21, p. 167.
23. G. Formicka-Kozlowska, W. Maret, and M. Zeppezauer, in Reference 21, p. 579.
24. I. Bertini and C. Luchinat, in Reference 30, p. 101.
25. I. Bertini and C. Luchinat, *Adv. Inorg. Biochem.* **6** (1984), 71.
26. H. Sigel, ed., *Metal Ions in Biological Systems*, Dekker, **12** (1981).
27. T. G. Spiro, ed., *Copper Proteins*, Wiley, 1981.
28. J. E. Coleman and P. Gettins, in Reference 21, p. 77.
29. J. E. Coleman and D. P. Giedroc, in Reference 1, **25** (1989).
30. H. Sigel, ed., *Metal Ions in Biological Systems*, Dekker, **15** (1983).
31. N. U. Meldrum and F. J. W. Roughton, *J. Physiol.* **75** (1932), 4.
32. D. Keilin and T. Mann, *Biochem. J.* **34** (1940), 1163.

33. S. Lindskog, in T. G. Spiro, ed., *Zinc Enzymes: Metal Ions in Biology,* Wiley, **5** (1983), 78; D. N. Silverman and S. Lindskog, *Acc. Chem. Res.* **21** (1988), 30.

34. J.-Y. Liang and W. N. Lipscomb, *J. Am. Chem. Soc.* **108** (1986), 5051.

35. K. M. Merz, *J. Am. Chem. Soc.* **112** (1990), 7973.

36. (a) G. Sanyal, *Ann. N. Y. Acad. Sci.* **429** (1984), 165; (b) G. Sanyal and T. H. Maren, *J. Biol. Chem.* **256** (1981), 608.

37. T. Kararli and D. N. Silverman, *J. Biol. Chem.* **260** (1985), 3484.

38. A. Liljas *et al.*, *Nature* **235** (1972), 131.

39. K. K. Kannan *et al.*, *Proc. Natl. Acad. Sci. USA,* **72** (1975), 51.

40. E. A. Eriksson *et al.*, in Reference 21, p. 317.

41. S. Lindskog in Reference 21, p. 307.

42. E. Clementi *et al.*, *FEBS Lett.* **100** (1979), 313.

43. B. P. N. Ko *et al.*, *Biochemistry* **16** (1977), 1720.

44. A. Ikai, S. Tanaka, and H. Noda, *Arch. Biochem. Biophys.* **190** (1978), 39.

45. I. Bertini, C. Luchinat, and A. Scozzafava, *Struct. Bonding* **48** (1982), 45.

46. A useful review by W. W. Cleland on pH-dependent kinetics can be found in *Methods Enzymol.*, 1987, 390.

47. Y. Pocker and S. Sarkanen, *Adv. Enzymol.* **47** (1978), 149.

48. I. Bertini and C. Luchinat, *Acc. Chem. Res.* **16** (1983), 272.

49. S. Lindskog and I. Simonsson, *Eur. J. Biochem.* **123** (1982), 29.

50. B.-H. Jonsson, H. Steiner, and S. Lindskog, *FEBS Lett.* **64** (1976), 310.

51. H. Steiner, B.-H. Jonsson, and S. Lindskog, *Eur. J. Biochem.* **59** (1975), 253.

52. D. N. Silverman *et al.*, *J. Am. Chem. Soc.* **101** (1979), 6734.

53. S. H. Koenig *et al.*, *Pure Appl. Chem.* **40** (1974), 103.

54. I. Simonsson, B.-H. Jonsson, and S. Lindskog, *Eur. J. Biochem.* **93** (1979), 4.

55. T. J. Williams and R. W. Henkens, *Biochemistry* **24** (1985), 2459.

56. I. Bertini, C. Luchinat, and M. Monnanni, in Reference 99, p. 139.

57. I. Bertini *et al.*, in Reference 21, p. 371.

58. I. Bertini *et al.*, *Inorg. Chem.* **24** (1985), 301.

59. I. Bertini *et al.*, *J. Am. Chem. Soc.* **100** (1978), 4873.

60. R. C. Rosenberg, C. A. Root, and H. B. Gray, *J. Am. Chem. Soc.* **97** (1975), 21.

61. B. Holmquist, T. A. Kaden, and B. L. Vallee, *Biochemistry* **14** (1975), 1454, and references therein.

62. M. W. Makinen and G. B. Wells, in H. Sigel, ed., *Metal Ions in Biological Systems,* Dekker, **22** (1987).

63. M. W. Makinen *et al.*, *J. Am. Chem. Soc.* **107** (1985), 5245.

64. A. E. Eriksson, Uppsala Dissertation, Faculty of Science, n. 164, 1988.

65. A. E. Eriksson, A. T. Jones, and A. Liljas, *Proteins* **4** (1989), 274.

66. A. E. Eriksson *et al.*, *Proteins* **4** (1989), 283.

67. I. Bertini *et al.*, *Inorg. Chem.,* **31** (1992), 3975.

68. P. H. Haffner and J. E. Coleman, *J. Biol. Chem.* **248** (1973), 6630.

69. I. Bertini and C. Luchinat, *NMR of Paramagnetic Molecules in Biological Systems,* Benjamin/Cummings, 1986.

70. R. D. Brown III, C. F. Brewer, and S. H. Koenig, *Biochemistry* **16** (1977), 3883.

71. I. Bertini, C. Luchinat, and M. Messori, in Reference 62, vol. 21.

72. I. Solomon, *Phys. Rev.* **99** (1955), 559.

73. I. Bertini *et al.*, *J. Magn. Reson.* **59** (1984), 213.

74. L. Banci, I. Bertini, and C. Luchinat, *Magn. Res. Rev.*, **11** (1986), 1.

75. I. Bertini *et al.*, *J. Am. Chem. Soc.*, **103** (1981), 7784.

76. T. H. Maren, A. L. Parcell, and M. N. Malik, *J. Pharmacol. Exp. Theor.* **130** (1960), 389.

77. J. E. Coleman, *J. Biol. Chem.* **243** (1968), 4574.

78. S. Lindskog, *Adv. Inorg. Biochem.* **4** (1982), 115.

79. G. Alberti *et al.*, *Biochim. Biophys. Acta* **16** (1981), 668.

80. J. I. Rogers, J. Mukherjee, and R. G. Khalifah, *Biochemistry* **26** (1987), 5672.

81. C. Luchinat, R. Monnanni, and M. Sola, *Inorg. Chim. Acta* **177** (1990), 133.

82. L. Morpurgo *et al.*, *Arch. Biochem. Biophys.* **170** (1975), 360.

83. I. Bertini and C. Luchinat, in K. D. Karlin and J. Zubieta, eds., *Biological and Inorganic Copper Chemistry*, vol. 1, Adenine Press, 1986.

84. I. Bertini *et al.*, *J. Chem. Soc., Dalton Trans.* (1978), 1269.

85. P. H. Haffner and J. E. Coleman, *J. Biol. Chem.* **250** (1975), 996.

86. I. Bertini *et al.*, *J. Inorg. Biochem.* **18** (1983), 221.

87. I. Bertini, E. Borghi, and C. Luchinat, *J. Am. Chem. Soc.* **102** (1979), 7069.

88. I. Bertini *et al.*, *J. Am. Chem. Soc.* **109** (1987), 7855.

89. P. Yeagle, Y. Lochmüller, and R. W. Henkens, *Proc. Natl. Acad. Sci. USA* **48** (1975), 1728.

90. P. J. Stein, S. T. Merrill, and R. W. Henkens, *J. Am. Chem. Soc.* **99** (1977), 3194.

91. P. S. Hubbard, *Proc. Roy. Soc. London* **291** (1966), 537.

92. A. Lanir and G. Navon, *Biochemistry* **11** (1972), 3536.

93. J. J. Led and E. Neesgard, *Biochemistry* **26** (1987), 183.

94. N. B.-H. Johnsson *et al.*, *Proc. Natl. Acad. Sci. USA* **77** (1980), 3269.

95. J. L. Evelhoch, D. F. Bocian, and J. L. Sudmeier, *Biochemistry* **20** (1981), 4951.

96. J.-Y. Liang and W. N. Lipscomb, *Proc. Natl. Acad. Sci. USA* **87** (1990), 3675.

97. K. M. Merz, *J. Mol. Biol.* **214** (1990), 799.

98. C. A. Fierke, T. L. Calderone, and J. F. Krebs, *Biochemistry* **30** (1991) 11054.

99. M. Aresta and G. Forti, eds., *Carbon Dioxide as a Source of Carbon*, Reidel, 1987.

100. M. Aresta and J. V. Schloss, eds., *Enzymatic and Model Carboxylation and Reduction Reactions for Carbon Dioxide Utilization*, Kluwer, 1990.

101. C. K. Tu and D. N. Silverman, *J. Am. Chem. Soc.* **108** (1986), 6065.

102. E. Chaffee, T. P. Dasgupta, and J. M. Harris, *J. Am. Chem. Soc.* **95** (1973), 4169.

103. J. B. Hunt, A. C. Rutenberg, and H. Taube, *J. Am. Chem. Soc.* **74** (1983), 268.

104. R. S. Brown, N. J. Curtis, and J. Huguet, *J. Am. Chem. Soc.* **103** (1981), 6953.

105. I. Tabushi and Y. Kuroda, *J. Am. Chem. Soc.* **106** (1984), 4580.

106. I. Bertini *et al.*, *Gazz. Chim. Ital.* **118** (1988), 777.

107. L. Meriwether and F. H. Westheimer, *J. Am. Chem. Soc.* **78** (1956), 5119.

108. D. A. Buckingham, D. M. Foster, and A. M. Sargeson, *J. Am. Chem. Soc.* **92** (1970), 6151.

109. M. L. Bender, R. J. Bergeron, and M. Komiyama, *The Bioorganic Chemistry of Enzymatic Catalysis*, Wiley, 1984.

110. B. Anderson *et al.*, *J. Am. Chem. Soc.* **99** (1977), 2652.

111. M. L. Bender and B. W. Turnquest, *J. Am. Chem. Soc.* **77** (1955), 4271.

112. D. L. Miller and F. H. Westheimer, *J. Am. Chem. Soc.* **88** (1966), 1514.

113. H. Kroll, *J. Am. Chem. Soc.* **74** (1952), 2036.

114. D. A. Buckingham, D. M. Foster, and A. M. Sargeson, *J. Am. Chem. Soc.* **90** (1968), 6032.

115. R. Breslow, in R. F. Gould, ed., *Bioinorganic Chemistry* (*Advances in Chemistry Series*, vol. **100**), American Chemical Society, 1971; Chapter 2.

116. A. Schepartz and R. Breslow, *J. Am. Chem. Soc.* **109** (1987), 1814.

117. J. T. Groves and R. R. Chambers, Jr., *J. Am. Chem. Soc.* **106** (1984), 630.

118. M. A. Wells and T. C. Bruice, *J. Am. Chem. Soc.* **99** (1977), 5341.

119. H. Sigel, ed., *Metal Ions in Biological Systems*, Dekker, **5** (1976).

120. N. E. Dixon and A. M. Sargeson, in Reference 33.

121. R. W. Hay, G. Wilkinson, R. D. Gillard, and J. A. McCleverty, eds., in *Comprehensive Coordination Chemistry*, Pergamon Press, 1987.

122. D. C. Rees, M. Lewis, and W. N. Lipscomb, *J. Mol. Biol.* **168** (1983), 367.

123. D. C. Rees *et al.*, in Reference 21, p. 155.

124. D. S. Auld, K. Larson, and B. L. Vallee, in Reference 21, p. 133.

125. W. J. Rutter, personal communication.

126. D. Hilvert *et al.*, *J. Am. Chem. Soc.* **108** (1986), 5298.

127. S. J. Gardell *et al.*, *J. Biol. Chem.* **262** (1987), 576.

128. D. S. Auld *et al.*, *Biochemistry,* **31** (1992), 3840; W. L. Mock and J. T. Tsay, *J. Biol. Chem.* **263** (1988), 8635.

129. G. Shoham, D. C. Rees, and W. N. Lipscomb, *Proc. Natl. Acad. Sci. USA* **81** (1984), 7767.

130. K. F. Geoghegan *et al.*, *Biochemistry* **22** (1983), 1847.

131. I. Bertini *et al.*, *J. Inorg. Biochem.* **32** (1988), 13.

132. C. Luchinat *et al.*, *J. Inorg. Biochem.* **32** (1988), 1.

133. R. Bicknell *et al.*, *Biochemistry* **27** (1988), 1050.

134. I. Bertini *et al.*, *Biochemistry* **27** (1988), 8318.

135. M. E. Sander and H. Witzel, in Reference 21, p. 207.

136. M. W. Makinen, in Reference 21, p. 215.

137. D. W. Christianson and W. N. Lipscomb, *Acc. Chem. Res.* **22** (1989), 62.
138. D. W. Christianson *et al.*, *J. Biol. Chem.* **264** (1989), 12849.
139. B. W. Matthews, *Acc. Chem. Res.* **21** (1988), 333, and references therein.
140. B. S. Cooperman, in Reference 119.
141. D. M. Blow, J. J. Birktoft, and B. S. Hartley, *Nature* **221** (1969), 337.
142. R. C. Nordlie, in P. D. Boyer, ed., *T he Enzymes*, Academic Press, 3d ed., **4** (1975), 543.
143. R. Breslow *et al.*, *Proc. Natl. Acad. Sci. USA* **80** (1983), 4585.
144. H. W. Wyckoff *et al.*, *Adv. Enzymol.* **55** (1983), 453.
145. E. E. Kim and H. W. Wyckoff, *J. Mol. Biol.* **218** (1991), 449.
146. L. Banci *et al.*, *J. Inorg. Biochem.* **30** (1987), 77.
147. P. Gettins and J. E. Coleman, *J. Biol. Chem.* **259** (1984), 11036.
148. I. Bertini *et al.*, *Inorg. Chem.* **28** (1989), 352.
149. A. Chaidaroglou *et al.*, *Biochemistry* **27** (1988), 8338.
150. E. C. Dinovo and P. D. Boyer, *J. Biol. Chem.* **246** (1971), 4586.
151. H. Dutler, A. Ambar, and J. Donatsch, in Reference 21, p. 471.
152. H. Eklund and C.-I. Bränden, in *Biological Macromolecules and Assemblies*, Wiley, 1985.
153. C.-I. Bränden *et al.*, *The Enzymes* **11** (1975), 104.
154. E. S. Cedergren-Zeppezauer, in Reference 21, p. 393.
155. H. Theorell, *Feder. Proc.* **20** (1961), 967.
156. M. W. Makinen and W. Maret, in Reference 21, p. 465.
157. J. Kvassman and G. Pettersson, *Eur. J. Biochem.* **103** (1980), 565.
158. P. F. Cook and W. W. Cleland, *Biochemistry* **20** (1981), 1805.
159. I. Bertini *et al.*, *J. Am. Chem. Soc.* **106** (1984), 1826.
160. W. Maret *et al.*, *J. Inorg. Biochem.* **12** (1980), 241.
161. H. B. Gray and E. I. Solomon, in Reference 27, p. 1.
162. J. S. Valentine and M. W. Pantoliano, in Reference 27, p. 291.
163. M. F. Dunn, A. K. H. MacGibbon, and K. Pease, in Reference 21, p. 486.
164. I. Bertini *et al.*, *Eur. Biophys. J.* **14** (1987), 431.
165. W. Maret and M. Zeppezauer, *Biochemistry* **25** (1986), 1584.
166. C. Sartorius, M. Zeppezauer, and M. F. Dunn, *Rev. Port. Quim.* **27** (1985), 256; C. Sartorius *et al.*, *Biochemistry* **26** (1987), 871.
167. G. Pettersson, in Reference 21, p. 451.
168. E. Garces and W. W. Cleland, *Biochemistry* **8** (1969), 633.
169. B. Edlund *et al.*, *Eur. J. Biochem.* **9** (1969), 451.
170. R. K. Crane, in M. Florkin and E. H. Stotz, eds., *Comprehensive Biochemistry*, Elsevier, **15** (1964), 200.
171. B. M. Babior and J. S. Krouwer, *CRC Crit. Rev. Biochem.* **6** (1979), 35.
172. C. Brink-Shoemaker *et al.*, *Proc. Roy. Soc. London, Ser. A* **278** (1964), 1.
173. B. T. Golding and P. J. Sellars, *Nature* (1983), p. 204.
174. D. Lexa and J. M. Saveant, *Acc. Chem. Res.* **16** (1983), 235.
175. R. A. Firth *et al.*, *Chem. Commun.* (1967), 1013.
176. G. N. Schrauzer and L. P. Lee, *J. Am. Chem. Soc.* **90** (1968), 6541.
177. R. A. Firth *et al.*, *Biochemistry* **6** (1968), 2178.
178. V. B. Koppenhagen and J. J. Pfiffner, *J. Biol. Chem.* **245** (1970), 5865.
179. H. A. O. Hill, in G. L. Eichhorn, ed., *Inorganic Biochemistry*, Elsevier, **2** (1973), 1067.
180. J. Halpern, *Pure. Appl. Chem.* **55** (1983), 1059.
181. J. Halpern, S. H. Kim, and T. W. Leung, *J. Am. Chem. Soc.* **106** (1984), 8317.
182. R. G. Finke and B. P. Hay, *Inorg. Chem.* **23** (1984), 3041; B. P. Hay and R. G. Finke, *Polyhedron* **4** (1988), 1469; R. G. Finke, in C. Bleasdale and B. T. Golding, eds., *Molecular Mechanisms in Bioorganic Processes,* The Royal Society of Chemistry; Cambridge, England (1990).
183. B. P. Hay and R. G. Finke, *J. Am. Chem. Soc.* **108** (1986), 4820.
184. B. P. Hay and R. G. Finke, *J. Am. Chem. Soc.* **109** (1987), 8012.
185. J. M. Pratt, *Quart. Rev.* (1984), 161.
186. J. Halpern, *Science* **227** (1985), 869.
187. B. M. Babier, *Acc. Chem. Res.* **8** (1975), 376.
188. B. T. Golding, in D. Dolphin, ed., B_{12}, Wiley, **2** (1982), 543.
189. N. Bresciani-Pahor *et al.*, *Coord. Chem. Rev.* **63** (1985), 1.
190. J. M. Pratt, *J. Mol. Cat.* **23** (1984), 187.

191. S. M. Chennaly and J. M. Pratt, *J. Chem. Soc., Dalton Trans.* (1980), 2259.
192. *Ibid.*, p. 2267.
193. *Ibid.*, p. 2274.
194. L. G. Marzilli *et al.*, *J. Am. Chem. Soc.* **101** (1979), 6754.
195. J. Glusker, in Reference 188, **1** (1982), 23.
196. G. De Alti *et al.*, *Inorg. Chim. Acta* **3** (1969), 533.
197. K. Geno and J. Halpern, *J. Am. Chem. Soc.* **109** (1987), 1238.
198. C. Mealli, M. Sabat, and L. G. Marzilli, *J. Am. Chem. Soc.* **109** (1987), 1593.
199. *Nickel and its Role in Biology*, vol. 23 of Reference 62.
200. C. T. Walsh and W. H. Orme-Johnson, *Biochemistry* **26** (1987), 4901.
201. J. F. Riordan and H. Hayashida, *Biochem. Biophys. Res. Commun.* **41** (1970), 122.
202. R. G. Khalifah, J. I. Rogers, and J. Mukherjee, in Reference 21, p. 357.
203. A. Chaidaroglou *et al.*, *Biochemistry* **27** (1988), 8338.
204. C. Forsman *et al.*, *FEBS Lett.* **229** (1988), 360; S. Lindskog *et al.*, in *Carbonic Anhydrase*, F. Botrè, G. Gros, and B. T. Storey, eds., VCH, 1991.
205. C. A. Fierke, J. F. Krebs, and R. A. Venters, in *Carbonic Anhydrase*, F. Botrè, G. Gros, and B. T. Storey, eds., VCH, 1991.
206. K. Wüthrich, *NMR in Biological Research*, Elsevier-North Holland, 1976.
207. K. Wüthrich, *NMR of Proteins and Nucleic Acids*, Wiley, 1986.
208. I. Bertini, H. Molinari, and N. Niccolai, eds., *NMR and Biomolecular Structure*, Verlag Chemie, 1991.
209. J. T. J. LeComte, R. D. Johnson, and G. N. La Mar, *Biochim. Biophys. Acta* **829** (1985), 268.
210. Recently, ^{67}Zn has been used as a relaxing probe to monitor the binding of ^{13}C-enriched cyanide to zinc in carbonic anhydrase (see Section IV.C).
211. Recent work on HCA II has improved the resolution to 1.54 Å (K. Håkan *et al.*, *J. Mol. Biol.* **227** (1993), 1192). Mutants at positions 143 (R. S. Alexander, S. K. Nair, and D. W. Christianson, *Biochemistry* **30** (1991), 11064) and 200 (J. F. Krebs *et al.*, *Biochemistry* **30** (1991), 9153; Y. Xue *et al.*, *Proteins* **15** (1993), 80) also have been characterized by x-ray methods.
212. An x-ray study of the cyanate and cyanide derivatives of the native enzyme has shown that the anions sit in the cavity without binding to the metal ion (M. Lindahl, L.A. Svensson, and A. Liljas, *Proteins* **15** (1993), 177). Since NCO$^-$ has been shown to interact with the paramagnetic cobalt(II) center, and ^{13}C-enriched cyanide has been shown to interact with ^{67}Zn-substituted CA (see Reference 67), it appears that the structures in the solid state and solution are strikingly different.
213. Recent x-ray data on the adduct of 1,2,4-triazole with HCA II confirm H-bonding with Thr-200 (S. Mangani and A. Liljas, *J. Mol. Biol.* **232** (1993), 9).
214. An HCO$_3^-$-complex of the His-200 mutant of HCA II has been studied by x-ray methods. The data are consistent with the coordinated oxygen being protonated and H-bonded to Thr-199 (Y. Xue *et al.*, *Proteins* **15** (1993), 80).

Calcium in Biological Systems

STURE FORSÉN AND JOHAN KÖRDEL

Physical Chemistry 2,
Chemical Centre,
University of Lund

I. INTRODUCTION

Calcium, like many other "inorganic elements" in biological systems, has during the last decade become the subject of much attention both by scientists and by the general public.[1] The presence and central role of calcium in mammalian bones and other mineralized tissues were recognized soon after its discovery as an element by Davy in 1808. Much later, the insight arrived that Ca^{2+} ions could play an important role in other tissues as well. Experiments of great historical influence were performed by the British physiologist Sidney Ringer a little over a century ago.[2] He was interested in the effects of various cations on frog-heart muscle and somewhat serendipitously discovered that Ca^{2+} ions, everpresent in the tap water distributed in central London, in millimolar concentrations were necessary for muscle contraction and tissue survival.

Today it is widely recognized that Ca^{2+} ions are central to a complex intracellular messenger system that is mediating a wide range of biological processes: muscle contraction, secretion, **glycolysis** and **gluconeogenesis**, ion transport, cell division and growth (for definitions of terms in boldface, see Appendix A in Section IX). The detailed organization of this messenger system is presently the subject of considerable scientific activity, and some details are already known. One of the links in the system is a class of highly homologous Ca^{2+}-binding proteins, to be discussed later on in this chapter, that undergo Ca^{2+}-dependent conformational changes and respond to transitory increases in intracellular Ca^{2+}-ion concentrations. A prerequisite for the proper function of the calcium messenger system in higher organisms is that the **cytosolic** Ca^{2+} concentration in a "resting" cell be kept very low, on the order of 100 to 200 nM. Transitory increases in the Ca^{2+} concentration that may result from hormonal action on a membrane receptor must rapidly be reduced. Several transport proteins, driven either by ATP hydrolysis or by gradients of some other ion like Na^+, are involved in this activity.

Ca^{2+} ions are also known to play various roles outside cells. In the plant kingdom Ca^{2+} ions often form links between individual cells and are required for maintaining the rigidity of whole plants; some seaweeds are typical examples. In the blood plasma of mammals, in which the Ca^{2+} concentration exceeds the intracellular by a factor of about 10^4, Ca^{2+} ions are instrumental in joining certain proteins in the blood-clotting system with membrane surfaces of circulating cells. Many extracellular enzymes also contain Ca^{2+} ions, sometimes at the active site but most often at other locations. It is generally believed that Ca^{2+} ions confer on proteins an increased thermal stability, and indeed proteins in heat-tolerant microorganisms often hold many such ions.

Vertebrates require much calcium in their food; in the USA the recommended daily allowance (RDA) for adult humans is 800 mg, and most other countries have comparable recommendations. During gestation in mammals, calcium must be transported across the placenta into the fetus, in particular during those phases of pregnancy when bone formation is most rapid. Interestingly, there appear to be some parallels between intestinal and placental transport that will be discussed further below. The role of calcium in biominerals is a vast subject that we can treat only superficially in this chapter.

To provide a background to the more biologically oriented sections that follow, we begin with a brief recapitulation of some basic facts about calcium. Then we continue with an outline of calcium distribution in biological tissues and organelles, and of the methods that can be used to obtain this information. After this follows a brief section on Ca^{2+} transport, and an account of the mechanism of intracellular Ca^{2+} release as it is presently understood. A discussion of some selected Ca^{2+}-binding proteins of general interest, both intracellular and extracellular, then follows. Before we conclude the chapter, we will summarize some recent observations on Ca^{2+}-binding proteins in prokaryotes.

II. BASIC FACTS ABOUT CALCIUM: ITS COMPOUNDS AND REACTIONS

A. Basic Facts

Calcium was first recognized as an element in 1808 by Humphry Davy, and the name was given after the Latin for lime: *calx*. Several isotopes of calcium are known. The stable isotopes are, in order of decreasing natural abundance, ^{40}Ca (96.94%), ^{44}Ca (2.1%), ^{42}Ca (0.64%), and ^{43}Ca (0.145%). ^{43}Ca is the only isotope with a nuclear spin ($I = \frac{7}{2}$) different from zero, which makes it amenable to NMR studies. ^{45}Ca is a radioactive isotope of some importance (β^- decay; 8.8 min half life).[3] It has been used in studies of calcium localization and transport in biological systems.

Calcium constitutes about 3 percent by weight of the Earth's crust, mostly in the form of sedimentary rocks of biological origin dating back some three

Table 3.1
Ca^{2+} concentrations in fluids and tissues.[6–9]

Specimen	Units are mM if not otherwise stated
Sea water	10
Fresh water	0.02–2
Rain water	0.002–0.02
"Hard" tap water	1.5
"Good" beer	4
Adult human serum	2.45 ± 0.05
Serum of other vertebrates	1.5–5
Nematote body fluids	6
Molluscan serum—marine	9–15
—fresh water	1.5–7.8
—land	3.3–12.3
Milk	70
Bone	0.8–1.0
Mitochondria from rat liver	0.8 ± 0.1 mmol/kg
Endoplasmatic reticulum	8–10 mmol/kg
Cytoplasm of a resting mammalian cell	0.0001
Cytoplasm of *E. coli*	0.0001

billion years. In sea water the total concentration of calcium ranges from 5 to 50 times higher than in fresh water, which, in turn, has a calcium concentration ten times that of rain water (see Table 3.1). This explains the pleasant feeling when ordinary soaps are used in rain water. The calcium concentration in ordinary tap water varies with location; calcium is usually added to water in distributing networks in order to prevent corrosion of iron pipes. Tap water with a calcium concentration above 1.5 mM is usually classified as "hard." Interestingly, the taste of beer seems related to the calcium concentration, and it is claimed that "good" beer should have a concentration higher than that of "hard" tap water.

In the body fluids of higher organisms the total calcium concentration is usually on the order of a few millimolar (see Table 3.1). In adult human serum, the concentration is observed to be, within narrow limits, 2.45 mM.

B. Essentials of Ca^{2+} Chemistry

Since the Ca^{2+} ion accomplishes its biological tasks in an environment with 1 to 3 mM Mg^{2+}, it is of particular interest to compare the properties of these two ions in order to understand how a discrimination is made in biological systems. In addition, the coordination chemistry of Ca^{2+} is closely related to that of Mg^{2+} (as well as Cd^{2+}), though there are several obvious differences. First of all, the ionic radius of a Ca^{2+} ion with a given coordination number (CN) is always higher than that of an Mg^{2+} or Cd^{2+} ion with the same CN.

At CN = 6, the ionic radii of Ca^{2+}, Cd^{2+}, and Mg^{2+} are 1.00, 0.95, and 0.72 Å, respectively, whereas at CN = 8 they are 1.12, 1.10, and 0.89 Å, respectively.[4]

Ligand preferences of Ca^{2+} depend on the fact that it is a hard metal ion. Thus Ca^{2+} strongly prefers oxygen ligands over nitrogen or sulfur ligands; Ca^{2+}····N bonds are about 0.25—0.3 Å longer than Ca^{2+}····O bonds.[5,6,10] Large differences in coordination number and geometry have been observed for Ca^{2+} complexes. In a study of 170 x-ray structures of Ca^{2+} complexes involving carboxylate groups,[11] binding was found to be either (i) unidentate, in which the Ca^{2+} ion interacts with only one of the two carboxylate oxygens, (ii) bidentate, in which the Ca^{2+} ion is chelated by both carboxylate oxygens, or (iii) mixed (''α-mode'') in which the Ca^{2+} ion is chelated by one of the carboxylate oxygens and another ligand attached to the α-carbon (see Figure 3.1). The Ca^{2+}-oxygen distances span a range from 2.30 to 2.50 Å, with the average distance being 2.38 Å in the unidentate and 2.53 Å in the bidentate mode, respectively.[11] Observed coordination numbers follow the order $8 > 7 > 6 > 9$. By contrast, Mg^{2+} nearly always occupies the center of an octahedron of oxygen atoms (CN = 6) at a fixed Mg^{2+}-oxygen distance of 2.05 ± 0.05 Å.

In Table 3.2, stability constants for the binding of Ca^{2+} and Mg^{2+} to various ligands are collected. We may note that selectivity of Ca^{2+} over Mg^{2+} is not very great for simple carboxylate ligands, but that it tends to increase for large multidentate ligands, such as EDTA and in particular EGTA. The Ca^{2+} sites in many intracellular proteins with ''EF-hand'' binding sites (see Section V. C) bind Ca^{2+} about 10^4 times more strongly than Mg^{2+}.

Another difference in ligand-binding properties of Mg^{2+} and Ca^{2+} can be seen by comparing the rates of substitution of water molecules in the inner

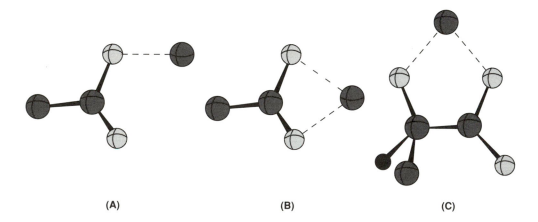

(A)   **(B)**   **(C)**

Figure 3.1

The three commonly observed modes of calcium carboxylate ligation. (A) The unidentate mode, in which the calcium ion interacts with only one of the two carboxylate oxygens. (B) The bidentate mode, in which the calcium ion is chelated by both oxygen atoms. (C) The α-mode, in which the calcium ion is chelated by one carboxylate oxygen, and another ligand is attached to the α-carbon. Adapted from Reference 11.

Table 3.2
Ca^{2+} and Mg^{2+} (where available) stability constants (log K) for different organic and biochemical ligands. Most values are at ionic strength 0.1 and 25°C.[5,6,12–15]

Ligand	Ca^{2+}	Mg^{2+}
Acetate	0.5	0.5
Lactate	1.1	0.9
Malonate	1.5	2.1
Aspartate	1.6	2.4
Citrate	3.5	3.4
Nitrilotriacetate	6.4	5.5
EGTA[a]	10.9	5.3
EDTA[b]	10.6	8.8
Glycine (Gly)	1.4	3.4
γ-Carboxyglutamic acid (Gla)	1.3	
Gly-Gly dipeptide	1.2	
Gla-Gla dipeptide	3.2	
Macrobicyclic amino cryptate [2.2.2]	4.5	
Fluo-3	6.2	2.0
Fura-2	6.9	2.0
BAPTA	7.0	1.8
Quin-2	7.1	2.7
Phospholipase A2	3.6	
Thrombin fragment 1	3.7	3.0
Trypsinogen	3.8	
Chymotrypsinogen	3.9	
Chymotrypsin	4.1	
Calmodulin, N-terminal	4.5	3.3
Trypsin	4.6	
Calmodulin, C-terminal	5.3	
Protein kinase C	~7	
α-Lactalbumin	~7	
Rabbit skeletal muscle Troponin C, Ca^{2+}/Mg^{2+} sites	7.3	3.6
Carp parvalbumin	~8.5	4.2
Bovine calbindin D_{9k}	8.8	~4.3

[a] EGTA: ethylenebis(oxyethylenenitrilo)tetraacetate
[b] EDTA: ethylenedinitrilotetraacetate

hydration sphere by simple ligands, according to

$$M(H_2O)_n^{2+} + L \xrightarrow{k} ML(H_2O)_{n-1}^{2+} + H_2O$$

This rate (log k, with k in s^{-1}) has been determined to be 8.4 for Ca^{2+} and 5.2 for Mg^{2+}.[16]

The formation of biominerals is a complex phenomenon. In order to obtain a feeling for the conditions under which inorganic solid phases in biological systems are stable, it is of some interest to look at solubility products. Solubility products, K_{sp}°, have a meaning only if the composition of the solid phase is specified. For a solid compound with the general composition $(A)_k(B)_l(C)_m$ the solubility product is defined as

$$K_{sp}^{\circ} = [A]^k[B]^l[C]^m, \tag{3.1}$$

where $[A]$, $[B]$, etc., denote activities of the respective species, usually ionic, in equilibrium with the solid. Activities are concentrations multiplied by an activity coefficient, γ, nearly always less than unity. Activity coefficients for ions in real solutions can be estimated from Debye-Hückel theory [17] if the ionic strength of the solution is known. In human blood plasma, the ionic strength, I, is about 0.16, and the activity coefficient for Ca^{2+} at 37°C is 0.34. In many discussions it may be sufficient to equate concentrations with activities.

The solid phase involved is essentially assumed to be an infinitely large, defect- and impurity-free crystal with a well-defined structure. Microscopic crystals have higher solubilities than large crystals, a well-known phenomenon that leads to "aging" of precipitates, in which larger crystals grow at the expense of smaller ones.

Many anionic species appearing in the solubility products may also be involved in protonation equilibria in solution, such as those of phosphoric acid: $H_2PO_4^- \rightleftharpoons H^+ + HPO_4^{2-}$; $HPO_4^{2-} \rightleftharpoons PO_4^{3-} + H^+$; etc. When the prospects for the formation of a solid phase under certain solution conditions are investigated, the activity, or concentration, of the particular anionic species specified in the solubility product must be known, not only "total phosphate" or "total calcium," etc. The data in Table 3.3 show that, at pH > 5, the most stable (i.e., insoluble) solid calcium phosphate is hydroxyapatite.

Table 3.3
Solubility products, at pH 5 and 25°C, for solid calcium phosphates.

Solid phase	$-\log K_{sp}^{\circ}$	$-\log K_{sp}^{\circ}$ of corresponding Mg^{2+} compound where applicable
$CaSO_4 \cdot 2H_2O$ (sulfate, "gypsum")	5.1	< 1.0
$Ca(OH)_2$ (hydroxide)	5.3	10.7
$CaHPO_4 \cdot 2H_2O$ (hydrogen phosphate)	6.6	—
$CaCO_3$ (carbonate, "calcite," "aragonite")	8.5	7.5
$CaC_2O_4 \cdot H_2O$ (oxalate, "whewellite")	10.5	5.0
$\beta\text{-}Ca_3(PO_4)_2$ (β-phosphate)	29	—
$Ca_5(PO_4)_3OH$ (hydroxyapatite)	58	—

III. CALCIUM IN LIVING CELLS: METHODS FOR DETERMINING CONCENTRATIONS AND SPATIAL DISTRIBUTIONS

Much of our present knowledge of the biological role of Ca^{2+} ions in the regulation and modulation of cellular activities rests on the development of analytical techniques in three different areas: our ability to measure the low concentration levels in the cytoplasm of resting cells, follow the concentration changes, both temporally and spatially, that may occur as a result of an external stimulus, and measure the distribution of Ca^{2+} in various compartments of a cell. The last decade has seen the emergence of many such new techniques, and the improvement of old ones, which has had a major impact on our understanding of the detailed molecular mechanisms and dynamics of the Ca^{2+} messenger system. In this section, we will survey some of the most important techniques and results obtained using these. Broadly speaking there are two main groups of experimental techniques: those that aim at measuring the concentration of *"free"* (or uncomplexed) *Ca^{2+}-ion concentrations* (or activities), and those that measure *total calcium*.

A. Measurements of "Free" Calcium Concentrations

1. Ca^{2+}-selective microelectrodes

Ion-selective electrodes can be made from a micropipette (external diameter $0.1–1\,\mu m$) with an ion-selective membrane at the tip.[18,19] For Ca^{2+} the membrane can be made of a polyvinyl chloride gel containing a suitable Ca^{2+}-selective complexing agent soluble in the polymer gel. A commonly used complexing agent is "ETH 1001" (see Figure 3.2A). An additional "indifferent" reference electrode is needed. For measurements inside cells, the reference electrode can also be made from a micropipette filled with an electrolyte gel. Often the ion-selective and reference electrodes are connected in a double-barrelled combination microelectrode.[21] The whole assembly can then be inserted, using a micromanipulator, into a single cell typically 30–50 μm across. The arrangement is depicted in Figure 3.2B. With proper care, Ca^{2+} microelectrodes can be used to measure Ca^{2+}-ion concentrations down to 10^{-8} M.[19, 21] One limitation of the technique is that the response time is usually in seconds or even minutes, making rapid concentration transients difficult to follow.

2. Bioluminescence

Several living oganisms are able to emit light. The light-emitting system in the jellyfish (Aequorea) is a protein called aequorin ($M_r \simeq 20$ kDa). The light is emitted when a high-energy state involving a prosthetic group (coelenterazine) returns to the ground state in a chemical reaction that is promoted by Ca^{2+}

114

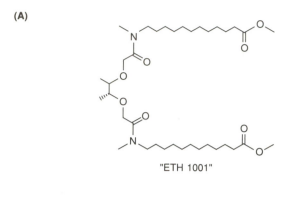

(A)

"ETH 1001"

(B)

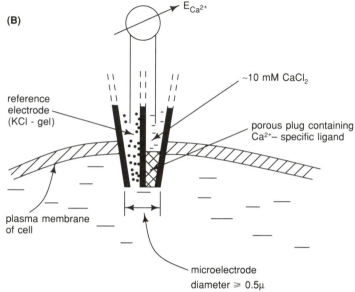

$E_{Ca^{2+}}$

reference
electrode
(KCl - gel)

~10 mM CaCl$_2$

porous plug containing
Ca^{2+}– specific ligand

plasma membrane
of cell

microelectrode
diameter ⩾ 0.5μ

Figure 3.2
(A) Structure of a commonly used neutral Ca^{2+} chelator in Ca^{2+}-selective electrodes, "ETH 1001" (N,N'-di[11-(ethoxycarbonyl) undecul]-N,N'-4,5-tetramethyl-3,6-dioxaoctan-1,8-diam- ide).[20] (B) Schematic arrangement for the measurement of the activity (or concentration) of Ca^{2+} ions in cells using a Ca^{2+}-selective double-barreled microelectrode. Frequently the mi- croelectrode is supplemented by a third, indifferent, electrode inserted into the bathing medium surrounding the cell.[21]

ions. At Ca^{2+} concentrations below ~0.3 μM the emission is weak, but in the range 0.5–10 μM the emission is a very steep function of the concentration (roughly as [Ca^{2+}]$^{2.5}$).[18,19,22] The response to a Ca^{2+}-concentration transient is rapid ($\tau_{1/2} \approx 10$ ms at room temperature), and the light emitted can be accu- rately measured even at very low light levels by means of image intensifiers and/or photon counting. For measurements of Ca^{2+} concentrations inside cells, aequorin has usually been introduced either through microinjection or through some other means. A novel idea, however, is to utilize recombinant aequorin reconstituted within the cells of interest, thus circumventing the often difficult injection step.[174]

3. Complexing agents with Ca^{2+}-dependent light absorption or fluorescence

An important advance in the field of Ca^{2+}-ion determination was made by R. Y. Tsien, who in 1980 described[23] the synthesis and spectroscopic properties of several new tetracarboxylate indicator dyes that had high affinity and reasonable selectivity for Ca^{2+}. All these dye molecules have a high UV absorbance that is dependent on whether Ca^{2+} is bound or not; a few also show a Ca^{2+}-dependent fluorescence. Tsien has also demonstrated that these anionic chelators can be taken up by cells as tetraesters, which, once inside the cells, are rapidly enzymatically hydrolyzed to give back the Ca^{2+}-binding anionic forms. Fluorescent tetracarboxylate chelators with somewhat improved Ca^{2+} selectivity such as "BAPTA," "Quin-2," and "Fura-2" (Figure 3.3) were later described.[24] These chelators are very suitable for measurement of Ca^{2+}-ion concentrations in the range 1 μM to 10 nM in the presence of 1 mM Mg^{2+} and 100 mM Na^+ and/or K^+—i.e., conditions typically prevailing in animal cells. Recently a new set of chelators that are more suitable for measurements of calcium concentrations above 1 μM was presented.[25] The most interesting of these is "Fluo-3," with a calcium-binding constant of 1.7×10^6.

Whereas the emission spectrum for Fura-2 (Figure 3.3B), which peaks at 505–510 nm, hardly shifts wavelength when Ca^{2+} is bound, the absorption spectrum shifts toward shorter wavelengths. In studies of free Ca^{2+} concentrations where internal referencing is necessary, for example, in studies of single cells, it is therefore advantageous to excite alternately at ~350 and 385 nm, and to measure the ratio of fluorescence intensity at ~510 nm.

The use of fluorescent chelators has recently permitted studies in single cells of rapid fluctuations or oscillations of free Ca^{2+} and the formation of Ca^{2+} concentration gradients. Using a fluorescence microscope coupled to a low-light-level television camera feeding a digital image processor, Tsien et al.[26] have been able to reach a time resolution of about 1 s in single-cell studies. The results of some highly informative studies made using this instrument are shown in Figure 3.4. (*See color plate section, page C-7.*) The concentration of free Ca^{2+} is presented in pseudocolor, and the Fura-2 concentration inside cells is 50–200 μM, as indicated in the figures. We see a Ca^{2+} gradient diffusing through an entire sea-urchin egg (~120 μ across) in 30 s. The free Ca^{2+} concentration of the resting egg (~100 nM) is increased to about 2 μM as Ca^{2+} diffuses through the egg. The mechanism of propagation is believed to be a positive feedback loop with inositol trisphosphate releasing Ca^{2+} and vice versa (see Section V).

A pertinent question concerning the uses of intracellular Ca^{2+} chelators is whether or not the chelator significantly perturbs the cell. The chelator will obviously act as a Ca^{2+} buffer in addition to all other Ca^{2+}-binding biomolecules in the cell. The buffer effect is probably not of any major consequence, since the cell may adjust to the new situation by an increase in *total* Ca^{2+}, especially if the chelator concentration is in the μM range. The chelators could, however, interact with and inhibit intracellular enzymes or other molecules, an

116

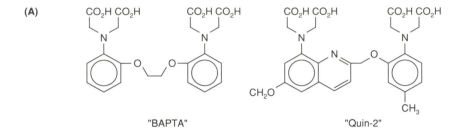

"BAPTA"   "Quin-2"

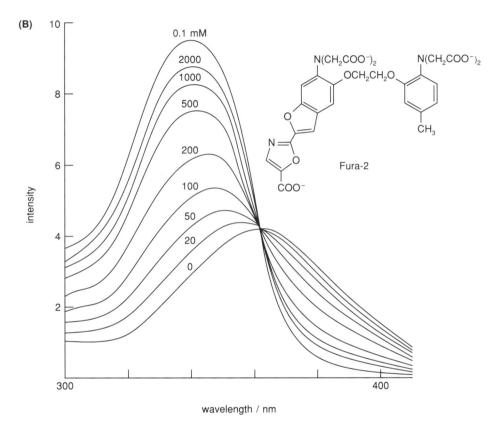

Fura-2

wavelength / nm

Figure 3.3

Molecular structure of three chelators frequently used in measurements of "free" Ca^{2+}-ion concentrations. They may all be regarded as aromatic analogues of the classical chelator "EDTA"; their optical spectroscopic properties change upon binding Ca^{2+} ions. (A) For "BAPTA" the spectral changes are confined to the absorption spectrum, whereas "Quin-2" and the "Fura-2" in (B) show Ca^{2+}-dependent changes in their fluorescence spectra.[23, 24] (B) The relative fluorescence intensity of "Fura-2" at 505 nm as a function of the wavelength of the excitation light at different Ca^{2+} concentrations. The data[26] refer to a solution containing 115 mM KCl, 20 mM NaCl, and 1 mM Mg^{2+} at 37°C and pH 7.05. At increasing Ca^{2+} concentration, the excitation efficiency at $\sim$350 nm is increased, but that at $\sim$385 nm is decreased. In order to eliminate (as much as possible) variations in fluorescence intensity in biological samples due to slight variations in dye concentrations and/or cell thickness, it is often advantageous to measure the intensity *ratio* at 345 and 385 nm excitation wavelengths.

effect that could result in aberrant cellular behavior. It is not unlikely that BAPTA will bind to certain proteins.[27]

4. Complexing agents with Ca^{2+}-dependent NMR spectra

A series of symmetrically substituted fluorine derivatives of BAPTA (see Figure 3.3A) has been synthesized.[28,29] One of these chelators is 5F-BAPTA (Figure 3.5A), which has a binding constant for Ca^{2+}, K_B^{Ca}, of 1.4×10^6 M^{-1} and a ^{19}F NMR chemical shift, δ, that in the free ligand is different from that in the complex with Ca^{2+} ($\Delta\delta_{Ca^{2+}} \approx 6$ ppm). The rate of Ca^{2+} dissociation, k_{off}, is 5.7×10^2 s^{-1}, which gives the rate of association, k_{on}, as 8×10^8 M^{-1} s^{-1} according to

$$K_B = k_{on}/k_{off}. \tag{3.2}$$

This exchange rate means that we are approaching the slow exchange limit in ^{19}F NMR, and in subsaturating concentrations of Ca^{2+} two ^{19}F signals are seen (see Figure 3.5B).

Since the areas of the NMR signals from the bound (B) and free (F) forms of the ligand are proportional to their concentration, the free Ca^{2+} concentration

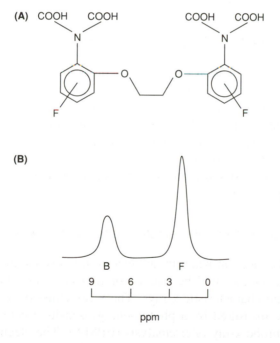

Figure 3.5
(A) Molecular structure of the calcium chelator 5F-BAPTA whose ^{19}F chemical shift changes upon calcium binding. (B) ^{19}F NMR spectrum of a solution containing 5F-BAPTA and Ca^{2+} in a molar ratio of 3:1. Signal B originates from the Ca^{2+}-5F-BAPTA complex, and F from free 5F-BAPTA. Adapted from Reference 29.

is obtained simply as

$$[Ca^{2+}]_{\text{free}} = \frac{B}{F} \cdot \frac{1}{K_B}. \qquad (3.3)$$

An additional beneficial property of 5F-BAPTA and other fluorinated analogues of BAPTA is that they will also bind other metal ions with a ^{19}F chemical shift of the complex that is characteristic of the metal ion.[29] Under favorable circumstances, it is thus possible to measure simultaneously the concentrations of several cations.

For 5F-BAPTA the selectivity for Ca^{2+} over Mg^{2+} is very good ($K_B^{Mg^{2+}} \approx 1 \text{ M}^{-1}$). In applications of 5F-BAPTA to intracellular studies, the same protocol is used as with the parent compound and its fluorescent derivatives: some esterified derivative, e.g., the acetoxymethyl ester, is taken up by the cells and allowed to hydrolyze in the cytoplasm. The intracellular concentrations of 5F-BAPTA needed to get good ^{19}F NMR signals depend on the density of cells in the sample tube and the number of spectra accumulated. With accumulation times on the order of ten minutes (thus precluding the observation of concentration transients shorter than this time), Ca^{2+} concentrations of the order of $1\mu M$ have been studied in perfused rat hearts using 5F-BAPTA concentrations of about $20\mu M$.[34]

B. Measurements of Total Calcium Concentrations

The measurement of total calcium in a biological sample can be made by any method sensitive only to the element and not to its particular chemical form. Atomic absorption spectroscopy is excellent as such a method. Obviously, the spatial resolution that can be obtained with this method is limited, and it is hard to imagine its application to elemental mapping of single cells. The techniques discussed in this subsection have been limited to those that permit a spatial resolution of at least 1 μm on samples usually prepared by sectioning the frozen biological specimens.

1. Electron probe and electron energy-loss techniques

When the electron beam in an electron microscope hits a thin sample, some atoms in the sample will be excited or ionized, and returning to their ground state will emit characteristic x-rays. The x-ray emission at different wavelengths may then be measured by a photon-energy-sensitive detector. This is the basis of *electron probe* x-ray *microanalysis* (EPMA). The electrons that pass through the sample, and that give the transmission image in electron microscopy, will suffer energy losses that depend on the nature (to some extent also, the chemical state) and distribution of different elements. The outcome of these phenomena forms the basis of *electron energy-loss spectroscopy* (EELS; see Figure 3.6).

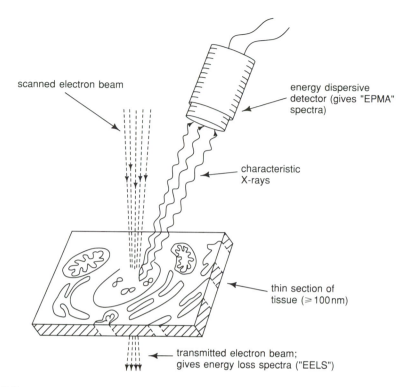

scanned electron beam

energy dispersive
detector (gives "EPMA"
spectra)

characteristic
X-rays

thin section of
tissue (≥ 100 nm)

transmitted electron beam;
gives energy loss spectra ("EELS")

Figure 3.6

Principles of *electron probe microanalysis* ("EPMA") and *electron energy loss spectroscopy* ("EELS").

A thin slice of a freeze-dried sample is exposed to a collimated beam of electrons that may be scanned across the sample. Atoms ionized by electron collisions will emit x-radiation at wavelengths characteristic of their nuclear charge (i.e., characteristic of each element). With the use of an energy dispersive x-ray detector, signals from different elements can be differentiated. Apart from the characteristic x-rays, a broad-spectrum background x-ray emission is also present because of inelastic scattering of the incident electrons. Some of the electrons that pass through the sample will have lost energy because of ionization of atoms in the sample. The energy loss is again characteristic for each element, and if the energy distribution of the transmitted electrons is analyzed, this will have "peaks" at certain characteristic energies. If the energy resolution is pushed far enough (<1 eV), the loss peaks even become sensitive to the chemical state of the element.

The EPMA technique as applied to calcium has been improved by Somlyo in particular.[30] Typically samples are rapidly frozen and sectioned at low temperatures ($-130°C$) to preserve the *in vivo* localization of diffusible ions and molecules. Spatial resolutions of 10 nm or better have been attained on ≥100 nm thick freeze-dried cryosections. The minimal detectable concentration, which requires some signal averaging, is approximately 0.3 mmol Ca per kg dry specimen (i.e., 10 ppm). The calcium content of **mitochondria** and **endoplasmic reticulum** in rat liver cells has been studied by EPMA (see Table 3.1).[8]

The high calcium content of endoplasmic reticulum (ER) is consistent with the view that this organelle is the major source of intracellular Ca^{2+} released through the messenger inositol trisphosphate (see Section IV.C). Other EPMA studies have shown mitochondria to have a large capacity for massive calcium accumulation in cells where cytoplasmic Ca^{2+} concentrations have been abnormally high, for example, as a result of damage of the cell membrane.[30]

EELS is presently less well-developed than EPMA. Two of the major difficulties in the use of EELS for quantitative analysis of calcium and other elements are (i) large background, since it is a difference technique, and (ii) sensitivity to specimen thickness. The major advantage of EELS is that the spatial resolution is potentially much better than in EPMA, and can be 1 to 2.5 nm in favorable specimens.

2. Proton-induced x-ray emission (PIXE)

A specimen exposed to a beam of high-energy (1 to 4 MeV) protons will also emit characteristic x-rays just as in EPMA. The advantage of using protons instead of electrons is that protons are more likely to collide with an atom, thus producing excited atoms emitting x-rays. The sensitivity in detecting a particular element is therefore much higher in PIXE than in EPMA or EELS. The PIXE technique, which was developed at the University of Lund, Sweden, in the late 1960s, was originally used mainly for studies of fairly large objects.[9]

In 1980 a group at Oxford University succeeded in focusing the proton beam to a diameter of 1 μm with sufficient energy (4 MeV) and beam intensity (100 pA/μm^2) to allow elemental mapping at ppm concentrations.[31] Similar beam performances ($\sim 0.5 \mu$m diameter) are now also available at the University of Lund and other laboratories. Beam diameters of 0.1 μm are likely to be achieved in the near future. Like EPMA, the PIXE method allows the simultaneous observation of several elements in the same sample. The biological applications of the microbeam PIXE technique are limited, but it is clear that its potential is great. Some representative results obtained with the Oxford microbeam are shown in Figure 3.7. (*See color plate section, page C-8.*)

3. Ion microscopy

Ion microscopy is another technique capable of detecting all elements at the ppm level. The basic idea is to expose a freeze-fixed, cryofractured, and freeze-dried sample, which has been put onto a conducting substrate in a vacuum chamber, to a beam of ions (e.g., D_2^+ or Ar^+). These ions will remove the top two or three atomic layers of the sample surface by sputtering. A certain fraction of the removed atoms will leave as ions. This secondary ion beam is accelerated into a double-focusing mass spectrometer, where the ions are separated according to their mass-to-charge ratio. The ion optics are designed to preserve the spatial distribution of the emitted secondary ions, and an element image of the sample can thus be produced with a spatial resolution of ~ 0.5 μm.[32] The ion-microscope technique can form images of *a particular isotope* of an element. In principle, then, one could perform isotope labeling or "isotope chase" studies and follow, say, the fate of isotope-enriched ^{43}Ca externally applied to a cell. The ion-microscope technique has not yet come into widespread use, but the quality of element (or ion) images obtained on single cells is impressive.[33]

C. Summary

Much of our present knowledge about the biological role of Ca^{2+} rests on detailed measurements of the concentration, distribution, and chemical nature of Ca^{2+} and its complexes. Concentrations of uncomplexed, or "free," Ca^{2+} can be measured by Ca^{2+}-selective microelectrodes, bioluminescence and complexing agents with Ca^{2+}-dependent light absorption, fluorescence, or NMR spectra. An outcome of such studies is that the "free" Ca^{2+} concentration in resting **eukaryotic** cells generally is very low, on the order of 100 to 200 nM. Total Ca^{2+} concentrations, uncomplexed and complexed, can be measured by a variety of physical techniques. Some techniques, like atomic absorption, are sensitive but give poor spatial resolution. Others involve the bombardment of the sample with electrons or charged atoms, and can yield spatial resolutions of the order of a few nm; however, there is a trade-off between detectability and resolution.

IV. THE TRANSPORT AND REGULATION OF Ca^{2+} IONS IN HIGHER ORGANISMS

All living organisms need calcium, which must be taken up from the environment. Thus, Ca^{2+} ions have to be distributed throughout the organism and made available where needed. In higher organisms, such as humans, the blood-plasma level of total calcium is kept constant (≈ 2.45 mM) within narrow limits, and there must be a mechanism for regulating this concentration. On a cellular level we have already seen in the preceding section that the basal cytoplasmic Ca^{2+} concentration, at least in eukaryotic cells, is very low, on the order of 100 nM. At the same time the concentrations of Ca^{2+} in certain organelles, such as endoplasmic (or sarcoplasmic) reticulum or mitochondria, may be considerably higher. If Ca^{2+} ions are to be useful as intracellular "messengers," as all present evidence has it, Ca^{2+} levels in the cytoplasm would have to be raised transitorily as a result of some stimulus. Ca^{2+} ions may enter the cytoplasm either from the extracellular pool or from the Ca^{2+}-rich organelles inside the cell (or both). We could imagine Ca^{2+} channels being regulated by chemical signaling, perhaps by a hormone acting directly on the channel, or by a small molecule released intracellularly when a hormone is attached to a membrane-bound receptor. Some channels may be switched on by voltage gradients, and both these mechanisms may operate concurrently.

Increased intracellular Ca^{2+} levels must eventually be brought back to the basal levels, in some cells very quickly. The ions could be transported out of the cell or back into the Ca^{2+}-rich organelles. This transport will be against an electrochemical potential gradient, and thus requires energy. There are many possibilities for different forms of Ca^{2+} transport and regulation in living systems, and we still know fairly little about the whole picture. Detailed studies are also complicated by the fact that, in higher organisms, cells are differentiated. Nature is multifarious, and what is valid for one type of cell may not be relevant for another. With these words of caution we will start out on a macroscopic level and continue on toward molecular levels.

A. Ca^{2+} Uptake and Secretion

The uptake of Ca^{2+} from food has mostly been studied in typical laboratory animals, such as rats, hamsters, chickens, and humans. In humans, uptake occurs in the small intestine, and transport is regulated by a metabolite of vitamin D, calcitriol (1,25-dihydroxy vitamin D_3).[34] The uptake process is not without loss; roughly 50 percent of the calcium content in an average diet is not absorbed. To maintain homeostasis and keep the calcium level in blood plasma constant, excess Ca^{2+} is excreted through the kidney. The main factor controlling this phenomenon in vertebrates is the level of the parathyroid hormone that acts on kidney (increases Ca^{2+} resorption), on bone, and, indirectly, via stimulated production of calcitriol, on the intestinal tract (increases Ca^{2+} uptake). Calcium enters the cells from the outside world, i.e., the intestinal lumen, by traveling through the brush-border membrane of the intestinal **epithelial cells**, through the cytosolic interior of these cells, and into the body fluids through the **basal lateral membranes** of the same cells. The molecular events involved need to be studied further. Figure 3.8 outlines the Ca^{2+} transport processes known or thought to occur.

Transfer through the brush-border membrane is assumed to be "passive" although indirectly facilitated by calcitriol. The calcitriol effect may be due to synthesis of a carrier protein,[35] but could also be an effect of altered membrane lipid composition.[36] The fate of Ca^{2+} ions, once inside the epithelial cell, is a much-debated subject. What appears clear is that the Ca^{2+} ions entering through the brush-border membrane do not cause an increase of the low cytosolic Ca^{2+} concentration. It is thus quite likely that the Ca^{2+} ions are carried through the cell but the means of transportation is unknown. One plausible carrier is the intracellular low-molecular-weight Ca^{2+}-binding protein calbindin D_{9K} ($M_r \approx 9$ kDa) formerly known as ICaBP (see Section V.C).[35] Its synthesis is induced by vitamin D, and it is mainly found in mammalian intestines. The porcine and bovine calbindin D_{9K} has a Ca^{2+} binding constant of $K_B \approx 3 \times 10^8$ M^{-1} in low ionic strength media [37] and $K_B \approx 2 \times 10^6$ M^{-1} in the presence of 1 mM Mg^{2+} and 150 mM K^+.[38] The concentration of calbindin D_{9K} in epithelial cells can reach millimolar levels,[35] which could allow it to facilitate Ca^{2+} diffusion across the cytosol. This was first suggested by Williams, subsequently elaborated by Kretsinger et al. in 1982,[39] and later demonstrated in a model cell by Feher.[40] The basic idea is that, although the diffusion rate of Ca^{2+} ions ($\sim 10^{-5}$ cm^2 s^{-1}) is higher than for the $(Ca^{2+})_2$ calbindin complex

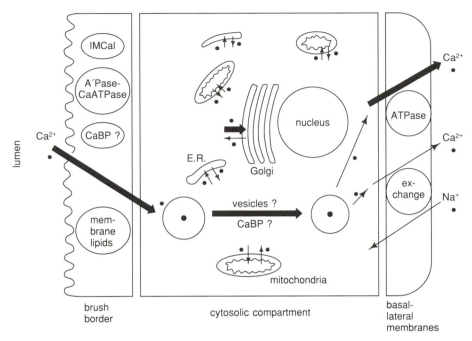

Figure 3.8

A scheme representing some of the known and hypothetical molecular participants in the transport of Ca^{2+} across intestinal epithelial cells. Transport across the brush-border membrane is generally assumed to be passive or to be facilitated by a carrier (IM Cal), and is also influenced by vitamin D. Transport through the cell may be in vesicles and/or in association with Ca^{2+}-binding proteins (CaBP), notably calbindins D_{9k} (mammals) or D_{28k} (avians). Temporary storage or buffering of Ca^{2+} may be through cytosolic CaBPs, mitochondria, endoplasmic reticula (ER), or other organelles. Transport of Ca^{2+} out of the cell through the basal-lateral membranes is energetically uphill, and appears primarily accomplished by a Ca^{2+}-ATPase and possibly to some extent by a Na^{2+}-Ca^{2+} antiport. Adapted from Reference 35.

($\sim 0.2 \times 10^{-5}$ cm^2 s^{-1}), the fact that the concentration of the latter complex may be about 10^3 times higher than that of free Ca^{2+} will result in an increased net calcium transport rate. Calbindin would, in fact, act very much like myoglobin in facilitating oxygen transport through muscle tissue.

Plausible as the above mechanism may seem, it may, however, not be the whole truth. An alternative mechanism is vesicular transport. In chicken intestine it has been shown that the only epithelial organelles that increased in Ca^{2+} content as a result of calcitriol treatment were the lysosomes.[41] The result lends support to a transport mechanism involving Ca^{2+} uptake across the brush-border membrane by **endocytic vesicles**, fusion of these vesicles with lysosomes, and possibly also delivery of Ca^{2+} to the basal lateral membrane of the epithelial cell by **exocytosis**. This process would also explain the vitamin-D-induced alterations in brush-border-membrane lipid compositions as a consequences of preferential incorporation of certain types of lipids into the vesicles. Interestingly, the lysosomes in the chicken studies also contained high levels of calbindin D_{28k}—a type of vitamin-D-induced Ca^{2+}-binding protein found in avian intestines—making it conceivable that this protein acts as a "receptor" for Ca^{2+} at the brush-border membrane and upon Ca^{2+} binding could become internalized in endocytic vesicles.[41]

The basal lateral plasma membrane contains at least two types of Ca^{2+} pumps that also may play a role in Ca^{2+} uptake, one ATP-driven, one driven by a concurrent flow of Na^+ ions into the cytoplasm (i.e., a Na^+-Ca^{2+} **antiport**; see Figure 3.8). We discuss these types of transporting proteins in the next subsection.

There are some apparent analogies between intestinal Ca^{2+} transport and that occurring in the placenta. Transplacental movements of Ca^{2+} increase dramatically during the last trimester of gestation.[42] In mammalian placental **trophoblasts**, high concentrations of calbindin D_{9K} are found.[43,44] The protein synthesis also in this tissue appears to be under calcitriol regulation. Ca^{2+} ions have to be supplied by mammalian females, not only to the fetus during pregnancy, but also to the newborn child through the mother's milk. The molecular details of Ca^{2+} transport in the mammalian glands have not been extensively studied. In milk, Ca^{2+} is bound mainly to micelles of casein, and the average Ca^{2+} content is reported to be 2.5 g/liter (see Table 3.1).

B. Intracellular Ca^{2+} Transport

In order to provide a better understanding of the role of Ca^{2+} as an almost universal regulator of cellular function, we need to take a brief look at the many ways by which Ca^{2+} ions can be transported in or out of eukaryotic cells. Although various transport pathways have been elucidated, the present picture is probably not complete, since the molecular structures and properties of the transport proteins are only partially known. The major pathways for Ca^{2+} transport across cellular membranes involve three membrane systems: the plasma membrane, the inner mitochondrial membrane, and the membrane of the endoplasmic reticulum (ER) (or, in striated muscle cells, a specialized form of ER called the sarcoplasmic reticulum (SR): (Figure 3.9). Two of the membrane-bound transport systems are Ca^{2+}-ATPases, since they derive their main energy from the hydrolysis of ATP (1 and 2 in Figure 3.9). Their properties do, however, differ in many other respects, as we will see.

1. The Ca^{2+}-ATPases

The plasma membrane Ca^{2+}-ATPase (PM Ca^{2+}-ATPase) of **erythrocytes**—first recognized by Schatzmann in 1966[45]—was isolated in pure form by Niggli *et al.* in 1979, using an affinity column with an ATP-ase binding protein, calmodulin (see Section V.A), coupled to the gel.[46] Ca^{2+}-ATPases purified from other types of plasma membranes appear to be very similar. The schematic structure of the erythrocyte membrane Ca^{2+}-ATPase is presented in

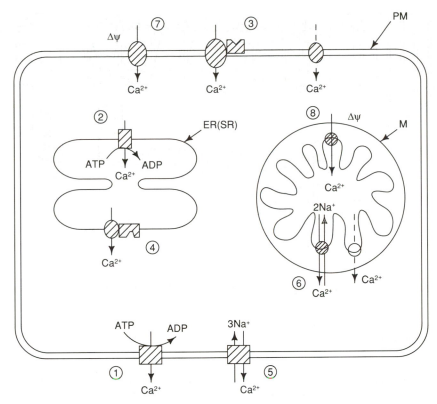

Figure 3.9

Schematic representation of the major pathways for the transport of Ca^{2+} across cellular membranes. PM, plasma membrane; ER(SR), endoplasmic reticulum (sarcoplasmic reticulum); M, mitochondria; $\Delta\Psi$, difference in membrane potential. The transport proteins shown are: 1 and 2, PM and ER(SR) Ca^{2+}-ATPases; 3 and 4, PM and ER(SR) receptor-mediated Ca^{2+} channels; 5 and 6, PM and M (inner-membrane) Na^+/Ca^{2+} exchangers; 7 and 8, PM and M voltage-sensitive Ca^{2+} channels. In addition, some not-well-defined "passive" transport pathways are indicated by dashed arrows.

Figure 3.10.[47] The sarcoplasmic reticulum in muscle cells is abundant in Ca^{2+}-ATPase. It is estimated that this protein constitutes more than 80 percent of the integral membrane proteins, and covers a third of the surface area.

The sarcoplasmic reticulum Ca^{2+}-ATPase (SR Ca^{2+}-ATPase) was first purified by MacLennan in 1970.[48] Presently it is the best characterized Ca^{2+}-ATPase. A schematic model and a summary of some properties are given in Figure 3.11.[49] Ten hydrophobic segments of about 20 amino-acid residues each are revealed by **hydropathy** plots, and these segments are assumed to span the membrane as α-helices. (For the one-letter codes for amino acids, see Appendix B in Section IX.) The phosphorylation site has been identified as Asp-351, and the nucleotide binding domain is following the phosphorylation domain. The Ca^{2+}-binding sites are located within the predicted trans-membrane domains

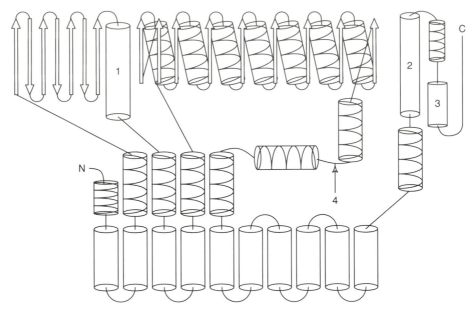

1 phospholipid sensitive domain
2 calmodulin binding domain
3 cAMP phosphorylation domain
4 hinge

Figure 3.10

Schematic structure of the calmodulin (CaM)-activated plasma membrane Ca^{2+}-ATPase of erythrocytes. Some molecular characteristics are: $M_r = 138,000$: transport rate (30°C), 20–70 Ca^{2+} ions per protein molecule per second; $K_M(Ca^{2+}) \approx 0.5 \mu M$ (cytoplasmic side in high-affinity form); Ca^{2+}/ATP ratio, 1(?); activated not only by CaM but also by acidic phospholipids and unsaturated fatty acids. Figure kindly provided by R. Moser and E. Carafoli.

(see Figure 3.11). This was shown through a series of site-directed mutations in which likely Ca^{2+}-liganding residues like Asp, Glu, and Thr were mutated into residues lacking possible side-chain ligands (e.g., Asn, Gln, and Ala).[50]

The presently accepted reaction cycle involves two main alternative conformations, E_1 and E_2, the former with two high-affinity sites $(K_m \lesssim 1 \ \mu M)^4$ on the cytoplasmic side, which in E_2 are open to the luminal side with $K_m \sim 1$ mM.[49,51] The mechanism suggested for Ca^{2+} transport (Figure 3.12) has many features similar to that suggested by Williams for H^+ translocation in the mitochondrial ATPase.[52]

It is instructive to consider briefly the thermodynamic limits of the transport. (The discussions about the thermodynamics behind Ca^{2+}/Na^+ transport pertain to Na^+/K^+ gradients in excitable tissues as well). Let us define an "inside" and an "outside" separated by a membrane, as shown in Figure 3.13, where $[Ca^{2+}]$ and ψ denote activities and membrane potentials, respectively. The difference in electrochemical potential, $\Delta\mu$, across the membrane for a Ca^{2+} ion is given by

$$\Delta\mu_{Ca^{2+}} = +RT\ln \frac{[Ca^{2+}]_o}{[Ca^{2+}]_i} + 2F\Delta\psi, \tag{3.4}$$

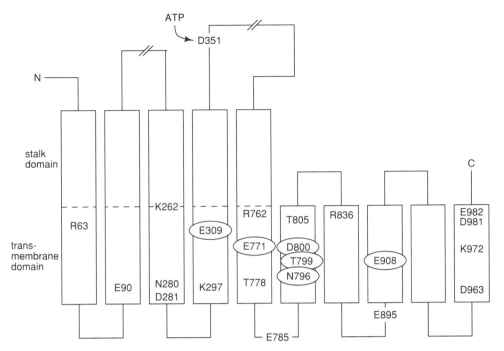

Figure 3.11
Schematic structure of the Ca^{2+}-ATPase of sarcoplasmic reticulum. Some molecular characteristics are: $M_r = 110,000$; $K_m < 1\mu M$ (two Ca^{2+} sites on cytoplasmic side in high-affinity form); Ca^{2+}/ATP ratio, 2; Mg^{2+} required for activity. The amino-acid residues labeled were mutated to a residue lacking side chains capable of binding Ca^{2+}. Mutations at the circled positions resulted in complete loss of Ca^{2+} transport activity, suggesting that the circled residues participate in Ca^{2+} binding. Adapted from Reference 50.

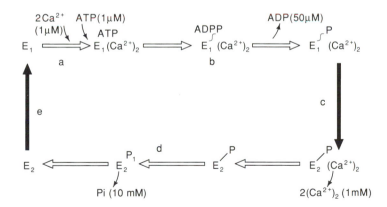

Figure 3.12
Simplified schematic reaction cycle of the Ca^{2+}-ATPase of sarcoplasmic reticulum (SR).[49,51] The transport protein is assumed to be in either of two states, E_1 on the cytoplasmic side, or E_2 on the side of the SR lumen. Starting from E_1 in the upper left corner, the reactions steps shown are: (a) binding of Ca^{2+} and ATP (approximate dissociation constants within parentheses); (b) rapid phosphorylation of Asp-351 of the protein ($E_1{}^{\sim P}$) and release of ADP; (c) transformation from an energy-rich, high-Ca^{2+}-affinity conformation ($E_1{}^{\sim P}$) $(Ca^{2+})_2$ to a low-energy, low-affinity conformation $(E_2{}^{-P})(Ca^{2+})_2$; (d) hydrolysis of the phosphorylated protein and release of the phosphate into the lumen; (e) return to the original state.

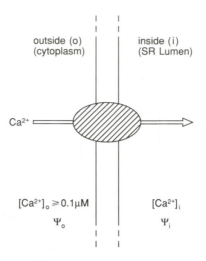

Figure 3.13
Schematic representation of Ca^{2+} transport through a membrane
by a Ca^{2+}-ATPase molecule. Ψ denotes membrane potentials.

where F is Faraday's constant, T the temperature, and R the gas constant. If we assume $\Delta\Psi = 0$, which appears reasonable for the SR membrane according to experimental evidence, we may calculate the free-energy change, ΔG, at 25°C for transferring Δn moles of Ca^{2+} across the membrane. This becomes $\Delta G = -\Delta n \times \Delta\mu_{Ca^{2+}} = \Delta n \times 4.1$ kcal/mol if $[Ca^{2+}]_o/[Ca^{2+}]_i = 10^{-3}$ and $\Delta G = \Delta n \times 5.4$ kcal/mol if $[Ca^{2+}]_o/[Ca^{2+}]_i = 10^{-4}$. Under the pertinent cellular conditions, the free-energy change associated with ATP hydrolysis to ADP and P_i has been calculated by Tanford to be $\Delta G = -13$ to -14 kcal/mol.[53] In the absence of a membrane potential, it is thus possible to transport two Ca^{2+} ions for every ATP molecule hydrolyzed against a concentration (or activity) gradient of 10^4 or more. This treatment says nothing, of course, about the molecular details of this transport. A more detailed model for the transport cycle has been proposed by Tanford.[53]

In the specialized cells of muscle tissue, the sarcoplasmic reticulum may contain much calcium, and if all were "free" Ca^{2+}, the concentration could be as high as 30 mM.[54] This value would cause an osmotic pressure difference across the membrane, as well as put a high demand on the SR Ca^{2+}-ATPase. A lowering of the free Ca^{2+} concentration inside the SR would clearly be beneficial. In the presence of oxalate or phosphate ions in the external medium, calcium oxalate or phosphate may precipitate inside the sarcoplasmic reticulum, but under normal circumstances it appears that Ca^{2+} ions inside the SR are bound to a very acidic protein, *calsequestrin*.[54] Each molecule ($M_r \approx 40$ kDa) is able to bind 40 to 50 Ca^{2+} ions with an effective dissociation constant of about 1 mM (at $I = 0.1$). The protein has a low cation specificity and behaves in many respects like a negatively charged polyelectrolyte. It has been crystallized[55] and we may soon have access to its x-ray structure.

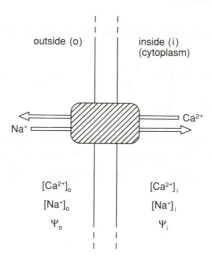

Figure 3.14
Schematic representation of the Ca^{2+}/Na^+ exchanger of the plasma membrane. Ψ denotes membrane potentials.

2. The Na^+/Ca^{2+} exchanger of the plasma membrane

Presently available information on the Na^+/Ca^{2+} exchanger has mainly been obtained from studies of the large cells of the giant squid axon and of plasma-membrane vesicles from various other tissues.[56,57] In heart plasma-membrane vesicles, the exchanger has the following characteristics: $K_m = 1.5–5~\mu M$ for Ca^{2+} and ~ 20 nM for Na^+; $V_{max} \approx 20$ nmol Ca^{2+}/mg protein.[58] The stoichiometry is at least 3:1 Na^+/Ca^{2+}. Very few molecular details of the exchanger are available at present. We may again briefly consider the thermodynamic framework for an Na^+/Ca^{2+} exchanger (Figure 3.14). The difference in electrochemical potential for Na^+ and Ca^{2+} across the membrane is:

$$\Delta\mu_{Ca^{2+}} = RT\ln\frac{[Ca^{2+}]_o}{[Ca^{2+}]_i} + 2F\Delta\Psi, \tag{3.5}$$

$$\Delta\mu_{Na^+} = RT\ln\frac{[Na^+]_o}{[Na^+]_i} + F\Delta\Psi. \tag{3.6}$$

The free-energy change, $\Delta G_t^{Ca^{2+}}$, associated with a transfer of $\Delta n_{Ca^{2+}}$ moles of Ca^{2+} from the inside to the outside is $\Delta G_t^{Ca^{2+}} = \Delta n_{Ca^{2+}} \times \Delta\mu_{Ca^{2+}}$, and the corresponding change associated with the movement of Δn_{Na^+} moles of Na^+ from the outside in is $\Delta G_t^{Na^+} = -\Delta n_{Na^+} \times \Delta\mu_{Na^+}$. If these free-energy changes are coupled via the exchanger, there will be a net flux of Ca^{2+} as long as the free-energy difference,

$$\Delta\Delta G = \Delta G_t^{Ca^{2+}} - \Delta G_t^{Na^+} = \Delta n_{Ca^{2+}} \times \Delta\mu_{Ca^{2+}} - \Delta n_{Na^+} \times \Delta\mu_{Na^+}, \tag{3.7}$$

is less than zero. We can write $\Delta\Delta G$ for the transport of 1 mol Ca^{2+} as

$$\Delta\Delta G = 2.303 \, RT \left[\log \frac{[Ca^{2+}]_o}{[Ca^{2+}]_i} - \Delta n_{Na^+} \times \log \frac{[Na^+]_o}{[Na^+]_i} \right] \tag{3.8}$$
$$+ (2 - \Delta n_{Na^+}) \times F\Delta\Psi.$$

Equating ion activities with concentrations, we note that in a typical mammalian cell $[Na^+]_o \approx 110\text{–}145$ mM, and $[Na^+]_i \approx 7\text{–}15$ mM, or $[Na^+]_o/[Na^+]_i \approx 10$. In the absence of a membrane potential difference ($\Delta\Psi = 0$), Equation (3.8) can thus be simplified to

$$\Delta\Delta G = 2.3 \, RT \left[\log \frac{[Ca^{2+}]_o}{[Ca^{2+}]_i} - \Delta n_{Na^+} \right]. \tag{3.9}$$

To pump one Ca^{2+} ion out of a cell against a concentration gradient of about 10^3 (1 μM $\rightarrow$ 1 mM) requires that at least 3 Na^+ ions pass in the opposite direction, thus maintaining $\Delta\Delta G < 0$. What then will be the effect of a membrane potential difference? Most animal cells, particularly excitable cells such as nerve and muscle cells, have resting potential differences, $\Delta\Psi$, over the plasma membrane of 30 to 90 mV (cytoplasm negative). For this value we find the change in free energy, $\Delta\Delta G$, for the transport of one mol Ca^{2+} to be

$$\Delta\Delta G = 2.3 \, RT \left[\log \frac{[Ca^{2+}]_o}{[Ca^{2+}]_i} - \Delta n_{Na^+} \right] + (2 - \Delta n_{Na^+}) \, 0.1 \, F. \tag{3.10}$$

Thus for $\Delta n_{Na^+} > 2$, we have $\Delta\Delta G < 0$, and the transport of Ca^{2+} against a concentration gradient of about 10^3 will be promoted. This is another good reason for having a Na^+/Ca^{2+} exchange stoichiometry of 3:1.

3. Mitochondrial Ca^{2+} transport: influx

Mitochondria isolated from various types of animal cells—but, interestingly, not those from plant cells—can rapidly accumulate exogenous Ca^{2+}.[59] The transporter is located in the inner membrane and the driving force behind the Ca^{2+} transport appears to be merely the high potential difference across this membrane ($\Delta\Psi \approx 150$ to 180 mV, negative in the inner matrix). This potential difference is fairly closely maintained by the pumping out of H^+ from the matrix by cell respiration. For the transport of 1 mol Ca^{2+} from the "outside" (= cytoplasm) to the "inside" (= inner mitchondrial matrix), we may deduce from Equation (3.4) that the free-energy change ΔG may be written ($\Delta n_{Ca^{2+}} = -1$)

$$\Delta G = - RT \cdot \ln \frac{[Ca^{2+}]_o}{[Ca^{2+}]_i} - 2F \, \Delta\Psi. \tag{3.11}$$

From this analysis it may be inferred that the limiting Ca^{2+} concentration (or activity) ratio that can be achieved by this **electrogenic** pump (i.e., $\Delta G = 0$) is

$$\frac{[Ca^{2+}]_o}{[Ca^{2+}]_i} = e^{-2F\Delta\Psi/RT} \tag{3.12}$$

With $\Delta\Psi = 150$ mV, this ratio is calculated to be 8.4×10^{-6} at 25°C. It is evident that, as long as the Ca^{2+} influx would not lower the membrane potential difference, the Ca^{2+} **uniporter** has a very high pumping potential. Measured values of the pumping rate, V_{max}, are indeed high (>10 nmol/mg protein[59]) and probably limited only by the rate of electron transport and H$^+$ extrusion in the mitochondria.

Mitochondria may accumulate large quantities of Ca^{2+}, probably to maintain electroneutrality. To prevent the buildup of high concentrations of free Ca^{2+} (and of osmotic pressure), phosphate ions are also transported into the inner matrix, where an amorphous calcium phosphate—or possibly a phosphocitrate[60]— is formed. The equilibrium concentration of free Ca^{2+} in the mitochondrial matrix may as a result be comparatively low, on the order of 1 μM.

The molecular nature of the mitochondrial Ca^{2+} uniporter continues to be elusive, and needs to be studied further.

4. Mitochondrial Ca^{2+} transport: efflux

Mitochondria, as well as SR, release Ca^{2+} ions by mechanisms other than "back leakage" through the pumps. In mitochondria from excitable cells, the efflux occurs mainly through an antiport, where 2 Na$^+$ ions are transported inward for every Ca^{2+} ion departing for the cytosolic compartment.[61] In other cells there is evidence for the dominance of a 2H$^+$−Ca^{2+} antiport.[59] In all likelihood the Ca^{2+} efflux is regulated, possibly by the redox state of pyridine nucleotides in the mitochondria. As with the Ca^{2+} uniporter, few details on the molecular nature of the antiporters are presently available.

5. Ca^{2+} efflux from non-mitochondrial stores

Release of Ca^{2+} from ER and SR presently appears to be the prime effect of the new intracellular messenger 1,4,5-triphosphoinositol (1,4,5-IP$_3$) released into the cytoplasm as a result of an external hormonal stimulus (see Section IV.C). It seems that receptors for 1,4,5-IP$_3$ have been established on ER, and that the binding of 1,4,5-IP$_3$ causes a release of Ca^{2+} stored in this organelle.[62,63,170,171] In addition to the receptor-controlled Ca^{2+} efflux, there may be other pathways for Ca^{2+} release, and Ca^{2+} mobilization may be regulated by other intracellular entities, the Ca^{2+} ions themselves included.

6. Other voltage-gated or receptor-activated Ca^{2+} channels

In addition to the transport pathways already discussed, some cells seem to have Ca^{2+} channels in the plasma membrane that can be opened by the action of an agonist on a receptor or that are gated in response to changes in membrane potential.[64] For example, Ca^{2+} channels can be opened by nicotinic cholinergic agonists[65] or by the excitatory amino acid N-methyl-D-aspartate (NMDA).[66] Endochrine cells and also some muscle and neuronal cells have voltage-sensitive Ca^{2+} channels.[67,68] We will not discuss these further, but merely point to their existence. We finally note that during the last few years knowledge about the mechanisms of Ca^{2+} entry and release to and from extracellular and intracellular pools has increased dramatically, and we refer the reader to recent reviews of the field.[175,176]

C. Inositol Trisphosphate and the Ca^{2+} Messenger System

A "second" messenger is an entity that inside a cell mediates the action of some hormone at the plasma membrane, the hormone being considered the "first" messenger. The first such second messenger to be discovered—in fact, the very molecule that led to the formulation of the whole concept—was cyclic AMP.[69] During the decade following the discovery of cAMP, it was gradually realized that intracellular release of Ca^{2+} ions also accompanied hormonal stimuli, and the Ca^{2+} ion slowly became regarded as a second messenger. This idea was first clearly enunciated by Rasmussen[70] as early as 1970, and gained general acceptance when the ubiquitous intracellular Ca^{2+}-binding protein calmodulin (see Section V.A) was discovered. In the mid-1970s this protein was shown to be a Ca^{2+}-dependent regulator of a large number of Ca^{2+}-dependent enzymes, transport proteins, etc., establishing a molecular basis for Ca^{2+} action in cells.

There were some puzzling facts, however. Although a transitory increase in intracellular Ca^{2+} concentration in response to the binding of a hormone or transmitter substance to a surface receptor could result from extracellular Ca^{2+} being released into the cytoplasm, there was compelling evidence for muscle cells that the main Ca^{2+} source was the sarcoplasmic reticulum (SR). This result led to the hypothesis of "Ca^{2+}-induced Ca^{2+} release," i.e., that upon stimulation of the cell, a small amount of Ca^{2+} entered into the cytoplasm and triggered the release of greater amounts of Ca^{2+} from the SR. For some cell types it could, however, be shown that transient increases in intracellular Ca^{2+} could occur even when extracellular Ca^{2+} was removed, although *prolonged* responses required the presence of extracellular Ca^{2+}. Although some specialized cells have gated plasma-membrane Ca^{2+} channels, release of Ca^{2+} into the cytoplasm from intracellular stores appears to be of at least equal importance. Furthermore, there is now overwhelming evidence[63,70–72] that intracellular Ca^{2+} is released in response to the formation of a new type of intracellular messenger: $1,4,5\text{-IP}_3$. Receptors for this messenger have recently been found in the membranes of intracellular organelles, and binding of $1,4,5\text{-IP}_3$ to these receptors results in the release of Ca^{2+} ions.[73]

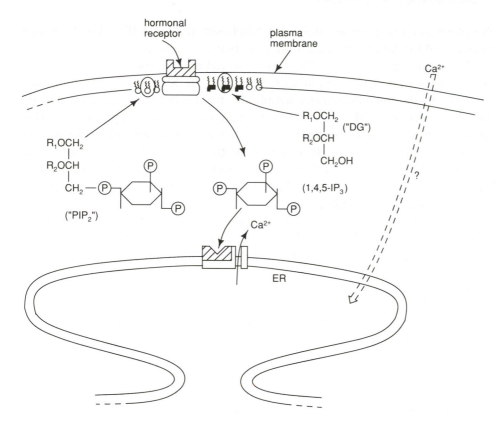

Figure 3.15
Outline of the presumed role of inositol phosphates in the intracellular mobilization of Ca^{2+}.
Upon binding of an agonist to a plasma-membrane receptor, phosphatidylinositol 4,5-bisphos-
phate (PIP_2) membrane lipids are hydrolyzed to give diacylglycerol (DG) and inositol 1,4,5-
trisphosphate (1,4,5-IP_3). The latter interacts with specific receptors on the endoplasmic reticulum (ER) membrane that trigger the release of Ca^{2+} into the cytosol. Ca^{2+} may be returned to
the ER through the Ca^{2+}-ATPase of the ER membrane (see Figure 3.9) and also by a direct
influx of Ca^{2+} from the extracellular medium.[70,81]

1,4,5-IP_3 is formed as a product in the hydrolysis of a special phospholipid
present in the cell membrane: phosphatidyl-inositol-4,5-bisphosphate. This reaction, then, is the initial receptor-stimulated event. The newly formed 1,4,5-
IP_3 is assumed to diffuse into the cytoplasm, and eventually reach intracellular
1,4,5-IP_3 receptors on the ER, thereby triggering the release of Ca^{2+}. A simplified reaction scheme is shown in Figure 3.15. A diacylglycerol (DG) is also
formed in the hydrolysis step. DG can also act as an intracellular messenger,
and stimulates the activity of a membrane-bound protein kinase, known as *protein kinase C* (PKC). As a result, PKC may phosphorylate certain key proteins
and influence their activity. Protein kinase C is also activated by Ca^{2+} ions, a
fact that illustrates Nature's knack in designing regulatory networks! 1,4,5-IP_3
is either directly degraded in a series of enzymatic steps back to inositol, which
is then used to resynthesize the phospholipid, or it may be further phosphorylated to inositol-1,3,4,5,-tetraphosphate (1,3,4,5-IP_4), which may undergo de-

phosphorylation to form inositol-1,3,4-trisphosphate (1,3,4-IP$_3$). The biological functions of the latter compounds are now being investigated.

The intracellular levels of Ca^{2+} are restored back to the normal low resting values (100 to 200 nM) via transport back into the SR, and/or into mitochondria, or out through the plasma membrane by the pumping mechanisms discussed in Section IV.B. As was briefly mentioned above, depriving a cell of extracellular Ca^{2+} will eventually make the cell incapable of prolonged responses to external stimuli. It appears that the intracellular Ca^{2+} stores may become depleted if not replenished. It has been suggested that the intracellular ER Ca^{2+} pool has a direct route of access to the extracellular pool, a route that is closed when the ER pool is full.[74]

In a sense, then, Ca^{2+} seems to have been downgraded by the inositolphosphates from a "second" to a "third" messenger; however, the pivotal role of Ca^{2+} as a regulator of cellular activities remains undisputed.

D. Summary

The fluxes of Ca^{2+} ions and their regulation in higher organisms, as well as in microorganisms, depend on several transport proteins in addition to vesicular and gated processes. An important class of transport proteins are the Ca^{2+}-ATPases, which are particularly abundant in muscle cells. These proteins translocate Ca^{2+} ions against large activity (or concentration) gradients through the expenditure of ATP. Transport of Ca^{2+} ions against activity gradients across membranes may also be accomplished by coupled transport of other ions, like Na$^+$, with a gradient in the opposite direction.

As a result of some external stimulus—the action of a hormone, for example—the "free" Ca^{2+}-ion concentrations in the cytoplasm of many cell types may transiently increase several orders of magnitude. This increase largely results from the release of Ca^{2+} from intracellular stores (ER, SR) in response to the initial formation of a new type of messenger, 1,4,5-IP$_3$. The activity of Ca^{2+}-transport proteins eventually restores the Ca^{2+} concentration levels to resting levels. This sequence of events forms the basis for Ca^{2+}'s role in the regulation of a wide variety of cellular activities (see Section V).

V. MOLECULAR ASPECTS OF Ca^{2+}-REGULATED INTRACELLULAR PROCESSES

So far we have mainly discussed the routes and means by which the concentration of Ca^{2+} ions in the cytoplasm can be transiently increased and brought back to resting levels. But changing the cytoplasmic Ca^{2+} concentration is not enough. In order to influence the cellular machinery, the Ca^{2+} ions must interact with different proteins, *intracellular Ca^{2+} receptors* if you like. These intracellular Ca^{2+}-receptor proteins must have certain properties in order to function.

(i) Their Ca^{2+}-affinity must be such that their Ca^{2+}-binding sites are essentially unoccupied at resting levels of free Ca^{2+} ($\sim 10^{-7}$ M) and occupied at levels reached upon stimulus (generally assumed to be 10^{-5} to 10^{-6} M). This means that the binding constants $K_B^{Ca^{2+}}$ should be $\sim 10^6 \, M^{-1}$.

(ii) We should also remember that Ca^{2+} must exert its function in the presence of a number of other ions; in mammalian cells the intracellular concentration of "free" Mg^{2+} ions is around 1 mM, and that of K^+ ions around 100 to 150 mM. The receptors must therefore have an adequate selectivity for Ca^{2+}.

(iii) In response to Ca^{2+} binding, a Ca^{2+} receptor must undergo some kind of conformation change that either alters its interaction with other molecules or changes its activity if it is an enzyme.

(iv) Finally, there are *kinetic* considerations. In many cells a rapid response is essential, and therefore the receptors must be able to interact swiftly—within milliseconds—with incoming Ca^{2+} ions, and the ions must also be able to depart almost as rapidly.

A few proteins have been discovered that qualify as intracellular Ca^{2+} receptors. The best known of these is *calmodulin* (CaM), which appears to be present in all eukaryotic cells. Most of the cellular responses elicited by Ca^{2+} appear to result from interactions between the Ca^{2+}-calmodulin complex and various other target enzymes and proteins.[75] Another important Ca^{2+}-receptor protein is *troponin C* (TnC), which occurs in muscle cells and is instrumental in mediating muscle contraction.[76] These two types of proteins are highly homologous, as we shall see, and may be considered members of a superfamily of closely related intracellular Ca^{2+}-binding proteins. This superfamily has been given the name "the calmodulin superfamily," and close to 200 distinct family members are presently known.[77] Not all members of the superfamily may qualify as Ca^{2+} receptors; some like *parvalbumins* and *calbindins* (see Section IV.A) appear to have a role in intracellular transport and/or Ca^{2+}-buffering. For others, such as the *S-100* proteins[78] found predominantly in brain tissue, and *calcimedins*,[79] isolated from smooth muscle, the biological function is still unclear.

One Ca^{2+} receptor with enzymatic activity is *protein kinase C*. Its activity is markedly increased in the presence of Ca^{2+}, and it has a high calcium-binding constant (see Table 3.2) in the presence of diacylglycerol or **phorbol esters**.[80]

During recent years, groups interested in the role of Ca^{2+} in secretion and in the control of membrane cytoskeleton have identified some intracellular Ca^{2+}/phospholipid-binding proteins that appear to be distinct from the calmodulin superfamily; these include *lipocortin, endonexin, calelectrin, p36*, and *calpactin*.[81–83] These membrane-binding proteins are collectively called *annexins*,[84] and contain repeated domains distinct from EF-hands. The Ca^{2+} sites are very similar to that observed in phospholipase A_2, as shown by the recently determined x-ray structure of annexin V.[172] A condensed overview of the interaction of Ca^{2+} with intracellular proteins is shown in Figure 3.16. We will now go on to discuss the molecular properties of some of the proteins mentioned above, starting with calmodulin.

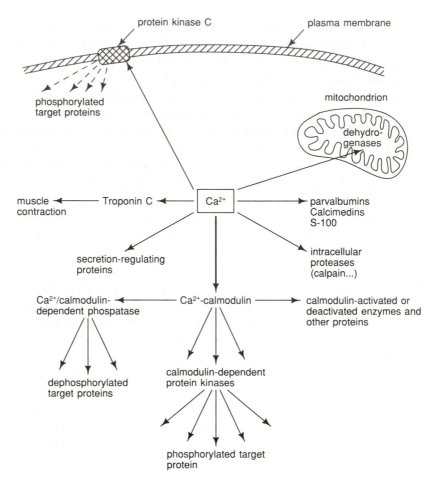

Figure 3.16
Condensed overview of the interaction of Ca²⁺ with intracellular proteins.

A. Calmodulin

Calmodulin is a small acidic protein ($M_r \approx 16{,}700$), the amino-acid sequence of which has been remarkably preserved during evolution. Early on, an analysis of its amino-acid sequence indicated that it should have four Ca^{2+}-binding sites, a deduction that proved to be correct. The three-dimensional x-ray structure of bovine brain calmodulin[85] has been solved to a resolution of 2.2 Å. A space-filling model is shown in Figure 3.17. (*See color plate section, page C-9.*) The molecule has a dumbbell-like shape, with two globular domains connected by an eight-turn α-helix—an unusual structural feature. In the crystal structure, there are no direct contacts between the two globular domains, each of which contains two Ca^{2+}-binding sites. The Ca^{2+} sites are all constructed in the same way: two α-helices separated by a calcium-binding loop, 12 amino acids long, and wrapped around the Ca^{2+} ion. This structural arrangement is nearly identical with that first observed in the x-ray structure of carp parvalbumin, and is colloquially termed *"the EF-hand."* [86] This structural unit is also observed in all available x-ray structures of proteins of the calmodulin superfamily (see Sections V.B and V.C). The Ca^{2+} ligands are all oxygen atoms, located approximately at the vertices of a pentagonal bipyramid.

The binding of Ca^{2+} and other cations to CaM has been extensively investigated.[87] The first two Ca^{2+} ions are bound in a cooperative manner, with an average binding constant of about $2 \times 10^5 \, M^{-1}$ in 150 mM KCl and 1 mM Mg^{2+}. The third and fourth Ca^{2+} ions are bound with binding constants of about $3 \times 10^4 \, M^{-1}$ under the same conditions. Spectroscopic evidence has shown that the first two Ca^{2+}-ions are bound in the C-terminal domain. Mg^{2+} has been shown to bind primarily to the N-terminal domain (see Table 3.2).[88]

The rates of dissociation of Ca^{2+} from the $(Ca^{2+})_4$ CaM complex have been studied by both stopped-flow and NMR techniques.[89,90] Fast and slow processes are observed, both corresponding to the release of two Ca^{2+} ions. At an ionic strength $I = 0.1$ and 25°C, the rates for the two processes differ by a factor of 30 (see Table 3.4).

A body of biophysical measurements, mostly made before the advent of x-ray structures, indicated that CaM is constructed from two largely independent domains.[87] This conclusion emanated from studies of the two tryptic fragments, TR_1C and TR_2C. The major site of cleavage is between Lys-77 and Asp-78 of the central helix, and results in N-terminal and C-terminal fragments of nearly equal size. To a good approximation, the biophysical properties of the intact CaM molecule—NMR, UV and CD spectra, kinetic properties, thermochemical data, etc.—are the sum of the same properties of the fragments TR_1C and TR_2C. This means that we may assign the slow dissociation process, k_{off}^s, to the C-terminal domain, and the fast, k_{off}^f, to the N-terminal domain of CaM. Combining binding constants and off-rates, we may calculate that the rates of Ca^{2+} binding to CaM are on the order of $10^7 \, M^{-1} \, s^{-1}$ at high ionic strength, and

Table 3.4
Rates of Ca^{2+}-dissociation and -association of some enzymes and proenzymes.

	k_{off} [s^{-1}]	k_{on} [M^{-1} s^{-1}]
Macrobicyclic amino cryptate [2.2.2]	0.3	10^4
Phospholipase A2	1.1×10^3	4×10^6
sTroponin C: Ca^{2+} sites	300	
Ca^{2+}-Mg^{2+} sites	5	
Trypsin	3	1.1×10^5
Trypsinogen	≤10	6×10^4
Chymotrypsin	70	$\sim 10^6$
Chymotrypsinogen	350	2.8×10^5
Calmodulin: N-terminal	300–500	10^7
Calmodulin: C-terminal	10–20	

10^8 M^{-1} s^{-1} or higher at low ionic strength. Recently the x-ray structure of the C-terminal fragment TR$_2$C was solved, and indeed showed a structure nearly identical with C-terminal domain of intact CaM.[91]

The structural changes occurring in CaM as Ca^{2+} ions are bound are associated with pronounced changes in ^{1}H NMR, UV, fluorescence, and CD spectra.[87] The observed changes in CD and fluorescence spectra in the presence of Mg^{2+} are only about 20 to 25 percent of those induced by Ca^{2+}. A comparison of the CD spectra of CaM and its **tryptic** fragments indicates that the structural changes induced by Ca^{2+} are substantially greater in the C-terminal than in the N-terminal half.[92] By and large, few structural details of the conformation changes have as yet been obtained. However, one aspect of the Ca^{2+}-induced conformation change is that *hydrophobic sites*, probably one on each domain of the molecule, become exposed. In the presence of excess Ca^{2+}, CaM will bind to other hydrophobic molecules, e.g., phenyl-Sepharose, a variety of drugs, many small peptides, and—last but not least—its target proteins. This brings us to the question of how CaM recognizes and interacts with the latter. We may suspect that the hydrophobic sites on each domain are somehow involved, but the role played by the central helix is still not clear. To explain small-angle x-ray scattering data, the interconnecting helix needs to be kinked, bringing the intact globular domains closer.[93]

A putative CaM-binding segment (27 amino acids long) of myosin light-chain kinase (MLCK), an enzyme activated by CaM, has been identified.[94] The interaction between the segment peptide ("M13") and CaM has been studied[95] by CD spectroscopy and ^{1}H NMR. From these studies it appears that a unique 1:1 complex is formed, and that secondary and tertiary structural changes occur not only in the peptide M13 but also in both halves of the CaM molecule. Further NMR studies[96,97] of the interaction between CaM and naturally occurring peptides (mellitin and mastoparan) that share some structural features of M13—clusters of basic residues, hydrophobic residues adjacent to the basic residues, and a predicted high α-helical content—show very much the same results. Based on these results, a model, shown in Figure 3.18, for the interaction between CaM and M13 has been proposed. In this model the central helix is kinked at position 81, allowing the two domains to wrap around the assumed α-helical M13. Preliminary structure calculations of calcium-loaded CaM, based on NMR data, indicate that the central helix in solution indeed is kinked and very flexible,[99] and comparisons[100] of chemical shifts in calmodulin with and without M13 complexed supports the model in Figure 3.18. Recent structural studies using NMR spectroscopy and x-ray diffraction have essentially confirmed the general features of this model, although the orientation of the peptide is found to be reversed.[173]

In conclusion, two important features of the protein should be recognized.

(i) The binding of Ca^{2+} to CaM (and to its complex with the target protein) is quite likely *cooperative*, meaning that the switch from inactive to active conformation may occur over a much more narrow Ca^{2+}-concentration interval than otherwise.

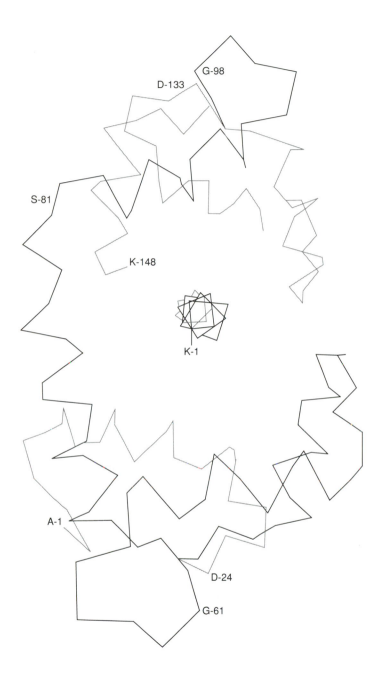

Figure 3.18
A model for the interaction between calmodulin (CaM) and the assumed α-helical ($\phi = -57$, $\psi = -47$) peptide M13. To produce this model, the backbone dihedral angles of Ser-81 in the central α-helix of CaM have been changed (to $\phi = -54$, $\psi = +98$), allowing the hydrophobic patches of both globular domains (green in Figure 3.17) of CaM to interact with the peptide simultaneously. Figure kindly provided by R. Kretsinger; see also Reference 98.

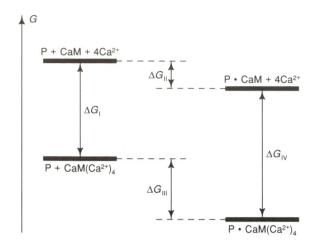

Figure 3.19
Scheme depicting the standard free energies of different states in a system consisting of Ca^{2+}, calmodulin (CaM), and a target protein (P). P·CaM denotes a complex between calcium-free CaM and P, P·CaM(Ca^{2+})$_4$ denotes a complex with Ca^{2+}-loaded CaM. If the affinity of the Ca^{2+}-loaded CaM with the target protein P is higher than that of the Ca^{2+}-free form— i.e., $|\Delta G_{III}| > |\Delta G_{II}|$—it follows that the Ca^{2+} affinity of the complex P·CaM is higher than that of CaM itself.

(ii) The effective Ca^{2+} affinity will be different in the presence of the target proteins. To illustrate this second point, consider the standard free energies in the minimum scheme depicted in Figure 3.19. If the affinity of the Ca^{2+}-cal-modulin complex (CaM(Ca)$_4$) for the target protein (P) is greater than that of Ca^{2+}-free calmodulin (CaM)—i.e., $|\Delta G_{III}| > |\Delta G_{II}|$—it follows that the Ca^{2+} affinity of the complex between P and CaM (P·CaM) must be higher than in CaM itself. This effect is also found experimentally in model systems.[101]

B. Troponin C

The contraction of striated muscle is triggered by Ca^{2+} ions. Muscle cells are highly specialized, and contain two types of filaments that may slide past each other in an energy-consuming process. One of the filaments, the thin filament, is built up by actin molecules ($M_r \approx 42$ kDa) polymerized end-to-end in a dou-ble helix. In the grooves of this helix runs a long rod-like molecule, tropomyo-sin; and located on this molecule at every seventh actin, is a complex of three proteins, *troponin*. The three proteins in the troponin complex are *troponin I* (TnI), *troponin T* (TnT), and *troponin C* (TnC). A schematic picture of the organization of the thin filament is shown in Figure 3.20.

Troponin C is the Ca^{2+}-binding subunit of troponin, and it is structurally highly homologous to calmodulin. Skeletal-muscle troponin C (sTnC; $M_r \approx 18$ kDa) can bind four Ca^{2+} ions, but cardiac-muscle troponin C (cTnC) has one of the four calcium sites modified, so that it binds only three Ca^{2+} ions. The

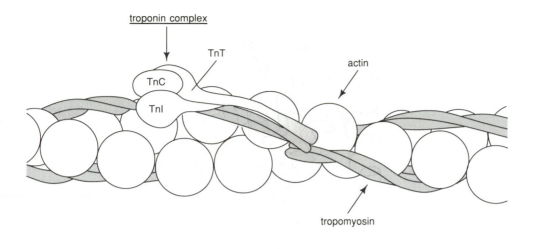

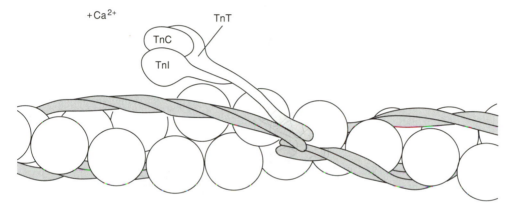

Figure 3.20
Schematic diagram of the organization of skeletal muscle thin filament, showing the position of tropo-myosin and the troponin complex on the actin filament. The binding of Ca^{2+} to TnC, the calcium-binding subunit of the troponin complex, removes TnI, the inhibitory subunit, from actin and thus permits an interaction with a specialized protein, myosin, on neighboring thick muscle filaments (not shown). An ATP-driven conformation change in the myosin head group makes the thick and thin filaments move relative to one another, so that muscle contraction occurs.

x-ray structures of sTnC from turkey and chicken skeletal muscle have been determined to resolutions of 2.8 and 3.0 Å, respectively.[102,103] The structure of turkey sTnC is shown in Figure 3.21. The similarity between the structures of CaM (Figure 3.17) and sTnC is obvious. In sTnC we again find two domains, each with two potential Ca^{2+} sites, separated by a 9-turn α-helix. The crystals were grown in the presence of Ca^{2+} at a low pH (pH $\approx$ 5), and only two Ca^{2+} ions are found in the C-terminal domain. The two Ca^{2+}-binding sites in this domain have the same helix-loop-helix motif that is found in CaM, and they both conform to the archetypal EF-hand structure. The interhelix angles between helices E and F and between G and H are close to 110°. By contrast, the helices in the N-terminal domain, where no Ca^{2+} ions are bound, are closer to being

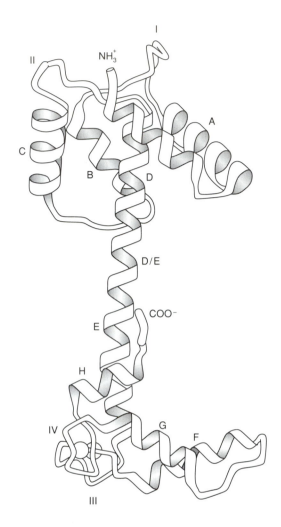

Figure 3.21
A ribbon backbone representation of the three-dimensional structure of turkey-skeletal-muscle troponin C according to Herzberg and James.[102] The crystals were grown at pH 5 in the presence of excess Ca^{2+}, and at this low pH only Ca^{2+} ions bound to the high-affinity domain (the C-terminal domain) are observed. Note the high structural homology with calmodulin (Figure 3.17).

ntiparallel, with interhelix angles of 133° (helices A and B) and 151° (helices and D).

Both sTnC and cTnC have two high-affinity Ca^{2+}-binding sites (see Table ?) that also bind Mg^{2+} ions competitively, although with a much lower affin-. These two sites are usually called "*the Ca^{2+}-Mg^{2+} sites.*"[76,104] In sTnC re are also two (in cTnC, only one) Ca^{2+}-binding sites of lower affinity $Ca^{2+} \approx 10^5$ M^{-1}) that bind Mg^{2+} weakly and therefore have been called "*the $^{2+}$-specific sites.*" Since Ca^{2+} binding to the latter sites is assumed to be the crucial step in the contractile event, they are often referred to as "*the regulatory sites*" (see below). The existence of additional weak Mg^{2+} sites ($K_B \approx 300\ M^{-1}$) on sTnC, not in direct competition with Ca^{2+}, has also been inferred.[76,104,105] Spectroscopic studies have shown that the two strong Ca^{2+}-Mg^{2+} sites are located in the C-terminal domain, and the weaker Ca^{2+}-specific sites in the N-terminal domain of sTnC.[106] This pattern is similar to that observed with CaM. NMR spectroscopic studies strongly suggest that binding of Ca^{2+} to both sTnC and cTnC is cooperative.[107] In sTnC the C-terminal domain binds Mg^{2+} much more strongly than the N-terminal domain, by contrast to CaM, where the reverse is true.

The rates of dissociation of Ca^{2+} and Mg^{2+} from sTnC have been measured by both stopped-flow and ^{43}Ca NMR techniques.[76,108] As with CaM, the actual numbers depend on the solution conditions, ionic strength, presence of Mg^{2+}, etc. (see Table 3.4). On the rate of Mg^{2+} dissociation from the Ca^{2+}-Mg^{2+} sites, quite different results have been obtained by stopped-flow studies[76] of fluorescence-labeled sTnC ($k_{off}^{Mg^{2+}} \approx 8\ s^{-1}$) and by ^{25}Mg NMR ($k_{off}^{Mg^{2+}} \simeq 800\text{--}1000\ s^{-1}$).[109] This apparent discrepancy seems to have been resolved by the observation that both binding and release of Mg^{2+} ions to the Ca^{2+}-Mg^{2+} sites occur *stepwise*, with $k_{off}^{Mg^{2+}} < 20\ s^{-1}$ for one of the ions, and $k_{off}^{Mg^{2+}} \geq 800\ s^{-1}$ for the other.[110] The rates of dissociation of the Mg^{2+} ions are important, since under physiological conditions the Ca^{2+}-Mg^{2+} sites of sTnC are likely to be predominantly occupied by Mg^{2+} ions, release of which determines the rate at which Ca^{2+} can enter into these sites.

Spectroscopic and biochemical data[111] indicate that upon binding Ca^{2+}, sTnC and cTnC undergo significant conformation changes. Comparisons of NMR spectroscopic changes on Ca^{2+} binding to intact sTnC, as well as to the two fragments produced by tryptic cleavage (essentially the N-terminal and C-terminal halves of the molecule, just as was the case with CaM), have shown that the conformation changes induced are mainly localized within the domain that is binding added ions.[110,112] Thus the central α-helix connecting the domains seems unable to propagate structural changes from one domain to the other. It has been suggested that the structural differences found in the x-ray structure of turkey sTnC between the C-terminal domain, which in the crystal contains two bound Ca^{2+} ions, and the N-terminal domain, in which no Ca^{2+} ions were found, may represent these conformational changes.[113] This rather substantial conformational change is schematically depicted in Figure 3.22. However, preliminary structure calculations[114] of the calcium-saturated and calcium-free forms of calbindin D_{9k} indicate that much more subtle conformational changes take place upon binding Ca^{2+} in calbindin D_{9k}. Interestingly, ^1H NMR spectroscopy has provided evidence for the concept that the structural change induced by Mg^{2+} binding to the C-terminal domain of sTnC must be very similar to that induced by Ca^{2+} ions. Another result obtained by ^{113}Cd NMR studies[108] is that the cadmium-loaded N-terminal domain of sTnC in solution undergoes a rapid interchange between two or more conformations, with an exchange rate on the order of $10^3\text{--}10^4\ s^{-1}$.

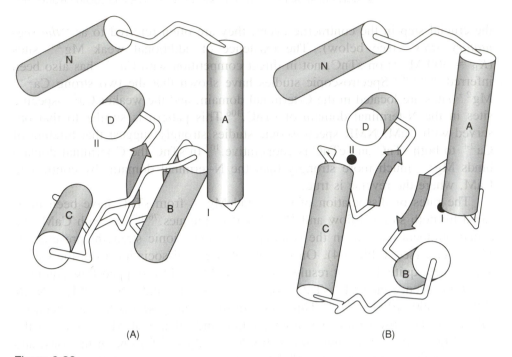

(A)   (B)

Figure 3.22

Diagrammatic representation of the proposed conformational changes in the N-terminal domain of troponin C upon Ca^{2+} binding.[113] Helices are depicted as cylinders, and I, II denote the Ca^{2+}-binding sites. Helices N, A, and D retain their relative positions, and the relative disposition of helices B and C are also kept constant. (A) Ca^{2+}-free conformation as determined by x-ray crystallography.[102] (B) Proposed Ca^{2+}-saturated conformation based on the structure of the highly sequence-homologous Ca^{2+}-loaded C-terminal. Figure kindly provided by N. C. J. Strynadka and M. N. G. James.

Just as CaM exerts its biological function in complexes with other proteins, TnC participates in the three-protein troponin complex. It presently appears that TnC and TnI form a primary complex that is anchored by TnT to a binding site on tropomyosin.[115] In the troponin complex the Ca^{2+} affinity is increased by a factor of about ten over that in isolated sTnC, both at the Ca^{2+}-Mg^{2+} sites and at the Ca^{2+}-specific sites. A similar increase in affinity is found for Mg^{2+}. Given the amounts of "free" Mg^{2+} inside muscle cells (1 to 3 mM), it seems likely that the Ca^{2+}-Mg^{2+} sites in the resting state of troponin are filled with Mg^{2+}, so that a transitory release of Ca^{2+} leads primarily to rapid Ca^{2+} binding to the Ca^{2+}-specific sites, and subsequently to conformation change and contraction.

C. Parvalbumin and Calbindins D_{9K} and D_{28K}

A few intracellular Ca^{2+}-binding proteins have been discovered that by sequence homology clearly belong to the CaM-TnC family with Ca^{2+} sites of the "EF-hand"-type, but that do not appear to exert a direct regulatory function. Parvalbumins ($M_r \approx 12$ kDa), calbindin D_{9K} ($M_r \approx 8.7$ kDa) and calbindin D_{28K}

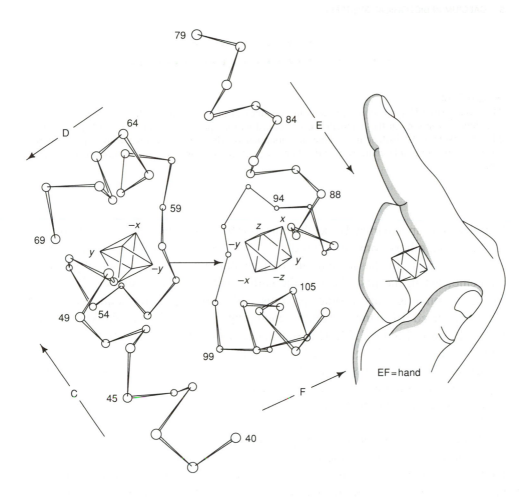

Figure 3.23
Structure of the Ca^{2+}-binding sites of carp parvalbumin. The Ca^{2+} ions are depicted as regular octahedra making six ligand contacts with oxygen atoms at each vertex, labeled x, y, z, $-x$, $-y$, $-z$. The helix-loop-helix structure that forms a Ca^{2+}-binding site can be regarded as a hand with the forefinger representing one helix (e.g., the E-helix) in the plane of the figure, the thumb oriented perpendicular to the plane representing the second helix (the F-helix), and the remaining fingers make up the Ca^{2+}-binding loop. After Kretsinger and Barry.[118]

$(M_r \approx 28$ kDa) belong to this group. *Parvalbumin(s)* exist in two main types, α and β, found in large quantities in the white muscle of fish, amphibia, and reptiles, but also in different mammalian tissues,[116,117] including neurons of the central and peripheral nervous system. The molecule has two fairly strong Ca^{2+}-binding sites (see Table 3.2). The x-ray structure of carp parvalbumin was solved in 1973 by Kretsinger *et al.*,[118] and for a decade provided the basis for all discussions on intracellular Ca^{2+}-binding proteins. The concept of the canonical "EF-hand" Ca^{2+}-binding site originated from the parvalbumin work, and the name "EF" derives from the labeling of the two helices that flank the second of the two Ca^{2+} sites in parvalbumin, as shown in Figure 3.23. If the first Ca^{2+}

x   y   z   −y   −x   −z

En • • nn • • n ♦ • ♦ • ♦ G • I ♦ • • En • • nn • • n

Figure 3.24
One consensus EF-hand sequence including residues in the flanking α-helices; x, y, z, $-x$, $-y$, $-z$ denote positions in the octahedral Ca^{2+} coordination sphere. E—glutamic acid residue, G—glycine residues, I—isoleucine residue, n—nonpolar residue, ♦ —a residue with a nonaromatic oxygen-containing side chain (i.e., Glu, Gln, Asp, Asn, Ser, or Thr), and •—nonspecific residue.

ligand in the approximately octahedral coordination sphere is given number 1 (or "x") the others come in the order 3("y"), 5("z"), 7("$-y$"), 9("$-x$"), and 12("$-z$"). In the second site of parvalbumin, "$-x$" is actually a H_2O molecule, but in the first site it is the carboxylate of a Glu. Studies[118] of putative Ca^{2+}-binding sites in other proteins with known primary sequences led to the generalized EF-hand structure—including residues in the flanking α-helices—shown in Figure 3.24. This sequence, with minor modifications, has been widely used in searching for "EF-hands" in libraries of amino-acid (or DNA) sequences of new proteins with unknown properties. In this way, *calbindin* D_{28k}, a protein with unknown function, initially discovered in chicken intestine, but later found also in brain, testes, and other tissue, has been shown to have four EF-hand sites.[119]

Recently two structures of carp parvalbumin, both with a resolution of 1.6 Å, were published.[120] One of these structures is the native calcium-loaded form of the protein; the second is the structure of parvalbumin in which Ca^{2+} has been replaced by Cd^{2+}. No significant differences are observed upon replacement of calcium by cadmium. ^{113}Cd has a nuclear spin of $I = \frac{1}{2}$, making it much more amenable to NMR studies than the quadrupolar ^{43}Ca ($I = \frac{7}{2}$). This study supports the use of ^{113}Cd NMR as a tool for the study of calcium-binding proteins.[121]

The function of parvalbumin has long been assumed to be that of buffering Ca^{2+} in muscle cells, i.e., taking up Ca^{2+} ions released from Ca^{2+}-troponin complexes, thereby ensuring that the cytoplasmic levels of free Ca^{2+} are always kept very low, even during short bursts of muscle activity.[122] The widespread occurrence of parvalbumin in non-muscle tissue indicates that it probably has other roles as well.

Calbindin D_{9k} ($M_r \approx 8.7$ kDa) is another intracellular Ca^{2+}-binding protein with unknown function. It was briefly mentioned in connection with Ca^{2+} uptake and transport in the intestine and placenta (Section IV.A). Like the avian calbindin D_{28k}, the D_{9k} calbindin has been observed in many types of tissue. The homology between the D_{9k} and D_{28k} calbindins is much less than the name suggests; both their syntheses are, however, regulated by vitamin D. The x-ray structure of bovine calbindin D_{9k} has been determined[123] and refined to a resolution of 2.3 Å, and a three-dimensional *solution* structure of porcine calbindin

D$_{9k}$ is also available.[124] The average solution structure calculated from NMR data is shown in Figure 3.25 (*See color plate section, page C-10.*)

The protein has four main α-helices and two Ca^{2+}-binding loops (I and II). The interior of the molecule shows a loose clustering of several hydrophobic side chains; in particular, three phenylalanine rings come very close in space. The Ca^{2+}-binding loops constitute the least-mobile parts of the molecule. The crystallographic temperature factors have pronounced minima in these regions, with the lowest overall B-factor observed in loop II. Both Ca^{2+} ions are roughly octahedrally coordinated with protein oxygen atoms. There are some striking differences between the two sites, however. Whereas the C-terminal site (II) has a general structure very similar to the archetypal "EF-hand," as observed in CaM, sTnC, and parvalbumin, the N-terminal site (I) has an extra amino-acid residue inserted between vertices x and y, and z and $-y$ (see Figure 3.24). As a consequence, the peptide fold in site I is different from that in site II. Three carboxylate groups are ligands in site II, but in site I there is only one.

Despite this marked difference in charge and peptide fold, the Ca^{2+} affinity of both Ca^{2+} sites is remarkably similar, as has been shown in a study in which site-directed mutagenesis was combined with different biophysical measurements.[37] Cooperative Ca^{2+} binding in the native calbindin D$_{9k}$ (the "wild type") was first demonstrated at low ionic strength by means of the values of the two stoichiometric Ca^{2+}-binding constants, K_1 and K_2, which could be measured with good accuracy ($K_1 = 4.4 \times 10^8$ M^{-1} and $K_2 = 7.4 \times 10^8$ M^{-1}). The effects of amino-acid substitutions in Ca^{2+} site I were primarily localized to this site, with virtually no effects on the structure or other biophysical properties pertinent to site II. The appearance of sequential Ca^{2+} binding in some of the calbindin mutants did allow the identification of ^{1}H NMR resonances that respond primarily to binding of Ca^{2+} to either one of the sites. This result in turn permitted an estimate of the ratio between the site-binding constants (K_A and K_B) in the wild-type protein and in one of the mutant proteins (Tyr-13 $\rightarrow$ Phe). In this way the researchers[125] could assess, to within narrow limits, the free energy of interaction, $\Delta\Delta G$, between the two Ca^{2+} sites as 7.7 kJ/mol at low ionic strength and 4.6 kJ/mol in the presence of 0.15 M KCl. How this site-site interaction is transmitted on a molecular level is still unknown.

Through a combination of site-specific mutations and biophysical measurements, it has recently been demonstrated that carboxylate groups at the surface of the protein, but not directly ligated to the bound Ca^{2+} ions, have a profound effect on the Ca^{2+} affinity.[126] Neutralization of the surface charges reduces affinity and increases the stability of the protein toward unfolding by urea.[127]

A surprising discovery about the structure of bovine calbindin D$_{9k}$ in solution has also been made recently.[128] Detailed analysis of the 2D ^{1}H NMR spectrum of wild-type calbindin has revealed that it exists as a 3:1 equilibrium mixture of two forms, corresponding to a *trans* and *cis* conformation around the Gly-42-Pro-43 peptide bond. The global fold appears essentially the same in the two forms, and structural differences are primarily located in the inter-domain loop in which Pro-43 is located.

D. Sarcoplasmic Calcium-Binding Protein from *Nereis diversicolor*

The calmodulin superfamily of proteins also includes *sarcoplasmic Ca^{2+}-binding proteins* (SCPs) that can be found in both vertebrate and invertebrate muscle.[129] The function of SCPs is not yet known, but their sequence homology with Ca^{2+}-binding proteins of known tertiary structure suggests that they originally contained four helix-loop-helix Ca^{2+}-binding domains. Ca^{2+} binding has been preserved in the first and third domains of all known SCPs, but only one, if any, of domains II and IV is functional. The three-dimensional crystal structure of an SCP from the sandworm *Nereis diversicolor* analyzed at 3.0 Å resolution[130] can be seen in Figure 3.26. (*See color plate section, page C-11.*) The C-terminal half (domains III and IV) of the molecule contains two Ca^{2+}-binding EF-hands (green and red in Figure 3.26) similar to calbindin D$_{9k}$ and the globular domains of troponin C and calmodulin. The N-terminal half is, on the contrary, markedly different from the normal helix-loop-helix geometry. Domain I binds Ca^{2+} with a novel helix-loop-helix conformation, whereas domain II lacks calcium-binding capacity. The two halves are packed closely together, and are not, as in troponin C or calmodulin, connected by a solvent-exposed α-helix.

E. Membrane Cytoskeleton and Phospholipid Binding Proteins

It has long been suspected that Ca^{2+} ions are somehow involved in exocytosis. Recently several groups[131] have isolated intracellular proteins that associate with membranes, and/or membrane cytoskeleton proteins, in a Ca^{2+}-dependent manner, and that seem able to mediate vesicle fusion or aggregation at Ca^{2+} concentrations above 200 μM. These proteins—endonexin, calelectrin, p36, and pII—have stretches of consensus amino-acid sequences that are also found in a phospholipase A$_2$ inhibitor protein, lipocortin.[132] It appears that further studies of this new class of proteins, known as annexins, will lead to new insights into cell-signaling pathways. Multiple functions have been proposed for the annexins, but no cellular role has yet been defined.[133]. The first crystal structure of an annexin, human annexin V—which *in vitro* will form voltage-gated Ca^{2+} channels—has been determined recently.[172] In annexin, the three Ca^{2+}-binding sites are located on the side of the molecule that is involved in membrane binding.

F. Ca^{2+}-Dependent Proteases

An interesting Ca^{2+}-activated intracellular protease, sometimes called *calpain*, was discovered during the last decade.[134] The ending -pain refers to its relation with other proteolytic enzymes like papain. It may seem dangerous to have a proteolytic enzyme loose inside a cell, and it must have rather specialized functions and be under strict control. The complete primary structure of the calcium protease ($M_r \approx 80,000$) in chicken tissues has recently been deduced from the nucleotide sequence of cloned DNA. [135,136] The findings are quite unexpected.

The protein contains four distinct domains. The first and third domains have no clear sequence homologies with known protein sequences, but the second domain has a high homology with the proteolytic enzyme papain, and the fourth domain is highly homologous to calmodulin. This fourth domain thus has four EF-hand-type Ca^{2+}-binding sites, although the third site has a somewhat unusual loop sequence. Here we apparently are faced with an unusual invention by Nature: by fusing the gene for a protease with that of the canonical Ca^{2+} receptor, she has created a molecule in which a regulatory protein is covalently linked to its target enzyme!

G. Protein Kinase C

Before we leave our brief survey of intracellular Ca^{2+}-binding proteins, we must write a few lines about an important Ca^{2+}-regulated kinase (a phosphorylating enzyme), i.e., *protein kinase C* (PKC). The activity of this enzyme, or rather family of enzymes,[137] appears to be regulated by three factors: phospholipids, in particular phosphatidylserine; diacyl-glycerols, one of the products of inositol lipid breakdown; and Ca^{2+} ions. The high-activity form of PKC, which appears responsible for much of the phosphorylation activity of many cells, is presumably membrane-bound, whereas the low-activity form may be partly cytosolic (Figure 3.27). The schematic structure of rabbit PKC ($M_r \approx 77$ kDa)

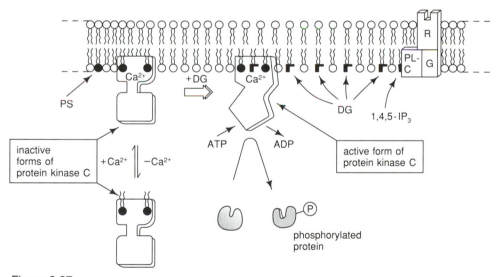

Figure 3.27
Outline of the cellular events that result in the activation of protein kinase C (PKC). The enzyme apparently exists in at least two states. Recent sequence work indicates that it has a Ca^{2+}-binding site of the EF-hand type. When no Ca^{2+} ion is bound, and when the "concentration" of diacylglycerol (DG) in the inner layer of the plasma membrane is low, the kinase exists in a low-activity form, possibly dissociated from the membrane. When a hormone binds to a plasma-membrane receptor (R), cleavage of phosphoinositol into 1,4,5-IP₃ and DG is induced. The latter lipid may bind to and activate the calcium-loaded form of PKC. The active form of protein kinase C will now phosphorylate other cytoplasmic proteins, and in this way modify their biochemical properties. R = receptor; PL-C = phospholipase C; G = a GTP-binding protein that is assumed to act as an intermediary between the receptor and the membrane bound PL-C.

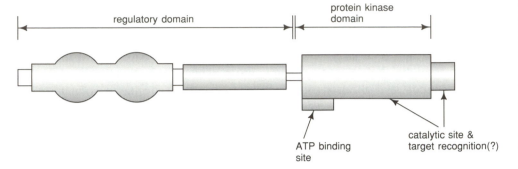

Figure 3.28
Schematic representation of the structure of rabbit protein kinase C.[138] Three highly homologous protein kinases C were actually identified with $M_r \approx 76{,}800$. The kinase region shows clear similarity with other kinases. The regulatory domain should contain binding sites for Ca^{2+}, phosphatidyl serine (PS), and diacylglycerol (DG).

according to Ohno *et al.*[138] is shown in Figure 3.28. The Ca^{2+} site(s) are presumably in the regulatory domain. No typical "EF-hand" pattern has been found in the amino-acid sequence. A protein kinase that requires Ca^{2+} but not phospholipids nor calmodulin for activity has been purified from soybean. From the amino-acid sequence the protein appears to have a calmodulin-like Ca^{2+}-binding domain, very much as in calpain.[139]

H. Summary

Many different biological processes in eukaryotic cells are regulated by intracellular Ca^{2+} concentration levels. Examples of such processes are muscle contraction, transport processes, cell division and growth, enzyme activities, and metabolic processes. A link in this regulatory chain is a number of intracellular Ca^{2+} receptors with Ca^{2+} affinities such that their binding sites are largely unoccupied at resting Ca^{2+} concentration levels, but are occupied at Ca^{2+} levels reached as a result of some external stimulus. This class of Ca^{2+} receptors is often called the "calmodulin superfamily" and includes the well-known members troponin C (regulating muscle contraction in striated muscle) and calmodulin (playing an important role in the regulation of many cellular processes). Amino-acid sequence determinations as well as x-ray and 2D ^1H NMR studies have revealed a strong homology between the regulatory Ca^{2+}-binding proteins. The Ca^{2+}-binding sites are located in a loop flanked by two helices, and the Ca^{2+} ions are ligated with approximately octahedral or pentagonal bipyramidal symmetry. The ligands are six or seven oxygen atoms that are furnished by side-chain carboxylate or hydroxyl groups, backbone carbonyls, and water molecules. Pairs of these Ca^{2+} sites, rather than individual sites, appear to be the functional unit, and a common consequence of their arrangement is cooperative Ca^{2+} binding. Ca^{2+} binding to the intracellular receptor proteins is accompanied by structural changes that expose hydrophobic patches on their surfaces, thereby enabling them to bind to their target proteins.

VI. EXTRACELLULAR Ca²⁺-BINDING PROTEINS

The Ca^{2+} concentration in extracellular fluids is usually orders of magnitude higher than intracellular concentrations. In mammalian body fluids, the "free" Ca^{2+} concentration is estimated to be 1.25 mM (total Ca^{2+} is ~ 2.45 mM) with only minor variations.[140] We would thus expect that Ca^{2+} ions in extracellular fluids play a very different role from that inside cells. To ensure Ca^{2+} binding the macromolecular binding sites need have only a modest Ca^{2+} affinity ($K_B^{Ca^{2+}} \approx 10^3$ to 10^4 M^{-1}), and since extracellular Ca^{2+} does not seem to have a signaling function, the rates of Ca^{2+} association or dissociation in protein-binding sites need not be very high.

One particularly important aspect of Ca^{2+} in mammals is its role in the blood coagulation system. Here we will meet a new type of amino acid, γ-carboxyglutamic acid ("Gla")—see Figure 3.29, that seems to have been de-

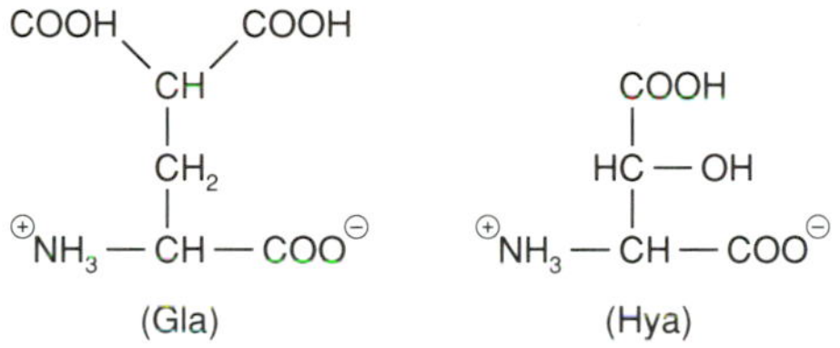

Figure 3.29
Chemical structures of two novel amino acids believed to bind calcium in, e.g., blood-clotting proteins.

signed by Nature as a Ca^{2+} ligand with rather special functions. Gla-containing proteins are also encountered in some mineralized tissues. The formation of bone, teeth, and other calcified hard structures is an intriguingly complicated phenomenon that will be dealt with in Section VII. We start, however, with a brief discussion of the role of Ca^{2+} in some extracellular enzymes.

A. Ca²⁺-Binding in Some Extracellular Enzymes

Several extracellular enzymes have one or more Ca^{2+} ions as integral parts of their structure. In a very few of them the Ca^{2+} ion is bound at or near the active cleft, and appears necessary for maintaining the catalytic activity (phospholipase A$_2$, α-amylase, nucleases), whereas other enzymes show catalytic activity even in the absence of Ca^{2+} (trypsin and other serine proteases). In the latter proteins, the Ca^{2+} ion is usually ascribed a "structural" role, although its function may be rather more related to "dynamics" and so be more subtle and complex.

Trypsin has one Ca^{2+}-binding site with four ligands (two side-chain and two backbone oxygens) donated by the protein (Glu-70, Asn-72, Val-75, and Glu-80) and two ligating water molecules, making the site roughly octahedral.[141] The binding constant of Ca^{2+} to trypsin and its inactive precursor "proenzyme," *trypsinogen*, has been measured (see Table 3.2). The binding constant is slightly smaller for the precursor, as is also true for *chymotrypsin* and *chymotrypsinogen*.[142] The Ca^{2+} affinities of the serine proteases and their proenzymes are such that their Ca^{2+} sites will be largely occupied in extracellular fluids, but would be unoccupied inside a cell. It has been suggested that this phenomenon constitutes a safeguard against unwanted conversion of the proenzymes into the active enzymes as long as they still are inside the cells where they are synthesized.

The rates of Ca^{2+} dissociation of the above enzymes and proenzymes have been measured by ^{43}Ca NMR and stopped-flow techniques,[142] and are collected in Table 3.4. We note that the values of k_{on} and k_{off} are generally much smaller than in the intracellular regulatory EF-hand proteins discussed in Section VI. Whereas the latter have dynamic and equilibrium properties similar to those of flexible low-molecular- weight chelators such as EDTA and EGTA, the serine proteases are more similar to the more-rigid cryptates, such as the macrobicyclic amino cryptate [2.2.2] (see Tables 3.2 and 3.4).

As mentioned above, there are a few enzymes in which a Ca^{2+} ion is present in the active cleft and essential for activity. Pancreatic *phospholipase* A_2 ($M_r \approx 14$ kDa) is an enzyme of this type. The x-ray structure is known to high resolution, and a single Ca^{2+} ion is found to be surrounded by six ligands, four presented by the protein (Tyr-28, Glu-30, Glu-32, and Asp-49) and two water molecules.[143] A mechanism for the action of phospholipase A_2 has been proposed[144] and is shown in Figure 3.30. This mechanism is based on three high-resolution x-ray crystal structures of phospholipase A_2 with and without transition-state analogues bound. The binding constant for Ca^{2+} together with the rate of dissociation found from variable-temperature ^{43}Ca NMR studies[145] can be used to calculate $k_{on} \approx 4 \times 10^6$ M^{-1} s^{-1}, again lower than in EF-hand proteins. Recent ^1H NMR studies indicate that the global structure of the lipase is very much the same in the Ca^{2+}-free and the Ca^{2+}-bound forms. Structural changes upon Ca^{2+} binding appear primarily located in the region of the binding site.[112,146]

The mammary glands produce, among other substances, a Ca^{2+}-binding enzyme activator, *α-lactalbumin*, that has about 40 percent sequence identity with lysozyme. This protein , which is involved in the conversion of glucose into lactose, is secreted in large quantities, and in human milk constitutes some 15 percent of total protein. The Ca^{2+}-binding constant of bovine or human α-lactalbumin is on the order of 10^7 M^{-1} under physiological conditions. In addition to Ca^{2+}, the enzyme also binds Zn^{2+}. It appears that Ca^{2+}-ion binding affects enzymatic activity, and somehow controls the secretion process, but the biological role of metal-ion binding to α-lactalbumin needs to be studied further. The x-ray structure of α-lactalbumin from baboon milk ($M_r \approx 15$ kDa) has been

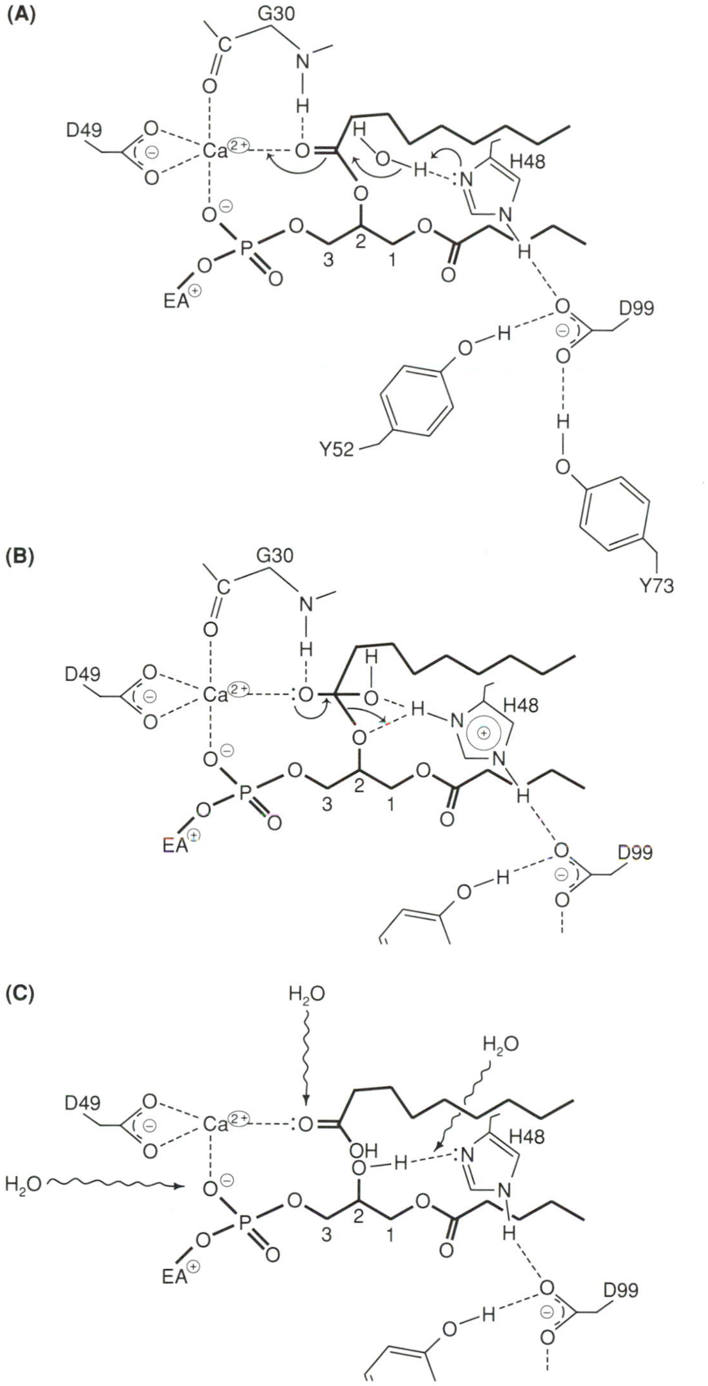

Figure 3.30
Catalytic mechanism[144] of phospholipase A₂. (A) Catalytic attack on substrate bound in a productive mode. (B) The tetrahedral intermediate as it collapses into products. (C) Products
formed by ''productive collapse'' in which three water molecules move into the active site to
replace the products. Two of these water molecules will coordinate the calcium ion. Figure
kindly provided by P. B. Sigler.

determined[147] to a high resolution ($\sim$1.7 Å). The Ca^{2+}-binding site has an interesting structure. The ion is surrounded by seven oxygen ligands, three from the carboxylate groups of aspartyl residues (82, 87, and 88), two carbonyl oxygens (79 and 84), and two water molecules. The spatial arrangement is that of a slightly distorted pentagonal bipyramid with the carbonyl oxygens at the apices, and the five ligands donated by the proteins are part of a tight "elbow"-like turn. The α-lactalbumin site has a superficial structural similarity to an "EF-hand," although the enzyme presumably has no evolutionary relationship with the intracellular Ca^{2+}-binding regulatory proteins.

Blood clotting proceeds in a complicated cascade of linked events involving many enzymes and proenzymes. About a decade ago it was shown that several of these proteins contained a previously unknown amino acid, γ-carboxyglutamic acid (Gla), and more recently yet another new amino acid, β-hydroxyaspartic acid (Hya), has been discovered (see Figure 3.29). The former is formed postribosomally by a vitamin-K-dependent process in the liver.[148] Presently the most-studied Gla protein in the blood-clotting system is *prothrombin* ($M_r \approx 66$ kDa). Ten Gla residues are clustered pairwise in the N-terminal region, essentially lining one edge of the molecule, forming a highly negatively charged region.[149] A small (48 residues) proteolytic fragment (F1) that contains all ten Gla amino acids can be prepared. Prothrombin can bind about 10 Ca^{2+} ions, but F1 binds only 7. Binding studies to F1 show that the Ca^{2+} ions bind at three high-affinity cooperative sites and four noninteracting sites,[150] and that this binding takes places in conjunction with a spectroscopically detectable conformational change (see Table 3.1).

In the presence of Ca^{2+} ions, prothrombin and other vitamin-K-dependent proteins in the blood-coagulation system will bind to cell membranes containing acidic phospholipids, in particular, the platelet membrane, which is rich in phosphatidylserine. A proposed model for the prothrombin-membrane interaction is shown in Figure 3.31.

It has long been known that calcium ions are involved in cell-to-cell and cell-to-extracellular matrix interactions, but the molecular details largely remain to be unraveled. In the late 1980s a large, adhesive, calcium-binding matrix glycoprotein ($M_r \sim 420$ kDa) named thrombospondin was characterized. This multifunctional adhesion molecule is composed of three polypeptide chains, each with 38 amino-acid-long repeats that are homologous with the calcium-binding helix-loop-helix sites of the calmodulin superfamily.[152] Each thrombospondin molecule is reported to bind 12 calcium ions with an affinity of about $10^4 \ M^{-1}$, and the removal of calcium is accompanied by a conformational change.[153,154]

B. Summary

In higher organisms, the Ca^{2+} concentration in extracellular fluids generally is considerably higher than the intracellular concentrations. In mammalian body fluids, the Ca^{2+} concentration is typically on the order of a few mM. The

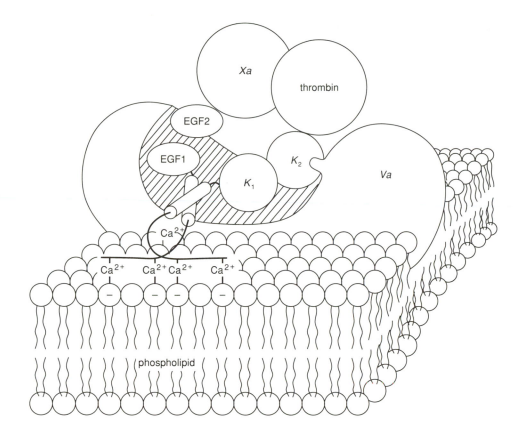

Figure 3.31
Proposed features of the interaction between the prothrombinase complex and a membrane lipid bilayer.[151] K_1 and K_2 are the kringle domains of prothrombin, and EGF1 and EGF2 are the two epidermal growth factor units of factor X_a. Prothrombin and factor X_a form a heterodimer complex harbored within the membrane protein factor V_a. The proposed interaction between prothrombin and factor X_a involves hydrophobic interactions between two helices and bridging by a Ca^{2+} ion between two Gla residues. The N-terminal Gla residues attach the heterodimer complex to the phospholipid surface. Figure kindly provided by C. C. F. Blake.

extracellular concentration levels are highly regulated and undergo only minor variations. A consequence of these high levels of Ca^{2+} in extracellular fluids is that the binding constant need be only 10^3 to 10^4 M^{-1} in order for a protein site to be highly occupied by Ca^{2+}. Several extracellular enzymes and enzyme activators have one or more Ca^{2+} ions as integral parts of their structures. Some Ca^{2+} ions are bound at, or near, the active cleft and may take part in the enzymatic reactions (e.g., phospholipase A_2, α-amylase). In other molecules, for example, serine proteases like trypsin and chymotrypsin, the Ca^{2+} ion is not essential for enzymatic activity, and may play more of a structural role. Ca^{2+} ions are involved in the cascade of enzymatic events that results in blood clotting in mammals. Several of the proteins in this system contain two new amino acids, γ-carboxyglutamic acid (Gla) and β-hydroxyaspartic acid (Hya), which are strongly suspected to be involved as ligands in Ca^{2+} binding. In the presence of Ca^{2+} ions, prothrombin and other Gla-containing proteins will bind to cell membranes containing acidic phospholipids, in particular, the platelet membrane. It appears likely that Ca^{2+} ions form a link between the protein and the membrane surface.

VII. CALCIUM IN MINERALIZED TISSUES

The formation of calcified tissue—shells, bone, and teeth—is a very complex process that is under strict regulatory control. Despite the obvious importance of this field, relatively little research has been directed toward elucidation of the underlying mechanisms, perhaps because the field spans a broad range of subjects, from inorganic solution and solid-state chemistry to cellular physiology.[155]

Historically, it was long held that formation of biological minerals such as bone was simply the nucleation and growth of calcium hydroxyapatite within an extracellular matrix of collagen. Many proteins other than collagen have now been discovered in appreciable quantities in bone and other biological minerals. It is also apparent that the pattern of calcification differs in shells, bone, teeth, and other mineralized tissues; so it is not likely that there is only one underlying mechanism. Considering the immensity of the subject, we will here only make a few brief comments, mainly about bone and teeth.

As was briefly mentioned earlier in this chapter, the inorganic matter of bone and teeth in many ways resembles apatite minerals $(Ca_5(OH)(PO_4)_3)$. Table 3.5 summarizes inorganic solid components of other biominerals. A detailed

Table 3.5
A summary of the main inorganic solid component[a] of the most-common biominerals in living systems.[148]

Anion	Formula	Crystal form	Occurrence	Main function
Carbonate	$CaCO_3$	Calcite Aragonite Valerite	Sea corals, molluscs, and many animals and plants	Exoskeleton; Ca-store; eye lens
Oxalate	$Ca(COO)_2 \cdot H_2O$ $Ca(COO)_2 \cdot 2H_2O$	Whewellite Weddellite	Insect eggs; vertebrate stones	Deterrent; cytoskeleton; Ca store
Phosphate	$(Ca)_{10}(PO_4)_6(OH)_2$ (unit cell comp.)	Hydroxyapatite	Bones; teeth; shells; intracellular in some bacteria	Skeletal; Ca storage; pressure-transducer (piezo-electric)
Sulfate	$CaSO_4 \cdot H_2O$	Gypsum	Jelly fish; plants	Gravity device; S and Ca store

[a] Most real biominerals are actually nonstoichiometric, and contain a number of additional cations (e.g., Mg^{2+}) or anions (e.g., F^-). In addition, the inorganic phase may be interpenetrated by a biopolymer.

analysis[156] shows that, apart from Ca^{2+} and PO_4^{3-}, many other cations and anions occur in bone, e.g., Mg^{2+}, Na^+, K^+, Sr^{2+}, CO_3^{2-}, F^-, Cl^-, and citrate. X-ray diffraction patterns and electron-microscope pictures of bone show that the inorganic phase is made up of many very small and imperfect crystals. By contrast, dental enamel is made up of much larger and uniform thin crystals. Although the solubility product of calcium hydroxyapatite (see Section II) is such that the equilibrium Ca^{2+} concentration should be in the low micromolar range, bone mineral appears to be in equilibrium with much higher Ca^{2+} concentrations (0.8–1.0 mM).[157] This discussion brings us to the question of how the inorganic crystallites are formed. Obviously both Ca^{2+} and PO_4^{3-} ions must be concentrated in cells or organelles bordering on the regions where mineralization is to take place. Fresh layers of bone matrix are formed by a continuously replenished layer of cells called *osteoblasts* (Figure 3.32A), which, in addition to apatite crystallites, also secrete collagen, and large specific proteins called *osteonectin, osteocalcin* (a Gla protein), proteoglycans, and phosphoproteins. In tissues undergoing rapid mineral deposition, the crystallites appear to be formed in vesicles that may have peeled off from the adjacent cell layers. These vesicles seem able to concentrate calcium and phosphate in a manner not well understood.

Bone, unlike diamond, is not forever. It can be remodeled and dissolved. A serious medical problem, which affects some women after menopause, is **osteoporosis**, i.e., the decalcification of bone. This loss of bone mass, which occurs with increasing age, makes bones more susceptible to breaking under stress. About 50 percent of American women, and 25 percent of American men, over 45 years of age are affected by osteoporosis.[158] Whereas osteoblast cells handle bone formation, another type of cells, *osteoclasts*, can erode it (Figure 3.32B). These macrophage-like cells can form deep tunnels in a bone matrix, and the cavities left behind are rapidly invaded by other cells forming blood vessels and new layers of osteoblasts. The *modus operandi* of osteoclast cells is not well understood at present. They may secrete calcium-chelating organic anions, such as citrate, to assist in the solubilization of the bones, as well as extracellular proteases that degrade the organic part of the matrix.

Summary

Calcium is, along with iron, silicon, and the alkaline earth metals, an important constituent of mineralized biological tissues. Some Ca^{2+}-based biominerals, like bone or mother-of-pearl, can be regarded as complex composites with microscopic crystallites embedded in a protein matrix. The formation of calcified biominerals is a highly regulated process, and human bone, for instance, is constantly being dissolved and rebuilt. When the rates of these two counteracting processes are not in balance, the result may be decalcification, or *osteoporosis*, which seriously reduces the strength of the bone.

158

(A)

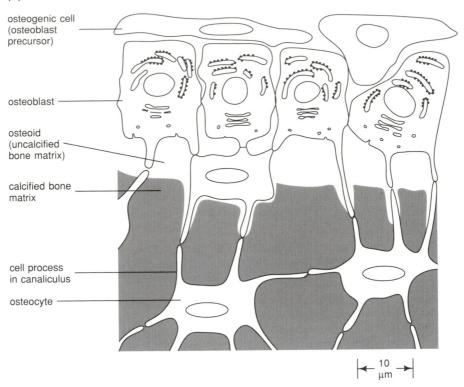

osteogenic cell
(osteoblast
precursor)

osteoblast

osteoid
(uncalcified
bone matrix)

calcified bone
matrix

cell process
in canaliculus

osteocyte

10
μm

(B)

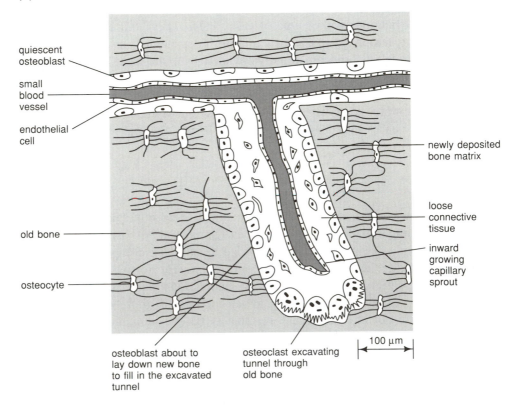

quiescent
osteoblast

small
blood
vessel

endothelial
cell

old bone

osteocyte

newly deposited
bone matrix

loose
connective
tissue

inward
growing
capillary
sprout

osteoblast about to
lay down new bone
to fill in the excavated
tunnel

osteoclast excavating
tunnel through
old bone

100 μm

Figure 3.32 *(facing page)*
Schematic diagram depicting the roles of the most important cell types in bone formation. (A) The *osteoblast* cells line the bone surface and secrete the inorganic and organic components (collagen, etc.) that will form new bone. Some osteoblast cells gradually become embedded in their own secretion. A particular secreted bone-specific protein, osteonectin, forms strong links between calcium hydroxyapatite and collagen. The bone-forming cells that become trapped in the bone matrix are now called *osteocytes*. (B) The osteoclast cells function to remodel compact bone. A group of cells acting together excavate a tunnel through old bone at a rate of about 50 μm per day. Behind the advancing osteoclasts follow a contingent of osteoblasts that line the wall of the tunnel and start to form new bone. Concurrently a capillary vessel is formed along the center of the tunnel and provides the cells with nutrients. Eventually the tunnel will become filled with concentric layers of new bone with only a narrow canal remaining. It is apparent that bone is far from a dull inorganic deposit, and very much a site of continuous activity. It is estimated that 5 to 10 percent of the bone in an adult mammal is replaced per year. Adapted from Reference 159.

VIII. Ca^{2+}-BINDING PROTEINS IN MICROORGANISMS: THE SEARCH FOR A PROKARYOTIC CALMODULIN

Since Ca^{2+} ions evidently play an important role in regulating a variety of cellular responses in animals and higher organisms, one may ask whether this use of Ca^{2+} is a recent discovery of Nature, or if it was invented early in evolution. It now appears well-established that the key intracellular "Ca^{2+}-receptor" protein calmodulin (CaM; see Section V.A) is present in all eukaryotic cells. Even in a unicellular eukaryote like common yeast (*Saccharomyces cerevisiae*), Ca^{2+} has an important regulatory role, and recently yeast CaM, as well as the single-gene encoding for it, was isolated.[160]

The amino-acid sequence of the yeast CaM (147 a.a.; $M_r = 16.1$ kDa) is 60 percent identical with the sequences of all other CaMs known. In fact, if generally accepted conservative amino-acid replacements are allowed, the homology increases to 80 percent or more, the most highly conserved portions being the four putative Ca^{2+}-binding sites. Sites I and III match the EF-hand test sequence (see Figure 3.24) very well; in site II, a His occurs after the "z"-ligand instead of the archetypal Gly; and in site IV there is no amino acid between the residues that usually make up ligands "x" and "y." The effect of these alterations on the Ca^{2+} affinity of yeast CaM is not yet known.

That CaM is essential for the growth of yeast cells was shown by deletion or disruption of the gene. This constitutes, in fact, the first demonstration in any organism that CaM is an essential protein. (Deletions of genes in mammals are ethically questionable research procedures!)

In the biochemically less sophiscated (than eukaryotes) **prokaryotic** cells, a regulatory role of Ca^{2+} is not well-established. What is known is that calcium is massively accumulated during sporulation in many bacteria, for example, in strains of *Bacillus*, *Streptomyces*, and *Myxococcus*. In *Myxococcus xanthus* a development-specific protein called protein S assembles at the surface of myxospores in the presence of Ca^{2+}. The DNA sequence of the gene that encodes this protein has been deciphered.[161] The primary sequence of protein S (175 a.a., M_r = 19.2 kDa) turns out to closely resemble mammalian CaM. It has four internally homologous regions with putative Ca^{2+} sites. At least two of these are partly similar to the typical EF-hand, but uncharacteristically there are many more prolines in the *M. xanthus* protein than in bovine CaM (12 versus 2); so it is questionable if the bacterial protein really has the repeated helix-loop-helix structure found in mammalian CaM.[162]

One candidate for a prokaryotic CaM was reported by Leadlay *et al.*[163] in *Streptomyces erythreaus*, the bacterium that produces the well-known antibiotic "erythromycin." The amino-acid sequence of a low-molecular-weight Ca^{2+}-binding protein, as determined from the gene encoding it, revealed a high homology with mammalian CaM. The protein is made up of 177 amino acids (M_r = 20.1 kDa), and has four regions that are predicted to have the helix-loop-helix secondary structure typical of EF-hand proteins. The aligned sequences of the 12 residues in each of the four potential calcium-binding loops in the *S. erythreaus* protein are compared with those of human calmodulin in Table 3.6. The pattern of residues in the *S. erythraeus* protein is typical of an EF-hand at least in sites I, III, and IV. Site II is unusual in having Gly at both positions 1 and 3. ^{113}Cd NMR studies show that the bacterial protein binds three metal ions strongly ($K \gtrsim 10^5$ M^{-1}) with chemical shifts close to those expected for EF-hands, and ^1H NMR studies show that it undergoes a Ca^{2+}- dependent conformational change.[164]

Although the *S. erythraeus* protein has a homology with eukaryotic CaM, it has been pointed out that the protein has an even higher homology with a group of eukaryotic sarcoplasmic Ca^{2+}-binding proteins[165] (see Section V.D). The search for a prokaryotic CaM analogue continues, and the prospect of success has been improved after recent reports of a 21-amino-acid-long polypeptide from an *E. coli* heat-shock protein[166] that shows the typical structural features of CaM-binding domains in other eukaryotic proteins.[167]

Table 3.6
Aligned EF-hand sequences for the prokaryotic and human calmodulins.

Ligands		1	3	5	7	9		12
S. erythraeus protein	I	D F	D G	N G	A L	E R	A	D
	II	G V	G S	D G	S L	T E	E	Q
	III	D K	N A	D G	Q I	N A	D	E
	IV	D T	N G	N G	E L	S L	D	E
Human calmodulin	I	D K	D G	D G	T I	T T	K	E
	II	D A	D G	N G	T I	D F	P	E
	III	D K	D G	N G	Y I	S A	A	E
	IV	D I	D G	D G	Q V	N Y	E	E

Summary

The role of Ca^{2+} ions in the regulation of biological activities of procaryotic organisms is still largely unsettled. Over the last decade, however, evidence has gradually accumulated that calcium ions are involved in diverse bacterial activities, such as chemotaxis and substrate transport, sporulation, initiation of DNA replication, phospholipid synthesis, and protein phosphorylation.[168] An important landmark is the recent demonstration that the intracellular Ca^{2+} concentration in *E. coli* is tightly regulated to about 100 nM, a level similar to that typical of resting eukaryotic cells.[169] Furthermore, increasing numbers of calcium-binding proteins, some of which also have putative EF-hand Ca^{2+} sites characteristic of the calmodulin superfamily of intracellular regulatory proteins, have been isolated in bacteria.[168]

IX. APPENDIXES

A. Definition of Biochemical Terms

Antiport	A transport protein that carries two ions or molecules in opposite directions across a membrane.
Basal lateral membrane	The membrane in intestinal epithelial cells that is located on the base of the cells, opposite the microvilli that face the intestinal lumen.
Cytosol	The unstructured portion of the interior of a cell—the cell nucleus excluded—in which the organelles are bathed.
Electrogenic	A biological process driven by electric field gradients.
Endocytosis	The process by which eukaryotic cells take up solutes and/or particles by enclosure in a portion of the plasma membrane to (temporarily) form cytoplasmic vesicles.
Endoplasmic reticulum (ER)	Sheets of folded membranes, within the cytoplasm of eukaryotic cells, that are the sites for protein synthesis and transport.
Epithelial cells	Cells that form the surface layer of most, if not all, body cavities (blood vessels, intestine, urinary bladder, mouth, etc.).
Erythrocytes	Red-blood corpuscles.
Eukaryotic cells	Cells with a well-defined nucleus.

Exocytosis	The process by which eukaryotic cells release packets of molecules (e.g., neurotransmitters) to the environment by fusing vesicles formed in the cytoplasm with the plasma membrane.
Gluconeogenesis	Metabolic synthesis of glucose.
Glycolysis	Metabolic degradation of glucose.
Hydropathy	A measure of the relative hydrophobic or hydrophilic character of an amino acid or amino-acid side chain.
Lamina propria mucosae	The layer of connective tissue underlying the epithelium of a mucous membrane.
Mitochondrion	A double-membrane organelle in eukaryotic cells that is the center for aerobic oxidation processes leading to the formation of energy-rich ATP.
Organelle	A structurally distinct region of the cell that contains specific enzymes or other proteins that perform particular biological functions.
Osteoporosis	Brittle-bone disease.
Phorbol esters	Polycyclic organic molecules that act as analogues to diacylglycerol and therefore are strong activators of protein kinase C.
Prokaryotic cells	Cells lacking a well-defined nucleus.
Sarcoplasmic reticulum	The ER of muscle cells.
Trophoblasts	The cells between the maternal and fetal circulation systems.
Tryptic digest	Fragmentation of proteins as a result of treatment with the proteolytic enzyme trypsin.
Uniporter	A transport protein that carries a particular ion or molecule in one direction across a membrane.

B. One-Letter Code for Amino-Acid Residues

A—alanine, C—cysteine, D—aspartate, E—glutamate, F—phenylalanine, G—glycine, H—histidine, I—isoleucine, K—lysine, L—leucine, M—methionine, N—asparagine, P—proline, Q—glutamine, R—arginine, S—serine, T—threonine, V—valine, W—tryptophan, Y—tyrosine.

C. The Activity of a Transport Protein

This is usually described in terms of the classical Michaelis-Menten scheme:

$$V \ (= \text{transport rate}) \ = \ V_{\max} \cdot \frac{[S]}{[S] + K_{\mathrm{m}}},$$

where [S] is the concentration of the solute to be transported and $K_{\mathrm{m}} = (k_{-1} + k_2)/k_1$ is the Michaelis constant (dimension ''concentration'') for the reaction

$$\mathrm{E + S \underset{k_{-1}}{\overset{k_1}{\rightleftarrows} ES \overset{k_2}{\rightarrow} P.}}$$

Approximated as the reciprocal ratio between on- and off-rate constants relevant to the solute-protein complex, $1/K_{\mathrm{m}} = k_1/k_{-1}$ may be taken as a lower limit of the affinity of the protein for the solute.

X. REFERENCES

1. See, for example, ''Going Crazy over Calcium'' in the Health and Fitness section of *Time*, February 23, 1987, p. 49, or C. Garland *et al.*, *The Calcium Diet*, Penguin Books, 1990.
2. S. Ringer, *J. Physiol.* **3** (1883), 195.
3. *Handbook of Chemistry and Physics*, 64th ed., CRC Press, 1984.
4. R. D. Shannon, *Acta Cryst.* A, **32** (1976), 751.
5. R. J. P. Williams, in *Calcium in Biological Systems*, Cambridge Univ. Press, 1976, p. 1.
6. B. A. Levine and R. J. P. Williams, in L. J. Anghileri and A. M. T. Anghileri, eds., *Role of Calcium in Biological Systems*, CRC Press, 1982, pp. 3–26.
7. A. K. Campbell, *Intracellular Calcium: Its Universal Role as Regulator*, Wiley, 1983.
8. A. P. Somlyo, M. Bond, and A. V. Somlyo, *Nature* **314** (1985), 622.
9. T. B. Johansson, R. Akselsson, and S. A. E. Johansson, *Nucl. Inst. Methods* **84** (1970), 141.
10. R. B. Martin, in H. Sigel, ed., *Metal Ions in Biological Systems*, Dekker, **17** (1984), 1.
11. H. Einspahr and C. E. Bugg, in *ibid.*, pp. 52–97.
12. L. G. Sillén and A. E. Martell, eds., *Stability Constants of Metal Ion Complexes*, Chemical Soc., London, 1964.
13. A. E. Martell and R. M. Smith, eds., *Critical Stability Constants*, Plenum Press, **1**, 1975.
14. J. D. Potter and J. Gergely, *J. Biol. Chem.* **250** (1975), 4628.
15. W. Märki, M. Oppliger, and R. Schwyzer, *Helvetica Chemica Acta* **60** (1977), 807.
16. C. M. Frey and J. Stuehr, in H. Sigel, ed., *Metal Ions in Biological Systems*, Dekker, **1** (1974), 51.
17 W. J. Moore, *Physical Chemistry*, Longman, 5th ed., 1972, Chapter 10.
18. M. V. Thomas, *Techniques in Calcium Research*, Academic Press, London, 1982.
19. T. J. Rink, *Pure Appl. Chem.* **55** (1988), 1977.
20. W. Simon *et al.*, *Ann. N.Y. Acad. Sci.* **307** (1987), 5269.
21. R. C. Thomas, *Ion-Sensitive Intracellular Microelectrodes*, Academic Press, London, 1987.
22. J. R. Blinks *et al.*, *Prog. Biophys. Mol. Biol.* **4** (1983), 1.
23. R. Y. Tsien, *Biochemistry* **19** (1980), 2396.
24. G. Grynkiewicz, M. Poenie, and R. Y. Tsien, *J. Biol. Chem.* **260** (1985), 3440.
25. A. Minta, J. P. Y. Kao, and R. Y. Tsien, *J. Biol. Chem* **264** (1989), 8171.
26. R. Y. Tsien and M. Poenie, *Trends Biochem. Sci.* **11** (1986), 450.
27. E. Chiancone *et al.*, *J. Biol. Chem.* **26** (1986), 16306.
28. G. A. Smith *et al.*, *Proc. Natl. Acad. Sci. USA* **80** (1983), 7178.
29. J. C. Metcalfe, T. R. Hesketh, and G. A. Smith, *Cell Calcium* **6** (1985), 183.

30. A. P. Somlyo, *Cell Calcium* **6** (1985), 197.
31. G. W. Grime *et al.*, *Trends Biochem. Sci.* **10** (1985), 6.
32. G. H. Morrison and G. Slodzian, *Anal. Chem.* **47** (1975), 932A.
33. S. Chandra and G. H. Morrison, *Science* **228** (1985), 1543.
34. E. Murphy *et al.*, *Circulation Research* **68** (1991), 1250.
35. R. H. Wasserman and C.S. Fullmer, in W. Y. Cheung, ed., *Calcium and Cell Function*, Academic Press, **2** (1982), 176.
36. H. Rasmussen, O. Fontaine, and T. Matsumoto, *Ann. N. Y. Acad. Sci.* **77** (1981), 518.
37. S. Linse *et al.*, *Biochemistry* **26** (1987), 6723.
38. D. T. W. Bryant and P. Andrews, *Biochem. J.* **219** (1984), 287.
39. R. H. Kretsinger, J. E. Mann, and J. G. Simmonds, in A. W. Norman *et al.*, eds., *Vitamin D, Chemical, Biochemical, and Clinical Endocrinology of Calcium Metabolism*, W. de Gruyter, pp. 232–248.
40. J. J. Feher, *Am. J. Physiol.* **244** (1983), 303.
41. I. Nemere, V. Leathers, and A. W. Norman, *J. Biol. Chem.* **261** (1986), 16106.
42. G. E. Lester. *Feder. Proc.* **45** (1986), 2524.
43. P. Marche, C. LeGuern, and P. Cassier, *Cell Tissue Res.* **197** (1979), 69.
44. M. Warembourg, C. Perret, and M. Thomasset, *Endocrinol.* **119** (1986), 176.
45. H. J. Schatzman, *Experientia* **22** (1966), 364.
46. V. Niggli, J. T. Penniston, and E. Carafoli, *J. Biol. Chem.* **254** (1979), 9955.
47. E. Carafoli, M. Zurini, and G. Benaim, in *Calcium and the Cell*, CIBA Foundation Symposium no. 122, Wiley, 1986, pp. 58–65.
48. D. MacLennan, *J. Biol. Chem.* **245** (1970), 4508.
49. N. M. Green *et al.*, in Reference 47, pp. 93.
50. D. M. Clarke *et al.*, *Nature* **339** (1989), 476.
51. G. Inesi, *Annu. Rev. Physiol.* **47** (1985), 573.
52. R. J. P. Williams, *Eur. J. Biochem.* **150** (1985), 231.
53. C. Tanford, *Proc. Natl. Acad. Sci. USA* **79** (1982), 6527.
54. D. H. MacLennan, K. P. Campbell, and R. A. Reithmeier, in Reference 35, pp. 151–173.
55. A. Maurer *et al.*, *Proc. Natl. Acad. Sci. USA* **82** (1985), 4036.
56. M. P. Blaustein and M. T. Nelson, in E. Carafoli, ed., *Membrane Transport of Calcium*, Academic Press, 1982, pp. 217–236.
57. P. F. Baker, in Reference 47, pp. 73–84.
58. E. Carafoli, G. Inesi, and B. P. Rosen, in Reference 10, pp. 140–143.
59. G. Fiskum, in Reference 10, pp. 187–214.
60. A. L. Lehninger *et al.*, in F. Bronner and M. Peterlik, eds., *Calcium and Phosphate Transport Across Membranes*, Academic Press, 1981, pp. 73–78.
61. E. Carafoli, in Reference 56, pp. 109–139.
62. M. J. Berridge, in Reference 47, pp. 39–49.
63. R. F. Irvine, *Brit. Med. Bull.* **42** (1986), 369.
64. E. W. McCleskey *et al.*, *J. Exp. Biol.* **124** (1986), 177.
65. D. J. Adams, T. Dwyer, and B. Hille, *J. Gen. Physiol.* **75** (1980), 493.
66. A. B. MacDermott *et al.*, *Nature* **321** (1986), 519.
67. R. J. Miller, *Science* **235** (1987), 46.
68. R. W. Tsien, *Annu. Rev. Physiol.* **45** (1983), 341.
69. E. W. Sutherland, *Science* **177** (1972), 401.
70. M. J. Berridge and R. F. Irvine, *Nature* **312** (1984), 315.
71. M. J. Berridge, *Biochem. J.* **212** (1983), 849.
72. M. J. Berridge, *J. Exp. Biol.* **124** (1986), 323.
73. C. C. Chadwick, A. Saito, and S. Fleischer, *Proc. Natl. Acad. Sci. USA* **87** (1990), 2132.
74. A. R. Hughes and J. W. Putney, *Envir. Health Perspec.* **84** (1990), 141.
75. A. S. Manalan and C. B. Klee, in *Adv. Cycl. Nucl. Prot. Phosph. Res.* (1984), 227.
76. J. D. Potter and J. D. Johnson, in Reference 35, pp. 145.
77. (a) R. H. Kretsinger, *Cold Spring Harbor Symp. Quant. Biol.* **52** (1987), 499; (b) C. W. Heizmann and K. Braun, *Trends in Neurosciences* **15** (1992), 259.
78. A. Marks *et al.*, *J. Neurochem.* **41** (1983), 107.
79. P. B. Moore and J. R. Dedman, in H. Hidaka and P. J. Hartshorne, eds., *Calmodulin Antagonists and Cellular Physiology*, Academic Press, 1985, pp. 483–494.
80. Y. Nishizuka, *Phil. Trans. Roy. Soc. Lond.* (B) **302** (1983), 101.

81. M. J. Geisow, *FEBS Lett.* **203** (1986) 99.
84. M. J. Crumpton and J. R. Dedman, *Nature* **345** (1990), 212.
85. Y. S. Babu, C. E. Bugg, and W. J. Cook, *J. Mol. Biol.* **204** (1988), 191.
86. R. H. Kretsinger and C. E. Nockolds, *J. Biol. Chem.* **248** (1973), 3313.
87. S. Forsén, H. J. Vogel, and T. Drakenberg, in Reference 35, **6** (1986) 113.
88. M. D. Tsai *et al.*, *Biochemistry* **26** (1987), 3635.
89. S. R. Martin *et al.*, *Eur. J. Biochem.* **151** (1985), 543.
90. A. Teleman, T. Drakenberg, and S. Forsén, *Biochim. Biophys. Acta* **873,** (1986), 204.
91. L. Sjölin, *Acta Cryst.* **B46** (1990), 209.
92. S. R. Martin and P. M. Bayley, *Biochem. J.* **238** (1986), 485.
93. D. B. Heidorn and J. Trewhella, *Biochemistry* **27** (1988), 909.
94. D. K. Blumenthal *et al.*, *Proc. Natl. Acad. Sci. USA* **82** (1985), 3187.
95. R. E. Klevit *et al.*, *Biochemistry* **24** (1985), 8152.
96. S. Linse, T. Drakenberg, and S. Forsén, *FEBS Lett.* **199** (1986), 28.
97. S. H. Seeholzer *et al.*, *Proc. Natl. Acad. Sci. USA* **83** (1986), 3634.
98. A. Persechini and R. H. Kretsinger, *J. Cardiovasc. Pharm.* **12** (suppl. 5, 1988), S1.
99. G. Barbato *et al.*, *Biochemistry,* in press.
100. M. Ikura *et al.*, *Biochemistry* **30** (1991), 5498.
101. C. H. Keller *et al.*, *Biochemistry* **21** (1982), 156.
102. O. Herzberg and M. N. G. James, *Nature* **313** (1985), 653.
103. K. A. Satyshur *et al.*, *J. Biol. Chem.* **263** (1988), 1628.
104. E. D. McCubbin and C. M. Kay, *Acc. Chem. Res.* **13** (1980), 185.
105. M. T. Hincke, W. D. McCubbin, and C. M. Kay, *Can. J. Biochem.* **56** (1978), 384.
106. P. C. Leavis *et al.*, *J. Biol. Chem.* **253** (1978), 5452.
107. O. Teleman *et al.*, *Eur. J. Biochem.* **134** (1983), 453.
108. H. J. Vogel and S. Forsén, in L. J. Berliner and J. Reuben, eds., *Biological Magnetic Resonance,* Plenum Press, **7** (1987), 247.
109. S. Forsén *et al.*, in B. de Bernard *et al.*, eds., *Calcium Binding Proteins 1983,* Elsevier, 1984, pp. 121–131.
110. T. Drakenberg *et al.*, *J. Biol. Chem.* **262** (1987), 672.
111. K. B. Seamon and R. H. Kretsinger, in T. G. Spiro, ed., *Calcium in Biology,* Wiley, **6** (1983), 1; review of the literature.
112. B. A. Levine and D. C. Dalgarno, *Biochim. Biophys. Acta* **726** (1983), 187.
113. O. Herzberg, J. Moult, and M. N. G. James, *J. Biol. Chem.* **261** (1986), 2638.
114. W. Chazin *et al.*, personal communication.
115. M. L. Greaser and J. Gergely, *J. Biol. Chem.* **246** (1971), 4226.
116. W. Wnuk, J. A. Cox, and E. A. Stein, in Reference 35, pp. 243–278.
117. C. W. Heizmann, in Reference 109, pp. 61–63.
118. R. H. Kretsinger, *CRC Crit. Rev. Biochem.* **8** (1980), 119; R. H. Kretsinger and C. D. Barry, *Biochim. Biophys. Acta* **405** (1975), 4051.
119. W. Hunziker, *Proc. Natl. Acad. Sci. USA* **83** (1986), 7578.
120. A. L. Swain, R. H. Kretsinger, and E. L. Amma, *J. Biol. Chem.* **264** (1988), 16620.
121. M. F. Summers, *Coord. Chem. Rev.* **86** (1988), 43; review of the literature.
122. J. M. Gillis *et al.*, *J. Muscl. Res. Cell. Mot.* **3** (1982), 377.
123. D. M. E. Szebenyi and K. Moffat, *J. Biol. Chem.* **261** (1986), 8761.
124. M. Akke, T. Drakenberg, and W. J. Chazin, *Biochemistry* **31** (1992), 1011.
125. S. Linse *et al.*, *Biochemistry* **30** (1991), 154.
126. S. Linse *et al.*, *Nature* **335** (1988), 651.
127. M. Akke and S. Forsén, *Proteins* **8** (1990), 23.
128. W. Chazin *et al.*, *Proc. Natl. Acad. Sci. USA* **86** (1989), 2195.
129. W. Wnuk, J. A. Cox, and E. A. Stein, in Reference 35, pp. 243–278.
130. W. J. Cook *et al.*, *J. Biol. Chem.* **266** (1991), 652.
131. J. J. Geisow and H. H. Walker, *Trends Biochem. Sci.* **11** (1986), 420.
132. R. H. Kretsinger and C. E. Creutz, *Nature* **320** (1986), 573.
133. M. R. Crompton, S. E. Moss, and M. J. Crumpton, *Cell* **55** (1988), 1.
134. J. Kay in A. Heidland and W. H. Horl, eds., *Protease Role in Health and Disease,* 1984, pp. 519–570.
135. S. Ohno *et al.*, *Nature* **312** (1984), 566.

136. D. E. Croall and G. N. DeMartino, *Physiol. Rev.* **71** (1991), 813.

137. D. Carpenter, T. Jackson, and M. J. Hanley, *Nature* **325** (1987), 107.

138. S. Ohno *et al.*, *Nature* **325** (1987), 161.

139. J. F. Harper *et al.*, *Science* **252** (1991), 951.

140. H. Rasmussen, *Am. J. Med.* **50** (1971), 567.

141. W. Bode and P. Schwager, *J. Mol. Biol.* **98** (1975), 693.

142. E. Chiancone *et al.*, *J. Mol. Biol.* **185** (1985), 201.

143. B. W. Dijkstra *et al.*, *J. Mol. Biol.* **147** (1983), 97.

144. D. L. Scott *et al.*, *Science* **250** (1990), 1541.

145. T. Drakenberg *et al.*, *Biochemistry* **23** (1984), 2387.

146. R. J. P. Williams, in Reference 47, pp. 144–159.

147. D. I. Stuart *et al.*, *Nature* **324** (1986), 84.

148. G. L. Nelsestuen, in Reference 10, pp. 354–380.

149. M. Soriano-Garcia *et al.*, *Biochemistry* **28** (1989), 6805.

150. D. W. Deerfield *et al.*, *J. Biol. Chem* **262** (1987), 4017.

151. K. Harlos *et al.*, *Nature* **330** (1987), 82.

152. J. Lawler and R. O. Hynes, *J. Cell. Biol.* **103** (1986), 1635.

153. V. M. Dixit *et al.*, *J. Biol. Chem.* **261** (1986), 1962.

154. J. M. K. Slane, D. F. Mosher, and C.-S. Lai, *FEBS Lett.* **229** (1988), 363.

155. R. E. Wuthier, in Reference 10, pp. 411–472.

156. F. G. E. Pautard and R. J. P. Williams, in *Chemistry in Britain*, pp. 188–193 (1982).

157. W. E. Brown and L. C. Chow, *Annu. Rev. Material Sci.* **6** (1976), 213.

158. O. R. Bowen, *Publ. Health Reports Suppl.* Sept.–Oct. 1989, pp. 11–13.

159. B. Alberts *et al.*, *Molecular Biology of the Cell,* Garland, 1983, pp. 933–938.

160. T. N. Davis *et al.*, *Cell* **47** (1986), 423.

161. S. Inouye, T. T. Franceschini, and M. Inouye, *Proc. Natl. Acad. Sci. USA* **80** (1983), 6829.

162. G. Wistow, L. Summers, and T. L. Blundell, *Nature* **315** (1985), 771.

163. P. F. Leadlay, G. Roberts, and J. E. Walker, *FEBS Lett.* **178** (1984), 157; D. G. Swan *et al.*, *Nature* **329** (1987), 84.

164. N. Bylsma *et al.*, *FEBS Lett.* **299** (1992), 44.

165. J. A. Cox and A. Bairoch, *Nature* **331** (1988), 491.

166. M. A. Stevenson and S. K. Calderwood, *Mol. Cell Biol.* **10** (1990), 1234.

167. K. T. O'Neil and W. F. DeGrado, *Trends Biochem. Sci.* **170** (1990), 59.

168. V. Norris *et al.*, *Mol. Microbiol.* **5** (1991), 775.

169. M. R. Knight *et al.*, *FEBS Lett.* **282** (1991), 405.

170. S. Supattapone *et al.*, *J. Biol. Chem.* **263** (1988), 1530.

171. C. D. Ferris *et al.*, *Nature* **342** (1989), 87.

172. R. Huber *et al.*, *J. Mol. Biol.* **223** (1992), 683.

173. (a) M. Ikura *et al.*, *Science* **256** (1992), 632; (b) W. E. Meador, A. R. Means, and F. A. Quiocho, *Science* **257** (1992), 1251.

174. M. R. Knight *et al.*, *FEBS Lett.* **282** (1991), 405.

175. R. W. Tsieu and R. Y. Tsien, *Annu. Rev. Cell Biol.* **6** (1990), 715.

176. T. E. Gunter and D. R. Pfeiffer, *Am. J. Physiol.* **258** (1990), c755.

177. The authors would like to express their warm gratitude to the many students, colleagues, and coworkers who, during the preparation of this chapter, have supplied helpful comments, preprints of unpublished work, background material for figures, etc. Their encouragement is much appreciated. Special thanks are due to Drs. R. J. P. Williams and G. B. Jameson, who critically read and commented on an early version of the chapter.

4

Biological and Synthetic Dioxygen Carriers

GEOFFREY B. JAMESON
Department of Chemistry
Georgetown University

JAMES A. IBERS
Department of Chemistry
Northwestern University

I. INTRODUCTION: BIOLOGICAL DIOXYGEN TRANSPORT SYSTEMS

Most organisms require molecular oxygen in order to survive. The dioxygen is used in a host of biochemical transformations, although most is consumed in the reaction

$$O_2 + 4H^+ + 4e^- \longrightarrow 2H_2O \tag{4.1}$$

that is the terminal (or primary) step of oxidative phosphorylation (Chapters 5 and 6). For some small animals and for plants, where the surface-to-volume ratio is large, an adequate supply of dioxygen can be obtained from simple diffusion across cell membranes. The dioxygen may be extracted from air or water; for plants that produce dioxygen in photosynthesis, it is also available endogenously. For other organisms, particularly those with non-passive life-styles, from scorpions to whales, diffusion does not supply sufficient dioxygen for respiration.

An elegant three-component system has evolved to transport dioxygen from regions of high abundance—water (at least if free of pollutant reductants) and air—to regions of relatively low abundance and high demand—the interior cells of the organism. This process is illustrated in Figure 4.1.[1-3] The central component is a dioxygen-carrier protein. In the three chemically distinct carriers that have evolved and are found today, the dioxygen-binding site in the protein, that is, the so-called "active site," is a complex either of copper or of iron.[4-6] For hemoglobins, the most widely distributed family of dioxygen carriers, the active

167

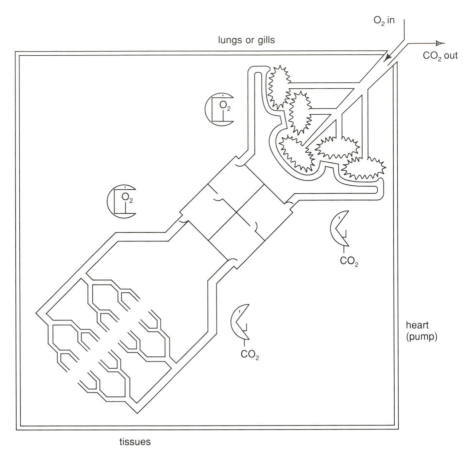

Figure 4.1
Oxygen sequestration and transport in the generalized organism *Squarus squorur*. The surface area of lungs or gills is typically 1–2 orders of magnitude greater than the external surface area of the organism.

site has long been known to consist of an iron porphyrin (heme) group embedded in the protein. Almost all hemoglobins share the basic structure illustrated in Figure 4.2.[7–12] Hemocyanin[13–15] and hemerythrin,[16–18] the other two biological dioxygen carriers, feature pairs of copper atoms and iron atoms, respectively, at the active sites.* Some basic properties of these metalloproteins are summarized in Table 4.1.[4–6]

The second component of the dioxygen-transport system facilitates the sequestration of dioxygen by the dioxygen-carrier protein. Specialized organs, such as lungs in air-breathing creatures or gills in fish, offer a very large surface area to the outside environment in order to facilitate diffusion. The third component is the delivery system. The oxygen carrier is dissolved or suspended in a fluid, called blood plasma or hemolymph, that is pumped throughout the animal by

---

* The use of the prefix *hem-* is confusing. In this context *hem* connotes blood. Thus, since hemocyanin and hemerythrin lack a hem*e* group [an iron(II) porphyrin], they are nonheme metalloproteins.

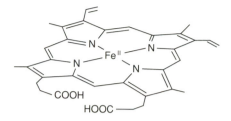

(A) Fe^{II} protoporphyrin IX (heme b)
[hemoglobins and erythrocruorins]

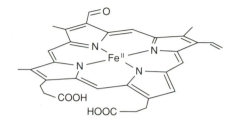

(B) Chloroheme
[chlorocruorin]

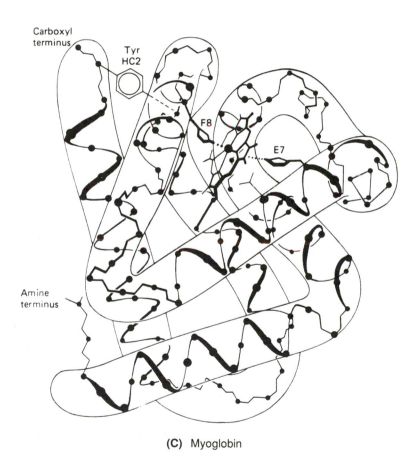

(C) Myoglobin

Figure 4.2
Heme groups used in hemoglobin: (A) Protoporphyrin IX (heme *b*), (hemoglobins and erythro-cruorins); (B) Chloroheme (chlorocruorin); (C) The encapsulation of the heme molecule in myoglobin.[11a] Reproduced with permission from M. F. Perutz, *Nature* **228** (1970), 726-737.

Table 4.1
General features of dioxygen-carrier proteins.

Metalloprotein	Active site of deoxy	Color change deoxy → oxy	MW (Dalton)	# Subunits	Average MW subunit (Dalton)
Hemoglobins					
Vertebrate					
Human A	heme Fe^{II}	purple → red	64,000	4	16,000
Invertebrate					
Erythrocruorin (*Lumbricus terrestris*, earthworm)	heme Fe^{II}	purple → red	up to 3.3×10^6	192	17,000
Chlorocruorin (*Eudistylia vancouveri*)	chloroheme Fe^{II}	purple → green	3.1×10^6	192	15,000
Hemocyanins					
Mollusc (*Helix pomatia*-α, edible snail)	$Cu^I \ldots Cu^I$	colorless → blue	$\sim 9 \times 10^6$	160	52,700
Arthropod (*Cancer magister*, crab)	$Cu^I \ldots Cu^I$	colorless → blue	$\sim 9 \times 10^5$	12	76,600
Hemerythrins					
(*Phascolopsis* syn. *golfingia gouldii*)	$Fe^{II} \ldots Fe^{II}$	colorless → burgundy	108,000	8	13,500

another specialized organ, the heart, through a network of tubes, the blood vessels. In many organisms an additional dioxygen-binding protein, which stores dioxygen, is located in tissues that are subject to sudden and high dioxygen demand, such as muscles. These dioxygen-storage proteins are prefixed *myo-* (from the Greek root *mys* for muscle). Thus for the dioxygen-transport protein hemerythrin there exists a chemically similar dioxygen-storage protein myohemerythrin. For the hemoglobin family the corresponding storage protein is called myoglobin. Interestingly, some organisms that use hemocyanin as the dioxygen-*transport* protein use myoglobin as the dioxygen-*storage* protein.

At the center of biological dioxygen transport are transition-metal complexes of iron or copper. To model such systems, chemists have prepared several synthetic oxygen carriers, especially of iron and cobalt porphyrins. In this chapter the structures and properties of biological and nonbiological oxygen carriers are described, with particular attention to the hemoglobin family. This family has been studied in more detail than any other group of proteins, and as a result a deeper understanding of the relationships among structure, properties, and biological function (i.e., physiology) exists. The central focus of this chapter is to delineate chemical features that determine the affinity of an active site, especially an iron porphyrin, for molecular oxygen. In order to develop this theme, macroscopic (thermodynamic and kinetic) factors associated with dioxygen binding and release are summarized first. The nonbiological chemistry of iron and copper in the presence of dioxygen is described briefly to elucidate the key role that the protein plays in supporting oxygen transport by preventing irreversible oxidation of the binding site or of its ligands. The macroscopic behavior of the biological systems is related to the microscopic picture that has been developed over the last 30 years from x-ray crystallographic studies and a miscellany of spectroscopic probes of the oxygen-binding site. Relationships between the geometry and charge distribution in the metal-dioxygen moiety and the nature of the interactions between this moiety and its surroundings are examined. Nonbiological dioxygen carriers have proved particularly useful in providing precise and accurate structural information as well as thermodynamic and kinetic data against which the corresponding data from biological oxygen carriers can be contrasted.

The bioinorganic chemistry of the hemoglobin family of oxygen binders is particularly amenable to study by means of small-molecule model systems: four of the five ligands that make up the active site are provided by a square-planar tetradentate ligand, the protoporphyrin IX dianion (Figure 4.2). One axial ligand in hemoglobin, imidazole from a histidine residue, is provided by the protein, and the remaining sixth coordination site is available for the exogenous ligand, e.g., dioxygen or carbon monoxide. Thus a model system that approximates the stereochemistry of the active site in hemoglobin may be assembled from an iron(II) porphyrin and a ligand, such as imidazole or pyridine. On the other hand, in hemocyanin and hemerythrin most of the ligands are supplied by the protein. Thus the assembly of a model system that provides appropriate ligands correctly disposed around the pair of metal atoms poses a major synthetic challenge, especially for hemocyanin, where details on the number, type, and arrangement of ligands have been difficult to establish. Many aspects of the physical, inorganic, and structural chemistry underlying biological oxygen transport and utilization (Chapter 5) have been clarified through model systems.

A. Requirements for Effective Oxygen Carriers

In order for dioxygen transport to be more efficient than simple diffusion through cell membranes and fluids, it is not sufficient that a metalloprotein merely binds dioxygen. Not only is there an optimal affinity of the carrier for dioxygen, but also, and more importantly, the carrier must bind *and* release dioxygen at a rapid rate. These thermodynamic and kinetic aspects are illustrated in Figure 4.3, a general diagram of energy vs. reaction coordinate for the process

$$M + O_2 \underset{k_{-1}}{\overset{k_1}{\rightleftharpoons}} MO_2 \qquad (4.2)$$

where M is an oxygen carrier, for example hemocyanin or a simple nonbiological metal complex. Thermodynamic or equilibrium aspects are summarized

172

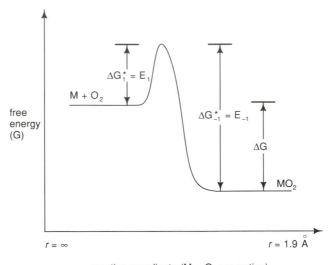

Figure 4.3
Schematic diagram of energy changes in dioxygen binding.

by ΔG in Figure 4.3. As illustrated there, ΔG is negative, and thus the forward reaction, dioxygen binding, is spontaneous. The equilibrium constant (K) is given by

$$K = \frac{a(MO_2)}{a(M)a(O_2)},\tag{4.3}$$

where a is the activity (crudely, concentration) of the component. The equilibrium constant is related to the change in free energy by

$$\Delta G^{\circ} = -RT\ln K.\tag{4.4}$$

The rate of the forward reaction (k_1) is related to ΔG_1^{*}; the rate of the reverse reaction (k_{-1}) is related to ΔG_{-1}^{*}. Provided that oxygen binding is effectively a single-step process, then

$$K = \frac{k_1}{k_{-1}}.\tag{4.5}$$

Usually the rates of the forward and reverse reactions are related by the empirical Arrhenius expression to quantities termed the activation energies (E_1 and E_{-1}) of the reactions, where

$$k_1 = A_1 \exp\left(-E_1/RT\right) \quad \text{and} \quad k_{-1} = A_{-1} \exp\left(-E_{-1}/RT\right).\tag{4.6}$$

These quantities are experimentally accessible through the change in rate as a function of temperature.

1. Thermodynamic factors [19-20]

The equilibrium constant K was defined in Equation (4.3) in terms of the activity a_i of component i. The a_i may be expressed as a function of concentration as

$$a_i = \gamma_i[i], \tag{4.7}$$

where for species i, γ_i is its activity coefficient and $[i]$ is its concentration (strictly molality, but usually as molarity in mol L^{-1}). At infinite dilution $\gamma_i = 1$. Provided that the charge and size of species M and MO_2 are similar and that O_2 forms an ideal solution, then the activities of Equation (4.3) may be approximated by concentrations to give the expression

$$K_c = \frac{[MO_2]}{[M][O_2]}. \tag{4.8}$$

However, Equation (4.8) does not permit a direct comparison of the oxygen-binding behavior of one species in some solvent with that of a second in some other solvent. First, for a given partial pressure of dioxygen, the concentration of O_2 in the solution varies considerably with temperature and from one solvent to another. Second, reliable measurements of oxygen solubilities are not always available, and it is only relatively recently that oxygen electrodes have been developed to measure directly oxygen concentrations (strictly, activities). However, oxygen-binding measurements are normally made with a solution of M in equilibrium with gaseous dioxygen. At equilibrium the molar Gibbs' free energies (chemical potentials) of the dissolved and gaseous dioxygen are identical—if they are not, gaseous O_2 would dissolve, or dissolved O_2 would be released. Thus the solvent-dependent quantity $[O_2]$ in Equation (4.8) may be replaced by the solvent-independent quantity $P(O_2)$, the partial pressure of dioxygen. Under almost all experimental conditions the quantity $P(O_2)$ is a very good approximation to the gas-phase activity (fugacity) of dioxygen; hence we obtain for the equilibrium constant*

$$K_p = \frac{[MO_2]}{[M]P(O_2)}. \tag{4.9}$$

It is very convenient to express the affinity as the partial pressure of dioxygen required for half-saturation of the species M, $P_{1/2}(O_2)$. Under such conditions, $[M] = [MO_2]$, one obtains

$$P_{1/2}(O_2) = 1/K_p, \tag{4.10}$$

---

* There has been considerable discussion as to whether K_c (4.8) or K_p (4.9) should be used to compare dioxygen binding under different solvent conditions.[21-23] We believe that the latter is more appropriate, since for a system at equilibrium, the chemical potential of gaseous O_2 must be identical with that of dissolved O_2.[19] On the other hand, the concentration of O_2 varies from one solvent to another.

where $P_{1/2}(O_2)$ is usually given in Torr or mm Hg.* As will be detailed shortly, values for $P_{1/2}(O_2)$ are typically in the range 0.5 to 40 Torr.

The dioxygen affinity is composed of enthalpic (ΔH) and entropic (ΔS) components, with

$$\Delta G° = -RT\ln K = \Delta H° - T\,\Delta S°. \tag{4.11}$$

Within a family of oxygen carriers the values of $\Delta S°$ and $\Delta H°$ are usually similar. Large deviations (such as a change of sign) are therefore indicative of a change in the nature of the oxygen-binding process.

a. Non-cooperative Dioxygen Binding   If the oxygen-binding sites M are mutually independent and noninteracting, as in moderately dilute solutions of monomeric molecules, then the concentration of species MO_2 as a function of the partial pressure of O_2 is generally well fit by a Langmuir isotherm.[20] Here a plot of the fractional saturation of dioxygen binding sites, θ, where

$$\theta = \frac{[MO_2]}{[M] + [MO_2]} = \frac{K_p\,P(O_2)}{1 + K_p\,P(O_2)} \tag{4.12}$$

versus $P(O_2)$ gives the hyberbolic curve labeled "non-cooperative" in Figure 4.4A.[9] Alternatively,[24] a plot of log ($\theta/(1 - \theta)$) versus log ($P(O_2)$), the so-called "Hill plot," gives a straight line with a slope of unity and an intercept of $-\log P_{1/2}(O_2)$ (Figure 4.4B). A differential form is shown as the dotted line in Figure 4.4C. Such binding, where the dioxygen sites are independent of each other, is termed *non-cooperative*.

b. Cooperative Dioxygen Binding   Many dioxygen-binding proteins are not independent monomers, with only one dioxygen-binding site, but oligomeric species with the protein comprising two or more similar subunits. The subunits may be held together by van der Waals' forces or by stronger interactions, such as hydrogen bonds or salt bridges, or even by covalent bonds. For example, most mammalian hemoglobins are tetramers, consisting of two pairs $[\alpha\beta]_2$ of myoglobin-like subunits denoted as α and β. Either none, one, two, three, or all four sites may be occupied by dioxygen. This situation is illustrated schematically in Figure 4.5, which also shows the statistical weighting of each level of saturation, treating the α and β subunits as identical. Thus the binding or release of dioxygen at one site *may* affect the affinity and kinetics of ligand binding and release at a neighboring site. As a result, the saturation curve becomes sigmoidal in shape, as illustrated in Figure 4.4A. The dioxygen binding is *cooperative*. When cooperativity is positive, the affinity of a vacant site is increased by occupancy of an adjacent one.

This behavior, where the binding of one molecule influences the binding of successive molecules *of the same kind*, is referred to as a *homotropic allosteric*

---

* Many authors use the symbol P_{50} (corresponding to 50% saturation) for $P_{1/2}$.

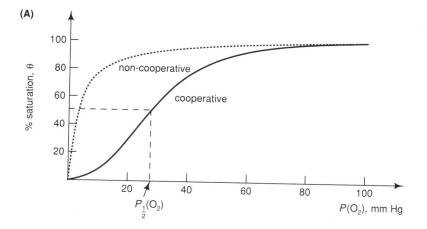

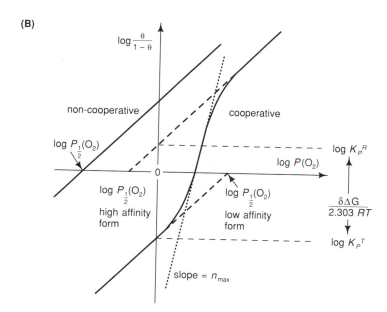

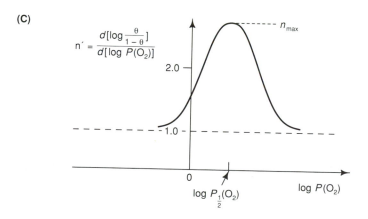

Figure 4.4
Cooperative and non-cooperative binding of dioxygen:[9] (A) Binding curves;
(B) Hill plot of binding curves; (C) First derivative (slope) of the Hill plots.

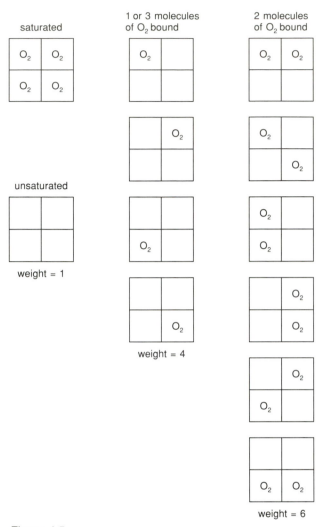

Figure 4.5
Diagram of tetrameric hemoglobin, showing statistical weights
of different saturations (see text).

interaction. A *heterotropic allosteric* interaction occurs when the interaction with
the protein of a second *unlike* molecule, for instance, an organic polyphosphate
for human hemoglobins, influences the binding of the first molecule (e.g., dioxygen). Such molecules are often termed *allosteric effectors*. A commonly observed heterotropic allosteric interaction is the Bohr effect, named after the biologist Christian Bohr, father of physicist Niels Bohr. This effect, which relates
the change in partial pressure of O_2 to a change in pH at constant saturation of
binding sites (θ), is related thermodynamically to the Haldane effect, which
relates the number of protons released ($\#H^+$) with a change in θ at constant
pH (Equation 4.13). A very large Bohr effect, where O_2 affinity decreases sharply
with pH, is often called the Root effect.[25a] It is physiologically important for

fish such as trout, probably in maintaining buoyancy, but its molecular basis in trout hemoglobin IV remains to be discovered.[25b]

$$\left[\frac{\partial(\#H^+)}{\partial \theta} \right]_{pH} = \left[\frac{\partial(\log P(O_2))}{\partial pH} \right]_{\theta} \qquad (4.13)$$

The degree of cooperativity can be characterized in a number of ways. By means of a Hill plot of $\log(\theta/(1 - \theta))$ versus $\log(P(O_2))$, the limiting slopes (which should be unity) at high O_2 pressure and low O_2 pressure may be extrapolated as shown in Figure 4.4B to $\log(\theta/(1 - \theta)) = 0$, where $\theta = 0.5$. Two limiting values for $P_{1/2}(O_2)$ are obtained, one characterizing the regime of high partial pressure of dioxygen, where the O_2 affinity is high (for the case illustrated of positive cooperativity). The other $P_{1/2}(O_2)$ value characterizes the regime of low partial pressure of dioxygen, where affinity is relatively low. This difference in affinities can be converted into a difference between the free-energy change upon O_2 binding in the low-affinity state (K_p^T) and the high-affinity state (K_p^R) [the designations T and R will be described in subsection d]:

$$\delta \Delta G^\circ = -RT \ln(K_p^T / K_p^R). \qquad (4.14)$$

A second way to characterize cooperativity involves fitting the oxygen-binding data at intermediate saturation ($0.2 < \theta < 0.8$)—that is, about the inflection point in a Hill plot—to the Hill equation

$$\theta/(1 - \theta) = K_p P^n(O_2)$$

or $$\log(\theta/(1 - \theta)) = -\log(P_{1/2}(O_2)) + n \log(P(O_2)). \qquad (4.15)$$

The Hill coefficient (n) is an empirical coefficient that has a value of unity for non-cooperative binding, where Equation (4.15) reduces to the Langmuir iso-therm, Equation (4.12). Any number greater than unity indicates positive coop-erativity. If O_2 binding is an all-or-nothing affair, where dioxygen binding sites are either all occupied or all vacant, n equals the number of subunits in the molecule. The fit is only approximate, since the Hill plot is only approximately linear about the inflection point, as may be seen in Figure 4.4B. A more precise value of n may be obtained by plotting the slope in the Hill plot (n') as a function

$$n' = \frac{d[\log(\theta/(1 - \theta))]}{d[\log(P(O_2))]} \qquad (4.16)$$

of $\log(P(O_2))$ (Figure 4.4C). The maximum value of n' is taken as the Hill coefficient n.[9] Note that the maximum in this first-derivative plot of the binding curve will occur at $P_{1/2}(O_2)$ only if the Hill plot is symmetric about its inflection point. For tetrameric hemoglobins, a maximum Hill coefficient of around 3.0 is seen, and for hemocyanins n may be as high as 9. These values, like $P_{1/2}(O_2)$ values, are sensitive to the nature and concentrations of allosteric effectors.

c. Benefits of Cooperative Ligand Binding   In general, oxygen-carrier proteins, being oligomeric, coordinate dioxygen cooperatively, whereas oxygen-storage proteins, being monomeric, do not. Oligomerization and cooperative binding confer enormous physiological benefits to an organism. The first benefit derives directly from oligomerization. Oxygen carriers either form small oligomers that are encapsulated into cells or erythrocytes (such hemoglobins are referred to as intracellular hemoglobins) or associate into large oligomers of 100 or more subunits. Such encapsulation and association reduce by orders of magnitude the number of independent particles in the blood, with consequent reductions in the osmotic pressure of the solution and in strain on vascular membranes.

The second benefit derives from cooperative binding of ligands and the abilities of heterotropic allosteric effectors to optimize exquisitely the oxygen-binding behavior in response to the external and internal environment. The situation is illustrated in general terms in Figure 4.6.[9] Most organisms that require O_2 live in an environment where the activity of O_2 corresponds to about 21 percent of an atmosphere, that is, to about 160 Torr, although usually the effective availability, because of incomplete exchange of gases in the lungs, for example, is around 100 Torr. The concentration of O_2 in vertebrate tissues at rest is equivalent to a partial pressure of about 35–40 Torr dioxygen; lower values obtain at times of exertion. Now consider a noncooperative oxygen binder with an affinity expressed as $P_{1/2}(O_2)$ of 60 Torr (Figure 4.6, curve a). Then, at 100 Torr the fractional saturation θ is 0.625. In other words, in a realm of high O_2 availability, only 62.5 percent of the oxygen-binding capacity is used, which is not particularly efficient if the organism wished to climb Mt. Everest, where the partial pressure of O_2 is less than half that at sea level. In the tissues, where $P(O_2) \approx 40$ Torr, the fractional saturation is about 40 percent. Thus, only about one third of the coordinated dioxygen is released to the tissues, and total effi-

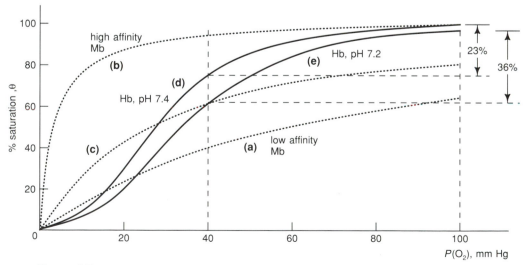

Figure 4.6
Physiological benefits of cooperativity and heterotropic allosteric effectors.[9]

ciency is only 22.5 percent. Consider now a noncooperative oxygen carrier with a much higher affinity, $P_{1/2}(O_2) = 1.0$ Torr (Figure 4.6, curve b). If we assume the same ambient pressure of O_2 in the tissues, the fractional saturation is 97.6 percent. Note that at 100 Torr of O_2 the carrier is 99.0 percent saturated. In other words, only about 1.4 percent of the available oxygen is delivered.

With an oligomeric protein that binds dioxygen cooperatively, the problem of inefficient and inflexible oxygen delivery disappears. For example, the tetrameric protein hemoglobin has a mean affinity for O_2 of $P_{1/2}(O_2) \approx 26$ Torr at 37°C and pH 7.4. If hemoglobin bound O_2 noncooperatively, then the hyberbolic binding curve (c) in Figure 4.6 would represent the O_2 binding. Instead, the observed binding follows curve (d). Since the partial pressure of dioxygen in the lungs and arterial blood of vertebrates is around 100 Torr, but in the tissues and venous blood it is around 40 Torr, then at these pressures a typical myoglobin ($P_{1/2}(O_2) \approx 1$ Torr) remains effectively saturated. On the other hand, about 25 percent of the available dioxygen can be delivered, even in the absence of myoglobin. With venous blood remaining 75 percent oxygenated, hemoglobin has substantial capacity to deliver more O_2 at times of exertion or stress when $P(O_2)$ in the tissues falls below 40 Torr.

The net result is that whole blood, which contains about 15 g of hemoglobin per 100 mL, can carry the equivalent of 20 mL of O_2 (at 760 Torr) per 100 mL, whereas blood plasma (no hemoglobin) has a carrying capacity of only 0.3 mL of O_2 per 100 mL.[9]

Oxygen binding *in vivo* is modulated by allosteric effectors that through interaction with the protein change the affinity and degree of cooperativity. For hemoglobin A (adult human hemoglobin), naturally occurring allosteric effectors include the proton, carbon dioxide, and 2,3-diphosphoglycerate (2,3-DPG). Increasing concentrations of these species progressively lower the affinity of free hemoglobin A, thereby enhancing the release of coordinated O_2 (Figure 4.6, curve e). For example, 2,3-DPG is part of a subtle mechanism by which dioxygen is transferred from mother to fetus across the placenta. The subunits comprising fetal hemoglobin and adult hemoglobin are slightly different. In the absence of allosteric effectors (referred to as stripped hemoglobin), the oxygen-binding curves are identical. However, 2,3-DPG binds less strongly to fetal hemoglobin than to adult hemoglobin. Thus fetal hemoglobin has a slightly higher affinity for dioxygen, thereby enabling dioxygen to be transferred. The proton and carbon dioxide are part of a short-term feedback mechanism. When O_2 consumption outpaces O_2 delivery, glucose is incompletely oxidized to lactic acid (instead of CO_2). The lactic acid produced lowers the pH, and O_2 release from oxyhemoglobin is stimulated (Figure 4.6, curve e). The CO_2 produced in respiration forms carbamates with the amino terminals, preferentially of deoxy hemoglobin.

$$\text{R-NH}_2 + CO_2 \rightleftharpoons \text{R-NH-COO}^- + H^+$$

Thus hemoglobin not only delivers O_2 but also facilitates removal of CO_2 to the lungs or gills, where CO_2 is exhaled.

d. Models for Cooperativity The binding of O_2 to hemoglobin can be described as four successive equilibria:

$$Hb + O_2 \xrightleftharpoons{K^{(1)}} Hb(O_2) \qquad P_{1/2}{}^{(1)}(O_2) = 123\ [46]\ \text{Torr}$$

$$Hb(O_2)_1 + O_2 \xrightleftharpoons{K^{(2)}} Hb(O_2)_2 \qquad P_{1/2}{}^{(2)}(O_2) = 30\ [16]\ \text{Torr}$$

$$Hb(O_2)_2 + O_2 \xrightleftharpoons{K^{(3)}} Hb(O_2)_3 \qquad P_{1/2}{}^{(3)}(O_2) = 33\ [3.3]\ \text{Torr} \tag{4.17}$$

$$Hb(O_2)_3 + O_2 \xrightleftharpoons{K^{(4)}} Hb(O_2)_4 \qquad P_{1/2}{}^{(4)}(O_2) = 0.26\ [0.29]\ \text{Torr}$$

(0.6 mM hemoglobin A, bis(Tris) buffer, pH 7.4, 0.1 M Cl$^-$, 2 mM 2,3-DPG, 25°C. The values in square brackets are affinities in Torr measured in the absence of 2,3-DPG.)

This simple scheme proposed by Adair[26] assumes that each of the four binding sites is identical. The $P_{1/2}(O_2)$ values given come from fitting the binding curve to this scheme.[27] When 2,3-DPG is removed, the affinity of hemoglobin for the first three molecules of O_2 is substantially increased, and the degree of cooperativity is lowered (values in square parentheses). For progressively stronger binding, the following inequalities, reflecting the proper statistical weighting illustrated in Figure 4.5, should hold:

$$\tfrac{1}{4}K^{(1)} > \tfrac{4}{6}K^{(2)} > \tfrac{6}{4}K^{(3)} > \tfrac{4}{1}K^{(4)} \tag{4.18}$$

The $\tfrac{6}{4}$ ratio, for example, reflects the six equivalent forms of the doubly and the four equivalent forms of the triply ligated species. In other words, relative to a noncooperative system, at low O_2 availability dioxygen *release* is facilitated; at high O_2 availability dioxygen *binding* is facilitated. The scheme is readily extended to higher orders of oligomerization.

A simple model for analyzing cooperative ligand binding was proposed by Monod, Wyman, and Changeux in 1965, and is usually referred to as the MWC two-state concerted model.[28] Molecules are assumed to be in equilibrium between two conformations or quaternary structures, one that has a low ligand affinity and a second that has a high ligand affinity. The low-affinity conformation is often designated the T or tense state, and the high-affinity conformation the R or relaxed state. The equilibrium between the two conformations is characterized by the allosteric constant

$$L_0 = [R_0]/[T_0] \tag{4.19}$$

where the subscript denotes the unliganded R and T states. The free-energy change upon binding a ligand to the R state, irrespective of saturation, is assumed to be a constant, and the associated equilibrium constant is designated K_R; a third constant, K_T, characterizes binding to the T state. Figure 4.7 illustrates this model, and introduces the terminology conventionally used. To a reasonable approximation, the cooperative binding of dioxygen can be summa-

(A)

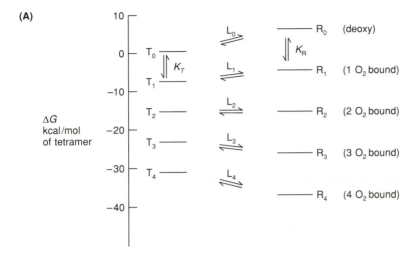

(B)

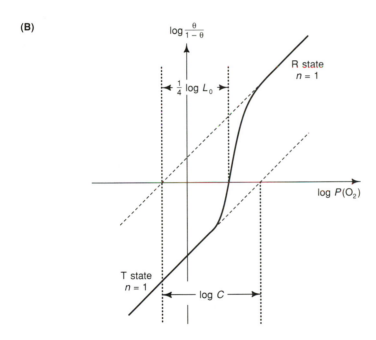

Figure 4.7
The MWC two-state model for cooperative ligand binding:[4] (A) Free-energy relationships among R and T states; (B) Calculation of the allosteric constants from the binding curve.

rized by these three parameters, L_0, K_R, and K_T. The Adair constants may be expressed in terms of these parameters:

$$K^{(1)} = \frac{(1 + L_0C)K_T}{1 + L_0}, \qquad K^{(2)} = \frac{(1 + L_0C^2)K_T}{1 + L_0C},$$

$$K^{(3)} = \frac{(1 + L_0C^3)K_T}{1 + L_0C^2}, \qquad K^{(4)} = \frac{(1 + L_0C^4)K_T}{1 + L_0C^3}, \tag{4.20}$$

where $C = K_R/K_T$. The fractional saturation is given as

$$\theta = \frac{\alpha(1 + \alpha)^3 + L_0\alpha C(1 + \alpha C)^3}{(1 + \alpha)^4 + L_0\alpha C(1 + \alpha C)^4}, \tag{4.21}$$

where $\alpha = K_T[X]$, and $[X]$ is the concentration of the free ligand (e.g., O_2) in the same units (M or Torr) in which K_T is expressed. Figure 4.7B illustrates how the allosteric parameters, $C = K_R/K_T$ and $L_0 = [R_0]/[T_0]$, are extracted from a plot of saturation (as log $[\theta/(1 - \theta)]$) versus partial pressure of dioxygen (as log $[P(O_2)]$). Notice how the two-state model (Figure 4.7B) matches very closely the form of the binding curve for hemoglobin (Figure 4.4B). Equations (4.20) and (4.21) may be generalized to an oligomer with n subunits. In the case of hemoglobin, Perutz and coworkers,[11] through the determination of the crystal structures of a variety of hemoglobin derivatives, have given subsequently a sound structural basis to the MWC model of two basic quaternary states (see below).

A more exact treatment of ligand-binding data would allow for different affinities for different binding sites (called subunit heterogeneity) and different intrinsic affinities for ligand binding to the R-state conformation compared with the T-state conformation, for each level of ligand saturation—that is, for tertiary structure change within subunits upon ligation. This more exact treatment requires 25 separate equilibrium constants. Statistical thermodynamical approaches exist.[29] These explicitly incorporate the different types of subunit interactions that structural studies have revealed, and give improved fits to oxygen-binding data and to the Bohr effect. The key element of two basic quaternary states is preserved, at least for dioxygen binding.[29b]

For some modified hemoglobins, for example $[\alpha\text{-Fe(II)}]_2[\beta\text{-Mn(III)}]_2$, where in the β subunits the heme iron is replaced by Mn(III), there is now strong evidence for three quaternary states,[29c] with the singly and several of the doubly ligated species having an energy state intermediate between the T (unliganded) and R (fully, triply, and the other doubly liganded) states.

2. Kinetic factors

It is of little benefit to the organism if its dioxygen carrier, such as hemoglobin, binds and releases O_2 at such slow rates that O_2 is not delivered faster than it would be by simple diffusive processes. Thus, a binding rate within a couple of orders of magnitude of the rate of diffusion, together with the high carrying capacity of O_2 that high concentrations of oxygen carrier enable (noted earlier), and a pumping system ensure adequate O_2 supplies under all but the most physiologically stressful conditions.

Whereas measurements of equilibrium give little or no molecular information, rather more molecular information may be inferred from kinetic data. The processes of binding and release can be examined by a variety of techniques, with timescales down to the picosecond range. The temperature behavior of the rates gives information on the heights of energy barriers that are encountered as dioxygen molecules arrive at or depart from the binding site. The quantitative interpretation of kinetic data generally requires a molecular model of some sort. It is because of this multibarrier pathway that the equilibrium constant measured as k_1/k_{-1} (Equation 4.5) may differ substantially from the thermodynamically measured value (Equation 4.3).

The simple Adair scheme outlined above is readily adapted to cater to kinetic data.

3. Dioxygen reactions

Most biological conversions involving dioxygen require enzymatic catalysis. It is reasonable then that metals found in the proteins involved in the transport and storage of O_2 also frequently appear, with minor modification of ligands, in enzymes that incorporate oxygen from dioxygen into some substrate. Dioxygen, in this case, is not only coordinated, but also activated and made available to the substrate. In the family of proteins with heme groups, hemoglobin is a dioxygen carrier and cytochrome P-450 is an oxygenase. A similar differentiation in function is also found for hemocyanin and tyrosinase from the family of proteins with a dinuclear copper complex at the active site. Note that not all enzymes that mediate the incorporation of oxygen from O_2 into some substrate coordinate and activate dioxygen. For example, lipoxygenase probably catalyzes the conversion of a 1,4-diene to a 1,3-diene-4-hydroxyperoxy species by activation of the organic substrate. The active site does not resemble that of any known oxygen-carrier protein. This topic is discussed more fully in Chapter 5.

B. Biological Oxygen Carriers

As noted earlier, three solutions to the problem of dioxygen transport have evolved: hemoglobin (Hb), hemocyanin (Hc), and hemerythrin (Hr). Their remarkable distribution over plant and animal kingdoms is shown in Figure 4.8.[15] The hemoglobins and myoglobins found in plants, snails, and vertebrates all appear to share a common, very ancient ancestor. There is some evidence now for a common ancestral hemocyanin.[42c] The appearance of hemerythrin in a few annelid worms is an evolutionary curiosity. These few words and the diagram will

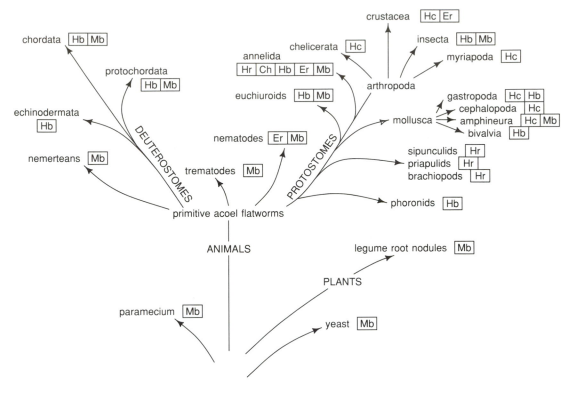

Figure 4.8
Phylogenetic distribution of oxygen-carrier proteins: Hb, hemoglobin; Mb, myoglobin; Er, ery-throcruorin; Ch, chlorocruorin; Hc, hemocyanin; Hr, hemerythrin.[15a] Reproduced with permission from K. E. van Holde and K. I. Miller, *Quart. Rev. Biophys.* **15** (1982), 1–129.

suffice to give some hints about how respiratory proteins evolved, a subject that is outside the scope of this book.

1. The hemoglobin family

Hemoglobins are the most evolutionarily diverse family of dioxygen carriers. They are found in some plants (e.g., leghemoglobin in the nitrogen-fixing nodules of legumes), many invertebrates (including some insect larvae), crustaceans, molluscs (especially bivalves and snails), almost all annelid worms, and in *all* vertebrates with one possible exception, the Antarctic fish *Cyclostomata*.

With few exceptions the monomeric and oligomeric hemoglobins all share a basically similar building block: a single heme group is embedded in a folded polypeptide with a molecular weight of about 16 kDa (see Figure 4.2), and is anchored to the protein by coordination of the iron center to an imidazole ligand from a histidine residue. Mammalian myoglobin is often taken as the archetypical myoglobin (see Table 4.1). Sperm whale, bovine, or equine myoglobin are specific examples; the muscle tissue from which they may be extracted is more available than that from *Homo sapiens*. The archetypical oligomeric hemoglobin that shows cooperative binding of O_2 is the tetrameric hemoglobin A. It is readily available from the blood of human donors.* In some invertebrate hemoglobins, especially those of annelids, aggregates may contain as many as 192 binding sites, to give a molecular weight of about 3×10^6 Dalton. These and other high-molecular-weight hemoglobins of arthropods are often referred to as erythrocruorins (Er). In a few annelid worms, the otherwise ubiquitous heme b or protoheme is replaced by chloroheme (see Figure 4.2) to give chlorocruorins (Ch), which turn green upon oxygenation (*chloros*, Greek for green). Some organisms, for example the clam *Scapharca equivalvis*, feature a dimeric hemoglobin.

The only known anomalous hemoglobin is Hb *Ascaris*, which comes from a parasitic nematode found in the guts of pigs. It has a molecular weight of about 39 kDa per heme; this value is not a multiple of the myoglobin building block.[31] Moreover, presumably in response to the low availability of O_2 in pigs' guts, Hb *Ascaris* has an extraordinarily high affinity for dioxygen, in large part owing to an extremely slow rate of dioxygen release.[32] Leghemoglobin is another carrier with a high affinity for dioxygen, in this case because of a high rate of O_2 binding. Since O_2 is a poison for the nitrogenase enzyme, yet the nodules also require dioxygen, diffusion of O_2 is facilitated, but the concentration of free dioxygen in the vicinity of nitrogen-fixing sites is minimized.[33]

Kinetic and thermodynamic data for dioxygen binding and release from a variety of hemoglobins are summarized in Table 4.2.[9,10,31,34–36] Notice that for the hemoglobin tetramer, which comprises two pairs of slightly dissimilar subunits, the α and β chains bind O_2 with significantly different affinities and rate constants, especially in the T state. Isolated chains behave like monomeric vertebrate hemoglobins, such as whale myoglobin, which have affinities close to those of R-state hemoglobin. The chlorocruorins have a low affinity compared to other erythrocruorins. Especially for proteins that bind O_2 cooperatively, a range of values is specified, since affinities and rates are sensitive to pH, ionic strength, specific anions and cations (allosteric effectors), and laboratory. For example, as we noted above, the O_2 affinity of hemoglobin A is sensitive to the concentration of 2,3-DPG and to pH (Bohr effect). Trout hemoglobin I is insensitive to these species, whereas a second component of trout blood, trout hemoglobin IV, is so sensitive to pH (Root effect) that at pH < 7 trout hemoglobin IV is only partially saturated at $P(O_2) = 160$ Torr.[4] Note that O_2 affinities span five orders of magnitude. Since heme catabolism produces carbon monoxide, and since in some environments CO is readily available exogenously, selected data for CO binding are also presented.

2. The hemocyanin family

Hemocyanins (Hc), the copper-containing dioxygen carriers, are distributed erratically in two large phyla, *Mollusca* (for example, octopi and snails) and

---

* Blood from human donors is also a source for a variety of abnormal hemoglobins, the most famous of which is HbS, the hemoglobin giving rise to sickle-cell anemia. It was Pauling and coworkers[30] who first found that HbS differs from HbA through the *single* substitution of valine for glutamic acid in each of two of the four subunits comprising Hb. Sickle-cell anemia was the first condition to be denoted a "molecular disease."

Table 4.2
Thermodynamics and kinetics of ligand binding to biological oxygen carriers (at 20–25°C and buffered at pH 6.5–8.5).

Carrier	Dioxygen binding					Carbon-monoxide binding				
	$P_{1/2}(O_2)$ Torr	ΔH kcal/mol	ΔS eu	k_{on} µM^{-1} s^{-1}	k_{off} s^{-1}	$P_{1/2}(CO)$ Torr	ΔH kcal/mol	ΔS eu	k_{on} µM^{-1} s^{-1}	k_{off} s^{-1}
Hemoglobins										
Hb *Ascaris*	0.0047	—	—	1.5	0.0041	0.063	—	—	0.21	0.018
Leg Hb	0.047	−18.9	—	156.	1.	0.00074	—	—	13.5	0.012
whale Mb	0.51	−14.9	—	14.	12.	0.018	−13.5	—	0.51	0.019
Whale Mb (E7His → Gly)	—	—	—	140.	1600.	—	—	—	—	—
HbA isolated chains α	0.74	−14.2	−21.	50.	28.	0.0025	—	—	4.0	0.013
HbA isolated chains β	0.42	−16.9	−29.	60.	16.	0.0016	—	—	4.5	0.008
HbA R α chain	0.15–1.5	−18.	−30.	29.	10.	0.001–0.004	—	—	3.2	0.005
β chain				100.	21.				9.8	0.009
HbA R α$_{E7His→Gly}$				220.	620.				19.	0.007
β$_{E7His→Gly}$				100.	3.				5.0	0.013
HbA T α chain	9–160	−12.	−35.	2.9	183.	0.10–2.8	—	—	0.099	0.09
β chain				11.8	2500.					
Chironimus Mb	0.40	—	—	300.	218.	0.0019	—	—	27.	0.095
Glycera Mb	5.2	—	—	190.	1800.	0.00089	—	—	27.	0.042
Aplysia Mb	2.7	−13.6	—	15.	70.	0.013	—	—	0.49	0.02
Spirographis chlorocruorin	16–78	−4.5	—	—	—	—	—	—	—	—

Hemocyanins[a]

Molluscan Hc

Helix pomatia R	2.7	−11.5	−12.6	3.8	10.	10.	−13.5	−24.	0.66	70.
Helix pomatia T	55.	−15.4	−31.1	1.3	300.	CO binding noncooperative since not measurable				
Levantina hierosohimia R	3.8	−7.5	−1.8	—	—	—	—	—	—	—
Levantina hierosohimia T	18.	+3.1	+31.	—	—	—	—	—	—	—
Arthropod Hc										
Panulirus interruptus R[b]	1.0	—	—	31.	60.	720.	−6.0	−2.7	4.1	8100.
P. interruptus monomer	9.3	—	—	57.	100.	CO binding noncooperative				
Leirus quinquestris R	1.7	−7.4	0.	—	—	—	—	—	—	—
Leirus quinquestris T	117.	+3.1	+27.	—	—	—	—	—	—	—

Hemerythrins

Phascolopsis gouldii	2.0	−12.4	−18.	7.4	56.	not known to bind CO				
Themiste zostericola 8-mer	6.0	—	—	7.5	82.					
T. zostericola monomer	2.2	—	—	78.	315.					

Solubility of O_2 in water: 1.86×10^{-6} M/Torr
Solubility of CO in water: 1.36×10^{-6} M/Torr
[a] 10 mM Ca^{2+} added: necessary for cooperativity.
[b] CO binding at pH 9.6.

Arthropoda (for example, lobsters and scorpions). The functional form of hemocyanin consists of large assemblies of subunits.[14,15,37] In the mollusc family the subunit has a molecular weight of about 50 kDa and contains two copper atoms. From electron-microscopic observations, hemocyanin molecules are cylindrical assemblies about 190 or 380 Å long and 350 Å in diameter comprising 10 or 20 subunits, respectively, for a molecular weight as high as 9×10^6 Dalton. In the arthropod family, the subunit has a molecular weight of about 70 kDa with two copper atoms. Molecular aggregates are composed of 6, 12, 24, or 48 subunits. Upon oxygenation the colorless protein becomes blue (hence cyanin from *cyanos*, Greek for blue). Spectral changes upon oxygenation, oxygen affinities, kinetics of oxygen binding (Table 4.2),[4,5,14,15,38] anion binding, and other chemical reactions show that the active site in the phylum *Arthropoda* and that in *Mollusca*, although both containing a pair of copper atoms, are not identical.[4,14]

No monomeric hemocyanins, analogous to myoglobin and myohemerythrin (next section), are known. For some hemocyanins the binding of dioxygen is highly cooperative, if calcium or magnesium ions are present, with Hill coefficients as high as $n \sim 9$. However, the free energy of interaction per subunit can be small in comparison with that for tetrameric hemoglobin; 0.9 to 2.5 kcal/mol compared to 3.0 kcal/mol. Allosteric effects, at least for a 24-subunit tarantula hemocyanin, can be separated into those within a dodecamer (12 subunits)—the major contributor to overall allostery—and those between dodecamers.[39c] This has been termed *nested allostery*. In contrast to the hemoglobin family, isolated chains have affinities typical of the T-state conformation for hemocyanin. The binding of CO, which binds to only one copper atom, is at best weakly cooperative.[39.]

As alluded to above, the distribution of hemocyanins is striking. Among the molluscs exclusive use of hemocyanin as the respiratory protein occurs only with the cephalopods (squid, octopi, and cuttlefish), and in the arthropods only among the decapod (ten-footed) crustaceans (lobsters, shrimp, and crabs). The bivalve molluscs (for example, oysters and scallops) all use small dimeric or octameric hemoglobins. The edible gastropod (snail) *Helix pomatia* uses hemocyanin, whereas the apparently closely related fresh-water snail *Planorbis* uses a high-oligomer hemoglobin. Both use a myoglobin as the oxygen-storage protein. The structure of the active site has been extensively probed by EXAFS methods,[40,41] and the x-ray crystal structure of a hexameric deoxyhemocyanin is known.[42] Each copper atom is coordinated to three imidazole groups from histidine residues. The pinwheel arrangement of the six subunits, the domain structure of a single subunit, and the domain containing the active site are shown in Figure 4.9.

3. The hemerythrin family

The biological occurrence of hemerythrins (Hr in Figure 4.8), the third class of dioxygen carriers, is relatively rare, being restricted to the sipunculid family (nonsegmented worms), a few members of the annelid (segmented worm) fam-

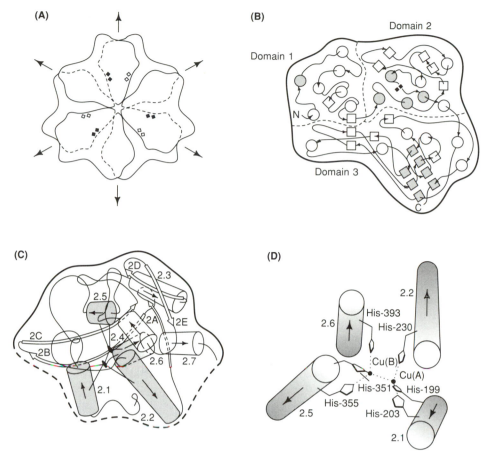

Figure 4.9
Diagram of the structure of deoxyhemocyanin from *Panulirus interruptus* at 3.2 Å resolution:[42c] (A) The hexameric arrangement of subunits; (B) The domain structure of one subunit; (C) The tertiary structure of domain 2, which contains the pair of copper atoms: α-helices are represented by cylinders; β-strands by arrows, and copper atoms by diamonds; (D) The active site and its histidine ligands. Reproduced with permission from B. Linzen, *Science* **229** (1985), 519–524.

ily, a couple of brachiopods (shrimps), and a couple of priapulids. The oxygen-binding site contains, like hemocyanin, a pair of metal atoms, in this case, iron. Upon oxygenation the colorless protein becomes purple-red. Monomeric (myo-hemerythrin), trimeric, and octameric forms of hemerythrin are known; all appear to be based on a similar subunit of about 13.5 kDa. When hemerythrin is extracted from the organism, its oxygen binding is at best only weakly cooperative, with Hill coefficients in the range 1.1 to 2.1.[18] In coelomic cells (the tissue between the inner membrane lining the digestive tract and the outer membrane of the worm—analogous to flesh in vertebrates), oxygen apparently binds with higher cooperativity ($n \sim 2.5$).[43] Perchlorate ions have been observed to induce cooperativity: since ClO_4^- has no biological role, it appears that in protein purifications the biological allosteric effector is lost. No Bohr effect occurs. Dioxygen binding data are accumulated in Table 4.2.[36,44]

The structure of hemerythrin in a variety of derivatives (oxy, azido, met, and deoxy) is now well-characterized. With three bridging ligands, a distinctive cofacial bioctahedral stereochemistry is seen (Figure 4.10).[45-48]

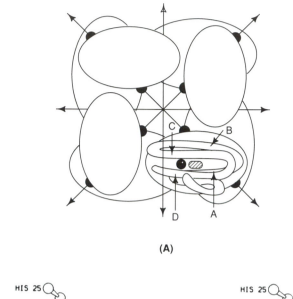

(A)

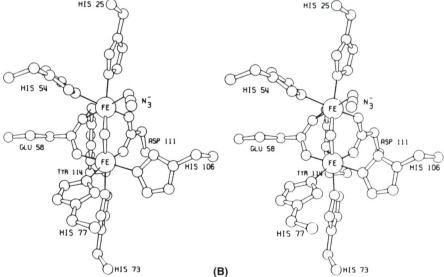

(B)

Figure 4.10
Structure of hemerythrin: (A) The tertiary structure of octameric hemerythrin[46b] with four α-helices (A, B, C, D) of one of the eight subunits. The filled half-circles denote anion binding sites (e.g., ClO_4^-); the filled circle the Fe_2 site; and the cross-hatched oval the N_3^- and SCN^- binding sites (Fe^{III})$_2$ and the O_2 binding sites (Fe^{II})$_2$. Reproduced with permission from R. E. Stenkamp, L. C. Sieker, and L. H. Jensen, *J. Mol. Biol.* **126** (1978), 457–466. (B) The structure of the active site of metazidomyohemerythrin,[48] showing the cofacial bioctahedral stereochemistry. The structure of oxyhemerythrin is very similar, including the orientations of the $(H)O_2^{II}$ ligand.[45] Reproduced with permission from S. Sheriff, W. A. Hendrickson, and J. L. Smith, *J. Mol. Biol.* **197** (1987), 273–296.

C. Hazards of Life with Dioxygen

The binding of dioxygen is normally a reversible process:

$$M + O_2 \rightleftharpoons MO_2 \qquad (4.22)$$

Under some circumstances, such as in the presence of added nucleophiles and protons, coordinated dioxygen is displaced as the superoxide anion radical, $O_2^{-\cdot}$, leaving the metal center oxidized by one electron and unreactive to dioxygen:[49,50]

$$MO_2 \rightleftharpoons M^+ + O_2^{-\cdot} \qquad (4.23)$$

For hemoglobin there exists a flavoprotein reductase system, comprising a reduced pyridine nucleotide (e.g., NADH), cytochrome b_5 reductase, and cytochrome b_5, that reduces the ferric iron back to the ferrous state, so that it may coordinate dioxygen again.[1,51] In addition, all aerobically respiring organisms and many air-tolerant anaerobes contain a protein, superoxide dismutase, that very efficiently catalyzes the dismutation of superoxide ion to dioxygen and hydrogen peroxide:[52–54]

$$2O_2^{-\cdot} + 2H^+ \rightarrow O_2 + H_2O_2 \qquad (4.24)$$

However, the physiological effects of the superoxide moiety remain controversial.[53,54] Finally, there is a third enzyme, the hemoprotein catalase, that converts the toxic hydrogen peroxide into water and dioxygen:[1]

$$2H_2O_2 \rightarrow O_2 + 2H_2O \qquad (4.25)$$

This topic is discussed further in Chapter 5.

II. SELECTED CHEMISTRY OF DIOXYGEN, IRON, COPPER, AND COBALT

Dioxygen is a powerful oxidant, capable of oxidizing all but the noble metals and of converting many low-valent metal complexes to higher-valent states. As will be detailed in this section, the binding of dioxygen to metals is most usefully considered as an oxidative addition process. The nature of the interaction is determined by the metal, its oxidation state, and its ligands that modulate the redox properties of the metal center. In biological and nonbiological oxygen carriers, several factors allow reversible binding of O_2 to occur, even though this process is metastable with respect to (irreversible) oxidation of the metal, or its ligands, or other species that may be present. Later in this section the bioinorganic chemistry of iron, copper, and cobalt is described. For a wider perspective on the coordination chemistry of these metals, see comprehensive texts on inorganic chemistry.[56–58]

Many techniques have been used to probe the metal-dioxygen moiety. A summary of these techniques, key concepts, and results is presented in Table 4.3.[59-61] UV-visible spectroscopy usually characterizes the oxidation state of the metal and in favorable cases the number, geometry, and ligand field strength of ligands. The O—O and M—O stretching modes may be investigated with infrared spectroscopy, provided that the complex is not a centrosymmetric dimer, for then the O—O stretch for the μ-dioxygen species is infrared-inactive. Resonance Raman techniques complement infrared spectroscopy. Not only are the selection rules different in Raman spectroscopy, but a suitable choice of the irradiating wavelength (to coincide approximately with an M-L electronic transition) can amplify those vibrational modes that are coupled, or in resonance, with the electronic transition. This technique is particularly suited as a probe of the metal-ligand environment of metalloproteins, since the many solely protein vibrational modes disappear into background noise. Geometric information on the orientation of the CO moiety with respect to the heme normal has been obtained by examining polarization behavior of infrared bands following photolysis of the Fe—CO bond by linearly polarized light.

Spin and oxidation states of mononuclear iron-porphyrin systems may be assigned directly from magnetic susceptibility measurements and indirectly from Mössbauer spectroscopy. Variable temperature susceptibility measurements are particularly useful for detecting dinuclear systems that share at least one ligand in common if there is antiferromagnetic (or ferromagnetic) coupling of the electron spin of one metal center with that of a second.

Definitive characterization of the stereochemistry is usually provided by x-ray diffraction data when single crystals are available. In general, the level of resolution and precision available from protein crystal structures leads to tantalizing uncertainties over the geometry of the M—O_2 species and of the structural changes occurring on oxygenation that are the origin of cooperativity. Precise structural data are more readily obtained from small-molecule model systems. The relevance of these to biological systems is established through congruence of spectroscopic and functional properties. X-ray diffraction techniques also provide important information on the environment beyond the immediate surroundings of the metal center: this information is usually unobtainable from other techniques, although recent developments in two-dimensional NMR spectroscopy can provide this information for diamagnetic systems. Limited information may be obtained with the use of spin labels or, if the metal center is paramagnetic, with EPR techniques.

Two other techniques that selectively probe the immediate environment of the metal center are EXAFS (Extended X-ray Absorption Fine Structure)[60] and XANES (X-ray Absorption Near-Edge Structure).[61] The former may yield information on the number and type of bonded atoms and their radial separation from the metal center. The latter technique may reveal the oxidation state and, in principle, may yield geometric information, although in its present state of development some interpretations are contentious. Both techniques have the ad-

Table 4.3
Techniques used to probe the active sites of oxygen carriers.

Technique	Abbrev.	Description of technique	Description of results
Nuclear magnetic resonance	NMR	Quantized orientation of nuclear spin in a magnetic field. Energy separations sampled with radio-frequency radiation.	Identification of histidine by deuterium exchange (N—H vs. N—D) at or near metal, especially if paramagnetic.
Electron paramagnetic resonance	EPR	Quantized orientation of electron spin in a magnetic field. Energy separations sampled with X- or Q-band microwave radiation.	Location of unpaired electron density from hyperfine splitting by metals or atoms with nuclear spin.
Magnetic susceptibility		Strength of interaction of sample with magnetic field. Solid state or solution state by Evans' NMR method.	Identification of spin state, spin-equilibria, and spin coupling (ferro- or antiferromagnetic); identification of Fe^{III}—O—Fe^{III} moiety.
Infrared spectroscopy	IR	Vibrational modes involving change in dipole moment.	Classification of O—O moiety (superoxo vs. peroxo). Identification of ν(M—O) and ν(M—O—M) modes, etc.
Raman and resonance Raman	R, RR	Vibrational modes involving a change in polarizability. For RR enhancement of modes coupled with electronic transition excited by laser light source.	Complementary to IR. ν(O—O) and ν(M—O) especially in metalloproteins. In porphyrins, oxidation and spin state.
UV-visible spectroscopy	UV-Vis	Valence electron transitions.	Electronic state of metal from d-d transitions. Identification of unusual ligands, e.g., Cu(II)—SR, Fe^{III}—OPh, Fe^{III}—O—Fe^{III}. Single crystals and polarized light give geometrical information.
X-ray photoelectron spectroscopy	XPS (ESCA)	Inner-shell electron transitions.	Oxidation state of metal.
Mössbauer spectroscopy		Excitation of nuclear spin by γ rays.	Oxidation and spin state. Antiferromagnetic coupling (Fe only).
X-ray single-crystal diffraction		Fourier transform of diffraction data reveals location of electron density.	Precise three-dimensional structure, bond distances and angles for small molecules. Lower resolution and precision for proteins.
Extended x-ray absorption fine structure	EXAFS	Backscattering of x-rays produces interference fringes on absorption curve at energies just greater than metal absorption edge ($K\beta$ transition)	Number, type, and radial distance of ligand donor atoms bonded to the metal.
X-ray absorption near-edge structure	XANES	Similar to EXAFS except that absorption is monitored at energies near and below the absorption edge.	As for EXAFS. May give geometric information.

vantage of not requiring crystalline material. The structural information is more reliable if definitive model systems are available for comparison.

X-ray (and, less frequently, neutron) diffraction techniques on single crystals give absolute structural information* and thus provide the basis for interpretation of data obtained from these other techniques that yield relative structural information.

A. General Aspects of the Chemistry of Dioxygen

1. Redox chemistry of free molecular dioxygen

Dioxygen has a rich redox chemistry that is not explicitly exploited in the oxygen carriers, but which is central to enzymes that coordinate and activate dioxygen for subsequent reaction with a substrate. On reduction of dioxygen by one electron, the superoxide anion radical $O_2^{-\cdot}$ is formed. Concomitant with a reduction in bond order from 2.0 to 1.5 is an increase in bond length from 1.21 to 1.30 Å. A second reduction step produces the peroxide anion O_2^{2-}; the bond order is one, and the O—O separation is 1.49 Å. Each of these reduced species, $O_2^{-\cdot}$ and O_2^{2-}, has a characteristic O—O stretching vibration in the infrared region. The free-energy changes and electrochemical potentials for the reduction of dioxygen at unit activity, pH = 1 ($E°$), are different from those at pH 7.0 ($E°'$), as shown in Figure 4.11.[58,62] The values at pH 7.0 are more relevant to physiological conditions. Note that the superoxide anion may function as either an oxidant or a reductant.

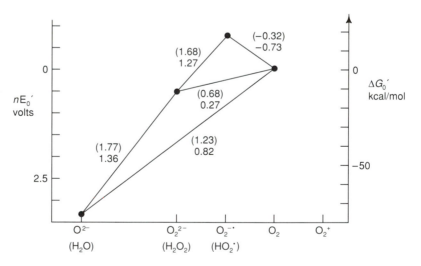

Figure 4.11
Free-energy changes in the aqueous redox chemistry of dioxygen. Standard state $P(O_2) = 1$. Electrode potentials are at pH 7; those in parentheses are at unit activity.

---

* In favorable situations, sophisticated NMR techniques have been applied successfully to determine the polypeptide folding (e.g., in metallothionein).[55]

2. Geometry and electronic structure of coordinated dioxygen

In coordinating to metals, dioxygen shows a great variety of geometries and two formal oxidation states. Many complexes have $\nu(O—O)$ values in the range 740 to 930 cm^{-1}, and, where known, an O—O separation in the range 1.40 to 1.50 Å. By analogy with the peroxide anion, these species are designated peroxo, O_2^{II-}. Similarly, the designation superoxo O_2^{I-} is applied to those complexes where $\nu(O—O)$ values are in the range 1075 to 1200 cm^{-1}, and the O—O separation is around 1.30 Å.[63] Although such O—O separations and vibrations are consistent with coordinated peroxide or superoxide moieties, the net amount of charge transferred onto the dioxygen ligand from the metal and its other ligands is difficult to measure experimentally and is probably variable. Thus the oxidation state of the dioxygen ligand and that of the metal are best considered in a formal sense rather than literally—hence the use of the terminology O_2^{I-} to indicate oxidation state I— for the O_2 moiety as a unit (not each O atom). Because of the high degree of covalency in the M—O bond, a more sensible comparison, at least for the peroxo class of compounds, is with organic peroxides, ROOH or ROOR. The clear separation of coordinated dioxygen into either the superoxo or the peroxo class is shown in Figure 4.12.[63-66] *Only* those compounds for which both stretching frequencies ($\nu(O—O)$) and O—O separations ($r(O—O)$) are available are shown; for the purpose of the plot, noncoordinated anions and cations, replacement of ethylenendiamine by two ammonia ligands, and replacement of triphenylphosphine by alkylphenylphosphines are assumed not to perturb significantly $\nu(O—O)$ or $r(O—O)$.

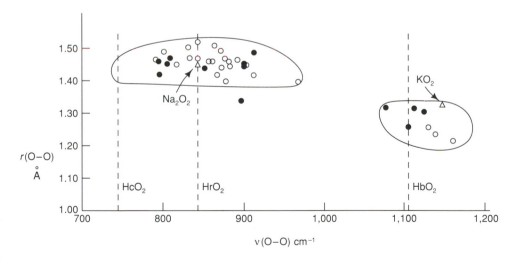

Figure 4.12
Scatter diagram showing the distribution of O—O stretching frequencies and separations in ionic superoxides and peroxides (△) and in coordination compounds. An open circle denotes O_2 coordinated to one metal; a filled circle denotes O_2 bridging two metals. The O—O stretching frequencies of oxyhemoglobin, oxyhemocyanin, and oxyhemerythrin are marked by dashed lines.

196

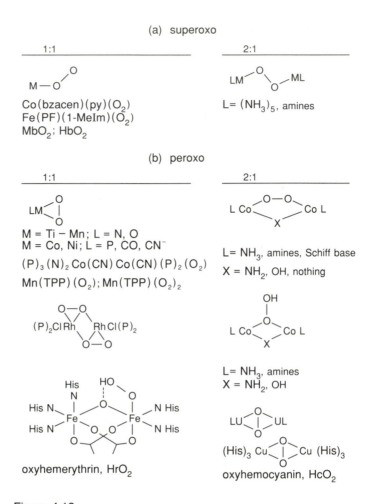

(a) superoxo

<u>1:1</u>

M—O / O

Co(bzacen)(py)(O₂)
Fe(PF)(1-MeIm)(O₂)
MbO₂; HbO₂

<u>2:1</u>

LM—O\O—ML

L= (NH₃)₅, amines

(b) peroxo

<u>1:1</u>

LM< O | O

M = Ti – Mn; L = N, O
M = Co, Ni; L = P, CO, CN⁻

(P)₃(N)₂Co(CN) Co(CN)(P)₂(O₂)

Mn(TPP)(O₂); Mn(TPP)(O₂)₂

O—O
(P)₂ClRh   RhCl(P)₂
O—O

oxyhemerythrin, HrO₂

<u>2:1</u>

L Co O—O Co L
X

L= NH₃, amines, Schiff base
X = NH₂, OH, nothing

OH
|
L Co O Co L
X

L= NH₃, amines
X = NH₂, OH

LU O | O UL

(His)₃ Cu O | O Cu (His)₃

oxyhemocyanin, HcO₂

Figure 4.13
Modes of attachment of O₂ to metals. P, N, and O denote phosphine, amine, and oxygen ligands, respectively.

At least seven different geometries have been observed for the coordination of dioxygen (Figure 4.13),[63–66] only three or four of which are currently known to be biologically relevant—the superoxo M—O (for oxyhemoglobin), the peroxo M—O...M or M (for oxyhemocyanin), and the hydroperoxo M—O—M OOH (for oxyhemerythrin). The geometry is a function of the metal, its oxidation state, and its ligands. For the late transition metals of the cobalt and nickel triads, with soft π-acid ligands, such as phosphines and carbonyls, and with an initially low oxidation state of the metal, triangular coordination of a peroxo species with covalent M—O₂ bonds is common.[63] Concomitant with the formal reduction of dioxygen, the metal center undergoes a formal two-electron oxidation:

$$M^n + O_2 \longrightarrow M^{n+II} \underset{O}{\overset{O}{\diagdown}} {}^{II-} \tag{4.26}$$

In this example, where the metal has undergone, at least formally, a two-electron oxidation, the UV-visible properties of the metal-dioxygen complex tend to resemble those of bona fide M^{n+II} rather than M^n species.

Early transition metals (Ti, V, Cr triads) often coordinate several peroxo species, leaving the metal in formally a very high oxidation state (e.g., $Cr(O_2)_4{}^{3-}$, a Cr^V ion).[63] The M—O_2 links have more ionic character, with the η-peroxo groups acting as bidentate ligands. Titanium and molybdenum(II) porphyrins bind, respectively, one and two dioxygen molecules in this manner.[65]

With harder σ-donor ligand systems, such as those containing nitrogen and oxygen donors, and the metal center in a normal oxidation state, a formal one-electron reduction to an end-on coordinated superoxo species occurs with a bent M—$O{\overset{O}{\diagup}}$ bond. Metal-dioxygen species can also be formed by adding the superoxide anion to the oxidized species:[64]

$$M^n + O_2 \longrightarrow M^{n+I} - O{\overset{O^{I-}}{\diagup}} \longleftarrow M^{n+I} + O_2{}^{-\cdot} \tag{4.27}$$

In the absence of steric constraints, dimerization to a (bridging) μ-peroxo species frequently occurs, especially for cobalt-dioxygen complexes:

$$Co^{III} - O{\overset{O^{I-}}{\diagup}} + Co^{II} \longrightarrow Co^{III} - O{\overset{O-Co^{III}}{\diagup}} {}^{II-} \tag{4.28}$$

There are several neutral dicobalt species formulated as μ-peroxo systems that contain Schiff-base ligands, and for which the O—O separation is anomalously short (e.g., 1.31(2) and 1.338(6) Å).[66] Few infrared or Raman data for $\nu(O—O)$ are available to check whether these compounds, as a result of their delocalized Co-Schiff base system and their neutrality, fall between the superoxo and peroxo classes. The isolated circle in Figure 4.12 is one such example.

These dicobalt species (right-hand side of Equation 4.28) may be oxidized by one electron to give a μ-superoxo moiety. A clear shortening of the O—O bond and concomitant increase in the value of $\nu(O—O)$ are observed in several superoxo-peroxo pairs. These and other modes of O_2 attachment are illustrated in Figure 4.13. Some geometries are represented by only one or two examples, and some geometries, for example, a linear M—O—O species, have never been observed.

In binding to metals, O_2 effectively functions both as a π acid, accepting into its π^* orbitals electron density from the filled d orbitals of the metal, and as a σ donor, donating electron density into an empty metal d orbital. Thus other σ donor or π acceptor ligands, such as nitric oxide (NO), alkyl isocyanides (R—NC), alkyl nitroso (R—NO), and carbon monoxide (CO), are often observed to bind to the same metal complexes that bind O_2. The nature of the metal-dioxygen linkage in biological oxygen carriers and their models will be examined in more detail later.

B. General Aspects of the Chemistry of Iron

1. Irreversible oxidation

In the presence of dioxygen, iron(II) species are readily oxidized to iron(III) species. In the presence of water, iron(III) species frequently associate into μ-oxodiiron(III) dimers. For iron(II)-porphyrin complexes this process may take only milliseconds at room temperature. The following mechanism was proposed in 1968 for the irreversible oxidation of iron(II)-porphyrinato species;[67,68] subsequent work has largely confirmed it.[69–71]

$$Fe^{II} + O_2 \rightleftarrows Fe^{III}-O_2^{I-} \tag{4.29a}$$

$$Fe^{III}-O_2^{I-} + Fe^{II} \rightleftarrows Fe^{III}-O_2^{II-}-Fe^{III} \tag{4.29b}$$

$$Fe^{III}-O_2^{II-}-Fe^{III} \longrightarrow 2Fe^{IV}=O \tag{4.29c}$$

$$Fe^{IV}=O + Fe^{II} \longrightarrow Fe^{III}-O-Fe^{III} \tag{4.29d}$$

In particular, the dimerization reaction (4.29b) may be rendered less favorable by low temperatures ($< -40°C$) or by sterically preventing the bimolecular contact of an $Fe^{III}-O_2^{I-}$ moiety with an Fe^{II} moiety. In the latter case, sterically bulky substituents on the equatorial ligand surround the coordinated O_2 ligand and the other axial position, trans to the coordinated dioxygen ligand, is protected with a nitrogenous base, such as imidazole, or with additional bulky substituents on the equatorial ligand (Figure 4.14).[72] The protein effectively provides such protection and thus plays a key role in preventing the bimolecular contact of two hemes. The first observation of reversible binding of dioxygen to an iron(II)-porphyrin in the absence of protein was made in 1958.[73] In that pioneering study, a heme group was immobilized on a polymer support specially modified to contain imidazole functions. The structurally characterizable hemoglobin or myoglobin species was replaced by a noncrystalline structurally uncharacterized polymer.

Why does this irreversible oxidation not occur analogously for cobalt systems? Step (4.29c) involves cleavage of the O—O bond, which in H_2O_2 has a bond energy of 34.3 kcal/mol or in Na_2O_2 of 48.4 kcal/mol. By way of comparison, for O_2 the bond energy is 117.2 and for $HO_2\cdot$ it is 55.5 kcal/mol.[64] A simple molecular orbital picture gives insight into why an $Fe^{IV}=O$ species is stabilized relative to the analogous $Co^{IV}=O$ species.[74] From Figure 4.15 we see that for metals with electronic configuration d^n, where $n \leq 5$, no electrons

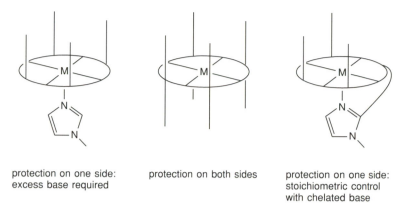

protection on one side: excess base required

protection on both sides

protection on one side: stoichiometric control with chelated base

Figure 4.14
Stylized representation of steric hindrances preventing irreversible oxidation.

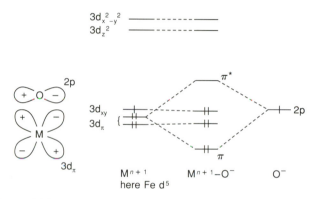

Figure 4.15
Orbital scheme showing the differing stabilities of M—O species, M = Co(d^6), Fe(d^5); σ-bonding of 2p$_z$ with 3d$_{z^2}$ not shown.[74]

occupy the antibonding orbital π^* for Fe^{III}—$O^{1-}\cdot$ or Fe^{IV}=O moieties. For Co^{III} (d^6) the extra electron goes into the antibonding orbital π^*. As predicted by the model, Mn^{III} is observed indeed to behave like Fe^{III}.

A second oxidation pathway does not require the bimolecular contact of two iron(II)-porphyrins. Coordinated dioxygen may be released not as O_2, as in normal dioxygen transport, but, as noted in Section I.C, as a superoxide radical anion $O_2^{-}\cdot$ in a process called *autoxidation*:

$$Fe^{III}—O_2^{1-} \rightleftharpoons Fe^{III} + O_2^{-}\cdot \tag{4.30}$$

This process is assisted by the presence of other nucleophiles that are stronger than the superoxide anion, such as chloride, and by protons that stabilize the $O_2^{-}\cdot$ anion as $HO_2^{\bullet}$:

$$Fe^{III} + Cl^- \rightleftharpoons Fe^{III}—Cl^- \tag{4.31}$$

The formation of methemoglobin occurs *in vivo*, probably by the above mechanism, at the rate of $\sim$ 3 percent of total hemoglobin per day.

If exogenous reductants are present, then further reduction of dioxygen can occur:

$$2H^+ + Fe^{III}-O_2^{I-} + e^- \rightarrow Fe^V=O + H_2O \qquad (4.32)$$

Such processes are important, for example, in the cytochrome P-450 system. With suitably small reductants, oxygenase activity also has been observed for hemoglobin A. This has led to the characterization of hemoglobin as a "frustrated oxidase."[75] Note the formal similarity between this process (Equation 4.32) and the bimolecular irreversible oxidation of iron(II) porphyrins: the second Fe(II) complex in Reaction (4.29b) functions like the electron in Reaction (4.32).

2. Spectroscopy of the $Fe^{III}-O-Fe^{III}$ moiety

The end products of the irreversible bimolecular oxidation of Fe^{II} species contain the $Fe^{III}-O-Fe^{III}$ fragment. Given the facile formation of μ-oxodiiron(III) species, it is not surprising that the Fe—O—Fe motif is incorporated into a variety of metalloproteins, including the oxygen-carrier hemerythrin (Figure 4.10),[16–18] the hydrolase purple acid phosphatase,[76] the oxidoreductases ribonucleotide reductase[77] and methane monooxygenase,[78] an iron-sulfur protein rubrerythrin,[79a] and the iron-transport protein ferritin.[79b] In ferritin higher-order oligomers are formed.

This μ-oxodiiron(III) moiety has a distinctive fingerprint that has made it easy to identify this motif in proteins.[80] Regardless of the number (4, 5, 6, or 7), geometry (tetrahedral, square pyramidal, tetragonally distorted octahedral, or pentagonal bipyramidal), and type of ligands (halide, RO^-, $RCOO^-$, aliphatic N, or aromatic N) around the iron center, and of the Fe—O—Fe angle, the magnetic susceptibility at room temperature lies in the range 1.5 to 2.0 Bohr magnetons per $Fe^{III}-O-Fe^{III}$ group, equivalent to about one unpaired electron.[81,82] In other words, the high-spin ($S = \frac{5}{2}$) iron centers are strongly antiferromagnetically coupled. Other bridging groups, such as OH^-, Cl^-, carboxylate, alkoxide, or phenoxide, give very weak coupling.[83–86]

The asymmetric Fe—O stretch, ν_{as}(Fe—O), lies in the range 730 to 880 cm^{-1}; in multiply bridged complexes this mode is weak in the infrared region. The symmetric vibration, ν_s(Fe—O), forbidden in the infrared region for linear, symmetric Fe—O—Fe groups, occurs in the range 360 to 545 cm^{-1}. The symmetric mode is usually,[87a] but not always,[87b] observed by resonance Raman techniques upon irradiating on the low-energy side of the Fe—O charge-transfer band that occurs at about 350 nm.

Few dinuclear iron(II) complexes are known where the ligands approximately resemble those believed or known to occur in the family of μ-oxodiiron(III) proteins.[88] The dioxygen-binding process in hemerythrin has no close nonbiological analogue. Although spectroscopically similar to oxyhemerythrin,

the unstable monomeric purple peroxo complex formed by the addition of hydrogen peroxide to basic aqueous Fe^{III}(EDTA) solutions remains structurally
uncharacterized.[89,90]

3. Oxidation and spin states of iron porphyrins

Iron porphyrins, the active sites of the hemoglobin family, have a rich mag-
netochemistry.[91] Iron porphyrins may be octahedral (two axial ligands), square
pyramidal (one axial ligand), or square planar (no axial ligand). The metal d
orbitals, now having partial porphyrin π^* character, are split, as shown in Figure 4.16. The radius of the metal atom is much greater when it is high spin
($S = 2$ for Fe^{II}, $S = \frac{5}{2}$ for Fe^{III}) than when it is low spin ($S = 0$ for Fe^{II},
$S = \frac{1}{2}$ for Fe^{III}). This difference influences $Fe—N_{porph}$ separations, porphyrin
conformation, and the displacement of the iron center with respect to the porphyrin plane. For iron(II)-porphyrins, two strong-field axial ligands, such as a
pair of imidazoles or an imidazole and carbon monoxide, lead to diamagnetic
complexes ($S = 0$) with the six 3d electrons occupying those orbitals of approximate t_{2g} symmetry. In a classic experiment in 1936, Pauling and Coryell
proved that oxyhemoglobin and carbonmonoxyhemoglobin are diamagnetic.[92]*

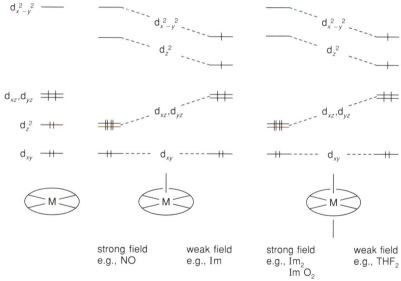

Figure 4.16
d-Orbital splitting in metalloporphyrins as a function of number and ligand-field strength
of ligands.[91] Orbital occupancy is illustrated for a d^6 species (Fe^{II} or Co^{III}).

---

* There was a considerable flurry of interest when an Italian group, using a SQUID (Superconducting
Quantum Mechanical Interference Device), reported that at room temperature oxyhemoglobin was significantly
paramagnetic.[93] Not surprisingly, several theoretical papers followed that "proved" the existence of low-lying
triplet and quintet excited states.[94–96] Subsequently, the residual paramagnetism was doubted[97] and shown to
arise from incomplete saturation of hemoglobin by O_2; in other words, small amounts of deoxy hemoglobin
remained.[98] Since oxygen affinity increases with decreased temperature, the concentration of paramagnetic
impurity decreased with decreasing temperature.

No axial ligands at all may lead to a spin state of $S = 1$, with unpaired electrons in the d_{xy} and d_{z^2}orbitals. Five-coordinate iron(II)-porphyrinato complexes are commonly high spin, $S = 2$, although strong σ-donor π-acceptor ligands, such as phosphines, carbon monoxide, nitric oxide, and benzyl isocyanide,[99] enforce a low-spin state. Five-coordinate iron(II)-porphyrinato complexes with aromatic nitrogenous axial ligands, such as pyridine or 1-methylimidazole, bind a second such axial ligand 10 to 30 times more avidly than the first to give the thermodynamically and kinetically (d^6, $S = 0$) stable hemochrome species, a process that is avoided by hemoglobins. That is, the equilibrium constant for the following disproportionation reaction is greater than unity,

$$\text{Fe—N} + \text{Fe—N} \rightleftharpoons \text{N—Fe—N} + \text{Fe} \tag{4.33}$$

except for bulky ligand N, such as 2-methylimidazole and 1,2-dimethylimidazole, for which the five-coordinate species predominates at room temperature even with a mild excess of ligand:[100]

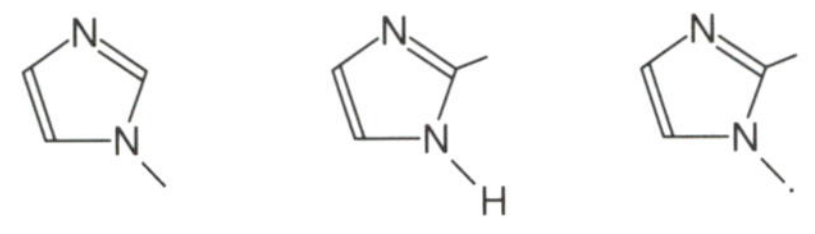

$$\tag{4.34}$$

1-MeIm   2-MeIm   1,2-Me$_2$Im

For iron(III)-porphyrinato complexes, strong-field ligands lead to low-spin ($S = \frac{1}{2}$) complexes. A pair of identical weak-field ligands, such as tetrahydrofuran, leads to intermediate-spin ($S = \frac{3}{2}$) species. Five-coordinate species are, with few exceptions, high-spin ($S = \frac{5}{2}$), with all five 3d electrons in separate orbitals. Spin equilibria $S = \frac{1}{2} \rightleftharpoons S = \frac{5}{2}$ and $S = \frac{3}{2} \rightleftharpoons S = \frac{5}{2}$ are not unusual. Specific examples of these spin systems are given in Table 4.4.[65,91] Higher oxidation states are found in some other hemoproteins. Fe(V)-porphyrin systems actually occur as Fe(IV)-porphyrin cation radical species, and Fe(I)-porphyrin systems exist as Fe(II)-porphyrin anion radical species.

Substantial structural changes occur upon the addition of ligands and upon changes in spin state. In one mechanism of cooperativity these changes are the "trigger" (metrical details are deferred until the next section). Spectral changes in the UV-visible region are observed also (Figure 4.17)[10] and may be monitored conveniently to evaluate the kinetic and thermodynamic parameters of ligand binding to hemoglobin.

Figure 4.17 *(facing page)*
Spectral changes accompanying the oxygenation and carbonylation of myoglobin.[10] Reproduced with permission from E. Antonini and M. Brunori, *Hemoglobin and Myoglobin in Their Reactions with Ligands*, North Holland, 1971.

Table 4.4
Oxidation and spin states of iron porphyrins and their biological occurrences.

State	Fe^{II}		Fe^{III}	
High spin	Fe(PF)(2-MeIm)	} 5-coord	Fe(TPP)Cl	} 5-coord
	Hb		Cytochrome P-450(ox)	
Fe^{II} $S = 2(d^6)$	Fe (TPP)(THF)$_2$	} 6-coord	$[Fe(TPP)(H_2O)_2]^+$	
Fe^{III} $S = \frac{5}{2}(d^5)$	Hb(H$_2$O)?		$[Fe(OEP)(3\text{-}ClPy)_2]^+$ (293 K)	} 6-coord
			Fe(OEP)(Py)(NCS)	
			MetHb(H$_2$O)	
Intermediate spin	Fe(TPP)	} 4-coord	$[Fe(TPP)(C(CN)_3)]_n$	
Fe^{II} $S = 1$ (d^6)	no biol. occurrence		Fe(TPP)(OClO$_3$) $(S = \frac{3}{2}, \frac{5}{2})$	
Fe^{III} $S = \frac{3}{2}(d^5)$			cytochrome c'	
Low spin	Fe(TPP)(NO)	}^{a} 5-coord	$[Fe(TPP)(Ph)]$	5-coord
Fe^{II} $S = 0(d^6)$	Hb(NO)·DPG(T)			
Fe^{III} $S = \frac{1}{2}(d^5)$	Fe(PF)(2-MeIm)(O$_2$)^b		$[Fe(TPP)(Im)_2]^+$	
$(S = \frac{1}{2}$, Fe—NO adducts)	Fe(TPP)(1-MeIm)(NO)		$[Fe(OEP)(3\text{-}Cl\text{-}Py)_2]^+$ (98 K)	
	Fe(TPP)(Py)(CO)		Fe(TPP)(Py)(CN)	} 6-coord
	Fe(TPP)(1-MeIm)$_2$	} 6-coord	Fe(TPP)(Py)(NCS)^c	
	Hb(CO), Hb(NO),^{a} Hb(O$_2$)^b		cytochrome b_5, c, c_3	
	cytochrome b_5, c, c_3		metHb (CN)	

^{a} Could be placed in Fe^{III} column.
^{b} Could be placed in Fe^{III} column with spin = 0.
^{c} Non-linear Fe—NCS moiety.

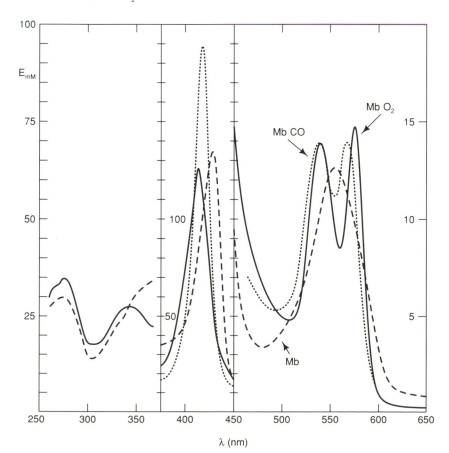

C. General Aspects of the Chemistry of Copper

The chemistry of copper in biological systems is limited to oxidation states I and II. The Cu^I state has electronic configuration d^{10}. Unless there are ligand bands or strong ligand-to-copper charge-transfer bands, diamagnetic Cu^I species are colorless. Complexes of Cu^{II} (d^9) are often blue in color. The single unpaired electron makes Cu^{II} amenable to electron paramagnetic resonance (EPR) techniques, at least if the electron spins of Cu^{II} centers are independent of one another. In oxyhemocyanin the spins are so strongly coupled ($-J > 600$ cm^{-1}) that at room temperature and below the system is effectively diamagnetic and the pair of Cu^{II} ions is EPR silent.[14]

In aqueous solutions the Cu^I ion is unstable with respect to disproportionation to Cu metal and Cu^{II} ion:[62]

$$Cu^{II+} + 2e^- \rightleftharpoons Cu \qquad E° = 0.3402 \text{ V}$$
$$Cu^+ + e^- \rightleftharpoons Cu \qquad E° = 0.522 \text{ V} \qquad (4.35)$$
$$2Cu^+ \rightleftharpoons Cu + Cu^{II+} \qquad E° = 0.182 \text{ V}$$

The Cu^I state may be stabilized by ligands, especially sulfur-containing ones, or by immobilization as afforded by a protein matrix, or in nonaqueous solvents, such as acetonitrile, in the absence of dioxygen. Whereas Cu^I thiolate species are stable, Cu^{II} thiolate species usually are unstable with respect to the disproportionation:[101]

$$2Cu^{II}—SR \longrightarrow 2Cu^I + R—S—S—R \qquad (4.36)$$

Again, immobilization may give kinetic stability to Cu^{II} thiolate species, as occurs in the blue-copper family of electron-transport proteins.

Copper(I) complexes are often two-coordinate with a linear arrangement of ligands. Three-, four-, and possibly five-coordinate complexes are known.

In the presence of O_2, nonbiological copper(I) [and iron(II)] complexes are often susceptible to ligand degradation, which may give the illusion of O_2 binding.[102] The mechanisms by which this reaction occurs remain essentially unknown. Iron-porphyrin systems are rather more robust. Nonetheless, there are now several well-characterized copper(I) systems that reversibly bind dioxygen,[15b,103] at least at low temperature. One that has been structurally characterized features a $Cu\diagdown^{O}\diagdown_{O}\diagup^{Cu}$ dicopper(II)-peroxo moiety,[103f] while a second, with more properties in common with oxyhemocyanin, features a

$Cu\diagup\diagdown^{O}_{O}\diagdown\diagup Cu$ moiety.[103g]

D. General Aspects of the Chemistry of Cobalt

Many parallels exist between the chemistry of Fe^{II}- and Co^{II}-porphyrinato systems. Dioxygen binds to many Co^{II} complexes to give mononuclear 1:1 $Co:O_2$ complexes with a bent geometry

$$Co-O \diagup^O \diagup \qquad (4.37)$$

and dinuclear 2:1 $Co:O_2$ complexes,[64,66] analogous to those described for Fe^{II} systems in Reactions (4.29a) and (4.29b). Indeed, these dinuclear systems were the first nonbiological oxygen carriers to be isolated. The geometry of the dioxygen moiety, spanning two metals, may be *cis* or *trans*:

$$\diagup^{O-O}\diagdown \qquad \diagup^{O}\diagdown\diagup^{Co} \qquad (4.38)$$
$$Co \qquad Co \qquad Co \qquad O$$

However, whereas these dinuclear cobalt species are invariably octahedral, dinuclear copper-peroxo species are tetrahedral or distorted square pyramidal.[40,41]

In the late 1960s, 1:1 $Co:O_2$ species were first isolated by use of a combination of low temperatures and specific Schiff-base ligands.[104] It was found that cobalt corrins, such as vitamin B_{12r}, also formed 1:1 dioxygen adducts,[105] although this chemistry is not known to be utilized by living systems.[103] Cobalt(II) porphyrins also form 1:1 adducts but with low O_2 affinity, especially in nonpolar, aprotic solvents. Thus hemoglobin and myoglobin may be reconstituted from a cobaltoheme with preservation not only of dioxygen-binding capabilities but also of cooperativity.[106] The synthetic 1:1 $Co:O_2$ complexes have proven to be very useful in increasing our understanding of factors that determine oxygen affinity for cobalt systems and by extrapolation for iron systems. Two important differences make Co^{II} systems more accessible. First, in contrast to iron systems, the cleavage reaction (4.29c) and redimerization to a μ-oxo species (Reaction 4.29d) do not occur (see Figure 4.15). Thus Co^{II} complexes of O_2 are stable in solution at room temperature without the need for protection illustrated in Figure 4.14. Second, for Co^{II}-porphyrinato systems, the equilibrium constant for the addition of a second axial base, such as pyridine or 1-methylimidazole, is small. Thus the disproportionation to four-coordinate and six-coordinate species that occurs for corresponding Fe^{II} systems (Reaction 4.33) does not occur. This difference simplifies the interpretation of spectral changes that are used to obtain thermodynamic and kinetic parameters of which there are now voluminous examples.[66]

Moreover, the 1:1 $Co-O_2$ complexes are paramagnetic. From the small ^{59}Co hyperfine splitting, it is deduced that the single unpaired electron resides primarily on the dioxygen moiety.[104a,105] From other experiments[107] it is apparent that net transfer of electron density from the metal onto the dioxygen varies considerably, from about $0.1e^-$ to about $0.8e^-$. For example, it is found for a given Co^{II} Schiff base, Co(bzacen), that the redox potential of the cobalt-Schiff-base center LCo, measured by cyclic voltammetry, $E_{1/2}$,

$$B_2LCo^{III} + e^- \rightleftharpoons B_2LCo^{II} \qquad (4.39)$$

B = substituted pyridine

is a linear function of log $K(O_2)$ as the axial base B is varied. The more easily the Co^{II} center may be oxidized, the higher is the O_2 affinity,[103] as illustrated in Figure 4.18A. The dioxygen affinity also increases as the basicity of the axial nitrogenous ligand increases.[104a] This effect is illustrated in Figure 4.18B. Because of differing steric requirements, dimethylformamide (DMF), substituted imidazole, and piperidine (pip) ligands do not fall on the correlation defined by the series of substituted pyridine species. Note the synergistic nature of dioxygen binding: in general, the more electron density that is pumped onto the metal by the axial base, the more electron density is available for donation into the π^* orbitals of the dioxygen ligand. $E_{1/2}$ and log $K(O_2)$ are also correlated, although more weakly, for a number of hemoglobins (Figure 4.18C).[108] Here the porphyrin and axial base remain constant, but presumably the surroundings of the heme group and O_2 binding site vary in a manner that is less well-defined than in the model systems of Figure 4.18A and B. Notwithstanding these various perturbations to the metal center, the O—O stretch occurs at about 1140 cm^{-1}, placing all 1:1 cobalt and iron-dioxygen complexes of nitrogenous and other hard ligands into the superoxo class.*

Cobalt(II) porphyrins and their adducts with diamagnetic molecules invariably have spin $S = \frac{1}{2}$. (See Figure 4.16, but add one electron.) Thus the structural changes are less pronounced than for corresponding iron(II) systems.[110,111] From the similarities in geometries and differences in electronic structures between cobalt- substituted and native hemoglobins and their models, many insights have been gained about the factors that determine oxygen affinity as well

---

* Because the O—O stretch may be coupled with other ligand modes,[109] its value should not be used to estimate superoxo character, although in a series of μ-superoxo and μ-peroxo complexes of carefully controlled stereochemistry, small changes in ν(O—O) have been correlated with the pK_a of the suite of ligands.[66]

Figure 4.18 *(facing page)*
Linear free-energy relationships: (A) Correlation of the O_2 affinity at $-21°C$ of Co(bzacen)L with the $Co^{III} \rightleftharpoons Co^{II}$ cyclic voltammetric wave of Co(bzacen)L$_2$ species.[104a] (B) Correlation of ligand affinity with pK_a. Squares (□) pertain to the binding of L to Co(PPIX) at 23°C. Circles (○) pertain to the O_2 affinity at $-45°C$ of Co(PPIX)L species. Filled shapes pertain to substituted pyridines; the least-squares lines shown are calculated from these data only.[104a] (C) Correlation of $E°'$ (from cyclic voltammetry) with affinity for a miscellany of hemoglobins:[108] 1, *Aphysia limacina* Mb; 2, *Physeter macrocephalus* Mb; 3, *Candida mycoderma* Hb; 4, *Chironomus thummi* Hb; 5, Hb$_M$-Hyde Park; 6, Hb$_M$-Iwate; 7, Hb α chain; 8, Hb β chain; 9, Hb F; 10, Hb γ chain; 11, HbA; 12, *Glycera dibranchiata*; 13, legHb-a; 14, legHb-c. Reproduced with permission from A. W. Addison and S. Burman, *Biochim. Biophys. Acta* **828** (1985), 362–368.

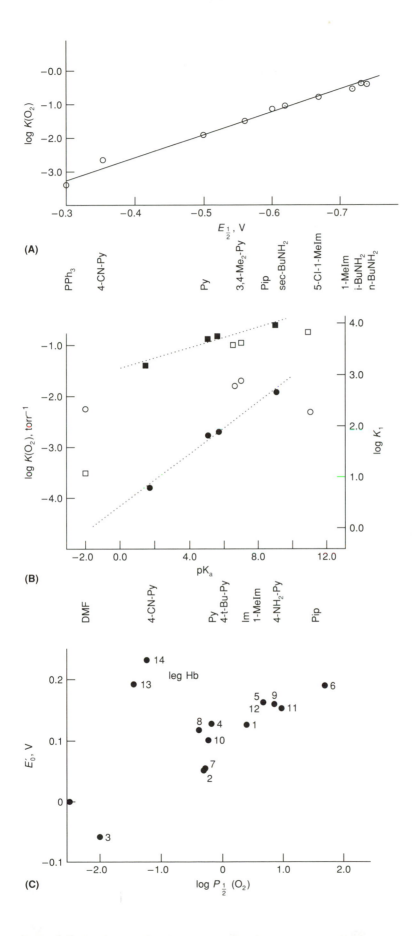

as how cooperativity might, or might not, work at the molecular level.[112,113] The mechanism of cooperativity has also been probed by the substitution of other metalloporphyrins into the globin: for example, zinc porphyrins have been used for their excited triplet-state properties,[114] manganese porphyrins for their EPR activity,[115] and ruthenium porphyrins as a member of the iron triad.[116]

E. Other Ligands for Biological Oxygen Carriers

As noted above, a variety of other σ-donor or π-acceptor ligands will bind to the active sites of biological oxygen carriers.

1. Carbon monoxide

As documented in Table 4.2, carbon monoxide (CO) generally binds more strongly to hemoglobin than does dioxygen, hence causing carbon-monoxide poisoning. In addition to being readily available from car exhausts and tobacco smoke to convert oxyhemoglobin to carbonmonoxyhemoglobin, CO is produced in the catabolism of heme molecules.[117] Thus under even the most favorable of conditions, about 3 percent of human hemoglobin is in the carbonmonoxy form. When CO binds to a single metal atom in nonbiological systems, *without exception* it does so through the carbon atom and in a linear manner:[56]

$$\text{Fe—C}\equiv\text{O} \longleftarrow \text{Fe}^{\text{II}} + {}^{-}\text{C}\equiv\text{O}^{+} \tag{4.40}$$

Model systems for carbonmonoxy (also called carbonyl) hemoglobin show a geometry similar to that of the Fe—C≡O group, linear or nearly so and essentially perpendicular to the porphyrin plane.[110,118–121] The biochemical literature is littered with reports that this is *not* the geometry adopted by CO in binding to hemoglobins.[122–128] We will return to this topic later in this chapter, since the physiological consequences are potentially important.

Carbon monoxide binds weakly as a σ-donor ligand to four-coordinate cobalt(II) systems.[129] Despite a bout of artifactual excitement,[130] CO has never been observed to bind significantly to five-coordinate Co^{II} systems with a nitrogenous axial base to yield octahedral six-coordinate species.[131] The sulfur analogue thiocarbonyl (CS), although not stable as a free entity, binds very strongly to iron-porphyrin species in a linear manner.[132]

2. Nitric oxide

Nitric oxide (NO) binds to hemes even more strongly than CO (and hence O_2),[10] so strongly, in fact, that the Fe—N_{Im} bond is very weak and easily ruptured.[11,111,133] Attachment to the metal is via the nitrogen atom; however, the geometry of attachment is sensitive to the π basicity of the metalloporphyrin, and ranges from linear to strongly bent. In binding to Co^{II} the NO ligand is effectively reduced to NO^{-}, with concomitant oxidation of Co^{II} to Co^{III}:[111]

$$
\begin{array}{cccc}
& \underset{\substack{\| \\ N \\ | \\ N \equiv O \cdot}}{O} & \underset{\substack{N^{140} \\ | \\ Mn}}{\overset{O}{\diagup}} & \underset{\substack{N^{120} \\ | \\ Co}}{\overset{O}{\diagup}} \\
v(N-O): & 1860\ cm^{-1} & 1740\ cm^{-1}\ \ 1625\ cm^{-1} & 1689\ cm^{-1}
\end{array}
\tag{4.41}
$$

In much the same way that cobalt-dioxygen systems are paramagnetic ($S = \frac{1}{2}$) and amenable to EPR studies, iron-nitric oxide (also called iron nitrosyl) species are also paramagnetic and isoelectronic with cobalt-dioxygen species. The unpaired spin is localized mostly on the NO group.

3. Isocyanide and nitroso species

In contrast to the dioxygen, carbon-monoxide, and nitric-oxide ligands, the isocyanide and nitroso functions bear an organic tail. Moreover, nitroso ligands are isoelectronic with dioxygen.

$$
R-N\equiv C \quad R = \text{alkyl or aryl}, \quad R-N\diagdown_O \quad R = \text{aryl, alkyl} \tag{4.42}
$$

Thus, in principle, not only may the steric bulk of the ligand be varied, in order to probe the dimensions[35] of the dioxygen-binding pocket,* but also the σ-donor/π-acceptor properties of the ligands may be varied by appropriate substituents on the aryl ring.

Isocyanide groups may bind to metals in a variety of ways. For 1:1 adducts (Figure 4.19), the isocyanide group is approximately linear, although some flexibility seems to exist in a bis(t-butylisocyanide)iron(II)tetraphenylporphyrinato complex.[135] For zerovalent metals with much electron density available for do-

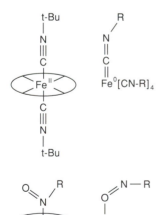

Figure 4.19
Modes of coordination of isocyanide and nitroso species.

---

* In fact, this classic experiment of St. George and Pauling established, for the first time and before any crystallographic data were available, that the heme group and the ligand-binding site in hemoglobin reside at least partway inside the protein, rather than on the surface.[134]

nation into ligand π^* orbitals, the isocyanide ligand has been observed to bend at the N atom.[136] One prediction exists that an isocyanide ligand binds in this manner to hemoglobin.[137]

For 1:1 adducts of nitroso ligands, side-on,[138] O-, and N-ligated modes are possible (Figure 4.19). No O-nitroso complexes have been definitively characterized by diffraction methods. For hemoglobin the N-nitroso mode is likely, since this is the mode found for the nitrosoalkane in Fe(TPP)(amine)(RNO).[139]

To date isocyanide ligands have not achieved their potential as probes of the geometry of the ligand-binding pocket in hemoglobin, partly because we lack structural data on the preferred geometry of attachment of these ligands in a sterically uncongested environment.

F. Nature of the Metal-Dioxygen Linkage in Biological Systems

Many techniques have been used to probe the geometry and electronic structure of the metal-dioxygen moiety in biological systems and in synthetic models. The results form the basis of any understanding of the factors that determine and modulate oxygen affinity.

1. Oxyhemocyanin and oxyhemerythrin

By resonance Raman techniques, the O—O stretch is observed at 744 cm^{-1} in oxyhemocycanin and at 844 cm^{-1} in oxyhemerythrin.[140–142] Dioxygen is therefore coordinated as peroxo species. By use of unsymmetrically labeled dioxygen, ^{18}O—^{16}O, it was established that dioxygen coordinates symmetrically:[141]

$$\text{Cu}^{II}\diagdown\overset{\displaystyle O-O}{}\diagdown\text{Cu}^{II} \quad \text{or} \quad \text{Cu}^{II}\diagdown\overset{\displaystyle O}{\underset{\displaystyle O}{}}\diagdown\text{Cu}^{II} \quad \text{or} \quad \text{Cu}^{II}\diagup\overset{\displaystyle O}{\underset{\displaystyle O}{}}\diagdown\text{Cu}^{II} \qquad (4.43)$$

Carbon monoxide binds to only one of the Cu^I centers in deoxyhemocyanin, through the C atom,* apparently blocking the second Cu^I site.[15] Similar behavior is also seen for the nitrosyl adduct.

On the other hand, in an experiment parallel to that just described for hemocyanin, dioxygen was found to bind asymmetrically in oxyhemerythrin:[142]

$$\text{Fe}^{III}\diagup\overset{\displaystyle O(H)}{\underset{\displaystyle |}{O}}\diagdown\text{Fe}^{III} \quad \text{or} \quad \text{Fe}^{III}\ldots\overset{\displaystyle (H)O}{\underset{\displaystyle |}{O}}\diagdown\text{Fe}^{III} \qquad (4.44)$$

---

* A report[143] that CO binds to the copper center through the O atom, an unprecedented mode, has been challenged.[144]

From single-crystal UV-visible spectroscopy with polarized radiation, an Fe · · · Fe—O_2 angle of approximately 90° was inferred, which is inconsistent with the bridging μ-1,1-(hydro)peroxo geometry.[145] Subsequently, this conclusion was confirmed in a single-crystal x-ray diffraction study of oxyhemerythrin that revealed, among other things, end-on coordination of dioxygen to only one iron center (see Figure 4.13).[45] This mode of dioxygen coordination remains unobserved to date in small-molecule synthetic systems. The O_2-binding process is formally very similar to the combinations of Reactions (4.29a) and (4.29b), except that dioxygen attaches to only one Fe center (bridging carboxylato groups omitted in Reaction 4.45):

$$Fe^{II} \overset{OH}{\diagup} Fe^{II} + O_2 \rightleftharpoons Fe^{III} \overset{HO-O}{\underset{O}{\diagup}} Fe^{III} \tag{4.45}$$

The existence and location of the proton, which cannot be proven in the crystal structure of deoxyhemerythrin or of oxyhemerythrin, are inferred from a model system for the former that contains a hydroxo group bridging two high-spin, weakly antiferromagnetically coupled Fe^{II} centers ($J = -10$ cm^{-1}).[88] For oxyhemerythrin (and for one conformation of methydroxohemerythrin), a small change in the position of the symmetric Fe—O—Fe mode is observed when H_2O is replaced by D_2O.[146] The strong antiferromagnetic coupling observed for methemerythrin and oxyhemerythrin ($-J \sim 100$ cm^{-1})[147] is uniquely consistent with a bridging oxo moiety between a pair of Fe^{III} centers.[80] Finally, a Bohr effect (release or uptake of protons) is absent in oxygen binding to hemerythrin.[16,18] These observations are consistent with a μ-oxo group slightly perturbed by hydrogen bonding to a coordinated hydroperoxo species. An important role for the protein in hemerythrin is to assemble an asymmetric di-iron(II) species. Only a few of the myriad of known μ-oxodiiron(III) complexes are asymmetric,[81] and the synthesis of realistic asymmetric models remains a challenge.

Deoxyhemocyanin (Cu^I d^{10}) and deoxyhemerythrin (Fe^{II} d^6) are colorless. In the oxygenated derivatives there is considerable charge transfer between the coordinated peroxo groups and the metal centers. This phenomenon makes the essentially d-d metal transitions more intense than those for the simple aquated Fe^{3+} or Cu^{2+} ions, and permits facile measurement of oxygen-binding curves. The spectral changes accompanying oxygenation are shown in Figures 4.20 and 4.21.[5]

Nitric oxide binds to deoxyhemocyanin, to deoxyhemerythrin, and to the mixed valence Fe^{III} · · · Fe^{II} semimethemerythrin.[148] Carbon monoxide binds to neither form of hemerythrin: apparently the other ligands have insufficiently strong fields to stabilize the low-spin state for which electron density would be available for back donation into the CO π^* orbitals.

212

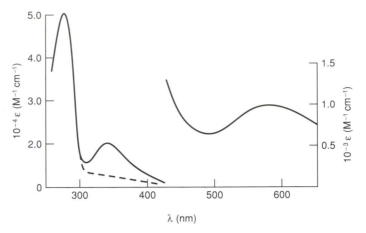

Figure 4.20
Spectral changes accompanying oxygenation of hemocyanin:
deoxyhemocyanin (– – –); oxyhemocyanin (————).[5] Reproduced
with permission from A. G. Sykes, in A. G. Sykes, ed.,
Advances in Irorganic and Bioinorganic Mechanisms,
Academic Press, **1** (1985), 121–178.

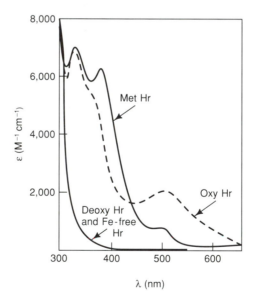

Figure 4.21
Spectral changes accompanying oxygenation of hemerythrin.[5]
Reproduced with permission from A. G. Sykes, in A. G. Sykes,
ed., *Advances in Inorganic and Bioinorganic Mechanisms*,
Academic Press, **1** (1985), 121–178.

2. Oxyhemoglobin

The O—O stretch that is observed by difference infrared techniques at around 1105 cm^{-1} for oxyhemoglobin and oxymyoglobin[149] clearly categorizes the dioxygen moiety as a superoxo species; that is, the order of the O—O bond is about 1.5. Considerable ink has been spilled about the nature of the Fe—O$_2$ fragment since Pauling's original suggestion[150] in 1948 that dioxygen binds to iron in an end-on bent fashion:

$$\text{Fe—O} \diagup^{\textstyle O} \qquad\qquad (4.46)$$

He subsequently reaffirmed this geometry, and proposed that hydrogen bonding between the coordinated dioxygen and the distal imidazole H—N group was important in stabilizing the Fe—O$_2$ species.[137] In an alternative model Weiss proposed that a low-spin Fe^{III} center ($S = \frac{1}{2}$) was very strongly antiferromagnetically coupled to a superoxide anion radical ($S = \frac{1}{2}$).[151] A triangular peroxo mode has also been advanced.[152,153] The problem has been how to resolve the observed diamagnetism of oxyhemoglobin[92,98] with UV-visible, x-ray absorption, and resonance Raman spectroscopic characteristics[154] that are distinctly different from those of Fe^{II} systems (such as carbonmonoxyhemoglobin and low-spin six-coordinated hemochromes, such as Fe(Porph)(Py)$_2$) and from unambiguously Fe^{III} systems (such as chloromethemoglobin or cyanomethemoglobin).

Any adequate theoretical treatment must also explain how iron-porphyrin systems can bind not only O$_2$, but also CO, NO, alkyl isocyanides, and alkyl-nitroso moieties. A simple qualitative model presented by Wayland and coworkers[129,155] conveniently summarizes ligand-binding geometries of cobalt and iron porphyrins. Although a reasonable quantitative theoretical consensus exists for 1:1 cobalt-dioxygen species, the same cannot be said yet for iron-dioxygen systems.

3. A simple model for the electronic structure of liganded hemoglobins

Why does dioxygen bind to iron and cobalt porphyrins in an end-on bent-bond fashion as in (4.37) and (4.46)? Why does carbon monoxide bind in a linear manner (Equation 4.40)? Why are six-coordinate dioxygen and carbon-monoxide adducts more stable than five-coordinate ones? A unified picture of ligand binding that addresses these questions is important in understanding properly the specific case of dioxygen binding to hemoglobin and related systems.

The splitting of the metal d orbitals for a four-coordinate metalloporphyrin is shown in the center of Figure 4.22. These orbitals contain some porphyrin character and are antibonding with respect to metal-porphyrin bonds. As shown in Figure 4.16, the primary effect of a single σ-donor axial ligand, such as pyridine or 1-methylimidazole, is to elevate the energy of the antibonding d$_{z^2}$ and lower the energy of the d$_{x^2-y^2}$ orbital and hence lead to a high-spin species in place of the intermediate-spin four-coordinate one. Thus, for simplicity in highlighting interaction of the metal center with the diatomic σ-donor: π-acid ligands CO, NO, and O_2, the perturbations wrought by primarily σ-donor ligands, such as 1-methylimidazole, are omitted. For the corresponding cobalt(II) compound, there is an additional electron. The diatomic ligands of interest share a qualitatively similar molecular orbital scheme. The filling of orbitals for CO is shown on the left-hand side. Dioxygen, which is shown on the right-hand side, has two more electrons than CO; these occupy the doubly degenerate π^* orbitals. Quantitative calculations show that the energy of the π^* orbitals decreases monotonically from CO to NO to O_2, indicating increasing ease of reduction of the coordinated molecule, a feature that has not been included in the diagram. Only those interactions of molecular orbitals that have appropriate symmetry and energy to interact significantly with the metal d orbitals are shown.

Two extremes are shown in Figure 4.22 for the interaction of a diatomic molecule A-B with the metal center: a linear geometry on the left and a bent geometry on the right. A side-on geometry is omitted for the binding of O_2 to a Co^{II} or Fe^{II} porphyrin, since this would lead to either an M^{III} side-on superoxo or an M^{IV} peroxo species; both these modes of coordination to these metals are currently without precedent.

Linear diatomic metal bonding maximizes the metal-d_π to ligand-p_{π^*} bonding. When a ligand coordinates in a bent manner, axial symmetry is destroyed, and the degeneracy of the ligand p_{π^*} orbitals is lifted. One p_{π^*} orbital is now oriented to combine with the metal d_{z^2} orbital to form a σ bond, and the other is oriented to combine with d_{xz} and d_{yz} orbitals to form a π bond. A bent geometry for the diatomic molecule will result when either or both of the metal d_{z^2} or the ligand p_{π^*} orbitals are occupied, since this geometry stabilizes the occupied d_{z^2} orbital in the five-coordinate complex. Thus O_2 binds in a strongly bent manner to Co^{II} and Fe^{II} porphyrins; NO binds in a strongly bent manner to Co^{II} porphyrins; CO binds in a linear fashion to Fe^{II} porphyrins.

The interaction of NO with Fe^{II} porphyrins and CO with Co^{II} porphyrins—the resultant species are formally isoelectronic—is more complicated. The degree of bending seen in $Fe^{II}(TPP)(NO)$ is midway between the two extremes.[111] For CO the higher-energy p_{π^*} orbitals lead to a greater mismatch in energy between the d_{z^2} and p_{π^*} orbitals, and less effective σ bonding. In EPR experiments the odd electron is found to be localized in a molecular orbital with about 0.87 metal d_{z^2} character for the five-coordinate Co—CO adduct, as expected for a nearly linear geometry.[129] On the other hand, for the Fe—NO adduct the metal d_{z^2} character of the odd electron is about 0.4 to 0.5;[155] a somewhat bent geometry (140°) is observed in the crystal structure of Fe(TPP)(NO). Because the CO ligand is a very weak σ donor, the Co—CO species exists only at low temperatures.

Only qualitative deductions can be made from this model about the extent of electron transfer, if any, from the metal onto the diatomic ligand, especially for dioxygen. The higher in energy the metal d_{z^2} orbital is with respect to the dioxygen p_{π^*} orbitals, the closer the superoxo ligand comes to being effectively a coordinated superoxide anion. With an additional electron, the dioxygen li-

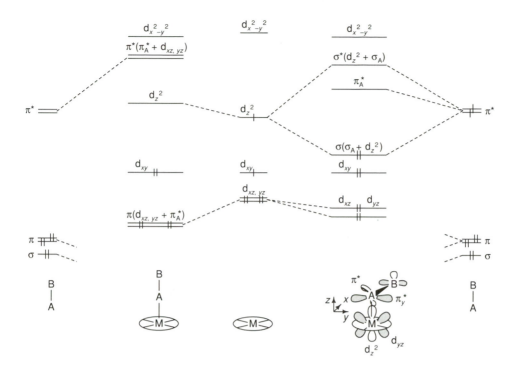

Figure 4.22
A simple general model for ligand binding to metalloporphyrins:[129] AB $=$ CO($\pi*^0$), NO ($\pi*^1$), O_2($\pi*^2$); M$=$Co^{II}(d^7), Fe^{II}(d^6).

gand in Co—O_2 complexes can acquire greater electron density than it can in Fe—O_2 complexes.

From the diagram it may be inferred that a ligand with very strong π-acceptor properties will lower the energy of the d_{xz} and d_{yz} orbitals through strong (d_{xz}, d_{yz})-$\pi*$ interaction. The resultant energy gap between these two orbitals and the other three metal d orbitals may be sufficient to overcome the energy involved in spin-pairing, and hence lead to five-coordinate low-spin species, as happens for complexes containing phosphines and carbon monosulfide.[99,132]

G. Role of the Protein in Effecting Biological Oxygen Transport

In our survey of the dioxygen chemistry of iron and copper species in earlier subsections, three general functions for the protein matrix became apparent: provision of ligand(s) in an appropriate stereochemistry; protection of the metal-dioxygen moiety from oxidation and competitive ligands; and modulation of dioxygen affinity through nonbonded interactions with distal groups.

1. Provision of ligands to the metal

In the hemoglobin family the heme group is anchored in a cleft in the globin chain by an imidazole ligand from a histidine residue (the proximal histidine).

The other (distal) side of the heme plane is more or less open to accommodate a small sixth ligand (see Figure 4.2).

For hemerythrin and hemocyanin the requirements of the protein chain are more severe. In contrast to the hemoglobin family, all but two of the ligands are provided by the protein chain, and in addition the metal ions are encapsulated as a pair. The exogenous ligands for hemerythrin are a μ-(hydr)oxo moiety and dioxygen or anions (depending on oxidation state); for hemocyanin, the identity of the second exogenous ligand, if there is one at all, is still unclear. Although hemerythrin has a distinctive cofacial bioctahedral structure (Figure 4.10)[46-48] that would appear to be very difficult to assemble in the absence of the protein, it turns out that with a variety of tridentate ligands the (μ-oxo)bis(μ-carboxylato)diiron(III) core may be assembled rather easily.[82,156,157] Thus, this core appears to be a thermodynamically very stable structural motif. Such a synthesis has been termed "self-assembly" and appears to be a common phenomenon in biological systems.[158] The low-temperature assembly of bis-copper(II)-μ-peroxo complexes (models for oxyhemocyanin) from mononuclear copper(I) compounds provides other examples of this phenomenon.[103f,g]

2. Protection of the metal-dioxygen moiety

The immobilization of the heme group inside the protein prevents (i) the bimolecular contact of an FeO_2 species with an Fe^{II} species (Reaction 4.29b), the key step in the irreversible oxidation of Fe^{II} porphyrins; (ii) the facile access of nucleophiles that would cause autoxidation (Reactions 4.30 and 4.31); (iii) the oxygenase activity (Reaction 4.32) that is the normal function of other hemoproteins, such as cytochrome P-450, horseradish peroxidase, catalase, etc.; and (iv) the self-oxygenase activity that has been observed in some iron(II) systems that bind dioxygen, activating it for destruction of the ligand itself. Avoiding these last two fates also appears to be very important in the active site of hemocyanin. Finally (v), the globin chain serves to restrain the binding of the distal histidine to give a six-coordinate hemochrome (Reaction 4.33), at least at room temperature.[159] Thus, unoxygenated hemoglobin is held in a five-coordinate state, allowing a rapid rate of oxygen binding and greater oxygen affinity—hemochromes such as $Fe(TPP)(Py)_2$ are impervious to oxygenation and subsequent oxidation.

3. Modulation of ligand-binding properties

The protein chain in hemoglobin may place restraints on the iron-to-proximal histidine bond. On the other side of the heme, the distal histidine and occluded water molecules may hydrogen-bond to the coordinated dioxygen and force ligands to adopt geometries that are different from those observed in the absence of steric hindrances. The conformation of the porphyrin skeleton may also be perturbed by the protein chain. Clearly, it is the protein chain that bestows the property of cooperativity on oligomeric oxygen carriers.

H. Requirements for a Model System for Hemoglobin

For hemoglobin, the majority of ligands around the iron center are provided by a fairly rigid macrocycle, the protoporphyrin IX dianion (Figure 4.1), that by itself enforces a square-planar stereochemistry. Thus, the task of assembling synthetic analogues of the active site in hemoglobin is simplified. Essentially, any square-planar, tetradentate ligand containing at least a couple of nitrogen atoms will suffice: to this end a variety of other porphyrins have been used, as well as Schiff-base and non-porphyrinato nitrogen-containing macrocycles that serve to delineate the role of the porphyrin in dioxygen binding. Tetraphenyl-porphyrin, in place of the naturally occurring porphyrins, has served as the basis of numerous model systems. It is easily synthesized and derivatized (see below). Its fourfold symmetry precludes formation of chemical isomers that may arise if substitution on the asymmetric, naturally occurring porphyrins is attempted. Moreover, its derivatives can be crystallized. Finally, with the porphyrin *meso* positions occupied by phenyl groups, the molecule is less susceptible to photoinduced oxidation.

In order to obtain five-coordinate species, putative models for a deoxy hemoglobin, access to one side of the porphyrin must be blocked to the coordination of a second axial base (Reaction 4.33), but still must be accessible for the binding of small molecules (O_2, NO, CO, etc). Or second, an axial base may be attached covalently to the porphyrin, to give so-called ''tail-under'' or ''chelated'' porphyrins. Here the chelate effect ensures an effectively 100 percent five-coordinate complex with only a 1:1 stoichiometric ratio of axial base to porphyrin. These two approaches are illustrated in Figure 4.14. A third means is to incorporate a sterically bulky substituent in the 2-position of the base, such as a methyl group, to give 2-methylimidazole (4.34). The formation of the hemochrome Fe(porph)(2-MeIm)$_2$, where the iron atom is in the center of the porphyrin ring, is strongly disfavored relative to the 1-methylimidazole analogue, because the 2-methyl substituents clash with the porphyrin ring. The axial base needs to be a strong σ donor, such as imidazole or pyridine, in order to increase affinity at the iron (or cobalt) center for dioxygen (Figure 4.18).

Steric hindrance on one side, or on both, provides a pocket for small molecules to bind and, for O_2, prevents the bimolecular contact of two iron(II)-porphyrinato species that would lead to irreversible oxidation (Reaction 4.29). A picturesque collection of substituted porphyrins has been synthesized. Some of these are illustrated in Figure 4.23.[31,72,160–164] The only system that has led to crystalline dioxygen complexes stable at room temperature is the ''picket-fence'' porphyrin.[72]* A derivative of this, the ''pocket'' porphyrin,[165] and var-

---

* The pivalamido pickets (—NH—CO—C(CH$_3$)$_3$) of the picket-fence porphyrin are sufficiently bulky that their free rotation is sterically hindered. Thus the various *atropisomers*—$\alpha\alpha\alpha\alpha$, $\alpha\alpha\alpha\beta$, $\alpha\beta\alpha\alpha$, $\alpha\alpha\beta\beta$, where α denotes picket ''up'' and β denotes picket ''down''—can be separated chromatographically on the basis of their different polarities. The tail-under picket-fence porphyrin is derived from the $\alpha\alpha\alpha\beta$ atropisomer.[72]

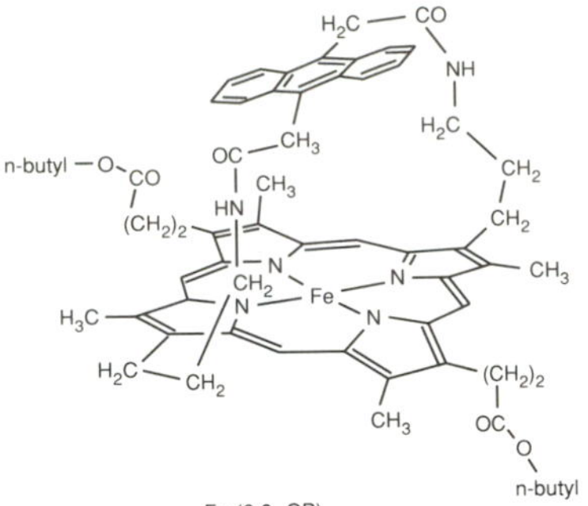

218

Fe (6,6 - CP)

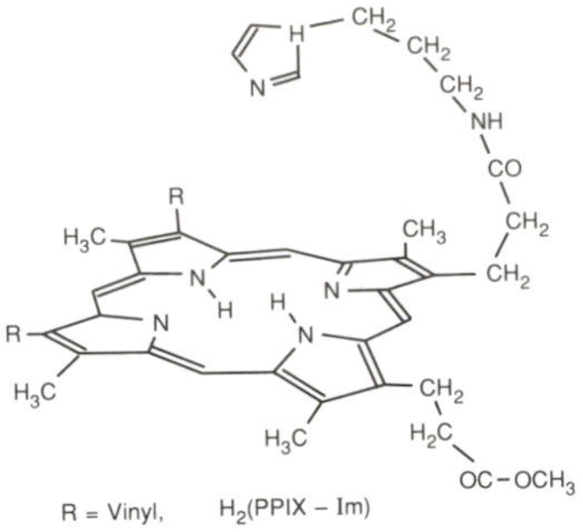

R = Vinyl,   H₂(PPIX – Im)

R = Ethyl,   H₂(MPPIX – B),   B = Im, 3 – Py

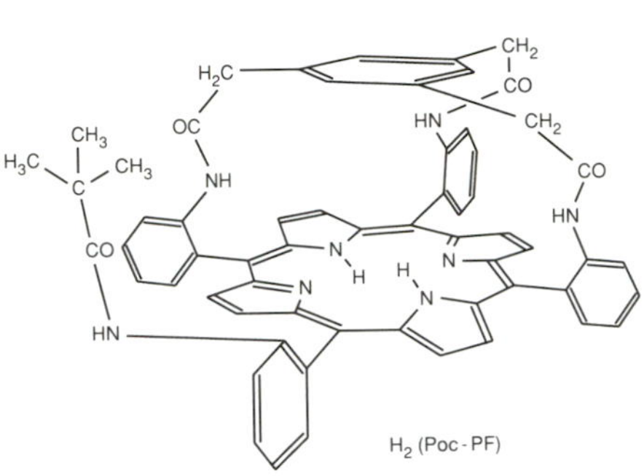

H₂ (Poc - PF)

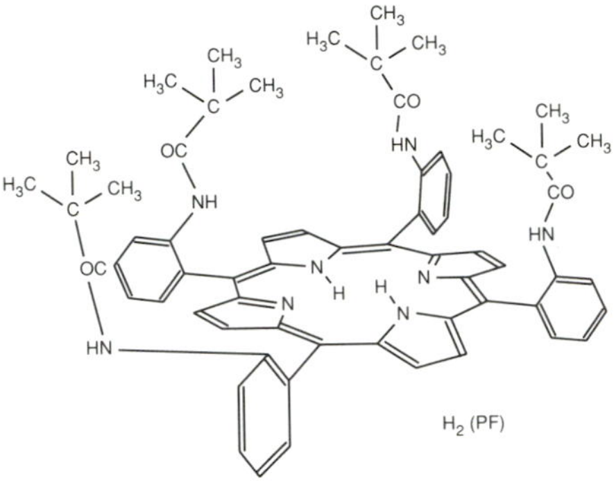

H₂ (PF)

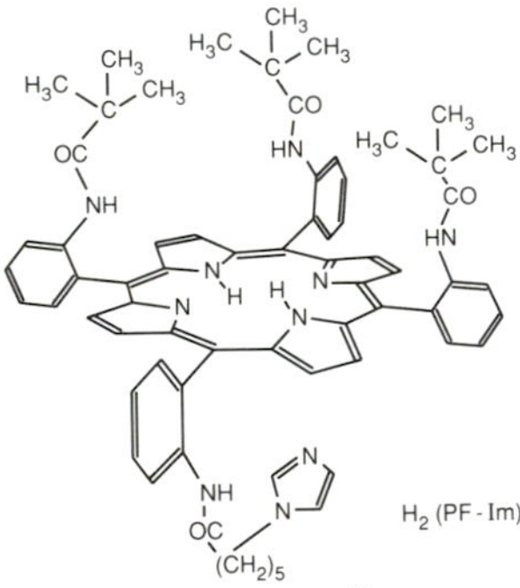

H₂ (PF - Im)

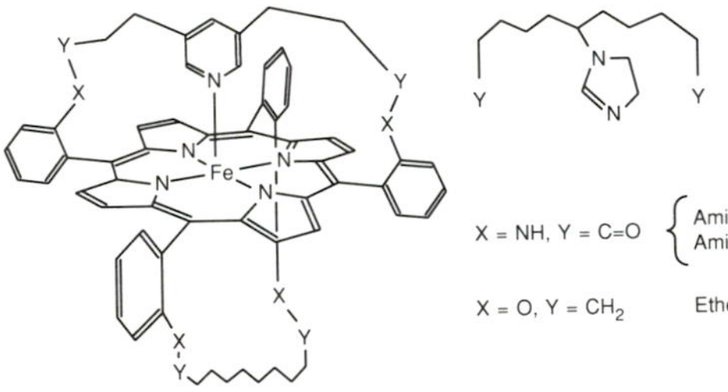

X = NH, Y = C=O { Amide - Im
                   Amide - Py

X = O, Y = CH₂   Ether - Py

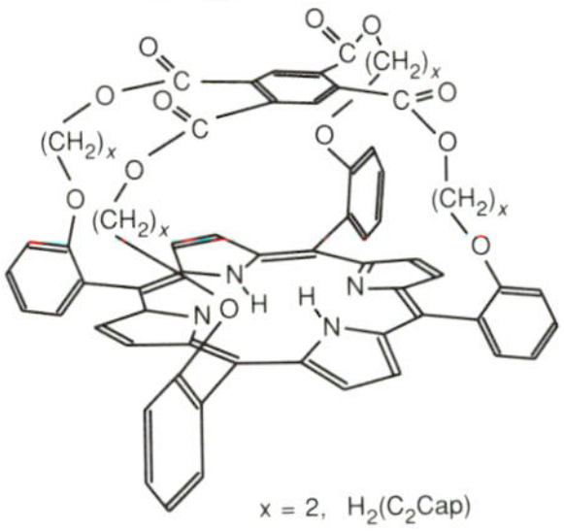

x = 2, H₂(C₂Cap)

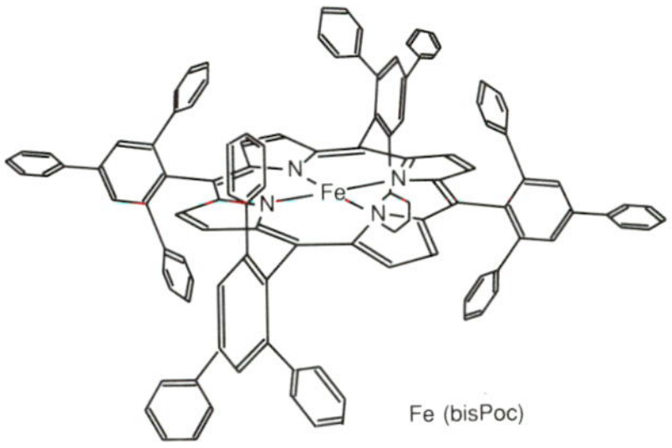

Fe (bisPoc)

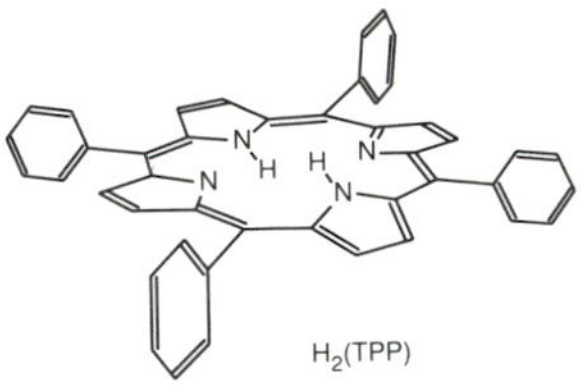

H₂(TPP)

ious "capped" porphyrins,[166] provide binding sites with steric hindrance even to small diatomic molecules.

In the next section the structures of various derivatives of hemoglobin and its models are presented, and the relationship of structure to ligand-binding properties is examined. Although there is now a wealth of thermodynamic data available from model systems, attention is focused primarily on those for which structural data are also available.

III. STRUCTURAL BASIS OF LIGAND AFFINITIES OF OXYGEN CARRIERS

The interaction of ligands, such as dioxygen, with metal complexes, such as iron-porphyrinato systems, and the means by which this interaction is characterized, have been covered in broad outline in the previous sections. As noted earlier, the affinities of hemoglobins for carbon monoxide and dioxygen span a wide range (see Table 4.2 and Figure 4.24). In this section the active site is examined in much finer detail than before in order to develop relationships between perturbations in structure and affinity (and hence function)—so called structure-function relationships. The reference point is the somewhat hypothetical situation where the dioxygen binder is in the gas phase and independent of interactions with solvent molecules, solute molecules, and itself, and where dioxygen, carbon monoxide, and other small molecules may bind without steric constraints—in other words, a state where intrinsic affinity is measured. In this section attention is focused exclusively on the hemoglobin family and on iron- and cobalt-porphyrinato systems. In recent years structural data on hemoglobin, myoglobin, and their derivatives have become available with a precision that permits meaningful comparison with the more precisely determined model or synthetic systems. In addition, the various hemoglobins and myoglobins, and especially the naturally occurring mutants of hemoglobin A (human Hb), have provided a sort of poor man's site-directed mutagenesis. Now the techniques of molecular biology permit the site of mutation to be selected, the altered gene to be inserted into *E. coli*, and the mutant protein to be expressed in large (mg) quantities. With the conditions for crystallization of hemoglobins now well-established, we can discover quite rapidly what structural perturbations are caused by the substitution of one amino acid for another, and can relate these to the perturbations in properties, such as cooperativity, dioxygen affinity, and kinetics of ligand binding.

The principles enunciated here are applicable generally to hemerythrin and hemocyanin; however, we currently lack the thermodynamic and especially structural data we would like to have for these systems.

Figure 4.23 *(facing page)*
A selection of synthetic porphyrins used to probe the dioxygen-binding process.

A. Ligand Affinities in Hemoglobins and Their Models

The O_2 affinities in biological carriers span five orders of magnitude, which at room temperature corresponds to a difference in the free energy of oxygen binding

$$\delta \Delta G = -RT \ln (K_{max}/K_{min}) = -RT \ln (P_{1/2min}/P_{1/2max}) \qquad (4.47)$$

of about 6.0 kcal/mol. This wide range of O_2 and CO affinities has not yet been paralleled in synthetic systems; the values for O_2 affinity do not exceed those for R-state human hemoglobin. A selection of values from model systems is given in Table 4.5.[23,31,160-165] For the flat-open porphyrin system (Figure 4.23) the dioxygen ostensibly binds in an unconstrained manner, but is actually subject to solvent influences. In order to obtain thermodynamic constants on these "unhindered" systems, one must gather data at several low temperatures and then extrapolate to room temperature, or obtain them from kinetic measurements, $K = k_{on}/k_{off}$, at room temperature.

For the picket-fence porphyrins, dioxygen binds in a protected pocket that is deep enough to accommodate it and to prevent the dimerization that leads to irreversible oxidation, provided that there is a slight excess of base to ensure full saturation of the coordination sites on the unprotected face of the porphyrin.[72] Thus the picket-fence, the capped, and the bis-pocket porphyrins reversibly bind dioxygen at room temperature with little oxidation over many cycles. This stability facilitated isolation of crystals of a synthetic iron-dioxygen species of the picket-fence porphyrin. The capped porphyrin offers a more highly protected site. The low affinity these latter systems have for dioxygen indicates that the binding cavity is so small that repulsive steric interactions between coordinated dioxygen and the cap are unavoidable. The left-hand side of Figure 4.24 depicts on a logarithmic scale the range of O_2 affinities. Each power of 10 corresponds to around 1.2 kcal/mol at 25°C.

The right-hand side of Figure 4.24 illustrates the range of affinities for CO binding. For many synthetic systems the CO affinities are orders of magnitude greater than in the biological systems that have an O_2 affinity similar to the synthetic; for example, see the entries for the picket-fence porphyrin. Comparison of the left- and right-hand sides of Figure 4.24 reveals that the strongest O_2 binder, hemoglobin *Ascaris*, is one of the weakest CO binders. The O_2 affinity of the picket-fence porphyrins is very similar to that of myoglobin, but, as will be detailed shortly, one cannot infer from this that the binding sites are strictly comparable. Indeed, similar affinities have been observed with a non-porphyrin iron complex.[121,162] Moreover, if the CO affinity of myoglobin paralleled that of the picket-fence porphyrins, some 20 percent of myoglobin (and hemoglobin) would be in the carbonmonoxy form (in contrast to the approximately 3 percent that occurs naturally), a level that could render reading this section while chewing gum physically taxing.[117]

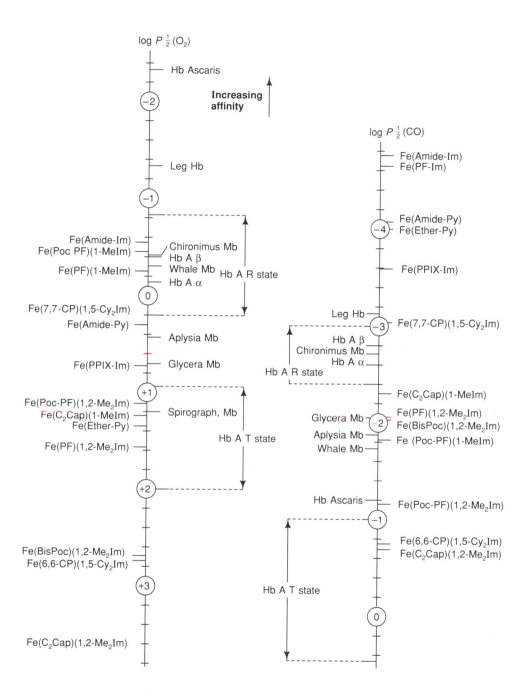

Figure 4.24
CO and O_2 affinities of a selection of hemoglobins and model systems. Affinities are given as $P_{1/2}$, and the scale is logarithmic. One order of magnitude corresponds to 1.2 kcal/mol at 25°C.

Table 4.5
Thermodynamics and kinetics of ligand binding to synthetic oxygen carriers at 20–25°C.

Carrier	Dioxygen binding					Carbon-monoxide binding				
	$P_{1/2}(O_2)$ Torr	ΔH kcal/mol	ΔS eu	k_{on} $\mu M^{-1} s^{-1}$	k_{off} s^{-1}	$P_{1/2}(CO)$ Torr	ΔH kcal/mol	ΔS eu	k_{on} $\mu M^{-1} s^{-1}$	k_{off} s^{-1}
Toluene/benzene solvent										
Picket fence, pocket										
Fe(PF-Im)	0.58	−16.3	−40	430	2,900	0.000022	—	—	36	0.0078
Fe(PF) (1, 2-Me₂Im)	38	−14.3	−42	106	46,000	0.0089	—	—	1.4	0.14
Fe(Poc-PF) (1-MeIm)	0.36	—	—	2.2	9	0.0015	—	—	0.58	0.0086
Fe(Poc-PF) (1, 2-Me₂Im)	12.6	−13.9	−28	1.9	280	0.067	—	—	0.098	0.055
Fe(Bis-Poc) (1, 2-Me₂Im)	508	−14.4	−47	—	—	0.0091	—	—	—	—
Cap										
Fe(C₂Cap) (1-MeIm)	23	−10.5	−28	—	—	0.0054	—	—	0.95	0.05
Fe(C₂Cap) (1, 2-Me₂Im)	4,000	−9.7	−36	—	—	0.20	—	—	—	—
Strapped										
Fe(7, 7-CP) (1, 5-Cy₂Im)	1.4	—	—	65	1,000	0.00091	—	—	6	0.05
Fe(6, 6-CP) (1, 5-Cy₂Im)	700	—	—	0.1	800	0.17	—	—	0.03	0.05
Flat open										
Fe(PPIX-Im)	5.6	—	—	62	4,200	0.00025	—	—	11	0.025
Bis-strapped										
Fe(Amide-Im)	0.29	—	—	310	620	0.000017	—	—	40	0.067
Fe(Amide-Py)	2.0	—	—	360	5,000	0.00009	—	—	35	0.03
Fe(Ether-Py)	18	—	—	300	40,000	0.0001	—	—	68	0.069
H₂O, alkylammonium micelles, pH 7.3										
Fe(PPIX-Im)	1.0	−14.0	−3.5	26	4.7	0.002	−17.5	−34	3.6	0.009
Fe(MPIX-Im)	0.57	—	—	22	23	0.0013	—	—	11	0.019
Fe(MPIX-Py)	12.2	—	—	1	380	0.0021	—	—	12	0.035

[a] When available $P_{1/2}$ are from thermodynamic measurements, otherwise from k_{on}/k_{off}, where solubility of O_2 in toluene is 1.02×10^{-5} M/Torr and of CO in toluene is 1.05×10^{-2}; solubilities in benzene are very similar.

[b] Some k_{off} are calculated from $K(O_2)$, $k_{on}(CO)$, and M.

There is a convenient index to summarize the extent to which CO (or O_2) binding is discriminated against for a given iron-porphyrin system. M is defined as the ratio of O_2 affinity (as $P_{1/2}$) to CO affinity for a particular system and experimental conditions:

$$M = \frac{P_{1/2}(O_2)}{P_{1/2}(CO)}. \qquad (4.48)$$

From Figure 4.24 and from Tables 4.2 and 4.5 the M values calculated may be somewhat arbitrarily divided into three classes: those where $M > 2 \times 10^4$ (good CO binder); those where $2 \times 10^2 < M < 2 \times 10^4$; and those where $M < 2 \times 10^2$ (good O_2 binder). An analogous parameter, N, may be defined to summarize the differences in the O_2 affinity between an iron-porphyrin system and its cobalt analogue:

$$N = \frac{P_{1/2}(O_2\text{—Co})}{P_{1/2}(O_2\text{—Fe})}. \qquad (4.49)$$

For the picket-fence porphyrins and for vertebrate hemoglobins N is in the range 10 to 250, whereas for the flat-open porphyrins and for some hemoglobins that lack a distal histidine (e.g., hemoglobin *Glycera* and hemoglobin *Aplysia*), N is at least an order of magnitude larger, indicating for these latter species that the cobalt analogue binds O_2 relatively poorly [167,168] (see Table 4.6).

Note that whereas the O_2 binding of the picket-fence porphyrins is similar to that for myoglobin, the kinetics of the process are very different; the synthetic system is more than an order of magnitude faster in k_1 and k_{-1} (often also referred to as k_{on} and k_{off}). On the other hand, O_2 binding to the pocket porphyrin is similar to that for the biological system. The factors by which ligand affinities are modulated, generally to the benefit of the organism, are subtle and varied, and their elucidation requires the *precise* structural information that is currently available only from x-ray diffraction experiments. Figure 4.25 shows the structural features of interest that will be elaborated upon in the next subsections. [110,169]

B. General Structural Features that Modulate Ligand Affinity

There are many ways in which ligand affinity may be perturbed (Figure 4.25). It is convenient to divide these into two groups, referred to as distal and proximal effects. [163] *Proximal* effects are associated with the stereochemistry of the metalloporphyrinato moiety and the coordination of the axial base, and thus their influence on O_2 and CO affinity is indirect. *Distal* effects pertain to *noncovalent interactions* of the metal-porphyrinato skeleton and the sixth ligand (O_2, CO, etc.) with neighboring solvent molecules, with substituents, such as pickets or caps, on the porphyrin, and with the surrounding protein chain. The distal groups that hover over the O_2-binding site engender the most important

Table 4.6
Relative affinities (M) of iron-porphyrinato systems for O_2 and CO, and relative affinities (N) for O_2 of iron and cobalt-porphyrinato systems.

Compound	$P_{1/2}$(Fe—CO) Torr	$P_{1/2}$(Fe—O$_2$) Torr	M $P_{1/2}$(Fe—O$_2$)/$P_{1/2}$(Fe—CO)	$P_{1/2}$(Co—O$_2$) Torr	N $P_{1/2}$(Co—O$_2$)/$P_{1/2}$(Fe—O$_2$)
H$_2$O, pH 7					
Whale Mb	0.018	0.51	28	57	110
Whale Mb (E7His→Gly)	0.0049	6.2	1,300	—	—
Aplysia Mb	0.013	2.7	200	50 × CoMb	>1,000
Glycera Mb	0.00089	5.2	5,800	50 × CoMb	>1,000
Fe(PPIX-Im)	0.002	1.0	500	—	—
Toluene/Benzene					
Fe(PF-Im)/ Co(PF) (1-MeIm)	0.000022	0.58	27,000	140	240
M(PF)(1, 2-Me$_2$Im)	0.0089	38	4,300	900	24
M(Bis-Poc)- (1, 2-Me$_2$Im)	0.0091	508	55,800	—	—
Fe(PPIX-Im)/ Co(PPIX) (1-MeIm)	0.00025	5.6	22,000	18,000	3,200
M(C$_2$-Cap)(1-MeIm)	0.0054	23	4,200	140,000	6,100

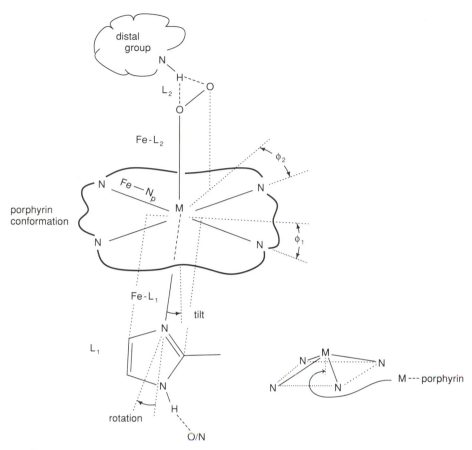

Figure 4.25
Structural parameters and features that determine the affinity of oxygen carriers.

distal effects. For convenience, the effects of crystal packing and the protein matrix on porphyrin conformation will be discussed among the proximal effects, although as nonbonded interactions they properly are distal effects.

To a first approximation, the effects of substituents on the porphyrin ring, as transmitted through bonds to the metal center, do not perturb the ligand binding properties as much as do distal effects.[170] Thus substituents, such as vinyl and propionic-acid groups on protoporphyrin IX and o-pivalamidophenyl pickets, are ignored; one porphyrin is much like another. At the end of this subsection the various ways ligand affinity may be modulated will be summarized in an augmented version of Figure 4.3.

1. Proximal effects

Few molecules have had their conformational properties characterized as exhaustively as have metalloporphyrins.

a. Porphyrin Conformation and M-N_p Separations   The cyclic aromatic 24-atom porphyrinato skeleton offers a tightly constrained metal-binding site. The conformation of least strain is planar, and the radius of the hole of the dianion is close to 2.00 Å,[110] leading to metal-porphyrinato nitrogen-atom separations, M-N_p, of 2.00 Å if the metal is centered in the square plane defined by the four porphyrinato nitrogen atoms. Small deviations from planarity are generally observed and attributed to crystal packing effects; large deviations may be induced by bulky substituents on the porphyrin skeleton, especially at the *meso* positions, by the crystal matrix,[65] or by the highly anisotropic protein matrix. The 2.00 Å radius hole neatly accommodates low-spin ($S = 0$) and intermediate-spin ($S = 1$) iron(II), low-spin ($S = \frac{1}{2}$) iron(III), and cobalt(II) and cobalt(III) ions.[91]* With few exceptions the metal is centered in or above the central hole for mononuclear porphyrin species; only rarely do M-N_p bonds show a significant (though still small) scatter about their mean value.

b. M $\cdots$ Porph Displacement   For five-coordinate complexes the magnitude of the displacement of the metal from the plane of the four nitrogen atoms, M $\cdots$ porph, is a consequence of the electronic configuration of ML$_5$ complexes. Of course, the effect is augmented if the $3d_{x^2-y^2}$ orbital (directed along M-N_p bonds, Figure 4.16) is occupied. Compare a displacement of 0.14 Å for Co(TPP)(1,2-Me$_2$Im) (no $3d_{x^2-y^2}$ occupancy)[111] with 0.43 Å for Fe(PF)(2-MeIm) ($3d_{x^2-y^2}$ occupied).[169] For six-coordinate complexes where the two axial ligands, L$_1$ and L$_2$, are different, the M $\cdots$ porph displacement usually reflects relative *trans* influences.

Generally, displacement of the metal from the plane of the porphyrinato-nitrogen atoms is within 0.04 Å of the displacement from the 24-atom mean plane of the entire porphyrin skeleton. On occasions this second displacement may be much larger, for example in Fe(TPP)(2-MeIm), where it is 0.15 Å larger[110] than it is for Fe(PF)(2-MeIm). This effect is called *doming*, and it is usually attributed to crystal packing forces. Interaction of the porphyrin with protein side chains leads to considerable doming or folding of the heme in vertebrate hemoglobins.

c. M-L Separations   The metal-axial ligand separations, M-L (when more than one, L$_1$ denotes the heterocyclic axial base), are dependent on the nature of the ligand, L. When L$_1$ and L$_2$ are different, the M-L separations are sensitive to the relative *trans* influences of L$_1$ and L$_2$ as well as to steric factors. For

---

* In order to accommodate smaller ions, such as nickel(II), the porphyrin skeleton may contract by ruffling, with little loss of aromaticity; like a pleated skirt the pyrrole rings rotate alternately clockwise and counterclockwise about their respective M-N_p vectors. This distortion leaves the four porphyrinato nitrogen atoms, N_p, still coplanar, Alternatively, the porphyrin skeleton may buckle to give a saddle conformation; the N_p atoms may acquire a small tetrahedral distortion in this process. M-N_p bonds as short as 1.92 Å have been observed. Metals with one or two electrons in their $3d_{x^2-y^2}$ orbital have a radius larger than 2.00 Å. In order to accommodate them in the plane of the porphyrin, the porphyrin skeleton expands. M-N_p separations as long as 2.07 Å may occur with the metal still centered in the plane of the N_p atoms.[110]

example, for Fe(TPP)(1-MeIm)$_2$, the Fe—N$_{Im}$ bond length is 2.016(5) Å,[110] whereas for Fe(TPP)(1-MeIm)(NO) it is 2.180(4) Å.[111] For sterically active ligands, such as 2-methylimidazole compared to 1-methylimidazole (4.34), the longer Co—N$_{Im}$ bond occurs for the 2-MeIm ligand because of steric clash between the 2-methyl group and the porphyrin.[111]

It is possible that combinations of intrinsic bonding and steric factors may give rise to a double minimum and two accessible axial ligand conformations (see Figure 4.26). This situation seems to occur in the solid state for Fe(PF)(2-

Fe(PF)(2-MeIm)(O$_2$)   Fe(PF)(1,2-Me$_2$Im)(O$_2$)
· EtOH

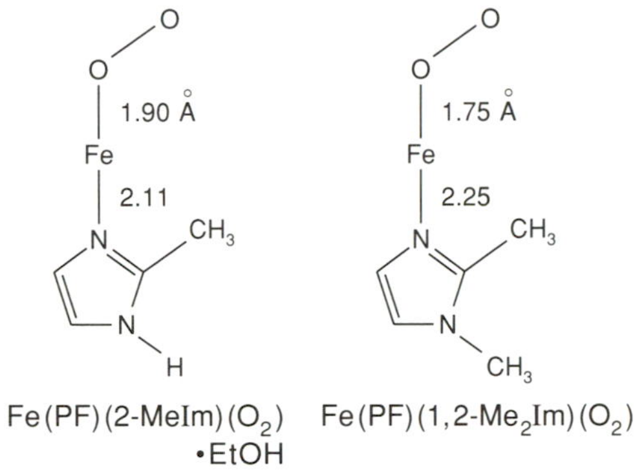

Figure 4.26
Two arrangements of axial ligands. The right-hand side features a short L$_2$ and long L$_1$; the left-hand side the opposite.

MeIm)(O$_2$) · EtOH, where a short Fe—N$_{Im}$ and a long Fe—O bond are observed both from the structure revealed by single-crystal x-ray diffraction methods and by EXAFS data. On the other hand, for solvate-free Fe(PF)(2-MeIm)(O$_2$) and for Fe(PF)(1,2-Me$_2$Im)(O$_2$), the EXAFS patterns are interpreted in terms of a short Fe—O and long Fe—Im bond.[171]

d. The Angle φ   This parameter is the minimum angle that the plane of the axial base (e.g., pyridine, substituted imidazole, etc.) makes with a plane defined by the N_p, M, and L_1 atoms (Figure 4.25).[65] If there are two axial ligands, e.g., 1-methylimidazole and O$_2$, then, as before, the angle the axial base makes is denoted ϕ_1 and the other angle ϕ_2. For a linear CO ligand bound perpendicularly to the porphyrin plane, ϕ_2 is undefined. Note that the orientation of the second ligand is influenced by distal effects.

When $\phi = 0$, the axial base eclipses a pair of M-N_p bonds; contacts with the porphyrin are maximized. When $\phi = 45°$, contacts are minimized. Unless the axial base has a 2-substituent, however, the contacts are not excessively close for any value of ϕ. With a 2-methyl substituent, the contacts are sufficiently severe that the M-N_{L_1} vector is no longer perpendicular to the porphyrin plane, and the imidazole group is rotated so that the M-N_{L_1} vector no longer approximately bisects the imidazole C—N—C bond angle, as illustrated in Figure 4.25.[110,172]

2. Distal effects

Distal effects arise from noncovalent interactions of the coordinated dioxygen, carbon monoxide, or other ligand with its surroundings. The protein matrix, the pickets, and the caps are functionally equivalent to an anisotropic solvent matrix that contains a variety of solutes. The limits of this simplification are illustrated in the following example. The electronically similar cobalt meso-, deutero-, and protoporphyrin IX complexes bind dioxygen with similar affinities under identical solvent conditions. When they are embedded in globin, larger differences in affinity and changes in cooperativity are observed.[170] These effects are attributed to the slightly different nestling of the porphyrin molecules in the cleft in hemoglobin or, in the generalization introduced, to slightly different solvation effects.

Interaction of the coordinated O_2 or CO molecule with solvent molecules or with the protein has a profound influence on kinetics and thermodynamics (see Figure 4.24, and Tables 4.2 and 4.5). As discussed earlier, there is accumulation of negative charge on the dioxygen ligand. The possibility then arises for stabilization of coordination through hydrogen bonding or dipolar interactions with solute molecules,[175] porphyrin substituents (such as amide groups in the picket-fence porphyrins[176] and some species of strapped porphyrins[161]), or with protein residues* (such as histidine).[167,177–179]

Destabilization of coordinated ligands and lowered affinity can result if the coordinated ligand is unable, through steric clash, to achieve its optimum stereochemistry or if the closest neighboring groups are electronegative, as are the ether and ester linkages on capped porphyrins.[31,180] We will describe in detail in the next subsection (III.C) the fascinating variety of means by which ligand binding is modulated by distal amino-acid residues.

3. Approximate contributions of proximal and distal effects to ligand affinity

Dissimilar systems may show similar affinities for a ligand as a result of a different mix of the proximal and distal effects enumerated above. These effects are not all of equal magnitude, and an attempt is made here to show the increment in free energy that occurs if the effect is manifest in the deoxy or liganded state of Figure 4.3. Increasing the free energy of the deoxy state while holding that of the liganded state constant leads to increase in affinity. The reference state is gaseous Fe(TPP)(1-MeIm). The magnitude and sign of these effects are shown in Figure 4.27. For the coordination of alkylisocyanide molecules to

---

* For *Glycera* CoMbO$_2$ no change in EPR parameters occurs on substituting D_2O for H_2O.[168] No hydrogen bond between O_2 and a distal group comparable in strength to that in whale CoMbO$_2$ was inferred.

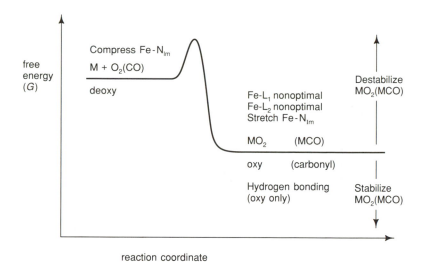

Figure 4.27
Proximal and distal effects on the ligand affinities.

hemoglobin, the steric effects of different alkyl groups have been quantified.[35] Lowered affinity occurs with increasing alkyl chain length, with the exception of methyl isocyanide.

C. Detailed Structures of Hemoglobins and Model Systems

With thermodynamic background and general structural features relevant to ligand affinity enumerated, attention may now be turned to the detailed structural aspects of the active site and its surroundings. As was shown crudely in Figure 4.3, the ligand affinity of an iron porphyrin may be perturbed either by modulating the structure of the deoxy material or by modulating the structure and surroundings of the liganded material or both. The model systems provide the reference points against which the protein structures may be compared.

1. Structures relevant to deoxy hemoglobins

The structure of the picket-fence porphyrin compound, Fe(PF)(2-MeIm), is shown in Figure 4.28.[172] Minus the pickets, it is essentially a magnified view of the active site of deoxymyoglobin, shown in Figure 4.29.[181] Some metrical details of these structures, of a very similar unsubstituted tetraphenylporphyrin,[110] and of several other deoxyhemoglobins[11c,182–185] are listed in Table 4.7. In general they are all similar, but important differences exist.

In all structures, except deoxyerythrocruorin,[183] the iron atom is displaced about 0.4 to 0.5 Å from the plane of the porphyrin toward the axial base. For deoxyerythrocruorin the displacement is less than half this, perhaps because the water molecule is weakly coordinated to the iron center.

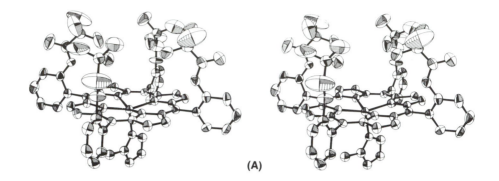

(A)

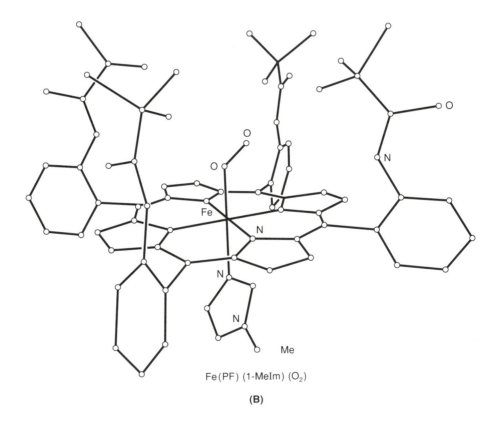

Fe(PF) (1-MeIm) (O$_2$)

(B)

Figure 4.28
(A) Stereodiagram of the structure of Fe(PF)(2-MeIm).[172]
(B) Structure of Fe(PF)(1-MeIm)(O$_2$).[187]

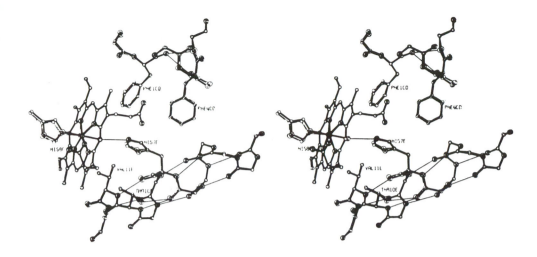

Figure 4.29
Structure of metmyoglobin and deoxymyoglobin at 2.0 Å resolution near the heme.[181a] Solid bonds are for metmyoglobin; open bonds for deoxymyoglobin. Note the water molecule coordinated to the iron center and hydrogen bonded to the distal imidazole group in metmyoglobin. Reproduced with permission from T. Takano, *J. Mol. Biol.* **110** (1977), 569–584.

An imidazole group from a histidine residue—the distal histidine E7 in position 7 on helix labeled E—hovers over the binding site for most vertebrate hemoglobins, except for genetically engineered mutants of human hemoglobin (βE7His → Gly), pathological mutant hemoglobins, such as hemoglobin Zürich (βE7His → Arg), and some others, such as elephant hemoglobin. Long believed to be noncoordinating, this distal histidine may, in fact, coordinate weakly to the Fe center at low temperature.[159] In the α chains of human deoxyhemoglo-

Table 4.7
Metrical details of deoxyhemoglobins and their models[a]

Compound	Resol. (Å)	Fe-N_p (Å)	Fe · · · Porp (Å)	Doming (Å)	Fe—N_{Im} (Å)	ϕ (deg)	Tilt (deg)
Fe(PF)(2-MeIm)	—	2.072(5)	0.43	0.03	2.095(6)	22.8	9.6
Fe(TPP)(2-MeIm)	—	2.086(6)	0.40	0.13	2.161(5)	7.4	10.3
Mb	1.4	2.03(10)	0.42	0.08	2.22	19	11
Er · · · H_2O	1.4	2.02	0.17	−0.06	2.25	7	3
HbA (α · · · H_2O)	1.74	2.08(3)	0.40(5)	0.16(6)	2.16(6)	18(1)	12(2)
(β)		2.05(3)	0.36(5)	0.10(6)	2.09(6)	24(1)	11(2)
CoHb[b]	2.5	—	0.14(5)	0.13	2.24(6)	—	—
Co(TPP)(1-MeIm)	—	1.977(6)	0.13	0.01	2.157(3)	3.8	0
Co(TPP)(1, 2-Me$_2$Im)	—	1.985(3)	0.15	0.05	2.216(2)	10	—

[a] See Figure 4.25 for definition of symbols.
[b] From a difference refinement of CoHb vs. Hb, where the difference in metal-to-porphyrin-plane separation was 0.24(2) Å and the difference in M-N_{Im} was 0.13(4) Å. Doming is similar to Hb.

bin, hemoglobin A, a water molecule is found in the binding cavity.[182] For many years the binding cavity has been referred to as the hydrophobic pocket—literally, water-hating. Although many hydrophobic groups, such as valine, leucine, isoleucine, and phenylalanine are positioned over the porphyrin, the immediate environment around the binding site is, in fact, polar, with the distal histidine and associated water molecules, as well as the heme group itself. As will be shown in the next section, the label "hydrophobic pocket" becomes more misleading when the interaction of coordinated ligands with distal groups is examined.

The orientation of the axial base, angle ϕ_1, is similar for Fe(PF)(2-MeIm) and for several vertebrate deoxyhemoglobins. On the other hand, Fe(TPP)(2-MeIm) and deoxyerythrocruorin have a similar eclipsed axial-base orientation. At least for five-coordinate species, where the iron center is substantially out of the porphyrin plane, orientation of the axial base does not invariably induce structural perturbations, e.g., doming, in the porphyrin skeleton.

The conformation of the protein chain is such that the proximal histidine in deoxyhemoglobin coordinates in a slightly tilted manner,[182,186] comparable to the tilt that the sterically active 2-methyl substituent induces in the synthetic systems.[172] Clearly, coordination of the histidine to the heme in a symmetric manner, as would be expected in the absence of the protein constraints, does *not* produce the conformation of lowest free energy for the *whole* molecule.

2. Structures relevant to liganded hemoglobins

a. Stereochemistry of the Active Site   Before the advent of techniques that enabled the preparation and stabilization of oxyhemoglobin crystals, key information on the probable structure of oxyhemoglobin and thence on the mechanism of cooperativity was extrapolated from structures of methemoglobin derivatives[11] and from various five- and six-coordinate cobalt-porphyrinato complexes.[110,112,113] The structures of these met derivatives have proved to be similar to that of oxyhemoglobin, at least in the stereochemistry of the metalloporphyrinato species and for the protein tertiary and quaternary structure as well.

Two synthetic iron-dioxygen adducts built from the picket-fence porphyrins have been structurally characterized.[172,187] The high, effectively fourfold symmetry of the binding pocket in these systems results in fourfold disorder of the angularly coordinated dioxygen molecule, and precludes the precise and accurate measurements of the Fe—O—O angle and O—O separation that are grist to the theoretical mills.* Figure 4.28B illustrates the stereochemistry for one conformer. Subsequently, the structures of several dioxygen adducts of biolog-

---

* On the other hand, very precisely described mononuclear 1:1 cobalt-dioxygen complexes invariably have been crystallized from dipolar aprotic solvents. Protic species, especially water, promote dimerization and, additionally for iron systems, irreversible oxidation. Although several Co—O$_2$ structures with tetradentate square-planar Schiff bases are very precisely known, the environment around the dioxygen bears little resemblance to that in the quasibiological system, Co hemoglobin.

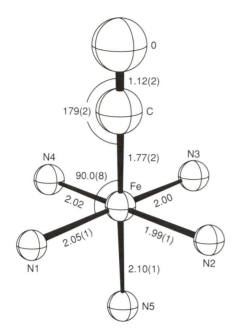

Figure 4.30
Molecular structure of Fe(TPP)(py)(CO), a model for unhindered coordination of CO.[118]

ical oxygen carriers have been determined.[183,188–191] Although the dioxygen moiety is usually ordered, the precision is tantalizingly just less than that needed to decide whether the apparently more-linear geometry seen for oxyerythrocruorin[183] and oxyhemoglobin[190] is significantly different from that for oxymyoglobin[188] and therefore attributable to the water molecule or imidazole that is hydrogen-bonded to the coordinated dioxygen ligand. Nonetheless, several interesting differences emerge.

The axial base in oxymyoglobin and oxyhemoglobin is almost eclipsed; that is, $\phi_1 \approx 0°$. The axial base has moved from a tilted position in deoxyhemoglobin to a symmetric one in oxyhemoglobin. In the absence of steric constraints, the iron atom is essentially in the center of the porphyrin plane for Fe(PF)(1-MeIm)(O$_2$), oxymyoglobin, and oxyhemoglobin. For the 2-methyl analogue, Fe(PF)(2-MeIm)(O$_2$), the iron remains significantly out of the plane, as also appears to occur for oxyerythrocruorin.

In the structure of Fe(TPP)(Py)(CO), Figure 4.30, a model for carbonyl hemoglobins, the iron atom is in the plane and the Fe—C≡O bond is linear and perpendicular, as expected.[118] Not so for carbonyl hemoglobins, where the blob of electron density that is identified with the coordinated carbon monoxide lies substantially off the normal to the porphyrin. We return to this point shortly. In general, with the exception of the coordinated ligand, the structures of six-coordinate low-spin hemoglobins, whether Fe^{II} or Fe^{III}, are similar. Indeed, the refined structures of oxy- and carbonmonoxyhemoglobin are superimposable within experimental uncertainties, except in the immediate vicinity of the diatomic ligand. Some metrical details are given in Table 4.8.[11c,118,121,172,183,187,188,190–192]

Table 4.8
Metrical details of selected liganded hemoglobins and their models [a]

Compound	Resol. Å	Fe-N$_p$ Å	Fe-porph Å	Doming Å	Fe-L$_1$ Å	Fe-L$_2$ Å	Fe-XY deg.	ϕ_1 deg.	ϕ_2 deg.	Tilt deg.
Dioxygen adducts										
Fe(PF)(1-MeIm)(O$_2$)	—	1.98(1)	−0.03	0.02	2.07(2)	1.75(2)	131(2)	20	45	0
Fe(PF)(2-MeIm)(O$_2$)	—	1.996(4)	0.09	0.02	2.107(4)	1.898(7)	129(1)	22	45	7
MbO$_2$	1.6	1.95(6)	0.18(3)	0.01	2.07(6)	1.83(6)	115(4)	1	~0	4
	EXAFS	2.02(2)			2.06(2)	1.80(2)	123(4), 148(8)[b]			
ErO$_2$	1.4	2.04	0.38	−0.08	2.1	1.8	150	7	3	—
HbO$_2$ α	2.1	1.99(5)	0.12(8)	0.04	1.94(9)	1.66(8)	153(7)	11	0	3
β		1.96(6)	−0.11(8)	0.11	2.07(9)	1.87(3)	159(12)	27	45	5
	EXAFS	1.99(2)			2.05(2)	1.82(2)	122(4), 143(8)[b]			
[α-FeO$_2$]$_2$[β-Fe]$_2$ α	2.1	2.04(4)	0.19(5)	0.17(5)	2.24(10)	1.82(4)	153(4)	6	—	11
β			0.3		2.2					
Carbonmonoxy adducts										
MbCO	1.5	1.97(3)	0.00	0.03	2.2	1.9	140	—	30	—
	EXAFS	2.01(2)			2.20(2)	1.93(2)	127(4), 145(8)[b]			
ErCO	1.4	2.01	−0.11	−0.10	2.1	2.2	161(9)	7	—	1
HbCO α	2.2	2.02	−0.10		1.95	1.83	175(15)			
β		2.03	−0.10		2.20	1.70	171(15)			
[α-Ni] [β-FeCO] α	2.6	—			2.23(5)	—	—			
β			0.15	0.12						
Fe(TPP)(Py)(CO)		2.02(3)	−0.02	−0.02	2.10(2)	1.77(2)	179(2)	45	—	~0
	EXAFS	2.02(2)			2.09(2)	1.81(2)	138(6), 180(11)[b]			
Fe(poc)(1,2-MeIm)(CO)		1.973(8)	0.001		2.079(5)	1.768(7)	172.5(6)			
Fe(C$_2$Cap)(1-MeIm)(CO)		1.990(7)	0.01		2.043(6)	1.742(7)	172.9(6)			
		1.988(13)	0.02		2.041(5)	1.748(7)	175.9(6)			

[a] See Figure 4.25 for definition of symbols.

b. Interactions of Coordinated Ligands with Distal Groups   Without exception to date (but see footnote 1 in Reference 168), in structurally characterized oxyhemoglobins, the coordinated dioxygen ligand is hydrogen-bonded to the distal histidine or to a water molecule—even though theoretical calculations show that hydrogen bonding would destabilize M—O_2 moieties.[192] This universal observation of hydrogen bonding in these biological systems is consistent with notions that electron density accumulates on the dioxygen molecule upon coordination. Given the errors associated with atomic positions (at best, $\pm$ 0.20 Å) the x-ray crystallographic evidence could be equivocal, since hydrogen atoms on the distal imidazole are not observed. There are at least three lines of evidence that support the existence of a specific $O_2 \cdots$ HN interaction. First, the EPR spectrum of cobalt oxyhemoglobin indicates that the coordinated dioxygen is hydrogen-bonded to something.[177-179] Second, and more directly, in the neutron-diffraction structure of oxymyoglobin,[189] where hydrogen and especially deuterium nuclides scatter strongly, the imino hydrogen or deuteron was located on the nitrogen atom closest to the coordinated dioxygen, as illustrated in Figure 4.31A. In contrast, in the neutron-diffraction structure of carbonmonoxymyoglobin, the alternative imidazole tautomer was observed (Figure 4.31B).[125,189] The absence of hydrogen bonding of the distal imidazole residue with the coordinated CO molecule is consistent with other lines of evidence that there is little accumulation of electron density on the carbonyl ligand. Third, but less directly, genetically engineered mutants have been produced in which the distal histidine has been replaced by glycine—sperm whale Mb E7His→Gly, and HbA αE7His→Gly and HbA βE7His→Gly.[35b,192] For the myoglobin mutant, the O_2 binding rate constant at room temperature increases by an order of magnitude, but the dissociation rate constant increases by two orders of magnitude, leading to a decrease in affinity of more than an order of magnitude, as derived from k_{on}/k_{off}. This leads to an estimate of the free energy associated with hydrogen bonding of

$$\Delta G = -RT\log\left[\frac{P_{1/2}(O_2)\text{-Native}}{P_{1/2}(O_2)\text{-Mutant}}\right] = 1.5 \text{ kcal/mol.}$$

In addition, this mutant myoglobin autooxidizes rapidly compared to the native one. On the other hand, the affinity for CO is greatly increased, leading to a value of M for the mutant of 1300, compared to 16 for the native. Thus the distal histidine stabilizes a coordinated O_2 ligand by hydrogen bonding and destabilizes a coordinated CO ligand by steric clash.

A similar discrimination is seen for the α chains of the hemoglobin mutant in the binding of the fourth O_2 or CO molecule. For the β chains little difference is seen relative to the native protein: hydrogen bonding between the distal histidine and the coordinated dioxygen ligand appears to be much weaker in β chains, as evidenced by longer N(H) $\cdots$ O separations than those seen in the α chains. Comparison of the crystal structures of the native and mutant $\alpha_2(\beta$E7His→Gly$)_2$ structures reveals negligible changes in the distal environ-

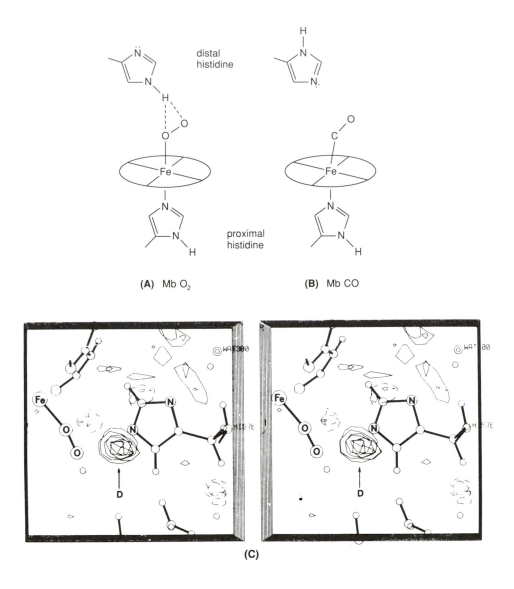

Figure 4.31
(A) The hydrogen bonding interaction between coordinated dioxygen and the distal histidine.
(B) The interaction of coordinated carbon monoxide and the distal histidine, showing the absence of hydrogen bonding. (C) Stereodiagram of a neutron-difference density map for oxymyoglobin. The refined structure showing the FeO_2 and distal histidine moieties is superimposed. The imidazole deuterium atom is arrowed.[189] Reproduced with permission from S. E. V. Phillips and B. P. Schoenborn, *Nature* **292** (1981), 81–82.

ment, except for that occasioned by the replacement of —CH_2—$C_3N_2H_3$ (histidine side chain) by —H (glycine side chain).

Studies of hemoglobin mutants where the nonpolar distal residue βValE11 (—$CH(CH_3)_2$) is replaced by alanine (—CH_3), isoleucine (—$CH(CH_3)CH_2CH_3$), and leucine (—$CH_2CH(CH_3)_2$) reveal that this valine offers steric hindrance to oxygen binding in the *T* state.

Whereas the angularly coordinated O_2 ligand fits comfortably around the distal histidine, a perpendicular and linear CO moiety cannot. Either the distal histidine rotates out of the way, or the CO tilts off axis, or the $Fe—C\equiv O$ group bends, or some combination of these occurs. Notwithstanding the absence of bent M—CO moieties in the inorganic literature, reports of strongly bent M—CO groups appear in the biochemical literature.[122-127] The controversy is illustrative of the synergistic interplay of data from models and proteins, and the importance of examining a problem with a miscellany of techniques. The molecular orbital model of ligand-metal interactions presented in Figure 4.22 does not preclude a bent $M—C\equiv O$ moiety on symmetry grounds. Groups related to CO can bend; the normally linearly coordinated SCN^- moiety has been observed[194] to become strongly bent under severe steric stress, with an Fe—N—CS of 140°. Unfortunately, the resolution in protein crystal structures is not sufficient to distinguish unequivocally a linear tilted stereochemistry from a bent one or from a combination of tilt and bend. Studies by the XANES technique have been interpreted in terms of a bent $Fe—C\equiv O$ moiety (150°)

$$\begin{array}{ccc} & \overset{\displaystyle O}{\underset{\displaystyle \parallel}{}} & \\ & \overset{C}{\diagup} & \\ N—Fe & \text{or} & N—Fe—C \overset{\displaystyle O}{\diagup} \end{array} \qquad (4.50)$$

both in MbCO[127] and in the CO adduct of a chelated heme in micelles, the latter being an especially surprising result. From EXAFS data on a number of carbonyl adducts, *two* interpretations were offered: linear or moderately bent (150°) FeCO moieties for unhindered model systems, and moderately bent or strongly bent (130°) FeCO moieties for hindered synthetic and biological systems.[195] In the crystal structure of MbCO at 1.5 Å resolution,[122] the CO group is disordered, and $Fe—C\equiv O$ angles of 120° and 140° were proposed, although the alternative model of tilted, nearly linear $Fe—C\equiv O$ stereochemistry could not be eliminated, and is indeed far more likely to account for the off-axis nature of the oxygen position. Vibrational spectroscopy confirms the existence of two major configurations, and indicates a third minor configuration of the $Fe—C\equiv O$ moiety in MbCO.[196] An elegant infrared study of the polarization of reattached carbon-monoxide molecules following photolysis of MbCO by linearly polarized light at 10 K gave tilt angles of the CO vector with respect to the heme normal of 15(3)°, 28(2)°, and 33(4)° for the three conformational substates;[196b] the former two values were confirmed in a similar study at room temperature.[196c] Note that these studies do not yield the tilt of the Fe—C bond to the heme normal.

In three synthetic compounds with severe steric hindrance, the extent of bending and tilting of the Fe—CO moiety is small. In one nonporphyrinic system the $Fe—C\equiv O$ group is bent by 9.4(5)° and tilted by 4.2°.[121a] In Fe(Poc-PF)(1,2-Me$_2$Im)(CO) the $Fe—C\equiv O$ angle is 172.5(6)° and modest tilting of the Fe—CO group and substantial buckling of the porphyrin ring are apparent.[121b] In Fe(C$_2$Cap)((1-MeIm)(CO) the two independent $Fe—C\equiv O$ angles are 172.9(6)° and 175.9(6)° and modest tilting of the Fe—CO group is again apparent. [121d]

From a detailed analysis of the force constants describing the vibrational spectroscopy for the Fe—CO moiety, values of 171° for the Fe—CO angle, 9.5° for the tilt, and 11° for porphyrin buckling were calculated for MbCO.[121c] These results are particularly important, for in a model complex very closely related to Fe(Poc-PF)(1,2-Me$_2$Im)(CO), just mentioned, an EXAFS study[195] suggested an Fe—C≡O bond angle of 127(4)°; that same study ascribed an Fe—C≡O bond angle of around 130° to MbCO. The structure of carbonmonoxyhemoglobin, Hb(CO)$_4$, now is interpreted in terms of a nearly linear tilted geometry.[192] Clearly the geometry of attachment of CO to hemoglobins is perturbed by the surroundings of the ligand-binding site and hence the affinity of hemoglobins for CO is also perturbed. Unfortunately, a clear resolution of the geometry of the Fe—CO moiety in MbCO does not exist yet.

D. Stereochemical Changes Upon Ligation

Upon binding a second axial ligand, the iron center together with the axial base move toward the plane of the porphyrin, initiating a change in spin state from high-spin to low-spin when the sixth ligand is O$_2$ or CO or any other strong ligand with an even number of valence electrons. Given these general features, what are the structural differences between systems that bind O$_2$ with high affinity and those that bind O$_2$ with low affinity? The answers to this question are relevant to understanding at the molecular level the mechanism of cooperativity, where a low-affinity conformation, the T state, and a high-affinity conformation, the R state, are in dynamic equilibrium in one tetrameric molecule. In looking at crystallographic data one sees a particular conformation frozen in the crystal, usually the one of lowest free energy among many in equilibrium in the solution state. The $R \rightleftarrows T$ equilibrium for hemoglobin is moderately rapid, at 4×10^3 s^{-1}; hemocyanin also switches quaternary conformations with a similar rate constant.[4]

Human hemoglobins are a heterogeneous group. Many mutants are known, and several have been structurally characterized. A structural alteration that affects the equilibrium between R and T states has a marked effect on ligand affinity and cooperativity in hemoglobin. If a specific amino-acid substitution destabilizes the T state, then the transition to the R state will occur earlier in the ligation process, and the hemoglobin will have an increased oxygen affinity. Hemoglobin Kempsey is an example. In this mutant an aspartic acid on the β chain is replaced by asparagine. Conversely, if the R state is destabilized, then the hemoglobin will have a lowered oxygen affinity. Hemoglobin Kansas is an example. Here an asparagine on the β chain has been replaced by threonine.[9]

1. Structural changes in normal-affinity systems

It was proposed earlier that the molecule Fe(PF)(1-MeIm)(O$_2$) in the solid or solution state was a fair approximation to the reference gas-phase molecule. The axial base, although not oriented for minimization of contacts with the porphyrin (i.e., $\phi = 45°$), is well-removed from an eclipsing orientation where $\phi = 0 \pm 10°$; the Fe atom is centered in the plane of a highly planar porphyrin;

the O_2 ligand is oriented·for minimization of contacts with the porphyrin, and its geometry is largely unconstrained by distal groups (the pickets); no groups are hydrogen-bonded to the axial base. The major difference from the reference state is that there is a significant attractive interaction between the electronegative dioxygen moiety and the amide groups on the pickets, and a smaller repulsive interaction with the picket t-butyl groups.[176]

For the CO adduct, contacts with the pickets are all at ideal van der Waals' separations and the Fe—CO moiety is free to assume its normal linear geometry. For CO binding the reference molecule is again the carbonyl adduct of the iron picket-fence porphyrinato molecule. In contrast to O_2 binding, there are no specific distal effects, such as hydrogen bonding, by which CO affinity may be increased; there remain many ways, as with O_2 binding, by which CO affinities may be reduced. Thus, the CO binder with highest affinity is the iron picket-fence porphyrin.

The O_2 affinities of myoglobin, R-state hemoglobin, and the Fe(PF)(1-MeIm) system are similar. However, the means by which this is achieved are different, and this difference is reflected most clearly in the kinetics of binding and release of O_2, which for Mb are much slower. The similarities and differences are summarized in Table 4.9, which is culled from Tables 4.2, 4.4, and 4.5.

Table 4.9
Comparison of the picket-fence porphyrin system with Mb.

Characteristic	Mb	Fe(PF-Im)[a] Fe(PF)(1-MeIm)	Fe(Poc-PF) (1-MeIm)
$P_{1/2}(O_2)$, Torr	0.7	0.58	0.36
$P_{1/2}(CO)$, Torr	0.018	0.000022	0.0015
$k_{on}(O_2)$, $\mu M^{-1} s^{-1}$	15	430	2.2
$k_{off}(O_2)$, s^{-1}	10	2,900	9
$k_{on}(CO)$, $\mu M^{-1} s^{-1}$	0.50	36	0.58
$k_{off}(CO)$, s^{-1}	0.015	0.0078	0.0086
Solvent	H_2O/PO_4^{3-}	toluene	toluene
Local environment	His (H_2O)	H-N(amide)	H-N (amide)
	polar, protic	polar, aprotic	phenyl
$O_2 \cdots$ distal, Å	2.97(18)	NH 4.06(5)	
		CH_3 2.67(6)	
$CO \cdots$ distal, Å (calc.)	2.7	NH 5.0	
		CH_3 3.3	
Fe-N_{Im}(deoxy), Å	2.22	2.095(6)	
Fe-N_{Im}(oxy), Å	2.07(6)	2.07(2)	
Fe-O, Å	1.83(6)	1.75(2)	
ϕ_{Im}(deoxy), deg	19°	22.8°	
ϕ_{Im}(oxy), deg	1°	20°	
ϕ_{O2}, deg	~0°	45°	
Fe $\cdots$ porph(deoxy), Å	0.42	0.43	
Fe $\cdots$ porph(oxy), Å	0.18	−0.03	

[a] Solution state studies on Fe(PF-Im). Structural details on Fe(PF)(1-MeIm)(O_2) and Fe(PF)(2-MeIm).

For the cobalt-dioxygen derivative, the putative hydrogen bonding between the dioxygen and the amide groups of the pickets assumes greater importance because the coordinated dioxygen is substantially more negative. Again the picket-fence porphyrin, being structurally characterized, is the reference system. Although no Co picket-fence porphyrin structures have been determined, the structures may be predicted with confidence from the iron analogues together with related structures of Co^{II} and Co^{III} tetraphenylporphyrinato systems.*

2. Structural changes in low-affinity systems

The 2-methyl substituent on 2-methylimidazole is not sterically active in the five-coordinate structures Fe(PF)(2-MeIm) and Fe(TPP)(2-MeIm), since the iron atom is displaced from the plane of the porphyrin by the expected amount and the Fe—N_{Im} bond is unstretched and similar to that in deoxyhemoglobin (low O_2 affinity) and deoxyMb (higher O_2 affinity). Moreover, resonance Raman measurements also indicate little strain[†] in this bond.[200] In other words, there is no "tension at the heme," a key concept in early discussions of cooperativity before structures on model systems and high-resolution, refined protein structures became available.[11a] On moving into the plane of the porphyrin upon oxygenation, the 2-methyl substituent prevents the Fe-imidazole group from achieving its optimum geometry with the iron at the center of the porphyrin hole, as seen in the structure of Fe(PF)(1-MeIm)(O_2). Thus, the sterically active 2-methyl substituent leads to lowered O_2 (and CO) affinity relative to the 1-methyl analogue. In metrical terms the lowered affinity is reflected in an increase in the sum of the axial bond lengths from $1.75 + 2.07 = 3.82$ Å to $1.90 + 2.11 = 4.01$ Å.

In the crystal structure of Fe(C_2Cap)(1-MeIm)(CO) the cap is about 5.6 Å from the porphryin plane.[121d] Hence, in the crystal structures of the free base $H_2(C_2Cap)$[201a] and FeCl(C_2Cap)[201b] species, in which the cap is screwed down to approximately 4.0 Å from the porphyrin plane, considerable conformational rearrangement of the cap and the four chains attaching it to the porphyrin is needed to provide room for a small ligand such as CO. This is even more pronounced in a Co(C_3Cap) complex where the cap is only 3.49 Å from the mean porphyrin plane.[202] Thus not only is affinity for CO lowered, but some additional discrimination against it is induced, since a linear, perpendicular coordination creates considerable strain energy elsewhere in the molecule.

For the pocket porphyrin (Figure 4.23), structural data are available on the

---

* For $CoHbO_2$, single-crystal EPR spectra have been interpreted in terms of a nearly triangularly coordinated O_2,[197] although a crystal structure of $CoMbO_2$ shows a bent CoOO group.[198] There is no precedent for this triangular arrangement in any Co^{III}-superoxo (O_2^{1-}) system, whereas there are many for angularly coordinated O_2 in electronically not dissimilar square-planar Schiff-base systems.[66] Regardless of geometry, the picket amide $\cdots O_2$ contacts do not change substantially.

† Shortly (10^{-9}—10^{-12} s) after a ligand dissociates, a large difference in v(Fe—N_{Im}) between R and T structures is observed, prior to relaxation to the equilibrium R and T conformations.[199]

carbonyl adduct.[121b] The CO ligand is unable to achieve the linear perpendicular geometry seen in the high-affinity picket-fence porphyrin derivative, Fe(PF)(1-MeIm)(CO),[110] and distortion of the porphyrin core is greater. In the pocket-porphyrin system, O_2 affinity is unaffected, but CO affinity is lowered.

The crystal structure of partially oxygenated hemoglobin, $[\alpha\text{-FeO}_2]_2[\beta\text{-Fe}]_2$,[191a] reveals that the quaternary structure, except in the immediate vicinity of the α hemes, which have O_2 coordinated, resembles that of T-state deoxyhemoglobin rather than R-state liganded hemoglobin. In accord with the low affinity of T-state hemoglobin, the Fe—N_{Im} bonds for the six-coordinate α-hemes at 2.37 Å are significantly longer than those in fully oxygenated R-state oxyhemoglobin, $[\alpha\text{-FeO}_2]_2[\beta\text{-FeO}_2]_2$ in the notation above, (1.94 (α-hemes) and 2.06 Å (β hemes)) and that found in oxymyoglobin (2.07 Å). In contrast to the R-state structure and oxyMb, the α-hemes are folded as seen in the deoxy parent, leaving the Fe still substantially displaced (0.2 Å) from the plane of the four pyrrole nitrogen atoms. The deoxyhemoglobin T-state quaternary structure also has been observed in two other partially liganded hybrid hemoglobins, $[\alpha\text{-FeCO}]_2[\beta\text{-Mn(II)}]_2$[203] and $[\alpha\text{-Ni}]_2[\beta\text{-FeCO}]_2$.[191d] Again, structural changes upon coordination do not propagate beyond the immediate vicinity of the liganded heme to the critical $\alpha_1\beta_2$ interfaces.

Note that although the crystal structure of hemoglobin A reveals that access to the binding site for the β chains is blocked by groups at the entrance to the cavity above the iron center, this does not prevent facile access to the binding site; the rate of O_2 binding is slowed by a factor of only five. A similar situation occurs also for vertebrate myoglobins.

The large structural differences that exist between deoxy (T) and oxy (R) hemoglobin and the much smaller differences between deoxy (T) and partially liganded (T) hybrid hemoglobin are shown in Figure 4.32.[203]

Because of the steric hindrance afforded by the distal histidine, all biological systems have low affinity for CO relative to the picket-fence porphyrins, with the exception of mutants where the distal histidine has been replaced by glycine. Thus low affinity to CO is associated primarily with the inability of the Fe-CO group to achieve its preferred linear geometry perpendicular to the porphyrin.

Low-affinity O_2 binding in the hemoglobins appears to be associated with the inability of the Fe-proximal histidine unit to move into the plane of the porphyrin and less so to distal effects, such as a cavity too small to accommodate the coordinated ligand. The blocked access to the site affects the kinetics but not necessarily the thermodynamics of ligand binding, as evidenced by the structure of T-state $[\alpha\text{-Ni}]_2[\beta\text{-FeCO}]_2$.[191d] Some similarities between the structures and properties of partially oxygenated (T-state) $[\alpha\text{-FeO}_2]_2[\beta\text{-Fe}]_2$ hemoglobin and Fe(PF)(2-MeIm)(O_2) are provided in Table 4.10. In the synthetic systems low O_2 affinity can be induced by 2-methyl substituents—a restraint on the movement of the Fe-imidazole moiety analogous to that provided by the protein chain. A second means is by distal effects, such as caps and straps.

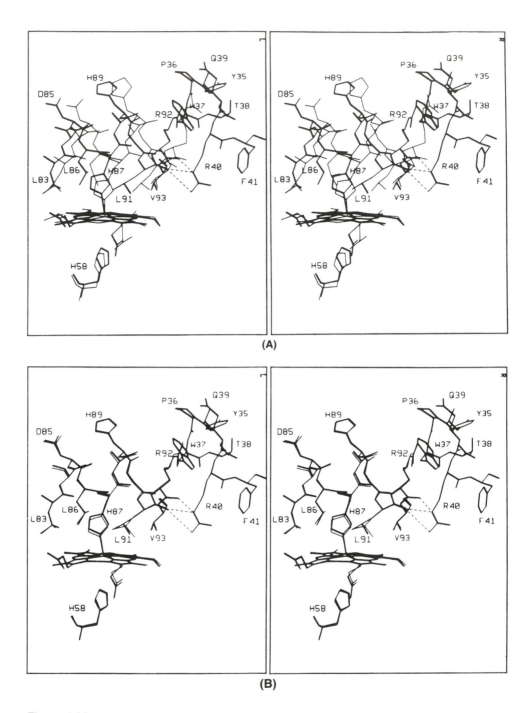

Figure 4.32
The effects of changes in ligation and in quaternary structure on stereochemistry in the vicinity of α hemes:[203] (A) Stereodiagram of the large structural differences between deoxy T-state (heavy lines) and oxy R-state hemoglobin (light lines). (B) Stereodiagram of the small structural differences between deoxy T-state (heavy lines) and partially liganded T-state $[\alpha\text{-FeCO}]_2[\beta\text{-Mn}]_2$ hemoglobin (light lines). Reproduced with permission from A. Arnone *et al.*, *J. Mol. Biol.* **188** (1986), 693–706.

Table 4.10

Comparison of the low-affinity picket-fence porphyrin system with low-affinity (*T*-state) partially liganded hemoglobin.

Characteristic	HbA^T [α-FeO$_2$]$_2$[β-Fe]$_2$	Fe(PF)(2-MeIm) Fe(PF)(1,2-Me$_2$Im)^a
$P_{1/2}(O_2)$, Torr	46, first O$_2$	38
$P_{1/2}(CO)$, Torr	~0.7, 1st CO	0.0089
$k_{on}(O_2)$, $\mu M^{-1} s^{-1}$	2.9(α)	106
$k_{off}(O_2)$, s^{-1}	183(β)	46,000
$k_{on}(CO)$, $\mu M^{-1} s^{-1}$	0.099	1.4
$k_{off}(CO)$, s^{-1}	0.09	0.14
Solvent	Tris buffer, pH 7, no 2,3-DPG	toluene
Local environment	histidine polar	H-N(amide) polar, aprotic
O$_2$ · · · distal, Å	?	NH 3.88 CH$_3$ 2.77(3)
CO · · · distal, Å		NH 4.9 CH$_3$ 3.5
Fe-N$_{Im}$(deoxy), Å	2.13(6) (average of α & β)	2.095(6)
Fe-N$_{Im}$(oxy), Å	2.24(10)	2.107(4)
Fe-O, Å	1.82(4)	1.898(7)
ϕ_{Im}(deoxy), deg	21(3) (average of α & β)	22.8
ϕ_{Im}(oxy), deg	~6	22.2
ϕ_{O_2}, deg	?	45
Fe · · · porph(deoxy), Å	0.38(5) (average of α & β)	0.43
Fe · · · porph(oxy), Å	0.19(5)	0.09

^{a} Ligand binding to Fe(PF)(1,2-Me$_2$Im)

3. Structural changes in high-affinity systems

Few structural data are available for high-affinity oxygen carriers. The crystal structures of two leghemoglobin derivatives, a monomeric myoglobin-like oxygen carrier found in the nitrogen-fixing nodules of legumes, are known at 2.0 and 3.3 Å.[204,205] The binding pocket appears more open, perhaps allowing H$_2$O to enter and partake in stronger hydrogen bonding than that offered by the distal imidazole. Consistent with this notion is the more rapid rate of autoxidation observed for oxyleghemoglobin. *Aplysia* oxymyoglobin, which lacks a distal histidine, also autoxidizes rapidly,[204] although a distal arginine further along the helix E, E10Arg, fulfills the role of the distal histidine by hydrogen bonding to the sixth ligand, at least in the fluoride derivative, met-MbF.

Although no structural data are available, a tenfold increase in O$_2$ affinity was observed between an ester-strapped porphyrin, offering no hydrogen-bonding possibilities, and its conformationally very similar amide analogue. O$_2$ · · · amide hydrogen bonding was demonstrated by means of NMR shift data (Zn and Fe—CO complexes vs. the Fe—O_2 complex) and from infrared spectroscopy, which showed shifted amide N—H absorptions.[166]

The *specific* structural features that lead to the extraordinarily high affinity for O_2 and low affinity for CO in hemoglobin *Ascaris* remain unidentified. This high affinity is due to an *extremely* slow dissociation rate of O_2 of only 0.1 s^{-1}; in most hemoglobins the rate is about 10 to 2,500 s^{-1} (Table 4.2). Dioxygen binding is thus close to irreversible. Figure 4.27 shows that hydrogen bonding to the coordinated dioxygen ligand, unrestrained motion of the Fe-proximal histidine group into the plane of the porphyrin, hydrogen bonding to the proximal histidine, and, in the deoxy form, compression of the Fe—N_{Im} bond and decrease in the out-of-plane displacement of the Fe atom will all increase O_2 affinity over that of a system where these effects are absent.

When hydrogen bonding is impossible, as in various synthetic systems (Table 4.5) as well as hemoglobin *Glycera* and Mb(E7His→Gly), O_2 affinity is much lower than when hydrogen bonding can occur (see Table 4.6), especially for the cobalt analogues. But caution is needed in the absence of complete structural information: the lowered affinity of *Aplysia* hemoglobin had been attributed to the lack of a distal histidine and its attendant hydrogen-bonding capabilities. However, the crystal structure reveals that an arginine residue, normally directed out into the solution, is capable of folding back into the ligand-binding pocket and of hydrogen bonding to ligands at the sixth site. In oxyhemerythrin the hydrogen bonding of the coordinated hydroperoxy group to the oxo bridge linking the two iron atoms (Figure 4.10B), described in Section II.F.1, may not only increase the stability of oxyhemerythrin,[146] but also facilitate electron transfer that occurs in dioxygen binding.[205]

IV. DIOXYGEN CARRIERS AND BIOINORGANIC CHEMISTRY

To the student the subject of biological and synthetic molecular-oxygen carriers offers unusual insights into how bioinorganic chemistry works and what its aims and uses are. First, consider another bioinorganic problem, that of the nature of the blue copper proteins. When bioinorganic chemists entered the scene, the nature, function, and structure of copper blues were largely unknown. To take a Cu(II) solution, add a nitrogen base, and obtain a spectrum that "resembled" that of the proteins was not a contribution to the solution of the biological puzzle, although it was the activity of some bioinorganic chemists. The difficulty here was that too little was known about the protein system. But the challenge did not diminish once the structure of a blue copper protein was known, for that structure allowed definition of the active site, one that contained (in the oxidized form) a Cu(II) center surrounded by two imidazole, one cysteine, and one methionine residue. Now the bioinorganic chemist was faced with the formidable (and still incompletely solved) synthetic problem of designing a tetradentate ligand that (i) would present two N atoms, one thiolate S atom, and one thioether SR_2 group to a Cu(II) center and (ii) would remain intact if the Cu(II) center were reduced to Cu(I). If we could prepare such complexes, we would be in a position to examine in some detail the effects on physical properties, such as redox potentials or spectra, of chemical substitution. In other words, we could learn about structure-function relationships in the copper blues. But the risks involved included the possibility that the specially designed ligand, even if it could be synthesized, might not bind Cu(II) or Cu(I) in the desired manner.

Contrast such a situation with that of the oxygen carriers. Hemoglobin[208] and myoglobin[209] were the first crystallized proteins to have their structures determined. Their functions were well-known. They had been studied by a wide variety of physical techniques, in part because their structures were known, and even before that because of their role in human health. The central tetradentate ligand of the heme group, namely, the porphyrin, was well-defined and much porphyrin chemistry was known. The structural puzzles that intrigued chemists and biologists were not answered in the initial, early structural studies of the proteins; for example, how is O_2 bound and why is CO not bound more firmly? Model chemistry in this area looked as though it would be easy; after all, the metalloporphyrins were readily synthesized, and all one needed was an axial base, some spectroscopic equipment, perhaps some single crystals, and then the structure-function relationships in the biological oxygen carriers would be understood! Indeed, as often happens, the situation was more complicated than it appeared. The irreversible oxidation of iron porphyrins was a major stumbling block to simple modeling. This obstacle was overcome in solution studies through the use of low temperatures and aprotic solvents; some very useful measurements of O_2 and CO binding were made on model systems in such solutions. But in order to isolate oxygen complexes so that they could be studied by diffraction methods, another approach, that of synthesizing elaborated porphyrins, such as those in Figure 4.23, was necessary. This task entailed difficult organic chemistry that ultimately led to successful models that proved to be stable under ambient conditions. From such models we have learned much about local stereochemistry and, through spectroscopic congruence, about the biological systems. In short, bioinorganic chemistry has made a major contribution to the understanding of biological molecular-oxygen carriers, primarily because knowledge of the biological systems was advanced, the systems "self-assemble," and the goals of the studies were well-defined.

The complementarity of the two approaches continues. There are several unanswered questions, including:

(1) What is the structural basis for cooperativity? Indeed, is there a structural basis at all, or is the ~6 kcal/mol that represents the effect spread over many interactions, so that there is no obvious structural effect to be modeled?

(2) Can one design a model where hydrogen bonding to the bound O_2 molecule can be demonstrated by diffraction experiments? How will the oxygen uptake properties depend on the strength of the hydrogen bond?

(3) Can one design a "high-affinity" model system? What will this tell us about the largely ill-defined high-affinity systems that are found in Nature?

There remain many intriguing questions about biological molecular-oxygen carriers, questions that will be answered by complementary studies on the biological and model systems. To make and study such model systems is an example of the challenge and excitement of this aspect of bioinorganic chemistry.

V. REFERENCES

1. C. K. Mathews and K. E. van Holde, *Biochemistry*, Benjamin/Cummings, 1990.
2a. M. Nikinmaa, *Vertebrate Red Blood Cells: Adaptations of Function to Respiratory Requirements*, Springer-Verlag, 1990.
2b. D. Hershey, ed., *Blood Oxygenation*, Plenum, 1970.
2c. D. W. Lübbers, H. Acker, E. Leniger-Follert, and T. K. Goldstick, eds., *Oxygen Transport to Tissues V*, Plenum, 1984.
2d. F. Kreuzer, S. M. Cain, Z. Turek, and T. K. Goldstick, eds., *Oxygen Transport to Tissues VII*, Plenum, 1984.
3a. P. Astrup and M. Rørth, eds., *Oxygen Affinity of Hemoglobin and Red Cell Acid Base Status*, Munksgaard, 1972.
3b. G. L. Eichhorn, ed., *Inorganic Biochemistry*, Elsevier, 2 vols., 1973.
4. M. Brunori, A. Coletta, and B. Giardina, in P. M. Harrison, ed., *Topics in Molecular and Structural Biology*, Verlag Chemie, **7**, Part 2 (1985), 263–331.
5. A. G. Sykes, in A. G. Sykes, ed., *Advances in Inorganic and Bioinorganic Mechanisms*, Academic Press, **1** (1985), 121–178.
6. M. Brunori, B. Giardina, and H. A. Kuiper, in H. A. O. Hill, ed., *Inorganic Biochemistry*, Royal Society of Chemistry **3** (1982), 126–182.
7. R. E. Dickerson and I. Geis, *Hemoglobin: Structure, Function, Evolution, and Pathology*, Benjamin/Cummings, 1983.
8. H. F. Bunn, B. G. Forget, and H. M. Ranney, *Human Hemoglobins*, Saunders, 1977.
9. K. Imai, *Allosteric Effects in Haemoglobin*, Cambridge University Press, 1982.
10. E. Antonini and M. Brunori, *Hemoglobin and Myoglobin in Their Reactions with Ligands*, North Holland, 1971.
11a. M. F. Perutz, *Nature* **228** (1970), 726–739.
11b. M. F. Perutz, *Annu. Rev. Biochem.* **48** (1979), 327–386.
11c. M. F. Perutz *et al.*, *Acc. Chem. Res.* **20** (1987), 309–321.
12. M. C. M. Chung and H. D. Ellerton, *Prog. Biophys. Mol. Biol.* **35** (1979), 53–102.
13. J. Lamy and J. Lamy, eds., *Invertebrate Oxygen-Binding Proteins: Structure, Active Site and Function*, Dekker, 1981.
14. H. D. Ellerton, N. F. Ellerton, and H. A. Robinson, *Prog. Biophys. Mol. Biol.* **41** (1983), 143–248.
15a. K. E. van Holde and K. I. Miller, *Quart. Rev. Biophys.* **15** (1982), 1–129.
15b. K. D. Karlin and J. Zubieta, eds., *Biological and Inorganic Chemistry of Copper*, Academic Press, 2 vols., 1986.
16. P. C. Wilkins and R. G. Wilkins, *Coord. Chem. Rev.* **79** (1987), 195–214; in this see references to the original literature.
17. I. M. Klotz and D. M. Kurtz, Jr., *Acc. Chem. Res.* **17** (1984), 16–22.
18. J. Sanders Loehr and T. M. Loehr, *Adv. Inorg. Biochem.* **1** (1979), 235–252.
19. K. Denbigh, *The Principles of Chemical Equilibrium*, Cambridge University Press, 4th ed., 1981.
20. P. W. Atkins, *Physical Chemistry*, Freeman, 3d ed., 1986.
21. J. P. Collman, J. I. Brauman, and K. M. Doxsee, *Proc. Natl. Acad. Sci. USA* **76** (1979), 6035–6039.
22. D. Lexa *et al.*, *Inorg. Chem.* **25** (1986), 4857–4865.
23. K. S. Suslick, M. M. Fox, and T. J. Reinert, *J. Am. Chem. Soc.* **106** (1984), 4522–4525.
24. A. V. Hill, *J. Physiol.* **40** (1910), iv–vii.

25a. R. W. Root, *Biol. Bull.* (Woods Hole, Mass.) **61** (1931), 427–456.

25b. G. G. Dodson *et al.*, *J. Mol. Biol.* **211** (1990), 691–692.

26. G. S. Adair, *J. Biol. Chem.* **63** (1925), 529–545.

27. Reference 9, 114.

28. J. Monod, J. Wyman, and J.-P. Changeux, *J. Mol. Biol.* **12** (1965), 88–118.

29a. M. L. Johnson, B. W. Turner, and G. K. Ackers, *Proc. Natl. Acad. Sci. USA* **81** (1984), 1093–1097.

29b. M. Straume and M. L. Johnson, *Biochemistry* **27** (1988), 1302–1310.

29c. G. K. Ackers and F. R. Smith, *Annu. Rev. Biophys. Chem.* **16** (1987), 583–609.

30. L. Pauling *et al.*, *Science*, **110** (1949), 543–548.

31. G. B. Jameson and J. A. Ibers, *Comments Inorg. Chem.* **2** (1983), 97–126; in this see references to the original literature.

32. Q. H. Gibson and M. H. Smith, *Proc. Roy. Soc., Ser. B, Biol. Sci.* **163** (1965), 206–214.

33a. T. Imamura, A. Riggs, and Q. H. Gibson, *J. Biol. Chem.* **247** (1972), 521–526.

33b. J. B. Wittenberg, C. A. Appleby, and B. A. Wittenberg, *J. Biol. Chem.* **247** (1972), 527–531.

34. L. J. Parkhurst, *Annu. Rev. Phys. Chem.* **30** (1979), 503–546; in this see references to the original literature.

35a. M. P. Mims *et al.*, *J. Biol. Chem.* **258** (1983), 14219–14232; in this see references to the original literature.

35b. J. S. Olson *et al.*, *Nature* **336** (1988), 265–266.

36. G. D. Armstrong and A. G. Sykes, *Inorg. Chem.* **25** (1986), 3135–3139.

37. R. Lontie and R. Winters, in H. Sigel, ed., *Metal Ions in Biological Systems*, Dekker, **13** (1981), 229–258.

38a. M. Brunori *et al.*, in Reference 13, 693-701.

38b. E. Antonini *et al.*, *Biophys. Chem.* **18** (1983), 117–124.

39a. B. Richey, H. Decker, and S. J. Gill, *Biochemistry* **24** (1985), 109–117.

39b. M. Brunori *et al.*, *J. Mol. Biol.* **153** (1981), 1111–1123.

39c. H. Decker *et al.*, *Biochemistry* **27** (1988), 6901–6908.

40a. G. L. Woolery *et al.*, *J. Am. Chem. Soc.* **106** (1984), 86–92.

40b. J. M. Brown *et al.*, *J. Am. Chem. Soc.* **102** (1980), 4210–4216.

41. M. S. Co. *et al.*, *J. Am. Chem. Soc.* **103** (1981), 984–986.

42a. W. P. J. Gaykema *et al.*, *Nature* **309** (1984), 23–29.

42b. W. P. J. Gaykema, A. Volbeda, and W. G. J. Hol, *J. Mol. Biol.* **187** (1985), 255–275.

42c. B. Linzen *et al.*, *Science* **229** (1985), 519–524.

43a. H. A. DePhillips, Jr., *Arch. Biochem. Biophys.* **144** (1971), 122–126.

43b. D. E. Richardson, R. C. Reem, and E. I. Solomon, *J. Am. Chem. Soc.* **105** (1982), 7780–7781.

44. D. J. A. de Waal and R. G. Wilkins, *J. Biol. Chem.* **251** (1976), 2339–2343.

45. R. E. Stenkamp *et al.*, *Proc. Natl. Acad. Sci. USA* **82** (1985), 713–716.

46a. R. E. Stenkamp, L. C. Sieker, and L. H. Jensen, *J. Am. Chem. Soc.* **106** (1984), 618–622.

46b. Ibid., *J. Mol. Biol.* **126** (1978), 457–466.

47. S. Sheriff *et al.*, *Proc. Natl. Acad. Sci. USA* **82** (1985), 1104–1107.

48. S. Sheriff, W. A. Hendrickson, and J. L. Smith, *J. Mol. Biol.* **197** (1987), 273–296.

49. H. P. Misra and I. Fridovich, *J. Biol. Chem.* **247** (1972), 6960–6962.

50. W. J. Wallace, J. C. Maxwell, and W. S. Caughey, *FEBS Lett.* **43** (1974), 33–36; *Biochem. Biophys. Res. Comm.* **57** (1974), 1104–1110.

51a. D. E. Hultquist, L. J. Sannes, and D. A. Juckett, *Curr. Top. Cell. Regul.* **24** (1984), 287–300.

51b. M. R. Mauk and A. G. Mauk, *Biochemistry* **21** (1982), 4730–4734.

52. I. Fridovich, *Acc. Chem. Res.* **5** (1972), 321–326.

53. D. T. Sawyer and J. S. Valentine, *Acc. Chem. Res.* **14** (1981), 393–400.

54. I. Fridovich vs. D. T. Sawyer and J. S. Valentine, *Acc. Chem. Res.* **15** (1982), 200 (correspondence).

55. W. Braun *et al.*, *J. Mol. Biol.* **187** (1986), 125–129.

56. F. A. Cotton and G. Wilkinson, *Advanced Inorganic Chemistry: A Comprehensive Text*, Wiley, 4th ed., 1980.

57. A. F. Wells, *Structural Inorganic Chemistry*, Oxford University Press, 5th ed., 1984.

58. J. E. Huheey, *Inorganic Chemistry: Principles of Structure and Reactivity*, Harper and Row, 3rd ed., 1983.

59. S. A. Fairhurst and L. H. Sutcliffe, *Prog. Biophys. Mol. Biol.* **34** (1978), 1–79.

60. B. K. Teo, *EXAFS: Basic Principles and Data Analysis*, Springer-Verlag, 1986.

61. A. Bianconi *et al.*, *Phys. Rev. B* **26** (1982), 6502–6508.

62. *Handbook of Chemistry and Physics*, CRC Press, 68th ed., 1987-88, D151-D155.
63. L. Vaska, *Acc. Chem. Res.* **9** (1976), 175–183; in this see references to the original literature.
64. R. D. Jones, D. A. Summerville, and F. Basolo, *Chem. Rev.* **79** (1979), 139–179; in this see references to the original literature.
65. W. R. Scheidt and Y. J. Lee, *Structure and Bonding* **64** (1987), 1–70; in this see references to the original literature.
66. E. C. Niederhoffer, J. H. Timmons, and A. E. Martell, *Chem. Rev.* **84** (1984), 137–203; in this see references to the original literature.
67. J. O. Alben *et al.*, *Biochemistry* **7** (1968), 624–635.
68. G. S. Hammond and C.-S. Wu, *Adv. Chem. Ser.* **77** (1968), 186–207.
69. D.-H. Chin *et al.*, *J. Am. Chem. Soc.* **99** (1977), 5486–5488.
70. J. E. Penner-Hahn *et al.*, *J. Am. Chem. Soc.* **108** (1986), 7819–7825.
71. A. L. Balch *et al.*, *J. Am. Chem. Soc.* **106** (1984), 7779–7785.
72. J. P. Collman, *Acc. Chem. Res.* **10** (1977), 265–272.
73. J. H. Wang, *J. Am. Chem. Soc.* **80** (1958), 3168–3169.
74. E.-I. Ochiai, *Inorg. Nucl. Chem. Lett.* **10** (1974), 453–457.
75. C. C. Winterbourn and J. K. French, *Biochem. Soc. Trans.* **5** (1977), 1480–1481; J. K. French, C. C. Winterbourn, and R. W. Carrell, *Biochem. J.* **173** (1978), 19–26.
76. B. C. Antanaitis and P. Aisen, *Adv. Inorg. Biochem.* **5** (1983), 111–136.
77. B.-M. Sjöberg and S. A. Gräslund, *Adv. Inorg. Biochem.* **5** (1983), 87–110.
78a. H. Toftlund *et al.*, *J. Chem. Soc. Chem. Comm.* (1986), 191–192.
78b. M. P. Woodland and H. Dalton, *J. Biol. Chem.* **259** (1984), 53–59.
79a. J. LeGall *et al.*, *Biochemistry* **27** (1988), 1636–1642.
79b. E. C. Theil, *Adv. Inorg. Biochem.* **5** (1983), 1–38.
80. K. S. Murray, *Coord. Chem. Rev.* **12** (1974), 1–35.
81. P. Gomez-Romero, G. C. DeFotis, and G. B. Jameson, *J. Am. Chem. Soc.* **108** (1986), 851–853.
82. W. H. Armstrong *et al.*, *J. Am. Chem. Soc.* **106** (1984), 3653–3667.
83. C. C. Ou *et al.*, *J. Am. Chem. Soc.* **100** (1978), 2053–2057.
84. W. M. Reiff, G. J. Long, and W. A. Baker, Jr., *J. Am. Chem. Soc.* **90** (1968), 6347–6351.
85. J. A. Bertrand and P. G. Eller, *Inorg. Chem.* **13** (1974), 927–934.
86. B. F. Anderson *et al.*, *Nature* **262** (1976), 722–724.
87a. R. S. Czernuszewicz, J. E. Sheats, and T. G. Spiro, *Inorg. Chem.* **26** (1987), 2063–2067.
87b. B. A. Averill *et al.*, *J. Am. Chem. Soc.* **109** (1987), 3760–3767.
88a. P. Chaudhuri *et al.*, *Angew. Chem. Intl. Ed. Engl.* **24** (1985), 778–779.
88b. J. A. R. Hartman *et al.*, *J. Am. Chem. Soc.* **109** (1987), 7387–7396.
89. C. Bull, G. J. McClune, and J. A. Fee, *J. Am. Chem. Soc.* **105** (1983), 5290–5300.
90a. R. E. Hester and E. M. Nour, *J. Raman Spectrosc.* **11** (1981), 35–38.
90b. S. Ahmad *et al.*, *Inorg. Chem.* **27** (1988), 2230–2233.
91. W. R. Scheidt and C. A. Reed, *Chem. Rev.* **81** (1981), 543–555; in this see references to the original literature.
92. L. Pauling and C. D. Coryell, *Proc. Natl. Acad. Sci. USA* **22** (1936), 210–216.
93. M. Cerdonio *et al.*, *Proc. Natl. Acad. Sci. USA* **74** (1977), 398–400.
94. Z. S. Herman and G. H. Loew, *J. Am. Chem. Soc.* **102** (1980), 1815–1821.
95. A. Dedieu, M.-M. Rohmer, and A. Veillard, in B. Pullman and N. Goldblum, eds., *Metal Ligand Interactions in Organic Chemistry and Biochemistry*, Part 2, Reidel, 1977, 101–130.
96. W. A. Goddard III and B. D. Olafson, *Ann. N. Y. Acad. Sci.* **367** (1981), 419–433.
97. B. Boso *et al.*, *Biochim. Biophys. Acta* **791** (1984), 244–251.
98. J. P. Savicki, G. Lang, and M. Ikeda-Saito, *Proc. Natl. Acad. Sci. USA* **81** (1984), 5417–5419.
99. P. E. Ellis, Jr., R. D. Jones, and F. Basolo, *J. Chem. Soc. Chem. Comm.* (1980), 54–55.
100. M. Rougee and D. Brault, *Biochemistry* **14** (1975), 4100–4106.
101. J. S. Thompson, T. J. Marks, and J. A. Ibers, *J. Am. Chem. Soc.* **101** (1979), 4180–4192.
102. M. G. Burnett *et al.*, *J. Chem. Soc. Chem. Comm.* (1980), 829–831; M.G. Burnett, V. McKee, and S. M. Nelson, *loc. cit.* (1980), 599–601.
103a. J. S. Thompson, *J. Am. Chem. Soc.* **106** (1984), 8308–8309.
103b. C. L. Merrill *et al.*, *J. Chem. Soc., Dalton Trans.* (1984), 2207–2221.
103c. L. Casella, M. S. Silver, and J. A. Ibers, *Inorg. Chem.* **23** (1984), 1409–1418.
103d. Y. Nishida *et al.*, *Inorg. Chim. Acta* **54** (1981), L103–L104.
103e. K. D. Karlin *et al.*, *J. Am. Chem. Soc.* **110** (1988), 1196–1207.

103f. R. R. Jacobson *et al.*, *J. Am. Chem. Soc.* **110** (1988), 3690–3692.

103g. N. Kitajima *et al.*, *J. Am. Chem. Soc.* **114** (1992), 1277–1291.

104a. F. Basolo, B. M. Hoffman, and J. A. Ibers, *Acc. Chem. Res.* **8** (1975), 384–392; in this see references to the original literature.

104b. G. A. Rodley and W. T. Robinson, *Nature* **235** (1972), 438–439.

105. T. D. Smith and J. R. Pilbrow, *Coord. Chem. Rev.* **39** (1981), 295–383; in this see references to the original literature, and a critical reanalysis of results in References 107a and 197.

106. B. M. Hoffman and D. H. Petering, *Proc. Natl. Acad. Sci. USA* **67** (1970), 637–643.

107a. B. S. Tovrog, D. J. Kitko, and R. S. Drago, *J. Am. Chem. Soc.* **98** (1976), 5144–5153.

107b. R. S. Drago and B. B. Corden, *Acc. Chem. Res.* **13** (1980), 353–360.

108. A. W. Addison and S. Burman, *Biochim. Biophys. Acta* **828** (1985), 362–368.

109. L. M. Proniewicz, K. Nakamoto, and J. R. Kincaid, *J. Am. Chem. Soc.* **110** (1988), 4541–4545.

110. J. L. Hoard, in K. M. Smith, ed., *Porphyrins and Metalloporphyrins*, Elsevier, 1975, 317–380.

111. W. R. Scheidt, *Acc. Chem. Res.* **10** (1977), 339–345.

112. R. G. Little and J. A. Ibers, *J. Am. Chem. Soc.* **96** (1974), 4452–4463.

113. J. L. Hoard and W. R. Scheidt, *Proc. Natl. Acad. Sci. USA* **70** (1973), 3919–3922, and **71** (1974), 1578.

114. S. E. Peterson-Kennedy *et al.*, *J. Am. Chem. Soc.* **108** (1986), 1739–1746.

115. N. V. Blough and B. M. Hoffman, *J. Am. Chem. Soc.* **104** (1982), 4247–4250.

116a. D. R. Paulson *et al.*, *J. Biol. Chem.* **254** (1979), 7002–7006.

116b. T. S. Srivastava, *Biochim. Biophys. Acta* **491** (1977), 599–604.

117. R. B. Frydman and B. Frydman, *Acc. Chem. Res.* **20** (1987), 250–256.

118. S.-M. Peng and J. A. Ibers, *J. Am. Chem. Soc.* **98** (1976), 8032–8036.

119. V. L. Goedken and S.-M. Peng, *J. Am. Chem. Soc.* **96** (1974), 7826–7827.

120. V. L. Goedken *et al.*, *J. Am. Chem. Soc.* **98** (1976), 8391–8400.

121a. D. H. Busch *et al.*, *Proc. Natl. Acad. Sci. USA* **78** (1981), 5919–5923.

121b. K. Kim *et al.*, *J. Am. Chem. Soc.* **111** (1989), 403–405.

121c. X.-Y. Li and T. G. Spiro, *J. Am. Chem. Soc.* **110** (1988), 6024–6033.

121d. K. Kim and J. A. Ibers, *J. Am. Chem. Soc.* **113** (1991), 6077–6081.

122. J. Kuriyan *et al.*, *J. Mol. Biol.* **192** (1986), 133–154.

123. J. M. Baldwin, *J. Mol. Biol.* **136** (1980), 103–128.

124. E. J. Heidner, R. C. Ladner, and M. F. Perutz, *J. Mol. Biol.* **104** (1976), 707–722.

125a. J. C. Norvell, A. C. Nunes, and B. P. Schoenborn, *Science* **190** (1975), 568–570.

125b. J. C. Hanson and B. P. Schoenborn, *J. Mol. Biol.* **153** (1981), 117–146.

126. E. A. Padlan and W. E. Love, *J. Biol. Chem.* **249** (1974), 4067–4078.

127. A. Bianconi *et al.*, *Nature* **318** (1985), 685–687.

128. A. Bianconi *et al.*, *Biochim. Biophys. Acta* **831** (1985), 114–119.

129. B. B. Wayland, J. V. Minkiewicz, and M. E. Abd-Elmageed, *J. Am. Chem. Soc.* **96** (1974), 2795–2801.

130. B. S. Tovrog and R. S. Drago, *J. Am. Chem. Soc.* **96** (1974), 6765–6766.

131. B. M. Hoffman, T. Szymanski, and F. Basolo, *J. Am. Chem. Soc.* **97** (1975), 673–674.

132. W. R. Scheidt and D. K. Geiger, *Inorg. Chem.* **21** (1982), 1208–1211.

133. J. C. Maxwell and W. S. Caughey, *Biochemistry* **15** (1976), 388–396.

134. R. C. C. St. George and L. Pauling, *Science* **114** (1951), 629–634.

135. G. B. Jameson and J. A. Ibers, *Inorg. Chem.* **18** (1979), 1200–1208.

136. J.-M. Bassett *et al.*, *J. Chem. Soc. Chem. Comm.* (1977), 853–854.

137. L. Pauling, *Nature* **203** (1964), 182–183.

138. L. S. Liebeskind *et al.*, *J. Am. Chem. Soc.* **100** (1978), 7061–7063.

139. D. Mansuy *et al.*, *J. Am. Chem. Soc.* **105** (1983), 455–463.

140. T. B. Freedman, J. S. Loehr, and T. M. Loehr, *J. Am. Chem. Soc.* **98** (1976), 2809–2815.

141. T. J. Thamann, J. S. Loehr, and T. M. Loehr, *J. Am. Chem. Soc.* **99** (1977), 4187–4189.

142. D. M. Kurtz, Jr., D. F. Shriver, and I. M. Klotz, *J. Am. Chem. Soc.* **98** (1976), 5033–5035.

143. L. Y. Fager and J. O. Alben, *Biochemistry* **11** (1972), 4786–4792.

144. M. Munakata, S. Kitagawa, and K. Goto, *J. Inorg. Biochem.* **16** (1982), 319–322.

145. R. R. Gay and E. I. Solomon, *J. Am. Chem. Soc.* **100** (1978), 1972–1973.

146. A. K. Shiemke, T. M. Loehr, and J. Sanders-Loehr, *J. Am. Chem. Soc.* **108** (1986), 2437–2443.

147a. J. W. Dawson *et al.*, *Biochemistry* **11** (1972), 461–465.

147b. M. J. Maroney *et al.*, *J. Am Chem. Soc.* **108** (1986), 6871–6879.

148. J. M. Nocek *et al.*, *J. Am. Chem. Soc.* **107** (1985), 3382–3384.

149. C. H. Barlow *et al.*, *Biochem. Biophys. Res. Comm.* **55** (1973), 91–95.

150. L. Pauling, *Stanford Med. Bull.* **6** (1948), 215–222.

151. J. J. Weiss, *Nature* **202** (1964), 83–84.

152. J. S. Griffith, *Proc. Roy. Soc. A* **235** (1956), 23–36.

153. H. B. Gray, *Adv. Chem. Ser.* **100** (1971), 365–389.

154. T. G. Spiro, in A. B. P. Lever and H. B. Gray, eds., *Iron Porphyrins*, Addison-Wesley (1983), Part II, 89–159.

155. B. B. Wayland and L. W. Olson, *J. Am Chem. Soc.* **96** (1974), 6037–6041.

156. K. Wieghardt, K. Pohl, and W. Gebert, *Angew. Chem. Intl. Ed. Engl.* **22** (1983), 727.

157. P. Gomez-Romero *et al.*, *J. Am. Chem. Soc.* **110** (1988), 1988–1990.

158. J. A. Ibers and R. H. Holm, *Science* **209** (1980), 223–235.

159. A. Levy and J. M. Rifkind, *Biochemistry* **24** (1985), 6050–6054.

160. K. S. Suslick and M. M. Fox, *J. Am. Chem. Soc.* **105** (1983), 3507–3510.

161a. M. Momenteau *et al.*, *J. Chem. Soc. Chem. Comm.* (1983), 962–964.

161b. J. Mispelter *et al.*, *J. Am. Chem. Soc.* **105** (1983), 5165–5166.

162. N. Herron *et al.*, *J. Am. Chem. Soc.* **105** (1983), 6585–6596.

163. T. G. Traylor, *Acc. Chem. Res.* **14** (1981), 102–109.

164. T. G. Traylor, N. Koga, and L. A. Deardurff, *J. Am Chem. Soc.* **107** (1985), 6504–6510.

165. J. P. Collman *et al.*, *J. Am. Chem. Soc.* **105** (1983), 3052–3064.

166. J. Almog *et al.*, *J. Am. Chem. Soc.* **97** (1975), 226–227.

167. M. Ikeda-Saito M. Brunori, and T. Yonetani, *Biochim. Biophys. Acta* **533** (1978), 173–180.

168. M. Ikeda-Saito *et al.*, *J. Biol. Chem.* **252** (1977), 4882–4887.

169. G. B. Jameson, W. T. Robinson, and J. A. Ibers, in C. Ho, ed., *Hemoglobin and Oxygen Binding*, Elsevier, 1982, 25–35.

170. T. Yonetani, H. Yamamoto, and G. V. Woodrow III, *J. Biol. Chem.* **249** (1974), 682–690.

171. G. L. Woolery *et al.*, *J. Am. Chem. Soc.* **107** (1985), 2370–2373.

172. G. B. Jameson *et al.*, *J. Am. Chem. Soc.* **102** (1980), 3224–3237.

173. W. Byers *et al.*, *Inorg. Chem.* **25** (1986), 4767–4774.

174. M. M. Doeff, D. A. Sweigart, and P. O'Brien, *Inorg. Chem.* **22** (1983), 851–852.

175. R. S. Drago, J. P. Cannady, and K. A. Leslie, *J. Am. Chem. Soc.* **102** (1980), 6014–6019.

176. G. B. Jameson and R. S. Drago, *J. Am. Chem. Soc.* **107** (1985), 3017–3020.

177. F. A. Walker and J. Bowen, *J. Am. Chem. Soc.* **107** (1985), 7632–7635.

178. T. Yonetani, H. Yamamoto, and T. Iizuka, *J. Biol. Chem.* **249** (1974), 2168–2174.

179. T. Kitagawa *et al.*, *Nature* **298** (1982), 869–871.

180. J. E. Linard *et al.*, *J. Am. Chem. Soc.* **102** (1980), 1896–1904.

181a. T. Takano, *J. Mol. Biol.* **110** (1977), 569–584.

181b. Deoxymyoglobin at 1.6 Å resolution has been determined (S. E. V. Phillips). See Reference 182.

182. G. Fermi *et al.*, *J. Mol. Biol.* **175** (1984), 159–174.

183. W. Steigemann and E. Weber, *J. Mol. Biol.* **127** (1979), 309–338.

184. G. Fermi *et al.*, *J. Mol. Biol.* **155** (1982), 495–505.

185. E. A. Padlan, W. A. Eaton, and T. Yonetani, *J. Biol. Chem.* **250** (1975), 7069–7073.

186. J. Baldwin and C. Chothia, *J. Mol. Biol.* **129** (1979), 175–220.

187. G. B. Jameson *et al.*, *Inorg. Chem.* **17** (1978), 850–857.

188. S. E. V. Phillips, *J. Mol. Biol.* **142** (1980), 531–554.

189. S. E. V. Phillips and B. P. Schoenborn, *Nature* **292** (1981), 81–82.

190. B. Shaanan, *J. Mol. Biol.* **171** (1983), 31–59.

191a. A. Brzozowski *et al.*, *Nature* **307** (1984), 74–76.

191b. Z. Derewenda *et al.*, *J. Mol. Biol.* **211** (1990), 515–519.

191c. R. Liddington *et al.*, *Nature* **331** (1988), 725–728.

191d. B. Luisi *et al.*, *J. Mol. Biol.* **214** (1990), 7–14.

192. K. Nagai *et al.*, *Nature* **329** (1987), 858–860.

193. A. Dedieu *et al.*, *J. Am. Chem. Soc.* **98** (1976), 3717–3718.

194a. J. C. Stevens *et al.*, *J. Am. Chem. Soc.* **102** (1980), 3283–3285.

194b. P. J. Jackson *et al.*, *Inorg. Chem.* **25** (1986), 4015–4020.

195. L. Powers *et al.*, *Biochemistry* **23** (1984), 5519–5523.

196a. D. Braunstein *et al.*, *Proc. Natl. Acad. Sci. USA* **85** (1988), 8497–8501.

196b. P. Ormos *et al.*, *Proc. Natl. Acad. Sci. USA* **85** (1988), 8492–8496.

196c. J. N. Moore, P. A. Hansen, and R. M. Hochstrasser, *Proc. Natl. Acad. Sci. USA* **85** (1988), 5062–5066.

197. J. C. W. Chien and L. C. Dickinson, *Proc. Natl. Acad. Sci. USA* **69** (1972), 2783–2787; L. C. Dickinson and J. C. W. Chien, *Proc. Natl. Acad. Sci. USA* **77** (1980), 1235–1239.

198. G. A. Petsko *et al.*, in P. L. Dutton, J. S. Leigh, and A. Scarpa, eds., *Frontiers in Bioenergetics*, Academic Press, 1978, 1011–1017.

199. E. W. Findsen *et al.*, *Science* **229** (1985), 661–665.

200. K. Nagai and T. Kitagawa, *Proc. Natl. Acad. Sci. USA* **77** (1980), 2033–2037.

201a. G. B. Jameson and J. A. Ibers, *J. Am. Chem. Soc.* **102** (1980), 2823–2831.

201b. M. Sabat and J. A. Ibers, *J. Am. Chem. Soc.* **104** (1982), 3715–3721.

202. J. W. Sparapany *et al.*, *J. Am. Chem. Soc.* **110** (1988), 4559–4564.

203. A. Arnone *et al.*, *J. Mol. Biol.* **188** (1986), 693–706.

204. É. G. Arutyunyan *et al.*, *Sov. Phys. Crystallogr.* **25** (1980), 43–58.

205. D. L. Ollis *et al.*, *Austral. J. Chem.* **36** (1983), 451–468.

206a. K. Shikama and A. Matsuoka, *Biochemistry* **25** (1986), 3898–3903.

206b. M. Bolognesi *et al.*, *J. Mol. Biol.* **213** (1990), 621–625.

207. S. J. Lippard, *Angew. Chem. Intl. Ed. Engl.* **27** (1988), 344–361.

208. M. F. Perutz and F. S. Mathews, *J. Mol. Biol.* **21** (1966), 199–202.

209. J. C. Kendrew *et al.*, *Nature* **185** (1960), 422–427.

210. GBJ gratefully acknowledges support from the National Institutes of Health (DK 37702) and sabbatical-leave support from the Department of Chemistry and Biochemistry, Massey University, Palmerston North, New Zealand. JAI is pleased to acknowledge the support of the National Institutes of Health (HL 13157).

VI. ABBREVIATIONS

1-MeIm	1-methylimidazole
1, 2-Me$_2$Im	1, 2-dimethylimidazole
2-MeIm	2-methylimidazole
2,3-DPG	2,3-diphosphoglycerate
3,4-Me$_2$-Py	3,4-dimethylpyridine
4-t-Bu-Py	4-t-butylpyridine
4-CN-Py	4-cyanopyridine
4-NH$_2$-Py	4-aminopyridine
a_i	activity of component i
Arg	arginine
B	general axial base ligand that binds to a metalloporphyrin
Ch	chlorocruorin
E_1, E_{-1}	activation energy
EDTA	ethylenediaminetetraacetic acid
E°	electrochemical potential at unit activity and fugacity
E$^{0'}$	electrochemical potential at physiological pH (7.4) and unit pressure (1 atm = 760 Torr)
Er	erythrocruorin
EtOH	ethanol
EXAFS	extended x-ray absorption fine structure
G	Gibbs free energy
H	enthalpy
H$_2$(6, 6-CP)	cyclophane-strapped porphyrin (see Figure 4.23)
H$_2$(Poc-PF)	pocket picket-fence porphyrin
H$_2$Amide-Im	basket-handle porphyrin: imidazole base covalently attached to porphyrin with amide straps; amide straps on other side of porphyrin (see Figure 4.23)

H_2Bis-Poc	meso-tetrakis (2, 4, 6-triphenylphenyl) porphyrin
H_2bzacen	N, N-ethylenebis (benzoylacetoninime)
H_2C_2Cap	capped porphyrin; see Figure 4.23
H_2Ether-Py	basket-handle porphyrin: as for H_2Amide-Im except ether straps and pyridine base (see Figure 4.23)
H_2MPIX-Im	mesoporphyrin IX dimethylester with imidazole covalently attached to porphyrin (see Figure 4.23)
H_2PF	picket fence porphyrin, meso-tetrakis (α, α, α, α-o-pivalamidephenyl) porphyrin (see Figure 4.23)
H_2PPIX	protoporphyrin IX dimethylester
H_2PPIX-Im	protoporphyrin IX dimethylester with imidazole covalently attached to porphyrin (see Figure 4.23)
Hb	hemoglobin
HbA	human adult hemoglobin
HbF	human fetal hemoglobin
Hc	hemocyanin
His	histidine
His^{203}	position on polypeptide chain (203) of a histidine residue
Hr	hemerythrin
Im	imidazole
k	rate constant
K_p, K_c	equilibrium constant: concentration of gas expressed in terms of pressure (P) and molarity (M), respectively
L_i	allosteric constant: equilibrium constant for conformational change of protein with i ligands bound
M	general metalloporphyrinato species
M	molarity, moles/L
M	general metal complex
Mb	myoglobin
Me	methyl group
met	oxidized (e.g. met Hb)
OEP	2,3,7,8,12,13,17,18-octaethylporphyrinato
P	pressure, usually in Torr or mm Hg
THF	tetrahydrofuran
TPP	5,10,15,20-tetraphenylporphyrinato

5

Dioxygen Reactions

JOAN SELVERSTONE VALENTINE

Department of Chemistry and Biochemistry
University of California, Los Angeles

I. INTRODUCTION[1]

The major pathway of dioxygen use in aerobic organisms is four-electron reduction to give two molecules of water per dioxygen molecule:[2]

$$O_2 + 4H^+ + 4e^- \rightarrow 2H_2O \qquad E° = +0.815 \text{ V} \qquad (5.1)$$

This reaction represents the major source of energy in aerobic organisms when coupled with the oxidation of electron-rich organic foodstuffs, such as glucose. Biological oxidation of this type is called respiration, and has been estimated to account for 90 percent or more of the dioxygen consumed in the biosphere. It is carried out by means of a series of enzyme-catalyzed reactions that are coupled to ATP synthesis, and the ATP produced is the major source of energy for the organism. The actual site of the reduction of dioxygen in many organisms is the enzyme cytochrome c oxidase.[2]

Another use of dioxygen in aerobic organisms is to function as a source of oxygen atoms in the biosynthesis of various molecules in metabolic pathways, or in conversions of lipid-soluble molecules to water-soluble forms for purposes of excretion. These reactions are also enzyme-catalyzed, and the enzymes involved are either monooxygenase or dioxygenase enzymes, depending on whether one or both of the oxygen atoms from dioxygen are incorporated in the final organic product. Many of these enzymes are metalloenzymes.[2–4]

The advantages of life in air are considerable for an aerobic organism as compared to an anaerobic organism, mainly because the powerful oxidizing power of dioxygen can be controlled and efficiently converted to a form that can be stored and subsequently used.[5] But aerobic metabolism has its disadvantages as well. The interior of a living cell is a reducing environment, and many of the components of the cell are fully capable thermodynamically of reacting directly with dioxygen, thus bypassing the enzymes that control and direct the beneficial reactions of dioxygen.[6] Luckily, for reasons that are discussed below, these

253

reactions generally are slow, and therefore represent minor pathways of biological dioxygen consumption. Otherwise, the cell would just burn up, and aerobic life as we know it would be impossible. Nevertheless, there are small but significant amounts of products formed from nonenzymatic and enzymatic reactions of dioxygen that produce partially reduced forms of dioxygen, i.e., superoxide, O_2^-, and hydrogen peroxide, H_2O_2, in aerobic cells. These forms of reduced dioxygen or species derived from them could carry out deleterious reactions, and enzymes have been identified that appear to protect against such hazards. These enzymes are, for superoxide, the superoxide dismutase enzymes, and, for peroxide, catalase and the peroxidase enzymes. All of these enzymes are metalloenzymes.[2-4]

Much of the fascination of the subject of biological reactions of dioxygen stems from the fact that the mechanisms of the biological, enzyme-catalyzed reactions are clearly quite different from those of the uncatalyzed reactions of dioxygen or even those of dioxygen reactions catalyzed by a wide variety of nonbiological metal-containing catalysts.[7] Investigators believe, optimistically, that once they truly understand the biological reactions, they will be able to design synthetic catalysts that mimic the biological catalysts, at least in reproducing the reaction types, even if these new catalysts do not match the enzymes in rate and specificity. To introduce this topic, therefore, we first consider the factors that determine the characteristics of nonbiological reactions of dioxygen.

II. CHEMISTRY OF DIOXYGEN

A. Thermodynamics

The reduction potential for the four-electron reduction of dioxygen (Reaction 5.1) is a measure of the great oxidizing power of the dioxygen molecule.[8] However, the reaction involves the transfer of four electrons, a process that rarely, if ever, occurs in one concerted step, as shown in Reaction (5.2).

$$O_2 \xrightarrow{e^-} O_2^- \xrightarrow{e^-, 2H^+} H_2O_2 \xrightarrow{e^-, H^+} H_2O + OH \xrightarrow{e^-, H^+} 2H_2O \qquad (5.2)$$

dioxygen   super-   hydrogen   water + hydroxyl water
            oxide   peroxide   radical

Since most reducing agents can transfer at most one or two electrons at a time to an oxidizing agent, the thermodynamics of the one- and two-electron reductions of dioxygen must be considered in order to understand the overall mechanism.

In aqueous solution, the most common pathway for dioxygen reduction in the absence of any catalyst is one-electron reduction to give superoxide. But this is the least favorable of the reaction steps that make up the full four-electron reduction (see Table 5.1) and requires a moderately strong reducing agent. Thus

Table 5.1
Standard reduction potentials for dioxygen species in water.[8]

Reaction	E°, V vs. NHE, pH 7, 25°C
$O_2 + e^- \rightarrow O_2^-$	-0.33[a]
$O_2^- + e^- + 2H^+ \rightarrow H_2O_2$	$+0.89$
$H_2O_2 + e^- + H^+ \rightarrow H_2O + OH$	$+0.38$
$OH + e^- + H^+ \rightarrow H_2O$	$+2.31$
$O_2 + 2e^- + 2H^+ \rightarrow H_2O_2$	$+0.281$[a]
$H_2O_2 + 2e^- + 2H^+ \rightarrow 2H_2O$	$+1.349$
$O_2 + 4H^+ + 4e^- \rightarrow 2H_2O$	$+0.815$[a]

[a] The standard state used here is unit pressure. If unit activity is used for the standard state of O_2, the redox potentials for reactions of that species must be adjusted by $+0.17$ V.[8,9]

if only one-electron pathways are available for dioxygen reduction, the low reduction potential for one-electron reduction of O_2 to O_2^- presents a barrier that protects vulnerable species from the full oxidizing power of dioxygen that comes from the subsequent steps. If superoxide is formed (Reaction 5.3), however, it disproportionates quite rapidly in aqueous solution (except at very high pH) to give hydrogen peroxide and dioxygen (Reaction 5.4). The stoichiometry of the overall reaction is therefore that of a net two-electron reduction (Reaction 5.5). It is thus impossible under normal conditions to distinguish one-electron and two-electron reaction pathways for the reduction of dioxygen in aqueous solution on the basis of stoichiometry alone.

$$2O_2 + 2e^- \longrightarrow 2O_2^- \tag{5.3}$$

$$2O_2^- + 2H^+ \longrightarrow H_2O_2 + O_2 \tag{5.4}$$

$$O_2 + 2e^- + 2H^+ \longrightarrow H_2O_2 \tag{5.5}$$

The thermodynamics of dioxygen reactions with organic substrates is also of importance in understanding dioxygen reactivity. The types of reactions that are of particular interest to us here are hydroxylation of aliphatic and aromatic C—H bonds and epoxidation of olefins, since these typical reactions of oxygenase enzymes are ones that investigators are trying to mimic using synthetic reagents. Some of the simpler examples of such reactions (plus the reaction of H_2 for comparison) are given in the reactions in Table 5.2. It is apparent that all these reactions of dioxygen with various organic substrates in Table 5.2 are thermodynamically favorable. However, *direct* reactions of dioxygen with organic substrates in the absence of a catalyst are generally very slow, unless the substrate is a particularly good reducing agent. To understand the sluggishness of dioxygen reactions with organic substrates, we must consider the kinetic barriers to these reactions.

Table 5.2
Examples of hydroxylation and epoxidation reactions.

Reaction	ΔH in kcal/mol	Reference
$CH_{4(g)} + \frac{1}{2}O_{2(g)} \rightarrow CH_3OH_{(g)}$	-30	10
$C_6H_{6(g)} + \frac{1}{2}O_{2(g)} \rightarrow C_6H_5OH_{(g)}$	-43	11,12
$C_6H_5OH_{(g)} + \frac{1}{2}O_{2(g)} \rightarrow C_6H_4(OH)_{2(g)}$	-42	12,13
$C_2H_{4(g)} + \frac{1}{2}O_{2(g)} \rightarrow C_2H_4O_{(g)}$	-25	10
$C_5H_5N_{(g)} + \frac{1}{2}O_{2(g)} \rightarrow C_5H_5NO_{(g)}$	-13	14
$H_{2(g)} + \frac{1}{2}O_{2(g)} \rightarrow H_2O_{(g)}$	-58	10

B. Kinetics

The principal kinetic barrier to direct reaction of dioxygen with an organic substrate arises from the fact that the ground state of the dioxygen molecule is triplet, i.e., contains two unpaired electrons.[15,16] Typical organic molecules that are representative of biological substrates have singlet ground states, i.e., contain no unpaired electrons, and the products resulting from their oxygenation also have singlet ground states. Reactions between molecules occur in shorter times than the time required for conversions from triplet to singlet spin. Therefore the number of unpaired electrons must remain the same before and after each elementary step of a chemical reaction. For these reasons, we know that it is impossible for Reaction (5.6) to go in one fast, concerted step.

$$\frac{1}{2} \, {}^3O_2 + {}^1X \longrightarrow {}^1XO \qquad (5.6)$$

The arrows represent electron spins: $\downarrow \uparrow$ represents a singlet molecule with all electron spins paired; $\uparrow \uparrow$ represents a triplet molecule with two unpaired electrons; and $\uparrow$ (which we will see in Reaction 5.13) represents a doublet molecule, also referred to as a free radical, with one unpaired electron. The pathways that do not violate the spin restriction are all costly in energy, resulting in high activation barriers. For example, the reaction of ground-state triplet dioxygen, i.e., ${}^3O_2$, with a singlet substrate to give the excited triplet state of the oxygenated product (Reaction 5.7) is spin-allowed, and one could imagine a mechanism in which this process is followed by a slow spin conversion to a singlet product (Reaction 5.8).

$$\frac{1}{2} \, {}^3O_2 + {}^1X \longrightarrow {}^3XO \qquad (5.7)$$

$$ {}^3XO \xrightarrow{\text{slow}} {}^1XO \qquad (5.8)$$

But such a reaction pathway would give a high activation barrier, because the excited triplet states of even unsaturated molecules are typically 40–70 kcal/mol less stable than the ground state, and those of saturated hydrocarbons are much higher.[17]

Likewise, a pathway in which O_2 is excited to a singlet state that then reacts with the substrate would be spin-allowed (Reactions 5.9 and 5.10). The high reactivity of singlet dioxygen, generated by photochemical or chemical means, is well-documented.[18,19] However, such a pathway for a reaction of dioxygen, which is initially in its ground triplet state, would also require a high activation energy, since the lowest-energy singlet excited state of dioxygen is 22.5 kcal/mol higher in energy than ground-state triplet dioxygen.[15,16]

$$^3O_2 + 22.5 \text{ kcal/mol} \longrightarrow {}^1O_2 \qquad (5.9)$$

$$\tfrac{1}{2} \, {}^1O_2 + {}^1X \longrightarrow {}^1XO \qquad (5.10)$$

Moreover, the products of typical reactions of singlet-state dioxygen with organic substrates (Reactions 5.11 and 5.12, for example) are quite different in character from the reactions of dioxygen with organic substrates catalyzed by oxygenase enzymes (see Section V):

$$\text{(ring)} + {}^1O_2 \longrightarrow \text{(endoperoxide)} \qquad (5.11)$$

$$\text{(alkene)} + {}^1O_2 \longrightarrow \text{OOH} \qquad (5.12)$$

One pathway for a direct reaction of triplet ground-state dioxygen with a singlet ground-state organic substrate that can occur readily without a catalyst begins with the one-electron oxidation of the substrate by dioxygen. The products of such a reaction would be two doublets, i.e., superoxide and the one-electron oxidized substrate, each having one unpaired electron (Reaction 5.13). These free radicals can diffuse apart and then recombine with their spins paired (Reaction 5.14).

$$^3O_2 + {}^1X \longrightarrow {}^2O_2{}^- + {}^2X^+ \qquad (5.13)$$

$$^2O_2{}^- + {}^2X^+ \longrightarrow {}^2O_2{}^- + {}^2X^+ \longrightarrow {}^1XO_2 \qquad (5.14)$$

Such a mechanism has been shown to occur for the reaction of dioxygen with reduced flavins shown in Reaction (5.15).[20]

$$(5.15)$$

However, this pathway requires that the substrate be able to reduce dioxygen to superoxide, a reaction that requires an unusually strong reducing agent (such as a reduced flavin), since dioxygen is not a particularly strong one-electron oxidizing agent (see Table 5.1 and discussion above). Typical organic substrates in enzymatic and nonenzymatic oxygenation reactions usually are not sufficiently strong reducing agents to reduce dioxygen to superoxide; so this pathway is not commonly observed.

The result of these kinetic barriers to dioxygen reactions with most organic molecules is that uncatalyzed reactions of this type are usually quite slow. An exception to this rule is an oxidation pathway known as free-radical autoxidation.

C. Free-Radical Autoxidation

The term free-radical autoxidation describes a reaction pathway in which dioxygen reacts with an organic substrate to give an oxygenated product in a free-radical chain process that requires an initiator in order to get the chain reaction started.[21] (A free-radical initiator is a compound that yields free radicals readily upon thermal or photochemical decomposition.) The mechanism of free radical autoxidation is as shown in Reactions (5.16) to (5.21).

Initiation:	$X_2 \longrightarrow 2X\cdot$	(5.16)
	$X\cdot + RH \longrightarrow XH + R\cdot$	(5.17)
Propagation:	$R\cdot + O_2 \longrightarrow ROO\cdot$	(5.18)
	$ROO\cdot + RH \longrightarrow ROOH + R\cdot$	(5.19)
Termination:	$R\cdot + ROO\cdot \longrightarrow ROOR$	(5.20)
	$2ROO\cdot \longrightarrow ROOOOR \longrightarrow O_2 + ROOR$	(5.21)

(plus other oxidized products, such as ROOH, ROH, RC(O)R, RC(O)H). This reaction pathway results in oxygenation of a variety of organic substrates, and is not impeded by the spin restriction, because triplet ground-state dioxygen can react with the free radical $R\cdot$ to give a free-radical product $ROO\cdot$ in a spin-allowed process (Reaction 5.18). It is a chain reaction, since $R\cdot$ is regenerated in Reaction (5.19), and it frequently occurs with long chain lengths prior to the termination steps, resulting in a very efficient pathway for oxygenation of some organic substrates, such as, for example, the oxidation of cumene to give phenol and acetone (Reaction 5.22).[22]

$$+ \ (CH_3)_2CO \qquad\qquad (5.22)$$

When free-radical autoxidation is used for synthetic purposes, initiators are intentionally added. Common initiators are peroxides and other compounds capable of fragmenting readily into free radicals. Free-radical autoxidation reactions are also frequently observed when no initiator has been intentionally added, because organic substrates frequently contain peroxidic impurities that may act as initiators. Investigators have sometimes been deceived into assuming that a metal-complex catalyzed reaction of dioxygen with an organic substrate occurred by a nonradical mechanism. In such instances, the reactions later proved, upon further study, to be free-radical autoxidations, the role of the metal complex having been to generate the initiating free radicals.

Although often useful for synthesis of oxygenated derivatives of relatively simple hydrocarbons, free-radical autoxidation lacks selectivity and therefore, with more complex substrates, tends to give multiple products. In considering possible mechanisms for biological oxidation reactions used *in vivo* for biosynthesis or energy production, free-radical autoxidation is not an attractive possibility, because such a mechanism requires diffusion of highly reactive free radicals. Such radicals, produced in the cell, will react indiscriminately with vulnerable sites on enzymes, substrates, and other cell components, causing serious damage.[6] In fact, free-radical autoxidation is believed to cause certain deleterious reactions of dioxygen in biological systems, for example the oxidation of lipids in membranes. It is also the process that causes fats and oils to become rancid (Reaction 5.23).[23,24]

$$\qquad\qquad (5.23)$$

$$R_1 = -C_5H_{11}$$

$$R_2 = -(CH_2)_7COOH$$

D. How Do Enzymes Overcome These Kinetic Barriers?

We see then the reasons that uncatalyzed reactions of dioxygen are usually either slow or unselective. The functions of the metalloenzymes for which dioxygen is a substrate are, therefore, to overcome the kinetic barriers imposed by spin restrictions or unfavorable one-electron reduction pathways, and, for the oxygenase enzymes, to direct the reactions and make them highly specific. It is instructive to consider (1) how these metalloenzymes function to lower the kinetic barriers to dioxygen reactivity, and (2) how the oxygenase enzymes redirect the reactions along different pathways so that very different products are obtained. The first example given below is cytochrome *c* oxidase. This enzyme catalyzes the four-electron reduction of dioxygen. It overcomes the kinetic barriers to dioxygen reduction by binding dioxygen to two paramagnetic metal ions at the dioxygen binding site, thus overcoming the spin restriction, and by reducing dioxygen in a two-electron step to peroxide, thus bypassing the unfavorable one-electron reduction to form free superoxide. The reaction occurs in a very controlled fashion, so that the energy released by dioxygen reduction can be used to produce ATP. A second example is provided by the catechol dioxygenases, which appear to represent substrate rather than dioxygen activation, and in which dioxygen seems to react with the substrate while it is complexed to the paramagnetic iron center. Another example given below is the monooxygenase enzyme cytochrome P-450, which catalyzes the reaction of dioxygen with organic substrates. It binds dioxygen at the paramagnetic metal ion at its active site, thus overcoming the spin restriction, and then carries out what can be formally described as a multielectron reduction of dioxygen to give a highly reactive high-valent metal-oxo species that has reactivity like that of the hydroxyl radical. Unlike a free hydroxyl radical, however, which would be highly reactive but nonselective, the reaction that occurs at the active site of cytochrome P-450 can be highly selective and stereospecific, because the highly reactive metal-oxo moiety is generated close to a substrate that is bound to the enzyme in such a way that it directs the reactive oxygen atom to the correct position. Thus, metalloenzymes have evolved to bind dioxygen and to increase while controlling its reactivity.

III DIOXYGEN TOXICITY

A. Background

Before we consider the enzymatically controlled reactions of dioxygen in living systems, it is instructive to consider the uncontrolled and deleterious reactions that must also occur in aerobic organisms. Life originally appeared on Earth at a time when the atmosphere contained only low concentrations of dioxygen, and was reducing rather than oxidizing, as it is today. With the appearance of photosynthetic organisms approximately 2.5 billion years ago, however, the conversion to an aerobic, oxidizing atmosphere exposed the existing anaerobic organisms to a gradually increasing level of oxidative stress.[25,26] Modern-day anaerobic bacteria, the descendants of the original primitive anaerobic organisms, evolved in ways that enabled them to avoid contact with normal atmospheric concentrations of dioxygen. Modern-day aerobic organisms, by contrast, evolved by developing aerobic metabolism to harness the oxidizing power of dioxygen and thus to obtain usable metabolic energy. This remarkably successful adaptation enabled life to survive and flourish as the atmosphere became aerobic, and also allowed larger, multicellular organisms to evolve. An important aspect of dioxygen chemistry that enabled the development of aerobic metabolism is the relatively slow rate of dioxygen reactions in the absence of catalysts. Thus, enzymes could be used to direct and control the oxidation of substrates either for energy generation or for biosynthesis. Nevertheless, the balance achieved between constructive and destructive oxidation is a delicate one, maintained in aerobic organisms by several means, e.g.: compartmentalization of oxidative reactions in mitochondria, peroxisomes, and chloroplasts; scavenging or detoxification of toxic byproducts of dioxygen reactions; repair of some types of oxidatively damaged species; and degradation and replacement of other species.[6]

The classification "anaerobic" actually includes organisms with varying degrees of tolerance for dioxygen: strict anaerobes, for which even small concentrations of O_2 are toxic; moderate anaerobes, which can tolerate low levels of dioxygen; and microaerophiles, which require low concentrations of O_2 for growth, but cannot tolerate normal atmospheric concentrations, i.e., 21 percent O_2, 1 atm pressure. Anaerobic organisms thrive in places protected from the atmosphere, for example, in rotting organic material, decaying teeth, the colon, and gangrenous wounds. Dioxygen appears to be toxic to anaerobic organisms largely because it depletes the reducing equivalents in the cell that are needed for normal biosynthetic reactions.[6]

Aerobic organisms can, of course, live in environments in which they are exposed to normal atmospheric concentrations of O_2. Nevertheless, there is much evidence that O_2 is toxic to these organisms as well. For example, plants grown in varying concentrations of O_2 have been observed to grow faster in lower than normal concentrations of O_2.[27] E. coli grown under 5 atm of O_2 ceased to grow unless the growth medium was supplemented with branched-chain amino acids or precursors. High concentrations of O_2 damaged the enzyme dihydroxy acid dehydratase, an important component in the biosynthetic pathway for those amino acids.[28] In mammals, elevated levels of O_2 are clearly toxic, leading first to coughing and soreness of the throat, and then to convulsions when the level of 5 atm of 100 percent O_2 is reached. Eventually, elevated concentrations of O_2 lead to pulmonary edema and irreversible lung damage, with obvious damage to other tissues as well.[6] The effects of high concentrations of O_2 on humans is of some medical interest, since dioxygen is used therapeutically for patients experiencing difficulty breathing, or for those suffering from infection by anaerobic organisms.[6]

B. Biological Targets

The major biochemical targets of O_2 toxicity appear to be lipids, DNA, and proteins. The chemical reactions accounting for the damage to each type of target are probably different, not only because of the different reactivities of these three classes of molecules, but also because of the different environment for each one inside the cell. Lipids, for example, are essential components of membranes and are extremely hydrophobic. The oxidative damage that is observed is due to free-radical autoxidation (see Reactions 5.16 to 5.21), and the products observed are lipid hydroperoxides (see Reaction 5.23). The introduction of the hydroperoxide group into the interior of the lipid bilayer apparently causes that structure to be disrupted, as the configuration of the lipid rearranges in order to bring that polar group out of the hydrophobic membrane interior and up to the membrane-water interface.[6] DNA, by contrast, is in the interior of the cell, and its exposed portions are surrounded by an aqueous medium. It is particularly vulnerable to oxidative attack at the base or at the sugar, and multiple products are formed when samples are exposed to oxidants *in vitro*.[6] Since oxidation of DNA *in vivo* may lead to mutations, this type of damage is potentially very serious. Proteins also suffer oxidative damage, with amino-acid side chains, particularly the sulfur-containing residues cysteine and methionine, appearing to be the most vulnerable sites.[6]

C. Defense and Repair Systems

The biological defense systems protecting against oxidative damage and its consequences are summarized below.

1. Nonenzymatic oxidant scavengers

Some examples of small-molecule antioxidants are α-tocopherol (vitamin E; 5.24), which is found dissolved in cell membranes and protects them against lipid peroxidation, and ascorbate (vitamin C; 5.25) and glutathione (5.26), which are found in the cytosol of many cells. Several others are known as well.[6,29]

$$\alpha\text{-tocopherol} \tag{5.24}$$

ascorbic acid

$$(5.25)$$

$$\text{(5.26)}$$

glutathione

2. Detoxification enzymes

The enzymatic antioxidants are (a) catalase and the various peroxidases, whose presence lowers the concentration of hydrogen peroxide, thereby preventing it from entering into potentially damaging reactions with various cell components (see Section VI and Reactions 5.82 and 5.83), and (b) the superoxide dismutases, whose presence provides protection against dioxygen toxicity that is believed to be mediated by the superoxide anion, O_2^- (see Section VII and Reaction 5.95).

Some of the enzymatic and nonenzymatic antioxidants in the cell are illustrated in Figure 5.1.

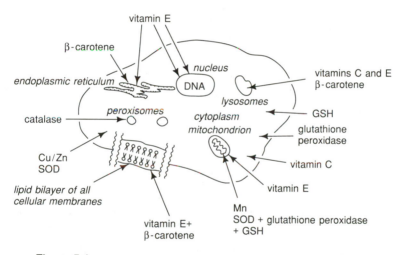

Figure 5.1
Cartoon showing some of the antioxidant agents inside the cell.[29]

3. Systems for sequestration of redox-active metal ions

Redox-active metal ions are present in the cell in their free, uncomplexed state only in extremely low concentrations. They are instead sequestered by metal-ion storage and transport proteins, such as ferritin and transferrin for iron (see Chapter 1) and ceruloplasmin for copper. This arrangement prevents such metal ions from catalyzing deleterious oxidative reactions, but makes them available for incorporation into metalloenzymes as they are needed.

In vitro experiments have shown quite clearly that redox-active metal ions such as $Fe^{2+/3+}$ or $Cu^{+/2+}$ are extremely good catalysts for oxidation of sulfhydryl groups by O_2 (Reaction 5.27).[30]

$$4RSH + O_2 \xrightarrow{M^{n+}} 2RSSR + 2H_2O \qquad (5.27)$$

In addition, in the reducing environment of the cell, redox-active metal ions catalyze a very efficient one-electron reduction of hydrogen peroxide to produce hydroxyl radical, one of the most potent and reactive oxidants known (Reactions 5.28 to 5.30).[31]

$$M^{n+} + Red^- \longrightarrow M^{(n-1)+} + Red \qquad (5.28)$$
$$M^{(n-1)+} + H_2O_2 \longrightarrow M^{n+} + OH^- + HO\cdot \qquad (5.29)$$
$$\overline{Red^- + H_2O_2 \longrightarrow Red + OH^- + HO\cdot} \qquad (5.30)$$

(Red$^-$ = reducing agent)

Binding those metal ions in a metalloprotein usually prevents them from entering into these types of reactions. For example, transferrin, the iron-transport enzyme in serum, is normally only 30 percent saturated with iron. Under conditions of increasing iron overload, the empty iron-binding sites on transferrin are observed to fill, and symptoms of iron poisoning are not observed *in vivo* until after transferrin has been totally saturated with iron.[32] Ceruloplasmin and metallothionein may play a similar role in preventing copper toxicity.[6] It is very likely that both iron and copper toxicity are largely due to catalysis of oxidation reactions by those metal ions.

4. Systems for the repair or replacement of damaged materials

Repair of oxidative damage must go on constantly, even under normal conditions of aerobic metabolism. For lipids, repair of peroxidized fatty-acid chains is catalyzed by phospholipase A_2, which recognizes the structural changes at the lipid-water interface caused by the fatty-acid hydroperoxide, and catalyzes removal of the fatty acid at that site. The repair is then completed by enzymatic reacylation.[6] Although some oxidatively damaged proteins are repaired, more commonly such proteins are recognized, degraded at accelerated rates, and then replaced.[6] For DNA, several multi-enzyme systems exist whose function is to repair oxidatively damaged DNA.[6] For example, one such system catalyzes recognition and removal of damaged bases, removal of the damaged part of the strand, synthesis of new DNA to fill in the gaps, and religation to restore the DNA to its original, undamaged state. Mutant organisms that lack these repair enzymes are found to be hypersensitive to O_2, H_2O_2, or other oxidants.[6]

One particularly interesting aspect of oxidant stress is that most aerobic organisms can survive in the presence of normally lethal levels of oxidants if they have first been exposed to lower, nontoxic levels of oxidants. This phenomenon has been observed in animals, plants, yeast, and bacteria, and suggests that low levels of oxidants cause antioxidant systems to be induced *in vivo*. In certain bacteria, the mechanism of this induction is at least partially understood. A DNA-binding regulatory protein named OxyR that exists in two redox states has been identified in these systems.[33] Increased oxidant stress presumably increases concentration of the oxidized form, which then acts to turn on the transcription of the genes for some of the antioxidant enzymes. A related phenomenon may occur when bacteria and yeast switch from anaerobic to aerobic metabolism. When dioxygen is absent, these microorganisms live by fermentation, and do not waste energy by synthesizing the enzymes and other proteins needed for aerobic metabolism. However, when they are exposed to dioxygen, the synthesis of the respiratory apparatus is turned on. The details of this induction are not known completely, but some steps at least depend on the presence of heme, the prosthetic group of hemoglobin and other heme proteins, whose synthesis requires the presence of dioxygen.[34]

D. Molecular Mechanisms of Dioxygen Toxicity

What has been left out of the preceding discussion is the identification of the species responsible for oxidative damage, i.e., the agents that directly attack the various vulnerable targets in the cell. They were left out because the details of the chemistry responsible for dioxygen toxicity are largely unknown. In 1954, Rebeca Gerschman formulated the "free-radical theory of oxygen toxicity" after noting that tissues subjected to ionizing radiation resemble those exposed to elevated levels of dioxygen.[35] Fourteen years later, Irwin Fridovich proposed that the free radical responsible for dioxygen toxicity was superoxide, O_2^-, based on his identification of the first of the superoxide dismutase enzymes.[36] Today it is still not known if superoxide is the principal agent of dioxygen toxicity, and, if so, what the chemistry responsible for that toxicity is.[6]

There is no question that superoxide is formed during the normal course of aerobic metabolism,[121] although it is difficult to obtain estimates of the amount under varying conditions, because, even in the absence of a catalyst, superoxide disproportionates quite rapidly to dioxygen and hydrogen peroxide (Reaction 5.4) and therefore never accumulates to any great extent in the cell under normal conditions of pH.[37]

One major problem in this area is that a satisfactory chemical explanation for the purported toxicity of superoxide has never been found, despite much indirect evidence from *in vitro* experiments that the presence of superoxide can lead to undesirable oxidation of various cell components and that such oxidation can be inhibited by superoxide dismutase.[38] The mechanism most commonly proposed is production of hydroxyl radicals via Reactions (5.28) to (5.30) with $Red^- = O_2^-$, which is referred to as the "Metal-Catalyzed Haber-Weiss Reaction''. The role of superoxide in this mechanism is to reduce oxidized metal ions, such as Cu^{2+} or Fe^{3+}, present in the cell in trace amounts, to a lower oxidation state.[37] Hydroxyl radical is an extremely powerful and indiscriminate oxidant. It can abstract hydrogen atoms from organic substrates, and oxidize most reducing agents very rapidly. It is also a very effective initiator of free-radical autoxidation reactions (see Section II.C above). Therefore, reactions that produce hydroxyl radical in a living cell will probably be very deleterious.[6]

The problem with this explanation for superoxide toxicity is that the only role played by superoxide here is that of a reducing agent of trace metal ions. The interior of a cell is a highly reducing environment, however, and other reducing agents naturally present in the cell such as, for example, ascorbate anion can also act as Red^- in Reaction (5.28), and the resulting oxidation reactions due to hydroxyl radical are therefore no longer inhibitable by SOD.[39]

Other possible explanations for superoxide toxicity exist, of course, but none has ever been demonstrated experimentally. Superoxide might bind to a specific enzyme and inhibit it, much as cytochrome oxidase is inhibited by cyanide or hemoglobin by carbon monoxide. Certain enzymes may be extraordinarily sensitive to direct oxidation by superoxide, as has been suggested for the enzyme aconitase, an iron-sulfur enzyme that contains an exposed iron atom.[122] Another possibility is that the protonated and therefore neutral form of superoxide, HO_2, dissolves in membranes and acts as an initiator of lipid peroxidation. It has also been suggested that superoxide may react with nitric oxide, NO, in the cell producing peroxynitrite, a very potent oxidant.[123] One particularly appealing mechanism for superoxide toxicity that has gained favor in recent years is the "Site-Specific Haber-Weiss Mechanism."[40,41] The idea here is that traces of redox-active metal ions such as copper and iron are bound to macromolecules under normal conditions in the cell. Most reducing agents in the cell are too bulky to come into close proximity to these sequestered metal ions. Superoxide, however, in addition to being an excellent reducing agent, is very small, and could penetrate to these metal ions and reduce them. The reduced metal ions could then react with hydrogen peroxide, generating hydroxyl radical, which would immediately attack at a site near the location of the bound metal ion. This mechanism is very similar to that of the metal complexes that cause DNA cleavage; by reacting with hydrogen peroxide while bound to DNA, they generate powerful oxidants that react with DNA with high efficiency because of their proximity to it (see Chapter 8).

Although we are unsure what specific chemical reactions superoxide might undergo inside of the cell, there nevertheless does exist strong evidence that the superoxide dismutases play an important role in protection against dioxygen-induced damage. Mutant strains of bacteria and yeast that lack superoxide dismutases are killed by elevated concentrations of dioxygen that have no effect on the wild-type cells. This extreme sensitivity to dioxygen is alleviated when the gene coding for a superoxide dismutase is reinserted into the cell, even if the new SOD is of another type and from a different organism.[42,43]

E. Summary of Dioxygen Toxicity

In summary, we know a great deal about the sites that are vulnerable to oxidative damage in biological systems, about the agents that protect against such damage, and about the mechanisms that repair such damage. Metal ions are involved in all this chemistry, both as catalysts of deleterious oxidative reactions and as cofactors in the enzymes that protect against and repair such damage. What we still do not know at this time, however, is how dioxygen initiates the sequence of chemical reactions that produce the agents that attack the vulnerable biological targets *in vivo*.

IV. CYTOCHROME *c* OXIDASE

A. Background

Most of the O_2 consumed by aerobic organisms is used to produce energy in a process referred to as "oxidative phosphorylation," a series of reactions in which electron transport is coupled to the synthesis of ATP and in which the driving force for the reaction is provided by the four-electron oxidizing power of O_2 (Reaction 5.1). (This subject is described in any standard text on biochemistry and will not be discussed in detail here.) The next to the last step in the electron-transport chain produces reduced cytochrome *c*, a water-soluble electron-transfer protein. Cytochrome *c* then transfers electrons to cytochrome *c* oxidase, where they are ultimately transferred to O_2. (Electron-transfer reactions are discussed in Chapter 6.)

Cytochrome *c* oxidase is the terminal member of the respiratory chain in all animals and plants, aerobic yeasts, and some bacteria.[44-46] This enzyme is always found associated with a membrane: the inner mitochondrial membrane in higher organisms or the cell membrane in bacteria. It is a large, complex, multisubunit enzyme whose characterization has been complicated by its size, by the fact that it is membrane-bound, and by the diversity of the four redox metal sites, i.e., two copper ions and two heme iron units, each of which is found in a different type of environment within the protein. Because of the complexity of this system and the absence of detailed structural information, spectroscopic studies of this enzyme and comparisons of spectral properties with O_2-binding proteins (see Chapter 4) and with model iron-porphyrin and copper complexes have been invaluable in its characterization.

B. Spectroscopic Characterization

1. Models

Iron-porphyrin complexes of imidazole are a logical starting point in the search for appropriate spectroscopic models for heme centers in metalloproteins, since the histidyl imidazole side chain is the most common axial ligand bound to iron in such enzymes. Iron-porphyrin complexes with two axial imidazole ligands are known for both the ferrous and ferric oxidation states.[47]

$$(5.31)$$

These complexes are six coordinate, i.e., coordinatively saturated, and therefore do not bind additional ligands unless an imidazole is lost first. Six-coordinate heme centers with two axial imidazole ligands in metalloproteins are known to act as electron-transfer redox centers, e.g., in cytochrome b_5,[48] but are not observed to coordinate dioxygen or any other substrate under normal conditions, because they are coordinatively saturated. In the ferrous state, the iron-porphyrin complex with two axial imidazole ligands is low-spin d^6, and thus diamagnetic. In the ferric state, it is low-spin d^5, and has one unpaired electron. Ferric porphyrin complexes with two axial imidazole ligands have EPR spectra characteristic of low-spin d^5 complexes with axial symmetry,[49,50] as shown in Figure 5.2.

$$(5.32)$$

Monoimidazole complexes of iron porphyrins are also known for both the ferrous and the ferric oxidation states. The design of these model complexes has been more challenging than for six-coordinate complexes because of the high affinity of the five-coordinate complexes for a sixth ligand. In the ferrous complex, five coordination has been achieved by use of 2-methylimidazole ligands, as described in Chapter 4. The ferrous porphyrin binds a single 2-methylimidazole ligand, and, because the Fe^{II} center is raised out of the plane of the porphyrin ring, the 2-methyl substituent suffers minimal steric interactions with the porphyrin. However, the affinity of the five-coordinate complex for another 2-methylimidazole ligand is substantially lower, because the Fe^{II} must drop down into the plane of the porphyrin to form the six-coordinate complexes, in which case the 2-methyl substitutents on both axial ligands suffer severe steric interactions with the porphyrin.[51] Using this approach, five-coordinate monoimidazole complexes can be prepared. They are coordinatively unsaturated, and will

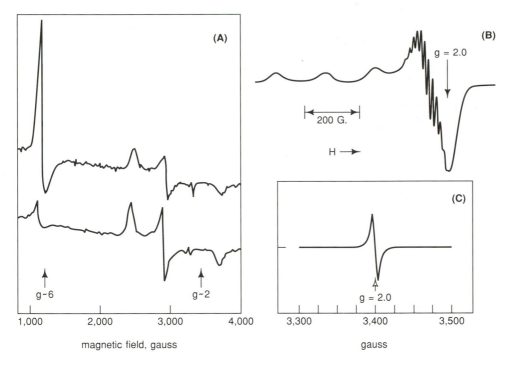

Figure 5.2
(A) Frozen solution EPR spectra of $Fe^{III}(TPP)(4\text{-MeIm})$ (top) and $Fe^{III}(TPP)(4\text{-MeIm})_2^-$ (bottom) prepared by addition of 4-methylimidazolate anion (4-MeIm$^-$) to a solution of $Fe(TPP)(SbF_6)$. The top spectrum is characteristic of a high-spin Fe^{III}-porphyrin complex, with a resonance at $g = 6$ ($g = 2.7$, 2.3, and 1.8 are due to formation of a small amount of $Fe^{III}(TPP)(4\text{-MeIm})_2^-$). The bottom spectrum is characteristic of a low-spin ferric-porphyrin bis(imidazole)-type complex.[52] (B) Frozen solution EPR spectrum of $Cu^{II}(ImH)_4^{2+}$ with $g_\parallel = 2.06$, $A_\parallel = 183$ G, and $g_\perp = 2.256$ (courtesy of Dr. J. A. Roe). This type of spectrum is typical of square-planar Cu^{II} complexes, except that the ligand hyperfine splitting of the $g_\perp$ feature is frequently unresolved, especially in copper proteins (for example, see Figure 5.20). (C) Simulated EPR spectrum of a typical organic free radical with no hyperfine interaction. Note the narrow linewidth and a g value close to 2.00, i.e., near the g value for a free electron.

bind a second axial ligand, such as O_2 and CO. They have been extensively studied as models for O_2- binding heme proteins such as hemoglobin and myoglobin. Monoimidazole ferrous porphyrins thus designed are high-spin d^6 with four unpaired electrons. They are even-spin systems and EPR spectra have not been observed.

Five-coordinate monoimidazole ferric-porphyrin complexes have also been prepared in solution[52] by starting with a ferric porphyrin complex with a very poorly coordinating anion, e.g., $Fe^{III}P(SbF_6)$. Addition of one equivalent of imidazole results in formation of the five-coordinate monoimidazole complex (Reaction 5.33).

$$Fe^{III}(TPP)(SbF_6) + ImH \longrightarrow [Fe^{III}(TPP)(ImH)]^+ + SbF_6^- \qquad (5.33)$$

When imidazole is added to ferric-porphyrin complexes of other anionic ligands, e.g., Cl$^-$, several equivalents of imidazole are required to displace the more strongly bound anionic ligand; consequently, only six-coordinate complexes are observed (Reaction 5.34).

$$Fe^{III}(TPP)Cl + 2ImH \longrightarrow [Fe^{III}(TPP)(ImH)_2]^+ + Cl^- \qquad (5.34)$$

Monoimidazole ferric porphyrins are coordinatively unsaturated, readily bind a second axial ligand, and thus are appropriate models for methemoglobin or metmyoglobin. The five-coordinate complexes are high-spin d^5, but usually become low-spin upon binding another axial ligand to become six-coordinate.

	$Fe^{II}P(ImH)$	$Fe^{III}P(ImH)^+$	$Fe^{II}P(ImH)(CO)$	$Fe^{III}P(ImH)L^+$
	$S = 2$	$S = 5/2$	$S = 0$	$S = 1/2$

(5.35)

heme *b*   heme *a*

Copper(I) complexes are typically two-, three-, or four-coordinate.[53] The metal center is d^{10}, diamagnetic, and thus has no EPR spectrum. Copper(II) complexes are typically four- or five-coordinate, d^9, and have one unpaired electron.[53] The interaction of the unpaired electron with the nuclear spin of copper ($I = \frac{3}{2}$) causes characteristic hyperfine splitting of $g_{//}$ into four components in the EPR, and the value of $g_\perp$ is found to be near to but greater than 2.00. The EPR signal due to Cu^{II} is thus normally distinguishable from that of an organic free radical for which $g_\perp = 2.00$ (see Figure 5.2).

2. Spectroscopy of the enzyme

The oxidized form of cytochrome c oxidase contains two Cu^{II} and two Fe^{III} heme centers. It can be fully reduced to give a form of the enzyme containing two Cu^{I} and two Fe^{II} heme centers.[44–46] The heme found in cytochrome c oxidase is different from that found in other heme proteins. It is heme a, closely related to heme b, which is found in hemoglobin, myoglobin, and cytochrome P-450, but has one of the vinyl groups replaced by a farnesyl substituent and one of the methyl groups replaced by a formyl substituent (see 5.35).

Each of the four metal centers has a different coordination environment appropriate to its function. Cytochrome a and Cu_A appear solely to carry out an electron-transfer function without interacting directly with dioxygen. Cytochrome a_3 and Cu_B appear to be part of a binuclear center that acts as the site for dioxygen binding and reduction. A schematic describing the probable nature of these four metal sites within cytochrome oxidase is given in Figure 5.3 and a description of the evidence supporting the formulation of each center then follows.

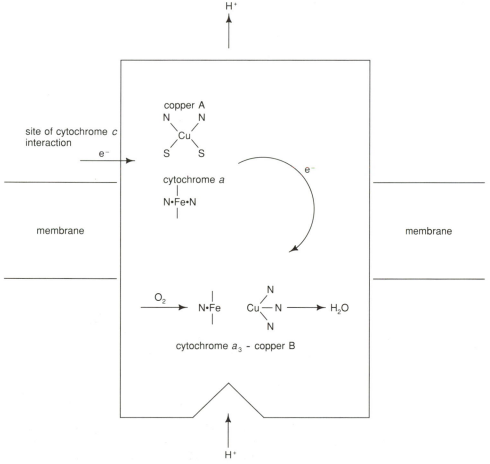

Figure 5.3
Schematic representation of the metal centers in cytochrome c oxidase.

Cytochrome *a* in both oxidation states has spectral characteristics that are entirely consistent with a low-spin ferric heme center with two axial imidazole ligands. In its oxidized form, it gives an EPR spectrum with *g* values[44-46] similar to those obtained with model ferric-porphyrin complexes with two axial imidazole ligands[49,50] (see above, Section IV.B.1). Moreover, addition of cyanide anion to the oxidized enzyme or CO to the reduced enzyme does not perturb this center, indicating that cyanide does not bind to the heme, again consistent with a six-coordinate heme. The absence of ligand binding is characteristic of six-coordinate heme sites found in electron-transfer proteins and suggests strongly that cytochrome *a* functions as an electron-transfer center within cytochrome *c* oxidase.

Cu$_A$ is also believed to act as an electron-transfer site. It has quite remarkable EPR spectroscopic characteristics, with *g* values at *g* = 2.18, 2.03, and 1.99, and no hyperfine splitting,[44-46] resembling more an organic free radical than a typical Cu^{II} center (see Figures 5.2 and 5.4A). ENDOR studies of yeast cytochrome *c* oxidase containing ^{2}H-cysteine or ^{15}N-histidine (from yeast grown with the isotopically substituted amino acids) showed shifts relative to the unsubstituted enzyme, indicating that both of these ligands are bound to Cu$_A$.[54] But the linear electric-field effect of Cu$_A$ did not give the patterns characteristic of Cu^{II}-histidine complexes, indicating that the unpaired electron is not on the copper ion.[55] The current hypothesis about this center is that copper is bonded

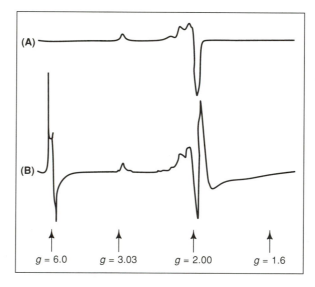

Figure 5.4
(A) EPR spectrum of oxidized cytochrome *c* oxidase. The signal at *g* = 2 is assigned to Cu$_A$^{II} and the signal at *g* = 3 is assigned to cytochrome *a* in the low-spin ferric state. No signals attributable to the other two metal centers, i.e., cytochrome *a$_3$* and Cu$_B$, are observed (see text). (B) Spectrum obtained after addition of NO. Binding of NO to Cu$_B$^{II} decouples the two metal ions, and thereby allows the high-spin ferric heme signal of cytochrome *a$_3$* at *g* = 6 to be observed. (From M. Brunori *et al.*, *Adv. Inorg. Biochem.* **7** (1988), 93–153, adapted from Reference 56.)

in a highly covalent fashion to one, or more likely two, sulfur ligands, and that the unpaired electron density is principally on sulfur, i.e., $[Cu^{II}\text{-}^-SR \leftrightarrow Cu^{I}\text{-} \cdot SR]$. Copper-thiolate model complexes with spectroscopic properties similar to Cu_A have never been synthesized, presumably because such complexes are unstable with respect to disulfide bond formation, i.e, $2\ RS \cdot \rightarrow RS\text{-}SR$. In the enzyme, $RS \cdot$ radicals are presumably constrained in such a way that they cannot couple to form disulfide bonds.

The other heme center, cytochrome a_3, does bind ligands such as cyanide to the Fe^{III} form and carbon monoxide to the Fe^{II} form, indicating that it is either five-coordinate or that it has a readily displaceable ligand. Reaction with CO, for example, produces spectral changes characteristic of a five-coordinate ferrous heme binding CO to give the six-coordinate carbonmonoxy product analogous to MbCO. The cytochrome a_3 site is therefore an excellent candidate for O_2 binding within cytochrome oxidase.

The EPR spectrum of fully oxidized cytochrome c oxidase might be expected to give signals corresponding to two Cu^{II} centers and two ferric heme centers. In fact, all that is observed in the EPR spectrum of the oxidized enzyme is the typical low-spin six-coordinate ferric heme spectrum due to cytochrome a and the EPR signal attributed to Cu_A (see Figure 5.4A). The fact that signals attributable to cytochrome a_3 and Cu_B are not observed in the EPR spectrum led to the suggestion that these two metal centers are antiferromagnetically coupled.[44-46] The measured magnetic susceptibility for the isolated enzyme was found to be consistent with this hypothesis, suggesting that these two metal centers consist of an $S = \frac{1}{2}$ Cu^{II} antiferromagnetically coupled through a bridging ligand to a high-spin $S = \frac{5}{2}$ Fe^{III} to give an $S = 2$ binuclear unit.[45] EXAFS measurements indicating a copper-iron separation of 3–4 Å as well as the strength of the magnetic coupling suggest that the metal ions are linked by a single-atom ligand bridge, but there is no general agreement as to the identity of this bridge.[45,46]

The cytochrome a_3-Cu_B coupling can be disrupted by reduction of the individual metal centers. In this fashion, a $g = 6$ ESR signal can be seen for cytochrome a_3 or $g = 2.053$, 2.109, and 2.278 signals for Cu_B. Nitric-oxide binding to Cu_B also decouples the metals, allowing the $g = 6$ signal to be seen[56] (see Figure 5.4B). Mössbauer spectroscopy also indicates that cytochrome a_3 is high-spin in the oxidized as well as the reduced state.[57] ENDOR studies suggest that Cu_B has three nitrogens from imidazoles bound to it with water or hydroxide as a fourth ligand.[58] Studies using ^{15}N-labeled histidine in yeast have demonstrated that histidine is a ligand to cytochrome a_3.[59] All of these features have been incorporated into Figure 5.3.

C. Mechanism of Dioxygen Reduction

1. Models

Before we consider the reactions of cytochrome c oxidase with dioxygen, it is instructive to review the reactions of dioxygen with iron porphyrins and copper complexes. Dioxygen reacts with ferrous-porphyrin complexes to make mononuclear dioxygen complexes (Reaction 5.36; see preceding chapter for discussion of this important reaction). Such dioxygen complexes react rapidly with another ferrous porphyrin, unless sterically prevented from doing so, to form binuclear peroxo-bridged complexes[60,61] (Reaction 5.37). These peroxo complexes are stable at low temperature, but, when the temperature is raised, the O—O bond cleaves and two equivalents of an iron(IV) oxo complex are formed (Reaction 5.38). Subsequent reactions between the peroxo-bridged complex and the Fe^{IV} oxo complex produce the μ-oxo dimer (see Reactions 5.39–5.40).

$$3Fe^{II}(P) + 3O_2 \longrightarrow 3Fe(P)(O_2) \tag{5.36}$$

$$3Fe(P)(O_2) + 3Fe^{II}(P) \longrightarrow 3(P)Fe^{III}-O-O-Fe^{III}(P) \tag{5.37}$$

$$(P)Fe^{III}-O-O-Fe^{III}(P) \longrightarrow 2Fe^{IV}(P)(O) \tag{5.38}$$

$$2Fe^{IV}(P)(O) + 2(P)Fe^{III}-O-O-Fe^{III}(P) \longrightarrow$$
$$2(P)Fe^{III}-O-Fe^{III}(P) + 2Fe(P)(O_2) \tag{5.39}$$

$$\underline{2Fe(P)(O_2) \longrightarrow 2Fe^{II}(P) + 2O_2} \tag{5.40}$$

$$4Fe^{II}(P) + O_2 \longrightarrow 2(P)Fe^{III}-O-Fe^{III}(P) \tag{5.41}$$

The reaction sequence (5.36) to (5.40) thus describes a four-electron reduction of O_2 in which the final products, two oxide, O^{2-}, ligands act as bridging ligands in binuclear ferric-porphyrin complexes (Reaction 5.41).

Copper(I) complexes similarly react with dioxygen to form peroxo-bridged binuclear complexes.[62] Such complexes do not readily undergo O—O bond cleavage, apparently because the copper(III) oxidation state is not as readily attainable as the Fe(IV) oxidation state in an iron-porphyrin complex. Nevertheless, stable peroxo complexes of copper(II) have been difficult to obtain, because, as soon as it is formed, the peroxo complex either is protonated to give free hydrogen peroxide or is itself reduced by more copper(I) (Reactions 5.42 to 5.46).

$$2Cu^{I} + O_2 \longrightarrow Cu^{II}-O-O-Cu^{II} \tag{5.42}$$

$$Cu^{II}-O-O-Cu^{II} + 2H^+ \longrightarrow 2Cu^{II} + H_2O_2 \tag{5.43}$$

$$2Cu^{I} + H_2O_2 + 2H^+ \longrightarrow 2Cu^{II} + 2H_2O \tag{5.44}$$

$$\text{or} \quad \underline{Cu^{II}-O-O-Cu^{II} + 2Cu^{I} + 4H^+ \longrightarrow 4Cu^{II} + 2H_2O} \tag{5.45}$$

$$4Cu^{I} + O_2 + 4H^+ \longrightarrow 4Cu^{II} + 2H_2O \tag{5.46}$$

Recently, however, examples of the long-sought stable binuclear copper(II) peroxo complex have been successfully synthesized and characterized, and interestingly enough, two entirely different structural types have been identified, i.e., μ-1,2 and μ-η^2:η^2 dioxygen complexes[63,64] (see 5.47)

$$\mu\text{-1,2} \qquad\qquad \mu\text{-}\eta^2:\eta^2 \tag{5.47}$$

2. Mechanistic studies of the enzyme

A single turnover in the reaction of cytochrome c oxidase involves (1) reduction of the four metal centers by four equivalents of reduced cytochrome c, (2) binding of dioxygen to the partially or fully reduced enzyme, (3) transfer of four electrons to dioxygen, coupled with (4) protonation by four equivalents of protons to produce two equivalents of water, all without the leakage of any substantial amount of potentially harmful partially reduced dioxygen byproducts such as superoxide or hydrogen peroxide.[44–46] At low temperatures, the reaction can be slowed down, so that the individual steps in the dioxygen reduction can be observed. Such experiments are carried out using the fully reduced enzyme to which CO has been bound. Binding of CO to the Fe^{II} heme center in reduced cytochrome c oxidase inhibits the enzyme and makes it unreactive to dioxygen. The CO-inhibited derivative can then be mixed with dioxygen and the mixture cooled. Photolysis of metal-CO complexes almost always leads to dissociation of CO, and CO-inhibited cytochrome c oxidase is no exception. Photolytic dissociation of CO frees the Fe^{II} heme, thereby initiating the reaction with dioxygen, which can then be followed spectroscopically.[44–46] Dioxygen reacts very rapidly with the fully reduced enzyme to produce a species that appears to be the dioxygen adduct of cytochrome a_3 (Reaction 5.48). Such a species is presumed to be similar to other mononuclear oxyheme derivatives. The dioxygen ligand in this species is then rapidly reduced to peroxide by the nearby Cu_B, forming what is believed to be a binuclear μ-peroxo species (Reaction 5.49). These steps represent a two-electron reduction of dioxygen to the peroxide level, and are entirely analogous to the model reactions discussed above (Reactions 5.36 to 5.46), except that the binuclear intermediates contain one copper and one heme iron. The μ-peroxo $Fe^{III} - (O_2^{2-}) - Cu^{II}$ species is then reduced by a third electron, resulting in cleavage of the O—O bond (Reaction 5.50). One of the oxygen atoms remains with iron in the form of a ferryl complex, i.e., an Fe^{IV} oxo, and the other is protonated and bound to copper in the form of a Cu^{II} aquo complex.[65] Reduction by another electron leads to hydroxo complexes of both the Fe^{III} heme and the Cu^{II} centers (Reaction 5.51).[65] Protonation then causes dissociation of two water molecules from the oxidized cytochrome a_3-Cu_B center (Reaction 5.52).

$$(cyt\ a_3)Fe^{II} \quad Cu_B^{I} + O_2 \longrightarrow (cyt\ a_3)Fe^{III}(O_2^-) \quad Cu_B^{I} \qquad (5.48)$$

$$(cyt\ a_3)Fe^{III}(O_2^-) \quad Cu_B^{I} \longrightarrow (cyt\ a_3)Fe^{III}\text{-}(O_2^{2-})\text{-}Cu_B^{II} \qquad (5.49)$$

$$(cyt\ a_3)Fe^{III}\text{-}(O_2^{2-})\text{-}Cu_B^{II} + e^- + 2H^+ \longrightarrow (cyt\ a_3)Fe^{IV}{=}O \quad H_2O\text{-}Cu_B^{II} \qquad (5.50)$$

$$(cyt\ a_3)Fe^{IV}{=}O \quad H_2O\text{-}Cu_B^{II} + e^- \longrightarrow (cyt\ a_3)Fe^{III}\text{-}(OH^-) \quad (HO^-)\text{-}Cu_B^{II} \qquad (5.51)$$

$$(cyt\ a_3)Fe^{III}\text{-}(OH^-) \quad (HO^-)\text{-}Cu_B^{II} + 2H^+ \longrightarrow (cyt\ a_3)Fe^{III} \quad Cu_B^{II} + 2H_2O \qquad (5.52)$$

Several important questions remain to be resolved in cytochrome c oxidase research. One is the nature of the ligand bridge that links cytochrome a_3 and

Cu_B in the oxidized enzyme. Several hypotheses have been advanced (imidazolate, thiolate sulfur, and various oxygen ligands), but then discarded or disputed, and there is consequently no general agreement concerning its identity. However, EXAFS measurements of metal-metal separation and the strength of the magnetic coupling between the two metal centers provide evidence that a single atom bridges the two metals.[45,46] Another issue, which is of great importance, is to find out how the energy released in the reduction of dioxygen is coupled to the synthesis of ATP. It is known that this occurs by coupling the electron-transfer steps to a proton-pumping process, but the molecular mechanism is unknown.[46] Future research should provide some interesting insights into the mechanism of this still mysterious process.

V. OXYGENASES

A. Background

The oxygenase enzymes catalyze reactions of dioxygen with organic substrates in which oxygen atoms from dioxygen are incorporated into the final oxidized product.[2-4] These enzymes can be divided into dioxygenases, which direct both atoms of oxygen into the product (Reaction 5.53), and monooxygenases, where one atom of oxygen from dioxygen is found in the product and the other has been reduced to water (Reaction 5.54):

$$\text{Dioxygenase: substrate} + {}^*O_2 \longrightarrow \text{substrate}({}^*O)_2 \tag{5.53}$$

$$\text{Monooxygenase: substrate} + {}^*O_2 + 2H^+ + 2e^- \longrightarrow \text{substrate}({}^*O) + H_2{}^*O \tag{5.54}$$

B. Dioxygenases

Dioxygenase enzymes are known that contain heme iron, nonheme iron, copper, or manganese.[66,67] The substrates whose oxygenations are catalyzed by these enzymes are very diverse, as are the metal-binding sites; so probably several, possibly unrelated, mechanisms operate in these different systems. For many of these enzymes, there is not yet much detailed mechanistic information. However, some of the intradiol catechol dioxygenases isolated from bacterial sources have been studied in great detail, and both structural and mechanistic information is available.[66,67] These are the systems that will be described here.

1. Intradiol catechol dioxygenases

The role of these nonheme iron-containing enzymes is to catalyze the degradation of catechol derivatives to give muconic acids (Reaction 5.55, for example). The enzymes are induced when the only carbon sources available to the bacteria are aromatic molecules. The two best-characterized members of this class are catechol 1,2-dioxygenase (CTD) and protocatechuate 3,4-dioxygenase (PCD).

R = H, cis,cis-muconic acid;
Enz = catechol 1,2-dioxygenase

$$\qquad\qquad\qquad\qquad\qquad\qquad (5.55)$$

R = COO⁻, β-carboxy cis,cis-muconic acid;
Enz = protocatechuate 3,4-dioxygenase

a. Characterization of the Active Sites Even before the x-ray crystal structure of PCD was obtained, a picture of the active site had been constructed by detailed spectroscopic work using a variety of methods. The success of the spectroscopic analyses of these enzymes is a particularly good example of the importance and usefulness of such methods in the characterization of metalloproteins. The two enzymes referred to in Reaction (5.55) have different molecular weights and subunit compositions,[66] but apparently contain very similar active-site structures and function by very similar mechanisms. In both, the resting state of the enzyme contains one Fe^{III} ion bound at the active site. EPR spectra show a resonance at $g = 4.3$, characteristic of high-spin Fe^{III} in a so-called rhombic (low symmetry) environment,[66] and the Mössbauer parameters are also characteristic of high-spin ferric.[66–68] Reactions with substrate analogues (see below) cause spectral shifts of the iron chromophore, suggesting strongly that the substrate binds directly to the iron center in the course of the enzymatic reaction.

It is straightforward to rule out the presence of heme in these enzymes, because the heme chromophore has characteristic electronic-absorption bands in the visible and ultraviolet regions with high extinction coefficients, which are not observed for these proteins. Likewise, the spectral features characteristic of other known cofactors or iron-sulfur centers are not observed. Instead, the dominant feature in the visible absorption spectrum is a band with a maximum near 460 nm and a molar extinction coefficient of 3000 to 4000 $M^{-1}cm^{-1}$ per iron (see Figure 5.5). This type of electronic absorption spectrum is characteristic of a class of proteins, sometimes referred to as iron-tyrosinate proteins, that contain tyrosine ligands bound to iron(III) in their active sites, and which consequently show the characteristic visible absorption spectrum due to phenolate-to-iron(III) charge-transfer transitions. This assignment can be definitively proven by examination of the resonance Raman spectrum, which shows enhancement of the characteristic tyrosine vibrational modes (typically ~1170, 1270, 1500, and 1600 cm^{-1}) when the sample is irradiated in the charge-transfer band described above. Ferric complexes of phenolate ligands may be seen to give almost identical resonance Raman spectra (see Figure 5.6). These bands have been assigned as a C—H bending vibration and a C—O and two C—C stretching vibrations of the phenolate ligand.[69] In addition, NMR studies of the relaxation rates of the proton spins of water indicate that water interacts with the paramagnetic Fe^{III} center in the enzyme. This conclusion is supported by the broadening of the Fe^{III} EPR signal in the presence of $H_2{}^{17}O$, due to interaction

278

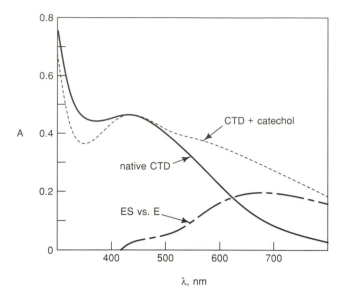

Figure 5.5
Visible absorption spectra of catechol
1,2-dioxygenase and its substrate
complex: E, native enzyme; ES,
enzyme-substrate complex:[66]

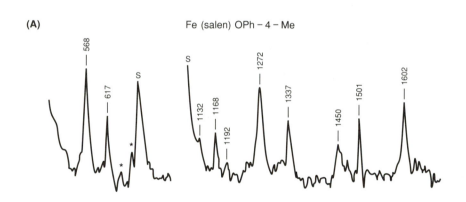

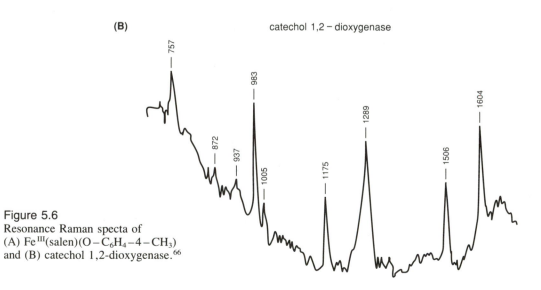

Figure 5.6
Resonance Raman specta of
(A) $Fe^{III}(salen)(O-C_6H_4-4-CH_3)$
and (B) catechol 1,2-dioxygenase.[66]

with the $I = \frac{5}{2}$ nuclear spin of ^{17}O. Thus numerous spectroscopic studies of the catechol dioxygenases led to the prediction that the high-spin ferric ion was bound to tyrosine ligands and water. In addition, EXAFS data, as well as the resemblance of the spectral properties to another, better characterized iron-tyrosinate protein, i.e., transferrin (see Chapter 1), suggested that histidines would also be found as ligands to iron in these proteins.[66,67]

Preliminary x-ray crystallographic results on protocatechuate 3,4-dioxygenase completely support the earlier predictions based on spectroscopic studies.[70] The Fe^{III} center is bound to two histidine and two tyrosine ligands and a water, the five ligands being arranged in a trigonal bipyramidal arrangement, with a tyrosine and a histidine located in axial positions, and with the equatorial water or hydroxide ligand facing toward a cavity assumed to be the substrate-binding cavity. The cavity also contains the positively charged guanidinium group of an arginine side chain, in the correct position to interact with the negatively charged carboxylate group on the protocatechuate substrate (see Figure 5.7).

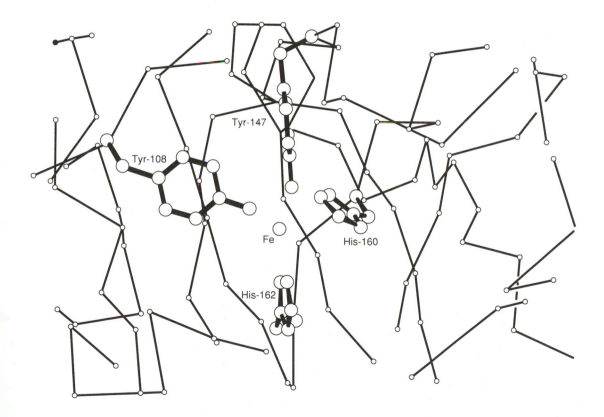

Figure 5.7
A view of the active site of protocatechuate 3,4-dioxygenase based on the results of the x-ray crystal structure.[70] The Fe^{III} center is approximately trigonal bipyramidal, with Tyr-147 and His-162 as axial ligands, and Tyr-108 and His-160 in the equatorial plane along with a solvent molecule (not shown) in the foreground.[66]

b. Mechanistic Studies   As mentioned above, substrates and inhibitors that are substrate analogues bind to these enzymes and cause distinct changes in the spectral properties, suggesting strongly that they interact directly with the Fe^{III} center. Nevertheless, the spectra remain characteristic of the Fe^{III} oxidation state, indicating that the ferric center has not been reduced. Catecholates are excellent ligands for Fe^{III} (see, for example, the catecholate siderophores, Chapter 1) and it might therefore be assumed that the catechol substrate would bind to iron using both oxygen atoms (see 5.56).

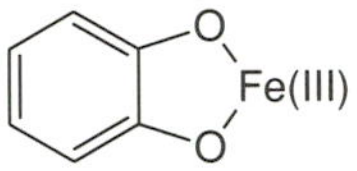

$$\text{(structure)} \quad Fe(III) \qquad\qquad (5.56)$$

However, the observation that phenolic inhibitors $p\text{-}X\text{-}C_6H_4\text{-}OH$ bind strongly to the enzymes suggested the possibility that the substrate binds to the iron center through only one oxygen atom (see 5.57).

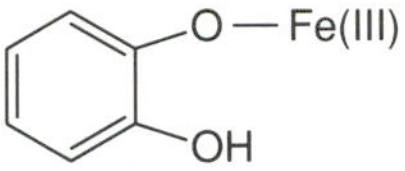

$$\text{(structure)} \quad O — Fe(III) \qquad\qquad (5.57)$$

Paramagnetic NMR studies have been invaluable in distinguishing these two possibilities.[67] The methyl group of the 4-methylcatecholate ligand is shifted by the paramagnetic ferric ion to quite different positions in the ^1H NMR spectra, depending on whether the catecholate is monodentate or bidentate. Comparison of the positions of the methyl resonances in the ^1H NMR spectra of the model complexes with those of the substrate 4-methylcatechol bound to the enzymes CTD and PCD indicates quite clearly that the substrate is bound to CTD in a monodentate fashion and to PCD in a bidentate fashion (see Figure 5.8). These

Figure 5.8 *(facing page)*
(A) Paramagnetically shifted ^{1}H-NMR spectra of Fe(salen) complexes with 4-methylphenolate and 4-methylcatecholate ligands. Starred resonances are assigned to methyl groups. The top spectrum is of the ferric-salen complex of the monodentate 4-methylphenolate ligand. The middle spectrum is of the ferric-salen complex of the monoprotonated 4-methylcatecholate ligand. Note that two isomers are present, because the two oxygen atoms on the ligand are inequivalent, and either may be used to make the monodentate complex. The bottom spectrum is of the ferric-salen complex of the fully deprotonated 4-methylcatecholate ligand, which binds to the iron in a bidentate fashion. Note the smaller span of isotropic shifts, which has been demonstrated to be diagnostic of bidentate coordination. (B) Paramagnetically shifted ^{1}H-NMR spectra of enzyme-substrate complexes of the dioxygenase enzymes catechol 1,2-dioxygenase (CTD) and protocatecholate 3,4-dioxygenase (PCD). Note the resemblance between the position of the methyl resonance in the complex of 4-methylcatechol with CTD and that of one of the methyl resonances in the middle spectrum in (A), indicating that 4-methylcatechol binds to the ferric center in CTD in a monodentate fashion through the O-1 oxygen only. By contrast, the spectrum of 4-methylcatechol with PCD resembles the bottom spectrum in (A), indicating that here the 4-methylcatecholate ligand is bidentate.[66]

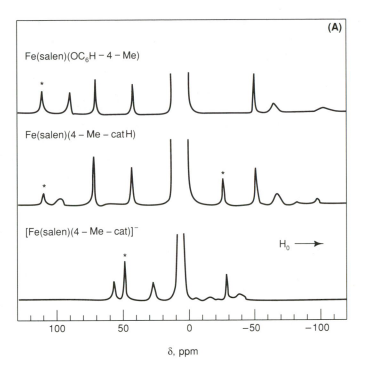

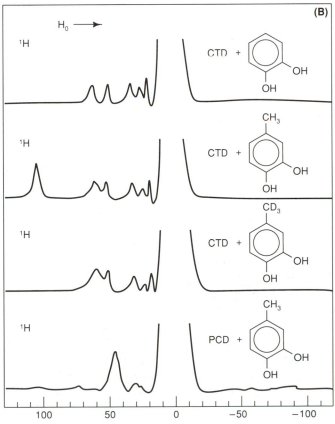

results contradict an early hypothesis that the mode of substrate binding, i.e., monodentate versus bidentate, might be a crucial factor in activating the substrate for reaction with dioxygen.[67]

Spectroscopic observations of the enzymes during reactions with substrates and substrate analogues have enabled investigators to observe several intermediates along the catalytic pathway. Such studies have led to the conclusion that the iron center remains high-spin Fe^{III} throughout the entire course of the reaction. This conclusion immediately presents a problem in understanding the nature of the interaction of dioxygen with the enzyme, since dioxygen does not in general interact with highly oxidized metal ions such as Fe^{III}. The solution seems to be that this reaction represents an example of *substrate* rather than *dioxygen* activation.

Studies of the oxidation of ferric catecholate coordination complexes have been useful in exploring mechanistic possibilities for these enzymes.[71] A series of ferric complexes of 3,5-di-*t*-butyl-catechol with different ligands L have been found to react with O_2 to give oxidation of the catechol ligand (Reaction 5.58).

$$(5.58)$$

The relative reactivities of these complexes appear to vary with the donating properties of the ligand L. Ligands that are the poorest donors of electron density tend to increase the reactivities of the complexes with O_2. These results suggest that the reactivity of these complexes with O_2 is increased by an increase in the contribution that the minor resonance form B makes to the ground state of the complex (see Figure 5.9) and that complexes of ligands that are poor donors tend to favor electron donation from catechol to Fe^{III}, thus increasing the relative amount of minor form B. It should be noted that the spectroscopic characteristics of these complexes are nevertheless dominated by the major resonance form A, regardless of the nature of L.

All these studies of the enzymes and their model complexes have led to the mechanism summarized in Figure 5.9.[66] In this proposed mechanism, the catechol substrate coordinates to the ferric center in either a monodentate or a bidentate fashion, presumably displacing the water or hydroxide ligand. The resulting catechol complex then reacts with dioxygen to give a peroxy derivative of the substrate, which remains coordinated to Fe^{III}. The subsequent rearrangement of this peroxy species to give an anhydride intermediate is analogous to well-characterized reactions that occur when catechols are reacted with alkaline hydrogen peroxide.[72] The observation that both atoms of oxygen derived from O_2 are incorporated into the product requires that the ferric oxide or hydroxide complex formed in the step that produces the anhydride does not exchange with external water prior to reacting with the anhydride to open it up to the product diacid.

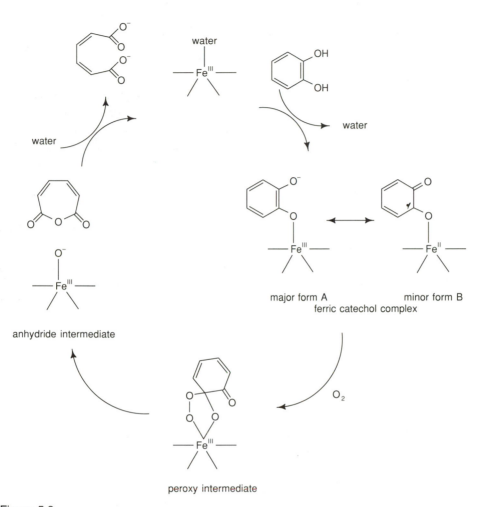

Figure 5.9
Proposed mechanism for catechol dioxygenases (modified from Reference 45). The major form A is shown here with the catechol bound in a monodentate fashion. It may sometimes be bound in a bidentate fashion as well (see text).

It is interesting to consider how the intradiol dioxygenase enzymes overcome the kinetic barriers to oxidations by dioxygen, and why this particular mechanism is unlikely to be applicable to the monooxygenase enzymes. The first point is that the ferric catechol intermediate is paramagnetic, with resonance forms that put unpaired electron density onto the carbon that reacts with dioxygen. The spin restriction is therefore not a problem. In addition, the catechol ligand is a very good reducing agent, much more so than the typical substrates of the monooxygenase enzymes (see next section). It is possible, therefore, that the reaction of dioxygen with the ferric catechol complex results in a concerted two-electron transfer to give a peroxy intermediate, thus bypassing the relatively unfavorable one-electron reduction of O_2.

C. Monooxygenases

Metal-containing monooxygenase enzymes are known that contain heme iron, nonheme iron, or copper at their active sites.[2] For most of these enzymes, there is only limited information about the nature of the active site and the mode of interaction with dioxygen or substrates. But there are three monooxygenase enzymes that strongly resemble well-characterized reversible dioxygen-carrying proteins (see preceding chapter), suggesting that dioxygen binding to the metalloenzyme in its reduced state is an essential first step in the enzymatic mechanisms, presumably followed by other steps that result in oxygenation of substrates. The enzymes are:

(1) cytochrome P-450,[73] a heme-containing protein whose active site resembles the dioxygen-binding sites of myoglobin or hemoglobin in many respects, except that the axial ligand to iron is a thiolate side chain from cysteine rather than an imidazole side chain from histidine;

(2) tyrosinase,[74] which contains two copper ions in close proximity in its active site and which has deoxy, oxy, and met states that closely resemble comparable states of hemocyanin in their spectroscopic properties; and

(3) methane monooxygenase,[75,76] which contains two nonheme iron ions in close proximity and which resembles hemerythrin in many of its spectroscopic properties.

In addition to these three, there are also monooxygenase enzymes containing single nonheme iron[77] or copper ions,[78] or nonheme iron plus an organic cofactor such as a reduced pterin at their active sites.[79] Just as with the dioxygenase enzymes, we do not know how similar the mechanisms of the different metal-containing monooxygenase enzymes are to one another. The enzyme for which we have the most information is cytochrome P-450, and we will therefore focus our discussion on that system. Speculations about the mechanisms for the other systems are discussed at the end of this section.

1. Cytochrome P-450

Cytochrome P-450 enzymes are a group of monooxygenase enzymes that oxygenate a wide variety of substrates.[73] Examples of such reactions are:

(1) hydroxylation of aliphatic compounds (Reaction 5.59);

(2) hydroxylation of aromatic rings (Reaction 5.60);

(3) epoxidation of olefins (Reaction 5.61);

(4) amine oxidation to amine oxides (Reaction 5.62);

(5) sulfide oxidation to sulfoxides (Reaction 5.63); and

(6) oxidative dealkylation of heteroatoms (for example, Reaction 5.64).

$$-\overset{|}{\underset{|}{C}}-H \longrightarrow -\overset{|}{\underset{|}{C}}-OH \qquad\qquad (5.59)$$

$$\text{(benzene)}-H \longrightarrow \text{(benzene)}-OH \qquad\qquad (5.60)$$

$$\searrow=\swarrow \longrightarrow \text{(epoxide)} \qquad\qquad (5.61)$$

$$\searrow N \longrightarrow \searrow N^+ - O^- \qquad\qquad (5.62)$$

$$-S- \longrightarrow -\overset{O}{\underset{||}{S}}- \qquad\qquad (5.63)$$

$$Ph-O-CH_3 \longrightarrow Ph-OH + HCHO \qquad\qquad (5.64)$$

Some of these reactions have great physiological significance, because they represent key transformations in metabolism, as in lipid metabolism and biosynthesis of corticosteroids, for example.[73] Cytochrome P-450 is also known to catalyze the transformation of certain precarcinogens such as benzpyrene into their carcinogenic forms.[73]

Many of the P-450 enzymes have been difficult to characterize, because they are membrane-bound and consequently relatively insoluble in aqueous solution. However, cytochrome P-450$_{cam}$, which is a component of the camphor 5-monooxygenase system isolated from the bacterium *Pseudomonas putida*, is soluble and has been particularly useful as the subject of numerous spectroscopic and mechanistic studies, as well as several x-ray crystallographic structure determinations.[80] This enzyme consists of a single polypeptide chain, mainly α-helical, with a heme *b* group (Fe-protoporphyrin IX) sandwiched in between two helices, with no covalent attachments between the porphyrin ring and the protein. One axial ligand complexed to iron is a cysteinyl thiolate. In the resting state, the iron is predominantly low-spin Fe^{III}, probably with a water as the other axial ligand. When substrate binds to the resting enzyme, the spin state changes to high-spin, and the non-cysteine axial ligand is displaced. The enzyme can be reduced to an Fe^{II} state, which is high-spin, and resembles deoxyhemoglobin or myoglobin in many of its spectroscopic properties. This ferrous form binds dioxygen to make an oxy form or carbon monoxide to make a carbonyl form. The CO derivative has a Soret band (high-energy π-π^* transition of the porphyrin ring) at 450 nm, unusually low energy for a carbonyl derivative of a heme protein because of the presence of the axial thiolate ligand. This spectroscopic feature aids in the isolation of the enzyme and is responsible for its name.

a. "Active Oxygen" Camphor 5-monooxygenase is a three-component system, consisting of cytochrome P-450$_{cam}$ and two electron-transfer proteins, a flavoprotein, and an iron-sulfur protein (see Chapters 6 and 7). The role of the electron-transfer proteins is to deliver electrons to the P-450 enzyme, but these may be replaced *in vitro* by other reducing agents. The reaction sequence is in Figure 5.10.

For cytochrome P-450, the question that is possibly of greatest current interest to the bioinorganic chemist is just what mechanism enables activation of dioxygen and its reaction with substrate. It seems clear that dioxygen binds to

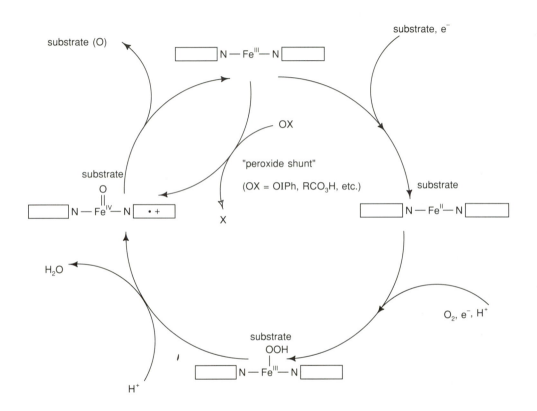

Figure 5.10
Proposed mechanism for cytochrome P-450. The resting enzyme is shown at the top of the cycle. Substrate binds to the enzyme at a position close to the iron center, but it is not directly coordinated to the metal ion. The enzyme-substrate complex is then reduced to the ferrous form. Dioxygen then binds to form an oxy complex (not shown). The oxy complex is then reduced by another electron and protonated, giving a ferric-hydroperoxy complex shown at the bottom of the cycle. The ligand bound here to the Fe^{III} center is HO_2^-, i.e. deprotonated hydrogen peroxide. The ferric hydroperoxy form of the enzyme-substrate complex then undergoes heterolytic O—O bond cleavage, giving a high-valent Fe^{IV} oxo center, with the porphyrin ligand oxidized by one equivalent (see text). This species then transfers a neutral oxygen atom to the bound substrate, which is then released, giving the oxygenated product and regenerating the resting form of the enzyme. The "peroxide shunt" refers to the mechanism proposed for the cytochrome P-450-catalyzed oxygenation of substrates by single-oxygen-atom donors (see text). It is believed that the same high-valent iron-oxo intermediate is generated in these types of reactions as well.

the ferrous state of the enzyme-substrate complex, and that the resulting oxy ligand, which presumably is similar to the oxy ligand in oxyhemoglobin and oxymyoglobin, is not sufficiently reactive to attack the bound substrate. The oxy form is then reduced and the active oxidant is generated, but the nature of the active oxidant has not been deduced from studies of the enzyme itself, nor has it been possible to observe and characterize intermediates that occur between the time of the reduction and the release of product. Three species are potential candidates for "active oxygen," the oxygen-containing species that attacks the substrate, in cytochrome P-450. They are:

(1) a ferric peroxo, **1a**, or hydroperoxo complex, **1b**, formed from one-electron reduction of the oxy complex (Reaction 5.65);
(2) an iron(IV) oxo complex, **2**, formed by homolytic O—O bond cleavage of a ferric hydroperoxo complex (Reaction 5.66); and
(3) a complex at the oxidation level of an iron(V) oxo complex, **3**, formed by heterolytic O—O bond cleavage of a ferric hydroperoxo complex (Reaction 5.67).

The hydroxyl radical, HO·, although highly reactive and capable of attacking P-450 substrates, is considered to be an unlikely candidate for "active oxygen" because of the indiscriminate character of its reactivity.

$$Fe^{II}P + O_2 \longrightarrow FePO_2 \xrightarrow{\ e^- \ } [Fe^{III}P(O_2^{2-})]^- \xrightarrow{\ H^+ \ } Fe^{III}P(O_2H^-) \qquad (5.65)$$
$$\phantom{Fe^{II}P + O_2 \longrightarrow FePO_2 \xrightarrow{\ e^- \ }}\textbf{1a}\textbf{1b}$$

$$Fe^{III}P(O_2H^-) \longrightarrow Fe^{IV}P(O) + HO· \qquad (5.66)$$
$$\textbf{1b}\textbf{2}$$

$$Fe^{III}P(O_2H^-) \longrightarrow [Fe^V(P^{2-})(O)^+ \longleftrightarrow Fe^{IV}(P^-)(O)^+] + HO^- \qquad (5.67)$$
$$\textbf{1b}\textbf{3}$$

(P^{2-} = porphyrin ligand; P^- = one-electron oxidized porphyrin ligand)

An iron(V) oxo complex (or a related species at the same oxidation level), **3**, formed via Reaction (5.67), is the favored candidate for "active oxygen" in cytochrome P-450.[81] This conclusion was initially drawn from studies of reactions of the enzyme with alkylhydroperoxides and single-oxygen-atom donors. Single-oxygen-atom donors are reagents such as iodosylbenzene, OIPh, and periodate, IO_4^-, capable of donating a neutral oxygen atom to an acceptor, forming a stable product in the process (here, iodobenzene, IPh, and iodate, IO_3^-). It was discovered that ferric cytochrome P-450 could catalyze oxygenation reactions using organic peroxides or single-oxygen-atom donors in place of dioxygen and reducing agents. Usually the same substrates would give the identical oxygenated product. This reaction pathway was referred to as the "peroxide shunt" (see Figure 5.10). The implication of this discovery was that the same form of "active oxygen" was generated in each reaction, and the fact that single-oxygen-atom donors could drive this reaction implied that this species contained only one oxygen atom, i.e., was generated subsequent to O—O bond cleavage. The mechanism suggested for this reaction was Reactions (5.68) and (5.69).

$$Fe(III)P^+ + OX \longrightarrow \mathbf{3} + X \qquad\qquad (5.68)$$

$$\mathbf{3} + substrate \longrightarrow Fe(III)P^+ + substrate(O) \qquad\qquad (5.69)$$

b. Metalloporphyrin Model Systems   Studies of the reactivities of synthetic metalloporphyrin complexes in oxygen-transfer reactions and characterization of intermediate species observed during the course of such reactions have been invaluable in evaluating potential intermediates and reaction pathways for cytochrome P-450. Logically, it would be most desirable if one could mimic the enzymatic oxygenation reactions of substrates using iron porphyrins, dioxygen, and reducing agents. However, studies of such iron-porphyrin-catalyzed reactions have failed to produce meaningful results that could be related back to the P-450 mechanism. This is perhaps not surprising, since the enzyme system is designed to funnel electrons into the iron-dioxygen-substrate complex, and thus to generate the active oxidant within the confines of the enzyme active site in the immediate proximity of the bound substrate. Without the constraints imposed by the enzyme, however, iron porphyrins generally will either (1) catalyze the oxidation of the reducing agent by dioxygen, leaving the substrate untouched, or (2) initiate free-radical autoxidation reactions (see Section II.C). A different approach was suggested by the observation of the peroxide shunt reaction (Reactions 5.68 and 5.69) using organic peroxides or single-oxygen-atom donors, and the earliest successful studies demonstrated that Fe(TPP)Cl (TPP = tetraphenylporphyrin) would catalyze the epoxidation of olefins and the hydroxylation of aliphatic hydrocarbons by iodosylbenzene[81](Reactions 5.70 and 5.71).

$$(5.70)$$

$$(5.71)$$

Reactions (5.70) and (5.71) were postulated to occur via an iron-bound oxidant such as **3** in Reaction (5.67). This hypothesis was tested by studying the reaction of dioctyl Fe(PPIX)Cl with iodosylbenzene, which resulted in 60 percent hydroxylation at positions 4 and 5 on the hydrocarbon tail (see 5.72), positions for which there is no reason to expect increased reactivity except for the fact that those particular locations are predicted from molecular models to come closest to the iron center when the tail wraps around the porphyrin molecule.[82]

$$(5.72)$$

The nature of the species produced when single-oxygen-atom donors react with Fe^{III}-porphyrin complexes has been deduced from studies of an unstable, bright-green porphyrin complex produced by reaction of $Fe^{III}(TMP)Cl$ (TMP = 5,10,15,20-tetramesitylporphyrin) with either iodosylbenzene or peroxycarboxylic acids in solution at low temperatures.[81,83] Titrations of this green porphyrin complex using I^- as a reducing agent demonstrated that this species is readily reduced by two electrons to give the ferric complex $Fe^{III}(TMP)^+$, i.e., that the green complex is two equivalents more oxidized than Fe^{III}. A logical conclusion would be that the green species is $Fe^V(P^{2-})(O^{2-})^+$. However, spectroscopic studies of this species have led to the conclusion that it is, in fact, an Fe^{IV} oxo porphyrin-radical complex, $Fe^{IV}(P^{\cdot-})(O^{2-})^+$, and that this formulation is the best description of **3**, the product formed from heterolytic cleavage of the hydroperoxy intermediate in Reaction (5.67).[81,83] EXAFS studies indicate that the green porphyrin complex contains iron bonded to an atom at an unusually short distance, i.e., 1.6 Å, in addition to being bonded to the porphyrin nitrogens at 2 Å. This short Fe—O distance is consistent with the formulation of the complex as a "ferryl" complex, i.e., $Fe^{IV}=O$. In such a complex, the oxo ligand, O^{2-}, is bonded to the Fe^{IV} center by a combination of σ and π bonding, the latter because of overlap of the filled ligand p-orbitals with the partially filled d_{xz} and d_{yz} orbitals of the metal. Confirmation that the oxidation state of iron is indeed Fe^{IV} comes from comparison of the Mössbauer parameters (δ_{Fe} = 0.06 and ΔE_Q = 1.62) with those of other known Fe^{IV}-porphyrin complexes (see Figure 5.11).[83]

Visible absorption spectra of porphyrin complexes are due largely to π-π^* transitions of the porphyrin ligand. The bright green color is unusual for iron-porphyrin complexes, which are usually red or purple. (However, this green color has been seen for compound I of catalase and peroxidases; see Section VI below.) The unusually long-wavelength visible absorption bands that account for the green color result from the fact that the porphyrin ring has been oxidized by one electron. Similar visible absorption bands can be seen, for example, in other oxidized porphyrin complexes, such as $Co^{III}(P^{\cdot-})^+$, formed by two-electron oxidation of $Co^{II}(P^{2-})$ (see 5.73).[84]

$$(5.73)$$

$Co^{II}(P^{2-})$ $Co^{II}(P^{\cdot-})X]^+$

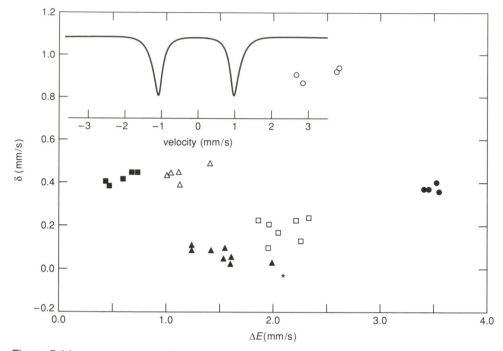

Figure 5.11

Comparative Mössbauer data for iron-porphyrin complexes: ■ high-spin Fe^{III}, □ low-spin Fe^{III}, ● intermediate-spin (admixed) Fe^{III}, ○ high-spin Fe^{II}, △ low-spin Fe^{II}, ▲ oxo- and imido-Fe^{IV}, * dimethoxyiron(IV)TMP. Insert shows the zero-field Mössbauer spectrum of ^{57}Fe-dimethoxyiron(IV)TMP at 4.2 K. (From Reference 83.)

Oxidized porphyrin ligands also give characteristic proton NMR spectra, which are seen for the green porphyrin complex as well.[81,83]

Magnetic measurements indicate that the green porphyrin complex contains three unpaired electrons. Detailed analysis of the Mössbauer spectra has indicated that the two unpaired electrons on the Fe^{IV} ion are strongly ferromagnetically coupled to the unpaired electron on the porphyrin, accounting for the resulting $S = \frac{3}{2}$ state.[81,83]

$$
\begin{array}{lll}
d_{x^2-y^2} & - & \\
d_{z^2} & - & \text{gives } S = 3/2 \\
d_{xz}, d_{yz} & \uparrow\ \uparrow & \uparrow \\
d_{xy} & \uparrow\downarrow & \\
\quad Fe^{IV} & & P^{\cdot -} \\
\quad S = 1 & & S = 1/2
\end{array}
\tag{5.74}
$$

Studies of the reactions of this species with P-450-type substrates demonstrate that this species is reactive enough to make it an attractive candidate for "active oxygen" in the enzymatic mechanism.[81,83]

Synthetic analogues for two of the other candidates for "active oxygen" have also been synthesized and their reactivities assessed. For example, Fe^{III}

and Mn^{III}-porphyrin peroxo complexes analogous to **1a** in Reaction (5.65) have been synthesized. The x-ray crystal structure of the Mn complex shows that the peroxo ligand is bound to the metal in a triangular, side-on fashion (see 5.75). The Fe complex is believed to have a similar structure.[85,86]

$$
\begin{array}{c}
\text{O} - \text{O} \\
\diagdown \ \diagup \\
\text{Fe} \\
\diagup \quad \diagdown
\end{array}
\qquad (5.75)
$$

Studies of this species indicate that **1a** in Reaction (5.65) would not have the requisite reactivity to be a candidate for "active oxygen" in the cytochrome P-450 mechanism, since it will not even oxidize triphenylphosphine, PPh_3, to triphenylphosphine oxide, $OPPh_3$, one of the more facile oxygenation reactions known.[87] Attempts to examine the protonated form, **1b** in Reaction (5.65), however, indicate that it is highly unstable, and its reactivity has not yet been thoroughly examined.[87] Fe^{IV}-oxo-porphyrin complexes analogous to **2** in Reaction (5.66) have also been prepared in solution and characterized by NMR.[60,61] Such complexes will react with PPh_3 to give $OPPh_3$, but are relatively unreactive with olefins and totally unreactive with saturated hydrocarbons. Thus **2** is also ruled out as a candidate for "active oxygen" in P-450 mechanisms.

These reactivity studies, and the observation of the peroxide shunt described above, indicate that $Fe^{V}(P^{2-})(O)^{+}$ or $Fe^{IV}(P^{-})(O)^{+}$ is the most likely candidate for "active oxygen." These two formulations are, of course, isoelectronic, and it is tempting to conclude that the latter is the more likely formulation of the enzymatic intermediate. However, it is important to remember that the model systems lack the axial cysteinyl ligand present in cytochrome P-450. The effect of the relatively easily oxidized sulfur ligand on the electron distribution within that intermediate is not known, since model systems for high-valent iron-oxo complexes containing axial thiolate ligands have not been synthesized.

The mechanism of reactions of the high-valent oxo complex **3** in Reaction (5.67) with a variety of substrates is an area of active interest.[81,88] Such studies are generally carried out by generation of the species *in situ* from the reaction of a ferric porphyrin with a single-oxygen-atom donor, such as a peracid or iodosylbenzene.[89] In hydroxylation reactions of aliphatic hydrocarbons, the initial step appears to be abstraction of a hydrogen atom from the substrate to form a substrate radical and an Fe^{IV} hydroxide complex held together in a cage created by the enzyme active site so that they cannot diffuse away from each other (Reaction 5.76). This step is then followed by recombination of the OH fragment with the substrate radical to make the hydroxylated product (Reaction 5.77). This mechanism is referred to as the "oxygen rebound mechanism."[83]

$$
\text{X} - \overset{|\,\cdot}{\underset{|}{\text{Fe}^{IV}}} = \text{O} + \text{H} - \text{C} \diagup \diagdown \longrightarrow \text{X} - \overset{|}{\underset{|}{\text{Fe}^{IV}}} - \text{OH} + \cdot \text{C} \diagup \diagdown \qquad (5.76)
$$

$$
\text{X} - \overset{|}{\underset{|}{\text{Fe}^{IV}}} - \text{OH} + \cdot \text{C} \diagup \diagdown \longrightarrow \text{X} - \overset{|}{\underset{|}{\text{Fe}^{III}}} + \text{HO} - \text{C} \diagup \diagdown \qquad (5.77)
$$

The radical character of the intermediates formed in this reaction is supported by the observation that such reactions carried out using synthetic porphyrins and single-oxygen-atom donors in the presence of $BrCCl_3$ give substantial amounts of alkyl bromides as products, a result that is consistent with radical intermediates and inconsistent with either carbanion or carbonium-ion intermediates.[83]

In the enzymatic reactions themselves, there is also strong evidence to support a stepwise mechanism involving free-radical intermediates. For example, cytochrome P-450$_{cam}$ gives hydroxylation of d-camphor only in the 5-exo position, but deuterium-labeling studies show that either the 5-exo or the 5-endo hydrogen is lost (Reaction 5.78).[88]

(5.78)

Such results are obviously inconsistent with a concerted mechanism in which the oxygen atom would be inserted into the 5-exo C—H bond in one step; so there would be no chance for the hydrogens in the two positions to exchange. (Remember that alcohol protons exchange rapidly with water and therefore are not expected to remain deuterated when the reaction is carried out in H_2O.)

The crystal structure of reduced cytochrome P-450$_{cam}$ with CO bound to the iron and the substrate camphor bound[90] adjacent to it has been examined and compared with the crystal structure of the oxidized enzyme with camphor bound. The former is expected to be similar in structure to the less-stable oxy complex. The comparison shows that the substrate camphor is closer to the iron center in the oxidized enzyme. It is therefore possible that a similar movement of the substrate occurs during the catalytic reaction after either a 5-exo or a 5-endo hydrogen is abstracted, and that the new position of the camphor molecule then restricts the hydroxylation step to the 5-exo position. It is interesting to note that the 5-exo position on the camphor that is hydroxylated is held in very close proximity to the Fe^{III} center, and therefore to the presumed location of the oxo ligand in the high-valent oxo intermediate in the structure of the ferric enzyme plus camphor derivative (Figure 5.12). Crystal structures of the ferric form of cytochrome P-450$_{cam}$ with norcamphor and adamantanone bound in place of camphor have also been determined.[90] These alternative substrates are smaller than camphor, and appear to fit more loosely than camphor. It is therefore reasonable to assume that they "rattle around" to a certain extent in the substrate

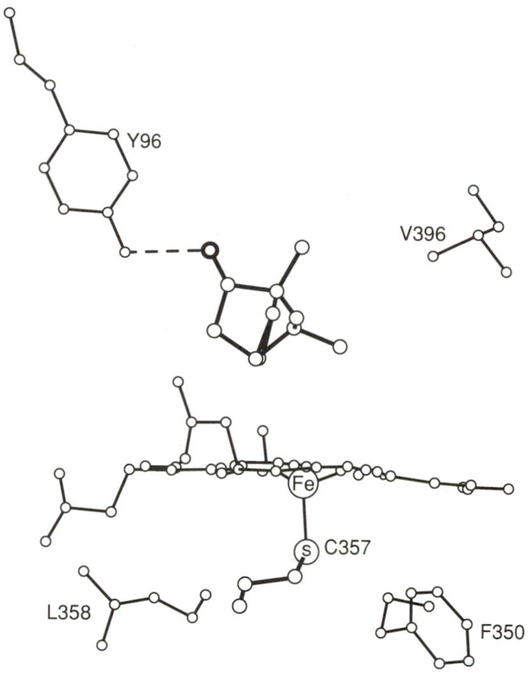

Figure 5.12
Edge-on view of the P-450 active-site region with the substrate camphor molecule bound.[80] The substrate camphor is located in a hydrophobic pocket directly above the heme and is oriented by a hydrogen bond between the carbonyl oxygen of the camphor and Tyr-96. The position that is hydroxylated in the oxygenation reaction, i.e., the 5-*exo* position, is the closest point of approach of the substrate to the expected position of the oxygen atom bound to iron in the high-valent iron-oxo intermediate.

binding site, which probably accounts for the less-specific pattern of hydroxylation observed for these alternative substrates.

Mechanisms for olefin epoxidations catalyzed either by the enzyme or by model porphyrin complexes are not as well understood as those for hydroxylation of aliphatic hydrocarbons. Some of the possibilities that have been proposed[88,91] are represented schematically in Figure 5.13.

c. *O—O Bond Cleavage*   The evidence is persuasive that the "active oxygen" species that attacks substrate in cytochrome P-450 is a high-valent iron-oxo complex. However, the mechanism of formation of that species in the catalytic reaction with dioxygen is less well-understood. Heterolytic O—O bond cleavage of a ferric porphyrin hydroperoxide complex, **1b** (Reaction 5.67), is the logical and anticipated route, but it has not yet been unequivocally demonstrated in a model complex.[92,93] The catalase and peroxidase enzymes catalyze heterolytic O—O bond cleavage in reactions of hydrogen peroxide, but in them the active sites contain amino-acid side chains situated to facilitate the devel-

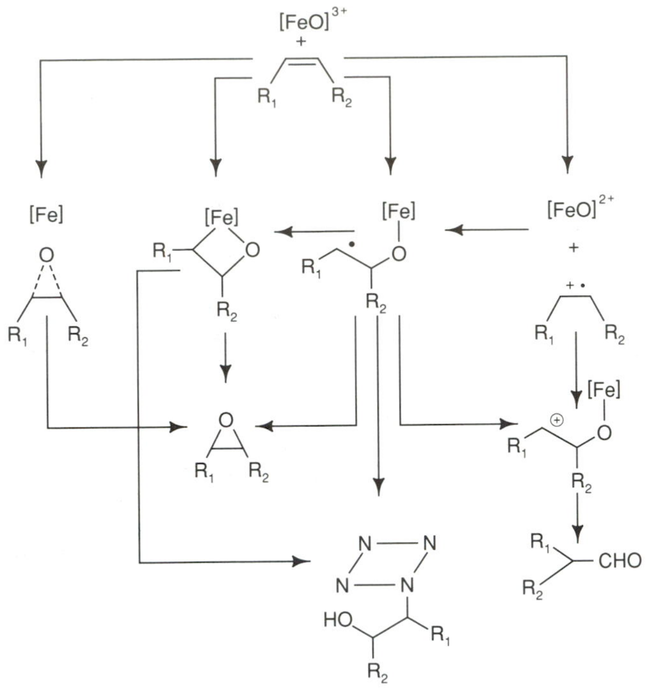

Figure 5.13
Schematic representation of possible mechanistic pathways for olefin epoxidation by **3**. The mechanisms described are, from right to left, concerted addition of oxygen to the double bond, reaction via a metallocylic intermediate, reaction via a ring-opened radical intermediate, and reaction proceeding via an initial electron-transfer step.[91]

oping charge separation that occurs in heterolytic cleavage (see Section VI). The crystal structure of cytochrome P-450$_{cam}$ shows no such groups in the active-site cavity, nor does it give any clue to the source of a proton to protonate the peroxide ligand when it is produced.[80] Also, we have little experimental evidence concerning possible roles that the cysteinyl sulfur axial ligand might play in facilitating O—O bond cleavage. These issues remain areas of active interest for researchers interested in cytochrome P-450 mechanisms.

2. Other metal-containing monooxygenase enzymes

As mentioned above, much less is known about the structural characteristics and mechanisms of the nonheme metal-containing monooxygenase enzymes. From the similarities of the overall stoichiometries of the reactions and the resemblance of some of the enzymes to dioxygen-binding proteins, it is likely that the initial steps are the same as those for cytochrome P-450, i.e., dioxygen binding followed by reduction to form metal-peroxide or hydroperoxide complexes. It is not obvious that the next step is the same, however (i.e., O—O bond cleavage to form a high-valent metal-oxo complex prior to attack on substrate). The problem is that such a mechanism would generate metal-oxo complexes that appear to contain metal ions in chemically unreasonable high-oxidation states, e.g., Fe^V, Cu^{III}, or Cu^{IV} (Reactions 5.79–5.81).

$$(Fe^{III} - OOH)^{2+} \longrightarrow (Fe^VO)^{3+} + OH^- \tag{5.79}$$

$$(Cu^{II} - OOH)^+ \longrightarrow (Cu^{IV}O)^{2+} + OH^- \tag{5.80}$$

$$(Cu^{II} - OO - Cu^{II})^{2+} \longrightarrow 2\,(Cu^{III}O)^+ \tag{5.81}$$

An alternative mechanism is for the peroxide or hydroperoxide ligand to attack the substrate directly; i.e., O—O bond cleavage could be concerted with attack on substrate. Another possibility is that the oxygen atom is inserted in a metal-ligand bond prior to transfer to the substrate. Neither of these alternative mechanisms has been demonstrated experimentally. These various possibilities remain to be considered as more information about the monooxygenase enzymes becomes available.

VI. CATALASE AND PEROXIDASE

A. Description of the Enzymes

Catalase and peroxidase are heme enzymes that catalyze reactions of hydrogen peroxide.[94,95] In catalase, the enzymatic reaction is the disproportionation of hydrogen peroxide (Reaction 5.82) and the function of the enzyme appears to be prevention of any buildup of that potentially dangerous oxidant (see the discussion of dioxygen toxicity in Section III).

$$2H_2O_2 \xrightarrow{\text{catalase}} 2H_2O + O_2 \tag{5.82}$$

Peroxidase reacts by mechanisms similar to catalase, but the reaction catalyzed is the oxidation of a wide variety of organic and inorganic substrates by hydrogen peroxide (Reaction 5.83).

$$H_2O_2 + AH_2 \xrightarrow{\text{peroxidase}} 2H_2O + A \tag{5.83}$$

(The catalase reaction can be seen to be a special case of Reaction 5.83 in which the substrate, AH_2, is hydrogen peroxide.) Some examples of peroxidases that have been characterized are horseradish peroxidase, cytochrome c peroxidase, glutathione peroxidase, and myeloperoxidase.[94,95]

X-ray crystal structures have been determined for beef-liver catalase[80] and for horseradish peroxidase[96] in the resting, high-spin ferric state. In both, there is a single heme b group at the active site. In catalase, the axial ligands are a phenolate from a tyrosyl residue, bound to the heme on the side away from the active-site cavity, and water, bound to heme within the cavity and presumably replaced by hydrogen peroxide in the catalytic reaction. In horseradish peroxidase, the axial ligand is an imidazole from a histidyl residue. Also within the active-site cavity are histidine and aspartate or asparagine side chains that appear to be ideally situated to interact with hydrogen peroxide when it is bound to the iron. These residues are believed to play an important part in the mechanism by facilitating O—O bond cleavage (see Section VI.B below).

Three other forms of catalase and peroxidase can be generated, which are referred to as compounds I, II, and III. Compound I is generated by reaction of the ferric state of the enzymes with hydrogen peroxide. Compound I is green and has spectral characteristics very similar to the $Fe^{IV}(P^{\cdot-})(O)^+$ complex prepared at low temperatures by reaction of ferric porphyrins with single-oxygen-atom donors (see Section V.C.1.a). Titrations with reducing agents indicate that it is oxidized by two equivalents above the ferric form. It has been proposed (see 5.84) that the anionic nature of the tyrosinate axial ligand in catalase may serve to stabilize the highly oxidized iron center in compound I of that enzyme,[80] and furthermore that the histidyl imidazole ligand in peroxidase may deprotonate, forming imidazolate,[52,97] or may be strongly hydrogen bonded,[98] thus serving a similar stabilizing function for compound I in that enzyme.

| tyrosinate | imidazolate | H-bonded imidazole |

$$(5.84)$$

Reduction of compound I by one electron produces compound II, which has the characteristics of a normal ferryl-porphyrin complex, analogous to **2**, i.e., $(L)Fe^{IV}(P)(O)$. Reaction of compound II with hydrogen peroxide produces compound III, which can also be prepared by reaction of the ferrous enzyme with dioxygen. It is an oxy form, analogous to oxymyoglobin, and does not appear to have a physiological function. The reactions producing these three forms and their proposed formulations are summarized in Reactions (5.85) to (5.88).

$$Fe^{III}(P)^+ + H_2O_2 \longrightarrow Fe^{IV}(P^{\cdot-})(O)^+ + H_2O \qquad (5.85)$$

ferric form   Compound I

$$Fe^{IV}(P^{\cdot-})(O)^+ + e^- \longrightarrow Fe^{IV}(P)(O) \qquad (5.86)$$

Compound I   Compound II

$$Fe^{IV}(P)(O) + H_2O_2 \longrightarrow Fe(P)O_2 + H_2O \qquad (5.87)$$

Compound II   Compound III

$$Fe^{II}(P) + O_2 \longrightarrow Fe(P)O_2 \qquad\qquad (5.88)$$

ferrous form   Compound III

B. Mechanism

The accepted mechanisms for catalase and peroxidase are described in Reactions (5.89) to (5.94).

$$Fe^{III}(P)^+ + H_2O_2 \longrightarrow Fe^{III}(P)(H_2O_2)^+ \longrightarrow Fe^{IV}(P^{\cdot-})(O)^+ + H_2O \qquad (5.89)$$

Compound I

catalase:

$$Fe^{IV}(P^{\cdot-})(O)^+ + H_2O_2 \longrightarrow Fe^{III}(P)^+ + H_2O + O_2 \qquad (5.90)$$

Compound I

peroxidase:

$$Fe^{IV}(P^{\cdot-})(O)^+ + AH_2 \longrightarrow Fe^{IV}(P)(O) + HA^{\cdot} + H^+ \qquad (5.91)$$

Compound I   Compound II

$$Fe^{IV}(P)(O) + \quad AH_2 \longrightarrow Fe^{III}(P)^+ + HA^{\cdot} + OH^- \qquad (5.92)$$

Compound II

$$2HA^{\cdot} \longrightarrow A + AH_2 \qquad\qquad (5.93)$$

or
$$2HA^{\cdot} \longrightarrow HA - AH \qquad\qquad (5.94)$$

In the catalase reaction, it has been established by use of $H_2^{18}O_2$ that the dioxygen formed is derived from hydrogen peroxide, i.e., that O—O bond cleavage does not occur in Reaction (5.90), which is therefore a two-electron reduction of compound I by hydrogen peroxide, with the oxo ligand of the former being released as water. For the peroxidase reaction under physiological conditions, it is believed that the oxidation proceeds in one-electron steps (Reactions 5.91 and 5.92), with the final formation of product occurring by disproportionation (Reaction 5.93) or coupling (Reaction 5.94) of the one-electron oxidized intermediate.[94,95]

C. Comparisons of Catalase, Peroxidase, and Cytochrome P-450

The proposal that these three enzymes all go through a similar high-valent oxo intermediate, i.e., **3** or compound I, raises two interesting questions. The first of these is why the same high-valent metal-oxo intermediate gives two very different types of reactions, i.e., oxygen-atom transfer with cytochrome P-450 and electron transfer with catalase and peroxidase. The answer is that, although the high-valent metal-oxo heme cores of these intermediates are in fact very similar, the substrate-binding cavities seem to differ substantially in how much access the substrate has to the iron center. With cytochrome P-450, the substrate is jammed right up against the location where the oxo ligand must reside in the high-valent oxo intermediate. But the same location in the peroxidase enzymes is blocked by the protein structure so that substrates can interact only with the heme edge. Thus oxidation of the substrate by electron transfer is possible for catalase and peroxidase, but the substrate is too far away from the oxo ligand for oxygen-atom transfer.[99,124]

The second question is about how the the high-valent oxo intermediate forms in both enzymes. For catalase and peroxidase, the evidence indicates that hydrogen peroxide binds to the ferric center and then undergoes heterolysis at the O—O bond. Heterolytic cleavage requires a significant separation of positive and negative charge in the transition state. In catalase and peroxidase, analysis of the crystal structure indicates strongly that amino-acid side chains are situated to aid in the cleavage by stabilizing a charge-separated transition state (Figure 5.14). In cytochrome P-450, as mentioned in Section V.C.1, no such groups

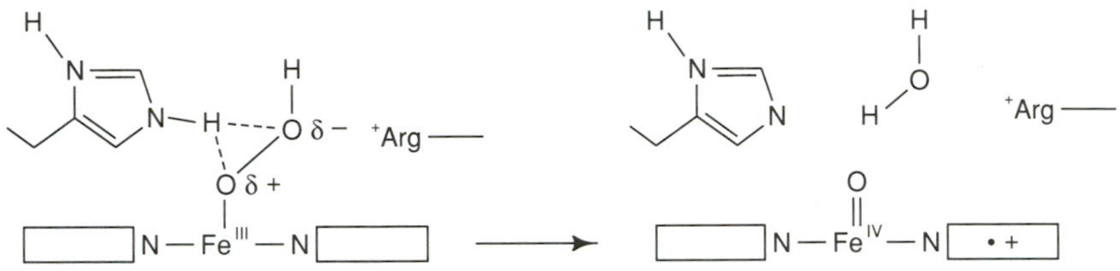

Figure 5.14
Schematic representation of the mechanism for heterolytic O—O bond cleavage to form Compound I in peroxidase, as proposed in Reference 96. The histidyl imidazole aids in transfer of a proton from the oxygen atom of hydrogen peroxide that is bound to the iron to the departing oxygen atom, while the positive charge of the arginyl side chain stabilizes the developing negative charge on the departing oxygen, thus facilitating heterolytic O—O bond cleavage.

are found in the hydrophobic substrate-binding cavity. It is possible that the cysteinyl axial ligand in cytochrome P-450 plays an important role in O—O bond cleavage, and that the interactions found in catalase and peroxidase that appear to facilitate such cleavage are therefore not necessary.

VII. COPPER-ZINC SUPEROXIDE DISMUTASE

A. Background

Two families of metalloproteins are excellent catalysts for the disproportionation of superoxide (Reaction 5.95).

$$2O_2^- + 2H^+ \xrightarrow{\text{SOD}} O_2 + H_2O_2 \qquad (5.95)$$

These are (1) the copper-zinc superoxide dismutases, CuZnSOD,[100-102] found in almost all eukaryotic cells and a very few prokaryotes, and (2) the manganese and iron superoxide dismutases, MnSOD and FeSOD, the former found in the mitochondria of eukaryotic cells, and both found in many prokaryotes.[103] Recent studies of bacterial[104] and yeast[105] mutants that were engineered to contain no superoxide dismutases demonstrated that the cells were unusually sensitive to dioxygen and that the sensitivity to dioxygen was relieved when an SOD gene was reintroduced into the cells. These results indicate that the superoxide dismutase enzymes play a critical role in dioxygen metabolism, but they do not define the chemical agent responsible for dioxygen toxicity (see Section III).

B. Enzymatic Activity

Several transition-metal complexes have been observed to catalyze superoxide disproportionation; in fact, aqueous copper ion, Cu^{2+}, is an excellent SOD catalyst, comparable in activity to CuZnSOD itself![37] Free aqueous Cu^{2+} would not itself be suitable for use as an SOD *in vivo*, however, because it is too toxic (see Section III) and because it binds too strongly to a large variety of cellular components and thus would not be present as the free ion. (Most forms of complexed cupric ion show much less superoxide dismutase activity than the free ion.) Aside from aqueous copper ion, few other complexes are as effective as the SOD enzymes.

Two mechanisms (Reactions 5.96 to 5.99) have been proposed for catalysis of superoxide disproportionation by metal complexes and metalloenzymes.[37]

Mechanism I:

$$M^{n+} + O_2^- \longrightarrow M^{(n-1)+} + O_2 \tag{5.96}$$

$$M^{(n-1)+} + O_2^- \longrightarrow M^{n+}(O_2^{2-}) \xrightarrow{2H^+} M^{n+} + H_2O_2 \tag{5.97}$$

Mechanism II:

$$M^{n+} + O_2^- \longrightarrow M^{n+}(O_2^-) \tag{5.98}$$

$$M^{n+}(O_2^-) + O_2^- \longrightarrow M^{n+}(O_2^{2-}) \xrightarrow{2H^+} M^n + H_2O_2 \\ + O_2 \tag{5.99}$$

In Mechanism I, which is favored for the SOD enzymes and most redox-active metal complexes with SOD activity, superoxide reduces the metal ion in the first step, and then the reduced metal ion is reoxidized by another superoxide, presumably via a metal-peroxo complex intermediate. In Mechanism II, which is proposed for nonredox metal complexes but may be operating in other situations as well, the metal ion is never reduced, but instead forms a superoxo complex, which is reduced to a peroxo complex by a second superoxide ion. In both mechanisms, the peroxo ligands are protonated and dissociate to give hydrogen peroxide.

Analogues for each of the separate steps of Reactions (5.96) to (5.99) have been observed in reactions of superoxide with transition-metal complexes, thereby establishing the feasibility of both mechanisms. For example, superoxide was shown to reduce $Cu^{II}(phen)_2^{2+}$ to give $Cu^I(phen)_2^+$ (phen = 1,10-phenanthro-line),[106] a reaction analogous to Reaction (5.96). On the other hand, superoxide reacts with $Cu^{II}(tet\ b)^{2+}$ to form a superoxo complex[107] (a reaction analogous to Reaction 5.98), presumably because $Cu^{II}(tet\ b)^{2+}$ is not easily reduced to the cuprous state, because the ligand cannot adjust to the tetrahedral geometry that Cu^I prefers.[53]

(5.100)

tet b

Reaction of superoxide with a reduced metal-ion complex to give oxidation of the complex and release of hydrogen peroxide (analogous to Reaction 5.97) has been observed in the reaction of $Fe^{II}EDTA$ with superoxide.[108] Reduction of a Co^{III} superoxo complex by free superoxide to give a peroxo complex (analogous to Reaction 5.99) has also been observed.[109]

If a metal complex can be reduced by superoxide and if its reduced form can be oxidized by superoxide, both at rates competitive with superoxide disproportionation, the complex can probably act as an SOD by Mechanism I. Mechanism II has been proposed to account for the apparent catalysis of superoxide disproportionation by Lewis acidic nonredox-active metal ions under certain conditions.[37] However, this mechanism should probably be considered possible for redox metal ions and the SOD enzymes as well. It is difficult to distinguish the two mechanisms for redox-active metal ions and the SOD enzymes unless the reduced form of the catalyst is observed directly as an intermediate in the reaction. So far it has not been possible to observe this intermediate in the SOD enzymes or the metal complexes.

C. Structure

The x-ray crystal structure of the oxidized form of CuZnSOD from bovine erythrocytes shows a protein consisting of two identical subunits held together almost entirely by hydrophobic interactions.[100–102] Each subunit consists of a flattened cylindrical barrel of β-pleated sheet from which three external loops of irregular structure extend (Figure 5.15). The metal-binding region of the protein binds Cu^{II} and Zn^{II} in close proximity to each other, bridged by the imidazolate ring of a histidyl side chain. Figure 5.16 represents the metal-binding region. The Cu^{II} ion is coordinated to four histidyl imidazoles and a water in a highly distorted square-pyramidal geometry with water at the apical position. The Zn^{II} ion is coordinated to three histidyl imidazoles (including the one shared with

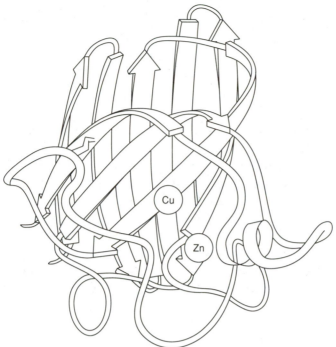

Figure 5.15
Schematic drawing of the polypeptide backbone of one of the two subunits of bovine CuZnSOD. The strands of the β structure are shown as arrows. The active-site channel provides access to the copper site from the direction of the viewer. (From J. A. Tainer *et al.*, *J. Mol. Biol.* **160** (1982), 181–217.)

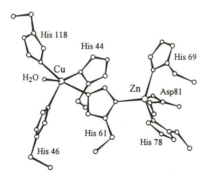

Figure 5.16
Representation of the metal-binding region of bovine CuZnSOD.

copper) and an aspartyl carboxylate group, forming a distorted tetrahedral geometry around the metal ion.

One of the most unusual aspects of the structure of this enzyme is the occurrence of the bridging imidazolate ligand, which holds the copper and zinc ions 6 Å apart. Such a configuration is not unusual for imidazole complexes of metal ions, which sometimes form long polymeric imidazolate-bridged structures.

$$M-N\diagdown N-M-N\diagdown N-M-N\diagdown N-M \qquad (5.101)$$

However, no other imidazolate-bridged bi- or polymetallic metalloprotein has yet been identified.

The role of the zinc ion in CuZnSOD appears to be primarily structural. There is no evidence that water, anions, or other potential ligands can bind to the zinc, so it is highly unlikely that superoxide could interact with that site. Moreover, removal of zinc under conditions where the copper ion remains bound to the copper site does not significantly diminish the SOD activity of the enzyme.[110] However, such removal does result in a diminished thermal stability, i.e., the zinc-depleted protein denatures at a lower temperature than the native protein, supporting the hypothesis that the role of the zinc is primarily structural in nature.[111]

The copper site is clearly the site of primary interaction of superoxide with the protein. The x-ray structure shows that the copper ion lies at the bottom of a narrow channel that is large enough to admit only water, small anions, and similarly small ligands (Figure 5.17). In the lining of the channel is the positively charged side chain of an arginine residue, 5 Å away from the copper ion

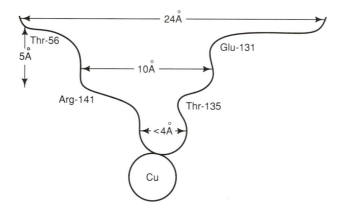

Figure 5.17
Schematic diagram of a cross section of the active-site channel in CuZnSOD. The diameter of the channel narrows as the Cu^{II} center is approached, and only small ligands can actually reach that site. In addition to the positively charged Cu^{II} ion at the bottom of the channel, the positively charged side chain of Arg-141 is part of the walls of the channel. Two positively charged lysine side chains, not shown in this diagram, are close to the mouth of the channel. (From E. D. Getzoff, R. A. Hallewell, and J. A. Tainer, in M. Inouye, ed., *Protein Engineering: Applications in Science, Industry, and Medicine*, Academic Press, 1986, pp. 41–69.)

and situated in such a position that it could interact with superoxide and other anions when they bind to copper. Near the mouth of the channel, at the surface of the protein, are two positively charged lysine residues, which are believed to play a role in attracting anions and guiding them into the channel.[112] Chemical modification of these lysine or arginine residues substantially diminishes the SOD activity, supporting their role in the mechanism of reaction with superoxide.[100–102]

The x-ray structural results described above apply only to the oxidized form of the protein, i.e., the form containing Cu^{II}. The reduced form of the enzyme containing Cu^{I} is also stable and fully active as an SOD. If, as is likely, the mechanism of CuZnSOD-catalyzed superoxide disproportionation is Mechanism I (Reactions 5.96–5.97), the structure of the reduced form is of critical importance in understanding the enzymatic mechanism. Unfortunately, that structure is not yet available.

D. Enzymatic Activity and Mechanism

The mechanism of superoxide disproportionation catalyzed by CuZnSOD is generally believed to go by Mechanism I (Reactions 5.96–5.97), i.e., reduction of Cu^{II} to Cu^{I} by superoxide with the release of dioxygen, followed by reoxidation of Cu^{I} to Cu^{II} by a second superoxide with the release of HO_2^- or H_2O_2. The protonation of peroxide dianion, O_2^{2-}, prior to its release from the enzyme is required, because peroxide dianion is highly basic and thus too unstable to be released in its unprotonated form. The source of the proton that protonates peroxide in the enzymatic mechanism is the subject of some interest.

Reduction of the oxidized protein has been shown to be accompanied by the uptake of one proton per subunit. That proton is believed to protonate the bridging imidazolate in association with the breaking of the bridge upon reduction of the copper. Derivatives with Co^{II} substituted for Zn^{II} at the native zinc site have been used to follow the process of reduction of the oxidized Cu^{II} form to the reduced Cu^{I} form. The Co^{II} in the zinc site does not change oxidation state, but acts instead as a spectroscopic probe of changes occurring at the native zinc-binding site. Upon reduction (Reaction 5.102), the visible absorption band due to Co^{II} shifts in a manner consistent with a change occurring in the ligand environment of Co^{II}. The resulting spectrum of the derivative containing Cu^{I} in the copper site and Co^{II} in the zinc site is very similar to the spectrum of the derivative in which the copper site is empty and the zinc site contains Co^{II}. This result suggests strongly that the imidazolate bridge is cleaved and protonated and that the resulting imidazole ligand is retained in the coordination sphere of Co^{II} (Reaction 5.102).[101]

$$\text{(5.102)}$$

The same proton is thus an attractive possibility for protonation of peroxide as it is formed in the enzymatic mechanism (Reactions 5.103 and 5.104).

$$\text{Cu(II)} - N \overset{\frown}{} N - \text{Zn(II)} + O_2^- \xrightarrow{H^+} \text{Cu(I)} \ H - N \overset{\frown}{} N - \text{Zn(II)} + O_2 \qquad (5.103)$$

$$\text{Cu(I)} \ H - N \overset{\frown}{} N - \text{Zn(II)} + O_2^- \longrightarrow \text{Cu(II)} - N \overset{\frown}{} N - \text{Zn(II)} + HO_2^- \qquad (5.104)$$

Attractive as this picture appears, there are several uncertainties about it. For example, the turnover of the enzyme may be too fast for protonation and deprotonation of the bridging histidine to occur.[113] Moreover, the mechanism proposed would require the presence of a metal ion at the zinc site to hold the imidazole in place and to regulate the pK_a of the proton being transferred. The observation that removal of zinc gives a derivative with almost full SOD activity is thus surprising and may also cast some doubt on this mechanism. Other criticisms of this mechanism have been recently summarized.[102]

Studies of CuZnSOD derivatives prepared by site-directed mutagenesis are also providing interesting results concerning the SOD mechanism. For example, it has been shown that mutagenized derivatives of human CuZnSOD with major differences in copper-site geometry relative to the wild-type enzyme may nonetheless remain fully active.[114] Studies of these and similar derivatives should provide considerable insight into the mechanism of reaction of CuZnSOD with superoxide.

E. Anions as Inhibitors

Studies of the interaction of CuZnSOD and its metal-substituted derivatives with anions have been useful in predicting the behavior of the protein in its reactions with its substrate, the superoxide anion, O_2^-.[101,102] Cyanide, azide, cyanate, and thiocyanate bind to the copper ion, causing dissociation of a histidyl ligand and the water ligand from the copper.[115] Phosphate also binds to the enzyme at a position close to the Cu^{II} center, but it apparently does not bind directly to it as a ligand. Chemical modification of Arg-141 with phenylglyoxal blocks the interaction of phosphate with the enzyme, suggesting that this positively charged residue is the site of interaction with phosphate.[116]

Electrostatic calculations of the charges on the CuZnSOD protein suggest that superoxide and other anions entering into the vicinity of the protein will be drawn toward and into the channel leading down to the copper site by the distribution of positive charges on the surface of the protein, the positively charged lysines at the mouth of the active-site cavity, and the positively charged arginine and copper ion within the active-site region.[112] Some of the anions studied, e.g.,

CN^-, F^-, N_3^-, and phosphate, have been shown to inhibit the SOD activity of the enzyme. The source of the inhibition is generally assumed to be competition with superoxide for binding to the copper, but it may sometimes result from a shift in the redox potential of copper, which is known to occur sometimes when an anion binds to copper.[100,101]

F. Metal-Ion Substitutions

1. SOD activity

In the example described above, studies of a metal-substituted derivative helped in the evaluation of mechanistic possibilities for the enzymatic reaction. In addition, studies of such derivatives have provided useful information about the environment of the metal-ion binding sites. For example, metal-ion-substituted derivatives of CuZnSOD have been prepared with Cu^{II}, Cu^I, Zn^{II}, Ag^I, Ni^{II}, or Co^{II} bound to the native copper site, and with Zn^{II}, Cu^{II}, Cu^I, Co^{II}, Hg^{II}, Cd^{II}, Ni^{II}, or Ag^I bound to the native zinc site.[100,101,117] The SOD activities of these derivatives are interesting; only those derivatives with copper in the copper site have a high degree of SOD activity, whereas the nature of the metal ion in the zinc site or even its absence has little or no effect.[100,101]

2. Spectroscopy

Derivatives of CuZnSOD are known with Cu^{II} ion bound either to the native copper site or to the native zinc site. The electronic absorption spectra of these derivatives indicate that the ligand environments of the two sites are very different. Copper(II) is a d^9 transition-metal ion, and its d-d transitions are usually found in the visible and near-IR regions of the spectrum.[53] Copper(II) complexes with coordinated nitrogen ligands are generally found to have an absorption band between 500 and 700 nm, with an extinction coefficient below $100\ M^{-1}cm^{-1}$. Bands in the absorption spectra of complexes with geometries that are distorted away from square planar tend to be red-shifted because of a smaller d-d splitting, and to have higher extinction coefficients because of the loss of centrosymmetry. Thus the optical spectrum of CuZnSOD with an absorption band with a maximum at 680 nm (14,700 cm^{-1}; see Figure 5.18A) and an extinction coefficient of $155\ M^{-1}cm^{-1}$ per Cu is consistent with the crystal structural results that indicate that copper(II) is bound to four imidazole nitrogens and a water molecule in a distorted square-pyramidal geometry. Metal-substituted derivatives with Cu^{II} at the native copper site but with Co^{II}, Cd^{II}, Hg^{II}, or Ni^{II} substituted for Zn^{II} at the native zinc site all have a band at 680 nm, suggesting that the substitution of another metal ion for zinc perturbs the copper site very little, despite the proximity of the two metal sites. The absorption spectra of native CuZnSOD and these CuMSOD derivatives also have a shoulder at 417 nm (24,000 cm^{-1}; see Figure 5.18A), which is at lower energy than normal imidazole-to-Cu^{II} charge-transfer transitions, and has been assigned to

306

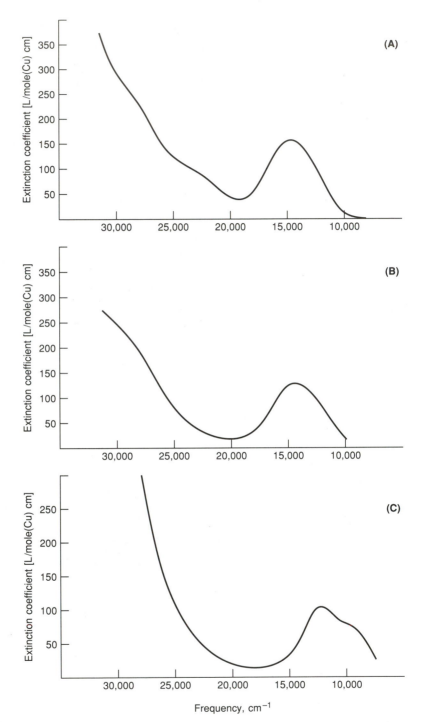

Figure 5.18
Vis-UV spectra of (A) CuZnSOD, (B) copper-only SOD (zinc site empty), and (C) Ag^{I}CuSOD.
In all three spectra, the low-energy band is the Cu^{II} d-d transition. Note that the spectrum of
CuZnSOD contains a shoulder at 24,000 cm^{-1} (417 nm) that is assigned to the imidazolate-to-
copper(II) charge-transfer transition. This shoulder is not present in the copper-only derivative.
For Ag^{I}CuSOD, the d-d transition is red-shifted because of the change in the ligand geometry
when Cu^{II} is moved from the copper site to the zinc site (see text). (From M. W. Pantoliano, L.
A. Nafie, and J. S. Valentine, *J. Am. Chem. Soc.* **104** (1982), 6310–6317.)

an imidazolate-to-Cu^{II} charge transfer, indicating that the imidazolate bridge between Cu^{II} and the metal ion in the native zinc site is present, as observed in the crystal structure of CuZnSOD. Derivatives with the zinc site empty, which therefore cannot have an imidazolate bridge, are lacking this 417 nm shoulder.

Small but significant changes in the absorption spectrum are seen when the metal ion is removed from the zinc site, e.g., in copper-only SOD (Figure 5.18B). The visible absorption band shifts to 700 nm (14,300 cm^{-1}), presumably due to a change in ligand field strength upon protonation of the bridging imidazolate. In addition, the shoulder at 417 nm has disappeared, again due to the absence of the imidazolate ligand.

The spectroscopic properties due to copper in the native zinc site are best observed in the derivative Ag^{I}CuSOD, which has Ag^{I} in the copper site and Cu^{II} in the zinc site (see Figure 5.18C), since the d^{10} Ag^{I} ion is spectroscopically silent. In this derivative, the d-d transition is markedly red-shifted from the visible region of the spectrum into the near-IR, indicating that the ligand environment of Cu^{II} in that site is either tetrahedral or five coordinate. The EPR properties of Cu^{II} in this derivative are particularly interesting (as discussed below).

The derivative with Cu^{II} bound at both sites, CuCuSOD, has a visible-near IR spectrum that is nearly a superposition of the spectra of CuZnSOD and Ag^{I}CuSOD (see Figure 5.19), indicating that the geometry of Cu^{II} in each of these sites is little affected by the nature of the metal ion in the other site.

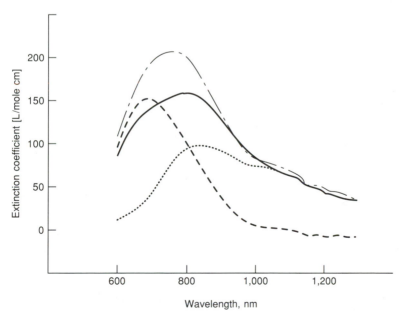

Figure 5.19

Comparison of the visible absorption spectrum of CuCuSOD, – – – – –, with that of Cu-ZnSOD, –——–, and of Ag^{I}CuSOD, ·········. A digital addition of the spectra of Cu-ZnSOD and of Ag^{I}CuSOD generated the other spectrum, –·–·–·–. Note that the spectrum of CuCuSOD, which has Cu^{II} ions in both the copper and the zinc sites, closely resembles a superposition of the spectra of CuZnSOD, which has Cu^{II} in the copper site, and Ag^{I}CuSOD, which has Cu^{II} in the zinc site. (From M. W. Pantoliano, L. A. Nafie, and J. S. Valentine, *J. Am. Chem. Soc.* **104** (1982), 6310–6317.)

EPR spectroscopy has also proven to be particularly valuable in characterizing the metal environments in CuZnSOD and derivatives. The EPR spectrum of native CuZnSOD is shown in Figure 5.20A. The $g_\parallel$ resonance is split by the hyperfine coupling between the unpaired electron on Cu^{II} and the $I = \frac{3}{2}$ nuclear spin of copper. The $A_\parallel$ value, 130 G, is intermediate between the larger $A_\parallel$

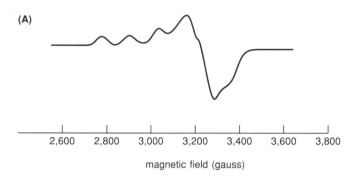

(A)

2,600   2,800   3,000   3,200   3,400   3,600   3,800

magnetic field (gauss)

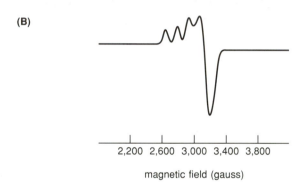

(B)

2,200 2,600 3,000 3,400 3,800

magnetic field (gauss)

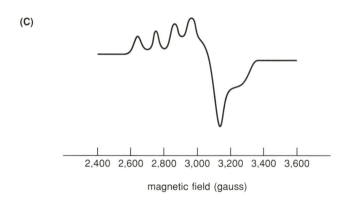

(C)

2,400 2,600 2,800 3,000 3,200 3,400 3,600

magnetic field (gauss)

Figure 5.20
Frozen solution EPR spectra of (A) CuZnSOD, (B) copper-only SOD (zinc site empty), and (C) Ag^{I}CuSOD. See text for discussion. (Adapted from References 100 and 101.)

typical of square-planar Cu^{II} complexes with four nitrogen donor ligands and the lower $A_\parallel$ observed in blue copper proteins (see Chapter 6). The large linewidth seen in the $g\perp$ region indicates that the copper ion is in a rhombic (i.e., distorted) environment. Thus, the EPR spectrum is entirely consistent with the distorted square-pyramidal geometry observed in the x-ray structure.

Removal of zinc from the native protein to give copper-only SOD results in a perturbed EPR spectrum, with a narrower $g_\perp$ resonance and a larger $A_\parallel$ value (142 G) more nearly typical of Cu^{II} in an axial N_4 environment (Figure 5.20B). Apparently the removal of zinc relaxes some constraints imposed on the geometry of the active-site ligands, allowing the copper to adopt to a geometry closer to its preferred tetragonal arrangement.

The EPR spectrum due to Cu^{II} in the native Zn^{II} site in the Ag^ICuSOD derivative indicates that Cu^{II} is in a very different environment than when it is in the native copper site (Figure 5.20C). The spectrum is strongly rhombic, with a low value of $A_\parallel$ (97 G), supporting the conclusion based on the visible spectrum that copper is bound in a tetrahedral or five-coordinate environment. This type of site is unusual either for copper coordination complexes or for copper proteins in general, but does resemble the Cu^{II} EPR signal seen when either laccase or cytochrome c oxidase is partially reduced (see Figure 5.21). Partial

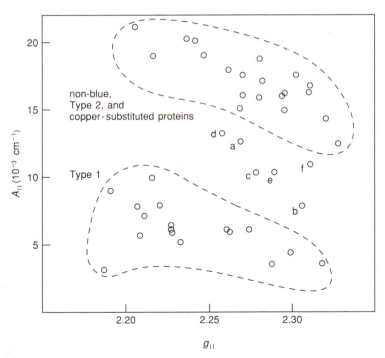

Figure 5.21
Relationship between $g_\parallel$ and $A_\parallel$ for naturally occurring copper proteins and copper-substituted metalloproteins. Points labeled a–f are for laccases (a and b), cytochrome c oxidase (c), Cu-ZnSOD (d), Ag^ICuSOD (e), and $Cu^ICu^{II}SOD$ (f). See text for discussion. (From M. W. Pantoliano, L. A. Nafie, and J. S. Valentine, *J. Am. Chem. Soc.* **104** (1982), 6310–6317.)

reduction disrupts the magnetic coupling between these Cu^{II} centers that makes them EPR-silent in the fully oxidized protein.

The EPR spectrum of CuCuSOD is very different from that of any of the other copper-containing derivatives (Figure 5.22) because the unpaired spins on

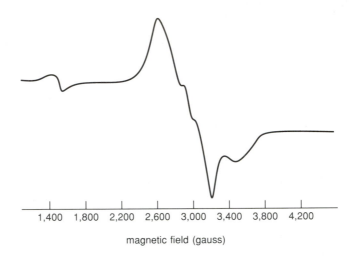

magnetic field (gauss)

Figure 5.22
Frozen-solution EPR spectrum of CuCuSOD. Note the very different appearance of this spectrum from those shown in Figure 5.20. These differences are due to the fact that the two Cu^{II} centers are magnetically coupled across the imidazolate bridge.[100]

the two copper centers interact and magnetically couple across the imidazolate bridge, resulting in a triplet EPR spectrum. This spectrum is virtually identical with that of model imidazolate-bridged binuclear copper complexes.[101]

Electronic absorption and EPR studies of derivatives of CuZnSOD containing Cu^{II} have provided useful information concerning the nature of the metal binding sites of those derivatives. ^{1}H NMR spectra of those derivatives are generally not useful, however, because the relatively slowly relaxing paramagnetic Cu^{II} center causes the nearby proton resonances to be extremely broad. This difficulty has been overcome in two derivatives, CuCoSOD and CuNiSOD, in which the fast-relaxing paramagnetic Co^{II} and Ni^{II} centers at the zinc site interact across the imidazolate bridge and increase the relaxation rate of the Cu^{II} center, such that well-resolved paramagnetically shifted ^{1}H NMR spectra of the region of the proteins near the two paramagnetic metal centers in the protein can be obtained and the resonances assigned.[118,119]

The use of ^{1}H NMR to study CuCoSOD derivatives of CuZnSOD in combination with electronic absorption and EPR spectroscopies has enabled investigators to compare active-site structures of a variety of wild-type and mutant CuZnSOD proteins in order to find out if large changes in active-site structure have resulted from replacement of nearby amino-acid residues.[120]

VIII. REFERENCES

1. The references in this chapter cite recent review articles or books when available; these are indicated by (R) or (B), respectively, in the citation, and the titles of review articles are given. Students should consult these sources if they want more detailed information about a particular topic or references to the original literature.

2. B. G. Malmström, "Enzymology of Oxygen" (R), *Annu. Rev. Biochem.* **51** (1982), 21–59.

3. L. L. Ingraham and D. L. Meyer, *Biochemistry of Dioxygen* (B), Plenum, 1985.

4. O. Hayaishi, ed., *Molecular Mechanisms of Oxygen Activation* (B), Academic Press, 1974, pp. 405–451.

5. P. George, "The Fitness of Oxygen" (R), in T. E. King, H. S. Mason, and M. Morrison, eds., *Oxidases and Related Redox Systems*, Wiley, **1** (1965), 3–36.

6. B. Halliwell and J. M. C. Gutteridge, *Free Radicals in Biology and Medicine* (B), Clarendon Press, 1989.

7. R. A. Sheldon and J. K. Kochi, *Metal-Catalyzed Oxidations of Organic Compounds* (B), Academic Press, 1981.

8. D. T. Sawyer, "The Chemistry and Activation of Dioxygen Species (O_2, $O_2^{\cdot -}$, and HOOH) in Biology" (R), in A. E. Martell and D. T. Sawyer, eds., *Oxygen Complexes and Oxygen Activation by Transition Metals*, Plenum, 1988, pp. 131–148.

9. P. M. Wood, *Trends in Biochem. Sci.* **12** (1987), 250–251.

10. R. H. Holm, "Metal-Centered Oxygen Atom Transfer Reactions" (R), *Chem. Rev.* **87** (1987), 1401–1449.

11. Calculated from data in D. D. Wagman *et al.*, *Selected Values of Chemical Thermodynamic Properties*, Institute for Basic Standards, NBS, 1968.

12. Calculated from data in M. Bertholon *et al.*, *Bull. Soc. Chim. Fr.* **9** (1971), 3180–3187.

13. Calculated from data in A. Finch, P. J. Gardner, and D. Wu, *Thermochim. Acta* **66** (1983), 333–342.

14. L. Shaofeng and G. Pilcher, *J. Chem. Thermodynamics* **20** (1988), 463–465.

15. G. A. Hamilton, "Chemical Models and Mechanisms for Oxygenases" (R), in Reference 4, pp. 405–451.

16. H. Taube, "Mechanisms of Oxidation with Oxygen" (R), *J. Gen. Physiol.* **49**, part 2 (1965), 29–52.

17. J. G. Calvert and J. N. Pitts, *Photochemistry*, Wiley, 1966.

18. H. H. Wasserman and R. W. Murray, eds., *Singlet Oxygen* (B), Academic Press, 1979.

19. A. A. Frimer, ed., *Singlet O_2* (B), CRC Press, 1985.

20. T. C. Bruice, "Chemical Studies and the Mechanism of Flavin Mixed Function Oxidase Enzymes (R), in J. F. Liebman and A. Greenberg, eds., *Mechanistic Principles of Enzyme Activity*, VCH Publishers, 1988, pp. 315–352.

21. Reference 7, p. 18.

22. Reference 7, p. 316.

23. Reference 3, p. 16.

24. N. A. Porter *et al.*, *J. Am. Chem. Soc.* **103** (1981), 6447–6455.

25. W. Day, *Genesis on Planet Earth: The Search for Life's Beginning*, Yale University Press, 2d ed., 1984.

26. D. L. Gilbert, ed., *Oxygen and Living Processes: An Interdisciplinary Approach* (B), Springer-Verlag, 1981.

27. B. Quebedeaux *et al.*, *Plant Physiol.* **56** (1975), 761–764.

28. O. R. Brown and F. Yein, *Biochem. Biophys. Res. Commun.* **85** (1978), 1219–1224.

29. L. J. Machlin and A. Bendich, *FASEB J.* **1** (1987), 441–445.

30. D. S. Tarbell, in N. Kharasch, ed., *Organic Sulfur Compounds*, Pergamon, **1** (1961), 97–102.

31. C. Walling, *Acc. Chem. Res.* **8** (1975), 125–132.

32. D. C. Harris and P. Aisen, in T. M. Loehr, ed., *Iron Carriers and Iron Proteins*, VCH Publishers, 1989, pp. 239–371.

33. G. Storz, L. A. Tartaglia, and B. N. Ames, *Science* **248** (1990), 189–194, and references therein.

34. T. Keng and L. Guarente, *Proc. Natl. Acad. Sci. USA* **84** (1987), 9113–9117, and references therein.

35. R. Gerschman *et al.*, *Science* **119** (1954), 623–626.

36. J. M. McCord and I. Fridovich, *J. Biol. Chem.* **244** (1969), 6049–6055.

37. D. T. Sawyer and J. S. Valentine, "How Super is Superoxide?" (R), *Acc. Chem. Res.* **14** (1981), 393–400.

38. L. W. Oberley, ed., *Superoxide Dismutase*, CRC Press, 2 vols., 1982, vol. 3, 1985.

39. J. A. Fee, "Is Superoxide Important in Oxygen Poisoning?" (R), *Trends Biochem. Sci.* **7** (1982), 84–86, and references therein.

40. P. Korbashi *et al., J. Biol. Chem.* **264** (1989), 8479–8482.

41. E. R. Stadtman, "Metal Ion-Catalyzed Oxidation of Proteins: Biochemical Mechanism and Biological Consequences" (R), *Free Radicals in Biology & Medicine* **9** (1990), 315–325.

42. D. O. Natvig *et al., J. Biol. Chem.* **262** (1987), 14697–14701.

43. C. Bowler *et al., J. Bacteriol.* **172** (1990), 1539–1546.

44. M. Wikström and G. T. Babcock, *Nature* **348** (1990), 16–17.

45. G. Palmer, "Cytochrome Oxidase: a Perspective" (R), *Pure Appl. Chem.* **59** (1987), 749–758.

46. T. Vanngard, ed., *Biophysical Chemistry of Dioxygen Reactions in Respiration and Photosynthesis, Chemica Scripta* **28A** (B), Cambridge University Press, 1988.

47. W. R. Scheidt and C. A. Reed, "Spin-State/Stereochemical Relationships in Iron Porphyrins: Implications for the Hemoproteins" (R), *Chem. Rev.* **81** (1981), 543–555.

48. F. S. Mathews, M. Levine, and P. Argos, *J. Mol. Biol.* **64** (1972), 449–464.

49. W. E. Blumberg and J. Peisach, *Adv. Chem. Ser.* **100** (1971), 271–291.

50. J. Peisach, W. E. Blumberg, and A. Adler, *Ann. N.Y. Acad. Sci.* **206** (1973), 310–327.

51. J. P. Collman, T. R. Halbert, and K. S. Suslick, "O$_2$ Binding to Heme Proteins and Their Synthetic Analogues" (R), in T. G. Spiro, ed., *Metal-Ion Activation of Dioxygen*, Wiley, 1980, pp. 1–72.

52. R. Quinn, M. Nappa, and J. S. Valentine, *J. Am. Chem. Soc.* **104** (1982), 2588–2595.

53. F. A. Cotton and G. Wilkinson, *Advanced Inorganic Chemistry*, Wiley, 5th ed., 1988, pp. 755–775.

54. T. H. Stevens *et al., J. Biol. Chem.* **257** (1982), 12106–12113.

55. W. B. Mims *et al., J. Biol. Chem.* **255** (1980), 6843–6846.

56. T. H. Stevens *et al., Proc. Natl. Acad. Sci USA* **76** (1979), 3320–3324.

57. T. A. Kent *et al., J. Biol. Chem.* **258** (1983), 8543–8546.

58. J. Cline *et al., J. Biol. Chem.* **258** (1983), 5124–5128.

59. T. H. Stevens and S. I. Chan, *J. Biol. Chem.* **256** (1981), 1069–1071.

60. D. H. Chin, G. N. La Mar, and A. Balch, *J. Am. Chem. Soc.* **102** (1980), 4344–4350.

61. A. L. Balch *et al., J. Am. Chem. Soc.* **106** (1984), 7779–7785.

62. K. D. Karlin and Y. Gultneh, "Binding and Activation of Molecular Oxygen by Copper Complexes" (R), *Prog. Inorg. Chem.* **35** (1987), 219–327.

63. R. R. Jacobson *et al., J. Am. Chem. Soc.* **110** (1988), 3690–3692.

64. N. Kitajima, K. Fujisawa, and Y. Moro-oka, *J. Am. Chem. Soc.* **111** (1989), 8975–8976.

65. S. Han, Y.-C. Chin, and D. L. Rousseau, *Nature* **348** (1990), 89–90.

66. L. Que, Jr., "The Catechol Dioxygenases" (R), in Reference 32, pp. 467–524.

67. L. Que, Jr., "Spectroscopic Studies of the Catechol Dioxygenases" (R), *J. Chem. Ed.* **62** (1985), 938–943.

68. J. W. Whittaker *et al., J. Biol. Chem.* **259** (1984), 4466–4475.

69. Y. Tomimatsu, S. Kint, and J. R. Scherer, *Biochemistry* **15** (1976), 4918–4924.

70. D. H. Ohlendorf, J. D. Lipscomb, and P. C. Weber, *Nature* **336** (1988), 403–405.

71. D. D. Cox and L. Que, Jr., *J. Am. Chem. Soc.* **110** (1988), 8085–8092.

72. Y. Sawaki and C. S. Foote, *J. Am. Chem. Soc.* **105** (1983), 5035–5040.

73. P. R. Ortiz de Montellano, ed., *Cytochrome P-450: Structure, Mechanism, and Biochemistry* (B), Plenum, 1986.

74. K. Lerch, "Copper Monooxygenases: Tyrosinase and Dopamine β-Monooxygenase" (R), *Metal Ions Biol. Syst.* **13** (1981), 143–186.

75. J. Green and H. Dalton, *J. Biol. Chem.* **264** (1989), 17698–17703, and references therein.

76. A. Ericson *et al., J. Am. Chem. Soc.* **110** (1988), 2330–2332.

77. J. E. Colbert, A. G. Katopodis, and S. W. May, *J. Am. Chem. Soc.* **112** (1990), 3993–3996, and references therein.

78. L. C. Stewart and J. P. Klinman, "Dopamine β-Hydroxylase of Adrenal Chromaffin Granules: Structure and Function" (R), *Annu. Rev. Biochem.* **57** (1988), 551–592.

79. T. A. Dix and S. J. Benkovic, "Mechanism of Oxygen Activation by Pteridine-Dependent Monooxygenases" (R), *Acc. Chem. Res.* **21** (1988), 101–107.

80. T. L. Poulos, "The Crystal Structure of Cytochrome P-450$_{cam}$" (R), in Reference 73, pp. 505–523.

81. T. J. McMurry and J. T. Groves, "Metalloporphyrin Models for Cytochrome P-450" (R), in Reference 73, pp. 1–28.

82. J. T. Groves, T. E. Nemo, and R. S. Myers, *J. Am. Chem. Soc.* **101** (1979), 1032–1033.

83. J. T. Groves, "Key Elements of the Chemistry of Cytochrome P-450: The Oxygen Rebound Mechanism" (R), *J. Chem. Ed.* **62** (1985), 928–931.

84. D. Dolphin *et al.*, *Ann. N.Y. Acad. Sci.* **206** (1973), 177–200.

85. E. McCandlish *et al.*, *J. Am. Chem. Soc.* **102** (1980), 4268–4271.

86. J. N. Burstyn *et al.*, *J. Am. Chem. Soc.* **110** (1988), 1382–1388.

87. J. S. Valentine, J. N. Burstyn, and L. D. Margerum, "Mechanisms of Dioxygen Activation in Metal-Containing Monooxygenases: Enzymes and Model Systems" (R), in Reference 8, pp. 175–187.

88. T. C. Bruice, "Chemical Studies Related to Iron Protoporphyrin-IX Mixed Function Oxidases" (R), in Reference 20, pp. 227–277.

89. T. G. Traylor, W.-P. Fann, and D. Bandyopadhyay, *J. Am. Chem. Soc.* **111** (1989), 8009–8010.

90. R. Raag and T. L. Poulos, *Biochemistry* **28** (1989), 7586–7592, and references therein.

91. P. R. Ortiz de Montellano, "Oxygen Activation and Transfer" (R), in Reference 73, pp. 217–271.

92. K. Murata *et al.*, *J. Am. Chem. Soc.* **112** (1990), 6072–6083, and references therein.

93. T. G. Traylor and J. P. Ciccone, *J. Am. Chem. Soc.* **111** (1989), 8413–8420, and references therein.

94. J. Everse, K. E. Everse, and M. B. Grisham, eds., *Peroxidases in Chemistry and Biology* (B), CRC Press, 1991.

95. H. B. Dunford, "Peroxidases" (R), *Adv. Inorg. Biochem.* **4** (1982), 41–68.

96. T. L. Poulos and J. Kraut, *J. Biol. Chem.* **225** (1980), 8199–8205.

97. M. Morrison and G. R. Schonbaum, *Annu. Rev. Biochem.* **45** (1976), 861–888.

98. R. Quinn *et al.*, *J. Am. Chem. Soc.* **106** (1984), 4136–4144.

99. P. R. Ortiz de Montellano *et al.*, *J. Biol. Chem.* **262** (1987), 11641–11646.

100. J. S. Valentine and M. W. Pantoliano, "Protein-Metal Ion Interactions in Cuprozinc Protein (Superoxide Dismutase)" (R), in T. G. Spiro, ed., *Copper Proteins*, 1981, pp. 291–358.

101. J. S. Valentine and D. Mota de Freitas, *J. Chem. Ed.* **62** (1985), 990–997.

102. J. V. Bannister, W. H. Bannister, and G. Rotilio, "Aspects of the Structure, Function, and Applications of Superoxide Dismutase" (R), *CRC Crit. Rev. Biochem.* **22** (1987), 111–180.

103. See Reference 38.

104. S. B. Farr, R. D'Ari, and D. Touati, *Proc. Natl. Acad. Sci. USA* **83** (1986), 8268–8272.

105. O. Bermingham-McDonogh, E. B. Gralla, and J. S. Valentine, *Proc. Natl. Acad. Sci. USA* **85** (1988), 4789–4793.

106. J. S. Valentine and A. B. Curtis, *J. Am. Chem. Soc.* **97** (1975), 224–226.

107. M. Nappa *et al.*, *J. Am. Chem. Soc.* **101** (1979), 7744–7746.

108. G. J. McClune *et al.*, *J. Am. Chem. Soc.* **99** (1977), 5220–5222.

109. P. Natarajan and N. V. Raghavan, *J. Am. Chem. Soc.* **102** (1980), 4518–4519.

110. M. W. Pantoliano *et al.*, *J. Inorg. Biochem.* **17** (1982), 325–341.

111. J. A. Roe *et al.*, *Biochemistry* **27** (1988), 950–958.

112. J. A. Tainer *et al.*, *Nature* **306** (1983), 284–289.

113. J. A. Fee and C. Bull, *J. Biol. Chem.* **261** (1986), 13000–13005.

114. I. Bertini *et al.*, *J. Am. Chem. Soc.* **111** (1989), 714–719.

115. L. Banci *et al.*, *Inorg. Chem.* **27** (1988), 107–109.

116. D. Mota de Freitas *et al.*, *Inorg. Chem.* **26** (1987), 2788–2791.

117. L.-J. Ming and J. S. Valentine, *J. Am. Chem. Soc.* **109** (1987), 4426–4428.

118. I. Bertini *et al.*, *J. Am. Chem. Soc.* **107** (1985), 4391–4396.

119. L.-J. Ming *et al.*, *Inorg. Chem.* **27** (1988), 4458–4463.

120. L. Banci *et al.*, *Inorg. Chem.* **29** (1990), 2398–2403, and references therein.

121. J. A. Imlay and I. Fridovich, *J. Biol. Chem.* **266** (1991), 6957–6965.

122. P. R. Gardner and I. Fridovich, *J. Biol. Chem.* **266** (1991), 19328–19333.

123. J. S. Beckman *et al.*, *Proc. Natl. Acad. Sci. USA* **87** (1990), 1620–1624.

124. P. R. Ortiz de Montellano, *Annu. Rev. Pharmacol.* **32** (1992), 89–107.

125. The author gratefully acknowledges research support from the National Science Foundation and the National Institutes of Health while this chapter was being written, editoral assistance from Dr. Bertram Selverstone, and patience and support from Dr. Andrew J. Clark.

126. It has been suggested recently that the Cu_A site in cytochrome c oxidase may contain two copper ions. See P. M. Kroneck *et al.*, *FEBS Lett.* **268** (1990), 274–276.

127. An excellent review relevant to this chapter has recently appeared. K. D. Karlin, *Science* **261** (1993), 701–708.

6

Electron Transfer

HARRY B. GRAY
Beckman Institute
California Institute of Technology

WALTHER R. ELLIS, JR.
Department of Chemistry
University of Utah

I. ELECTRON TRANSFERS IN BIOLOGY

A. Biological Redox Components

Three types of oxidation-reduction (redox) centers are found in biology: protein side chains, small molecules, and redox cofactors. The first class is frequently overlooked by mechanistic enzymologists. The sulfhydryl group of cysteine is easily oxidized to produce a dimer, known as cystine:

$$2R\text{-}SH \xrightarrow[-2H^+]{-2e^-} R\text{-}S\text{-}S\text{-}R \tag{6.1}$$

This type of interconversion is known to occur in several redox proteins, including xanthine oxidase, mercuric ion reductase, and thioredoxin. Other enzyme systems display spectral evidence pointing to the presence of a protein-based radical in at least one intermediate. EPR spectroscopy provides a powerful tool in studying such systems; the observation of a $g = 2.0$ signal that cannot be attributed to impurities or an organic redox cofactor is generally taken to be evidence for a protein-based radical. Radicals localized on tyrosine (e.g., in photosystem II and the B2 subunit of ribonucleotide reductase[1]) and tryptophan (e.g., in yeast cytochrome c peroxidase[2]) have been unambiguously identified using EPR techniques together with protein samples containing isotopically labeled amino acids (e.g., perdeuterated Tyr) or single amino-acid mutations (e.g., Trp $\rightarrow$ Phe).

A variety of small molecules, both organic and inorganic, can function as redox reagents in biological systems. Of these, only the nicotinamide and qui-

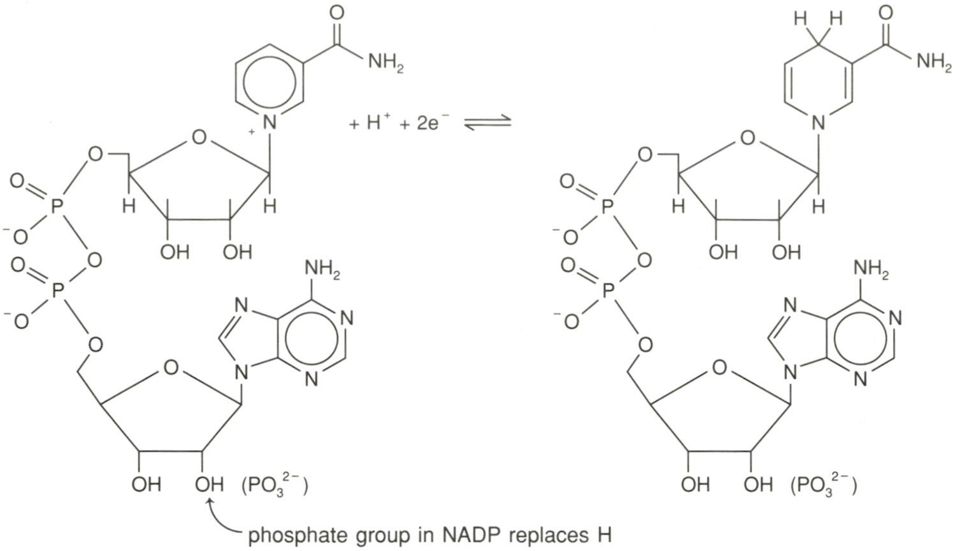

Figure 6.1
Reduction of NAD$^+$ to NADH.

none coenzymes are found throughout the biosphere. Nicotinamide adenine dinucleotide (NAD) and nicotinamide adenine dinucleotide phosphate (NADP) participate in a wide variety of biological redox reactions. The 4-position of the pyridine ring is the reactive portion of both molecules (Figure 6.1). Both typically function as 2-electron redox reagents.

In contrast, quinones may function as either 1- or 2-electron carriers:

$$Q \underset{\longleftarrow}{\overset{e^-, H^+}{\longrightarrow}} QH\cdot \underset{\longleftarrow}{\overset{e^-, H^+}{\longrightarrow}} QH_2 \tag{6.2}$$

Free-radical semiquinone (QH$\cdot$) intermediates have been detected by EPR spectroscopy in some electron transfers. Coenzyme Q, also called ubiquinone because it occurs in virtually all cells, contains a long isoprenoid tail that enables it to diffuse through membranes rapidly. This quinone derivative, which occurs in both free and protein-bound forms, is called ubiquinol when reduced (Figure 6.2). Other types of quinones are less frequently found in cells.

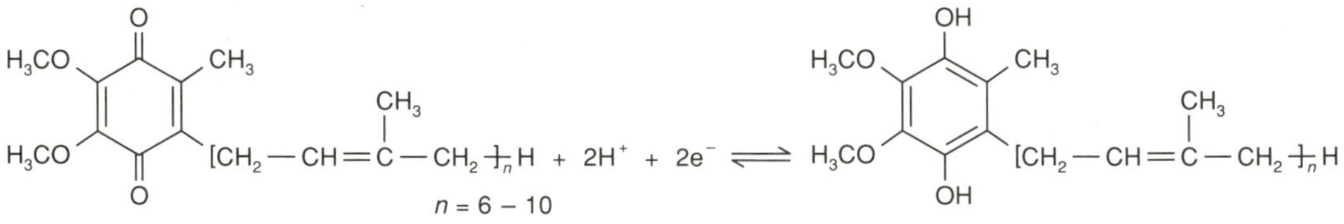

Figure 6.2
Reduction of coenzyme Q (ubiquinone) to ubiquinol.

Metalloproteins containing a single type of redox cofactor can be divided into two general classes: electron carriers and proteins involved in the transport or activation of small molecules. Adman[3] has identified some of the factors that seem to be characteristic of electron-transfer proteins (these proteins are sometimes called "electron transferases"): (a) possession of a suitable cofactor to act as an electron sink; (b) placement of the cofactor close enough to the protein surface to allow electrons to move in and out; (c) existence of a hydrophobic shell adjacent to, but not always entirely surrounding, the cofactor; (d) small structural changes accompanying electron transfer; and (e) an architecture that permits slight expansion or contraction in preferred directions upon electron transfer.

Proteins that function as electron transferases typically place their prosthetic groups in a hydrophobic environment and may provide hydrogen bonds (in addition to ligands) to assist in stabilizing both the oxidized and the reduced forms of the cofactor. Metal-ligand bonds remain intact upon electron transfer to minimize inner-sphere reorganization[4] (discussed in Section III). Many of the complex multisite metalloenzymes (e.g., cytochrome c oxidase, xanthine oxidase, the nitrogenase FeMo protein) contain redox centers that function as intramolecular electron transferases, shuttling electrons to/from other metal centers that bind exogenous ligands during enzymatic turnover.

There are four classes[3,5] of electron transferases, each of which contains many members that exhibit important structural differences: flavodoxins, blue copper proteins, iron-sulfur proteins, and cytochromes.

The flavodoxins[6] are atypical in that they contain an organic redox cofactor, flavin mononucleotide (FMN; see Figure 6.3). These proteins have molecular

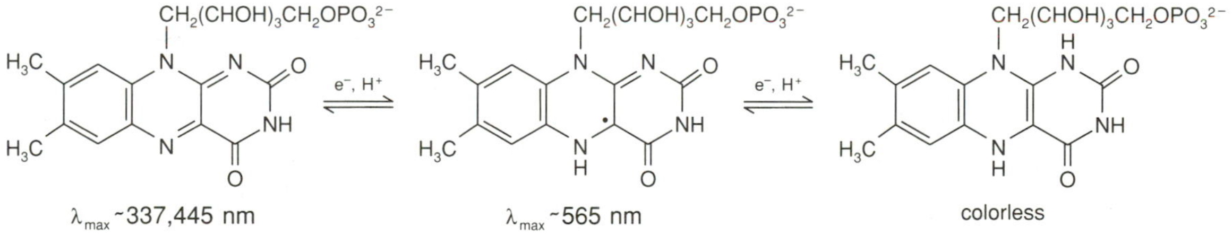

Figure 6.3
Reduction of FMN.

weights in the 8–13 kDa range, and are found in many species of bacteria and algae. The FMN cofactor is found at one end of the protein, near the molecular surface, but only the dimethylbenzene portion of FMN is significantly exposed to the solvent (Figure 6.4). FMN can act as either a 1- or a 2-electron redox center. In solution, the semiquinone form of free FMN is unstable, and disproportionates to the quinone (oxidized) and hydroquinone (reduced) forms. Hence, free FMN functions in effect as a 2-electron reagent. FMN in flavodoxins, on the other hand, can function as a single-electron carrier. This is easily discerned

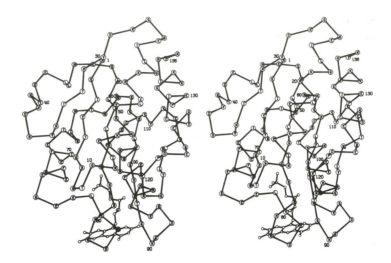

Figure 6.4
Stereo view of the structure of a clostridial flavodoxin. See R. D. Andersen *et al.*, *Proc. Natl. Acad. Sci. USA* **69** (1972), 3189–3191. Figure kindly provided by M. L. Ludwig.

by comparing reduction potentials for free and protein-bound FMN (Table 6.1). Clearly, the protein medium is responsible for this drastic alteration in oxidation-state stability. From an NMR study[7] of the *M. elsdenii* flavodoxin quinone/semiquinone and semiquinone/hydroquinone electron self-exchange rates, it was concluded that the latter is approximately 300 times faster than the former, in keeping with the view that the physiologically relevant redox couple is semiquinone/hydroquinone.

The blue copper proteins are characterized by intense S(Cys) → Cu charge-transfer absorption near 600 nm, an axial EPR spectrum displaying an unusually small hyperfine coupling constant, and a relatively high reduction potential.[4,8–10] With few exceptions (e.g., photosynthetic organisms), their precise roles in bacterial and plant physiology remain obscure. X-ray structures of several blue copper proteins indicate that the geometry of the copper site is approximately trigonal planar, as illustrated by the *Alcaligenes denitrificans* azurin structure (Figure 6.5).[11,12] In all these proteins, three ligands (one Cys, two His) bind tightly to the copper in a trigonal arrangement. Differences in interactions between the copper center and the axially disposed ligands may significantly contribute to variations in reduction potential that are observed[12] for the blue copper

Table 6.1
Reduction potentials of FMN couples.

	$E^{\circ\prime}Q/SQ$	$E^{\circ\prime}SQ/HQ$
Free FMN	− 238 mV	− 172 mV
C.M.P. flavodoxin	− 92 mV	− 399 mV

Abbreviations: Q, quinone; SQ, semiquinone; HQ, hydroquinone.

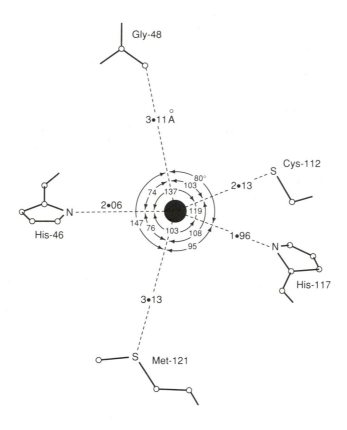

Figure 6.5
Structure of the blue copper center in azurin.[11]

electron transferases. For example, $E°' = 276$ mV for *A. denitrificans* azurin, whereas that of *P. vulgaris* plastocyanin is 360 mV. In *A. denitrificans* azurin, the Cu–S(Met) bond is 0.2 Å longer than in poplar plastocyanin, and there is a carbonyl oxygen 3.1 Å from the copper center, compared with 3.8 Å in plastocyanin. These differences in bond lengths are expected to stabilize Cu^{II} in azurin to a greater extent than in plastocyanin, and result in a lower $E°'$ value for azurin.

The iron-sulfur proteins play important roles[13,14] as electron carriers in virtually all living organisms, and participate in plant photosynthesis, nitrogen fixation, steroid metabolism, and oxidative phosphorylation, as well as many other processes (Chapter 7). The optical spectra of all iron-sulfur proteins are very broad and almost featureless, due to numerous overlapping charge-transfer transitions that impart red-brown-black colors to these proteins. On the other hand, the EPR spectra of iron-sulfur clusters are quite distinctive, and they are of great value in the study of the redox chemistry of these proteins.

The simplest iron-sulfur proteins, known as rubredoxins, are primarily found in anaerobic bacteria, where their function is unknown. Rubredoxins are small proteins (6 kDa) and contain iron ligated to four Cys sulfurs in a distorted tetrahedral arrangement. The $E^{\circ\prime}$ value for the $Fe^{III/II}$ couple in water is 770 mV; that of *C. pasteurianum* rubredoxin is -57 mV. The reduction potentials of iron-sulfur proteins are typically quite negative, indicating a stabilization of the oxidized form of the redox couple as a result of negatively charged sulfur ligands.

The [2Fe-2S] ferredoxins (10–20 kDa) are found in plant chloroplasts and mammalian tissue. The structure of *Spirulina platensis* ferredoxin[15] confirmed earlier suggestions, based on EPR and Mössbauer studies, that the iron atoms are present in a spin-coupled [2Fe-2S] cluster structure. One-electron reduction ($E^{\circ\prime} \sim -420$ mV) of the protein results in a mixed-valence dimer (Equation 6.3):

$$[Fe_2S_2(SR)_4]^{2-} \underset{-e^-}{\overset{+e^-}{\rightleftarrows}} [Fe_2S_2(SR)_4]^{3-} \qquad (6.3)$$

$$\begin{array}{cc} Fd_{ox} & Fd_{red} \\ 2Fe(III) & Fe(II) + Fe(III) \end{array}$$

The additional electron in Fd_{red} is associated with only one of the iron sites, resulting in a so-called trapped-valence structure.[16] The $[Fe_2S_2(SR)_4]^{4-}$ cluster oxidation state, containing two ferrous ions, can be produced *in vitro* when strong reductants are used.

Four-iron clusters [4Fe-4S] are found in many strains of bacteria. In most of these bacterial iron-sulfur proteins, also termed ferredoxins, two such clusters are present in the protein. These proteins have reduction potentials in the -400 mV range and are rather small (6–10 kDa). Each of the clusters contains four iron centers and four sulfides at alternate corners of a distorted cube. Each iron is coordinated to three sulfides and one cysteine thiolate. The irons are strongly exchange-coupled, and the [4Fe-4S] cluster in bacterial ferredoxins is paramagnetic when reduced by one electron. The so-called "high-potential iron-sulfur proteins" (HiPIPs) are found in photosynthetic bacteria, and exhibit anomalously high (~ 350 mV) reduction potentials. The *C. vinosum* HiPIP (10 kDa) structure demonstrates that HiPIPs are distinct from the [4Fe-4S] ferredoxins, and that the reduced HiPIP cluster structure is significantly distorted, as is also observed for the structure of the oxidized *P. aerogenes* ferredoxin. In addition, oxidized HiPIP is paramagnetic, whereas the reduced protein is EPR-silent.

This bewildering set of experimental observations can be rationalized in terms of a "three-state" hypothesis (i.e., $[4Fe-4S(SR)_4]^{n-}$ clusters exist in three physiological oxidation states).[17] This hypothesis nicely explains the differences in magnetic behavior and redox properties observed for these iron-sulfur proteins (Equation 6.4):

$$[4Fe-4S(SR)_4]^{-} \underset{-e^-}{\overset{+e^-}{\rightleftarrows}} [4Fe-4S(SR)_4]^{2-} \underset{-e^-}{\overset{+e^-}{\rightleftarrows}} [4Fe-4S(SR)_4]^{3-} \qquad (6.4)$$

$$\begin{array}{ccc} HiPIP_{ox} & HiPIP_{red} & Ferredoxin_{red} \\ & Ferredoxin_{ox} & \end{array}$$

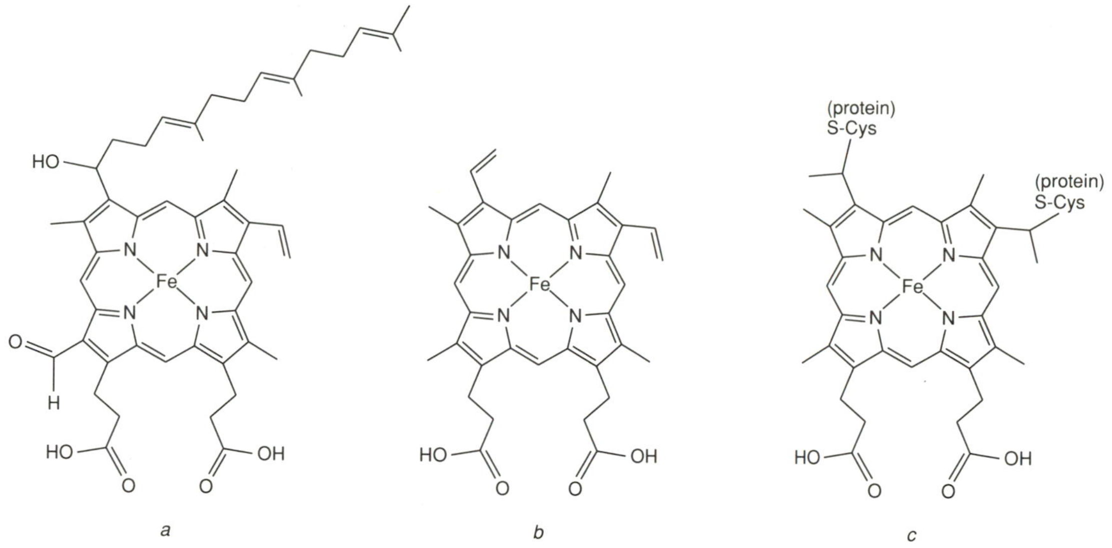

Figure 6.6
Structures of hemes a, b, and c.

The bacterial ferredoxins and HiPIPs all possess tetracubane clusters containing thiolate ligands, yet the former utilize the $-2/-3$ cluster redox couple, whereas the latter utilize the $-1/-2$ cluster redox couple.

The protein environment thus exerts a powerful influence over the cluster reduction potentials. This observation applies to *all* classes of electron transferases—the factors that are critical determinants of cofactor reduction potentials are poorly understood at present but are thought[18] to include the low dielectric constants of protein interiors (~4 for proteins vs. ~78 for H_2O), electrostatic effects due to nearby charged amino-acid residues, hydrogen bonding, and geometric constraints imposed by the protein.

As a class, the cytochromes[19-22] are the most thoroughly characterized of the electron transferases. By definition, a cytochrome contains one or more heme cofactors. These proteins were among the first to be identified in cellular extracts because of their distinctive optical properties, particularly an intense absorption in the 410–430 nm region (called the Soret band). Cytochromes are typically classified on the basis of heme type. Figure 6.6 displays the three most commonly encountered types of heme: heme a possesses a long phytyl "tail" and is found in cytochrome c oxidase; heme b is found in b-type cytochromes and globins; heme c is covalently bound to c-type cytochromes via two thioether linkages. Cytochrome nomenclature presents a real challenge! Some cytochromes are designated according to the historical order of discovery, e.g., cytochrome c_2 in bacterial photosynthesis. Others are designated according to the λ_{max} of the α band in the absorption spectrum of the reduced protein (e.g., cytochrome c_{551}).

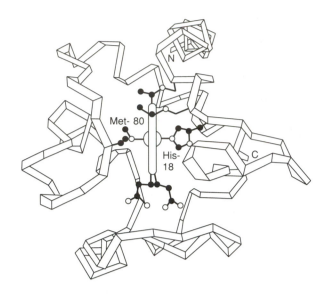

Figure 6.7
Structure of tuna cytochrome c.

Cytochromes c are widespread in nature. Ambler[23] divided these electron carriers into three classes on structural grounds. The Class I cytochromes c contain axial His and Met ligands, with the heme located near the N-terminus of the protein. These proteins are globular, as indicated by the ribbon drawing of tuna cytochrome c (Figure 6.7). X-ray structures of Class I cytochromes c from a variety of eukaryotes and prokaryotes clearly show an evolutionarily conserved "cytochrome fold," with the edge of the heme solvent-exposed. The reduction potentials of these cytochromes are quite positive (200 to 320 mV). Mammalian cytochrome c, because of its distinctive role in the mitochondrial electron-transfer chain, will be discussed later.

Class II cytochromes c ($E^{\circ\prime} \sim -100$ mV) are found in photosynthetic bacteria, where they serve an unknown function. Unlike their Class I cousins, these c-type cytochromes are high-spin: the iron is five-coordinate, with an axial His ligand. These proteins, generally referred to as cytochromes c', are four-α-helix bundles (Figure 6.8). The vacant axial coordination site is buried in the protein interior.

Finally, Class III cytochromes c, also called cytochromes c_3, contain four hemes, each ligated by two axial histidines. These proteins are found in a restricted class of sulfate-reducing bacteria and may be associated with the cytoplasmic membrane. The low molecular weights of cytochromes c_3 (~14.7 kDa) require that the four hemes be much more exposed to the solvent than the hemes of other cytochromes (see Figure 6.9), which may be in part responsible for their unusually negative (-200 to -350 mV) reduction potentials. These proteins possess many aromatic residues and short heme-heme distances, two properties that could be responsible for their anomalously large solid-state electrical conductivity.[24]

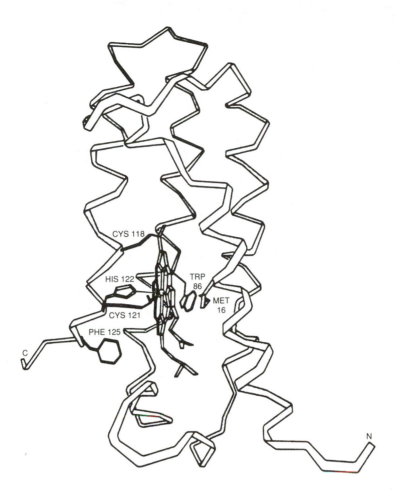

Figure 6.8
Structure of cytochrome c'.

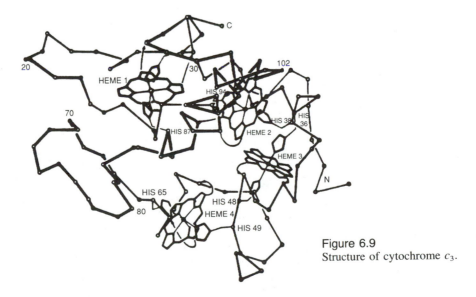

Figure 6.9
Structure of cytochrome c_3.

B. Energy Storage and Release

Electron-transfer reactions play key roles in a great many biological processes, including collagen synthesis, steroid metabolism, the immune response, drug activation, neurotransmitter metabolism, nitrogen fixation, respiration, and photosynthesis. The latter two processes are of fundamental significance—they provide most of the energy that is required for the maintenance of life. From the point of view of global bioenergetics, aerobic respiration and photosynthesis are complementary processes (Figure 6.10). The oxygen that is evolved by photosynthetic organisms is consumed by aerobic microbes and animals. Similarly, the end products of aerobic respiratory metabolism (CO_2 and H_2O) are the major nutritional requirements of photosynthetic organisms. The global C, H, and O cycles are thus largely due to aerobic respiration and photosynthesis.

The extraction of energy from organic compounds, carried out by several catabolic pathways (e.g., the citric-acid cycle), involves the oxidation of these compounds to CO_2 and H_2O with the concomitant production of water-soluble reductants (NADH and succinate). These reductants donate electrons to components of the mitochondrial electron-transfer chain, resulting in the reduction of oxygen to water:

$$\tfrac{1}{2}O_2 + NADH + H^+ \longrightarrow H_2O + NAD^+ \tag{6.5}$$

In aerobic organisms, the terminal oxidant is, of course, oxygen. However, some species of bacteria respire anaerobically and are able to use inorganic oxyanions (nitrate or sulfate) as terminal oxidants. The translocation of protons across the inner mitochondrial membrane accompanies the electron transfers that

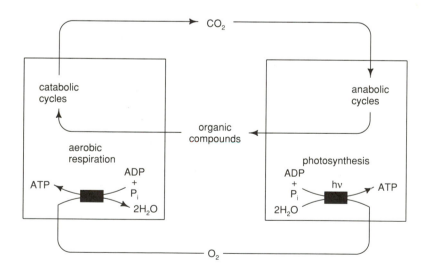

Figure 6.10
Aerobic respiration and photosynthesis.

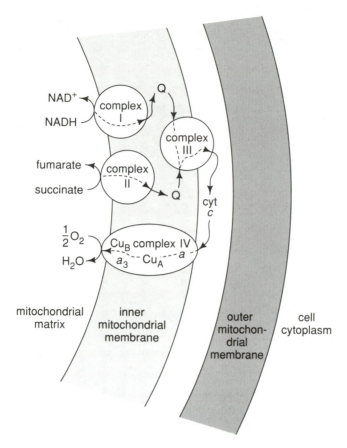

Figure 6.11
Redox components in mitochondria.

ultimately lead to the reduction of O_2; these protons, in turn, activate ATP synthase, which catalyzes the phosphorylation of ADP to ATP (a process known as oxidative phosphorylation). Because the hydrolysis of ATP is very exoergonic (i.e., $\Delta G < 0$), the newly synthesized ATP is used as a molecular energy source to drive thermodynamically unfavorable reactions to completion.

The rediscovery of cytochromes by Keilin[25] in 1925 led him to propose that the reduction of O_2 is linked to the oxidation of reduced substrates by a series of redox reactions, carried out by cellular components collectively referred to as the respiratory electron-transport chain. Progress toward a molecular understanding of these redox reactions has been painfully slow. Most of the components are multisubunit proteins that reside in the inner mitochondrial membrane (Figure 6.11). These proteins (Complexes I–IV) are quite difficult to purify with retention of *in vivo* properties, and they do not crystallize well.

The components[26–28] of the respiratory chain contain a variety of redox cofactors. Complex I (NADH-Q reductase; >600 kDa) contains five iron-sulfur clusters and FMN. Complex II (succinate-Q reductase; 150 kDa) contains sev-

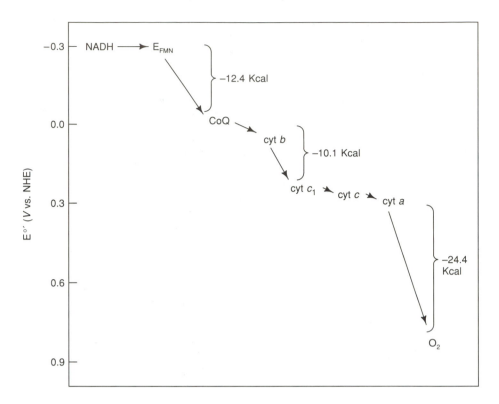

Figure 6.12
Electron flow from NADH to O_2 in the mitochondrial electron-transport chain.

eral iron-sulfur clusters, FAD (flavin adenine dinucleotide), and cytochrome b_{568}. Complex III (ubiquinol-cytochrome c reductase; 250 kDa) contains a [2Fe-2S] iron-sulfur center and cytochromes b_{562}, b_{566}, and c_1. Complex IV (cytochrome c oxidase; 200 kDa) contains at least two copper ions and cytochromes a and a_3; Q denotes coenzyme Q, which may be bound to hydrophobic subunits of Complexes I, II, and/or III *in vivo*. Cytochrome c (cyt c in Figure 6.11) is a water-soluble protein (12.4 kDa) that is only peripherally associated with the inner mitochondrial membrane; it has been so thoroughly studied that it is generally regarded as the prime example of an electron transferase.

More than 20 redox centers are involved in the electron-transport chain. Figure 6.12 depicts a simplified view of the flow of electrons from NADH to O_2 via this series of electron carriers. Electron flow through Complexes I, III, and IV is associated with the release of relatively large amounts of energy, which is coupled to proton translocation by these complexes (and therefore ATP production). The redox potentials of the electron carriers thus appear to play a role in determining the pathway of electron flow through the electron-transport chain.

Approximately 50 percent of the surface area of the inner mitochondrial membrane is lipid bilayer that is unoccupied by membrane proteins and through which these proteins, in principle, are free to diffuse laterally. Kinetic (laser photobleaching and fluorescence recovery) and ultrastructural (freeze-fracture electron microscopy) studies[29,30] indicate that Complexes I–IV diffuse independently and laterally over the inner membrane, whereas cytochrome c diffuses in three dimensions (i.e., through the intramembrane space). Respiratory electron transport has been shown to be a diffusion-coupled kinetic process.[29,30] The term "electron-transport chain" is thus somewhat misleading, because it implies a degree of structural order that does not exist beyond the level of a given protein complex.

In view of these observations, why are all of the electron transfers associated with mitochondrial respiration required? For example, why is cytochrome c needed to shuttle electrons in Figures 6.11 and 6.12 when the cofactor reduction potentials of Complex III are more negative than those of Complex IV? Evidently, factors other than $\Delta G°$ are of importance—these will be discussed in Sections III and IV.

Photosynthesis could be viewed as the most fundamental bioenergetic process. Biological reactions are driven by an energy flux, with sunlight serving as the energy source. Photosynthesis[31–36] is the process by which radiant solar energy is converted into chemical energy in the form of ATP and NADPH, which are then used in a series of enzymatic reactions to convert CO_2 into organic compounds. The photosynthetic algae that appeared on Earth two million years ago released oxygen into the atmosphere and changed the environment from a reducing to an oxidizing one, setting the stage for the appearance of aerobically respiring organisms.

Photosynthesis is initiated by the capture of solar energy, usually referred to as "light harvesting." A large number of organic pigments, including chlorophylls, carotenoids, phycoerythrin, and phycocyanin (in green plants and algae) are clustered together in pigment-protein complexes called photosystems. These pigments collectively absorb most of the sunlight reaching the Earth—their absorption spectra are displayed in Figure 6.13. Light is transformed into chemical energy in pigment-protein complexes called reaction centers. The concentration of reaction centers within a photosynthetic cell is too small to offer a suitable absorption cross section for sunlight. Hence, hundreds of these light-harvesting pigments function as molecular antennas; an x-ray structure[35] of one subunit of a bacteriochlorophyll-protein complex is displayed in Figure 6.14.

Absorption of a photon by an antenna pigment promotes the pigment into an electronically excited state, which can return to the ground state by a variety of relaxation processes, including fluorescence or resonance transfer of excitation energy to a nearby pigment at picosecond rates. As much as 100 ps may elapse between the photon absorption and the arrival of the light energy at a reaction center. During this time, the energy may "migrate" in a random-walk fashion among hundreds of pigments.

The energy of the excited state is converted into electrochemical potential energy at the reaction center, which contains a primary electron donor P that transfers an electron to a nearby acceptor A_1 within the same protein (and P

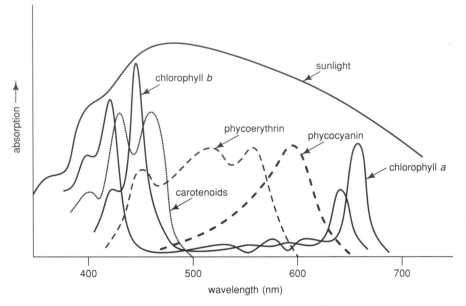

Figure 6.13
Absorption spectra of the photosystem pigments.

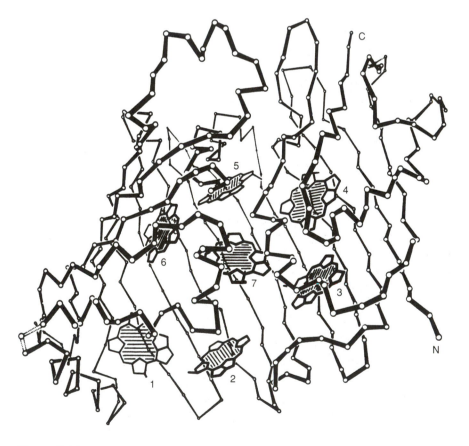

Figure 6.14
Structure of a subunit of a bacteriochlorophyll-protein complex. Reproduced with permission from Reference 35.

becomes oxidized to P^+):

$$PA_1A_2A_3 \cdots \xrightarrow{h\nu} P^*A_1A_2A_3 \cdots \longrightarrow P^+A_1{}^-A_2A_3 \cdots \qquad (6.5)$$

This *charge separation* is of paramount importance. The key problem is maintaining the charge separation, which involves minimization of the energy-wasting back reaction. Reaction centers contain an ordered array of secondary electron acceptors (A_1, A_2, $A_3 \cdots$) that optimize the $\Delta G°$ that occurs at each step:

$$P^+A_1{}^-A_2A_3 \cdots \longrightarrow P^+A_1A_2{}^-A_3 \cdots \longrightarrow P^+A_1A_2A_3{}^- \cdots \qquad (6.6)$$

Thus, the back reaction is circumvented by optimizing forward electron transfers that rapidly remove electrons from $A_1{}^-$. As the acceptors are separated by greater and greater distances from P^+, the probability of the back electron transfer to P^+ decreases. Put another way, the overlap of P^+ and each acceptor orbital decreases in the order $P^+/A_1{}^- > P^+/A_2{}^- > P^+/A_3{}^-$.

Photosynthetic bacteria contain only one type of reaction center (100 kDa). The solution of the x-ray structure (at 2.9 Å resolution) of the *Rps. viridis* reaction center was reported[36] in 1984, providing conclusive proof that electrons can "tunnel" over 10–20 Å distances through protein interiors. The reaction-center protein contains many cofactors (Figure 6.15): two bacteriochlorophylls (BChl) in close proximity (the so-called "special pair"), two further bacteriochlorophylls that are spectroscopically identical, two bacteriopheophytins (BPh), two quinones (Q_A and Q_B), and one iron center. (Q_B was lost during isolation of the *Rps. viridis* reaction center and thus does not appear in Figure 6.15.) The reaction center contains an approximate two-fold rotation axis. Despite this strikingly high symmetry in the reaction center, one pathway of electron flow predominates, as the cartoon in Figure 6.16 indicates.

C. Coupling Electron Transfers and Substrate Activation

Electron transfers are key steps in many enzymatic reactions involving the oxidation or reduction of a bound substrate. Relevant examples include cytochrome c oxidase ($O_2 \rightarrow 2H_2O$) and nitrogenase ($N_2 \rightarrow 2NH_3$). To reinforce the claim that electron-transfer steps are of widespread importance, several other redox systems, representative of diverse metabolic processes, will be mentioned here.

Xanthine oxidase (275 kDa; α_2 dimer) catalyzes the two-electron oxidation[37–39] of xanthine to uric acid (Equation 6.7).

$$\text{xanthine} + H_2O \longrightarrow \text{uric acid} + 2e^- + 2H^+ \qquad (6.7)$$

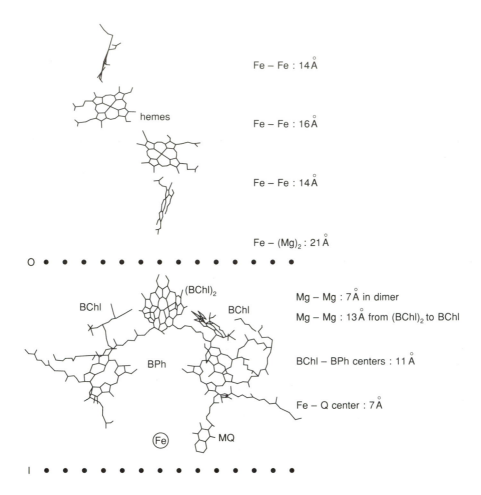

Fe – Fe : 14 Å

Fe – Fe : 16 Å

Fe – Fe : 14 Å

Fe – $(Mg)_2$: 21 Å

Mg – Mg : 7 Å in dimer
Mg – Mg : 13 Å from $(BChl)_2$ to BChl

BChl – BPh centers : 11 Å

Fe – Q center : 7 Å

Figure 6.15
Structure of the *Rps. viridis* photosynthetic reaction-center cofactors. The black dots delineate the outward (O)- and inward (I)-facing portions of the membrane. Adapted from Reference 36.

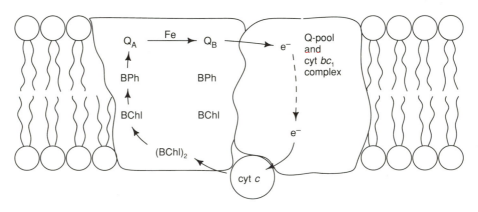

Figure 6.16
Electron flow in the bacterial photosynthetic reaction center.

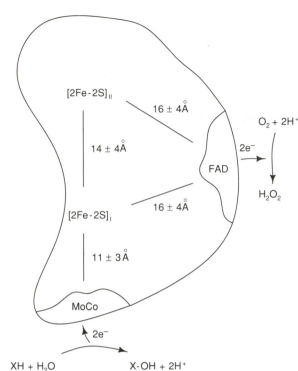

Figure 6.17
Representation of the cofactors in one subunit of xanthine oxidase.

This enzyme, which plays a prominent role in the biodegradation of purines, is the target of drugs administered to patients suffering from gout (joint inflammation, due to precipitation of sodium urate). Figure 6.17 displays the cofactors in a subunit: a Mo-pterin, termed MoCo; two [2Fe-2S] centers; and one FAD. The binuclear iron-sulfur sites serve to shuttle electrons between the reduced substrate (XH) and O_2.

The first step in the biosynthesis of DNA involves the reduction of ribonucleotides (Equation 6.8) catalyzed by ribonucleotide reductase.[40] The E. coli enzyme is an $\alpha_2\beta_2$ tetramer composed of a B1 protein (160 kDa) and a B2 protein (78 kDa). The B1 protein (a dimer) contains redox-active dithiol groups, binding sites for ribonucleotide substrates, and regulatory binding sites for nucleotide diphosphates. Protein B2, also a dimer, possesses a phenolate radical (Tyr-122) that is stabilized by an antiferromagnetically coupled binuclear iron center (Figure 6.18). This radical is essential for enzyme activity, and is ~10 Å from the protein-B1/protein-B2 interface. Hence it cannot directly participate in an H-atom abstraction from the substrate (bound to protein B1). Instead, the x-ray structure of the B2 protein[41] suggests that a long-range electron transfer from the Tyr radical to a residue (perhaps Trp-48) on the B1 protein is operative during enzyme turnover.

$$(6.8)$$

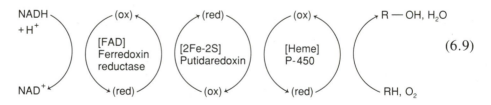

Figure 6.18
Schematic of the binuclear iron center and Tyr-122 radical in the B2 protein of *E. coli* ribonucleotide reductase.[41]

Most of the presently known metal-containing mono- and dioxygenases are multicomponent, requiring the involvement of additional proteins (electron transferases) to shuttle electrons from a common biological reductant (usually NADH or NADPH) to the metallooxygenase. Cytochrome P-450, whose substrate oxidation chemistry was discussed in detail in Chapter 5, serves as an excellent example. Figure 5.10 presented a catalytic cycle for cytochrome P-450-dependent hydroxylations[42] that begins with substrate (RH) binding to the ferric enzyme (RH is camphor for *Pseudomonas putida* cytochrome P-450). To hydroxylate the camphor substrate, the monooxygenase must be reduced via the electron-transport chain in Equation (6.9).

$$
\begin{array}{ccccccc}
\text{NADH} & & \text{(ox)} & & \text{(red)} & & \text{(ox)} & & R\!-\!OH, H_2O \\
+H^+ & & [FAD] & & [2Fe\text{-}2S] & & [Heme] & & \\
 & & \text{Ferredoxin} & & \text{Putidaredoxin} & & P\text{-}450 & & (6.9) \\
 & & \text{reductase} & & & & & & \\
\text{NAD}^+ & & \text{(red)} & & \text{(ox)} & & \text{(red)} & & RH, O_2
\end{array}
$$

The ferredoxin reductase receives two electrons from NADH and passes them on, one at at time, to putidaredoxin, a [2Fe-2S] iron-sulfur protein. Thus, two single-electron-transfer steps from reduced putidaredoxin to cytochrome P-450 are required to complete one enzyme turnover.

The activity of the enzyme appears to be regulated at the first reduction step.[43] In a 1:1 putidaredoxin-cytochrome P-450 complex, the reduction potential of putidaredoxin is −196 mV, but that of cytochrome P-450 is −340 mV in the absence of camphor; reduction of the cytochrome P-450 is thus thermodynamically unfavorable ($k \sim 0.22$ s^{-1}). Upon binding camphor, the reduction potential of cytochrome P-450 shifts to -173 mV, and the electron-transfer rate in the protein complex accordingly increases to 41 s^{-1}. "Costly" reducing equivalents are not wasted, and there are no appreciable amounts of noxious oxygen-reduction products when substrate is not present.

In the third step, molecular oxygen binds to the camphor adduct of ferrous cytochrome P-450. This species, in the presence of reduced putidaredoxin, accepts a second electron, and catalyzes the hydroxylation of the bound camphor substrate. The turnover rate for the entire catalytic cycle is 10–20 s^{-1}, and the second electron-transfer step appears to be rate-determining.[44]

The bulk of the interest in electron-transfer reactions of redox proteins has been directed toward questions dealing with long-range electron transfer and the nature of protein-protein complexes whose structures are optimized for rapid intramolecular electron transfer. Before we undertake a discussion of these issues, it is worth noting that studies of the reactions of redox proteins at electrodes are attracting increasing attention.[45–47] Direct electron transfer between a variety of redox proteins and electrode surfaces has been achieved. Potential applications include the design of substrate-specific biosensors, the development of biofuel cells, and electrochemical syntheses. An interesting application of bioelectrochemical technology is the oxidation of p-cresol to p-hydroxybenzaldehyde (Figure 6.19).[48]

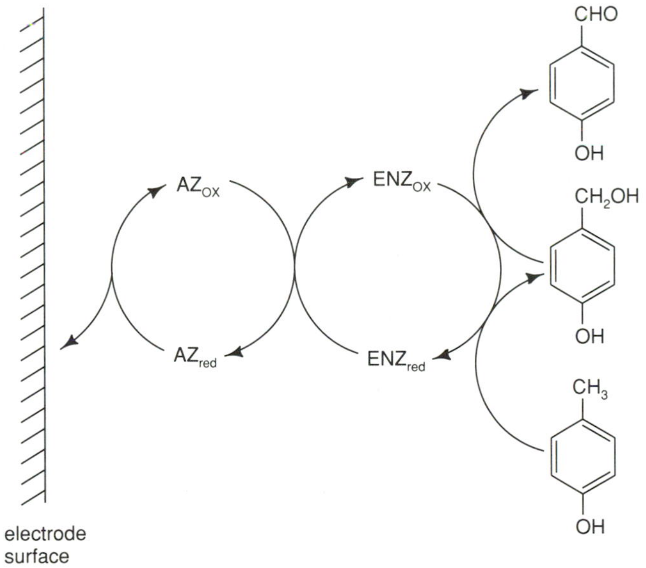

Figure 6.19
Enzyme-catalyzed electrochemical oxidation of p-cresol to p-hydroxybenzaldehyde. AZ is azurin, and ENZ is p-cresol methylhydroxylase.[48]

II. ELECTRON-TRANSFER RATES

A. Overview

Measurements of the rates of oxidation-reduction reactions began in the late 1940s. A great deal of the early experimental work was carried out by inorganic chemists, and by the 1970s the reactivity patterns of many complexes had been uncovered.[49-51] Chemists studying the mechanisms of metalloprotein electron-transfer reactions frequently seek parallels with the redox behavior of less-complicated inorganic complexes.

In examining biological electron transfers, it is important to remember that metalloproteins are more than just metal ions in disguise. Virtually every property of a protein (excluding its amino-acid sequence) depends on the solution pH. Redox proteins are very large polyelectrolytes whose redox prosthetic groups are typically buried in the protein interior. One important distinction between redox reactions of proteins and redox reactions of small transition-metal complexes is the magnitude of the electron donor-to-acceptor distance. The relevant distance for small molecules, unlike redox proteins, is generally taken to be van der Waals contact. Within the last ten years, it has been convincingly demonstrated that electrons can "tunnel" at significant rates across distances of 15 Å or more in protein interiors.[52-58]

Experimental investigation of the factors that control the rates of biological redox reactions has not come as far as the study of the electron transfers of metal complexes, because many more variables must be dealt with (e.g., asymmetric surface charge, nonspherical shape, uncertain details of structures of proteins complexed with small molecules or other proteins). Many experimental approaches have been pursued, including the covalent attachment of redox reagents to the surfaces of metalloproteins.

B. Self-Exchange and Cross Reactions

The simplest reactions in solution chemistry are electron self-exchange reactions (Equation 6.10), in which the reactants and products are the same (the asterisk is used to identify a specific isotope). The only

$$^*A_{ox} + A_{red} \longrightarrow {}^*A_{red} + A_{ox} \tag{6.10}$$

way to establish chemically that a reaction has taken place is to introduce an isotopic label. There is no change in the free energy ($\Delta G° = 0$) for this type of reaction. As will become evident later on, the reason why these types of reactions are studied is because self-exchange rates and activation parameters are needed to interpret redox reactions in which a net chemical change occurs. The experimental measurement[58] of self-exchange rates is tedious and usually only results in an order-of-magnitude estimate of the rate constant (as inferred from

Table 6.2
Experimental timescales in seconds.

Laser flash photolysis	$\geqslant 10^{-14}$
Pulse radiolysis	$\sim 10^{-9}$
Mössbauer spectroscopy (^{57}Fe)	10^{-9}–10^{-6}
EPR (transition metals)	10^{-9}–10^{-8}
Temperature-jump spectrometry	$\geqslant 10^{-8}$
NMR (^{1}H)	$\sim 10^{-5}$
Chemical mixing	$\geqslant 10^{-3}$

the experimental timescale; see Table 6.2). Most of the protein self-exchange rates reported to date have been measured by NMR line-broadening studies. Other potentially useful methods, such as Mössbauer spectroscopy and EPR, have not been widely used.

An elegant example of the measurement of an electron self-exchange rate of a redox protein was reported by Dahlin *et al.*[59] The copper ion of stellacyanin was removed and then replaced with either ^{63}Cu or ^{65}Cu. Oxidized [^{63}Cu] stellacyanin was allowed to react with reduced [^{65}Cu] stellacyanin for various times (10 ms to 7 min) at 20°C, after which the reaction was quenched by lowering the solution temperature to -120°C using a rapid-freeze apparatus:

$$^{63}\text{Cu}^{2+} + {}^{65}\text{Cu}^{+} \longrightarrow {}^{63}\text{Cu}^{+} + {}^{65}\text{Cu}^{2+} \qquad (6.11)$$

Subtle differences in the EPR spectra (Figure 6.20) of the two isotopic forms of stellacyanin (due to a small difference in the nuclear magnetic moments of the two isotopes) were used to monitor the progress of the reaction, yielding a rate constant of $1.2 \times 10^{5} \text{ M}^{-1}\text{s}^{-1}$.

Much more common are cross reactions (Equation 6.12), where A_{ox} is the oxidized reactant, B_{red} is the reduced reactant, A_{red} is the reduced product, and B_{ox} is the oxidized product.

$$A_{ox} + B_{red} \longrightarrow A_{red} + B_{ox} \qquad (6.12)$$

For these reactions, $\Delta G° \neq 0$. The experimental measurement of cross-reaction rates is generally more straightforward than the measurement of self-exchange rates. Either the reactants are simply mixed together, or a thermodynamically unstable system is generated rapidly (via pulse radiolysis, flash photolysis, or temperature-jump relaxation) to initiate the redox reaction. Absorption spectroscopy has almost always been used to monitor the progress of protein cross reactions. The primary goal of theory, as will become evident, is to provide a relationship between $\Delta G°$ and $\Delta G^{\ddagger}$ for cross reactions.

Both self-exchange and cross reactions can be broadly classified as inner-sphere or outer-sphere reactions. In an inner-sphere reaction, a ligand is shared between the oxidant and reductant in the transition state. An outer-sphere reac-

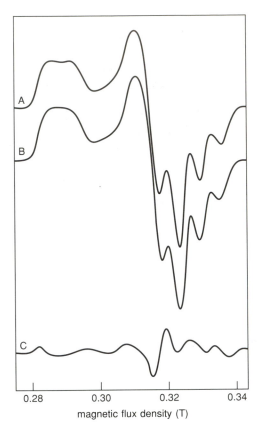

Figure 6.20
Frozen-solution EPR spectra of stellacyanin (20 K, 9.25 GHz): (A) ^{65}Cu; (B) ^{63}Cu; (C) Difference spectrum $(A - B)$.[59]

magnetic flux density (T)

tion, on the other hand, is one in which the inner coordination shells of both the oxidant and the reductant remain intact in the transition state. There is no bond breaking or bond making, and no shared ligands between redox centers. Long-range electron transfers in biology are all of the outer-sphere type.

III. ELECTRON-TRANSFER THEORY

A. Basic Concepts

The simplest electron transfer occurs in an outer-sphere reaction. The changes in oxidation states of the donor and acceptor centers result in a change in their equilibrium nuclear configurations. This process involves geometric changes, the magnitudes of which vary from system to system. In addition, changes in the interactions of the donor and acceptor with the surrounding solvent molecules will occur. The Franck-Condon principle governs the coupling of the electron transfer to these changes in nuclear geometry: during an electronic transition, the electronic motion is so rapid that the nuclei (including metal ligands and solvent molecules) do not have time to move. Hence, electron transfer occurs at a fixed nuclear configuration. In a self-exchange reaction, the energies of the donor and acceptor orbitals (hence, the bond lengths and bond angles of the donor and acceptor) must be the same before efficient electron transfer can take place.

The incorporation of the Franck-Condon restriction leads to the partitioning[60-65] of an electron-transfer reaction into reactant (precursor complex) and product (successor complex) configurations. The steps in Equations (6.13) to (6.15) go from reactants to products: K is the equilibrium constant for the formation of the precursor complex $[A_{ox}, B_{red}]$, and k_{et} is the forward electron-transfer rate to produce the successor complex $[A_{red}, B_{ox}]$.

$$A_{ox} + B_{red} \underset{\longleftarrow}{\overset{K}{\longrightarrow}} [A_{ox}, B_{red}] \tag{6.13}$$

$$[A_{ox}, B_{red}] \xrightarrow{k_{et}} [A_{red}, B_{ox}] \tag{6.14}$$

$$[A_{red}, B_{ox}] \xrightarrow{\text{fast}} A_{red} + B_{ox} \tag{6.15}$$

Marcus pioneered the use of potential energy diagrams as an aid in describing electron-transfer processes.[60] For the sake of simplicity, the donor and acceptor are assumed to behave like collections of harmonic oscillators. Instead of two separate potential energy surfaces being used for the reactants, they are combined into a single surface that describes the potential energy of the precursor complex as a function of its nuclear configuration (i.e., the sum of the translational, rotational, and vibrational degrees of freedom of the reactant molecules and the molecules in the surrounding solvent-3N coordinates, where N is the number of nuclei present). Similarly, a single potential energy (3N-dimensional) surface is used to describe the potential energy of the successor complex as a function of its nuclear configuration. It has become conventional to simplify such potential energy diagrams by using one-dimensional slices through the reactant and product surfaces in order to visualize the progress of a reaction, as illustrated in Figure 6.21.

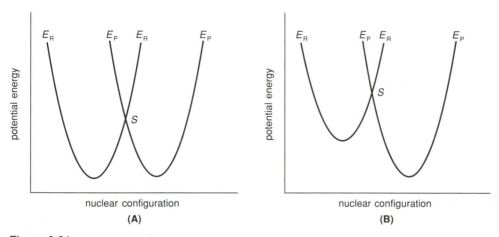

Figure 6.21
Potential energy diagrams: (A) self-exchange reaction; (B) cross reaction. Point S represents the activated complex. E_R and E_P are the reactant and product surfaces, respectively.

The intersection of the reactant and product surfaces (point S) represents the transition state (or "activated complex"), and is characterized by a loss of one degree of freedom relative to the reactants or products. The actual electron-transfer event occurs when the reactants reach the transition-state geometry. For bimolecular reactions, the reactants must diffuse through the solvent, collide, and form a precursor complex prior to electron transfer. Hence, disentangling the effects of precursor complex formation from the observed reaction rate can pose a serious challenge to the experimentalist; unless this is done, the factors that determine the kinetic activation barrier for the electron-transfer step cannot be identified with certainty.

The surfaces depicted in Figure 6.21 presume that the electrons remain localized on the donor and acceptor; as long as this situation prevails, no electron transfer is possible. Thus some degree of electronic interaction, or coupling, is required if the redox system is to pass from the precursor to the successor complex. This coupling removes the degeneracy of the reactant and product states at the intersection of their respective zero-order surfaces (points S in Figure 6.21) and leads to a splitting in the region of the intersection of the reactant and product surfaces (Figure 6.22). If the degree of electronic interaction is sufficiently small, first-order perturbation theory can be used to obtain the energies of the new first-order surfaces, which do not cross. The splitting at the intersection is equal to $2H_{AB}$, where H_{AB} is the electronic-coupling matrix element.

The magnitude of H_{AB} determines the behavior of the reactants once the intersection region is reached. Two cases can be distinguished. First, H_{AB} is very small; for these so-called "nonadiabatic" reactions, there is a high probability that the reactants will "jump" to the upper first-order potential energy surface, leading to very little product formation. If the electronic interaction is sufficiently large, as it is for "adiabatic" reactions, the reactants will remain on the lower first-order potential energy surface upon passage through the transition-state region.

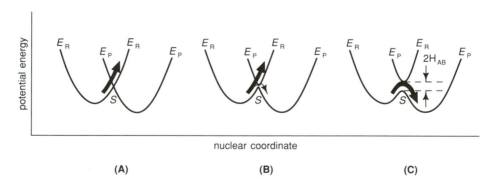

Figure 6.22
Potential energy diagrams: (A) $H_{AB} = 0$, $\kappa = 0$ (no transfer); (B) H_{AB} small, $\kappa \ll 1$ (nonadiabatic transfer); (C) H_{AB} large, $\kappa = 1$ (adiabatic transfer). The arrows indicate the relative probability of crossing to the product surface (E_R to E_P).

The term adiabatic (Greek: *a-dia-bainein*, not able to go through) is used in both thermodynamics and quantum mechanics, and the uses are analogous. In the former, it indicates that there is no heat flow in or out of the system. In the latter, it indicates that a change occurs such that the system makes no transition to other states. Hence, for an adiabatic reaction, the system remains on the same (i.e., lower) first-order electronic surface for the entire reaction. The probability of electron transfer occurring when the reactants reach the transition state is unity. The degree of adiabaticity of the reaction is given by a transmission coefficient, κ, whose value ranges from zero to one. For systems whose H_{AB} is sufficiently large ($>k_B T$, where k_B is the Boltzmann constant), $\kappa = 1$. This situation occurs when the reacting centers are close together, the orbital symmetries are favorable, and no substantial changes in geometry are involved. The transmission coefficient is generally very small ($\kappa < 1$) for electron-transfer reactions of metalloproteins, owing to the long distances involved.

B. Marcus Theory

In classical transition-state theory, the expression for the rate constant of a bimolecular reaction in solution is

$$k = \kappa \nu_n \exp\left(-\Delta G^*/RT\right), \tag{6.16}$$

where ν_n, the nuclear frequency factor, is approximately 10^{11} M^{-1} s^{-1} for small molecules, and ΔG^* is the Gibbs-free-energy difference between the activated complex and the precursor complex. This theoretical framework provides the starting point for classical electron-transfer theory. Usually the transmission coefficient κ is initially assumed to be unity. Thus, the problem of calculating the rate constant involves the calculation of ΔG^*, which Marcus partitioned into several parameters:

$$\Delta G^* = w^r + (\lambda/4)(1 + \Delta G^{\circ\prime}/\lambda)^2, \tag{6.17}$$

$$\Delta G^{\circ\prime} = \Delta G^\circ + w^p - w^r. \tag{6.18}$$

Here w^r is the electrostatic work involved in bringing the reactants to the mean reactant separation distance in the activated complex, and w^p is the analogous work term for dissociation of the products. These terms vanish in situations where one of the reactants (or products) is uncharged. ΔG° is the Gibbs-free-energy change when the two reactants and products are an infinite distance apart, and $\Delta G^{\circ\prime}$ is the free energy of the reaction when the reactants are a distance r apart in the medium; ΔG° is the standard free energy of the reaction, obtainable from electrochemical measurements (the quantity $-\Delta G^\circ$ is called the *driving force* of the reaction).

The reorganization energy λ is a parameter that contains both inner-sphere (λ_i) and outer-sphere (λ_o) components; $\lambda = \lambda_i + \lambda_o$. The inner-sphere reorganization energy is the free-energy change associated with changes in the bond lengths and angles of the reactants. The λ_i term can be evaluated within the simple harmonic-oscillator approximation:

$$\lambda_i = (\tfrac{1}{2}) \sum_j k_j \, (\Delta x_j)^2, \tag{6.19}$$

where k_j values are normal-mode force constants, and the Δx_j values are differences in equilibrium bond lengths between the reduced and oxidized forms of a redox center.

The outer-sphere reorganization energy reflects changes in the polarization of solvent molecules during electron transfer:

$$\lambda_o = e^2[(1/2r_A) + (1/2r_B) - (1/d)][(1/D_{op}) - (1/D_s)]; \tag{6.20}$$

d is the distance between centers in the activated complex, generally taken to be the sum of the reactant radii r_A and r_B; D_{op} is the optical dielectric constant of the medium (or, equivalently, the square of the refractive index); and D_s is the static dielectric constant. This simple model for the effect of solvent reorganization assumes that the reactants are spherical, and that the solvent behaves as a dielectric continuum. (Sometimes the latter approximation is so rough that there is no correspondence between theory and experiment.)

Variations in λ can have enormous effects on electron-transfer rates, Some of the possible variations are apparent from inspection of Equation (6.20). First, λ_o decreases with increasing reactant size. Second, the dependence of the reaction rate on separation distance attributable to λ_o occurs via the $1/d$ term. Third, λ_o decreases markedly as the solvent polarity decreases. For nonpolar solvents, $D_s \simeq D_{op} \simeq 1.5$ to 4.0. It is significant to note that protein interiors are estimated to have $D_s \simeq 4$, whereas, $D_s \simeq 78$ for water. An important conclusion is that metalloproteins that contain buried redox cofactors need not experience large outer-sphere reorganization energies.

The key result of Marcus theory is that the free energy of activation displays a quadratic dependence on $\Delta G°$ and λ (ignoring work terms). Hence, the reaction rate may be written as

$$k_{et} = \nu_n \kappa \exp{[-(\lambda + \Delta G°)^2/4\lambda RT]}. \tag{6.21}$$

For intramolecular reactions, the nuclear frequency factor (ν_n) is $\sim 10^{13}$ s^{-1}. One of the most striking predictions of Marcus theory follows from this equation: as the driving force of the reaction increases, the reaction rate increases, reaching a maximum at $-\Delta G° = \lambda$; when $-\Delta G°$ is greater than λ, the rate decreases as the driving force increases (Figure 6.23). Two free-energy regions, depending on the relative magnitudes of $-\Delta G°$ and λ, are thus distinguished. The normal free-energy region is defined by $-\Delta G° < \lambda$. In this region, ΔG^* decreases if $-\Delta G°$ increases or if λ decreases. If $-\Delta G° = \lambda$, there is no free-

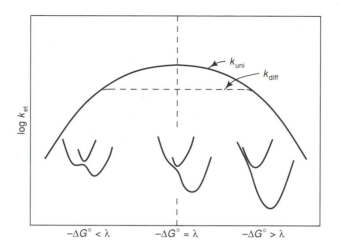

Figure 6.23
Plot of log k_{et} as a function of increasing driving force ($-\Delta G°$). Three $\Delta G°$ regions are indicated: normal ($-\Delta G° < \lambda$); activationless ($-\Delta G° = \lambda$); and inverted ($-\Delta G° > \lambda$). The corresponding two-well (E_R, E_P), diagrams also are shown. The dashed curve (k_{diff}) is for a bimolecular reaction. The predicted behavior of a unimolecular reaction (k_{uni}) is given by the solid curve; here the rate could be as high as 10^{13} s^{-1}, because it is not masked by diffusional processes.

energy barrier to the reaction. In the inverted region, defined by $-\Delta G° > \lambda$, ΔG^* increases if λ decreases or if $-\Delta G°$ increases.

Another widely used result of Marcus theory deals with the extraction of useful kinetic relationships for cross reactions from parameters for self-exchange reactions. Consider the cross reaction, Equation (6.22), for which the rate

$$A_1(ox) + A_2(red) \longrightarrow A_1(red) + A_2(ox) \tag{6.22}$$

and equilibrium constants are k_{12} and K_{12}, respectively. Two self-exchange reactions are pertinent here:

$$A_1(ox) + A_1(red) \longrightarrow A_1(red) + A_1(ox), \tag{6.23a}$$

$$A_2(ox) + A_2(red) \longrightarrow A_2(red) + A_2(ox). \tag{6.23b}$$

These reactions are characterized by rate constants k_{11} and k_{22}, respectively. The reorganization energy (λ_{12}) for the cross reaction can be approximated as the mean of the reorganization energies for the relevant self-exchange reactions:

$$\lambda_{12} = \tfrac{1}{2}(\lambda_{11} + \lambda_{22}). \tag{6.24}$$

Substitution of Equation (6.24) into Equation (6.17) leads to the relation

$$\Delta G_{12}^* = \tfrac{1}{2}(\Delta G_{11}^* + \Delta G_{22}^*) + \tfrac{1}{2}\Delta G_{12}^°(1 + \alpha), \tag{6.25a}$$

where

$$\alpha = \frac{\Delta G_{12}^{*}}{4(\Delta G_{11}^{*} + \Delta G_{22}^{*})}. \qquad (6.25b)$$

When the self-exchange rates k_{11} are corrected for work terms or when the latter nearly cancel, the cross-reaction rate k_{12} is given by the **Marcus cross relation**,

$$k_{12} = (k_{11}k_{22}K_{12}f_{12})^{1/2}, \qquad (6.26a)$$

where

$$\ln f_{12} = (\ln K_{12})^2/[4 \ln (k_{11}k_{22}/\nu_n^2)]. \qquad (6.26b)$$

This relation has been used to predict and interpret both self-exchange and cross-reaction rates (or even K_{12}), depending on which of the quantities have been measured experimentally. Alternatively, one could study a series of closely related electron-transfer reactions (to maintain a nearly constant λ_{12}) as a function of ΔG_{12}°; a plot of $\ln k_{12}$ vs. $\ln K_{12}$ is predicted to be linear, with slope 0.5 and intercept 0.5 $\ln (k_{11}k_{22})$. The Marcus prediction (for the normal free-energy region) amounts to a linear free-energy relation (LFER) for outer-sphere electron transfer.

1. Cross reactions of blue copper proteins

Given the measured self-exchange rate constant for stellacyanin ($k_{11} \sim 1.2 \times 10^5 \, M^{-1} \, s^{-1}$), the Marcus cross relation (Equation 6.26a) can be used to calculate the reaction rates for the reduction of Cu^{II}-stellacyanin by $Fe(EDTA)^{2-}$ and the oxidation of Cu^{I}-stellacyanin by $Co(phen)_3^{3+}$. $E^{\circ}(Cu^{2+/+})$ for stellacyanin is 0.18 V vs. NHE, and the reduction potentials and self-exchange rate constants for the inorganic reagents are given in Table 6.3.[66,67] For relatively small ΔE° values, f_{12} is ~ 1; here a convenient form of the Marcus cross relation is log $k_{12} = 0.5[\log k_{11} + \log k_{22} + 16.9\Delta E_{12}^{\circ}]$. Calculations with k_{11}, k_{22}, and ΔE_{12}° from experiments give k_{12} values that accord quite closely with the measured rate constants.

$$Cu^{II}St + Fe(EDTA)^{2-} \longrightarrow Cu^{I}St + Fe(EDTA)^{-}$$

$k_{12}(\text{calc.}) = 2.9 \times 10^5 \, M^{-1} \, s^{-1} \qquad (\Delta E_{12}^{\circ} = 0.06 \, V)$

$k_{12}(\text{obs.}) = 4.3 \times 10^5 \, M^{-1} \, s^{-1}$

$$Cu^{I}St + Co(phen)_3^{3+} \longrightarrow Cu^{II}St + Co(phen)_3^{2+}$$

$k_{12}(\text{calc.}) = 1.4 \times 10^5 \, M^{-1} \, s^{-1} \qquad (\Delta E_{12}^{\circ} = 0.19 \, V)$

$k_{12}(\text{obs.}) = 1.8 \times 10^5 \, M^{-1} \, s^{-1}$

The success of the Marcus cross relation with stellacyanin indicates that the copper site in the protein is accessible to inorganic reagents. The rate constants

Table 6.3
Reduction potentials and self-exchange rate
constants for inorganic reagents.

Reagent	E°(V vs. NHE)	$k_{22}(M^{-1} s^{-1})$
Fe(EDTA)$^{-/2-}$	0.12	6.9×10^4
Co(phen)$_3^{3+/2+}$	0.37	9.8×10^1

for the reactions of other blue copper proteins with inorganic redox agents show deviations from cross-relation predictions (Table 6.4).[68] These deviations suggest the following order of surface accessibilities of blue copper sites: stellacyanin > plastocyanin > azurin. Rate constants for protein-protein electron transfers also have been subjected to cross-relation analysis.[69]

Table 6.4
Reactions of blue copper proteins with inorganic reagents.

Protein	Reagent	k_{12}(obs.)[a]	$\Delta E_{12}°, V$	k_{11}(obs.)[a]	k_{11}(calc.)[a]
Stellacyanin	Fe(EDTA)$^{2-}$	4.3×10^5	0.064	1.2×10^5	2.3×10^5
	Co(phen)$_3^{3+}$	1.8×10^5	0.186	1.2×10^5	1.6×10^5
	Ru(NH$_3$)$_5$py^{3+}	1.94×10^5	0.069	1.2×10^5	3.3×10^5
Plastocyanin	Fe(EDTA)$^{2-}$	1.72×10^5	0.235	$\sim 10^3 - 10^4$	7.3×10^1
	Co(phen)$_3^{3+}$	1.2×10^3	0.009	$\sim 10^3 - 10^4$	1.1×10^4
	Ru(NH$_3$)$_5$py^{3+}	3.88×10^3	-0.100	$\sim 10^3 - 10^4$	4.9×10^4
Azurin	Fe(EDTA)$^{2-}$	1.39×10^3	0.184	2.4×10^6	2.8×10^{-2}
	Co(phen)$_3^{3+}$	2.82×10^3	0.064	2.4×10^6	7.0×10^3
	Ru(NH$_3$)$_5$py^{3+}	1.36×10^3	-0.058	2.4×10^6	1.1×10^3

[a]$M^{-1} s^{-1}$.

IV. LONG-RANGE ELECTRON TRANSFER IN PROTEINS

A. Electronic Coupling

The electron-transfer reactions that occur within and between proteins typically involve prosthetic groups separated by distances that are often greater than 10 Å. When we consider these distant electron transfers, an explicit expression for the electronic factor is required. In the nonadiabatic limit, the rate constant for reaction between a donor and acceptor held at fixed distance and orientation is:[70–73]

$$k_{et} = \frac{H_{AB}^2}{\hbar} \left[\frac{\pi}{\lambda RT} \right]^{1/2} \exp\left[\frac{-(\lambda + \Delta G°)^2}{4\lambda RT} \right]. \qquad (6.27)$$

The electronic (or tunneling) matrix element H_{AB} is a measure of the electronic coupling between the reactants and the products at the transition state. The magnitude of H_{AB} depends upon donor-acceptor separation, orientation, and the nature of the intervening medium. Various approaches have been used to test the validity of Equation (6.27) and to extract the parameters H_{AB} and λ. Driving-force studies have proven to be a reliable approach, and such studies have been emphasized by many workers.[73,74]

In the nonadiabatic limit, the probability is quite low that reactants will cross over to products at the transition-state configuration.[72] This probability depends upon the electronic hopping frequency (determined by H_{AB}) and upon the frequency of motion along the reaction coordinate.[75] In simple models, the electronic-coupling strength is predicted to decay exponentially with increasing donor-acceptor separation (Equation 6.28):[72,76]

$$H_{AB} = H_{AB}^{\circ} \exp\left[-\frac{\beta}{2}(\mathbf{d} - \mathbf{d}^{\circ})\right] \qquad (6.28)$$

In Equation (6.28), H_{AB}° is the electronic coupling at close contact ($\mathbf{d}^{\circ}$), and β is the rate of decay of coupling with distance ($\mathbf{d}$). Studies of the distance dependence of electron-transfer rates in donor-acceptor complexes, and of randomly oriented donors and acceptors in rigid matrices, have suggested $0.8 \leq \beta \leq 1.2$ Å^{-1}.[73,74,77,78]

Analysis of a large number of intramolecular electron-transfer rates has suggested a β value of 1.4 Å^{-1} for protein reactions (Figure 6.24).[79,80] Assigning a single protein β implies that the intervening medium is homogenous. At best this is a rough approximation, because the medium separating two redox sites in a protein is a heterogenous array of bonded and nonbonded interactions.[81–86] Beratan and Onuchic have developed a formalism that describes the medium in

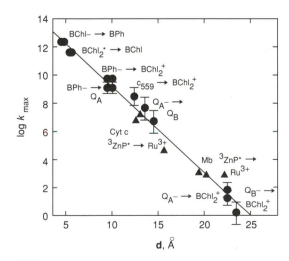

Figure 6.24
Maximum electron-transfer rate (k_{max}) vs. edge-to-edge distance (**d**) for proteins. Photosynthetic reaction center rates are shown as circles and $^3ZnP^*$ to Ru^{3+} rates in modified myoglobins and cytochromes c are shown as triangles. Adapted from Reference 80.

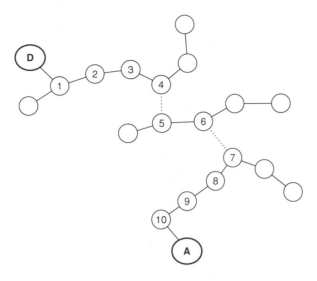

Figure 6.25
Example of a tunneling pathway.[79] The donor is coupled to the bonded pathway through bond 1 and the acceptor through bond 10. There are three bonded segments and two through-space jumps (between orbitals 4 and 5 and between orbitals 6 and 7).

terms of "unit blocks" connected together to form a tunneling pathway.[84-86] A unit block may be a covalent bond, a hydrogen bond, or a through-space jump, each with a corresponding decay factor. Dominant tunneling pathways in proteins are largely composed of bonded groups (e.g., peptide bonds), with less favorable through-space interactions becoming important when a through-bond pathway is prohibitively long (Figure 6.25).[84] The tunneling pathway model has been used successfully in an analysis of the electron-transfer rates in modified cytochromes c (Section IV.D.1).

1. Binding sites on the plastocyanin molecular surface

Plastocyanin cycles between the Cu^{II} and Cu^{I} oxidation states, and transfers electrons from cytochrome f to the P_{700} component of photosystem I in the chloroplasts of higher plants and algae.[87-89] The low molecular weight (10.5 kDa) and availability of detailed structural information[90] have made this protein an attractive candidate for mechanistic studies, which, when taken together,[87,91-94] point to two distinct surface binding sites (i.e., regions on the plastocyanin molecular surface at which electron transfer with a redox partner occurs). The first of these, the solvent-exposed edge of the Cu ligand His-87 (the adjacent site A in Figure 6.26), is ~6 Å from the copper atom and rather nonpolar. The second site (the remote site R in Figure 6.26) surrounds Tyr-83, and is much farther (~15 Å) from the copper center. Negatively charged carboxylates at positions 42–45 and 59–61 make this latter site an attractive one for positively charged redox reagents.

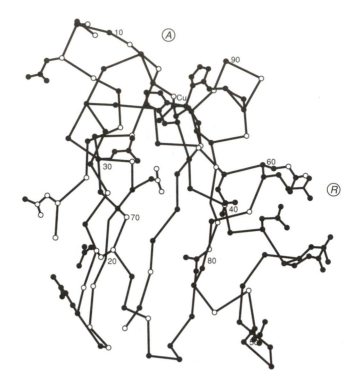

Figure 6.26
Structure of poplar plastocyanin illustrating the adjacent
(A) and remote (R) surface binding sites.

Bimolecular electron-transfer reactions are typically run under pseudo-first-order conditions (e.g., with an inorganic redox reagent present in $\sim$15-fold excess):

$$\text{Rate} = k[\text{plastocyanin}][\text{complex}] = k_{\text{obs}}[\text{plastocyanin}]. \qquad (6.29)$$

For some reactions [e.g., Co(phen)$_3^{3+}$ oxidation of plastocyanin (Cu^I)] the expected linear plot of k_{obs} vs. [complex] is not observed. Instead, the rate is observed to saturate (Figure 6.27).[95] A ''minimal'' model used to explain this behavior involves the two pathways for electron transfer shown in Equation (6.30).

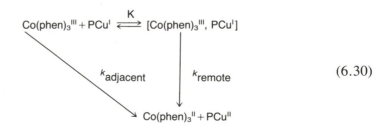

(6.30)

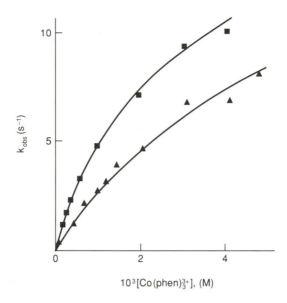

Figure 6.27
Dependence of first-order rate constants k_{obs} (25°C) on $[Co(phen)_3{}^{3+}]$ for the oxidation of plastocyanin PCu^I at pH 7.5. Key: ■, spinach; and ▲, parsley.[95]

Surprisingly, the rate ratio $k_{remote}/k_{adjacent}$ is 7.

Calculations[81] indicate that, despite the significant differences in distances, H_{AB} for the remote site is ~15 percent of H_{AB} for the adjacent site. This figure is much higher than would be expected from distance alone, suggesting that the value of the decay parameter β in Equation (6.28) depends strongly on the structure of the intervening medium.

B. Modified Metalloproteins

Chemical modification of structurally characterized metalloproteins by transition-metal redox reagents has been employed[52,53,96–98] to investigate the factors that control long-range electron-transfer reactions. In these semisynthetic multisite redox systems, the distance is fixed, and tunneling pathways between the donor and acceptor sites can be examined.

1. Ruthenium-modified myoglobin

Sperm-whale myoglobin can be reacted with $(NH_3)_5Ru(OH_2)^{2+}$ and then oxidized to produce a variety of ruthenated products,[52,99–101] including a His-48 derivative whose Ru ↔ Fe tunneling pathway is depicted in Figure 6.28. Electrochemical data (Table 6.5) indicate that the $(NH_3)_5Ru^{3+}$ group does not significantly perturb the heme center, and that equilibrium (i.e., $k_{obs} = k_1 + k_{-1}$) should be approached when a mixed-valent intermediate is produced by flash-

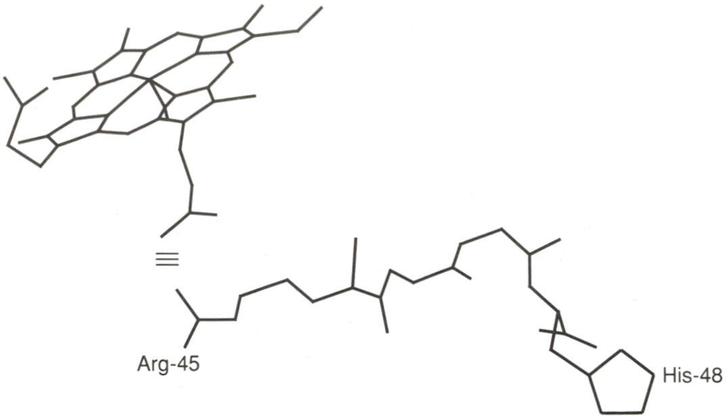

Figure 6.28
Electron-tunneling pathway for myoglobin modified at His-48.
The pathway moves along the protein backbone from His-48 to
Arg-45, and then to the heme via an H-bond ($\equiv$) to the heme
propionate. The His-48 to heme edge-edge distance is 12.7 Å.[101]

photolysis techniques:

$$(NH_3)_5Ru^{3+}\text{-}Mb(Fe^{3+}) \xrightarrow[e^-]{fast} (NH_3)_5Ru^{2+}\text{-}Mb(Fe^{3+}) \underset{k_{-1}}{\overset{k_1}{\rightleftarrows}} (NH_3)_5Ru^{3+}\text{-}Mb(Fe^{2+}) \quad (6.31)$$

This kinetic behavior was observed,[52] and both the forward (k_1) and reverse
(k_{-1}) reactions were found to be markedly temperature-dependent: $k_1 =$
0.019 s^{-1} (25°C), $\Delta H_1^\ddagger = 7.4$ kcal/mol, $k_{-1} = 0.041$ s^{-1} (25°C), $\Delta H_1^\ddagger =$
19.5 kcal/mol. X-ray crystallographic studies[102] indicate that the axial water
ligand dissociates upon reduction of the protein. This conformational change
does not control the rates, since identical results were obtained when a second
flash-photolysis technique[99] was used to generate $(NH_3)_5Ru^{3+}\text{-}Mb(Fe^{2+})$ in order to approach the equilibrium from the other direction.

Cyanogen bromide has been used[103] to modify the six-coordinate metmyoglobin heme site, causing the coordinated water ligand to dissociate. The CNBr-

Table 6.5
Thermodynamic parameters for the reduction of $(NH_3)_5Ru^{3+}$ and the heme
site in native and modified myoglobin (Mb).[a]

| | | Modified Mb | |
| | Native Mb | | |
Thermodynamic parameter	Fe$^{3+/2+}$	Fe$^{3+/2+}$	$(NH_3)_5Ru^{3+/2+}$
E°, mV vs. NHE (25°C)	58.8 ± 2	65.4 ± 2	85.8 ± 2
$\Delta G°$, kcal mol^{-1} (25°C)	−1.26 ± 0.05	−1.51 ± 0.05	−1.98 ± 0.05
$\Delta S°$, e.u.	−39.2 ± 1.2	−37.6 ± 1.2	4.2 ± 1.2
$\Delta H°$, kcal mol^{-1} (25°C)	−13.0 ± 0.4	−12.7 ± 0.4	−0.7 ± 0.4

[a] pH 7.0, $\mu = 0.1$ M phosphate buffer.

modified myoglobin heme site is thus five-coordinate in *both* oxidation states. As expected, the self-exchange rate increased from $\sim 1 \ M^{-1} \ s^{-1}$ to $\sim 10^4 \ M^{-1} \ s^{-1}$.

Recent efforts in modeling biological electron transfers using chemically modified redox proteins[104–106] point the way toward the design of semisynthetic redox enzymes for catalytic applications. An intriguing example, termed flavohemoglobin, was produced by reaction of hemoglobin with a flavin reagent designed to react with Cys-93 of the β-chain (i.e., the hemoglobin molecule was modified by two flavin moieties).[107] The resulting derivative, unlike native hemoglobin, accepts electrons directly from NADPH and catalyzes the *para*-hydroxylation of aniline in the presence of O_2 and NADPH.

C. Protein-Protein Complexes

In physiologically relevant precursor complexes, both redox centers are frequently buried in protein matrices. Characterization of such protein-protein complexes is clearly important, and several issues figure prominently:

(1) What are the ''rules'' that govern complex formation? How important are protein-dipole/protein-dipole interactions, intermolecular hydrogen bonding, and hydrophobic interactions?

(2) Are the water (and small solute) molecules associated with protein surfaces ''squeezed'' out of the interfacial region upon complex formation?

(3) Within a given complex, is there a high degree of structural order, or do the proteins retain some independent mobility?

Most of our knowledge about the structures of protein-protein complexes comes from crystallographic studies[108–110] of antigen-antibody complexes and multisubunit proteins; such systems generally exhibit a high degree of thermodynamic stability. On the other hand, complexes formed as a result of bimolecular collisions generally are much less stable, and tend to resist attempts to grow x-ray-quality crystals; the high salt conditions typically used in protein crystallizations often lead to dissociation of such complexes.

1. Cytochrome b_5-cytochrome c

One of the most widely studied protein-protein complexes is that formed between mammalian cytochrome b_5 and cytochrome c. Using the known x-ray structures of both proteins, Salemme[111] generated a static computer graphics model of this electron-transfer complex by docking the x-ray structures of the individual proteins. Two features of this model and its revision[112] by molecular dynamics simulations (Figure 6.29 *See color plate section, page C-12.*) are noteworthy: (1) several Lys residues on cytochrome c and carboxylate-containing groups on cytochrome b_5 form ''salt bridges'' (i.e., intermolecular hydrogen

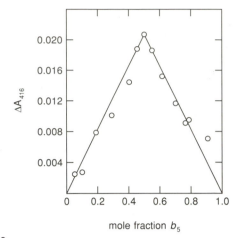

Figure 6.30
Job plot of the change in absorbance at 416 nm due to complex formation
between cytochrome b_5 and cytochrome c (25°C, pH 7.0 (phosphate),
$\mu = 1$ mM, 10.54 μM total protein concentration).[114]

bonds); and (2) the hemes are nearly coplanar and are ~17 Å (Fe-Fe) apart.
This distance was confirmed by an energy-transfer experiment[113] in which the
fluorescence of Zn-substituted cytochrome c was quenched by cytochrome b_5.
Spectroscopic studies[114,115] have verified the suggestion that these proteins form
a 1:1 complex at low ionic strength (Figure 6.30). In addition, chemical
modification[116] and spectroscopic analyses[117–119] are all in agreement with the
suggestion[111,112] that the complex is primarily stabilized by electrostatic interactions of the ($-NH_3^+ \cdots {}^-O_2C-$) type. The effect of ionic strength on the
reduction of cytochrome c by cytochrome b_5 is also in accord with this picture:[120] lowering the ionic strength increases the reaction rate, as expected for
oppositely charged molecules.

2. Hybrid hemoglobins

A common[52,55,57,121,122] experimental strategy for studying electron transfers
between proteins uses a metal-substituted heme protein as one of the reactants.
In particular, the substitution of zinc for iron in one of the porphyrin redox
centers allows facile initiation of electron transfer through photoexcitation of the
zinc porphyrin (ZnP). The excited zinc porphyrin, ^{3}ZnP* in Equation (6.32),
may decay back ($k_d \sim 10^2$ s^{-1}) to the ground state or transfer an electron to an
acceptor.

$$[^3ZnP^*, Acceptor_{ox}] \xrightarrow{k_f} [ZnP^+, Acceptor_{red}]$$

$$hv \uparrow \quad \updownarrow k_d \qquad \swarrow k_b \qquad (6.32)$$

$$[ZnP, Acceptor_{ox}]$$

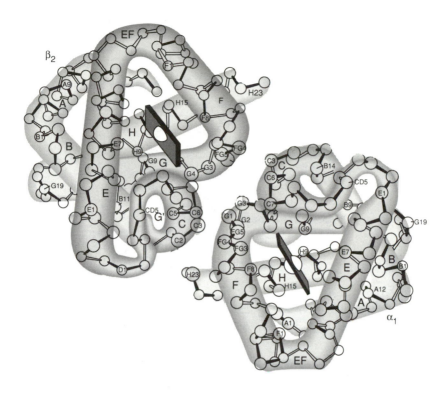

Figure 6.31
Structure of the α_1 and β_2 subunits of hemoglobin.
The edge-edge separation of the two hemes is 20 Å.

The ZnP$^+$ cation radical produced in the k_f step is a powerful oxidant; back electron transfer (k_b) will thus occur and regenerate the starting material.

The reactions shown in Equation (6.32) have been investigated in mixed-metal [Zn, Fe] hemoglobins.[123–125] A hemoglobin molecule can be viewed as two independent electron-transfer complexes, each consisting of an α_1-β_2 sub-unit pair (Figure 6.31), since the α_1-α_2, β_1-β_2, and α_1-β_1 distances are prohibitively long (> 30 Å).

Both [α(Zn), β(Fe)] and [α(Fe), β(Zn)] hybrids have been studied. The ZnP and FeP are nearly parallel, as in the cytochrome b_5-cytochrome c model complex. Long-range electron transfer (^{3}ZnP* $\rightarrow$ Fe^{3+}) between the α_1 and β_2 sub-units has been observed (the heme-edge/heme-edge distance is ~20 Å). The driving force for the forward electron-transfer step is ~0.8 eV, and k_f (see Equation 6.32) is ~100 s^{-1} at room temperature, but decreases to ~9 s^{-1} in the low-temperature region (Figure 6.32). Below 140–160 K the vibrations that induce electron transfer "freeze out"; nuclear tunneling is usually associated with such slow, temperature-independent rates. A complete analysis of the full temperature dependence of the rate requires a quantum-mechanical treatment[126,127] of λ_i rather than that employed in the Marcus theory. It is interesting to note that the heme b vinyl groups (see Figure 6.6) for a given [α_1(Fe), β_2(Zn)] hybrid point toward each other and appear[125] to facilitate electron transfer.

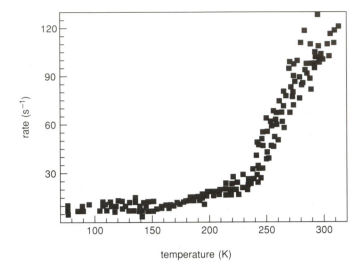

Figure 6.32
Temperature dependence of the forward electron-transfer rate, k_f, for [α(Zn), β(Fe^{III}H$_2$O)]; adapted from Reference 124.

D. Cytochrome *c*

Cytochrome *c* occupies a prominent place in the mitochondrial electron-transport chain. Its water solubility, low molecular weight (12.4 kDa), stability, and ease of purification have allowed many experiments, which, when taken together, present a detailed picture of the structure and biological function of this electron carrier.[128-133]

X-ray structures[134] of oxidized and reduced tuna cytochrome *c* are very similar; most of the differences are confined to changes in the orientations of the side chains of some surface-exposed amino acids and sub-Ångström adjustments of some groups in the protein interior. Upon reduction, the heme active site becomes slightly more ordered (Figure 6.33). Two-dimensional NMR studies[135-137] confirm this interpretation of the x-ray data, and further establish that the crystal and solution structures of cytochrome *c* differ in only minor respects.

Cytochrome *c* exhibits several pH-dependent conformational states. In particular, an alkaline transition with a $pK_a \sim 9.1$ has been observed for ferricytochrome *c*. This transition is believed to be associated with the dissociation of Met-80; the reduction potential decreases dramatically,[138] and the 695-nm absorption band, associated with a sulfur $\rightarrow$ iron charge-transfer transition, disappears. The ^{2}H NMR resonance due to (^{2}H$_3$C-) Met-80 in deuterium-enriched ferricytochrome *c* disappears from its hyperfine-shifted upfield position without line broadening, and reappears coincident with the (^{2}H$_3$C-)Met-65 resonance.[139] In contrast, ferrocytochrome *c* maintains an ordered structure over the pH range 4 to 11.[140] The heme iron in ferricytochrome *c* remains low-spin throughout this

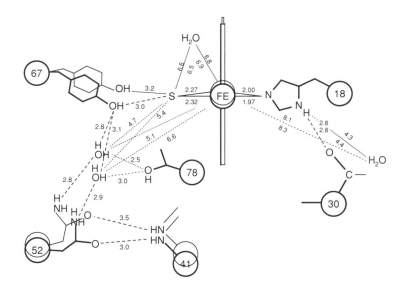

Figure 6.33
Side-chain motion in the vicinity of the heme of cytochrome c. Heavy lines
indicate the reduced molecule, and light lines the oxidized molecule.[134]

transition, and a new strong-field ligand must therefore replace Met-80. It has
been suggested that an ε-amino nitrogen of a nearby Lys provides the new donor
atom, but this has not been confirmed. However, it is clear that reduction of
ferricytochrome c at alkaline pH values below 11 causes a drastic conformational change at the heme site. The unknown sixth ligand must be displaced by
Met-80 in order for the reduced protein to assume a structure similar to the one
at neutral pH. This structural change is accompanied by a decrease in the rate
of reduction of ferricytochrome c by hydrated electrons,[141] as expected.

How does the protein control the reduction potential of the iron center in
cytochrome c? Factors that appear to play a role include the nature of the axial
ligands, the stability and solvent accessibility of the heme crevice, and the hy-
drophobicities of the amino acids that line the heme crevice. These issues have
been addressed theoretically[142,143] and experimentally[144-149] using cytochrome c
variants engineered by protein semisynthesis or site-directed mutagenesis. Results for horse heart cytochrome c are set out in Table 6.6. Point mutations at
either of positions 78 or 83 do not significantly alter $E^{0'}$; however, the double
mutant (Thr-78 $\rightarrow$ Asn-78; Tyr-83 $\rightarrow$ Phe-83) exhibits a substantially lower redox potential. Evidently, the results of such changes are not necessarily additive; great care must be taken in drawing conclusions about structure-function
relations in engineered proteins. Finally, the $\sim$310 mV difference between the
values for the heme octapeptide and the native protein (the axial ligands are the
same in both) provides a dramatic illustration of protein environmental effects
on the redox potential: shielding the heme from the solvent is expected to stabilize Fe^{II} and therefore result in an increase in $E^{0'}$.

Table 6.6
Reduction potentials[a] of horse heart cytochrome c.

Cytochrome	E°′(mV vs. NHE)	Reference
Native	262	138
Met-80 → His-80	41	144
Tyr-67 → Phe-67	225	145
Thr-78 → Asn-78	264	145
Tyr-83 → Pro-83	266	145
Thr-78, Tyr-83 → Asn-78, Pro-83	235	145
Heme octapeptide[b]	− 50	146

[a] pH 7.0, 25°C.
[b] In 2M N-acetyl-DL-methionine.

During the last fifteen years, much has been learned about the interaction of cytochrome c with its redox partners.[128–133] Cytochrome c is a highly basic protein (pI = 10.05); lysine residues constitute most of the cationic amino acids. Despite the indication from the x-ray structures that only ~1 percent of the heme surface is solvent-exposed, the asymmetric distribution of surface charges, particularly a highly conserved ring of Lys residues surrounding the exposed edge of the heme crevice, led to the suggestion that electron-transfer reactions of cytochrome c (and other Class I cytochromes as well) occur via the exposed heme edge.

Chemical modification of the surface Lys residues of cytochrome c has afforded opportunities to alter the properties of the surface ε-amino groups suspected to be involved in precursor complex formation. Margoliash and coworkers[133,150,151] used a 4-carboxy-2,6-dinitrophenol (CDNP) modification of the Lys residues to map out the cytochrome c interaction domains with various transition-metal redox reagents and proteins. These experiments have shown that cytochrome c interacts with inorganic redox partners near the exposed heme edge.

Numerous studies[129,152,153] of cytochrome c with physiological reaction partners are in accord with electrostatic interactions featured in the model cytochrome c/cytochrome b_5 complex discussed earlier. Similar types of interactions have been proposed for cytochrome c/flavodoxin[154] and cytochrome c/cytochrome c peroxidase complexes.[155] (Recent x-ray crystal structure work[155b] has shed new light on this problem.) Theoretical work[156] additionally suggests that electrostatic forces exert torques on diffusing protein reactants that "steer" the proteins into a favorable docking geometry. However, the domains on cytochrome c for interaction with physiological redox partners are not identical, as Figure 6.34 illustrates.

Reactions between cytochrome c and its physiological redox partners at low ionic strength generally are very fast, $\sim 10^8 \ M^{-1} \ s^{-1}$, even though the thermodynamic driving force may be as low as 20 mV, as it is for the reduction of

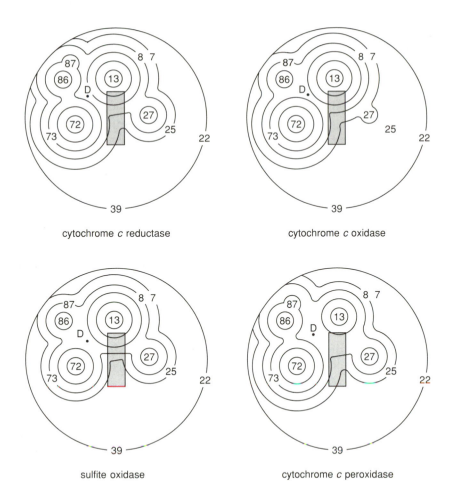

Figure 6.34
Domains on cytochrome c for interaction with physiological redox partners. The heme edge is represented by the shaded rectangle. The number of circles around a given Lys residue is proportional to the percentage of the observed inhibition in redox assays when the corresponding CDNP-modified cytochrome c is used.[151]

cytochrome a in cytochrome c oxidase. Such rates are probably at the diffusion-controlled limit for such protein-protein reactions.[157,158] A more detailed understanding of these reactions will require studies that focus on the dynamical (rather than static) features of complexes of cytochrome c with other proteins. For example, there is evidence[159] that a cytochrome c conformational change in the vicinity of the heme edge accompanies the formation of the complex with cytochrome c oxidase. Studies of the influence of geometry changes on activation energies[52,60,160] are of particular importance in elucidating the mechanisms of protein-protein reactions.

1. Ruthenium-modified cytochrome c

Intramolecular electron transfer in cytochrome c has been investigated by attaching photoactive Ru complexes to the protein surface.[98,161] $Ru(bpy)_2(CO_3)$ (bpy $= 2,2'$-bipyridine) has been shown to react with surface His residues to yield, after addition of excess imidazole (im), $Ru(bpy)_2(im)(His)^{2+}$. The protein-bound Ru complexes are luminescent, but the excited states ($*Ru^{2+}$) are rather short lived ($\tau \leq 100$ ns). When direct electron transfer from $*Ru^{2+}$ to the heme cannot compete with excited-state decay, electron-transfer quenchers (e.g., $Ru(NH_3)_6^{3+}$) are added to the solution to intercept a small fraction (1–10%) of the excited molecules, yielding (with oxidative quenchers) Ru^{3+}. If, before laser excitation of the Ru site, the heme is reduced, then the Fe^{2+} to Ru^{3+} reaction (k_{et}) can be monitored by transient absorption spectroscopy. The k_{et} values for five different modified cytochromes have been reported: (Ru(His-33), $2.6(3) \times 10^6$; Ru(His-39), $3.2(4) \times 10^6$; Ru(His-62), $1.0(2) \times 10^4$; Ru(His-72), $9.0(3) \times 10^5$; and Ru(His-79), $> 10^8$ s^{-1}).[162,163]

According to Equation (6.27), rates become activationless when the reaction driving force ($-\Delta G^\circ$) equals the reorganization energy (λ). The driving force (0.74 eV) is approximately equal to the reorganization energy (0.8 eV) estimated for the $Ru(bpy)_2(im)(His)$-cyt c reactions.[161] The activationless (maximum) rates (k_{max}) are limited by H_{AB}^2, where H_{AB} is the electronic matrix element that couples the reactants and products at the transition state. Values of k_{max} and H_{AB} for the Fe^{2+} to Ru^{3+} reactions are given in Table 6.7.

Calculations that explicitly include the structure of the intervening medium[81–86,164–169] have been particularly helpful in developing an understanding of distant electronic couplings. As discussed in Section IV.A, the couplings in proteins can be interpreted in terms of pathways comprised of covalent, H-bond, and through-space contacts. An algorithm has been developed[85,170] that searches a

Table 6.7
Electron-transfer parameters[163] for Ru(bpy)$_2$(im)(His-X)-cytochromes c.

X	k_{max} (s^{-1})	H_{AB} (cm^{-1}) [Fe^{2+}—Ru^{3+}]	$\mathbf{d}$ (Å)	n_{eff}[a]	σ l(Å)
79	$> 1.0 \times 10^8$	> 0.6	4.5	8 (8C)	11.2
39	3.3×10^6	0.11	12.3	14.0 (11C)	19.6
				(1H)	
33	2.7×10^6	0.097	11.1	13.9 (11C)	19.5
				(1H)	
72	9.4×10^5	0.057	8.4	17.6 (7C)	24.6
				(1S)	
62	1.0×10^4	0.006	14.8	20.6 (16C)	28.8
				(2H)	

[a] C = covalent bond, H = hydrogen bond, S = space jump.

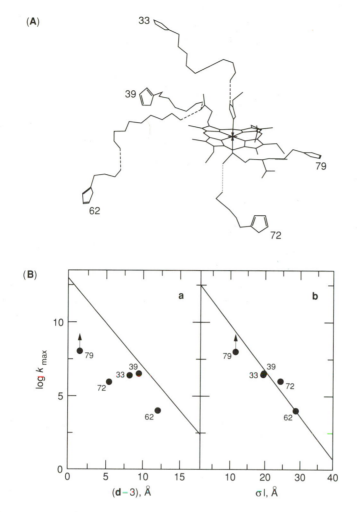

Figure 6.35
(A) Electronic coupling pathways to the heme from Ru-modified residues in native (His-33 horse heart, His-39 yeast), genetically engineered (His-62 yeast), and semisynthetic (His-72, His-79 horse heart) cytochromes c. Solid lines are covalent bonds; dashed lines are hydrogen bonds; and the dotted line (His-72 pathway) is a space jump. (B) The left half of the diagram (a) shows maximum electron-transfer rate vs. **d** minus 3 Å (van der Waals contact). Exponential-decay line with 1×10^{13} s^{-1} intercept and 1.4 Å^{-1} slope. The right half of the diagram (b) shows maximum rate vs. σl: 0.71 Å^{-1} slope; 3×10^{12} s^{-1} intercept. Adapted from References 162 and 163.

protein structure for the best pathways coupling two redox sites (the pathways between the histidines (33, 39, 62, 72, 79) and the heme are shown in Figure 6.35). A given coupling pathway consisting of covalent bonds, H-bonds, and through-space jumps can be described in terms of an equivalent covalent pathway with an effective number of covalent bonds (n_{eff}). Multiplying the effective number of bonds by 1.4 Å/bond gives σ-tunneling lengths (σl) for the five pathways (Table 6.7) that correlate well with the maximum rates (one-bond limit set at 3×10^{12} s^{-1}; slope of 0.71 Å^{-1}) (Figure 6.35). The 0.71 Å^{-1} decay accords closely with related distance dependences for covalently coupled donor-acceptor molecules.[73,77]

E. Bacterial Photosynthetic Reaction Centers

Photosynthetic bacteria produce only one type of reaction center, unlike green plants (which produce two different kinds linked together in series), and are therefore the organisms of choice in photosynthetic electon-transfer research.[171–176] As indicated in Section I.B, the original reaction center structure (Figure 6.15) lacked a quinone (Q_B). Subsequent structures for reaction centers from other photosynthetic bacteria[177,178] contain this quinone (Figure 6.36 *See color plate section, page C-13.*). The *Rps. sphaeroides* reaction center contains ten cofactors and three protein subunits. (Note that the *Rps. viridis* structure contains a cytochrome subunit as well.) The cofactors are arrayed so that they nearly span the 40-Å-thick membrane (Figure 6.37 *See color plate section, page C-13.*). The iron atom is indicated by the red dot near the cytoplasmic side of the membrane (bottom). In spite of the near two-fold axis of symmetry, electron transfer proceeds along a pathway that is determined by the A branch. In particular, $BChl_B$ and $BPhe_B$ do not appear to play an important role in the electron transfers.

It was demonstrated long ago that $(BChl)_2$ is the primary electron donor and that ubiquinone (or metaquinone) is the ultimate electron acceptor. Transient flash photolysis experiments indicate that several electron-transfer steps occur in order to translocate the charge across the membrane (Figure 6.38). Curiously, the high-spin ferrous iron appears to play no functional role in the Q_A to Q_B electron transfer.[179] In addition, the part played by $BChl_A$ is not understood—it may act to promote reduction of $BPhe_A$ via a superexchange mechanism.[180,181] Cytochromes supply the reducing equivalents to reduce the special pair $(BChl)_2^+$.

Estimated rate constants for the various electron-transfer steps, together with approximate reduction potentials, are displayed in Figure 6.39. For each step, the forward rate is orders of magnitude faster than the reverse reaction. The rapid rates suggest that attempts to obtain x-ray structures of intermediates (especially the early ones!) will not be successful. However, molecular dynamics methods are being explored in computer simulations of the structures of various intermediates.[182,183] Within a few years we may begin to understand why the initial steps are so fast.

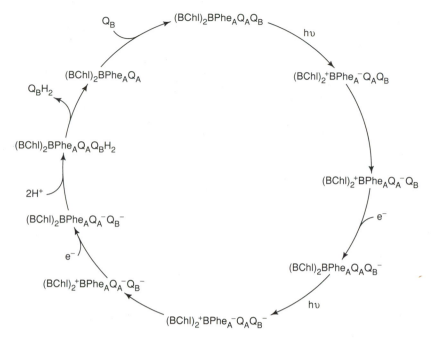

Figure 6.38
Forward electron transfer through the reaction center. Note that *two* charge translocations must occur in order for the (labile) quinol Q_BH_2 to be produced. Once Q_BH_2 dissociates from the RC and is replaced by another, *oxidized*, Q_B, the cycle can begin anew.

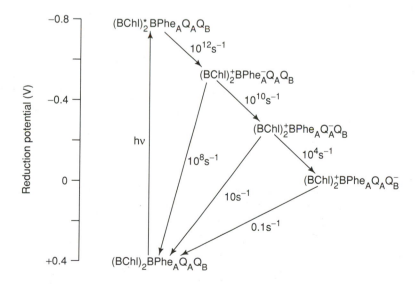

Figure 6.39
Electron-transfer rates (for forward and reverse reactions) and reduction potentials for RC intermediates.

V. REFERENCES

1. R. C. Prince, *Trends Biochem. Sci.* **13** (1988), 286–288.
2. R. C. Prince and G. N. George, *Trends Biochem. Sci.* **15** (1990), 170–172.
3. E. T. Adman, *Biochim. Biophys. Acta* **549** (1979), 107–144.
4. H. B. Gray and B. G. Malmström, *Comments Inorg. Chem.* **2** (1983), 203–209.
5. T. E. Meyer and M. A. Cusanovich, *Biochim. Biophys. Acta* **975** (1989), 1–28.
6. R. P. Simondsen and G. Tollin, *Mol. Cellular Biochem.* **33** (1980), 13–24.
7. C. T. W. Moonen and F. Müller, *Biochemistry* **21** (1982), 408–414.
8. H. B. Gray and E. I. Solomon, in T. G. Spiro, ed., *Copper Proteins*, Wiley, 1981, pp. 1–39.
9. E. T. Adman, in P. Harrison, ed., *Metalloproteins*, Macmillan, 1985, Part I, pp. 1–42.
10. O. Farver and I. Pecht, in R. Lontie, ed., *Copper Proteins and Copper Enzymes*, CRC Press, 1984, vol. I, pp. 183–214.
11. G. E. Norris, B. F. Anderson, and E. N. Baker, *J. Am. Chem. Soc.* **108** (1986), 2784–2785.
12. E. W. Ainscough *et al.*, *Biochemistry* **26** (1987), 71–82.
13. W. Lovenberg, ed., *Iron-Sulfur Proteins*, Academic Press, 1973–1977, 3 vols.
14. T. G. Spiro, ed., *Iron-Sulfur Proteins*, Wiley, 1982.
15. K. Fukuyama *et al.*, *Nature* **286** (1980), 522–524.
16. J. Rawlings, O. Siiman, and H. B. Gray, *Proc. Natl. Acad. Sci. USA* **71** (1974), 125–127.
17. C. W. Carter *et al.*, *Proc. Natl. Acad. Sci. USA* **69** (1972), 3526–3529.
18. G. R. Moore, G. W. Pettigrew, and N. K. Rogers, *Proc. Natl. Acad. Sci. USA* **83** (1986), 4998–4999.
19. T. E. Meyer and M. D. Kamen, *Adv. Protein Chem.* **35** (1982), 105–212.
20. G. R. Moore and G. W. Pettigrew, *Cytochromes c*, Springer-Verlag, 1984.
21. R. E. Dickerson, *Sci. Amer.* **242** (1980), 137–153.
22. F. S. Mathews, *Prog. Biophys. Mol. Biol.* **45** (1985), 1–56.
23. R. P. Ambler, in A. B. Robinson and N. D. Kaplan, eds., *From Cyclotrons to Cytochromes*, Academic Press, 1980, pp. 263–279.
24. K. Kimura *et al.*, *J. Chem. Phys.* **70** (1979), 3317–3323.
25. D. Keilin, *Proc. Roy. Soc. (London)* **B98** (1925), 312–339.
26. Y. Hatefi, *Annu. Rev. Biochem.* **54** (1985), 1015–1069.
27. B. P. S. N. Dixit and J. M. Vanderkooi, *Curr. Top. Bioenergetics* **13** (1984), 159–202.
28. C. W. Jones, *Biological Energy Conservation*, Chapman and Hall, 2d ed., 1981.
29. C. R. Hackenbrock, *Trends Biochem. Sci.* **6** (1981), 151–154.
30. S. Gupte *et al.*, *Proc. Natl. Acad. Sci. USA* **81** (1984), 2606–2610.
31. R. K. Clayton, *Photosynthesis: Physical Mechanisms and Chemical Patterns*, Cambridge Univ. Press, 1980.
32. F. K. Fong, ed., *Light Reaction Path of Photosynthesis*, Springer-Verlag, 1982.
33. R. K. Clayton and W. R. Sistrom, eds., *The Photosynthetic Bacteria*, Plenum, 1978.
34. *Photosynthesis, Vol. I: Energy Conversion by Plants and Bacteria*, Academic Press, 1982.
35. B. W. Matthews and R. E. Fenna, *Acc. Chem. Res.* **13** (1980), 309–317.
36. J. Deisenhofer *et al.*, *J. Mol. Biol.* **180** (1984), 385–398.
37. R. Hille and V. Massey, in T. G. Spiro, ed., *Molybdenum Enzymes*, Wiley, 1985, pp. 443–518.
38. R. Hille, W. R. Hagen, and W. R. Dunham, *J. Biol. Chem.* **260** (1985), 10569–10575.
39. A. Bhattacharyya *et al.*, *Biochemistry* **22** (1983), 5270–5279.
40. J. Stubbe, *Adv. Enzymology* **63** (1990), 349–419.
41. P. Nordlund, B.-M. Sjöberg, and H. Eklund, *Nature* **345** (1990), 593–598.
42. R. E. White and M. J. Coon, *Annu. Rev. Biochem.* **49** (1980), 315–356.
43. S. G. Sligar and R. I. Murray, in P. R. Ortiz de Montellano, ed., *Cytochrome P-450: Structure, Mechanism, and Biochemistry*, Plenum, 1986, pp. 429–503.
44. C. B. Brewer and J. A. Peterson, *J. Biol. Chem.* **263** (1988), 791–798.
45. F. A. Armstrong, H. A. O. Hill, and N. J. Walton, *Quart. Rev. Biophys.* **18** (1986), 261–322.
46. G. D. Hitchens, *Trends Biochem. Sci.* **14** (1989), 152–155.
47. F. A. Armstrong, *Structure and Bonding* **72** (1989), 137–221; A. Heller, *J. Phys. Chem.* **96** (1992), 3579–3587.
48. H. A. O. Hill *et al.*, *J. Chem. Soc. Chem. Commun.* (1985), 1469–1471.
49. H. Taube, *Angew. Chem. Int. Ed. Engl.* **23** (1984), 329–334.
50. R. D. Cannon, *Electron-Transfer Reactions*, Butterworths, 1980.

51. *Prog. Inorg. Chem.* **30** (1983).
52. S. L. Mayo *et al.*, *Science* **233** (1986), 948–952.
53. A. G. Sykes, *Chem. Brit.* **24** (1988), 551–554.
54. M. Faraggi, M. R. De Felippis, and M. H. Klapper, *J. Am. Chem. Soc.* **111** (1989), 5141–5145.
55. G. McLendon, *Acc. Chem. Res.* **21** (1988), 160–167.
56. O. Farver and I. Pecht, *Proc. Natl. Acad. Sci. USA* **86** (1989), 6968–6972.
57. S. E. Peterson-Kennedy *et al.*, *Coord. Chem. Rev.* **64** (1985), 125–133.
58. D. E. Richardson, *Comments Inorg. Chem.* **3** (1985), 367–384.
59. S. Dahlin, B. Reinhammar, and M. T. Wilson, *Biochem. J.* **218** (1984), 609–614.
60. R. A. Marcus, *Annu. Rev. Phys. Chem.* **15** (1964), 155–196.
61. W. L. Reynolds and R. W. Lumry, *Mechanisms of Electron Transfer*, Ronald Press, 1966.
62. J. Ulstrup, *Charge Transfer Process in Condensed Media*, Springer-Verlag, 1979.
63. N. Sutin, *Acc. Chem. Res.* **15** (1982), 275–282.
64. M. D. Newton and N. Sutin, *Annu. Rev. Phys. Chem.* **35** (1984), 437–480.
65. P. Bertrand, *Biochimie* **68** (1986), 619–628.
66. J. V. McArdle *et al.*, *J. Am. Chem. Soc.* **99** (1977), 2483–2489.
67. S. Wherland *et al.*, *J. Am. Chem. Soc.* **97** (1975), 5260–5262.
68. A. G. Mauk, R. A. Scott, and H. B. Gray, *J. Am. Chem. Soc.* **102** (1980), 4360–4363.
69. S. Wherland and I. Pecht, *Biochemistry* **17** (1978), 2585–2591.
70. B. S. Brunschwig and N. Sutin, *Comments Inorg. Chem.* **6** (1987), 209–235.
71. N. Sutin, in J. J. Zuckerman, ed., *Inorganic Reactions and Methods*, VCH Publishers, 1986, **XV,** 23–24.
72. R. A. Marcus and N. Sutin, *Biochim. Biophys. Acta* **811** (1985), 265–322.
73. J. R. Winkler and H. B. Gray, *Chem. Rev.* **92** (1992), 369–379.
74. H. B. Gray and J. R. Winkler, *Pure Appl. Chem.* **64** (1992), 1257–1262.
75. J. Jortner and M. Bixon, *J. Chem. Phys.* **88** (1988), 167–170.
76. J. J. Hopfield, *Proc. Natl. Acad. Sci. USA* **71** (1974), 3640–3644.
77. G. L. Closs and J. R. Miller, *Science* **240** (1988), 440–447.
78. H. Oevering *et al.*, *J. Am. Chem. Soc.* **109** (1987), 3258–3269.
79. C. C. Moser *et al.*, *Nature* **355** (1992), 796–802.
80. C. C. Moser and P. L. Dutton, *Biochim. Biophys. Acta* **1101** (1992), 171–176.
81. H. E. M. Christensen *et al.*, *Inorg. Chem.* **29** (1990), 2808–2816.
82. M. A. Ratner, *J. Phys. Chem.* **94** (1990), 4877–4883.
83. S. Larsson, *Chem. Scripta* **28A** (1988), 15–20.
84. D. N. Beratan and J. N. Onuchic, *Photosynthesis Res.* **22** (1989), 173–186.
85. D. N. Beratan *et al.*, *J. Am. Chem. Soc.* **112** (1990), 7915–7921.
86. J. N. Onuchic *et al.*, *Annu. Rev. Biophys. Biomol. Struct.* **21** (1992), 349–377.
87. A. G. Sykes, *Chem. Soc. Rev.* **14** (1985), 283–315.
88. H. C. Freeman, in J. L. Laurent, ed., *Coordination Chemistry-21*, Pergamon Press, 1981, pp. 29–51.
89. D. Boulter *et al.*, in D. H. Northcote, ed., *Plant Biochemistry*, Univ. Park Press, 1977, **II,** 1–40.
90. J. M. Guss *et al.*, *J. Mol. Biol.* **192** (1986), 361–387.
91. J. McGinnis *et al.*, *Inorg. Chem.* **27** (1988), 2306–2312.
92. O. Farver and I. Pecht, *Proc. Natl. Acad. Sci. USA* **78** (1981), 4190–4193.
93. O. Farver, Y. Shahak, and I. Pecht, *Biochemistry* **21** (1982), 1885–1890.
94. B. S. Brunschwig *et al.*, *Inorg. Chem.* **24** (1985), 3743–3749.
95. S. K. Chapman *et al.*, in M. Chisholm, ed., *Inorganic Chemistry: Toward the Twenty-first Century*, American Chemical Society, 1983, pp. 177–197.
96. H. B. Gray, *Chem. Soc. Rev.* **15** (1986), 17–30.
97. D. W. Conrad and R. A. Scott, *J. Am. Chem. Soc.* **111** (1989), 3461–3463.
98. L. P. Pan *et al.*, *Biochemistry* **27** (1988), 7180–7184.
99. C. M. Lieber, J. L. Karas, and H. B. Gray, *J. Am. Chem. Soc.* **109** (1987), 3778–3779.
100. J. L. Karas, C. M. Lieber, and H. B. Gray, *J. Am. Chem. Soc.* **110** (1988), 599–600.
101. J. A. Cowan *et al.*, *Ann. N.Y. Acad. Sci.* **550** (1988), 68–84.
102. T. Takano, *J. Mol. Biol.* **110** (1977), 537–568 and 559–584.
103. K. Tsukahara, *J. Am. Chem. Soc.* **111** (1989), 2040–2044.
104. E. T. Kaiser and D. S. Lawrence, *Science* **226** (1984), 505–511.
105. R. E. Offord, *Protein Eng.* **1** (1987), 151–157.
106. E. T. Kaiser, *Angew. Chem. Int. Ed. Engl.* **27** (1988), 913–922.

107. T. Kokubo, S. Sassa, and E. T. Kaiser, *J. Am. Chem. Soc.* **109** (1987), 606–607; J. Kuriyan *et al.*, *J. Am. Chem. Soc.* **110** (1988), 6261–6263.
108. C. Chothia and J. Janin, *Nature* **256** (1975), 705–708.
109. A. G. Amit *et al.*, *Science* **233** (1986), 747–753.
110. H. M. Geysen *et al.*, *Science* **235** (1987), 1184–1190.
111. F. R. Salemme, *J. Mol. Biol.* **102** (1976), 563–568.
112. J. J. Wendoloski *et al.*, *Science* **238** (1987), 794–797.
113. G. L. McLendon *et al.*, *J. Am. Chem. Soc.* **107** (1985), 739–740.
114. M. R. Mauk, L. S. Reid, and A. G. Mauk, *Biochemistry* **21** (1982), 1843–1846.
115. J. A. Kornblatt *et al.*, *J. Am. Chem. Soc.* **110** (1988), 5909–5911.
116. S. Ng *et al.*, *Biochemistry* **16** (1977), 4975–4978.
117. M. R. Mauk *et al.*, *Biochemistry* **25** (1986), 7085–7091.
118. P. W. Holloway and H. H. Mantsch, *Biochemistry* **27** (1988), 7991–7993.
119. A. M. Burch *et al.*, *Science* **247** (1990), 831–833; K. K. Rodgers and S. G. Sligar, *J. Mol. Biol.* **221** (1991), 1453–1460.
120. J. Stonehuerner, J. B. Williams, and F. Millett, *Biochemistry* **18** (1979), 5422–5427; L. D. Eltis *et al.*, *Biochemistry* **30** (1991), 3663–3674.
121. K. T. Conklin and G. McLendon, *Inorg. Chem.* **25** (1986), 4806–4807.
122. N. Liang *et al.*, *Science* **240** (1988) 311–313.
123. J. L. McGourty *et al.*, *Biochemistry* **26** (1987), 8302–8312.
124. S. E. Peterson-Kennedy *et al.*, *J. Am. Chem. Soc.* **108** (1986), 1739–1746.
125. D. J. Gingrich *et al.*, *J. Am. Chem. Soc.* **109** (1987), 7533–7534.
126. J. Jortner, *J. Chem. Phys.* **64** (1976), 4860–4867.
127. P. Siders and R. A. Marcus, *J. Am. Chem. Soc.* **103** (1981), 741–747.
128. F. R. Salemme, *Annu. Rev. Biochem.* **46** (1977), 299–329.
129. S. Ferguson-Miller, D. L. Brautigan, and E. Margoliash, in D. Dolphin, ed., *The Porphyrins*, Academic Press, 1979, **VII**, 149–240.
130. R. Timkovich, in Reference 129, pp. 241–294.
131. G. Williams, G. R. Moore, and R. J. P. Williams, *Comments Inorg. Chem.* **4** (1985), 55–98.
132. G. R. Moore, C. G. S. Eley, and G. Williams, *Adv. Inorg. and Bioinorg. Mech.* **3** (1984), 1–96.
133. E. Margoliash and H. R. Bosshard, *Trends Biochem. Sci.* **8** (1983), 316–320.
134. T. Takano and R. E. Dickerson, *J. Mol. Biol.* **153** (1981), 79–94 and 95–115.
135. G. Williams *et al.*, *J. Mol. Biol.* **183** (1985), 447–460.
136. Y. Feng, H. Roder, and S. W. Englander, *Biochemistry* **29** (1990), 3494–3504.
137. Y. Feng and S. W. Englander, *Biochemistry* **29** (1990), 3505–3509.
138. F. K. Rodkey and E. G. Ball, *J. Biol. Chem.* **182** (1950), 17–20.
139. J. B. Wooten *et al.*, *Biochemistry* **20** (1981), 5394–5402.
140. G. R. Moore and R. J. P. Williams, *Eur. J. Biochem.* **103** (1980), 513–521.
141. I. Pecht and M. Faraggi, *Proc. Natl. Acad. Sci. USA* **69** (1972), 902–906.
142. R. J. Kassner, *Proc. Natl. Acad. Sci. USA* **69** (1972), 2263–2267.
143. A. K. Churg and A. Warshel, *Biochemistry* **25** (1986), 1675–1681.
144. A. L. Raphael and H. B. Gray, *Proteins* **6** (1989), 338–340.
145. C. J. A. Wallace *et al.*, *J. Biol. Chem.* **264** (1989), 15199–15209.
146. H. A. Harbury *et al.*, *Proc. Natl. Acad. Sci. USA* **54** (1965), 1658–1664.
147. T. N. Sorrell, P. K. Martin, and E. F. Bowden, *J. Am. Chem. Soc.* **111** (1989), 766–767.
148. R. L. Cutler *et al.*, *Biochemistry* **28** (1989), 3188–3197.
149. A. L. Raphael and H. B. Gray, *J. Am. Chem. Soc.* **113** (1991), 1038–1040.
150. J. Butler *et al.*, *J. Am. Chem. Soc.* **103** (1981), 469–471.
151. G. D. Armstrong *et al.*, *Biochemistry* **25** (1986), 6947–6951.
152. W. H. Koppenol and E. Margoliash, *J. Biol. Chem.* **257** (1982), 4426–4437.
153. P. Nicholls, *Biochim. Biophys. Acta* **346** (1974), 261–310.
154. P. C. Weber and G. Tollin, *J. Biol. Chem.* **260** (1985), 5568–5573.
155. (a) T. L. Poulos and J. Kraut, *J. Biol. Chem.* **255** (1980), 10322–10330; (b) H. Pelletier and J. Kraut, *Science* **258** (1992), 1748–1755; D. N. Beratan *et al.*, *Science* **258** (1992), 1740–1741.
156. S. H. Northrup *et al.*, *J. Am. Chem. Soc.* **108** (1986), 8162–8170.
157. B. W. König *et al.*, *FEBS Lett.* **111** (1980), 395–398.
158. K. S. Schmitz and J. M. Schurr, *J. Phys. Chem.* **76** (1972), 534–545.
159. B. Michel *et al.*, *Biochemistry* **28** (1989), 456–462.

160. A. K. Churg *et al.*, *J. Phys. Chem.* **87** (1983), 1683–1694.
161. I.-J. Chang, H. B. Gray, and J. R. Winkler, *J. Am. Chem. Soc.* **113** (1991), 7056–7057.
162. D. S. Wuttke *et al.*, *Science* **256** (1992), 1007–1009.
163. D. S. Wuttke *et al.*, *Biochim, Biophys. Acta* **1101** (1992), 168–170.
164. A. Kuki, *Structure and Bonding* **75** (1991), 49–83.
165. P. Siddarth and R. A. Marcus, *J. Phys. Chem.* **94** (1990), 8430–8434; **96** (1992), 3213–3217.
166. A. Broo and S. Larsson, *J. Phys. Chem.* **95** (1991), 4925–4928.
167. C. Liang and M. D. Newton, *J. Phys. Chem.* **96** (1992), 2855–2866.
168. H. Sigel and A. Sigel, eds., *Metal Ions in Biological Systems,* Dekker, 1991, vol. 27.
169. D. N. Beratan, J. N. Betts, and J. N. Onuchic, *J. Phys. Chem.* **96** (1992), 2852–2855; *Science* **252** (1991), 1285–1288.
170. J. N. Betts, D. N. Beratan, and J. N. Onuchic, *J. Am. Chem. Soc.* **114** (1992), 4043–4046.
171. J. R. Norris and M. Schiffer, *Chem. Eng. News* **68** (July 30, 1990), 22–35.
172. G. Feher *et al.*, *Nature* **339** (1989), 111–116.
173. J. Deisenhofer and H. Michel, *Annu. Rev. Biophys. Biophys. Chem* **20** (1991), 247–266.
174. S. Kartha, R. Das, and J. R. Norris, *Metal Ions Biol. Syst.* **27** (1991), 323–359.
175. M. R. Wasielewski, *Metal Ions Biol. Syst.* **27** (1991), 361–430.
176. S. G. Boxer, *Annu. Rev. Biophys. Biophys. Chem.* **19** (1990), 267–299.
177. J. P. Allen *et al.*, *Proc. Natl. Acad. Sci. USA* **84** (1987), 5730–5734 and 6162–6166.
178. T. O. Yeates *et al.*, *Proc. Natl. Acad. Sci. USA* **84** (1987), 6438–6442.
179. R. J. Debus, G. Feher, and M. Y. Okamura, *Biochemistry* **25** (1986), 2276–2287.
180. S. G. Boxer *et al.*, *J. Phys. Chem.* **93** (1989), 8280–8294.
181. C.-K. Chan *et al.*, *Proc. Natl. Acad. Sci. USA* **88** (1991), 11202–11206.
182. M. Nonella and K. Schulten, *J. Phys. Chem.* **95** (1991), 2059–2067.
183. K. Schulten and M. Tesch, *Chem. Phys.* **158** (1991), 421–446.
184. The authors thank Deborah Wuttke for invaluable assistance with the preparation of the final draft of the manuscript and for many helpful discussions. We acknowledge the National Science Foundation, the National Institutes of Health, and the Arnold and Mabel Beckman Foundation for support of our work on biological electron-transfer reactions.

7

Ferredoxins, Hydrogenases, and Nitrogenases: Metal-Sulfide Proteins

EDWARD I. STIEFEL AND GRAHAM N. GEORGE

Exxon Research and Engineering Company

Transition-metal/sulfide sites, especially those containing iron, are present in all forms of life and are found at the active centers of a wide variety of redox and catalytic proteins. These proteins include simple soluble electron-transfer agents (the ferredoxins), membrane-bound components of electron-transfer chains, and some of the most complex metalloenzymes, such as nitrogenase, hydrogenase, and xanthine oxidase.

In this chapter we first review the chemistry of the Fe-S sites that occur in relatively simple rubredoxins and ferredoxins, and make note of the ubiquity of these sites in other metalloenzymes. We use these relatively simple systems to show the usefulness of spectroscopy and model-system studies for deducing bioinorganic structure and reactivity. We then direct our attention to the hydrogenase and nitrogenase enzyme systems, both of which use transition-metal-sulfur clusters to activate and evolve molecular hydrogen.

I. IRON-SULFUR PROTEINS AND MODELS

Iron sulfide proteins involved in electron transfer are called *ferredoxins* and *rubredoxins*.* The ferredoxins were discovered first, and were originally classified as bacterial (containing Fe_4S_4 clusters) and plant (containing Fe_2S_2 clusters) ferredoxins. This classification is now recognized as being not generally useful, since both Fe_2S_2 and Fe_4S_4 ferredoxins are found in plants,[14,15] animals,[2,6,16] and bacteria.[4] Ferredoxins are distinguished from rubredoxins by their possession of acid-labile sulfide; i.e., an inorganic S^{2-} ion that forms H_2S gas upon denaturation at low pH. Rubredoxins have no acid-labile sulfide, and generally have a single iron in a more or less isolated site. Despite their lack of acid-labile sulfide, rubredoxins are included in this chapter because they have se-

---

* For review articles, see References 1–11. For a discussion of nomenclature, see References 12 and 13.

quences much like those of the ferredoxins, and because their simple mononuclear Fe^{2+} and Fe^{3+} sites provide convenient illustrations of key structural and spectroscopic features.

In most ferredoxins, and in all rubredoxins, the protein ligands are cysteines, which provide four thiolate donors to the 1Fe, 2Fe, or 4Fe center. Additionally, the existence of 3Fe centers and of Fe-S sites that contain a second metal (i.e., heteronuclear clusters) make the Fe-S class a broad and multifunctional one.

Simple cytochromes and simple iron-sulfide proteins are similar, in that both can undergo one-electron transfer processes that are generally uncoupled from proton-, atom-, or group-transfer processes. Some of these proteins, such as cytochrome c_3 from *Desulfovibrio* with four hemes[17] or ferredoxin from *Clostridium pasteurianum* with two Fe_4S_4 centers,[6] can transfer more than one electron, because they have multiple copies of a one-electron transfer group. The cytochromes were discovered in 1886 by McMunn,[18] and their role in metabolism was discovered in the 1920s by Keilin (Chapter 6). The intense optical absorbance of these heme-containing proteins contributed singularly to their discovery and biochemical characterization. In contrast, the iron-sulfur proteins, although red to red-brown, absorb far more weakly in the visible region than do the cytochromes. Their presence is sometimes obscured by the cytochromes, and their frequent air instability made their initial recognition and isolation more difficult. It was not until the early 1960s that discoveries by several research groups[19] led to the isolation, recognition, and characterization of the ferredoxins. The use of EPR spectroscopy and its application to biological systems had a profoundly stimulating effect on the field (see below).

Although cytochromes were discovered first, the ferredoxins are likely to be the older proteins from an evolutionary perspective.[20] Ferredoxins have relatively low-molecular-weight polypeptide chains, require no organic prosthetic group, and often lack the more complex amino acids. In fact, the amino-acid composition in clostridial ferredoxin is close to that found in certain meteorites.[21]

The various Fe-S sites found in electron-transfer proteins (ferredoxins) are also found in many enzymes,[6,11,22,23] where these centers are involved in intra- or interprotein electron transfer. For example, sulfite reductase contains a siroheme and an Fe_4S_4 center,[24] which are strongly coupled and involved in the six-electron reduction of SO_3^{2-} to H_2S. Xanthine oxidase (see Figure 7.1) has two identical subunits, each containing two different Fe_2S_2 sites plus molybdenum and FAD sites. In xanthine oxidase, the Mo(VI) site carries out the two-electron oxidation of xanthine to uric acid, being reduced to Mo(IV) in the process.[25] The Mo(VI) site is regenerated by transferring electrons, one at a time, to the Fe_2S_2 and flavin sites, thereby readying the Mo site for the next equivalent of xanthine. Although the Fe_2S_2 sites do not directly participate in substrate reactions, they are essential to the overall functioning of the enzyme system. The Fe_2S_2 centers in xanthine oxidase play the same simple electron-transfer role as the Fe_2S_2 ferredoxins play in photosynthesis.

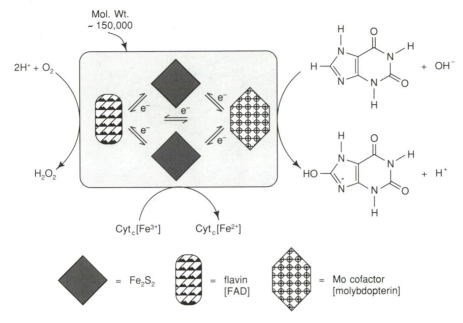

Figure 7.1
A schematic drawing of xanthine oxidase illustrating the Mo, flavin, and Fe_2S_2 sites and interaction of the enzyme with substrate and oxidant(s).

Structurally, all the iron-sulfur sites characterized to date are built up of (approximately) tetrahedral iron units (see Figure 7.2). In rubredoxins the single iron atom is bound in tetrahedral coordination by four thiolate ligands provided by cysteine side chains. In two-iron ferredoxins the Fe_2S_2 site consists of two tetrahedra doubly bridged through a pair of sulfide ions, i.e., $Fe_2(\mu_2\text{-}S)_2$, with the tetrahedral coordination of each Fe completed by two cysteine thiolates. In four-iron or eight-iron ferredoxins, the 'thiocubane' Fe_4S_4 cluster consists of four tetrahedra sharing edges with triply bridging S^{2-} ions, i.e., $Fe_4(\mu_3\text{-}S)_4$, with each Fe completing its tetrahedron by binding to a single cysteine thiolate. Finally, for Fe_3S_4 clusters, which are now being found in more and more proteins, the well-established structure has one triply bridging and three doubly bridging sulfide ions, $Fe_3(\mu_3\text{-}S)(\mu_2\text{-}S)_3$. The Fe_3S_4 unit can be thought of as derived from the 'thiocubane' Fe_4S_4 unit by the removal of a single iron atom.

In what follows we will introduce these structures in the order 1Fe, 2Fe, 4Fe, and 3Fe. For each, we will first discuss the physiological role(s) of the particular proteins, then the structural features, followed by the spectroscopic properties and model systems.

A. Rubredoxin: A Single-Fe Tetrathiolate Protein

The physiological role of rubredoxins (sometimes abbreviated as Rd) is not always known with certainty. In particular, although rubredoxin was first identified[26] in the anaerobe *Clostridium pasteurianum*, its role in anaerobic metabolism re-

368

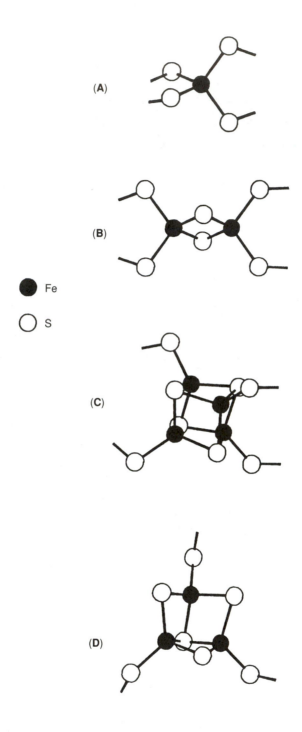

Figure 7.2
Structural systematics of Fe-S units found in proteins:
(A) rubredoxin single Fe center; (B) Fe₂S₂ unit;
(C) Fe₄S₄ unit; (d) Fe₃S₄ unit.

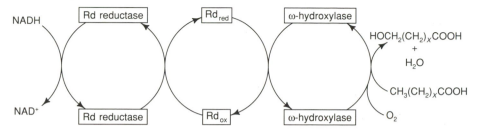

Figure 7.3
Diagram illustrating the redox changes that occur in the rubredoxin-dependent ω-hydroxylase of *Pseudomonas oleovorans*.[27]

mains obscure. Some rubredoxins, such as that from the aerobe *Pseudomonas oleovorans*, participate in fatty acid ω-hydroxylation, i.e., hydroxylation at the end of the hydrocarbon chain farthest from the carboxylic acid.[27] Like the Fe_2S_2 proteins putidaredoxin[28] and adrenodoxin,[29] the rubredoxin provides electrons to the hydroxylase, which acts as a monooxygenase forming the ω-alcohol product and water (see Figure 7.3). In a reaction catalyzed by rubredoxin reductase, rubredoxin is reduced by NADH to the ferrous state and reoxidized by the ω-hydroxylase to the ferric form during the catalytic cycle.

Most rubredoxins contain a single Fe atom, which can exist in the ferrous or ferric state. For the rubredoxin from *Clostridium pasteurianum*,[26] the $E^{0'}$ value is -57 mV, which is much more positive than that of ferredoxins from the same organism (see below). The 6-kDa clostridial protein has only 54 amino acids in its polypeptide chain, and has a very low isoelectric point of 2.93. The rubredoxin from *P. oleovorans*[27] has one or two iron atoms in a single polypeptide chain of MW ~ 20 kDa. Its redox potential is -37 mV for the $Fe^{3+/2+}$ couple. Rubredoxins as a class show considerable sequence identity, and the larger 2Fe members of the class show evidence, involving internal-sequence homology, that they may have evolved through gene duplication.

A protein from *Desulfovibrio gigas*, called desulforedoxin,[30,31] appears to resemble rubredoxins in some respects, but the two Fe atoms in the 7.6-kDa protein appear to be spectroscopically and structurally distinct from the Fe atoms in rubredoxins.[31] A protein from *Desulfovibrio vulgaris* called ruberythrin has a single rubredoxin site as well as a strongly coupled 2Fe site resembling that of hemerythrin. Its physiological function is unknown. Table 7.1 lists some of the known rubredoxins and their properties.

The x-ray crystal structures of the rubredoxins from *C. pasteurianum*[32] and *D. vulgaris*[33] have been determined.[33a] The *C. pasteurianum* protein structure is known to a resolution of 1.2 Å, placing it among the metalloproteins whose structures are known with greatest precision. The individual Fe and S atoms are clearly resolvable. As shown in Figure 7.4, the single iron is coordinated by four S ligands provided by Cys-6, Cys-9, Cys-39, and Cys-42. The sequence Cys-x-y-Cys is a common one in Fe-S proteins, because it allows both cysteine residues to bind to the same metal site or cluster. The Fe-S distances and angles

Table 7.1
Properties of some iron-sulfur proteins.

Protein source	Molecular weight (subunits)	Fe-S composition	Redox potential mV (pH)	EPR g values			References
Rubredoxins							
Clostridium pasteurianum	6,000	Fe	−58 (7)	9.3	4.3		26
Pseudomonas oleovorans	6,000	Fe		9.42	0.9	1.25[a]	44
				4.02	4.77	4.31	
Fe$_2$S$_2$ proteins							
Spinach ferredoxin	11,000	[2Fe-2S]	−420 (7.0)	2.05	1.96	1.89	350, 351
Parsley ferredoxin	11,000	[2Fe-2S]		2.05	1.96	1.90	352
Euglena ferredoxin	11,000	[2Fe-2S]		2.06	1.96	1.89	352
Adrenal cortex ferredoxin (pig) [Adrenodoxin]	16,000	[2Fe-2S]	−270 (7.0)	2.02	1.93	1.93	352, 353
Pseudomonas putida ferredoxin [Putidaredoxin]	12,500	[2Fe-2S]	−240 (7.0)	2.02	1.93	1.93	352, 353
Clostridium pasteurianum	25,000	[2Fe-2S]	−300 (7.5)	2.00	1.96	1.94	354
Xanthine Oxidase	280,000 (2)	2 × [2Fe-2S] I	−343 (8.2)	2.02	1.94	1.90	355, 356
		II	−303 (8.2)	2.12	2.01	1.91	
Thermus thermophilus Rieske	20,000	2 × [2Fe-2S]	+150 (7.8)	2.02	1.90	1.80	93,357
Fe$_4$S$_4$ proteins							
Clostridium pasteurianum	6,000	2 × [4Fe-4S]	−420 (8.2)	2.06	1.92	1.89[b]	115
Bacillus stearothermophilus	9,100	[4Fe-4S]	−280 (8.0)	2.06	1.92	1.89	358
Desulfovibrio gigas ferredoxin I	18,000 (3)	[4Fe-4S]	−455 (8.0)	2.07	1.94	1.92	359
Aconitase (beef heart) [active]	81,000	[4Fe-4S]		2.06	1.93	1.86	
Chromatium vinosum HiPIP	10,000	[4Fe-4S]	+356 (7.0)	2.12	2.04	2.04	353
Paracoccus sp.	10,000	[4Fe-4S]	+282 (7.0)				353

				g-values			Ref.
Azotobacter vinlandii Fd I	[3Fe-4S]	14,500					
	[4Fe-4S]		−645 (8.3)	2.06	1.93	1.89[c]	360
Thermus aquaticus	[3Fe-4S]	10,500					
	[4Fe-4S]		−550 (9.0)	2.06	1.93	1.92[c]	353, 361
Fe₃S₄ proteins							
Desulfovibrio gigas Fd II	[3Fe-4S]	6,000 (4)	−130 (8.0)	2.02			359
Azotobacter vinlandii Fd I	[3Fe-4S]	14,500	−450 (8.3)	2.01			360
	[4Fe-4S]						
Thermus aquaticus	[3Fe-4S]	10,500	−260 (9.0)	2.02	1.99	1.94[d]	353, 361
	[4Fe-4S]						
Aconitase (beef heart) [inactive]	[3Fe-4S]	81,000		2.01			

[a] g'-tensors for $\pm\frac{1}{2}$ and $\pm\frac{3}{2}$ Kramers doublets, respectively, of the $S = \frac{5}{2}$ system. The values of 0.9 and 1.25 are calculated (not observed).[44]

[b] The fully reduced protein has a complex spectrum due to magnetic coupling between the two identical Fe₄S₄ clusters. The g-values are those for partly reduced samples, and represent a magnetically isolated cluster.

[c] The reported spectrum is complex because of magnetic interaction with the reduced Fe₃S₄ cluster.

[d] Recent evidence suggests that *Thermus thermophilus* and *Thermus aquaticus* are actually the same species.[362] EPR parameters of the homologous *Thermus thermophilus* ferredoxin estimated from computer simulations.[361] In this protein a signal originating from the Fe₃S₄ cluster at $g' \simeq 12$, attributable to $\Delta M_s = \pm 4$ transitions, is observed for the reduced ($S = 2$) cluster.

[371]

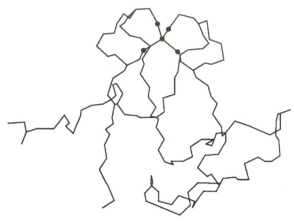

Figure 7.4
The x-ray crystal structure of rubredoxin from *Clostridium pasteurianum.*[32]

in the clostridial rubredoxin are shown in Table 7.2. The range of distances and angles reveals a slightly distorted tetrahedral structure.

The initial structural results on *C. pasteurianum* rubredoxin were reported at a slightly lower resolution than those displayed in Table 7.2. In fact, the early study[34] reported a range of Fe-S distances from 2.05 to 2.34 Å. Prior to the higher-resolution refinement, a synchrotron-radiation x-ray-absorption spectroscopy study of the iron-absorption edge of rubredoxin was reported.[35,36] Using the technique of *Extended X-ray Absorption Fine Structure,** EXAFS, the average Fe-S distance was found[35,36] to be 2.26 Å, in agreement with the average distance from the x-ray crystallographic study. However, the EXAFS indicated a much narrower permissible range of Fe-S distances than did the early crystallographic study. The later, more highly refined crystallographic treatment[32] agreed nicely with the EXAFS result, illustrating the importance of applying more than one technique to the elucidation of key parameters. Here, as with the 3Fe proteins we will discuss later, EXAFS proved a useful complementary technique to x-ray crystallography.

The tetrahedral iron sites in rubredoxins offer an interesting glimpse of ligand-field theory in action, and illustrate the use of various physical methods in

---

* X-ray absorption spectroscopy is most commonly (and conveniently) used with the K-edges of transition-metal ions, such as Fe or Mo. It can be split up into two distinct types; X-ray Absorption Near-Edge Structure (or XANES), and Extended X-ray Absorption Fine Structure (or EXAFS) spectroscopy. The former consists of features near the absorption edge itself, which are due to transitions of the photoelectron to bound states and also to other, more complex, phenomena (e.g., the so-called shape resonances). Although the spectra are highly dependent on the nature of the site, they are quite difficult to interpret, and most analyses are based upon simple comparisons with spectra from model compounds. The EXAFS are oscillations of the absorption coefficient at rather higher x-ray energies, and arise from scattering of the emitted photoelectron by surrounding atoms. In contrast to the XANES, EXAFS spectra are relatively simple to interpret in a quantitative manner, yielding a local radial structure. With proper interpretation of the spectra, very accurate interatomic distances (e.g., to ± 0.02 Å), plus more approximate ligand coordination numbers and atomic numbers can be obtained.

Table 7.2
Bond distances and bond angles around
Fe in rubredoxin from *Clostridium
pasteurianum* (W1).[32]

	Distance (Å)
Fe-S[Cys(6)]	2.333 (11)
Fe-S[Cys(9)]	2.288 (15)
Fe-S[Cys(39)]	2.300 (15)
Fe-S[Cys(42)]	2.235 (12)
	Angle (°)
S-Fe-S[Cys(6)-Fe-Cys(9)]	113.8 (4)
S-Fe-S[Cys(6)-Fe-Cys(39)]	109.0 (4)
S-Fe-S[Cys(6)-Fe-Cys(42)]	103.8 (4)
S-Fe-S[Cys(9)-Fe-Cys(39)]	103.7 (4)
S-Fe-S[Cys(9)-Fe-Cys(42)]	114.3 (5)
S-Fe-S[Cys(39)-Fe-Cys(42)]	112.4 (5)

deducing electronic structure and coordination geometry. The four sulfur ligands
are expected to split the iron 3d orbitals into e and t_2 sets, with the e set lower
as shown in Figure 7.5. The small tetrahedral splitting causes the $3d^5$ Fe^{3+} ion
to have five unpaired electrons, $(e)^2(t_2)^3$, $^6A_1$. Consistent with this configuration, the magnetic susceptibility of rubredoxin gives a μ_{eff} of 5.85 Bohr magnetons.[37] No spin-allowed ligand-field transitions are expected, and the red color
is caused by S $\rightarrow$ Fe charge-transfer transitions in the visible region.[38,39]

In contrast, the $3d^6$ Fe^{2+} state, with one additional electron, has four unpaired electrons, as confirmed by its magnetic moment of 5.05 Bohr magnetons.
In exact tetrahedral symmetry, a single, low-energy, low-intensity d-d absorp-
tion of designation $^5E \rightarrow {}^5T$ [$(e)^3(t_2)^3 \rightarrow (e)^2(t_2)^4$] is expected for the high-spin
ferrous site (Figure 7.5). Indeed, reduced rubredoxin displays a band in the
near-infrared region at 6,250 cm^{-1} that arises as a component of the $^5E \rightarrow {}^5T_2$
transition.[40] This band stands out particularly vividly in the low-energy circular
dichroism (CD) spectrum of reduced rubredoxin.[41] Moreover, magnetic circular
dichroism (MCD) has proven valuable in dissecting electronic transitions in several rubredoxins and metal-sulfide proteins.[38,39,42,43]

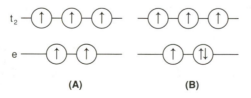

Figure 7.5
Splitting of the 3d orbitals of Fe by the tetrahedral ligand field
of four coordinated cysteine residues: (A) Fe^{3+}; (B) Fe^{2+}.

The EPR spectrum of oxidized rubredoxin (Figure 7.6) shows characteristic peaks at $g = 4.31$ and 9.42 (*P. oleovorans*), which have been assigned[44] to transitions within excited and ground-state Kramers doublets, respectively, of a nearly completely rhombic $S = \frac{5}{2}$ site, with $D = 1.8$ and $E = 0.5$ cm^{-1}. These values for the mononuclear Fe^{3+} ion stand in sharp contrast to those for other iron-sulfur proteins, which are usually $S = \frac{1}{2}$ (when reduced) and have g values close to 2. The even-electron Fe^{2+} state ($S = 2$) in reduced rubredoxin has no detectable EPR when conventional instruments are used.*

Mössbauer spectroscopy has proven to be a particularly powerful complementary tool to EPR in probing the iron sites in Fe-S proteins.[3,37,51,52] It is a nuclear spectroscopy that can give valuable information not available from other techniques.† Unlike EPR, where only *paramagnetic* centers are "seen," every ^{57}Fe atom in the sample will contribute to the Mössbauer spectrum. For rubredoxin, the high-spin nature of the ferric and ferrous sites are clearly seen in the Mössbauer spectra.[53] The high-spin Fe^{3+} sites show a small quadrupole splitting of roughly 0.7–0.8 mm/s due to the almost spherical distribution of the five d electrons in the five d orbitals (Figure 7.7A). In contrast, the high-spin Fe^{2+} ion with an additional d electron has a significant asymmetry, and thus displays

---

* But see Reference 45. EPR spectroscopy uses magnetic fields to split the electron spin states into levels that differ by energy in the microwave region of the spectrum. For an $S = \frac{1}{2}$ system, the g value (and its anisotropy) and the a values (hyperfine splitting from various nuclei and their anisotropy) are the major parameters reported. EPR spectroscopy has played a role in the development of Fe-S biochemistry akin to the role played by optical spectroscopy in the development of the biochemistry of the cytochromes,[46–49] particularly for mitochondria[47] and chloroplasts,[50] where the $g = 1.9$ EPR signal has facilitated the monitoring of electron flow through these redox systems. Although EPR has been a powerful tool, it does have some important limitations. A necessary but not sufficient condition for EPR is that the center to be observed must be in a paramagnetic state. Fortunately, this condition is met for at least one member of each one-electron redox couple, i.e., the odd-electron species. However, even when the even-electron species is paramagnetic, it is usually not observed in the EPR, because of the presence of large zero-field splittings. Moreover, relaxation effects and/or the population of excited states often cause the EPR of proteins to be unobservable at room temperature, necessitating the use of liquid N$_2$ or liquid He temperatures to observe the signals in the frozen state. The need to freeze samples prior to observation can lead to artifacts involving the observation of nonphysiological states and processes. On the positive side, the low temperature increases the signal intensity by altering the Boltzmann distribution of the spin population, and allows various quenching techniques to be used with EPR to evaluate kinetic and electrochemical parameters. Nevertheless, one cannot usually observe real-time kinetics or be certain that one is observing a physiologically relevant state. Despite these caveats, EPR has proven a valuable and, in some cases, indispensable tool for identification and monitoring of Fe-S sites. Recently, the advanced EPR techniques ENDOR (Electron Nuclear Double Resonance) and ESEEM (Electron Spin Echo Envelope Modulation) have allowed the extraction of additional information from the EPR signal.

† Mössbauer spectroscopy measures nuclear absorption of light at γ-ray energies, and can be used to probe nuclear energy levels (usually of ^{57}Fe). The splitting of these levels is influenced by the (s) electron density at the nucleus, and by the electric-field gradient that is set up by nearby atoms. These factors affect the isomer shift and the quadrupole splitting of the Mössbauer spectrum, respectively. Information on nuclear hyperfine couplings is also available when experiments are conducted in the presence of an external (usually applied) magnetic field. Fortunately, the nucleus most commonly (and easily) studied by this technique is present in all the proteins discussed in this chapter, although the level of ^{57}Fe (2 percent natural abundance) must be increased by isotopic enrichment to achieve a high-enough signal-to-noise ratio. For spectra containing one type of site, the spectra are relatively straightforward to interpret. For multisite systems deconvolution is required to get data on individual centers. When possible, selective labeling of sites with ^{57}Fe is extremely helpful in the deconvolution process.

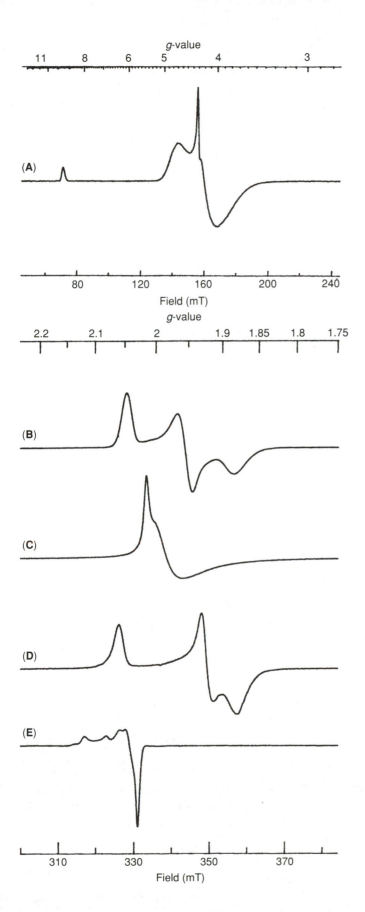

Figure 7.6
EPR spectra of various Fe-S proteins: (A) oxidized *Desulfovibrio gigas* rubredoxin; (B) reduced spinach ferredoxin $[Fe_2S_2]^+$; (C) reduced *Bacillus stearothermophilus* ferredoxin $[Fe_4S_4]^+$; (D) oxidized *Thermus aquaticus* ferredoxin $[Fe_3S_4]^+$; (E) oxidized *Chromatium vinosum* HiPIP $[Fe_4S_4]^{3+}$. (Spectra courtesy of S. J. George.)

376

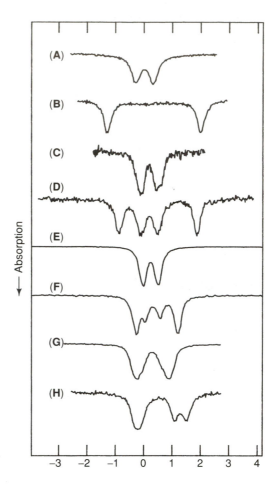

Figure 7.7
Mössbauer spectra of various Fe-S sites: (A) oxidized and (B) reduced $Fe(SR)_4$
(Fe^{3+} and Fe^{2+}, respectively) rubredoxin models (data from Reference 347);
(C) oxidized and (D) reduced *Scenedesmus* Fe_2S_2 ferredoxin ($2Fe^{3+}$ and [Fe^{3+},
Fe^{2+}], respectively, data from Reference 348); (E) oxidized and (F) reduced
Desulfovibrio gigas Fe_3S_4 ferredoxin II ($3Fe^{3+}$ and [$2Fe^{3+}$, Fe^{2+}], respectively,
data from Reference 158); (G) oxidized and (H) reduced *Bacillus sterother-
mophilus* Fe_4S_4 ferredoxin (data from Reference 349).

a large and quite characteristic quadrupole splitting of 3.1–3.4 mm/s (Figure
7.7B). The isotope shift also distinguishes between Fe^{2+} and Fe^{3+}, although
not as dramatically.[37] Finally, the observation[53] of magnetic hyperfine interaction in the Mössbauer spectrum at low temperature in the Fe^{3+} state directly
reveals the presence of unpaired electrons, i.e., magnetic coupling with a hyperfine field of 370 ± 3 kG. Although in rubredoxins with a single Fe atom,
this observation of magnetic coupling does not reveal any new information,
similar magnetic coupling is particularly useful in unraveling the Fe sites in
more complex multiiron proteins.

NMR studies on both the oxidized and the reduced states of rubredoxins have been reported. The strongly paramagnetic iron atoms have a profound effect on the NMR spectra of protons in the vicinity of the iron. The iron drastically affects the relaxation behavior of such protons, causing line-broadening, sometimes so much that the protons become nonobservable. If observed, the protons are shifted far from the values found in diamagnetic proteins by combinations of Fermi contact (through overlap/through bond) and pseudo-contact (through space/dipolar) coupling.[54,55] In the rubredoxins, the reduced state shows resolved spectra,[37,56] which can be assigned[56] with the help of data from model systems.*

Resonance Raman spectroscopy provides information involving molecular vibrations that is not dependent on either nuclear or magnetic properties. Electronic excitation of bands involving S $\rightarrow$ Fe charge transfer often leads to resonance enhancement of Fe-S stretching modes. In rubredoxin,[57,57a] the Fe-S stretching vibrations are located between 300–400 cm^{-1}. Deviations of the expected two-band tetrahedral pattern (T_2 and A_1 modes) are attributable to coupling of the Fe-S vibrations with S-C-C bending modes. This coupling makes for greater variability, and the detailed vibrational assignment is thus more difficult for bands involving the cysteinyl sulfur atoms. In contrast, for sites containing inorganic S^{2-}, the Fe-S vibrations involving the inorganic core are less variable and therefore more characteristic of the core type.[57]

In theory, each of the spectroscopic techniques applied to rubredoxins can give useful information about the other iron-sulfur proteins. In practice, some techniques have proven more useful than others in particular situations, and combined use of several techniques is necessary to draw meaningful conclusions.

Chemical studies of rubredoxins have led to the replacement of the Fe^{2+} with an Fe_4S_4 center,[58] with Co^{2+}, and with Ni^{2+}. The Co^{2+} replacement of Fe^{2+}, in *P. oleovorans* rubredoxin, leads to a stable protein that displays reduced (but not trivial) reactivity in the ω-hydroxylation reaction.[59,60] The spectral properties of the cobalt(II) site show the expected changes in d-d bands and the expected shifts in charge-transfer transitions.[59] Interestingly, when Ni^{2+} is substituted into rubredoxins from desulfovibrio species, the resultant proteins show hydrogenase activity.[61]

B. Rubredoxin Model Systems

The simple mononuclear tetrahedral site of Rd has been chemically modeled in both its reduced and its oxidized forms. The bidentate o-xylyl-α,α'-dithiolate ligand forms bis complexes of Fe(II) and Fe(III) that have spectroscopic features

---

* NMR is a technique whose great utility in the study of low-molecular-weight proteins and model systems has not (yet) carried over to the study of larger proteins. Slower tumbling rates, rapid electronic relaxation, multiple paramagnetic sites, large numbers of protons, and more dilute solutions conspire to make the observation and/or interpretation of NMR spectra a daunting task in multisite redox proteins of >50 kDa.

quite similar to those of the protein.[62,63] The preparative procedure is relatively straightforward (Equation 7.1).

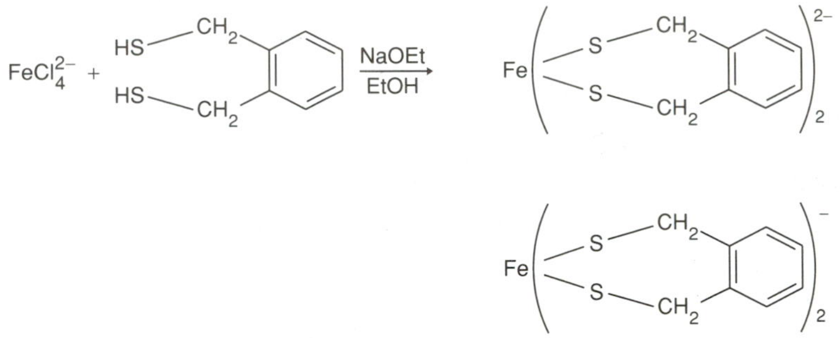

$$
(7.1)
$$

The UV-visible-NIR spectra, Mössbauer spectra, and magnetic susceptibility differ only slightly from those of oxidized and reduced rubredoxins.

The monodentate benzenethiolate (thiophenolate) ligand, $C_6H_5S^-$, similarly forms the ferrous $Fe(SC_6H_5)_4^{2-}$ complex.[64,65] Although for some time it was felt that the oxidized form, $Fe(SC_6H_5)_4^-$, was inherently unstable, the sterically hindered monothiolate ligand 2,3,5,6-tetramethylbenzenethiolate was found to form[66-68] a stable, quite symmetric Fe(III) tetrathiolate anion. Armed with this information, the preparation of the tetrakis(benzenethiolate) Fe(III) complex was reinvestigated, and the complex successfully synthesized[67] (Equation 7.2).

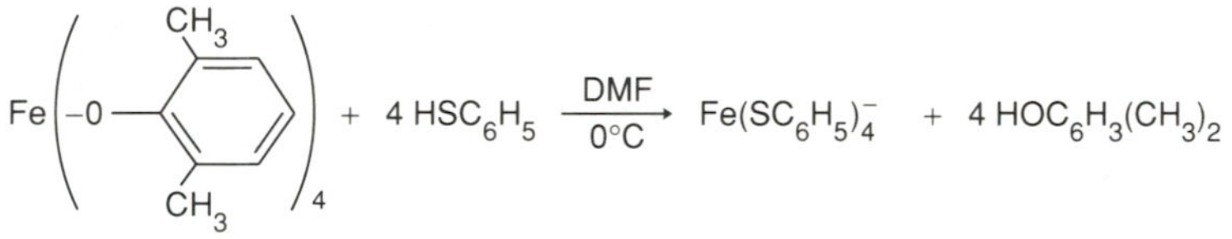

$$
Fe\left(-O-\underset{CH_3}{\overset{CH_3}{\bigcirc}}\right)_4 + 4\,HSC_6H_5 \xrightarrow[0°C]{DMF} Fe(SC_6H_5)_4^- + 4\,HOC_6H_3(CH_3)_2 \qquad (7.2)
$$

The Fe(III) and Fe(II) tetrathiolate species now serve as excellent structural models for the Fe sites of both oxidized and reduced Rd.[69]

The structural parameters for the oxidized rubredoxin analogues are very similar to those of the oxidized Rd iron site. The reduced complexes reveal a lengthening of the average Fe-S bond from 2.27 to 2.36 Å, consistent with the change in oxidation state from ferric to ferrous. The addition of an electron has a more profound structural effect in this single-iron center than in some of the multiiron clusters, where electrons are more delocalized.

Clearly, for the single-Fe sites, the dominant structural feature is their near-tetrahedral tetrathiolate coordination. The dominant electronic structural feature is the presence of high-spin Fe^{3+} and Fe^{2+} sites. The important mode of chemical reactivity is a simple one-electron transfer. Each of these features carries over to the 2Fe, 4Fe, and 3Fe sites discussed below.

C. Fe$_2$S$_2$ Ferredoxins

The simple 2Fe-2S proteins are sometimes referred to as "plant" or "plant-type" ferredoxins. The protein from spinach, which serves as an electron acceptor in the photosynthetic apparatus,[14,15,50,70] was among the first to be well-characterized and widely studied, and could be considered the prototypical 2Fe-2S ferredoxin. However, 2Fe-2S proteins are also well-known in bacteria.[4] The protein from the cyanobacterium (blue-green alga) *Spirulina platensis* has been structurally elucidated by x-ray crystallography.[47] Putidaredoxin, from *Pseudomonas putida*, which serves as a donor to the P-450 camphor monooxygenase system, has been extensively studied.[28] Fe_2S_2 centers are also well-established in mammalian proteins. Adrenodoxin[29] serves as the electron donor to the P-450 monooxygenase system that carries out the 11-β-hydroxylation of steroids. The so-called "Rieske proteins" are found in the bc_1 complex of mitochondria[47] as well as in the b_6f complex of the photosynthetic apparatus of plants.[71] In addition, Fe_2S_2 centers are well-known constituents of such redox proteins as xanthine oxidase,[25,72] CO oxidase,[25] succinate dehydrogenase,[73–75] and putidamonooxin.[76] Table 7.1 lists some of the Fe_2S_2 proteins and their properties.

The x-ray crystal structure of only the single 2Fe-2S protein mentioned above has been determined;[70a] the 2Fe-2S ferredoxin from the blue-green alga *Spirulina platensis*[6,22,47,77,78] shows significant sequence identity with chloroplast ferredoxins typical of higher plants.[79,80] As Figure 7.8 shows, the Fe_2S_2 unit in this 11-kDa protein is bound by Cys-41, Cys-46, Cys-49, and Cys-79. The binuclear iron cluster is found in a largely hydrophobic region of the protein, but is within 5 Å of the protein surface.[6] The sulfur atoms of the cluster, both inorganic and cysteinyl, are hydrogen-bonded to six peptide NH groups and one serine OH group, which presumably stabilize the cluster/protein complex. The serine involved in the H-bonding, Ser-40, is conserved in all plant and algal 2Fe-2S ferredoxins sequenced, which implies that it plays a crucial structural or functional role.

The structure of the 2Fe-2S core in Figure 7.2 reveals a tetrahedron of S ligands surrounding each Fe atom. The two tetrahedra share an edge defined by the two bridging sulfide ions, and the core structure is designated $Fe_2(\mu_2\text{-}S)_2$. Fe-S distances and angles cannot be measured accurately in the structure at the present 2.5-Å resolution;[70a] so we will later discuss these details in terms of model compounds.

The Fe_2S_2 center shows nicely how spectroscopy can be used to deduce the structure of an active site. Indeed, in this case the now well-established active-site structure was deduced by a combination of chemical, spectroscopic, and magnetic methods, and the site was successfully modeled long before the first protein crystallographic study was reported.

The presence of acid-labile, inorganic sulfide is a key feature of both the Fe_2S_2 and the Fe_4S_4 centers. The 1:1 stoichiometry between iron and acid-labile sulfide was eventually established analytically for Fe_2S_2 centers.[9–11] Care must be taken to ensure that both the protein and its active-site complement are homogeneous. Although protein homogeneity is usually established by electrophoretic methods, these methods may not distinguish between pure proteins and those with absent or incomplete active centers. Fortunately, absorption at 420 nm

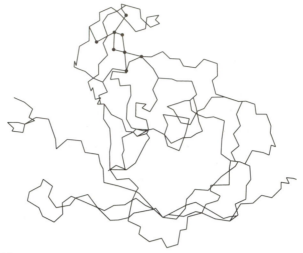

Figure 7.8
The x-ray crystal structure of the Fe_2S_2 ferredoxin from *Spirulina platensis*.[70a]

is due solely to the Fe_2S_2 cluster, whereas the 275-nm absorption is dominated by the protein. Therefore a good criterion for active-site saturation and homogeneity is the ratio of the absorbances at 420 and 275 nm, $A_{420 \text{ nm}}/A_{275 \text{ nm}}$, which is ~ 0.48 for pure spinach ferredoxin.[81] Once homogeneous protein is obtained, the Fe_2S_2 composition of the "plant" ferredoxins can be correctly deduced analytically.

The Fe_2S_2 center displays two redox states that differ by a single electron. The potential range for the couple is -250 to -420 mV, revealing the highly reducing nature of the ferredoxin. The correct structure of the Fe_2S_2 center was first proposed in 1966 based on EPR studies.[82] The reduced state of the cluster shows a rhombic EPR signal with g values of 1.88, 1.94, and 2.04 (Figure 7.6B) characteristic of an $S = \frac{1}{2}$ center. The oxidized state is EPR-silent. The weakness of the sulfur ligand field causes the iron atoms to be high-spin. But how can two sulfur-ligated iron atoms, each with a tendency to be high-spin, produce a state with a single unpaired electron?

The individual Fe atoms in the Fe_2S_2 cluster resemble those in rubredoxin quite closely. The two redox states of the Fe_2S_2 protein correspond to an Fe^{3+}-Fe^{3+} and an Fe^{3+}-Fe^{2+} pair, respectively, as shown in Figure 7.9. In the all-

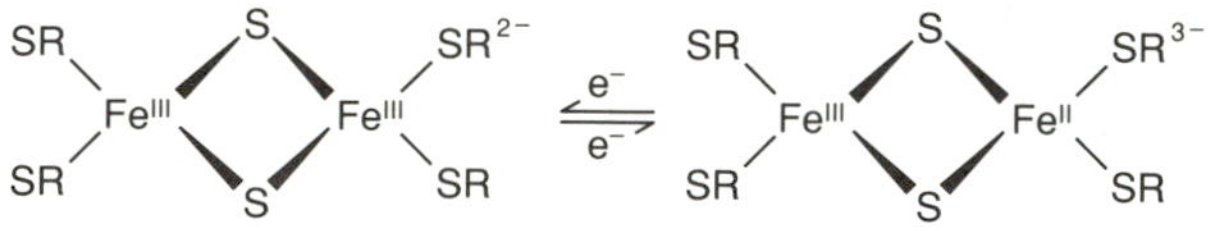

Figure 7.9
Redox states of Fe_2S_2 proteins: (A) reduced; (B) oxidized.

ferric oxidized state, the two Fe^{3+} sites are antiferromagnetically coupled; i.e., the spins of the five d electrons on the two iron atoms are oppositely aligned, such that their pairing produces an effective $S = 0$, diamagnetic ground state. In the reduced form, a single unpaired electron is present, because the $S = \frac{5}{2}$ Fe^{3+} and $S = 2$ Fe^{2+} sites are *antiferromagnetically* coupled, leaving one net unpaired spin and an $S = \frac{1}{2}$ ground state. The profound difference between the electronic properties of rubredoxin and Fe_2S_2 ferredoxin arises because the latter has two Fe atoms in close proximity, which allows for their magnetic coupling.

Strong support for the spin-coupling model in Fe_2S_2 ferredoxins comes from a detailed analysis of their absorption and circular dichroism spectra.[83] As with rubredoxin (see Figure 7.5), we expect no low-energy spin-allowed d-d bands for the ferric site in either the oxidized or the reduced state. Indeed, the oxidized state containing all Fe^{3+} shows no low-energy bands; the reduced state containing a single Fe^{2+} displays low-energy, low-intensity bands in the region 4,000–9,000 cm^{-1}, in close analogy to the situation in reduced rubredoxin. The combined EPR and optical spectra leave little doubt about the structural assignment: two coupled high-spin ferric ions in the oxidized state, and coupled high-spin ferric and ferrous ions in the reduced state. Moreover, the spectra are consistent only with a localized model, i.e., one in which the Fe(II) site is associated with a single iron.[83,83a] The Fe_2S_2 site is inherently asymmetric, and inequivalence of the Fe(III) sites is spectroscopically detectable in the all-ferric oxidized form.[84] In fact, the localized valence trapping is present in reduced model compounds that contain no ligand asymmetry.

Mössbauer spectra provide additional and striking confirmation of the structural assignment. The spectrum of the oxidized ferredoxin (Figure 7.7) resembles strongly that of oxidized rubredoxin, indicating the presence of high-spin Fe^{3+}, even though the net spin is zero. In the reduced form, the Mössbauer spectrum involves the superposition of signals from a high-spin Fe^{2+} and a high-spin Fe^{3+}, i.e., a reduced and an oxidized rubredoxin, respectively. Clearly, the simplest interpretation of this result consistent with the $S = \frac{1}{2}$ spin state required by the EPR is the localized Fe^{2+}-Fe^{3+} antiferromagnetic coupling model discussed above.

NMR studies of oxidized Fe_2S_2 proteins reveal broad isotropically shifted resonances for the CH_2 protons of the cysteine ligands.[85] Despite the coupling of the irons, the net magnetism at room temperature is sufficient to lead to large contact shifts (-30 to -40 ppm downfield from TMS). The assignment of the resonance was confirmed with the synthesis and spectroscopic analysis of model compounds.[86] Extensive NMR studies of the Fe_2S_2 proteins have been reported.[87,87a]

Resonance Raman spectra of Fe_2S_2 sites[88,89,90] reveal many bands attributable to Fe-S stretching. Detailed assignments have been presented for the four bridging and four terminal Fe-S modes. A strong band at ~390 cm^{-1}, which shifts on ^{34}S sulfide labeling, is assigned to the A_{1g} "breathing" mode; another band at 275 cm^{-1} is assigned to B_{3u} symmetry in point group D_{2h}.[57,88] Spectroscopic differences in the terminal, Fe-S(Cys) stretches between plant ferredoxins and adrenodoxin (which also differ somewhat in redox potential) seem to reflect different conformations of the cysteine ligands in the two classes. Evidence for asymmetry of the iron atoms is found in the intensity of the resonance enhancement of certain modes.

D. Rieske Centers

Within the class of Fe_2S_2 ferredoxins there is a subclass called the Rieske proteins, or the Rieske centers.[47,91,92] The Rieske iron-sulfur centers are found in proteins isolated from mitochondria and related redox chains.[47,92] In addition, the phthalate dioxygenase system from *Pseudomonas cepacia*[93,94] contains one Fe_2S_2 Rieske center as well as one additional nonheme Fe atom. Although the Rieske centers appear to contain an Fe_2S_2 core, there is extensive evidence for nonsulfur ligands coordinated to at least one of the Fe atoms. The proposed model in Scheme (7.3) has two imidazole ligands bound to one Fe atom. The nitrogen atoms are seen in ENDOR (*E*lectron *N*uclear *D*ouble *R*esonance) experiments,[93] and are manifest in EXAFS spectra, which are consistent with the presence of a low-Z (atomic number) ligand bound to iron.[94] The potentials for the Rieske proteins range from $+350$ to -150 mV,[47] in contrast to the plant-type Fe_2S_2 centers, which range from -250 to -450 mV. The strong dependence of redox potential on pH[95] suggests a possible role in coupling proton- and electron-transfer processes.

$$(7.3)$$

E. Fe_2S_2 Models

Although spectroscopic studies led to the correct deduction of the structure of the Fe_2S_2 core, the synthesis of model compounds containing this core provided unequivocal confirmation. The model compounds allowed detailed structural analysis unavailable for the proteins. Moreover, by using a uniform set of peripheral ligands, properties inherent to the Fe_2S_2 core could be discerned.

The Fe_2S_2 core has been synthesized by several routes[86,96,96a,b,c,d] (see Figure 7.10). For example, the reaction of $Fe(SR)_4^{2-}$, the ferrous rubredoxin model, with elemental sulfur produces the complex $Fe_2S_2(SR)_4^{2-}$. In this reaction the sulfur presumably oxidizes the Fe^{2+} to Fe^{3+}, being reduced to sulfide in the process. The Fe_2S_2 core has been prepared with a variety of peripheral S-donor ligands. Metrical details for $Fe_2S_2(SC_6H_4-p-CH_3)_4^{2-}$ are given in Table 7.3. Notable distances are the Fe-S (bridging) distance of 2.20 Å, the Fe-S (terminal) distance of 2.31 Å, and the Fe-Fe distance of 2.69 Å.

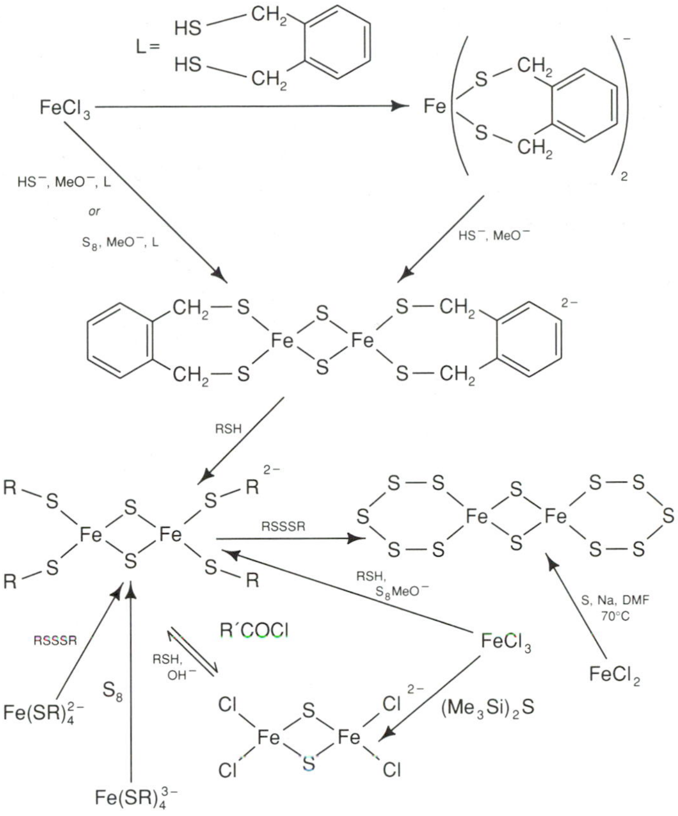

Figure 7.10
Preparative schemes leading to complexes containing the Fe_2S_2 core.[128]

Table 7.3
Structural parameters for
$Fe_2S_2(SC_6H_4\text{-}p\text{-}CH_3)_4{}^{2-}$.

Atoms[a]	Distance Å	Atoms[a]	Angle °
Fe-Fe	2.691 (1)	Fe-S-Fe	75.3
Fe-S1 (bridge)	2.200 (1)	S-Fe-S	104.6
Fe-S2 (bridge)	2.202 (1)	S-Fe-S	115.1
Fe-S3	2.312 (1)	S-Fe-S	105.4

[a] Data from Reference 211.

To date, all analogue systems structurally characterized contain the Fe^{3+}-Fe^{3+} fully oxidized form. Attempts to isolate the Fe^{3+}-Fe^{2+} form have so far failed. However, the mixed-valence Fe_2S_2 form can be generated and trapped by freezing for spectroscopic examination.[97,98] Mössbauer spectroscopy reveals the presence of distinct Fe^{2+} and Fe^{3+} ions, as found in the proteins, clearly showing that "trapped" valence states are an inherent characteristic of the $Fe_2S_2^{2+}$ core and are *not* enforced by the protein.[97,98]

The existence of noncysteine-bound Fe_2S_2 cores in Rieske-type proteins has led to attempts to synthesize complexes with oxygen and nitrogen ligands.[99-101] Characterized species include $Fe_2S_2(OC_6H_5)_4^{2-}$, $Fe_2S_2(OC_6H_4\text{-}p\text{-}CH_3)_4^{2-}$, $Fe_2S_2(C_4H_4N)_4^{2-}$, and $Fe_2S_2(L)_2^{2-}$, where L is a bidentate ligand.

$$(7.4)$$

The potentially tridentate ligand

$$(7.5)$$

acts in a bidentate fashion, binding through S and O but not N.

No Fe_2S_2 complexes containing mixed S,N terminal ligands, such as those suggested for the Rieske site, have been prepared. The Se^{2-} bridged analogue has been prepared for some of the complexes.[102,103]

F. Fe_4S_4 Ferredoxins (including HiPIPs)

We now turn our attention to proteins containing the Fe_4S_4 center. Historically, within this class a strong distinction was made between the "ferredoxins," which are low-potential (as low as -600 mV in chloroplasts) iron-sulfur proteins, and the "HiPIPs" = *H*igh *P*otential *I*ron *P*roteins, which have positive redox potentials (as high as $+350$ mV in photosynthetic bacteria). Although the HiPIP designation is still useful, proteins of both high and low potential are considered ferredoxins, whose key defining feature is the presence of iron and acid-labile sulfide.[13]

The Fe_4S_4 proteins participate in numerous electron-transfer functions in bacteria, and in some organisms (such as *Clostridium*) are the immediate electron donors for the nitrogenase and/or hydrogenase enzymes. The function of the HiPIPs seems obscure at present. In addition, Fe_4S_4 centers have been shown or postulated to occur in numerous microbial, plant, and mammalian redox enzymes, including nitrate reductase,[104] sulfite reductase,[24] trimethylamine dehydrogenase,[105] succinate dehydrogenase,[73,106] hydrogenase, and, possibly, in altered forms, nitrogenase. Table 7.1 lists some of the Fe_4S_4 ferredoxins and their properties.

In the Fe_2S_2 ferredoxins, combined spectroscopic, analytical, and model-system work led to an unequivocal assignment of the structural nature of the active site long before the crystallography was done. In contrast, for Fe_4S_4 systems and in particular the 8Fe-8S = $2Fe_4S_4$ systems from bacteria, the initial chemical suggestions were fallacious, and even the number and stoichiometry of the clusters were in doubt. In these cases, crystallography provided the definitive structural information.

The first indication of the presence of the "thiocubane" structure came in 1968, when a 4-Å resolution study[107] indicated a compact cluster of potentially tetrahedral Fe_4 shape in the HiPIP from *Chromatium vinosum*. This finding did not lead to much excitement, since it was not yet appreciated that HiPIPs and ferredoxins were structurally similar. In 1972, the high-resolution structure solution of both *Chromatium* HiPIP[108] and the 8Fe ferredoxin from *Peptococcus aerogenes* (formerly *Microbacter aerogenes*)[101] confirmed the presence of virtually identical thiocubane clusters in the two proteins.[108] Moreover, the structures for both oxidized and reduced HiPIP were deduced, and these revealed that the Fe_4S_4 cluster remained intact during the redox interconversion.[109] Subsequently, four-iron clusters have been crystallographically confirmed in an Fe_4S_4 ferredoxin from *Bacillus thermoproteolyticus*,[110,110a] in *Azotobacter vinelandii* ferredoxin I (also previously called Shethna Fe-S protein II), which also contains a 3Fe-4S cluster,[111,112] in the active form of aconitase,[113] and in sulfite reductase, where the cluster is probably bridged by cysteine sulfur to a siroheme.

In all the proteins characterized to date, the Fe_4S_4 clusters adopt the thiocubane structure,[108] which is discussed at greater length in the section on models. The clusters are usually bound to their proteins by four cysteine residues. As shown in Figure 7.11, in the *P. aerogenes* protein the two Fe_4S_4 clusters are bound by cysteines numbered 8, 11, 14, 18, 35, 38, 41, and 45.[101,114] The presence of the Cys-x-x-Cys unit is again apparent. However, this sequence seems prominent in all Fe-S proteins, and so is not specific for a particular Fe-S site. At first glance one might expect one cluster to be bound by cysteines 8, 11, 14, and 18, the other by cysteines 35, 38, 41, and 45. Actually, one cluster is bound by cysteines 8, 11, 14, and 45, the other by cysteines 35, 38, 41, and 18. The binding of a given cluster by cysteine residues from different portions of the polypeptide chain apparently helps stabilize the tertiary structure of the protein and brings the two clusters into relatively close proximity, the center-center distance being 12 Å.[114]

The *C. pasteurianum* protein displays weak magnetic coupling, which leads to an unusual EPR spectrum[115] consistent with the 12-Å cluster-cluster separation. However, the redox potentials for the two sites seem virtually identical at −412 mV, thus allowing the 8Fe ferredoxin to deliver two electrons at this low

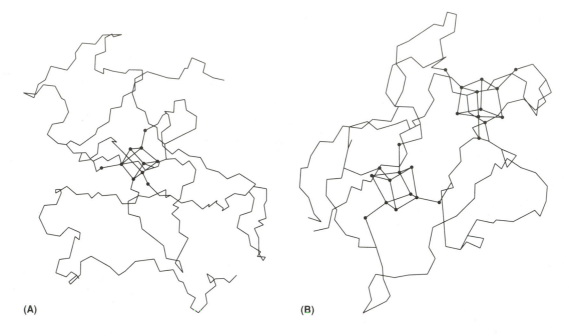

Figure 7.11
The x-ray crystal structures of (A) *Chromatium vinosum* HiPIP[108] and (B) the 8Fe-8S ferredoxin from *Peptococcus aerogenes*.[101]

redox potential.[115] Significant sequence identity indicates the likelihood that other 8Fe ferredoxins, such as the well-studied one from *C. pasteurianum*,[116-118] have quite similar structures.

The thiocubane unit of Fe_4S_4 proteins can exist in proteins in at least three stable oxidation states. This so-called three-state model[74,109,119,120] contrasts dramatically with the situation for Rd(1Fe), Fe_2S_2, and Fe_3S_4 systems, in which only two oxidation states are accessible through simple electron transfer for each center. For the thiocubane structure, the three accessible states can be designated $Fe_4S_4^{3+}$, $Fe_4S_4^{2+}$, and $Fe_4S_4^{+}$, corresponding to $[Fe(III)_3Fe(II)]$, $[Fe(III)_2Fe(II)_2]$, and $[Fe(III)Fe(II)_3]$ valence-state combinations, respectively. It is crucial to note that, in sharp contrast to the Fe_2S_2 and Fe_3S_4 sites, the oxidation states are not localized in the Fe_4S_4 clusters. In most cases, each Fe atom behaves as if it had the same average oxidation level as the other Fe atoms in the cluster. The redox interconversion of the Fe_4S_4 sites is shown in Figure 7.12. The $Fe_4S_4^{3+} \rightleftarrows Fe_4S_4^{2+}$ couple is the high-potential redox couple characteristic of HiPIPs; the $Fe_4S_4^{2+} \rightleftarrows Fe_4S_4^{+}$ couple is responsible for the low-potential process characteristic of the classical ferredoxins. In any given protein under physiological conditions, only one of the two redox couples appears to be accessible and functional.

Both the $Fe_4S_4^{+}$ and the $Fe_4S_4^{3+}$ states of the thiocubane cluster are paramagnetic and display characteristic EPR spectra (Figure 7.6C,D). The $Fe_4S_4^{3+}$

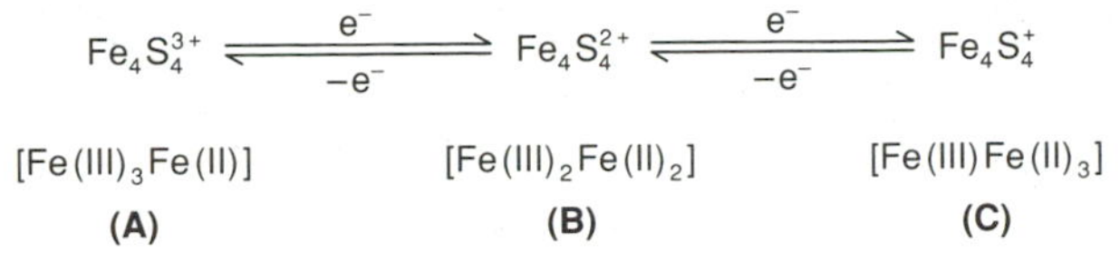

$$Fe_4S_4^{3+} \underset{-e^-}{\overset{e^-}{\rightleftharpoons}} Fe_4S_4^{2+} \underset{-e^-}{\overset{e^-}{\rightleftharpoons}} Fe_4S_4^+$$

$[Fe(III)_3Fe(II)]$ $[Fe(III)_2Fe(II)_2]$ $[Fe(III)Fe(II)_3]$

(A)   **(B)**   **(C)**

Figure 7.12
The redox interconversions of Fe_4S_4 sites illustrating the three-state model. The states are found in (A) oxidized HiPIP; (B) reduced HiPIP and oxidized ferredoxin; (C) reduced ferredoxin.

site in reduced ferredoxins[46,48,49,119] displays a rhombic EPR signal (Figure 7.6C) with $g = 1.88, 1.92$, and 2.06. The oxidized form of low-potential ferredoxins is EPR-silent, and attempts to "superoxidize" it to achieve the $Fe_4S_4^{3+}$ state invariably lead to irreversible cluster decomposition, probably through a 3Fe-4S structure. The $Fe_4S_4^{3+}$ signal is usually referred to as the HiPIP signal (Figure 7.6D) and shows distinct g values at $2.04(g_\perp)$ and $2.10(g_\parallel)$; it is present in oxidized HiPIP but absent in reduced HiPIP.[46,119] Reduction of HiPIP to a "super-reduced" state apparently occurs under partially denaturing conditions in aqueous DMSO.[108] The observed axial EPR signal with $g = 1.94$ and 2.05 is assigned to the $Fe_4S_4^+$ state characteristic of reduced ferredoxins. This result[108] is consistent with structural and spectroscopic identity of the HiPIP and Fd sites, as required by the three-state model of the Fe_4S_4 proteins (Figure 7.12).

In Fe_4S_4 centers at each level of oxidation, electronic transitions give rise to characteristic visible and UV spectra, although the delocalized nature of the electronic states makes detailed assignment difficult. MCD spectra of clusters in the three states of oxidation are clearly distinguishable from each other and from MCD of Fe_2S_2 clusters.[43,119] MCD, magnetic susceptibility, and Mössbauer spectra provide evidence that the $S = \frac{1}{2}$ state, whose EPR signal is so distinct in reduced ferredoxins, may coexist at higher T with $S = \frac{3}{2}$ and perhaps even higher spin states. Indeed, recent studies with model systems[121,122] and theoretical treatments[123,124] clearly support the ability of the Fe_4S_4 cluster to display a number of spin states that are in labile equilibria, which are influenced, perhaps quite subtly, by local structural conditions. The iron protein of nitrogenase also displays this behavior.

The Mössbauer spectra of Fe_4S_4 centers of ferredoxins reveal the equivalence of the Fe sites, and quadrupole splittings and isomer shifts at averaged values for the particular combination of oxidation states present.[3,51,52] Representative spectra are shown in Figure 7.7. Magnetic coupling is seen for the paramagnetic states.

Resonance Raman spectra (and IR spectra) have been extensively investigated in *C. pasteurianum* ferredoxin and in model compounds.[57,125] Selective labeling of either thiolate sulfur or sulfide sulfur with ^{34}S allows modes associated with the Fe_4S_4 core to be distinguished from modes associated with the Fe-SR ligands. The band at 351 cm^{-1} is assigned to Fe-SR stretching, and Fe_4S_4 modes occur at 248 and 334 cm^{-1} in reduced ferredoxin from *C. pasteurianum*. There is little difference between the oxidized and reduced spectra, although an extra band at 277 cm^{-1} seems present in the oxidized protein. The Fe_4Se_4 substituted protein has also been studied.[125]

As in the 1Fe and 2Fe proteins, ^1H NMR spectra reveal resonances from contact-shifted -CH_2- groups of cysteinyl residues.[125a] However, unlike the other proteins, where all states are at least weakly magnetic, only the reduced ferredoxin and the oxidized HiPIP states show contact shifts.[87a,125a,b,c]

EXAFS studies on proteins and on model compounds clearly identify the Fe-S distance of ~2.35 Å and an Fe-Fe distance of 2.7 Å. These distances, as expected, vary only slightly with state of oxidation.[125d]

G. Fe_4S_4 Models

Judging from the ease with which models of Fe_4S_4 are prepared under a variety of conditions and their relative stability, the $Fe_4S_4^{2+}$ core structure seems to be a relatively stable entity, a local thermodynamic minimum in the multitude of possible iron-sulfide-thiolate complexes. The initial preparation and structural characterization[126,127] of the models showed that synthetic chemistry can duplicate the biological centers in far-simpler chemical systems, which can be more easily studied in great detail.

The general synthetic scheme for Fe-S clusters is shown in Figure 7.13. Many different synthetic procedures can be used to obtain complexes with the Fe_4S_4 core.[126-138,138a,b] The multitude of preparative procedures is consistent with the notion that the $Fe_4S_4^{2+}$ core is the most stable entity present and "spontaneously self-assembles" when not limited by stoichiometric constraints.

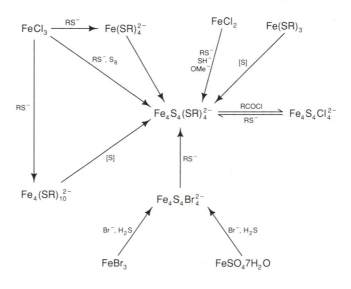

[S] = S_8 or RSSSR

Figure 7.13
Preparative schemes leading to complexes containing the Fe_4S_4 core.[128]

Table 7.4
Structural parameters for $Fe_4S_4(SCH_2C_6H_5)_4^{2-}$.

Atoms[a]	Average distances	Number of bonds	Type
Fe(1)-S(3)	2.310 (3)	8	Sulfide
Fe(1)-S(2)	2.239 (4)	4	Sulfide
Fe(1)-S(5)	2.251 (3)	4	Thiolate
Fe(1)-Fe(2)	2.776 (10)	2	
Fe-Fe(other)	2.732 (5)	4	

Atoms[a]	Average angle	Number of bonds	Type
Fe-S-Fe	73.8 (3)	12	
S-Fe-S	104.1 (2)	12	Sulfide-Fe-Sulfide
S-Fe-S	111.7–117.3	12	Sulfide-Fe-Thiolate

[a] Data from References 126 and 127.

The thiocubane structure can be viewed as two interpenetrating tetrahedra of 4Fe and 4S atoms. The 4S tetrahedra are the larger, since the S-S distance is ~3.5 Å, compared with the Fe-Fe distance of ~2.7 Å. The S_4 tetrahedron encloses ~2.3 times as much volume as does the Fe_4 tetrahedron.[128] Key distances and angles for $Fe_4S_4(SCH_2C_6H_5)_4^{2-}$ given in Table 7.4 are extremely similar to those found in oxidized ferredoxin and reduced HiPIP centers in proteins.[127]

The idealized symmetry of $Fe_4S_4^{2+}$ model systems is that of a regular tetrahedron, i.e., T_d. Though the distortion of the cube is quite pronounced, all known examples of the $Fe_4S_4^{2+}$ core show distortion, which lowers the symmetry at least to D_{2d}. In most $Fe_4S_4^{2+}$ core structures, this distortion involves a tetragonal compression, which leaves four short and eight long Fe-S bonds.

Complexes with non-S-donor peripheral ligands have been prepared and studied. The halide complexes $Fe_4S_4X_4^{2-}$ (X = Cl^-, Br^-, I^-) have been prepared, and serve as useful starting points for further syntheses.[129–133] The complex $Fe_4S_4(OC_6H_5)_4^{2-}$ can be prepared[134] from the tetrachloride (or tetrathiolate) thiocubane by reaction with $NaOC_6H_5$ (or HOC_6H_5). There are a few examples of synthetic $Fe_4S_4^{2+}$ cores in which the peripheral ligands are not identical. For example, $Fe_4S_4Cl_2(OC_6H_5)_2^{2-}$ and $Fe_4S_4Cl_2(SC_6H_5)_2^{2-}$ have structures characterized by D_{2d} symmetry.[135] The complexes $Fe_4S_4(SC_6H_5)_2[S_2CN(C_2H_5)_2]_2^{2-}$ and $Fe_4S_4(SC_6H_4OH)_4^{2-}$ are similarly asymmetric, containing both four- and five-coordinate iron.[136–138] The presence of five-coordinate iron in the Fe_4S_4 cluster is notable, since it offers a possible mode of reactivity for the cluster wherever it plays a catalytic role (such as in aconitase). Complexes with $Fe_4Se_4^{2+}$ and $Fe_4Te_4^{2+}$ cores have also been prepared.[138c,d]

One structural analysis of $Fe_4S_4(SC_6H_5)_4^{3-}$, which contains the reduced $Fe_4S_4^+$ core, revealed a tetragonal elongation[139] in the solid state. In contrast, analysis of $Fe_4S_4(SCH_2C_6H_5)_4^{3-}$ revealed a distorted structure possessing C_{2v} symmetry.[102] It would appear that the $Fe_4S_4^+$ clusters maintain the thiocubane structure, but are nevertheless highly deformable. Interestingly, when the solid-state C_{2v} structure, $Fe_4S_4(SCH_2C_6H_5)_4^{3-}$, is investigated in solution, its spectroscopic and magnetic behavior change to resemble closely those of the $Fe_4S_4(SC_6H_5)_4^{3-}$ cluster,[140] which does not change on dissolution. The simplest interpretation assigns the elongated tetragonal structure as the preferred form for $Fe_4S_4^+$ cores with deformation of sufficiently low energy that crystal packing (or, by inference, protein binding forces) could control the nature of the distortions in specific compounds.[128] The elongated tetragonal structure has four long and eight short bonds in the core structure. The terminal (thiolate) ligands are 0.03–0.05 Å longer in the reduced structure, consistent with the presence of 3Fe(II) and 1Fe(III) in the reduced form, compared to 2Fe(II) and 2Fe(III) in the oxidized form. There is no evidence for any valence localization.[128]

The oxidized $Fe_4S_4^{3+}$ core defied isolation and crystallization in a molecular complex prior to the use of sterically hindered thiolate ligands. With 2,4,6 tris(isopropyl)phenylthiolate, the $Fe_4S_4L_4^-$ complex could be isolated and characterized.[141] The structure is a tetragonally compressed thiocubane with average Fe-S and Fe-SR distances 0.02 and 0.04 Å shorter than the corresponding distances in the $Fe_4S_4L_4^{2-}$ complex. Again, there is no evidence for Fe inequivalence or more profound structural distortion in this 3Fe(III)-1Fe(II) cluster. Clearly, the Fe_4S_4 clusters have highly delocalized bonding.

Evidence from model systems using sterically hindered thiolate ligands indicates the existence of an $Fe_4S_4^{4+}$, i.e., all-ferric fully oxidized cube.[142] The existence of the complete series $Fe_4S_4[(Cy)_3C_6H_2S]_4^n$ (Cy = cyclohexyl; $n = 0$, -1, -2, -3) is implied by reversible electrochemical measurements. Clearly, five different states of the Fe_4S_4 core—including the (at least) transient fully oxidized state and the all-ferrous fully reduced state—may have stable existence. Although only the central three states have been shown to exist in biological contexts, one must not rule out the possible existence of the others under certain circumstances.

Recently, specifically designed tridentate ligands have been synthesized that bind tightly to three of the four Fe atoms in the thiocubane structure.[143,143a,b] The remaining Fe atom can then be treated with a range of reagents to produce a series of subsite-differentiated derivatives and variously bridged double-cubane units. These derivatives illustrate the potential to synthesize complexes that mimic the more unusual features of Fe_4S_4 centers that are bound specifically and asymmetrically to protein sites. The recently synthesized complex ion $[(Cl_3Fe_4S_4)_2S]^{4-}$, containing two Fe_4S_4 units bridged by a single S^{2-} ligand, illustrates the potential coupling of known clusters into larger aggregates.[143c]

The model-system work has made an important contribution to our understanding of the Fe_4S_4 centers. The existence of three states, the exchange of ligands, the redox properties, the metrical details of the basic Fe_4S_4 unit, and the subtleties of structural distortion can each be addressed through the study of models in comparison with the native proteins.

H. Core Extrusion/Cluster Displacement Reactions

Synthetic model-system work led to the realization that the cluster cores can exist outside the protein and undergo relatively facile ligand-exchange reactions.[128] This behavior of the purely inorganic complexes allowed core extrusion reactions[128,144] to be developed. The basic assumption behind these reactions is that the cluster core retains its integrity when it is substituted by low-MW thiolates, especially aryl thiolates, which replace the cysteinyl ligands that bind it to the protein. In order to free the cluster from the protein, one must at least partially denature the protein, usually by using ~80 percent aqueous solution of a polar aprotic solvent, such as DMSO or HMPA. The resulting inorganic clusters can be identified and quantified by measurement of their characteristic electronic absorption or NMR spectra. An alternative approach involves transferal of the unknown cluster in question to an apoprotein that binds a cluster of known type.[145]

Since the 1Fe, 2Fe, and 4Fe sites are each usually bound to the protein by four cysteine residues, it is perhaps not surprising that there have been reports[58] of interconversion of cluster types bound to a given protein. Specifically, in 90 percent aqueous DMSO, the single Fe site in rubredoxin from *C. pasteurianum* is converted to an Fe_4S_4 cluster by the addition of sodium sulfide, ferrous chloride, and ferric chloride in ratio $4:2:1$. Presumably the spacing and geometric disposition of the cysteines are suitable to bind a single Fe or the Fe_4S_4 cluster, which is readily formed under the reaction conditions. Another example of cluster rearrangement involves the three-iron center discussed below that does *not* extrude as an Fe_3S_4 center. Rather, at least under *certain* conditions, the Fe_3S_x center rearranges to form Fe_2S_2 centers.[146] The facile interconversion of the Fe clusters demonstrated the lability of Fe-S systems, and indicates that caution must be exercised in interpreting the results of cluster-displacement reactions.

I. Fe₃S₄ Centers

Three-iron centers are a comparatively recent finding,[119,146] and the full scope of their distribution is not yet known. Although they have now been confirmed in dozens of proteins, it often remains uncertain what physiological role these centers play. Indeed, since Fe_3S_4 centers can be produced as an artifact upon oxidation of Fe_4S_4 centers, it has been suggested that 3Fe centers may not be truly physiological, and could be side products of aerobic protein isolation. This caveat notwithstanding, the 3Fe sites are being found in more and more proteins and enzymes. Their physiological *raison d'être* may be more subtle than that of their 1, 2, and 4Fe cousins; we should certainly try to find out more about them. Some proteins containing Fe_3S_4 centers are listed in Table 7.1.

The 3Fe center was first recognized[147] in the protein ferredoxin I from the anaerobic nitrogen-fixing bacterium *Azotobacter vinelandii*. The protein is called *Av* FdI for short. It is instructive to sketch historically the evolution of our understanding of this protein. *Av* ferredoxin I was reported to have 6 to 8 Fe atoms and was first thought to resemble the clostridial 8Fe ferredoxins. However, unlike the clostridial protein, the *Av* FdI clusters appeared to have two quite different redox couples at $+320$ and -420 mV. Although it might have been thought that this protein contained one HiPIP-type and one Fd- or "ferredoxin"-type Fe_4S_4 cluster, the protein as isolated had an EPR signal with $g = 2.01$, which differed significantly (Figure 7.6) from that of an oxidized HiPIP or a reduced Fd.[148] Cluster extrusion reactions also seemed to indicate the presence of an unusual cluster type.[149]

Fortunately, the protein was crystallized, and could be studied by x-ray diffraction. Unfortunately, the initial conclusions[150] and subsequent revisions[151,152] of the crystal-structure analysis have proven to be wrong, teaching us in the process that protein x-ray crystallography, taken alone, does not always provide definitive results. Specifically, the first crystallographic report suggested the presence of a conventional Fe_4S_4 cluster and a smaller packet of electron density that was assigned as a 2Fe-2S center.[150] However, upon further refinement, and following the formulation of the 3Fe center by Mössbauer spectroscopy,[147] a 3Fe-3S center was identified and refined.[151,152] The "refined" Fe_3S_3 center was a six-membered alternating iron-sulfide ring with an open, almost flat, twisted-boat conformation (Figure 7.14). The Fe-Fe separation of 4.1 Å and the struc-

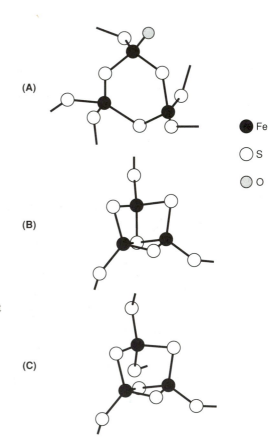

(A)

(B)

● Fe

○ S

◐ O

Figure 7.14
Fe_3S_n structures: (A) open Fe_3S_3 structure proposed from initial x-ray studies (now shown to be in error); (B) the thiocubane fragment structure believed present in most Fe_3S_4 proteins; (C) open Fe_3S_4 structure (not found to date).

(C)

tural type was unprecedented, and did not agree with the results of resonance Raman spectroscopy,[153] with x-ray absorption spectroscopy on the native protein or on samples from which the Fe_4S_4 center was removed,[103] or even with stoichiometry, which eventually led to the reformulation[154] of the cluster as Fe_3S_4. The x-ray absorption studies (EXAFS) clearly led to the assignment of a 2.7-Å Fe-Fe distance for the 3Fe cluster.[103]

In parallel with the studies on *Av* FdI, two additional proteins played key roles in the resolution of the nature of the Fe_3 cluster. These are FdII from *Desulfovibrio gigas* and aconitase from beef heart. Each contains (under certain conditions) only Fe_3S_4 sites, thus enabling more definitive structural, stoichiometric, and spectral information to be acquired. Studies on these proteins using EXAFS,[155] Mössbauer,[52,156,157] EPR,[146] and resonance Raman (to which we will return briefly) clearly favor the closed structure shown in Figure 7.14C. Indeed, x-ray crystallography on aconitase by the same group that did the initial x-ray work on *Av* FdII revealed the compact structure in agreement with the spectroscopy.[20]

In 1988 the structural error in the crystallography of *Av* FdI was found by two groups, and a new refinement in a corrected space group led to a structure in agreement with the spectroscopy.[111,112] The Fe_3S_4 cluster has the apoFe thiocubane structure, with each iron atom bound to the protein by a single cysteinyl thiolate. Clearly, even x-ray crystallography is potentially fallible, and its findings must be critically integrated with the data from other techniques in arriving at full structural definition of metalloenzyme sites.

The studies on the ferredoxins from *D. gigas* present an interesting lesson on the lability of the Fe-S cluster systems. Two distinct proteins from *D. gigas*, FdI and FdII, contain the same polypeptide chain (6 kDa) in different states of aggregation.[158,159] Whereas FdI is a trimer containing three Fe_4S_4 clusters, FdII is a tetramer that contains four Fe_3S_4 clusters. The ferredoxins differ in their redox potentials and appear to have different metabolic functions in *D. gigas*. The oxidation of *D. gigas* FdI with $Fe(CN)_6^{3-}$ leads to FdII, and treatment of FdII with iron salts leads to FdI. The *D. gigas* system reveals the lability and interconvertibility of Fe-S clusters. The recently reported[160] crystal structure of *D. gigas* FdII shown in Figure 7.15 confirms the partial (apoFe) thiocubane Fe_3S_4 center. The iron atoms are ligated by three cysteinyl residues from protein side chains. The cube missing an iron is now firmly established as a viable structural type.

The aconitase system presents yet another fascinating story.[159a] Aconitase is a key enzyme in the Krebs cycle, catalyzing the conversion of citrate and isocitrate through the intermediacy of *cis*-aconitate, as shown in Equation (7.6).

$$
\begin{array}{ccc}
\underset{\text{citrate}}{
\begin{array}{c}
H \\
| \\
H\text{—}C\text{—}COO^- \\
| \\
HO\text{—}C\text{—}COO^- \\
| \\
CH_2\text{—}COO^-
\end{array}
}
&
\underset{\text{\textit{cis}-aconitate}}{
\begin{array}{c}
H \quad COO^- \\
\diagdown\ / \\
C \\
\| \\
C \\
/\ \diagdown \\
H \quad CH_2 \\
| \\
COO^-
\end{array}
}
&
\underset{\text{isocitrate}}{
\begin{array}{c}
H \\
| \\
HO\text{—}C\text{—}COO^- \\
| \\
H\text{—}C\text{—}COO^- \\
| \\
CH_2\text{—}COO^-
\end{array}
}
\end{array}
\qquad (7.6)
$$

with the equilibrium steps labeled: $\xrightleftharpoons[H_2O]{-H_2O}$ between citrate and *cis*-aconitate, and $\xrightleftharpoons[H_2O]{-H_2O}$ between *cis*-aconitate and isocitrate.

Figure 7.15
The x-ray crystal structure of *D. gigas* FdII, illustrating the Fe_4S_4
and Fe_3S_4 centers in this protein.[160]

This is a hydrolytic nonredox process, and for some time it was thought that
aconitase was a simple Fe^{2+} protein wherein the ferrous iron was involved in
the Lewis-acid function of facilitating the hydrolytic reaction. Indeed, aconitase
is inactive when isolated from mitochondria, and requires the addition of Fe^{2+}
to achieve activity.

Surprisingly, the isolated aconitase was found by analysis and Mössbauer
spectroscopy to possess an Fe_3S_4 site in its inactive form.[161] Low-resolution
crystallographic study supports the presence of an apo-Fe thiocubane, Fe_3S_4
structure in aconitase.[162] Resonance Raman[163] and EXAFS[103,155] studies clearly
fingerprint the Fe_3S_4 cluster. The current hypothesis for aconitase activation involves the Fe_3S_4 thiocubane fragment reacting with Fe^{2+} to complete the cube,
which is the active form of the enzyme.[164] Recent crystallographic studies[113]
confirm the presence of a complete cube in the activated aconitase. The dimensions and positioning of the Fe_3S_4 and Fe_4S_4 centers in the cube are virtually
identical. The added Fe^{2+} iron atom is ligated by a water (or hydroxide) ligand,
consistent with the absence of any cysteine residues near the exchangeable iron.
Since this water (or a hydroxide) is also present in the Fe_3S_4 system, one won-
ders whether a small ion such as Na^+ might be present in the Fe_3S_4 aconitase
system.

Questions of detailed mechanism for aconitase remain open. ENDOR spectroscopy[165] shows that both substrate and water (or OH^-) can bind at the cluster. Does one of its Fe-atom vertices play the Lewis-acid role necessary for
aconitase activity? Is the $Fe_3S_4 \rightleftarrows Fe_4S_4$ conversion a redox- or iron-activated
switch, which works as a control system for the activity of aconitase? These
and other questions will continue to be asked. If aconitase is indeed an Fe-S
enzyme with an iron-triggered control mechanism, it may be representative of a

large class of Fe_3/Fe_4 proteins. Other hydrolytic enzymes containing similar Fe-S centers have recently been reported.[166,166a]

Spectroscopically, the Fe_3S_4 center is distinct and clearly distinguishable from 1Fe, 2Fe, and 4Fe centers. The center is EPR-active in its oxidized form, displaying a signal (Figure 7.6) with $g = 1.97, 2.00,$ and 2.06 (*D. gigas* FdII).[158] The Mössbauer spectrum (Figure 7.7) shows a single quadrupole doublet with $\Delta Q = 0.53$ nm/s and isomer shift of 0.27 nm, suggesting the now familiar high-spin iron electronic structure.[158,159] In its reduced form, the center becomes EPR-silent, but the Mössbauer spectrum now reveals two quadrupole doublets of intensity ratio 2:1. The suggestion of the presence of a 3Fe center was first made based on this observation.[147] The picture of the reduced Fe_3S_4 state that has emerged involves a coupled, delocalized Fe^{2+}/Fe^{3+} unit responsible for the outer doublet, with a single Fe^{3+} unit responsible for the inner doublet of half the intensity. The oxidized state contains all Fe^{3+} ions, which are coupled in the trinuclear center.

EXAFS studies were consistent and unequivocal in finding an Fe-Fe distance of $\sim$2.7 Å in all putative Fe_3S_4 proteins.[103,155,167] Resonance Raman spectra compared with those of other proteins and of model compounds with known structures[168,169] for other metals also favored the structure of Figure 7.14B. Clearly, what has been termed the spectroscopic imperative[170] has been crucial in the successful elucidation of the 3Fe structure.

An interesting excursion has led to isolation of what are presumed to be $ZnFe_3S_4$, $CoFe_3S_4$, and $NiFe_3S_4$ thiocubane structures by adding Zn^{2+}, Co^{2+}, or Ni^{2+}, respectively, to proteins containing reduced Fe_3S_4 cores.[170a,b,c] These modified proteins provide interesting electronic structural insights and, potentially, new catalytic capabilities.

J. Fe₃ Model Systems

To date, the Fe_3S_4 center is the only structurally characterized biological iron-sulfide center that does not have an analogue in synthetic Fe chemistry. In fact, the closest structural analogue[21,170d] is found in the $Mo_3S_4^{4+}$ or $V_3S_4^{3+}$ core in clusters such as $Mo_3S_4(SCH_2CH_2S)_3^{2-}$, whose structure is shown in Figure 7.16.

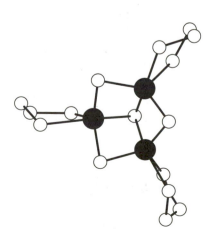

Figure 7.16
The x-ray crystal structure of $Mo_3S_4(SCH_2CH_2S)_3^{2-}$.
(From Reference 10a.)

The resonance Raman spectrum[169] of this complex bears a close resemblance to that of the *D. gigas* ferredoxin II. Since the vibrational bands responsible for the resonance Raman spectrum are not strongly dependent on the electronic properties, it is not surprising that an analogue with a different metal can be identified using this technique.

In synthetic Fe chemistry, although there is no precise structural analogue, it is instructive to consider three types of trinuclear and one hexanuclear center in relation to the three-iron biocenter.

The trinuclear cluster $Fe_3S[(SCH_2)_2C_6H_4]_3^{2-}$ is prepared[171,172] by the reaction of $FeCl_3$, $C_6H_4(CH_2SH)_2$, $Na^+OCH_3^-$, and p-CH_3-C_6H_4-SH in CH_3OH. As shown in Figure 7.17A, this cluster has, like the biocluster, a single triply bridging sulfide ion but, unlike the biocluster, it uses the sulfur atoms of the ethane 1,2-dithiolate as doubly bridging, as well as terminal, groups. The inorganic ring $Fe_3(SR)_3S_6^{3-}$ (X = Cl, Br) has a planar $Fe_3(SR)_3$ core, which resembles the now-discredited structure for *Av* FdI[173] (Figure 7.14A).

The complex $Fe_3S_4(SR)_4^{3-}$ is prepared[174,175] by reaction of $Fe(SR)_4^{2-}$ with sulfur. The x-ray-determined structure reveals two tetrahedra sharing a vertex with the linear Fe-Fe-Fe array shown in Figure 7.17B. This complex has distinctive EPR, Mössbauer, and NMR spectra that allow it to be readily identified.[174,175,175a] Interestingly, after the complex was reported, a study of (denatured) aconitase at high pH (>9.5) revealed that the thiocubane fragment Fe_3S_4 site in that enzyme rearranged to adopt a structure that was spectroscopically almost identical with that of the linear complex.[176] Although the state may not have any physiological significance, it does show that Fe-S clusters different

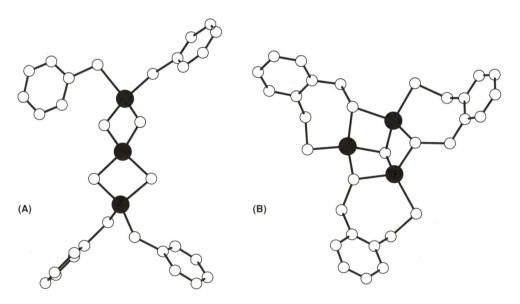

Figure 7.17
The x-ray crystal structures of trinuclear Fe complexes: (A) $Fe_3S_4(SR)_4^{3-}$; (B) $Fe_3S[(SCH_2)_2C_6H_4]_3^{2-}$. (Data from References 171, 172, 174, 175.)

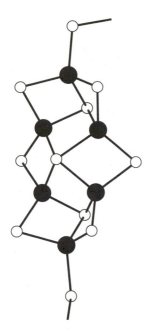

Figure 7.18
The x-ray crystal structure [176a,b] of $Fe_6S_9(SR)_2^{4-}$.

from the common (conventional) ones already discussed could be important in proteins under certain physiological conditions or in certain organisms; i.e., iron centers first identified synthetically may yet prove to be present in biological systems.

A synthetic cluster that displays features related to the biological Fe_3S_4 cluster is the hexanuclear cluster $Fe_6S_9(SR)_2^{4-}$ shown in Figure 7.18. This cluster contains two Fe_3S_4 units bridged through their diiron edges by a unique quadruply bridging S^{2-} ion (μ_4-S^{2-}) and by two additional μ_2-S^{2-} bridges. The inability of synthetic chemists to isolate an Fe_3S_4 analogue may indicate that in proteins this unit requires strong binding. Significant sequestration by the protein may be needed to stabilize the Fe_3S_4 unit against oligomerization through sulfide bridges or, alternatively, rearrangement to the stable Fe_4S_4 center.

K. Fe-S Chemistry: Comments and New Structures

The first successful model system for an iron-sulfur protein was an analogue of the Fe_4S_4 system, i.e., the system with the largest presently established biological Fe cluster. The reactions used to synthesize the cluster shown in Figure 7.13 are said to involve self-assembly, meaning that starting materials are simply mixed together, and thermodynamic control causes the cluster to assemble in its stable form. Interestingly, the Fe_4S_4-containing proteins, such as those of *C. pasteurianum*, are considered to be among the most ancient of proteins. Perhaps on the anaerobic primordial Earth, Fe-S clusters self-assembled in the presence of protein ligands to form the progenitors of the modern ferredoxins.

Much progress has been made in synthetic chemistry, and it is clear that both understanding and control of Fe-S chemistry are continuing to grow. New preparations for known clusters continue to be found, and new clusters continue to be synthesized. Although many of the new clusters appear to be abiological, we should not ignore them or their potential. They add to our understanding of Fe-S chemistry in general, and serve as starting points in the study of hetero-nuclear clusters. There is also the distinct possibility that one or more of these synthetic clusters represent an existing biological site that has not yet been iden-tified in an isolated system.

Among the "nonbiological" structures that have been synthesized are com-plexes with $Fe_6S_6^{3+/2+}$ cores [177,178] including the thioprismanes, [108,109,177-179] the octahedron/cube $Fe_6S_8^{3+}$ cores, [180,181] the $Fe_6S_9^{2-}$ cores [176a,b] discussed above, the adamantane-like $Fe_6(SR)_{10}^{4-}$ complexes related to Zn and Cu structures in metallothioneins, [182] basket $Fe_6S_6^{2+}/Fe_6S_6^{+}$ cores, [182a,b,c,d] monocapped pris-matic $Fe_7S_6^{3+}$ cores, [183] the cube/octahedron $Fe_8S_6^{5+}$ cores, [179] and the circular Na^+-binding $Fe_{18}S_{30}^{10-}$ unit. [183a,b] Representative ions are shown in Figure 7.19. Some of these cores are stabilized by distinctly nonbiological phosphine ligands. Nevertheless, one should not *a priori* eliminate any of these structures from a possible biological presence. Indeed, recently a novel, apparently six-iron pro-tein from *Desulfovibrio gigas* has been suggested [183c] to have the thioprismanc core structure first found in model compounds.

L. Detection of Fe-S Sites

Several recent reviews have concentrated on the ways in which the various Fe-S centers can be identified in newly isolated proteins. [5,6,184] It is instructive to summarize the central techniques used in the identification of active sites. Op-tical spectra are usually quite distinctive, but they are broad and of relatively low intensity, and can be obscured or uninterpretable in complex systems. MCD spectra can give useful electronic information, especially when the temperature dependence is measured. EPR spectra, when they are observed, are distinctive, and are usually sufficiently sharp to be useful even in complex systems. Möss-bauer and resonance Raman spectroscopies have each been applied with good effect when they can be deconvoluted, and NMR and magnetic susceptibility have given important information in some simple, lower-MW protein systems. X-ray absorption spectra, especially EXAFS, give accurate Fe-S and Fe-Fe dis-tances when a single type of Fe atom is present. Analytical and extrusion data complement the spectroscopic and magnetic information. Extrusion data must be viewed with considerable caution, because of possible cluster-rearrangement reactions. Even x-ray crystallography has led to incorrect or poorly refined structures. In general, no one technique can unequivocally identify a site except in the very simplest systems, and there is continued need for synergistic and collaborative application of complementary techniques to a given system.

M. Redox Behavior

Figure 7.20 shows the ranges of redox behavior known for Fe-S centers. Clearly, the Fe-S systems can carry out low-potential processes. The rubredoxins cover the mid-potential range, and the HiPIPs are active in the high-potential region.

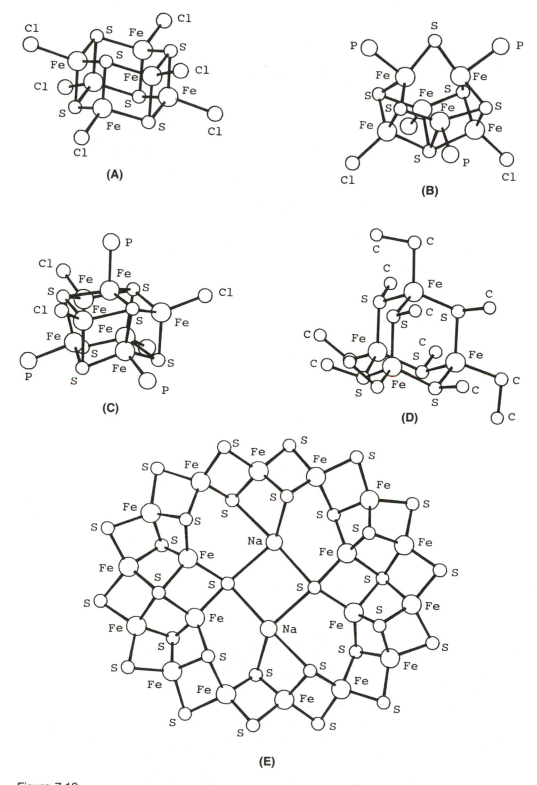

Figure 7.19

Structures of "to date nonbiological" Fe-S clusters: (A) the thioprismane structure;[177] (B) basket-handle structure;[182a] (C) monocapped prismatic structure;[183] (D) adamantane structure;[182] (E) circular $Fe_{18}S_{30}{}^{10-}$ core unit.[183a,b]

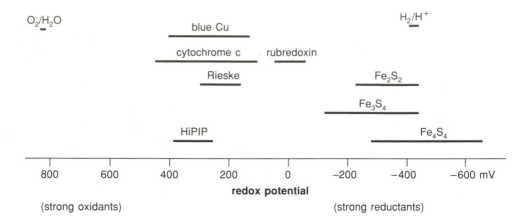

Figure 7.20
A schematic diagram of the redox potential of the various FeS centers in comparison with other known redox centers.

The lack of extensive Fe-S proteins in the positive potential region may reflect their instability under oxidizing conditions and their preemption by Mn, Cu, or heme-iron sites (such as in cytochrome c), which function in this region.

II. MULTISITE REDOX ENZYMES

In making the transition from the relatively simple electron-transfer proteins to the far more complex Fe-S-containing enzymes, we must recognize that the difference in our degree of understanding is enormous. Not only are the catalytic proteins ten to twenty times larger than the redox proteins, but they often also have several subunits and multiple copies of prosthetic groups. Moreover, very few crystal structures are known for the redox enzymes, and none is known at high resolution. In the absence of three-dimensional protein structural information, we do not know the arrangement, relative separation, or orientation of the prosthetic groups. Finally, studies on model systems have not yet approached the sophisticated state that they have for the structurally known centers.

In general, multicomponent redox enzyme systems appear to be organized in two distinct ways to effect their substrate reactions. In the first mode, the enzyme is designed to bring the oxidant and reductant together so that they may directly interact. For example, oxygenases bring O_2 and an organic molecule together, and activate one or both of these reactants to cause them to react directly with each other. This mode can be called *proximation*, as the reactants are brought near each other by the enzyme catalyst.

In contrast to proximation, many redox enzymes keep the oxidant and reductant well-separated, and use rapid (usually long-distance) internal electron transfer to bring electrons from the reductant to the oxidant. We can term their mode of action *electrochemical*. The oxidant and reductant are separated spatially. The enzyme provides the ''anode'' site to interact with the reductant, the ''cathode'' site to interact with the oxidant, and the wire to allow electronic flow between the ''anode'' and ''cathode'' sites. Hydrogenases and nitrogenases adopt the electrochemical mode of redox activation. In hydrogenases, the electron acceptor, even if it must formally take up hydrogen (e.g., $NAD^+ \rightarrow NADH$), does not interact at the same site as the H_2. There is no direct transfer of H^- from H_2 to NAD^+. Rather, H_2 reduces the enzyme at one site, and NAD^+ or other acceptors, such as methylene blue, retrieve the electrons at other sites following internal electron transfer. For nitrogenase, the redox partners are even more removed, as a separate protein, the Fe protein, delivers electrons to the FeMo protein, that eventually end up at the FeMoco site ready to reduce N_2 to NH_3. These enzymes work much like electrochemical cells.

A. Hydrogenase and Nitrogenase

Hydrogenase is the enzyme responsible for the uptake or evolution of H_2. Nitrogenase is the enzyme that catalyzes the ATP-dependent reduction of N_2 to NH_3, with concomitant evolution of H_2.

The relationship between H_2 and N_2 in biology is intricate.[184] Metabolically, H_2 use and N_2 use are tightly coupled in many nitrogen-fixing organisms, with H_2 serving indirectly as the reductant for N_2. Moreover, H_2 and N_2 react in related ways with various transition-metal complexes, which are at present the closest (albeit quite imperfect) models of the enzyme active sites. The biological fixation of molecular nitrogen is dependent on iron-sulfide proteins that also contain molybdenum or vanadium. The biological production or uptake of H_2 depends on the presence of iron-sulfide proteins, which often also contain nickel and sometimes selenium. Spectroscopic and model-system studies, which have played such a key role in advancing the understanding of simple Fe-S sites, are now helping to foster an understanding of these more complex enzyme sites, although we have much yet to learn about structure and mechanism in these enzymes. The remainder of this chapter seeks to convey the state of our rapidly evolving knowledge.

B. Hydrogenases

1. Physiological significance

Molecular hydrogen, H_2, is evolved by certain organisms and taken up by others. For either process, the enzyme responsible is called hydrogenase. The *raison d'être* for hydrogenases in particular organisms depends on the metabolic needs of the organism. Properties of some representative hydrogenases are given in Table 7.5.[184,184a-g]

Hydrogenases are found in a wide variety of anaerobic bacteria, such as the eubacterial *C. pasteurianum* and *Acetobacterium woodii* and the archaebacterial

Table 7.5
Properties of some representative hydrogenases.

Organism (designation)	MW (subunits)	Approximate composition	Reference
Clostridium pasteurianum (Hydrogenase I)	60,000 (1)	12Fe 22Fe	191 194
Clostridium pasteurianum (Hydrogenase II)	53,000 (1)	8Fe 17Fe	191 194
Acetobacterium woodii	15,000	Fe	192
Megasphaera elsdenii	50,000 (1)	12Fe	188
Desulfovibrio vulgaris (periplasmic)	56,000 (2)	12Fe	171
Desulfovibrio gigas	89,000 (2)	11Fe, 1Ni	363
Desulfovibrio africanus	92,000	11Fe, 1Ni	364
Methanobacterium thermoautotrophicum	200,000	Fe, 1Ni	145
Methanosarcina barkeri	60,000	8–10Fe, 1Ni	77
Methanococcus vanielli	340,000	Fe, Ni, 1Se, FAD	365
Desulfovibrio baculatus	85,000 (2)	12Fe, 1Ni, 1Se	365

Methanosarcina barkerii. Interestingly, *C. pasteurianum* sometimes evolves H_2 during its growth on sugars. This H_2 evolution is required for continued metabolism, since it allows the organism to recycle (reoxidize) cofactors that are reduced in the oxidation of sugars (or their metabolic descendants, lactate or ethanol). In effect, H^+ is acting as the terminal oxidant in clostridial metabolism, and H_2 is the product of its reduction. In contrast, methanogens such as methanosarcina take up H_2 and in effect use it to reduce CO_2 to CH_4 and other carbon products. Clearly, either H_2 uptake or H_2 evolution may be important in particular anaerobic metabolic contexts. The hydrogenases of the anaerobic sulfate-reducing bacteria of the genus *Desulfovibrio* have been particularly well-studied (see Table 7.5).

In nitrogen-fixing organisms, H_2 is evolved during the nitrogen-fixation process, and hydrogenase is present to recapture the reducing equivalents, which can then be recycled to fix more nitrogen. In N_2-fixing organisms, such "uptake" hydrogenases can make an important contribution to the overall efficiency of the nitrogen-fixation process. In fact, certain species of rhizobia lacking the hydrogen-uptake system (*hup⁻* strains) can be made more efficient by genetically engineering the *hup* activity into them.

Aerobic bacteria such as *Azotobacter vinelandii*, *Alcaligenes eutrophus*, and *Nocardia opaca*, and facultative anaerobes, such as *Escherichia coli* and various species of *Rhizobium* and *Bradyrhizobium* (the symbionts of leguminous plants), also contain hydrogenase, as do photosynthetic bacteria such as *Chromatium vinosum*, *Rhodobacter capsulatus* (formerly *Rhodopseudomonas capsulata*), and *Anabaena variabilis* (a filamentous cyanobacterium). The thermophilic hydrogen oxidizer *Hydrogenobacter thermophilus*, which grows in alkaline hot springs above 70°C, obviously has a critical requirement for hydrogenase.

In certain aerobic organisms, such as hydrogenomonas, H_2 and O_2 are caused to react (but not directly) according to the Knallgas reaction[185]

$$2H_2 + O_2 \longrightarrow 2H_2O \qquad \Delta G° = -54.6 \text{ kcal/mol} \qquad (7.7)$$

These organisms break up this thermodynamically highly favorable redox process by using intermediate carriers, thereby allowing the large negative free-energy change to be captured in biosynthetic capacity.

Hydrogenases seem to be especially prevalent in anaerobic, nitrogen-fixing, and photosynthetic organisms. However, although hydrogenases are obviously found widely among prokaryotes, unlike nitrogenase, their domain is not restricted to prokaryotes. Eukaryotic green algae such as *Chlorella fusca* and *Chlamydomonas reinhardtii* possess hydrogenase. The anaerobic protozoan *Trichomonas vaginalis*, which lacks typical aerobic organelles, such as mitochondria and peroxisomes, has an organelle called a hydrogenosome, whose function is to oxidize pyruvate to acetate, producing H_2, via hydrogenase, in the process.

The various hydrogenase enzymes are all transition-metal sulfide proteins. However, before we discuss these enzymes, we turn briefly to the dihydrogen molecule and its physical and chemical properties.

2. Dihydrogen: the molecule

Diatomic H_2 has a single H—H bond formed by overlap of the two 1s orbitals of the two hydrogen atoms. In molecular orbital terms, this overlap forms bonding σ and antibonding σ^* orbitals, shown in the energy-level diagram in Figure 7.21A and displayed spatially in Figure 7.21B. The H-H distance is 0.74 Å, and the bond dissociation energy is 103.7 kcal/mole. The isotopes of hydrogen $^1H_1$, $^2H_1 = {}^2D_1$, $^3H_1 = {}^3T_1$ are called protium (a designation seldom used), deuterium, and tritium, respectively. Deuterium, at natural abundance of 0.015 percent, is a stable isotope with nuclear spin $I = 1$, whereas both ^1H and ^3T have nuclear spin $I = \frac{1}{2}$. NMR has been fruitfully applied to all the hydrogen isotopes, including tritium. Tritium is radioactive, decaying to $^3He_2$ by β^- emission with a half-life of 12 years. The nuclear properties of deuterium and tritium make them useful as labels to probe structure and mechanism in hydrogen-containing compounds. "Exchange" reactions involving the formation of HD or HT have played a significant role in mechanistic studies of both hydrogenase and nitrogenase.

In molecular hydrogen, the existence of nuclear-spin energy levels is responsible for the distinction between *ortho* and *para* hydrogen, which correspond to the triplet and singlet (i.e., parallel and antiparallel) orientations, respectively, of the two nuclei in H_2. Because of the coupling of the rotational and spin levels, *ortho* and *para* hydrogen differ in specific heat and certain other properties. The correlated orientation of the nuclear spins in *para* H_2 has re-

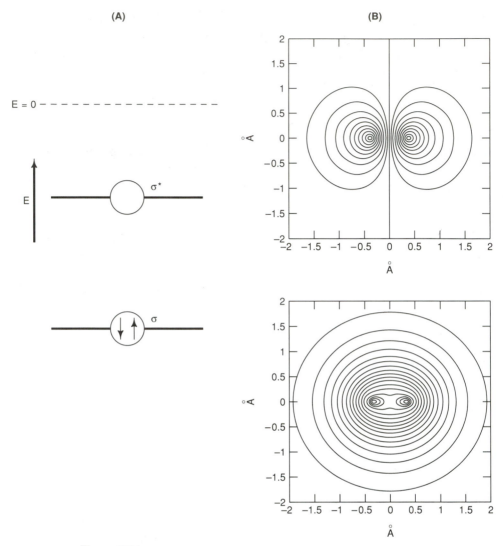

(A)

E = 0 – – – – – – – – – – – – – –

σ^*

σ

(B)

Figure 7.21
Molecular orbital scheme for binding in H_2: (A) energy-level diagram;
(B) bonding and antibonding orbitals.

cently been shown to constitute a powerful mechanistic probe, wherein NMR may be used to trace the relative fate of the two H nuclei in the original molecule.[186,187] Although this technique has not yet been applied to any enzyme systems, hydrogenase is known to catalyze the interconversion of the *ortho* and *para* forms of H_2 (as does the hydrogenase analogue Pd).

Dihydrogen is a reducing agent. The H_2/H^+ couple at $[H^+] = 1\ M$ defines the zero of the potential scale. At pH = 7 the hydrogen half-reaction

$$H_2 \longrightarrow 2H^+ + 2e^- \qquad (7.8)$$

has $E_o' = -420$ mV. Dihydrogen is therefore one of the strongest biological reductants.

Although many hydrogenases are reversible, some "specialize" in the uptake of H_2. One hydrogenase has been reported[188] to specialize in the evolution of H_2. This "specialization" seems curious, since it appears to contradict the notion of microscopic reversibility, and seems to violate the rule that catalysts increase the speed of both forward and backward reactions without changing the course (direction) of a reaction. In fact, there is no contradiction or violation, since the overall reactions catalyzed by the various types of hydrogenases are fundamentally different. The electron acceptor in uptake hydrogenases *differs* from the electron donor/acceptor in the reversible hydrogenases. The difference involves structure and, more importantly, redox potential. The reaction catalyzed by the uptake hydrogenase involves an acceptor of such high positive redox potential that its reaction with H_2 is essentially irreversible. The enzyme appears to be designed so that it can transfer electrons only to the high potential acceptor.

A selection of hydrogenases from various organisms is given in Table 7.5. All hydrogenases contain Fe-S centers. The hydrogenases from more than 20 organisms[189] have been found to contain Ni by analysis and/or spectroscopy. Many more Ni hydrogenases are likely to be found, given the nutritional requirements[189] for hydrogenase synthesis or growth on H_2. Hydrogenases may be cytoplasmic (as in *C. pasteurianum*), membrane-bound (as in *E. coli*), or located in the periplasmic space (as in *Desulfovibrio vulgaris*). The isolation of hydrogenases is sometimes complicated by their air sensitivity or membrane-bound nature. Many hydrogenases have now been isolated and studied in detail; they can be divided into two categories, the iron hydrogenases and the nickel-iron hydrogenases.

3. Iron hydrogenases

The iron hydrogenases[189a] generally have higher activities than the NiFe enzymes, with turnover numbers approaching 10^6 min^{-1}. Iron hydrogenases from four genera of anaerobic bacteria have been isolated: *Desulfovibrio*,[190] *Megasphaera*,[188] *Clostridium*,[191] and *Acetobacterium*.[192] Of these, the enzymes from *Desulfovibrio vulgaris*, *Megasphaera elsdenii*, *Clostridium pasteurianum* (which contains two different hydrogenases), and *Acetobacterium woodii*[192] have been well-characterized (especially the *D. vulgaris* and *C. pasteurianum* enzymes). Although *Acetobacterium* and *Clostridium* are closely related, the other organisms are only distant cousins.[193] Nevertheless, their hydrogenases display significant similarities; all contain two different types of iron-sulfur cluster, called F and H clusters,[194] and carbon monoxide is a potent inhibitor (although this has not been reported for the *M. elsdenii* enzyme). The F clusters are thought to be of the $Fe_4S_4{}^{+/2+}$ thiocubane type, and give $S = \frac{1}{2}$ EPR signals when the enzyme is in the reduced form. On the other hand, the H cluster, which is thought to be the hydrogen-activating site, gives an EPR signal only when the enzyme is in the oxidized form. The H-cluster EPR signals of all the enzymes are quite similar ($g = 2.09, 2.04, 2.00$), and are quite unlike the signals from other oxidized iron-sulfide clusters (such as Fe_3S_4 clusters and HiPIPs), in that they are observable at relatively high temperatures (>100 K). Inhibition of the *D. vulgaris* and both *C. pasteurianum* enzymes by carbon-monoxide yields a photosensitive species that has a modified H-cluster EPR signal.[195,196]

The two different hydrogenases of *C. pasteurianum*, called hydrogenase I and II, have both been quite extensively studied, and can be regarded as proto-typical iron-only hydrogenases. Hydrogenase I is active in catalyzing both H_2 oxidation and H_2 evolution, whereas hydrogenase II preferentially catalyzes H_2 oxidation.[188] The two enzymes differ in their iron contents: hydrogenase I contains about 20 iron atoms, 16 of which are thought to be involved in four F clusters,[194] while the remainder presumably constitute the H cluster, which may contain six Fe atoms.[194] Hydrogenase II contains about 14 iron atoms as two F clusters and one H cluster.[194] These estimates of iron content result from a recent reappraisal of the metal contents (based on amino-acid analysis) that in-dicated a rather higher Fe content than previously realized.[188,194] It is important to note that much of the spectroscopic work, which will be discussed below, was initially interpreted on the basis of the earlier, erroneous, iron analysis. Of particular interest is the possibility (first suggested[197] for the *D. vulgaris* en-zyme) that the H cluster contains six iron atoms.

Carbon-monoxide treatment of the *D. vulgaris* and both *C. pasteurianum* enzymes yields a photosensitive species that has a modified H cluster EPR sig-nal.[195,196] Interestingly, the *C. pasteurianum* enzymes also form complexes with O_2, in a process that is distinguishable from the *deactivation* of hydrogenase by O_2, which results from a much more prolonged exposure to O_2 than that re-quired to form the O_2 complex. The O_2 complexes have (photosensitive) EPR signals much like those of the CO complex.[196] It is important to note that al-though CO, when in excess, is a potent inhibitor of the enzymes, the hydroge-nase I-CO complex is actually quite active.[196] With hydrogenase II, the CO complex is dissociated on exposure to H_2, restoring the "active" enzyme.[196]

The EPR spectrum of reduced hydrogenase I is typical of (interacting) $Fe_4S_4^+$ clusters, and integrates to 3 or 4 spins/protein.[194,198] Electrochemical studies[199] show that these clusters possess indistinguishable reduction potentials. Recently, MCD and EPR spectroscopies have been used to demonstrate the presence of significant quantities of an $S = \frac{3}{2}$ species in reduced hydrogenase I. This signal apparently integrates to about one spin per molecule, and probably originates from an $S = \frac{3}{2}$ state of an Fe_4S_4 cluster.[198] No information is yet available on the reduction potential of the $S = \frac{3}{2}$ species. However, based on analogy with the nitrogenase iron protein,[200] we might expect the $S = \frac{3}{2}$ form to have electro-chemistry indistinguishable from the $S = \frac{1}{2}$ form. EPR signals with high g val-ues ($g = 6.1$ and 5.0) have also been observed in *C. pasteurianum* hydrogenase I, and in the *D. vulgaris* enzyme.[198,201] Since there is some uncertainty about the nature and origin of these signals,[198] we will not discuss them further. The F clusters of hydrogenase II, on the other hand, give two different EPR signals that integrate to one spin each per protein molecule, and that correspond to sites with different redox potentials.[194,198,199] This suggests that hydrogenase II con-

407

(A)

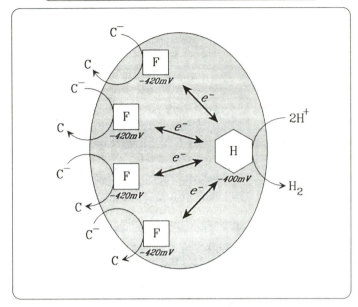

(B)

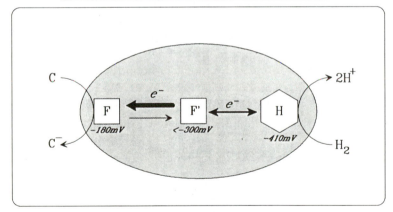

Figure 7.22
Redox schemes illustrating proposed action of Fe hydrogenases: (A) *Clostridium pasteurianum* hydrogenase I; (B) *Clostridium pasteurianum* hydrogenase II.[189a,199]

tains two different F clusters, called F and F′ (note that the presence of F′ was in fact first suggested by Mössbauer spectroscopy[202]). The EPR spectrum from the F′ cluster is unusually broad. The H-cluster EPR signals of active hydrogenase I and II are quite similar and have essentially identical redox potentials.[199]

The redox behavior of the F and H centers in hydrogenases I and II is nicely consistent with their respective modes of function. As shown in Figure 7.22, the F clusters are presumed to transfer electrons intermolecularly with the external electron carrier and intramolecularly with the H center. In the reversible hydrogenase I, the F and H centers have the same redox potentials (about -400 mV at pH 8), similar to that of the hydrogen electrode (-480 mV at pH 8). Thus, electrons may flow in either direction when a mediator such as methyl viologen ($E^{o\prime} = -440$ mV at pH 8) is used as the external electron acceptor (methyl viologen is the 4,4'-bipyridinium ion). On the other hand, for hydrogenase II, the F clusters have $E^{o\prime}$ (pH 8) $= -180$ mV and $E^{o\prime} > -300$ mV for F and F', respectively. In hydrogenase II, therefore, electrons can only move favorably from H_2 to the H cluster, through F' and F, and then to a lower-potential acceptor, such as methylene blue [for which $E^{o\prime}$ (pH 8) $= 11$ mV].

Mössbauer spectroscopic studies of both hydrogenase I and II have been reported.[202,203] Our discussion focuses primarily on the H cluster. The results are similar for the two enzymes, but better defined for hydrogenase II because of the smaller number of clusters. The H cluster apparently contains only two types of iron, in the ratio of 2:1, with quadrupole splittings reminiscent of Fe_3S_4 clusters. The oxidized cluster is confirmed to be an $S = \frac{1}{2}$ system, also reminiscent of Fe_3S_4 clusters, and the reduced H cluster is an $S = 0$ system; this contrasts with reduced Fe_3S_4 clusters, which have $S = 2$. In agreement with the Mössbauer studies, ENDOR spectroscopy of ^{57}Fe-enriched protein indicates at least two different types of iron in the H cluster, with metal hyperfine couplings of about 18 and 7 MHz in hydrogenase II. Rather-less-intense ENDOR features were also observed at frequencies corresponding to couplings of about 11 and 15 MHz.[204] The H-cluster EPR signals of hydrogenase II and hydrogenase I change on binding carbon monoxide. Although the signals of the uncomplexed enzymes are quite similar, the signals of the CO-bound species are very different (note, however, that *C. pasteurianum* CO-bound hydrogenase I has EPR similar to that of the CO-bound *D. vulgaris* enzyme). When produced with ^{13}C-enriched CO, the EPR signal of hydrogenase II shows resolved ^{13}C hyperfine coupling ($A_{av} = 33$ MHz) to a single ^{13}C nucleus, indicating that only a single CO is bound, presumably as a metal carbonyl. A slightly smaller coupling of 20 MHz was obtained using ENDOR spectroscopy for the corresponding species of hydrogenase I.[205]

Recent ESEEM spectroscopy of hydrogenase I indicates the presence of a nearby nitrogen, which may be a nitrogen ligand to the H-cluster. This nitrogen possesses an unusually large nuclear electric quadrupole coupling and a rather novel structure, involving an amide amino-acid side chain connected to an H-cluster sulfide via a bridging proton ligand, has been suggested for it.[206] Although the nitrogen in question must come from a chemically novel species, the proposed proton bridge might be expected to be exchangeable with solvent water. The ENDOR-derived result[205] that there are no strongly coupled exchangeable protons in the oxidized H cluster may argue against such a structure.

Rather weak MCD[198,207] and resonance Raman spectra[208] have also been reported for iron hydrogenases. The lack of an intense MCD spectrum[198,207] contrasts markedly with results for other biological FeS clusters. The resonance Raman spectra of hydrogenase I resemble, in some respects, spectra from Fe_2S_2

sites.[208] These results further emphasize that the hydrogenase H clusters are a unique class of iron-sulfur clusters.

Perhaps most tantalizing of all are the recent EXAFS results on hydrogenase II.[209] It is important to remember that for enzymes with multiple sites, the EXAFS represents the sum of all sites present (i.e., the iron of the F, F', and H clusters). Despite this complication, useful information is often forthcoming from these experiments. The EXAFS of oxidized hydrogenase I showed both iron-sulfur and iron-iron interactions; the latter, at about 2.7 Å, is at a distance typical of Fe_2S_2, Fe_3S_4, and Fe_4S_4 clusters, and thus is not unexpected. The reduced enzyme, however, gave an additional, long Fe-Fe interaction at 3.3 Å. This Fe-Fe separation is not found in any of the FeS model compounds reported to date.[209] The appearance of the 3.3-Å interaction indicates a change in structure on reduction of the H cluster, again revealing a cluster of unique structure and reactivity. The large structural change of this H cluster on H_2 reduction is likely to have significant mechanistic implications.

4. Nickel-iron hydrogenases

The presence of nickel in hydrogenases has only been recognized relatively recently. Purified preparations of the active enzymes were the subject of quite intensive studies for years before the Ni content was discovered by nutritional studies (see Reference 189 for a history). Some workers even tried (in vain) to purify out "impurity" EPR signals that were later found to be from the Ni. In contrast to the Fe hydrogenases discussed in the previous section, the Ni enzymes possess a variety of compositions, molecular weights, activation behavior, and redox potentials.[189,210,211] As Table 7.5 shows, some of the Ni hydrogenases contain selenium, likely in the form of selenocysteine, some contain flavin (FMN or FAD), and all contain iron-sulfur centers, but in amounts ranging from 4 to 14 iron atoms per Ni atom.

Among the different Ni hydrogenases there is a common pattern of protein composition, to which many, but not all, seem to conform (especially those enzymes originating from purple eubacteria). There are two protein subunits, of approximate molecular masses 30 and 60 kDa, with the nickel probably residing in the latter subunit. The hydrogenase of the sulfate-reducing bacterium *Desulfovibrio gigas* is among the best investigated, and we will concentrate primarily on this enzyme. *D. gigas* hydrogenase contains a single Ni, two Fe_4S_4 clusters, and one Fe_3S_4 cluster. Of primary interest is the Ni site, which is thought to be the site of H_2 activation.[189,210,211]

EPR signals attributable to mononuclear Ni [as shown by enrichment with ^{61}Ni ($I = \frac{3}{2}$)] have been used in numerous investigations of the role of Ni in hydrogenases.[189,210,211] Three major Ni EPR signals are known, which are called Ni-A, Ni-B, and Ni-C. The principal g values of these signals are: 2.32, 2.24, and 2.01 for Ni-A; 2.35, 2.16, and 2.01 for Ni-B; and 2.19, 2.15, and 2.01 for Ni-C. Of these, Ni-C is thought to be associated with the most active form of the enzyme (called active); the other two are thought to originate from less-active enzyme forms.[189,210]

In the enzyme as prepared (aerobically) the Ni-A EPR is characteristically observed. On hydrogen reduction the Ni-A EPR signal disappears, and the enzyme is converted into a higher-activity form (Ni-B arises from reoxidation of this form). Further progressive reduction of the enzyme gives rise to the Ni-C EPR signal, which also finally disappears. These redox properties show that Ni-C arises from an intermediate enzyme oxidation state. Although the Ni-A and Ni-B EPR signals[189,210–212] almost certainly originate from low-spin Ni(III), the formal oxidation state of Ni-C is rather less certain. Both an Ni(I) site[189,210] and an Ni(III) hydride[213] have been suggested, with the former alternative currently favored because of the apparent absence of the strong proton hyperfine coupling expected for the latter. In the fully reduced enzyme, Ni-C is converted to an EPR-silent species. This has variously been suggested to be Ni(0), Ni(II), or an Ni(II) hydride.[214] One possible reaction cycle[189,210] is shown in Figure 7.23.

Information on the coordination environment of the nickel has been obtained from both x-ray absorption spectroscopy and EPR spectroscopy. The Ni K-edge EXAFS of several different hydrogenases,[215–218] and EPR spectroscopy of ^{33}S enriched *Wolinella succinogenes* hydrogenase,[219] clearly indicate the presence of sulfur coordination to nickel. A recent x-ray absorption spectroscopic investigation of the selenium-containing *D. baculatus* hydrogenase, using both Ni and Se EXAFS, suggests selenocysteine coordination to Ni.[216]

ESEEM spectroscopy of the Ni-A and Ni-C EPR signals[220,221] indicate the presence of ^{14}N coupling, which probably arises from a histidine ligand to Ni.

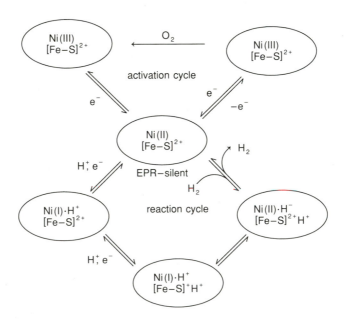

Figure 7.23
Proposed activation/reactivity scheme for Ni hydrogenases.[189]

Interestingly, Ni-C, but not Ni-A, shows coupling to a proton that is exchangeable with solvent water. Although this coupling is too small to suggest a nickel-hydride (consistent with conclusions drawn from EPR), the proton involved could be close enough to the Ni to play a mechanistic role.

Despite the extensive studies reported to date, there are still many unanswered questions about the mechanism of the NiFe hydrogenases, which remain as exciting topics for future research. Despite our lack of detailed knowledge of enzyme mechanism, it is nevertheless not premature to seek guidance from inorganic chemistry.

5. Insights from inorganic chemistry

Recent years have brought insights into the way dihydrogen can be bound at a transition-metal site. Unexpectedly, it has been shown that molecular H_2 forms simple complexes with many kinds of transition-metal sites.[222-224] This finding contrasts with the classical situation, in which H_2 interacts with a transition-metal site by oxidative addition to form a dihydride complex.[23] The H—H bond is largely maintained in the new/nonclassical structures. The dihydrogen and dihydride complexes can exist in simple equilibrium,[224] as in Equation (7.9).

$$M-N\equiv N + H_2 \longrightarrow M \leftarrow \begin{matrix} H \\ | \\ H \end{matrix} \rightleftarrows M \begin{matrix} H \\ \diagup \\ \diagdown \\ H \end{matrix} \qquad (7.9)$$

The bonding of dihydrogen to a metal occurs via the σ^b orbital of the H—H bond acting as a donor, with the σ^* level of H_2 acting as a weak acceptor. If the back donation is too strong, sufficient electron density will build up in the σ^* level to cause cleavage of the H—H bond, leading to the formation of a dihydride. Dihydrogen complexes therefore require a delicate balance, in which the metal coordination sphere facilitates some back-bonding, but not too much.

The proclivity of a metal center to form H_2 complexes can be judged by the stretching frequency of the corresponding N_2 complexes: N_2 can usually displace H_2 from the H_2 complex to form an N_2 complex without changing the remainder of the coordination sphere. If $\nu(N\equiv N)$ is between 2060 and 2160 cm^{-1}, H_2 complexes form upon replacement of N_2. If $\nu(N\equiv N)$ is less than 2060 cm^{-1}, indicative of electron back-donation from the metal center, a dihydride complex should form. For example, $\nu(N\equiv N) = 1950$ cm^{-1} in $MoN_2(PCy_3)_5$ and $MoH_2(PCy_3)_5$ is a dihydride complex, but $\nu(N\equiv N) = 2090$ cm^{-1} in $Mo(Ph_2PCH_2CH_2PPh_2)_2CO(N_2)$ and $Mo(Ph_2PCH_2CH_2PPh_2)H_2(CO)_2$ is a dihydrogen complex. By comparison, $Mo(Et_2PCH_2CH_2PEt_2)_2(CO)(N_2)$ has $\nu(N\equiv N) = 2050$ cm^{-1} and forms a dihydride complex, $Mo(Et_2PCH_2CH_2PEt_2)_2H_2(CO)$. The correlation between $\nu(N\equiv N)$ and the type of hydrogen complex formed seems quite useful.

Since Fe-S and Ni-S sites are implied for hydrogenase, the reactivity of transition-metal/sulfide systems with H_2 may also be relevant. Interestingly, H_2

can react with metal-sulfide systems at S instead of at the metal site. For example,[225] $(Cp')_2Mo_2S_4$ reacts with H_2 to form $(Cp')_2Mo_2(SH)_4$. Here the dihydrogen is cleaved without any evidence for direct interaction with the metal center, and the resulting complex contains bridging SH groups and no direct metal-H bonding.[225] In recent work,[226] the binuclear rhodium-sulfur complex $\{RhS[P(C_6H_5)_2CH_2CH_2]_3CH\}_2$ was reported to react with two equivalents of dihydrogen to yield the complex $\{Rh(H)(SH)[P(C_6H_5)_2CH_2CH_2]_3CH\}_2$, in which two SH groups bridge the two Rh centers, each of which contains a single hydrido ligand. Figure 7.24 illustrates the possibilities for hydrogen activation. Each of these types of reactivity must be considered as possibilities for the hydrogen activation process of hydrogenase.

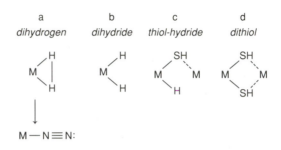

Figure 7.24
Possible modes of H_2/H bonding at a transition-metal sulfide site.

Recently, a great deal of attention has been given to the chemistry of nickel-sulfur systems, inspired in part by the results showing that many hydrogenases are nickel-sulfur proteins.[227–230] A particularly interesting finding is that Ni thiolates can react with O_2 to produce sulfinate complexes.[231,232] The oxygenated thiolate can be regenerated, thus providing a potential model for the O_2 inactivation of Ni hydrogenases.

C. Nitrogenases

Nitrogen fixation is a key reaction of the biological nitrogen cycle.[233] Fixed nitrogen, in which N is in molecules other than N_2, is frequently the limiting factor in plant growth.[234] Since natural systems often cannot provide enough fixed nitrogen for agriculture or animal husbandry, industrial processes have been developed to "fix nitrogen" chemically. The major process in use, often referred to as ammonia synthesis, is the Haber-Bosch process, in which N_2 and H_2 are reacted at temperatures between 300–500°C and pressures of more than 300 atm, using catalysts (usually) based on metallic iron.[235] Hundreds of massive chemical plants are located throughout the world, some producing more than 1,000 tons of NH_3/day. In contrast, in the biological process, N_2 is reduced locally as needed at room temperature and ~ 0.8 atm by the enzyme system called nitrogenase (variously pronounced with the accent on its first or second syllable).

1. The scope of biological nitrogen fixation

Biological nitrogen fixation occurs naturally only in certain prokaryotic organisms (sometimes called diazotrophs). Although the majority of bacterial species are not nitrogen fixers, the process of nitrogen fixation has been confirmed in at least some members of many important phylogenetic groups. Nitrogen fixation occurs in strict anaerobes such as *Clostridium pasteurianum*, in strict aerobes such as *Azotobacter vinelandii*, and in facultative aerobes such as *Klebsiella pneumoniae*. Much of the established biochemistry of N_2 reduction has been gleaned from studies of these three species. However, nitrogen fixation has a far broader range, occurring in archaebacterial methanogens such as *Methanobacillus omelianskii*, which produce methane, and eubacterial methanotrophs such as *Methylococcus capsulatus*, which oxidize methane. Photosynthetic organisms ranging from the purple bacterium *Rhodobacter capsulatus* (formerly *Rhodopseudomonas capsulata*) to the cyanobacterium (blue-green alga) *Anabaena cylindrica* fix nitrogen. Nitrogen fixation occurs mostly in mesophilic bacteria (existing between 15 and 40°C), but has been found in the thermophilic archaebacterial methanogen *Methanococcus thermolithotrophicus* at 64°C.

Many organisms fix N_2 in nature only symbiotically. Here the most studied systems are species of *Rhizobium* and *Bradyrhizobium*, which fix nitrogen in the red root nodules of leguminous plants such as soybeans, peas, alfalfa, and peanuts. The red color inside the nodules is due to leghemoglobin, a plant O_2-binding protein analogous to animal myoglobins and hemoglobins (Chapter 4). Other symbioses include that of blue-green algae such as *Anabaena azollae* with *Azola* (a water fern); actinomycetes such as *Frankia* with trees such as alder; and *Citrobacter freundii*, living in the anaerobic hind gut of termites. The distribution of nitrogenase clearly points to its adaptability as a metabolic option for species occupying widespread ecological niches.

The absence of nitrogen fixation in eukaryotes therefore seems somewhat puzzling. There would appear to be no fundamental limitation to the existence of nitrogen fixation in higher organisms. Indeed, *nif* genes have been transferred to yeast, where they work effectively under anaerobic conditions. Furthermore, the problem of the simultaneous presence of nitrogen fixation and aerobiosis has been solved effectively by aerobic bacteria such as *Azotobacter*, *Gleocapsa*, and *Anabaena*. Indeed, the lack of a fundamental limitation has encouraged researchers to propose the construction of nonsymbiotic nitrogen-fixing plants (whose niche to date is limited to the grant proposal).

Due to the mild conditions under which it occurs, the biological nitrogen-fixation process may seem inherently simpler than the industrial one. However, it is not; the biological process displays a complexity[235,235a] that belies the simplicity of the chemical conversion of $N_2 \rightarrow 2NH_3$. Genetic analysis reveals that

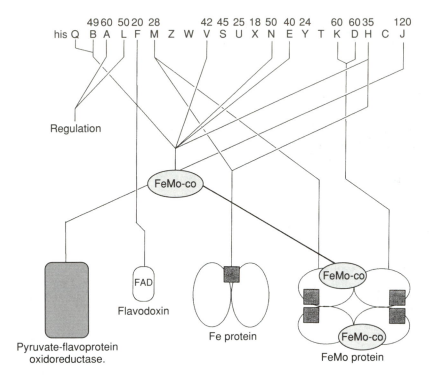

Figure 7.25

Nif genes required for nitrogen fixation as arranged in *Klebsiella pneumoniae*, and their respective gene products. The numbers on the top are the molecular weights in kilodaltons of the respective gene products.

at least twenty genes are required for nitrogen fixation in the bacterium *Klebsiella pneumoniae*.[236,237] These *nif* genes (illustrated in Figure 7.25) specify proteins that are involved in regulation (*nif* A and L), pyruvate oxidation/flavin reduction (*nif* J), electron transfer (*nif* F for flavodoxin), the subunits of the structural proteins of the nitrogenase (*nif* H, D, K), Fe-S cluster assembly (*nif* M) and biosynthesis of the iron-molybdenum cofactor, FeMoco (*nif* N, B, E, Q, V, H).[238] The last two functions specify proteins that are responsible for the incorporation of unusual transition-metal sulfide clusters into the nitrogenase proteins. These clusters have allowed nitrogenase to be studied by biophysical and bioinorganic chemists to establish aspects of its structure and mechanism of action.

We will first discuss the N_2 molecule and focus on its reduction products, which are the presumed intermediates or final product of nitrogen fixation. We then present what has been called[239,240] the "Dominant Hypothesis" for the composition, organization, and function of molybdenum-based nitrogenases. Until 1980, it was thought that molybdenum was essential for nitrogen fixation. However, work starting in 1980 led finally in 1986 to the confirmation of vanadium-based nitrogen fixation. The newly discovered vanadium-based nitrogenases differ in reactivity from the Mo-based enzyme in having "alternative" substrate specificity. The distinct reaction properties of the different nitrogenases point to the importance of the study of alternative substrate reactions in probing the mechanism of nitrogen fixation.

2. Dinitrogen: The molecule and its reduced intermediates

The N_2 molecule has a triple bond with energy 225 kcal/mole, a $\nu(N\equiv N)$ stretch of 2331 cm^{-1}, and an $N\equiv N$ distance of 1.098 Å. The stable isotopes of nitrogen are $^{14}N(I = 1)$ with natural abundance of 99.64 percent and $^{15}N(I = \frac{1}{2})$ with an abundance of 0.36 percent.

The challenge to which nitrogenase rises is to break and reduce at a reasonable rate the extremely strong $N\equiv N$ triple bond. The kinetic inertness of N_2 is highlighted by the fact that carrying out reactions "under nitrogen" is considered equivalent to doing the chemistry in an inert atmosphere. Despite this kinetic inertness, thermodynamically the reduction of N_2 by H_2 is a favorable process,

$$N_2 + 3H_2 \longrightarrow 2NH_3 \qquad \Delta G° = -3.97 \text{ kcal/mole} \tag{7.10}$$

and at pH = 7 the reaction

$$N_2 + 8H^+ + 6e^- \longrightarrow 2NH_4{}^+ \qquad E°' = -280 \text{ mV} \tag{7.11}$$

has an $E°'$ value that makes it easily accessible to biological reductants such as the low-potential ferredoxins discussed earlier in this chapter.

What, then, is the cause of the kinetic inertness of the N_2 molecule? The thermodynamically favorable reduction of N_2 to $2NH_3$ is a six-electron process. Unless a concerted $6e^-$, $6H^+$ process can be effected, intermediates between N_2 and NH_3 must be formed. However, all the intermediates on the pathway between N_2 and NH_3 are higher in energy than either the reactants or the products. The $E°'$ values for the formation of N_2H_2 (=diimine, diazene, diamide) or N_2H_4 (hydrazine) are estimated[241] as

$$N_2 + 2H^+ + 2e^- \longrightarrow N_2H_2 \qquad E°' \sim -1000 \text{ to } -1500 \text{ mV} \tag{7.12}$$
$$N_2 + 5H^+ + 4e^- \longrightarrow N_2H_5{}^+ \qquad E°' = -695 \text{ mV} \tag{7.13}$$

Clearly, these potentials are sufficiently negative that the normal biological reductants cannot effect the reaction. The difficulty of reaching these intermediates is indicated in Figure 7.26.

Several factors may allow this barrier to be overcome. First, the six-electron reduction might be carried out in a concerted or near-concerted manner to avoid the intermediates completely. Alternatively, the intermediates could be complexed at metal centers to stabilize them to a greater extent than either the reactants or products. Finally, the formation reaction for the unfavorable intermedi-

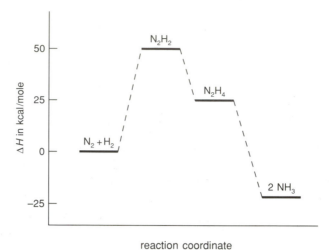

Figure 7.26

Energetics of N_2, NH_3, and some potential intermediates along the reaction pathway for their interconversion.

ate could be coupled with ATP hydrolysis or with the evolution of dihydrogen, each a favorable process, so that the overall process is favorable. Which of the above strategies is used by nitrogenase is unknown, but it seems likely that some combination of the last two of these is used to effect the difficult reduction of N_2 to NH_3. To probe the possibilities, a variety of complexes of N_2, diazenes, and hydrazines has been prepared and chemically characterized, and these are discussed toward the end of this section.

3. The Dominant Hypothesis for molybdenum nitrogenase[239,240,242,243]

The action of the Mo-nitrogenase enzyme involves the functioning of two separately isolatable component proteins, as sketched in Figure 7.27A. The larger of the two proteins, sometimes incorrectly[244] designated[245] dinitrogenase has, in the past, been called molybdoferredoxin, azofermo, or component I. More often this protein is called the MoFe or FeMo protein ([MoFe] or [FeMo]). The smaller protein, formerly called azoferredoxin or component II, is sometimes incorrectly[244] referred to[245] as dinitrogenase reductase.* This protein is properly

---

* The nomenclature proposal[245] that [FeMo] be designated as dinitrogenase and [Fe] as dinitrogenase reductase, although sometimes used in the literature, is incorrect or, at best, premature.[244] The suggested nomenclature implies that both [FeMo] and [Fe] are enzymes. However, neither protein can function catalytically in the absence of the other. [FeMo] will not reduce N_2 or C_2H_2 or evolve H_2 in the absence of [Fe]. The iron protein will not hydrolyze MgATP in the absence of [FeMo]. Nitrogen fixation requires the simultaneous presence of both proteins. Although mechanistic considerations[255] point to [FeMo] as the substrate binding and reducing protein, and [Fe] as the ATP binding locus, catalytic reactions characteristic of this enzyme system have never been consumated by one protein in the absence of the other (but see later for the uptake of H_2). In this chapter, we use the [FeMo] and [Fe] designations in accord with most workers in the field.

Table 7.6
Properties of some representative nitrogenases.

Organism	Component	MW	Metal content	Reference
Azotobacter vinelandii	[MoFe]	234,000	2Mo, 34–38Fe, 26–28S	366, 367
	[Fe]	64,000	3.4Fe, 2.8S	
Azotobacter chrococcum	[MoFe]	227,000	2Mo, 22Fe, 20S	366, 368
	[Fe]	65,400	4Fe, 3.9S	
Clostridium pasteurianum	[MoFe]	221,800	2Mo, 24Fe, 24S	366, 369
	[Fe]	55,000	4Fe, 4S	
Klebsiella pneumoniae	[MoFe]	229,000	2Mo, 32Fe, 24S	366
	[Fe]	66,800	4Fe, 3.8S	
Anabaena cylindrica	[MoFe]	223,000	2Mo, 20Fe, 20S	370
	[Fe]	60,000		
Rhodospirillum rubrum	[MoFe]	215,000	2Mo, 25–30Fe, 19–22S	371
	[Fe]	60,000		

called the Fe protein or [Fe]. A useful nomenclature for discussions of kinetics and comparative biochemistry designates the FeMo protein as *Xy1*, where *X* and *y* are the first letters of the first and second name of the bacterial source, respectively. For example, *Cp1* is the FeMo protein from *Clostridium pasteurianum*. Similarly, for the Fe protein the designation *Xy2* is given; for example, the Fe protein of *Azotobacter vinelandii* is called *Av2*. This system will be used where appropriate to distinguish the protein source. Properties of representative Mo nitrogenases are given in Table 7.6.

The schematic diagram in Figure 7.27 shows some of the compositional and functional relationships of the nitrogenase proteins. The iron protein contains two identical subunits of MW $\sim$ 30 kDa. The subunits are products of the *nif* H gene.[246] A single Fe_4S_4 center is present in the protein and appears to be bound between the subunits.[246a] A recent x-ray structure[246b] of the iron protein confirms this picture. As shown in Figure 7.27A, the single Fe_4S_4 center is located at one end of the molecule, in the only region of significant contact between the two subunits. *In vivo*, the Fe protein is reducible by flavodoxin or ferredoxin. *In vitro*, artificial reductants such as dithionite or viologens are generally used. The single Fe_4S_4 center undergoes a single one-electron redox process, wherein the reduced form is EPR-active and the oxidized form is diamagnetic. As such, this center resembles four-iron-cluster-containing ferredoxins. Its redox potential is dependent on the ATP or ADP level in the solution. For example, *Cp2* (the Fe protein from *Clostridium pasteurianum*) shows $E^{o\prime} = -294$ mV in the absence and -400 mV in the presence of MgATP.[247] Two equivalents of MgATP and MgADP each bind to [Fe].

Until recently there was a major mystery over the number of Fe_4S_4 centers in [Fe] as deduced by EPR quantitation of the Fe_4S_4 centers compared to the number derived analytically or by extrusion experiments.[239,248] However, it has now[9,200,249,250] been clearly established that the single Fe_4S_4 center can exist in

(A)

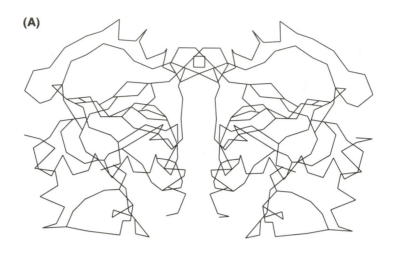

(B)

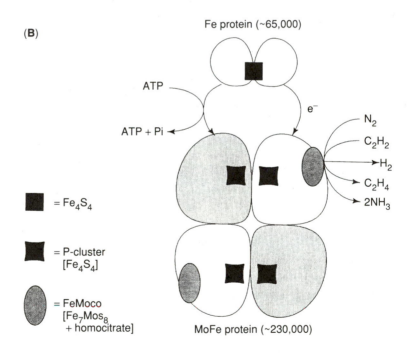

Fe protein (~65,000)

ATP

ATP + Pi

e⁻

N₂
C₂H₂
H₂
C₂H₄
2NH₃

= Fe₄S₄

= P-cluster
[Fe₄S₄]

= FeMoco
[Fe₇Mos₈
+ homocitrate]

MoFe protein (~230,000)

• *Alternative nitrogenase contains FeVco*

Figure 7.27
(A) Preliminary x-ray crystal structure of the *Azotobacter vinelandii* nitrogenase Fe protein.[246b]
(B) Schematic of the nitrogenase proteins illustrating their composition and mode of action.

this protein in two spin states, $S = \frac{1}{2}$ and $S = \frac{3}{2}$. Only that part of the EPR signal corresponding to the $S = \frac{1}{2}$ form, with its g values near 2, was considered in earlier spin quantitations. When the $S = \frac{3}{2}$ center, with g values between 4 and 6, is included, the EPR spin integration shows one paramagnetic site per Fe_4S_4 unit. Model systems[121,122] and theoretical studies[123,124,251] strongly support the ability of Fe_4S_4 to exist in various spin states. During enzyme turnover, the single Fe_4S_4 of the Fe-protein center transfers electrons to the FeMo protein in one-electron steps. There is no evidence for any difference in redox behavior between the $S = \frac{1}{2}$ and $S = \frac{3}{2}$ states of the protein.[200]

The Fe protein binds two molecules of MgATP.[252] The recent structure[246b,378] suggests that a cleft between the two subunits may serve as the ATP binding site. As the enzyme system turns over, a minimum of two molecules of MgATP are hydrolyzed to MgADP and phosphate in conjunction with the transfer of each electron to the FeMo protein.[253] The ATP/2e$^-$ ratio is generally accepted to have a minimum value of 4. Higher numbers represent decreased efficiency, often attributed to "futile cycling," where back electron transfer from [FeMo] to [Fe] raises the effective ratio.[248,253] Except for an as-yet-unconfirmed report of reduction by thermalized electrons produced by pulse radiolysis,[254] there is no evidence that the FeMo protein can be reduced to a catalytically active form without the Fe protein present.

Even though [Fe] must be present for catalysis to take place, the Dominant Hypothesis[239] designates [FeMo] as the protein immediately responsible for substrate reduction, and genetic/biochemical evidence supports this view. The FeMo protein contains an $\alpha_2\beta_2$ subunit structure, where α and β are coded by the *nif* D and *nif* K genes,[15,229] respectively. The overall molecular weight of about 230 kDa reflects the 50- to 60-kDa MW of each of its four subunits. In addition to protein, a total of 30 Fe, 2 Mo, and 30 S^{2-}, all presumed to be in the form of transition-metal sulfide clusters, add relatively little to the molecular weight, but are presumed to be major parts of the active centers of the protein. Figure 7.27, which is highly schematic, displays the cluster types in accord with the Dominant Hypothesis.

4. Protein purity and active sites

It has been almost 20 years since the first relatively pure preparations of nitrogenase became available. Indeed, homogeneous preparations are a *sine qua non* for progress in our understanding of the chemical nature and reactivity of any active site. In metalloproteins, there are two levels of homogeneity. The first involves purity with respect to the protein/subunit composition. This type of purity is achieved by conventional protein-purification techniques, and can be monitored by gel electrophoresis under native and denaturing conditions. In the language of polymer science, the macromolecular portion of the protein can be said to be monodisperse, corresponding to a single molecular weight for the polypeptide chain(s). However, even if the protein is homogeneous by this criterion, it may be inactive or only partially active because it does not have a full complement of active metal sites. The metal sites may be empty, filled with the wrong metals, or otherwise imperfect. Often the apo or inactive enzyme has chromatographic, electrophoretic, and centrifugal behavior very much like that of the holo protein, and therefore copurifies with it. Therefore, purification to electrophoretic homogeneity is only the first step. It is then necessary to ensure the chemical homogeneity of the active site. Very often activity is the major criterion for the approach toward such purity; i.e., the most homogeneous preparations are usually those in which the activity is highest. Several studies done on preparations that lacked active-site homogeneity were, as a result, not meaningful.

The two types of centers present in the nitrogenase FeMo protein are designated P clusters and FeMoco (or M) centers. Both types of centers display unique spectroscopic properties, but only FeMoco continues to display most of those properties when it is extracted from the protein.

5. FeMoco

The presence of the FeMo cofactor within the FeMo protein of nitrogenase, i.e., the M center, is revealed through spectroscopic and redox studies.[239] In the resting state of [FeMo], as isolated in the presence of dithionite, the M center has a distinct $S = \frac{3}{2}$ EPR signal, which is discussed below (see Figure 7.28).

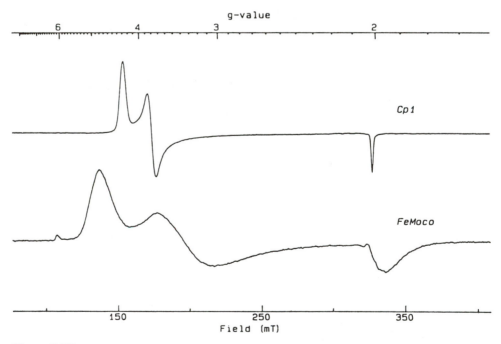

Figure 7.28
EPR spectra: (A) the $S = \frac{3}{2}$ M center in *Clostridium pasteurianum* nitrogenase FeMo protein; and (B) the FeMoco extracted into NMF from the protein. (Spectra courtesy of R. Bare and G. N. George.)

When the enzyme is turning over the EPR signal essentially disappears, leaving an EPR-silent state in which the FeMoco site is super-reduced to what is presumed to be its catalytically active form. In addition, a third state in which the $S = \frac{3}{2}$ EPR signal disappears is produced upon oxidation under non-turnover conditions. Thus the M center within the protein shows three states of oxidation, and these appear to have been reproduced in the FeMoco extracted from the protein:[255a]

$$\text{FeMoco (oxidized)} \xrightarrow{e^-} \text{FeMoco (reduced)} \xrightarrow{e^-} \text{FeMoco (super-reduced)} \quad (7.14)$$

The detailed characterization of the FeMoco site has involved parallel studies of the site within the protein and in its extracted form. The authentication of the extracted FeMoco involves the production and use of mutant organisms that make an inactive FeMo protein that contains all subunits and P clusters, but lacks the FeMoco sites.[172,256] A mutant of *Azotobacter vinelandii* called UW-45 (UW = *U*niversity of *W*isconsin) was first used to assay for isolated FeMoco.[257] Since several genes are involved in specifying FeMoco biosynthesis, mutants lacking these genes produce FeMo protein either lacking FeMoco or having a defective version of FeMoco. Mutants such as *Nif* B $^-$ of *Klebsiella pneumoniae*[172] lack cofactor, and an inactive "apo" protein can be isolated from them.

The breakthrough in this field[257] came in 1976, when FeMoco was extracted from [FeMo] into N-methylformamide[258] after the protein was acidified and then neutralized. The acidification removes most of the acid-labile P clusters, and partially denatures the protein. Reneutralization precipitates the protein (near its isoelectric point) and the precipitated denatured protein can then be extracted.

It has been shown that FeMoco can be extracted into many organic solvents,[10,257,259–259b] provided proper combinations of cations and anions[259a] are present in the solvent. The role of the cation is to balance the charge of the negatively charged cofactor. The role of the anion is to displace the cofactor from anion-exchange columns, such as DEAE cellulose or TEAE cellulose, to which the cofactor and/or its protein source had been adsorbed. The ability to dissolve cofactor in such solvents as CH_3CN, acetone, THF, and even benzene should facilitate attempts at further characterization and crystallization.[259,259a]

The biochemical authenticity of FeMoco has been assayed by its ability to activate the FeMo protein from the cofactor-less mutant organism.[258] The stoichiometry of the cofactor is $MoFe_{6-8}S_{7-10}$, with the variability likely due to sample inhomogeneity. The extracted cofactor resembles the M-center unit spectroscopically and structurally as shown in Table 7.7. The differences are presumed to result from differences in the peripheral ligands of the metal-sulfide center between the protein and the organic solvent.[260]

Strong evidence to support FeMoco as the site of substrate binding and reduction comes from the study of *nif* V mutants.[261–263] (The V designation is somewhat unfortunate, as *nif* V has *nothing* to do with vanadium.) The *Nif* V mutants do not fix nitrogen *in vivo*, and have altered substrate specificity *in vitro*. Dihydrogen evolution by isolated *nif* V nitrogenase is inhibited by CO, in contrast to the wild type, where H_2 evolution is insensitive to CO. FeMoco can be extracted from the *nif* V protein and used to reactivate the FeMoco-deficient mutants, such as *nif* B or UW-45. Remarkably, the reconstituted FeMo protein has CO-sensitive H_2 evolution, which is characteristic of *nif* V; i.e., the *nif* V phenotype is a property of FeMoco and not of the protein.[263] This result clearly implicates the FeMoco site as an important part of the substrate reactions of the nitrogenase enzyme complex.

Recently, a heat-stable factor called the V-factor has been discovered that restores the wild-type phenotype when added to *nif* V mutants during *in vitro* FeMoco assembly reactions.[264] The V-factor has been shown to be homocitrate (see Scheme 7.15) and ^{14}C labeling strongly suggests that homocitrate (or a part of it) is a component of the cofactor center. Interestingly, the far more metabolically common citrate appears to be present in the *nif* V mutant.[265a] Replacement of homocitrate by analogues that differ in structure or stereochemistry yields modified FeMoco sites that have altered substrate specificities.[265b] Thus, as is true for many cofactors (e.g., heme = porphyrin + iron; B_{12} = corrin + cobalt; F430 = corphin + nickel; Moco, the molybdenum cofactor = Mo + molybdopterin), both inorganic and organic components are present in FeMoco.

$$
\begin{array}{cc}
\begin{array}{l}
CH_2\!-\!CH_2\!-\!COOH \\
| \\
HO\!-\!C\!-\!COOH \\
| \\
CH_2\!-\!COOH \\
\\
\text{Homocitrate}
\end{array}
&
\begin{array}{l}
CH_2\!-\!COOH \\
| \\
HO\!-\!C\!-\!COOH \\
| \\
CH_2\!-\!COOH \\
\\
\text{Citrate}
\end{array}
\end{array}
\qquad (7.15)
$$

The biosynthesis of the cofactor and its insertion into [FeMo] apparently requires the presence of [Fe] and ATP in *A. vinelandii*.[266,266a] Whether this involves redox or conformational change in [FeMo] induced by [Fe] is unknown, but the fact that inactive versions of [Fe] are effective would seem to favor the nonredox mechanism. An attractive idea[266] is that [Fe]·MgATP binds to [FeMo], producing a state that is conformationally accessible for cofactor insertion.

Recently, site-directed mutagenesis studies[266b,c] have shown that cysteine residues are involved in binding FeMoco to the subunits of [FeMo]. Moreover, these studies again implicate FeMoco in the substrate-reducing site.

6. The P-clusters

Evidence has been presented[229] for the presence of four Fe_4S_4-like clusters (designated as P-clusters) in [FeMo]. The P-clusters are, however, by no means ordinary Fe_4S_4 clusters, and may not be Fe_4S_4 clusters at all. P-clusters are manifest[239,248] in electronic absorption and, especially, MCD and Mössbauer spectra of [FeMo]. These spectra are clearly not conventional; i.e., they are not like those found in ferredoxins and have not yet been seen in model compounds. In their oxidized forms, the P-clusters are high-spin, probably $S = \frac{7}{2}$, according to EPR studies.[267] Mössbauer spectra reveal decidedly inequivalent Fe populations,[268,269] indicating that the putative Fe_4S_4 clusters are highly distorted or asymmetric. The four P-clusters do not appear to behave identically under many circumstances, and it is clear that they form at least two subsets. There is open disagreement over the redox behavior of these sets.[239,270,271] Furthermore, an additional Mössbauer signal sometimes designated as S may also be part of the P-cluster signal.[268]

Although spectroscopic studies of the P-clusters do not unequivocally reveal their structural nature, extrusion of these clusters from the protein leads to the clear identification of three or four Fe_4S_4 clusters.[248,272] As discussed previously, the extrusion technique has inherent uncertainties, because it may be accompanied by cluster rearrangement. Nevertheless, the experimental result does support the Dominant Hypothesis, which designates the P centers as highly unusual Fe_4S_4 clusters.* The P-clusters are thought to be involved in electron storage and transfer, and presumably provide a reservoir of low-potential electrons to be used by the M center (FeMoco) in substrate reduction. Attractive as it may seem, there is no direct evidence to support this notion.

7. EPR, ENDOR, and ESEEM studies

The FeMoco or M center has been identified spectroscopically within the FeMo protein;[239,248,273] it has a distinctive EPR signal with effective g values of 4.3, 3.7, and 2.01, and originates from an $S = \frac{3}{2}$ state of the M center. The signal arises from transitions within the $\pm\frac{1}{2}$ ground-state Kramers doublet of the $S = \frac{3}{2}$ system ($D = +5.1$ cm^{-1}, $E/D = 0.04$). The isolated cofactor (FeMoco) gives a similar EPR signal, but with a rather larger rhombicity ($E/D = 0.12$). Spectra from the C. pasteurianum nitrogenase and cofactor are shown in Figure 7.28, and comparative data are given in Table 7.7. The M-center EPR signal has proved useful in characterizing the nature of the site, especially when more sophisticated magnetic resonance techniques, such as ENDOR or ESEEM, are used.

Extensive ENDOR investigations[274,275,275a] have been reported using protein samples enriched with the stable magnetic isotopes ^2H, ^{33}S, ^{57}Fe, ^{95}Mo, and ^{97}Mo. The ^{57}Fe couplings have been investigated in the most detail. Individual hyperfine tensors of five coupled ^{57}Fe nuclei are discernible, and were evaluated by simulation of the polycrystalline ENDOR spectrum.[275] The data from ^{33}S and ^{95}Mo were analyzed in less detail; ^{33}S gave a complex ENDOR spectrum, evidently with quite large hyperfine couplings, although no quantification was attempted because of the complexity of the spectrum.[274] On the other hand, ^{95}Mo was shown to possess a small hyperfine coupling, indicating that the molybde-

---

*Recent x-ray crystallographic results show that, if Fe_4S_4 clusters are present, they are very close together in two pairs,[379,380] which may account for their unusual properties.

Table 7.7
Comparison of the FeMo protein and isolated FeMoco.[a]

	FeMo protein (M center)	FeMoco (in NMF)
EPR		
g' values	4.27	4.8
	3.79	3.3
	2.01	2.0
EXAFS[a]		
Mo-S	2.36 (4)[b]	2.37 (3.1)[c]
Mo-Fe	2.69 (3)[b]	2.70 (2.6)[c]
Mo-O or N	2.18 (1)[b]	2.10 (3.1)[c]
Fe-S		2.25 (3.4)[d] 2.20 (3.0)[e]
Fe-Fe		2.66 (2.3)[d] 2.64 (2.2)[e]
		3.68 (0.8)[e]
Fe-Mo		2.76 (0.4)[d] 2.70 (0.8)[e]
Fe-O or N		1.81 (1.2)[d]
XANES		
MoO_3S_3 fits best[f]		

[a] Distance in Å with number of atoms in parentheses.
[b] From 287; earlier study reported in 286.
[c] Data from 373; earlier study reported in 372.
[d] Data from 290.
[e] Data from 291.
[f] Data from 288, 289.

num possesses very little spin density (although the quantitative aspects of the conclusions of the ^{95}Mo ENDOR study have recently been shown to be in error[276]).

Although no nitrogen splittings were reported in any of the ENDOR studies, evidence for involvement of nitrogen as a cluster component has been forthcoming from ESEEM spectroscopy.[277–279] ^{14}N modulations are observed in the ESEEM of the M center. The observed ^{14}N is not from the substrate (N_2), or from an intermediate or product of nitrogen fixation, because enzyme turnover using ^{15}N as a substrate does not change the ESEEM spectrum. The isolated cofactor (FeMoco) does not show the modulation frequencies observed for the M center in the protein. These experiments suggest that the M-center ^{14}N ESEEM arises from a nitrogen atom that is associated with the M center, and probably from an amino-acid side chain (most likely a histidine) ligated to the cluster.[279] Recent evidence from site-directed mutagenesis of the *Azotobacter vinelandii* protein[280] provides strong support for the presence of histidine ligation, and points specifically to His-195 of the α subunit as the N ligand.

8. Mössbauer studies

Extensive Mössbauer investigations of nitrogenase[271,281–283] and FeMoco[283a] have been reported. Unlike EPR and EPR-based spectroscopies, which can be used to investigate only the EPR-active $S = \frac{3}{2}$ oxidation state, all three available M-center oxidation states are accessible to Mössbauer spectroscopy. The fully reduced site was found to be diamagnetic with $S = 0$ (but see Reference 284), whereas the oxidized site was found to have $S \geq 1$. The zero-field spectrum of reduced *C. pasteurianum* nitrogenase is shown in Figure 7.29; the spectrum is comprised of four quadrupole doublets, one of which was concluded to originate from the M site.[282] Mössbauer spectra taken in the presence of applied magnetic fields were used to deduce the presence of four types of ^{57}Fe hyperfine coupling; these were called sites A1, A2, and A3, which have negative hyperfine couplings, and B sites, which have positive hyperfine couplings. The A sites were quantitated as a single Fe each; the B sites were estimated to contain three irons. These conclusions were largely confirmed and extended by later ENDOR investigations,[274] although the B sites were resolved as two inequivalent, rather than three equivalent, sites. ENDOR is rather more sensitive to the nature of the hyperfine couplings than Mössbauer, although it cannot usually be used to count numbers of exactly equivalent sites. Thus the number of iron atoms in the M center is minimally five, although larger numbers cannot be excluded. Note also that some of the quantitative aspects of the earlier Mössbauer investigations have been criticized.[285]

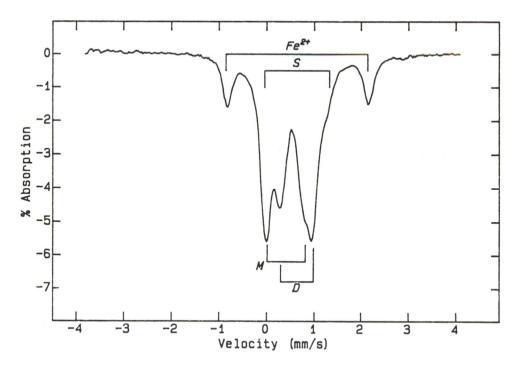

Figure 7.29
Mössbauer spectrum of *C. pasteurianum* nitrogenase FeMo protein,[282] indicating the various components (quadrupole doublets) and their assignments. The doublet labeled M is the cofactor signal; those labeled D, S, and Fe^{2+} are attributed to the P-clusters.

9. X-ray absorption studies

One of the early triumphs of biological x-ray absorption spectroscopy was the deduction that the nitrogenase M center is an Mo-Fe-S cluster.[286] (It is also worth noting that nitrogenase was the first *enzyme* to be studied by x-ray absorption spectroscopy.) Early work on lyophilized protein samples indicated the presence of two major contributions to the Mo K-edge EXAFS, which were attributed to Mo-S ligands, plus a more distant Mo-Fe contribution.[286] Subsequently, these conclusions have been confirmed and extended, using samples in solution and with much more sensitive detection systems.

Most EXAFS studies to date have been on the molybdenum K-edge of the protein or of FeMoco, and indicate a very similar Mo environment in both (Table 7.7, Figure 7.30). A consensus of the best available analyses[287] indicates that Mo is coordinated by three or four sulfur atoms at 2.4 Å, one to three oxygens or nitrogens at 2.2 Å, with approximately three nearby iron atoms at 2.7 Å. Of these, the EXAFS evidence for the oxygen/nitrogen contribution is weakest. However, comparison of Mo K-edge[288] and Mo L-edge XANES[289] spectra with model compounds indicates strong similarities with $MoFe_3S_4$ thio-cubane model compounds possessing MoS_3O_3 coordination, and provides some support for the presence of O/N ligands.

The iron EXAFS of FeMoco has been independently examined by two groups.[290,291] Both groups agree that the iron is coordinated largely to sulfur at about 2.2 Å, with more distant Fe-Fe interactions at about 2.6 Å. They differ, however, concerning the presence of short (1.8 Å) Fe-O interactions. Such interactions were apparently observed in the earlier study,[290] but not in the later study.[291] One possible explanation for this discrepancy is that the short Fe-O interactions of the earlier study were due to extraneous iron coordinated to solvent, contaminating the FeMoco preparation.[291] A final resolution of this discord must, however, await the results of further experiments. Interestingly, a long Fe-Fe interaction at 3.7 Å was also observed in the later study.[291]

Largely on the basis of the Mo K-edge EXAFS results and model studies discussed below, several proposals for the structure of the M center have been put forward. These are illustrated in Figure 7.31.

The MoFe proteins from *Clostridium pasteurianum*[292] and from *Azotobacter vinelandii*[293] have been crystallized. For the former protein, crystals of space group $P2_1$ are obtained, with two molecules per unit cell of dimensions $70 \times 151 \times 122$ Å. There is good evidence for a molecular two-fold axis, which presumably relates equivalent sites in the two $\alpha\beta$ dimers that make up the protein molecule.[294] Preliminary refinement reveals that the two FeMoco units per protein are about 70 Å apart and the four P clusters are grouped in two pairs.

Single crystal EXAFS studies[295] have provided important structural information on the molybdenum site. For different crystal orientations (relative to the polarized x-ray beam), the amplitude of the Mo-Fe EXAFS changes by a factor of 2.5, but the Mo-S EXAFS changes only slightly. Analysis of the an-

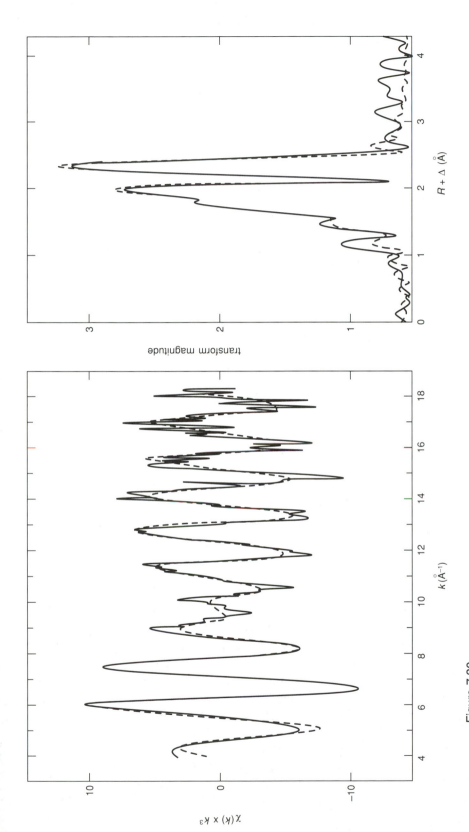

Figure 7.30

Mo K-edge EXAFS spectrum (left panel) and EXAFS Fourier transform (right panel) of *Klebsiella pneumoniae* nitrogenase MoFe protein. The solid line is the processed experimental spectrum and the dashed line a calculated one.[287]

[427]

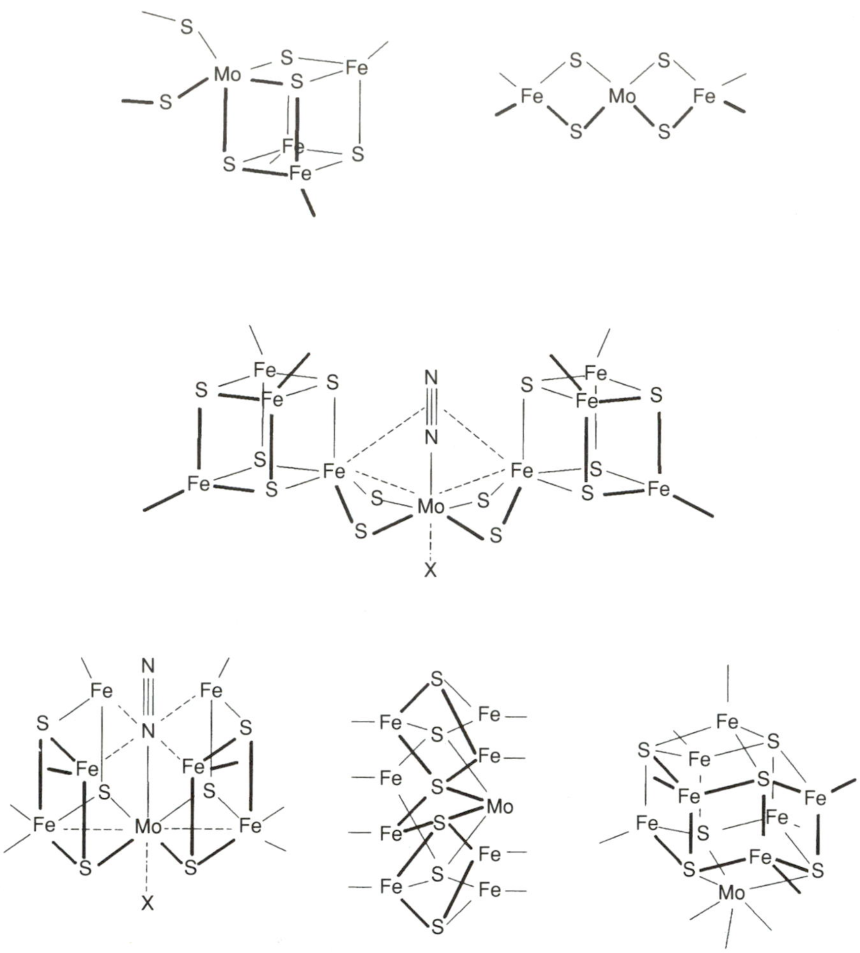

Figure 7.31
Proposed models for FeMoco. (Compare with the recent model from the x-ray structure on page 444.)

isotropy of the Mo-Fe EXAFS using the available crystallographic information[294] is consistent with either a tetrahedral $MoFe_3$ geometry such as that found in thiocubanes (Figure 7.32) or a square-based pyramidal $MoFe_4$ arrangement of metals. This interpretation tends to rule out some of the structural proposals shown in Figure 7.33. The observed orientation-dependence of the iron amplitudes is too small for clusters containing a linear or planar arrangement of iron and molybdenum (e.g., Figure 7.33B,C), and too large for arrangements that involve regular disposition of iron about molybdenum. Moreover, the lack of anisotropy of the sulfur EXAFS (which was apparently not considered in the original interpretation[295]) argues against an MoS_3 $(O/N)_3$ model that has molybdenum coordinated by sulfur atoms that bridge only to Fe atoms disposed to one side of the molybdenum. Significant anisotropy for the Mo-S EXAFS (of opposite polarization, and smaller than that for Mo-Fe) would be expected for

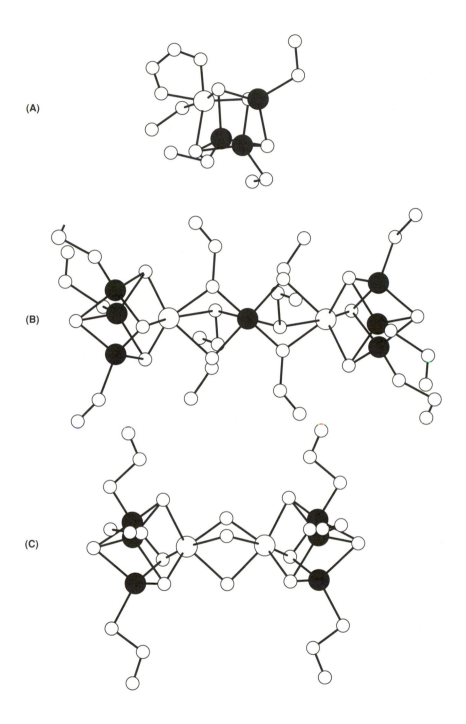

Figure 7.32
Structures of thiocubanes that display Mo-S and Mo-Fe distances similar to FeMoco:
(A) $(Fe_3MoS_4)_2(SR)_9{}^{3-}$; (B) $(MoFe_3S_4)_2Fe(SR)_{12}{}^{3-/4-}$; (C) $MoFe_3S_4(SEt)_3(cat)CN^{3-}$.
(Data on A and B from References 328, 330a; data on C from References 331, 332.)

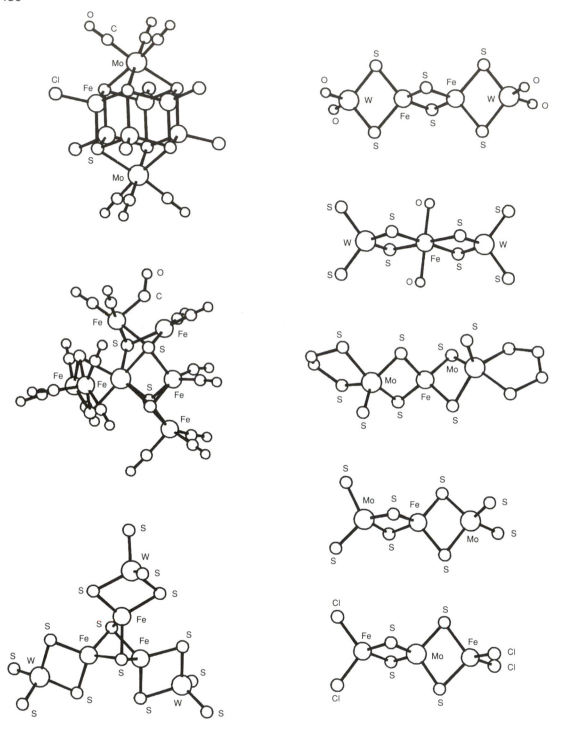

Figure 7.33
FeMoS and FeWS structures of potential interest with respect to nitrogenase.[331,332,332a-j]

such an arrangement of sulfur atoms. However, the cubane model of Figure 7.33, which provides the best model of both geometric and electronic structure, remains viable if one of the nonbridging ligands to molybdenum is a sulfur atom (rather than oxygen or nitrogen) with a bond length similar to that of the bridging sulfides.

10. Substrate reactions

The two-component Mo-nitrogenase enzyme catalyzes the reduction of N_2 to $2NH_4^+$ as its physiological reaction. Concomitant with the reduction of N_2, H_2 evolution occurs, with electrons supplied by the same reductants that reduce N_2. The limiting stoichiometry appears to be

$$N_2 + 10H^+ + 8e^- \longrightarrow 2NH_4^+ + H_2 \qquad (7.16)$$

If N_2 is omitted from the assay, all the electrons go to H_2 evolution. Indeed, to a first approximation the rate of electron flow through nitrogenase is independent of whether the enzyme is producing only H_2, producing both NH_4^+ and H_2, or reducing most of the alternative substrates.

As displayed in Table 7.8, many alternative substrates are known for this enzyme.[240,243,296] The most important of these from a practical perspective is

Table 7.8
Nitrogenase substrate reactions.[296,374-376]

Two-electron reductions
$2e^- + 2H^+ \rightarrow H_2$
$C_2H_2 + 2e^- + 2H^+ \rightarrow C_2H_4$
$N_3^- + 2e^- + 3H^+ \rightarrow NH_3 + N_2$
$N_2O + 2e^- + 2H^+ \rightarrow H_2O + N_2$

$$\underset{HC=CH}{\overset{CH_2}{\triangle}} + 2e^- + 2H^+ \rightarrow CH_2=CH-CH_3 + \underset{H_2C-CH_2}{\overset{CH_2}{\triangle}}$$

Four-electron reductions
$HCN + 4e^- + 4H^+ \rightarrow CH_3NH_2$
$RNC + 4e^- + 4H^+ \rightarrow RNHCH_3$

Six-electron reductions
$N_2 + 6e^- + 6H^+ \rightarrow 2NH_3$
$HCN + 6e^- + 6H^+ \rightarrow CH_4 + NH_3$
$HN_3 + 6e^- + 6H^+ \rightarrow NH_3 + N_2H_4$
$RNC + 6e^- + 6H^+ \rightarrow RNH_2 + CH_4$
$RCN + 6e^- + 6H^+ \rightarrow RCH_3 + NH_3$
$NCNH_2 + 6e^- + 6H^+ \rightarrow CH_3NH_2 + 2NH_3$
$NO_2^- + 6e^- + 6H^+ \rightarrow NH_3$

Multielectron reductions
$RNC \rightarrow (C_2H_6, C_3H_6, C_3H_8) + RNH_2$
$NCNH_2 + 8e^- + 8H^+ \rightarrow CH_4 + 2NH_3$

acetylene, C_2H_2, which is reduced by the Mo nitrogenase exclusively to ethylene, C_2H_4. Acetylene can completely eliminate H_2 evolution by nitrogenase. Many of the substrates in Table 7.8 have a triple bond. Indeed, the only triple-bonded molecule not reduced by nitrogenase is CO, which nevertheless inhibits all substrate reactions, but not H_2 evolution (in the wild type). Triple-bonded molecules such as acetylene ($H—C{\equiv}C—H$) are useful probe molecules for related reactivity as discussed below for simple inorganic systems. *All* substrate reductions involve the transfer of two electrons or multiples thereof (i.e., 4,6,8 . . .). Multielectron substrate reductions may involve the stepwise execution by the enzyme of two-electron processes. Further, about as many protons as electrons are usually transferred to the substrate. One way of viewing the nitrogenase active site is that it can add the elementary particles (H^+ and e^-) of H_2 to the substrate. This may have mechanistic implications.[297]

It is potentially fruitful to pursue the intimate connection between H_2 and the N_2 binding site in nitrogenase. It has been shown unequivocally[298,299] that one H_2 is evolved for each N_2 "fixed" even at 50 atm of N_2, a pressure of N_2 well above full saturation. Moreover, H_2 is a potent inhibitor of N_2 fixation, and under D_2, HD is formed, but only in the presence of N_2. These complex relationships between N_2 and $H_2(D_2)$ have elicited a variety of interpretations.[255,300-302]

Recently, it has been demonstrated that the FeMo protein alone acts as an uptake hydrogenase.[303] Dihydrogen in the presence of [FeMo] causes the reduction of oxidizing dyes such as methylene blue or dichlorophenolindophenol *in the absence of Fe protein*. This is the only known catalytic reaction displayed by the FeMo protein alone. The hydrogen evolution and uptake by [FeMo] suggest that understanding hydrogen interaction with transition-metal/sulfur centers may be crucial to understanding the mechanism of nitrogenase action.

11. The role of ATP

ATP hydrolysis appears to be mandatory, and occurs during electron transfer from [Fe] to [FeMo]. Dissociation of [Fe] and [FeMo] following electron transfer is probably the rate-limiting step in the overall turnover of the enzyme.[255] The fact that reductant and substrate levels do not affect turnover rates is consistent with this finding.

The role of ATP on a molecular level remains one of the great mysteries of the mechanism of nitrogen fixation. As discussed above, the overall thermodynamics of N_2 reduction to NH_3 by H_2 or by its redox surrogate flavodoxin or ferredoxin is favorable. The requirement for ATP hydrolysis must therefore arise from a kinetic necessity. This requirement is fundamentally different from the need for ATP in other biosynthetic or active transport processes, wherein the free energy of hydrolysis of ATP is needed to overcome a thermodynamic limitation.

What is the basis for the kinetic requirement of ATP hydrolysis in nitrogen fixation? To answer this question, we again look at the potential reduction products of the N_2 molecule. Of these, only N_2H_2 (diimide, three potential isomers), N_2H_4 (hydrazine and its mono and dications), and NH_3 (and its protonated form, NH_4^+) are isolable products. (In the gas phase, other species such as N_2H, N_2H_3, or NH_2 also have a "stable" existence.) In the presence of H_2, only the formation of ammonia is thermodynamically favored (Figure 7.26). Clearly, the formation of the intermediate species in the free state cannot occur to any reasonable extent. However, this does not mean that nitrogenase must form NH_3 directly without the formation of intermediates. It is possible for these reactive intermediates to be significantly stabilized by binding to a metal-sulfur center or centers.

Detailed kinetic studies[255,304] have suggested a scheme in which intermediates with bound and probably reduced nitrogen are likely to be present. Rapid quenching experiments in acid solution lead to the detection of hydrazine during nitrogenase turnover.[305] Likewise, studies of inhibition of N_2 fixation by H_2 and the formation of HD under D_2 have been interpreted in terms of a bound diimide intermediate.[306,307] Although a bound "dinitrogen hydride" is likely to be present, its detailed structure remains unknown.

D. The Alternative Nitrogenases

1. Vanadium nitrogenase

The "essentiality" of molybdenum for nitrogen fixation was first reported by Bortels in 1930.[308] This finding led ultimately to the characterization of the molybdenum nitrogenases discussed in the preceding section. Bortels' work has been cited many times, and is often referred to without citation. Following this seminal work, many other Mo-containing enzymes were subsequently sought and found.[25,309] At present more than a dozen distinct Mo enzymes are known, and new ones are continually being discovered.

In addition to the classic 1930 paper, Bortels[310] reported in 1935 that vanadium stimulated nitrogen fixation. In contrast to the 1930 paper, the 1935 paper languished in obscurity. Then, starting in the 1970s, attempts were made to isolate a vanadium nitrogenase. In 1971, two groups reported isolating a vanadium-containing nitrogenase from *A. vinelandii*.[311,312] The interesting notion at this time was that V might substitute for Mo in nitrogenase, *not* that there was a separate system. The isolated enzyme was reported to be similar to the Mo enzyme, but had a lower activity and an altered substrate specificity. One of the groups carefully reinvestigated their preparation, and found small amounts of molybdenum, which were presumed to be sufficient to account for the low activity, although the altered selectivity was not addressed.[313] The vanadium was suggested to play a stabilizing role for [FeMo], allowing the small amount of active Mo-containing protein to be effectively isolated. Apparently the possibility was not considered that a truly alternative nitrogenase system existed, whose protein and metal centers both differed from that of the Mo nitrogenase.

The unique essentiality of molybdenum for nitrogenase fixation went unchallenged until 1980, when it was demonstrated[314] that an alternative nitrogen-fixation system could be observed in *A. vinelandii* when this organism was starved for molybdenum.[315] Despite skepticism from the nitrogenase research community, it was eventually shown that even in a mutant from which the structural genes for the Mo nitrogenase proteins (*nif* H, D, and K) had been deleted, the alternative system was elicited upon Mo starvation. In 1986, two groups[316–321] isolated the alternative nitrogenase component proteins from different species of *Azotobacter,* and demonstrated unequivocally that one component contained vanadium and that neither component contained molybdenum.

One of the two components of the V-nitrogenase system is extremely similar to the Fe protein of nitrogenase. This similarity is evident in the isolated proteins from *A. vinelandii*[316] and in the genetic homology between *nif* H (the gene coding for the subunit of the Fe protein in the Mo-nitrogenase system) and *nif* H* (the corresponding gene in the V-based system). Both Fe proteins have an α_2 subunit structure, and contain a single Fe_4S_4 cluster that is EPR-active in its reduced state.

The FeV proteins from *Azotobacter vinelandii* and *Azotobacter chroococcum* each have an $\alpha_2\beta_2\delta_2$ subunit structure.[322] Metal composition and spectroscopic comparisons between the FeMo and FeV proteins are shown in Table 7.9. Although there is the major difference involving the presence of V instead of Mo in the FeV protein and in the probable presence of the small δ subunits (13 kDa), the two nitrogenase systems are otherwise quite similar.[322] In each, a system of two highly oxygen-sensitive proteins carries out an ATP-dependent N_2 reduction with concomitant H_2 evolution. The Fe proteins have the same subunit structure and cluster content, and are spectroscopically very similar. The V versions of the larger protein have somewhat lower molecular weights than their Mo analogues, and by MCD spectroscopy seem to contain P-like clusters.[318] The FeV site still may be an $S = \frac{3}{2}$ center (by EPR, although its EPR differs significantly from that of the FeMo center).[323] The V-S and V-Fe distances as measured by EXAFS[324,325] are similar to those in thiocubane VFe_3S_4 clusters and to Mo-S and Mo-Fe distances like those in [FeMo], which are in turn similar to those in $MoFe_3S_4$ thiocubanes. Likewise, XANES[324,325] indicates VS_3O_3 type coordination in [FeV] nitrogenase similar to the MoS_3O_3 coordination suggested by XANES for FeMoco. The "FeV cofactor" is extractable into NMF, and can reconstitute the *nif* B$^-$, FeMoco-deficient mutant of the Mo system.[326] Despite the substitution of V for Mo, the proteins and their respective M-Fe-S sites do not differ drastically. However, the compositional changes do correlate with altered substrate reactivity.

A major difference between the V and Mo enzymes lies in substrate specificity and product formation.[321] As is clearly shown in Table 7.9, the FeV nitrogenase has a much lower reactivity toward acetylene than does the Mo system. Furthermore, whereas the FeMo system exclusively produces ethylene from acetylene, the FeV system yields significant amounts of the four-electron reduction product, ethane.[321] The detection of ethane in the acetylene assay may

Table 7.9
Comparison of alternative nitrogenase proteins[a]

Property	$Av1$[47]	$Av1^{*}$[47]	$Ac1^{*}$[50]
Molecular weight	240,000	200,000	210,000
Molybdenum[b]	2	<0.05	<0.06
Vanadium[b]	—	0.7	2
Iron[b]	30–32	9.3	23
Activity[c]			
H[+]	2200	1400	1350
C_2H_2	2000	220	608
N_2	520	330	350
EPR g values	4.3	5.31	5.6
	3.7	4.34	4.35
	2.01	2.04	3.77
		1.93	1.93

[a] $Av1$ is the FeMo protein of *Azotobacter vinelandii*, $Av1^*$ is the FeV protein of *A. vinelandii* and $Ac1^*$ is the FeV protein of *A. chroococcum*. Data from References 317, 377, and 319, respectively.

[b] Atoms per molecule.

[c] nmol product/min/mg of protein.

prove a powerful technique for detecting the presence of the V nitrogenase in natural systems.[322] Moreover, this reactivity pattern is found in the *nif* B$^-$ mutant reconstituted with FeVco, indicating that the pattern is characteristic of the cofactor and not the protein.[326] The reactivity change upon going from Mo to V in otherwise similar protein systems clearly adds weight to the implication of the M-Fe-S center (M = V or Mo) in substrate reduction.

2. The all-iron nitrogenase[322]

The first sign that there is yet another alternative nitrogenase again came from genetic studies. A mutant of *A. vinelandii* was constructed with deletions in both *nif* HDK and *nif* H*D*K*, i.e., the structural genes for the Mo and V nitrogenases, respectively. Despite lacking the ability to make the two known nitrogenases, the mutant strain nevertheless was able to fix nitrogen, albeit poorly. Moreover, this mutant strain's nitrogenase activity was clearly inhibited when either Mo or V was present in the culture medium. Preliminary studies indicate that the nitrogenase proteins produced by this organism are closely related to those previously isolated. A 4Fe-4S Fe-protein *nif* H† and a protein due to *nif* D† was produced. The latter appeared to contain no stoichiometric metal other than iron. Symmetry of nomenclature would suggest calling this the FeFe protein and its cofactor FeFeco. Interestingly, this nitrogenase seems to be the poorest of the set in reducing N_2 and makes ethane from ethylene. The finding of the all-iron nitrogenase, if fully confirmed, will add significantly to the comparative biochemistry of nitrogen fixation. Speculatively, one might suggest that the concomitant absence of V and Mo suggests that nitrogen fixation need not directly involve the noniron heterometal in the cofactor cluster. This result may explain the lack of direct implication of Mo in the nitrogen fixation mechanism, despite many years of intense effort by workers in the field. (The above discussion should be taken *cum grano salis* until the existence of the all-iron nitrogenase is confirmed.)

3. Model systems

Three types of model systems for nitrogenase may be considered. First, there are transition-metal sulfide clusters that resemble the FeMoco or FeVco centers of the active proteins. Although there has been significant progress, there are not yet any definitive models (as there are for Fe_2S_2 and Fe_4S_4). A second approach uses the reactions of N_2 and related substrates or intermediates with metal centers in order to gain insights into the way in which transition-metal systems bind N_2 and activate it toward reduction. Here to date the most reactive systems bear little direct chemical resemblance to the nitrogenase active sites. Nevertheless, these systems carry out *bona fide* nitrogen fixation from which one may learn the various ways in which N_2 can be activated. Finally, there are other inorganic systems that display some of the structural and possibly some of the reactivity characteristics of the nitrogenase active sites without binding or reducing N_2 or precisely mimicking the active center. We may nevertheless be able to learn effectively about nitrogenase reactivity from these interesting chemical systems.

a. Transition-Metal Sulfide Models for Nitrogenase Sites   Although there has been great activity in synthetic Fe-S cluster chemistry, there is to date no example of a spectroscopic model for the P-cluster sites in nitrogenase. If the P-clusters are indeed asymmetrically bound high-spin Fe_4S_4 clusters, then the recent work on high-spin versions of Fe_4S_4 clusters[327] and site-selectively derivatized Fe_4S_4 centers[143] may hint that appropriate model systems are forthcoming.

b. Fe-Mo-S Cluster Models for FeMoco   Despite the importance of P-clusters, the modeling of the FeMoco center has properly received the most attention. The significant structural parameters that any model must duplicate are the Mo-S and Mo-Fe distances determined by EXAFS. Spectroscopically, the $S = \frac{3}{2}$ EPR signal provides a stringent feature that model systems should aspire to mimic.

Many FeMoS clusters have been prepared in the quest to duplicate the FeMoco center, but *none* of the chemically synthesized clusters can reactivate the (UW-45 or *Nif* B⁻) cofactor-less mutants, perhaps because of their lack of homocitrate, which only recently has been discovered as a key component of FeMoco. Undoubtedly, new FeMoS clusters containing homocitrate will be prepared, and perhaps these will activate the mutant proteins, thereby revealing a close or full identity with FeMoco.

Despite the absence of homocitrate, some interesting model systems have been investigated. It is beyond the scope of this chapter to give a comprehensive account of FeMoS chemistry. We concentrate on the so-called "thiocubane" model systems. Heterothiocubane models were first synthesized using self-assembly approaches analogous to those used for the simpler Fe-S model systems. The reaction[328–330a]

$$MoS_4^{2-} + Fe^{3+} + SR^- \longrightarrow (MoFe_3S_4)_2(SR)_9^{3-} \text{ and } (MoFe_3S_4)_2Fe(SR)_{12}^{3-,4-} \quad (7.17)$$

uses tetrathiomolybdate, MoS_4^{2-}, as the source of Mo, and leads to the double cubane structures shown in Figure 7.32A,B. The $Fe_7Mo_2S_8$ structure proved particularly interesting, since it was possible to complex the central ferric iron atom with substituted catecholate ligands[331,332] and eventually isolate a single thiocubane unit (Figure 7.32C). Significantly, the single unit has $S = \frac{3}{2}$ and Mo-S and Mo-Fe distances that match precisely those found by EXAFS for the M center of nitrogenase. Single cubes with VMo_3S_4 cores have also been prepared.[160,160a] Although the single thiocubanes display spectroscopic similarity and distance identity with FeMoco, they are not complete models. They are stoichiometrically Fe and S deficient, lack homocitrate, and most importantly, fail to activate the UW-45 and *Nif* B$^-$ mutants.

Other interesting FeMoS (and FeWS) clusters with structurally distinct properties are shown in Figure 7.33. These include the "linear" $(MoS_4)_2Fe^{3-}$ ion, the linear $(WS_4)_2Fe[HCON(CH_3)_2]_2^{2-}$ ion, the linear $Cl_2FeS_2MS_2Fe_2Cl_2^{2-}$ (M = Mo, W), the "linear" $(MoS_4)_2Fe_2S_2^{4-}$ ion, the trigonal $(WS_4)_3Fe_3S_2^{4-}$, the capped thioprismane $Fe_6S_6X_6[M(CO)_3]_2^{3-}$ (X = Cl, Br, I; M = Mo, W), and the organometallic clusters $MoFe_6S_6(CO)_{16}^{2-}$, $MoFe_3S_6(CO)_6(PEt_3)_3$, and $MoFe_3S_6(CO)_6^{2-}$. Structures suggested for FeMoco based on these and other chemically synthesized transition metal sulfides and on spectroscopic studies of the enzyme are shown in Figure 7.31.

E. N₂ and Related Complexes

The triple bond of N_2 has one σ and two π components. Each nitrogen atom has a lone pair oriented along the N-N direction. The two lone pairs allow N_2 to bind in an end-on fashion in either a terminal or a bridging mode. Both modes of binding are illustrated in the binuclear zirconium complex[333] shown in Figure 7.34. In this and in many other N_2 complexes, the N-N bond is not significantly lengthened and is therefore presumed to be insignificantly weakened in the complex. Interestingly, the complex in Figure 7.34, despite not having long N-N distances, forms hydrazine quantitatively upon protonation. Only one of the three N_2 molecules is reduced, and all four electrons required come from the two Zr(III) by presumed internal electron transfer. The related μ-N₂ complex $[W(\eta^5\text{-}C_5Me_5)Me_2(SC_6H_2Me_3)]_2(\mu\text{-}N_2)$ is one of the few dinitrogen complexes to contain an S donor ligand.[333a]

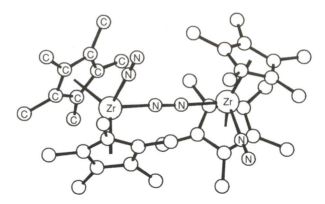

Figure 7.34
The x-ray crystal structure[333] of $(Cp')_2Zr(N_2)(\mu_2-N_2)Zr(N_2)(Cp')_2$.

In addition to the N lone pairs, the π components of the N≡N triple bond can serve as donor-acceptor orbitals in the Dewar-Chatt-Duncanson (olefin binding) manner. This less-common mode of N_2 binding is illustrated by the structure of the Ti complex[334] shown in Figure 7.35. Here, as in the few other known side-on bound N_2 complexes,[335] the N-N bond is significantly lengthened. The lengthened bond at 1.30 Å is presumed to be sufficiently weakened [ν(N-N) = 1280 cm^{-1}] that it is susceptible to further lengthening and reduction. As the N-N distance lengthens, it is more appropriate to consider the ligand as a deprotonated diimide or hydrazine.

Complexes that have proven particularly useful are bis(dinitrogen)phosphines of Mo(0) and W(0) such as $M(N_2)_2(Ph_2PCH_2CH_2PPh_2)_2$ and $M(N_2)_2(PPh_2Me)_4$.

Figure 7.35
The structure of a multiply bound dinitrogen titanium compound.[334]

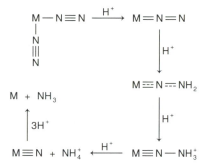

Figure 7.36
Structure and reactions of $M(N_2)_2(Ph_2PCH_2CH_2PPh_2)_2$ (M = Mo, W) and related complexes. Not all of the intermediates in this scheme have been isolated in any one particular system.[336,337,337a] The phosphine ligands are not shown.

As shown in Figure 7.36, treatment of the complexes[336–337a] with acid leads to the formation of the diazenido(-H) and hydrazido(-2H) complexes, and sometimes to the production of ammonia. The finding of a bound $N_2H_2^{2-}$ species is consistent with the proposed presence of similar bound species in nitrogenase. The complexes of reduced dinitrogen intermediates are stabilized by multiple M-N binding. Further protonation of these intermediates or treatment of the original complex with strong acid leads to the formation of NH_3 from the bound nitrogen. Here the Mo(0) starting complex has enough electrons [six from the Mo(0) → Mo(VI) conversion] to reduce one N_2 molecule in conjunction with its protonation from the external solution.

In a general sense this reaction may be telling us something about nitrogenase. The enzyme may be able to deliver six reducing equivalents to N_2, and protonation, perhaps carefully orchestrated by neighboring amino-acid or homocitrate groupings, may facilitate the process. However, it is virtually certain that the Mo in nitrogenase is not able to change its oxidation state by six units. In the enzyme the multimetal, multisulfur FeMoco site may serve the equivalent function, by providing multiple sites at which reduced intermediates can simultaneously bind.

Only a few of the known N_2 complexes contain S-donor ligands. One of these, $Mo(N_2)_2(S(CH_2C(CH_3)_2CH_2S)_3)$, shown in Figure 7.37, has four thioether S-donor atoms bound to Mo(0). This Mo(0) complex shows reactivity reminiscent of the related phosphine complexes.[337a] A remarkable complex (Figure 7.38) has been isolated[338] in which two lone pairs of *trans*-diimide bind to two Fe, concomitantly with H-binding of the two diimide hydrogen atoms to coordinated sulfur atoms. The ability of an Fe-S system to stabilize the very reactive trans-N_2H_2 grouping adds support to the notion that similar metal-sulfide sites of nitrogenase may stabilize related intermediates along the N_2 → $2NH_3$ reaction path.

Most of the model systems involving N_2 do not lead to NH_3 formation. Moreover, many systems that do form NH_3 are not catalytic. However, certain

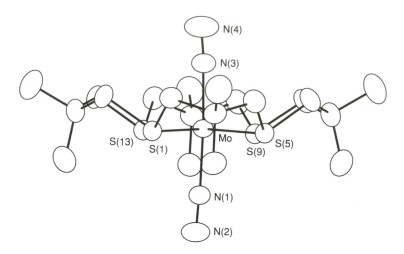

Figure 7.37
The structure of $Mo(N_2)_2L$ (L is a tetrathiacyclohexadecane).[337b,c]

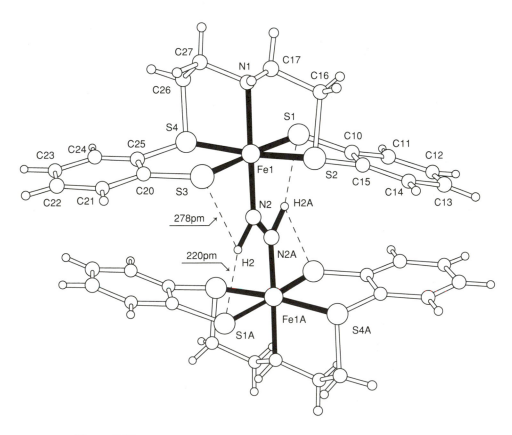

Figure 7.38
The structure[338] of $FeL(N_2H_2)FeL$ (L = $SC_6H_4SCH_2CH_2SCHCH_2SC_6H_4S$).

V-based and Mo-based systems can catalytically reduce N_2 to N_2H_4 or NH_3 using strong reducing agents.[339] Although kinetic studies indicate the possibility of intermediates, little structural information is available at present on these interesting systems.

F. Insights from Relevant Inorganic Reactivity

Certain studies on inorganic systems that do not model the nitrogen-fixation process can nevertheless potentially give insight into nitrogenase action. Two categories of relevant chemistry are acetylene binding/reactivity and dihydrogen binding/activation. Modes of dihydrogen activation on sulfide systems have previously been discussed in the section on hydrogenase.

Acetylene has long been known to bind to metal centers using its π and π^* orbitals as, respectively, σ-donor and π-acceptor orbitals. Even when the metal is predominantly sulfur-coordinated,[340,341] such side-on bonding of RC_2R is well known [340,341] as in $MoO(S_2CNR_2)_2(RC\equiv CR)$ and $Mo(S_2CNR_2)_2(RC\equiv CR)_2$. The direct interaction of acetylene with the metal center must be considered as a potential binding mode for nitrogenase substrates.

A totally different, sulfur-based mode of acetylene binding is now also well established. For example, $(Cp')_2Mo_2S_4$ reacts with acetylene[342,225] to produce

$$(Cp')_2Mo_2 \quad \text{(7.18)}$$

containing a bridging ethylene-1,2-dithiolate (dithiolene). The acetylene binds directly to the sulfur atoms by forming S—C bonds. Acetylenes or substituted (activated) acetylenes are able to displace ethylene from bridging or terminal 1,2-dithiolate ligands[225,341] to produce the 1,2-dithiolenes. In these reactions the sulfur rather than the metal sites of the cluster are reactive toward these small unsaturated molecules. Clearly, for nitrogenase, where we do not know the mode of binding, sulfur coordination might be a viable possibility. The $(Cp')_2Mo_2S_4$ systems that bind H_2 and C_2H_2, wherein bound C_2H_2 can be reduced to C_2H_4 and displaced by C_2H_2, are potential models for substrate reduction by nitrogenase.[225,342]

The versatility of transition-metal sulfur systems is further illustrated by the observation that activated acetylene can insert into a metal-sulfur bond in $Mo_2O_2S_2(S_2)_2{}^{2-}$, forming a vinyl-disulfide-chelating ligand

$$\text{(7.19)}$$

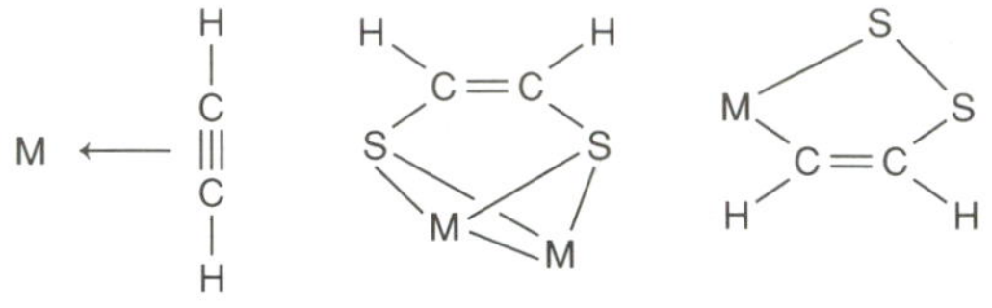

| conventional
π-bonding | dithiolene
coordination | vinyl
disulfide |

Figure 7.39
Possible modes of acetylene binding to metal-sulfur sites.

on an $Mo_2O_2S_2^{2-}$ core. In this case the acetylene is bound by both metal and sulfur atoms. Figure 7.39 shows three possible modes of C_2H_2 binding, each of which is possible for the nitrogenase system.

It has recently been suggested[343] that the presence of a dihydrogen complex is required for H_2 to be displaced by N_2 to form a dinitrogen complex. This reaction would explain the required stoichiometry of N_2 reduction and H_2 evolution. Such an explanation had been suggested previously with dihydride complexes acting as the N_2-binding and N_2-displacing site.[182] Clearly, this new suggestion is an interesting embellishment of potential N_2/H_2 relationships.

At present, the activation process that is at work in the enzyme is unknown. We need greater structural definition of the active site, which should be forthcoming through the continued application of sophisticated diffraction and spectroscopic probes. Diffraction alone, however, will be incapable of locating protons and possibly other low-molecular-weight ligands. Therefore, spectroscopic probes such as ENDOR[10] and ESEEM,[277-279,344] which are based on EPR spectroscopy, and x-ray-based techniques, such as EXAFS and XANES, will remain crucial in elucidating mechanistically significant structural details.

III. REPORT ON THE NITROGENASE CRYSTAL STRUCTURE[378-381]

A significant breakthrough has occurred in the crystallographic analysis of the iron-molybdenum protein of nitrogenase. The overall distribution of the metal clusters in the protein is shown in Figure 7.40. The distance between the two FeMoco units is fully consistent with each cofactor acting as an independent active site. On the other hand, the closeness of the P cluster and FeMoco centers in each unit is indicative of their likely cooperation in the N_2 fixation reaction.

The proposed structure of the P cluster, shown in Figure 7.41, involves a doubly bridged, double cubane unit consisting of one normally bound Fe_4S_4 cluster with all cysteine ligands and one Fe_4S_4 cluster that contains an unusual cysteine/serine (S/O) ligand pair on one of its two nonbridged Fe positions. Such five-coordinate iron in an Fe_4S_4 cluster is not unprecedented.[138] The two

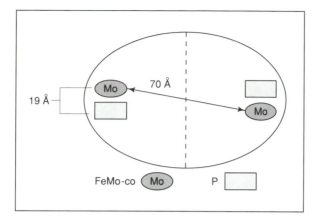

Figure 7.40
A schematic representation of the spatial arrangement of the metal sulfur clusters bound to *Cp1* as determined by the x-ray anomalous scattering studies described in the text. The representation of the large, "8-Fe" cluster with a P symbol indicates only that it must contain the Fe atoms normally assigned to P-clusters.

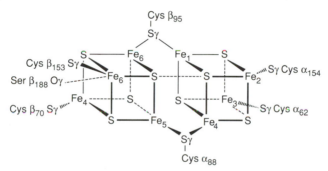

Figure 7.41
Proposed P-cluster pair in *A. vinelandii* FeMo protein.

Fe_4S_4 clusters are disposed to produce a face-sharing arrangement with two cysteine ligands bridging the two sets of Fe atoms. An interesting feature of the structure is a disulfide unit linking the two clusters; this unit potentially could be redox-active during nitrogenase turnover.

Most striking of the new results is the proposed structure of FeMoco shown in Figure 7.42. The cluster core of composition Fe_7MoS_8 can be viewed as two halves bridged by two S^{2-} ions and an unknown ligand (designated Y in the figure). The $MoFe_3S_3$ half of the core is in the shape of a thiocubane fragment missing one μ_3-S^{2-} ion. The Mo is six coordinate; the ligands are three μ_2-S^{2-} ions, which bridge to the three Fe ions, an α-His-442 nitrogen, and two oxygen donors (the hydroxyl and central carboxylate) of the homocitrate ligand. Interestingly, the second half of FeMoco is a similar thiocubane fragment, Fe_4S_3, also missing a μ_3-S^{2-} ion. This unit has a single noncore ligand, α-Cys-275, which is bound to the terminal Fe atom of the cluster. The two thiocubane

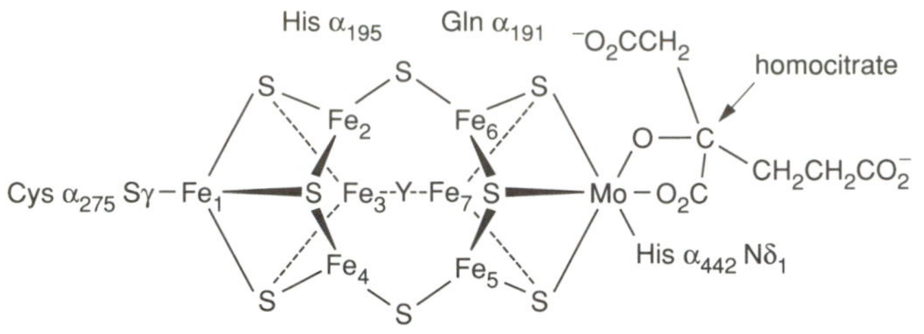

Figure 7.42
Proposed cofactor cluster in *A. vinelandii* FeMo protein.

fragments ($MoFe_3S_3$ and Fe_4S_3) are bridged by three ligands in a face-sharing mode with the two Fe_3 faces eclipsed with respect to each other. The eight metal ions display a bis(end-capped) trigonal prismatic arrangement with three bridges on the edges of the prism, which connect the two thiocubane fragments. The two sulfide bridges between the thiocubane halves are clearly defined in the structure, but the third bridge is not, suggesting the possibility that this is in fact part of the N_2-binding site. Interestingly, α-His-195, identified as essential for N_2 fixation by mutagenesis and ESEEM studies, does not appear to be covalently bound, although it is close to the FeMoco unit.

Clearly, this structure is not the same as any of those previously proposed (Figure 7.31), although it does possess many features that were identified in model studies. While it is tempting to speculate that the central bridge of the cluster (the Y ligand) is the site of N_2 reduction, this is in no way established at present.

The structural definition of the nitrogenase proteins is now progressing at a rapid rate. Many of the physical measurements will have to be reexamined in light of the new data. Through further experimentation involving physical methods, mutagenesis, and kinetic/mechanistic studies, much more information about the role of ATP, the activation of hydrogen, and the binding, activation, and reduction of N_2 and other nitrogenase substrates should be obtained.

IV. REFERENCES

1. F. Armstrong, in A. G. Sykes, ed., *Advances in Inorganic and Bioinorganic Mechanisms*, Vol. **I**, Academic Press, 1982.
2. H. Beinert and S. P. J. Albracht, *Biochim. Biophys. Acta* **683** (1982), 245.
3. A. V. Xavier, J. J. G. Moura, and I. Moura, in J. B. Goodenough *et al.*, eds., *Structure and Bonding*, Springer-Verlag, **43** (1981), 187–213.
4. D. C. Yoch and R. P. Carithers, *Microbiol. Rev.* **43** (1979), 384.
5. H. B. Dunford *et al.*, eds., *The Biological Chemistry of Iron: A Look at the Metabolism of Iron and Its Subsequent Uses in Living Organisms*, Reidel, 1981.
6. R. K. Thauer and P. Schönheit, in Reference 8, p. 329.
7. A. Bezkorovainy, *Biochemistry of Nonheme Iron*, Plenum, 1980, pp. 343–393.
8. T. G. Spiro, ed., *Iron-Sulfur Proteins*, Wiley-Interscience, 1985.
9. W. Lovenberg, ed., *Iron-Sulfur Proteins*, Vol. I, Academic Press, 1973.
10. Reference 9, Vol. II, Academic Press, 1973.

11. Reference 10, Vol. III, Academic Press, 1977.
12. Nomenclature, *Eur. J. Biochem.* **93** (1979), 427.
13. Nomenclature, *Biochim. Biophys. Acta* **549** (1979), 101.
14. C. F. Yocum, J. N. Sadow, and A. San Pietro, in Reference 9, p. 112.
15. B. B. Buchanan, in Reference 9, p. 129.
16. T. P. Singer and R. R. Ramsay, in A. N. Martonosi, ed., *The Enzymes of Biological Membranes*, Plenum, 1985, pp. 301–332.
17. T. Yagi, H. Inokuchi, and K. Kimura, *Acc. Chem. Res.* **16** (1983), 2; Y. Higuchi *et al.*, *J. Mol. Biol.* **172** (1984), 109.
18. R. Lemberg and J. Barrett, *Cytochromes*, Academic Press, 1973.
19. H. Beinert, in Reference 9, p. 1.
20. K. K. Rao and D. O. Hall, in G. V. Leigh, ed., *Evolution of Metalloenzymes, Metalloproteins, and Related Materials*, 1977, p. 39.
21. D. O. Hall, R. Cammack, and K. K. Rao, *Nature* **233** (1977), 136.
22. T. Ohnishi and J. C. Salerno, in Reference 8, p. 285.
23. F. A. Cotton and G. Wilkinson, *Advanced Inorganic Chemistry,* Wiley, 1980.
24. D. E. McRee *et al.*, *J. Biol. Chem.* **261** (1986), 10277.
25. S. J. N. Burgmayer and E. I. Stiefel, *J. Chem. Educ.* **62** (1985), 943.
26. W. Lovenberg and B. E. Sobel, *Proc. Natl. Acad. Sci. USA* **54** (1965), 193.
27. E. T. Lode and M. J. Coon, in Reference 9, pp. 173–191.
28. I. C. Gunsalus and J. D. Lipscomb, in Reference 11, p. 151.
29. R. W. Estabrook *et al.*, in Reference 9.
30. I. Moura *et al.*, *Biochem. Biophys. Res. Commun.* **75** (1977), 1037.
31. I. Moura *et al.*, *J. Biol. Chem.* **255** (1980), 2493.
31a. J. LeGall *et al.*, *Biochemistry* **27** (1988), 1636.
32. K. D. Watenpaugh, L. C. Sieker, and L. H. Jensen, *J. Mol. Biol.* **138** (1980), 615.
33. E. T. Adman *et al.*, *J. Mol. Biol.* **112** (1977), 113.
33a. C. D. Stout, in Reference 8, p. 97.
34. K. D. Watenpaugh *et al.*, *Acta Cryst.* **B29** (1973), 943.
35. R. G. Shulman *et al.*, *Proc. Natl. Acad. Sci. USA* **72** (1975), 4003.
36. R. G. Shulman *et al.*, *J. Mol. Biol.* **124** (1978), 305.
37. W. D. Phillips *et al.*, *Nature* **227** (1970), 574.
38. J. C. Rivoal *et al.*, *Biochim. Biophys. Acta* **493** (1977), 122.
39. D. E. Bennett and M. K. Johnson, *Biochim. Biophys. Acta* **911** (1987), 71.
39a. M. S. Gebhard *et al.*, *J. Am. Chem. Soc.* **112** (1990), 2217.
39b. K. D. Butcher, M. S. Gebhard, and E. I. Solomon, *Inorg. Chem.* **29** (1990), 2067.
40. M. C. W. Evans, in Reference 8, p. 249.
41. W. A. Eaton and W. J. Lovenberg, *J. Am. Chem. Soc.* **92** (1970), 7195.
42. M. K. Johnson, A. E. Robinson, and A. J. Thomson, in Reference 8, p. 367.
43. P. J. Stephens *et al.*, *Proc. Natl. Acad. Sci. USA* **75** (1978), 5273.
44. J. Peisach *et al.*, *J. Biol. Chem.* **246** (1971), 5877.
45. P. M. Champion and A. J. Siever, *J. Chem. Phys.* **66** (1977), 1819; D. Coucouvanis *et al.*, *J. Am. Chem Soc.* **101** (1979), 3392.
46. R. Cammack, D. S. Patil, and V. M. Fernandez, *Biochem. Soc. Trans.* **13** (1985), 572.
47. B. L. Trumpower, *Biochim. Biophys. Acta* **639** (1981), 129.
48. H. Beinert, *Biochem. Soc. Trans.* **13** (1985), 542.
49. G. Palmer, *Biochem. Soc. Trans.* **13** (1985), 548.
50. R. Malkin and A. J. Bearden, *Biochim. Biophys. Acta* **505** (1978), 147.
51. C. E. Johnson, *J. Inorg. Biochem.* **28** (1986), 207.
52. E. Münck and T. A. Kent, *Hyperfine Interactions* **27** (1986), 161.
53. K. K. Rao *et al.*, *Biochem. J.* **129** (1972), 1063.
54. I. Bertini and C. Luchinat, eds., *NMR of Paramagnetic Molecules in Biological Systems*, Benjamin/Cummings, 1986.
55. G. N. La Mar, W. D. Horrocks, Jr., and R. H. Holm, *NMR of Paramagnetic Molecules*, Academic Press, 1973.
56. M. T. Werth *et al.*, *J. Am. Chem. Soc.* **109** (1987), 273.
57. T. G. Spiro *et al.*, in Reference 8, p. 407.
57a. T. V. Long and T. M. Loehr, *J. Am. Chem. Soc.* **92** (1970), 6384.

58. G. Christou, B. Ridge, and H. N. Rydon, *J. Chem. Soc. Chem. Commun.* (1979), 20.
59. S. W. May and J.-Y. Kuo, *Biochemistry* **17** (1978), 3333.
60. S. W. May *et al.*, *Biochemistry* **23** (1984), 2187.
61. P. Saint-Martin *et al.*, *Proc. Natl. Acad. Sci. USA* **85** (1988), 9378.
62. R. W. Lane *et al.*, *Proc. Natl. Acad. Sci. USA* **72** (1975), 2868.
63. R. W. Lane *et al.*, *J. Am. Chem. Soc.* **99** (1977), 84.
64. D. G. Holah and D. Coucouvanis, *J. Am. Chem. Soc.* **97** (1975), 6917.
65. D. Coucouvanis *et al.*, *J. Am. Chem. Soc.* **93** (1976), 5721.
66. M. Millar *et al.*, *Inorg. Chem.* **21** (1982), 4105.
67. S. A. Koch and L. E. Madia, *J. Am. Chem. Soc.* **105** (1983), 5944.
68. M. Millar, S. A. Koch, and R. Fikar, *Inorg. Chim. Acta* **88** (1984), L15.
69. J. C. Deaton *et al.*, *J. Am. Chem. Soc.* **110** (1988), 6241.
70. D. B. Knaff, *Trends Biochem. Sci.* **13** (1988), 461.
70a. T. Tsukihara *et al.*, *J. Biochem.* **90** (1981), 1763.
71. D. R. Ort and N. E. Good, *Trends Biochem. Sci.* **13** (1988), 467.
72. V. Massey, in Reference 9, p. 301.
73. M. K. Johnson *et al.*, *J. Biol. Chem.* **260** (1985), 7368.
74. J. C. Salerno *et al.*, *J. Biol. Chem.* **254** (1979), 4828.
75. T. P. Singer, M. Gutman, and V. Massey, in Reference 9, p. 225.
76. H. Twilfer, F.-H. Bernhardt, and K. Gersonde, *Eur. J. Biochem.* **119** (1981), 595.
77. L. Petersson, R. Cammack, and K. Krishna Rao, *Biochim. Biophys. Acta* **622** (1980), 18.
78. B.-K. Teo and R. G. Shulman, in Reference 8, p. 343.
79. T. Tsukihara *et al.*, in K. Kimura, ed., *Molecular Evolution, Protein Polymorphism and the Neutral Theory*, Japan Scientific Societies Press and Springer-Verlag, 1982, p. 299.
80. T. Tsukihara *et al.*, *BioSystems* **15** (1982), 243.
81. D. Petering and G. Palmer, *Arch. Biochem. Biophys.* **141** (1970), 456.
82. J. F. Gibson *et al.*, *Proc. Natl. Acad. Sci. USA* **56** (1966), 987.
83. W. A. Eaton *et al.*, *Proc. Natl. Acad. Soc. USA* **68** (1971), 3015.
83a. L. B. Dugad *et al.*, *Biochemistry* **29** (1990), 2663.
84. J. Rawlings, O. Siiman, and H. B. Gray, *Proc. Natl. Acad. Sci. USA* **71** (1974), 125.
85. I. Salmeen and G. Palmer, *Arch. Biochem. Biophys.* **150** (1972), 767.
86. J. J. Mayerle *et al.*, *Proc. Natl. Acad. Sci. USA* **70** (1973), 2429.
87. J. L. Markley *et al.*, *Science* **240** (1988), 908.
87a. L. Banci, I. Bertini, and C. Luchinat, *Structure and Bonding* **72** (1990), 113.
88. V. K. Yachandra *et al.*, *J. Am. Chem. Soc.* **105** (1983), 6462.
89. S. Hwa, R. S. Czernuszewicz, and T. G. Spiro, *J. Am. Chem. Soc.* **111** (1989), 3496.
90. S. Hwa *et al.*, *J. Am. Chem. Soc.* **111** (1989), 3505.
91. J. R. Rieske, D. H. MacLennan, and R. Coleman, *Biochem. Biophys. Res. Commun.* **15** (1964), 338.
92. W. D. Bonner, Jr., and R. C. Prince, *FEBS Lett.* **177** (1984), 47.
93. J. F. Cline *et al.*, *J. Biol. Chem.* **260** (1985), 3251.
94. H.-T. Tsang *et al.*, *Biochemistry* **28** (1989), 7233.
95. R. C. Prince, S. J. G. Linkletter, and P. L. Dutton, *Biochem. Biophys. Acta* **635** (1981), 132.
96. J. G. Reynolds and R. H. Holm, *Inorg. Chem.* **19** (1980), 3257; **20** (1981), 1873.
96a. J. J. Mayerle *et al.*, *J. Am. Chem. Soc.* **97** (1977), 1032.
96b. Y. Do, E. D. Simhon, and R. H. Holm, *Inorg. Chem.* **22** (1983), 3809.
96c. S. Han, R. Czernuszewicz, and T. G. Spiro, *Inorg. Chem.* **25** (1986), 2276.
96d. H. Strasdeit, B. Krebs, and G. Henkel, *Inorg. Chim. Acta* **89** (1989), LII.
97. P. K. Mascharak *et al.*, *J. Am. Chem. Soc.* **103** (1981), 6110.
98. P. Beardwood *et al.*, *J. Chem. Soc. Dalton Trans.* (1982), 2015.
99. D. Coucouvanis *et al.*, *J. Am. Chem. Soc.* **106** (1984), 6081.
100. P. Beardwood and J. F. Gibson, *J. Chem. Soc. Chem. Commun.* **102** (1985), 490 and 1345.
101. L. C. Sieker, E. Adman, and L. H. Jensen, *Nature* **235** (1971), 40.
102. J. M. Berg, K. O. Hodgson, and R. H. Holm, *J. Am. Chem. Soc.* **101** (1970), 4586.
103. P. J. Stephens *et al.*, *Proc. Natl. Acad. Sci. USA* **82** (1985), 5661.
104. M. W. W. Adams and L. E. Mortenson, in Reference 235a.
105. L. W. Lim *et al.*, *J. Biol. Chem.* **261** (1986), 15 and 140.
106. J. C. Salerno *et al.*, *Biochem. Biophys. Res. Commun.* **73** (1976), 833.
107. G. Strahs and J. Kraut, *J. Mol. Biol.* **35** (1968), 503.

108. C. W. Carter, Jr., *et al.*, *Proc. Natl. Acad. Sci. USA* **69** (1972), 3526.
109. C. W. Carter, Jr., *et al.*, *J. Biol. Chem.* **249** (1974), 4212.
110. K. Fukuyama, *J. Mol. Biol.* **199** (1988), 183.
110a. K. Fukuyama *et al.*, *J. Mol. Biol.* **210** (1989), 383.
111. G. H. Stout *et al.*, *Proc. Natl. Acad. Sci. USA* **85** (1988), 1020.
112. C. D. Stout, *J. Biol. Chem.* **263** (1988), 9256.
113. A. H. Robbins and C. D. Stout, *Proc. Natl. Acad. Sci. USA* **86** (1989), 3639.
114. E. T. Adman, L. C. Sieker, and L. H. Jensen, *J. Biol. Chem.* **248** (1973), 3987.
115. R. C. Prince and M. W. W. Adams, *J. Biol. Chem.* **262** (1987), 5125.
116. L. E. Mortenson, R. C. Valentine, and J. E. Carnahan, *Biochem. Biophys. Res. Commun.* **7** (1962), 448.
117. W. Lovenberg, B. B. Buchanan, and J. C. Rabinowitz, *J. Biol. Chem.* **254** (1979), 4499.
118. R. Mathews *et al.*, *J. Biol. Chem.* **249** (1974), 4326.
119. A. J. Thomson, in P. M. Harrison, ed., *Metalloproteins, Part 1: Metal Proteins with Redox Roles*, Verlag Chemie, 1985, pp. 79–120.
120. R. Cammack, *Biochem. Biophys. Res. Commun.* **54** (1973), 548.
121. M. J. Carney *et al.*, *Inorg. Chem.* **27** (1988), 346.
122. M. J. Carney *et al.*, *J. Am. Chem. Soc.* **110** (1988), 6084.
123. L. Noodleman, D. A. Case, and A. Aizman, *J. Am. Chem. Soc.* **110** (1988), 1001.
124. L. Noodleman, *Inorg. Chem.* **27** (1988), 3677.
125. J.-M. Moulis, J. Meyer, and M. Lutz, *Biochemistry* **23** (1984), 6605.
125a. W. D. Phillips and M. Poe, in Reference 10, p. 255.
125b. E. L. Packer *et al.*, *J. Biol. Chem.* **252** (1977), 2245.
125c. I. Bertini *et al.*, *Inorg. Chem.* **29** (1990), 1874.
125d. B.-K. Teo *et al.*, *J. Am. Chem. Soc.* **101** (1979), 5624.
126. T. Herskovitz *et al.*, *Proc. Natl. Acad. Sci. USA* **69** (1972), 2437.
127. B. A. Averill *et al.*, *J. Am. Chem. Soc.* **95** (1973), 3523.
128. J. M. Berg and R. H. Holm, in Reference 8, p. 1.
129. G. B. Wang, M. A. Bobrick, and R. H. Holm, *Inorg. Chem.* **17** (1978), 578.
130. D. Coucouvanis *et al.*, *J. Am. Chem. Soc.* **104** (1982), 1874.
131. A. Müller and N. Schladerbeck, *Chimia* **39** (1985), 23.
132. A. Müller, N. Schladerbeck, and H. Bögge, *Chimia* **39** (1985), 24.
133. S. Rutchik, S. Kim, and M. A. Walters, *Inorg. Chem.* **27** (1988), 1513.
134. W. E. Cleland *et al.*, *J. Am. Chem. Soc.* **105** (1983), 6021.
135. M. G. Kanatzidis *et al.*, *J. Am. Chem. Soc.* **106** (1984), 4500.
136. M. G. Kanatzidis *et al.*, *Inorg. Chem.* **22** (1983), 179.
137. R. E. Johnson *et al.*, *J. Am. Chem. Soc.* **105** (1983), 7280.
138. M. G. Kanatzidis *et al.*, *J. Am. Chem. Soc.* **107** (1985), 4925.
138a. K. S. Hagen, J. G. Reynolds, and R. H. Holm, *J. Am. Chem. Soc.* **103** (1981), 4054.
138b. G. Christou and C. D. Garner, *J. Chem. Soc. Dalton Trans.* (1979), 1093.
138c. M. J. Carney *et al.*, *Inorg. Chem.* **27** (1988), 346.
138d. P. Barbaro *et al.*, *J. Am. Chem. Soc.* **112** (1990), 7238.
139. E. J. Laskowski *et al.*, *J. Am. Chem. Soc.* **100** (1978), 5322.
140. E. J. Laskowski *et al.*, *J. Am. Chem. Soc.* **101** (1979), 6562.
141. T. O'Sullivan and M. Millar, *J. Am. Chem. Soc.* **107** (1985), 4096.
142. M. Millar, private communication.
143. T. D. P. Stack and R. H. Holm, *J. Am. Chem. Soc.* **110** (1989), 2484.
143a. T. D. P. Stack, M. J. Carney, and R. H. Holm, *J. Am. Chem. Soc.* **111** (1989), 1670.
143b. S. Ciurli *et al.*, *J. Am. Chem. Soc.* **112** (1990), 2654.
143c. P. R. Challen *et al.*, *J. Am. Chem. Soc.* **112** (1990), 2455.
144. L. Que, Jr., R. H. Holm, and L. E. Mortenson, *J. Am. Chem. Soc.* **97** (1975), 463.
145. N. R. Bastian *et al.*, in Reference 10, p. 227.
146. H. Beinert and A. J. Thomson, *Arch. Biochem. Biophys.* **222** (1983), 333.
147. M. H. Emptage *et al.*, *J. Biol. Chem.* **255** (1980), 1793.
148. W. W. Sweeney, J. C. Rabinowitz, and D. C. Yoch, *J. Biol. Chem.* **250** (1985), 7842.
149. B. A. Averill, J. R. Bal, and W. H. Orme-Johnson, *J. Am. Chem. Soc.* **100** (1978), 3034.
150. C. D. Stout, *Nature* **279** (1979), 83.
151. D. Ghosh *et al.*, *J. Biol. Chem.* **256** (1981), 4185.

152. D. Ghosh *et al.*, *J. Mol. Biol.* **158** (1982), 73.

153. M. K. Johnson *et al.*, *J. Am. Chem. Soc.* **105** (1983), 6671.

154. H. Beinert and A. J. Thomson, *Arch. Biochem. Biophys.* **222** (1983), 333.

155. M. R. Antonio *et al.*, *J. Biol. Chem.* **257** (1982), 6646.

156. E. Münck, in Reference 8, p. 147.

157. C. E. Johnson, *J. Inorg. Biochem.* **207** (1986), 28.

158. B. H. Huynh *et al.*, *J. Biol. Chem.* **255** (1980), 3242.

159. J. J. G. Moura *et al.*, *J. Biol. Chem.* **257** (1982), 6259.

159a. H. Beinert and M. C. Kennedy, *Eur. J. Biochem.* **186** (1989), 1865.

160. C. R. Kissinger *et al.*, *J. Am. Chem. Soc.* **110** (1988), 8721.

160a. S. Ciurli and R. H. Holm, *Inorg. Chem.* **28** (1989), 1685.

161. T. A. Kent *et al.*, *J. Biol. Chem.* **260** (1985), 6871.

162. A. H. Robbins and C. D. Stout, *J. Biol. Chem.* **260** (1985), 2328.

163. M. K. Johnson *et al.*, *J. Biol. Chem.* **258** (1983), 12771.

164. M. C. Kennedy *et al.*, *J. Biol. Chem.* **258** (1983), 11098.

165. J. Telser *et al.*, *J. Biol. Chem.* **261** (1986), 4840.

166. D. H. Flint, M. H. Emptage, and J. R. Guest, *J. Inorg. Biochem.* **36** (1989), 306.

166a. R. L. Switzer, *BioFactors* **2** (1989), 77.

167. H. Beinert *et al.*, *Proc. Natl. Acad. Sci. USA* **80** (1983), 393.

168. M. K. Johnson *et al.*, *J. Biol. Chem.* **256** (1981), 9806.

169. T. R. Halbert *et al.*, *J. Am. Chem. Soc.* **106** (1984), 1849.

170. G. N. George and S. J. George, *Trends Biochem. Sci.* **13** (1988), 369.

170a. I. Moura *et al.*, *J. Am. Chem. Soc.* **108** (1986), 349.

170b. K. K. Surerus *et al.*, *J. Am. Chem. Soc.* **109** (1987), 3805.

170c. R. C. Conover *et al.*, *J. Am. Chem. Soc.* **112** (1990), 4562.

170d. J. K. Money, J. C. Huffman, and G. Christou, *Inorg. Chem.* **27** (1988), 507.

171. B. H. Huynh *et al.*, *Proc. Natl. Acad. Sci. USA* **81** (1984), 3728.

172. T. R. Hawkes and B. E. Smith, *Biochem. J.* **223** (1984), 783.

173. M. A. Whitener *et al.*, *J. Am. Chem. Soc.* **108** (1986), 5607.

174. K. S. Hagen and R. H. Holm, *J. Am. Chem. Soc.* **104** (1982), 5496.

175. K. S. Hagen, A. D. Watson, and R. H. Holm, *J. Am. Chem. Soc.* **105** (1983), 3905.

175a. J.-J. Girard *et al.*, *J. Am. Chem. Soc.* **106** (1984), 5941.

176. M. C. Kennedy *et al.*, *J. Biol. Chem.* **259** (1984), 14463.

176a. H. Strasdeit, B. Krebs, and G. Henkel, *Inorg. Chem.* **23** (1983), 1816.

177. M. G. Kanatzidis *et al.*, *J. Chem. Soc. Chem. Commun.* (1984), 356.

178. M. G. Kanatzidis, A. Salifoglou, and D. Coucouvanis, *J. Am. Chem. Soc.* **107** (1985), 3358; *Inorg. Chem.* **25** (1986), 2460.

179. S. Pohl and W. Saak, *Angew. Chem. Int. Ed. Engl.* **23** (1984), 907.

179a. S. A. Al-Ahmand *et al.*, *Inorg. Chem.* **29** (1990), 927.

180. F. Cecconi, C. A. Ghilardi, and S. Midolini, *J. Chem. Soc. Chem. Commun.* (1981), 640.

181. A. Agresti *et al.*, *Inorg. Chem.* **24** (1985), 689.

182. K. S. Hagen, J. M. Berg, and R. H. Holm, *Inorg. Chim. Acta* **45** (1980), L17.

182a. B. S. Snyder and R. H. Holm, *Inorg. Chem.* **27** (1988), 1816.

182b. B. S. Synder *et al.*, *Inorg. Chem.* **27** (1988), 595.

182c. M. S. Reynolds and R. H. Holm, *Inorg. Chem.* **27** (1988), 4494.

182d. B. S. Snyder and R. H. Holm, *Inorg. Chem.* **29** (1990), 274.

183. I. Noda, B. S. Snyder, and R. H. Holm, *Inorg. Chem.* **25** (1986), 3851.

183a. J.-F. You, B. S. Snyder, and R. H. Holm, *J. Am. Chem. Soc.* **110** (1988), 6589.

183b. J.-F. You *et al.*, *J. Am. Chem. Soc.* **112** (1990), 1067.

183c. W. R. Hagen, A. J. Pierik, and C. Veeger, *J. Chem. Soc., Faraday Trans. 1* **85** (1989), 4083.

184. E. I. Stiefel *et al.*, *Adv. Chem. Ser.* **162** (1977), 353.

184a. I. Moura and J. J. G. Moura, in Reference 5, p. 179.

184b. J. R. Lancaster, Jr., ed., *The Bioinorganic Chemistry of Nickel*, VCH Publishers, 1988.

184c. H. J. Grande *et al.*, in Reference 5, p. 193.

184d. M. W. W. Adams, L. E. Mortenson, and J.-S. Chen, *Biochim. Biophys. Acta* **594** (1981), 105.

184e. J. LeGall and H. D. Peck, Jr., in Reference 5, p. 207.

184f. J. LeGall *et al.*, in Reference 8, p. 177.

184g. S. P. Ballantine and D. H. Boxer, *Eur. J. Biochem.* **156** (1986), 276.

184h. W. H. Orme-Johnson and N. R. Orme-Johnson, in Reference 8, p. 67.

185. B. Bowien and H. G. Schlegel, *Annu. Rev. Microbiol.* **35** (1981), 401.

186. C. R. Bowers and D. P. Weitekamp, *J. Am. Chem. Soc.* **109** (1987), 5541.

187. T. C. Eisenschmid *et al.*, *J. Am. Chem. Soc.* **109** (1987), 8089.

188. M. W. W. Adams *et al.*, *Biochimie* **68** (1986), 35.

189. R. Cammack, V. M. Fernandez, and K. Schneider, in Reference 184b, p. 167.

189a. M. W. W. Adams, *Biochem. Biophys. Acta* **1020** (1990), 115.

190. H. J. Grande *et al.*, *Eur. J. Biochem.* **136** (1983), 201.

191. M. W. W. Adams and L. E. Mortenson, *J. Biol. Chem.* **259** (1984), 7045.

192. S. W. Ragsdale and L. G. L. Ljungdahl, *Arch. Microbiol.* **139** (1984), 361.

193. C. R. Woese, *Microbiol. Rev.* **S1** (1987), 221.

194. M. W. W. Adams, E. Eccleston, and J. B. Howard, *Proc. Natl. Acad. Sci. USA* **86** (1989), 4932.

195. D. S. Patil *et al.*, *J. Am. Chem. Soc.* **110** (1988), 8533.

196. A. T. Kowal, M. W. W. Adams, and M. K. Johnson, *J. Biol. Chem.* **264** (1989), 4342.

197. W. R. Hagen *et al.*, *FEBS Lett.* **203** (1986), 59.

198. I. C. Zambrano *et al.*, *J. Biol. Chem.* **264** (1989), 20974.

199. M. W. W. Adams, *J. Biol. Chem.* **262** (1987), 15054.

200. T. V. Morgan, R. C. Prince, and L. E. Mortenson, *FEBS Lett.* **206** (1986), 4.

201. W. R. Hagen *et al.*, *FEBS Lett.* **201** (1986), 158.

202. F. M. Rusnak *et al.*, *J. Biol. Chem.* **262** (1987), 38.

203. G. Wang *et al.*, *J. Biol. Chem.* **259** (1984), 14328.

204. J. Telser *et al.*, *J. Biol. Chem.* **262** (1987), 6589.

205. J. Telser *et al.*, *J. Biol. Chem.* **261** (1986), 15536.

206. H. Thomann, M. Bernardo, and M. W. W. Adams, *J. Am. Chem. Soc.* **113** (1991), 7044.

207. A. J. Thomson *et al.*, *Biochem. J.* **227** (1985), 333.

208. K. A. Macor *et al.*, *J. Biol. Chem.* **282** (1987), 9945.

209. G. N. George *et al.*, *Biochem. J.* **259** (1989), 597.

210. R. Cammack, *Adv. Inorg. Chem.* **32** (1988), 297.

211. J. J. G. Moura *et al.*, in Reference 5, p. 191.

212. J. R. Lancaster, *FEBS Lett.* **115** (1980), 285.

213. M. Teixeira *et al.*, *J. Biol. Chem.* **260** (1985), 8942.

214. J. W. Van der Zwaan *et al.*, *FEBS Lett.* **179** (1985), 271.

215. M. K. Eidsness, R. J. Sullivan, and R. A. Scott, in Reference 184b, p. 73.

216. M. K. Eidsness *et al.*, *Proc. Natl. Acad. Sci. USA* **86** (1989), 147.

217. P. A. Lindahl *et al.*, *J. Am. Chem. Soc.* **106** (1984), 3062.

218. R. A. Scott *et al.*, *J. Am. Chem. Soc.* **106** (1984), 6864.

219. S. P. J. Albracht *et al.*, *Biochim. Biophys. Acta* **874** (1986), 116.

220. A. Chapman *et al.*, *FEBS Lett.* **242** (1988), 134.

221. S. L. Tau *et al.*, *J. Am. Chem. Soc.* **106** (1984), 3064.

222. G. J. Kubas *et al.*, *J. Am. Chem. Soc.* **106** (1984), 451.

223. G. J. Kubas and R. R. Ryan, *Polyhedron* **5** (1986), 473.

224. G. J. Kubas *et al.*, *J. Am. Chem. Soc.* **108** (1986), 7000.

225. M. Rakowski DuBois *et al.*, *J. Am. Chem. Soc.* **102** (1980), 7456.

226. C. Bianchini *et al.*, *Inorg. Chem.* **25** (1986), 4617.

227. W. Tremel *et al.*, Inorg. Chem. **27** (1988), 3886.

228. W. Tremel and G. Henkel, *Inorg. Chem.* **27** (1988), 3896.

229. I. Dance, *Polyhedron* **5** (1986), 1037; P. J. Blower and J. R. Dilworth, *Coord. Chem. Rev.* **76** (1987), 121.

230. C. L. Coyle and E. I. Stiefel, in Reference 184b, p. 1.

231. M. Kumar *et al.*, *J. Am Chem. Soc.* **111** (1989), 5974.

232. M. Kumar *et al.*, *J. Am. Chem. Soc.* **111** (1989), 8323.

233. T. H. Blackburn, in W. E. Krumbein, ed., *Microbial Geochemistry*, Blackwell Scientific, 1983, p. 63.

234. J. R. Postgate, *Fundamentals of Nitrogen Fixation*, Cambridge University Press, 1982.

235. R. W. F. Hardy, *Treatise on Dinitrogen Fixation*, Wiley, 1979, Section I.

235a. T. G. Spiro, ed., *Molybdenum Enzymes*, Wiley-Interscience, 1985.

236. W. J. Brill, *NATO Adv. Sci. Inst., Ser. A* **63** (1983), 231.

237. R. Haselkorn, *Annu. Rev. Microbiol.* **40** (1986), 525.

238. A. C. Robinson, D. R. Dean, and B. K. Burgess, *J. Biol. Chem.* **262** (1987), 14327.

239. P. J. Stephens, in Reference 235a, p. 117.

240. A. H. Gibson and W. E. Newton, eds., *Current Perspectives in Nitrogen Fixation*, Australian Academy of Science, 1981.

241. E. I. Stiefel, in W. E. Newton and C. Rodriquez-Barrucco, eds., in *Recent Progress in Nitrogen Fixation*, Academic Press, 1977, p. 69.

242. C. Veeger and W. E. Newton, eds., *Advances in Nitrogen Fixation Research*, Nijhoff/Junk, 1984.

243. H. J. Evans, P. J. Bottomley, and W. E. Newton, eds., *Nitrogen Fixation Research Progress*, Martinus Nijhoff, 1985.

243a. A. Braaksma *et al.*, in Reference 5, p. 223.

243b. B. H. Huynh, E. Münck, and W. H. Orme-Johnson, in Reference 5, p. 241.

244. E. I. Stiefel, in Reference 240, p. 55.

245. R. V. Hageman and R. H. Burris, *Proc. Natl. Acad. Sci. USA* **75** (1978), 2699.

246. V. Sundaresan and F. M. Ausubel, *J. Biol. Chem.* **256** (1981), 2808.

246a. R. P. Hausinger and J. B. Howard, *J. Biol. Chem.* **258** (1983), 13486.

246b. M. M. Georgiadis, P. Chakrabarti, and D. C. Rees, *SSRL Annual Report* (1989), p. 94.

247. L. E. Mortenson, M. N. Walker, and G. A. Walker, in W. E. Newton and C. J. Nyman, eds., *Proceedings of the First International Conference on Nitrogen Fixation*, Washington State University Press (1976), p. 117.

248. W. H. Orme-Johnson *et al.*, in W. E. Newton, J. R. Postgate, and C. Rodriguez Barrucco, eds., *Recent Developments in Nitrogen Fixation*, Academic Press, 1977, p. 131.

249. G. D. Watt and J. W. McDonald, *Biochemistry* **24** (1985), 7226.

250. W. R. Hagen *et al.*, *FEBS Lett.* **189** (1986), 250.

251. L. Noodleman *et al.*, *J. Am. Chem. Soc.* **107** (1985), 3418.

252. G. D. Watt, Z.-C. Wang, and R. R. Knotts, *Biochemistry* **25** (1986), 8156; J. Cordewener *et al.*, *Eur. J. Biochem.* **148** (1985), 499.

253. L. E. Mortenson and R. N. F. Thorneley, *Annu. Rev. Biochem.* **48** (1979), 387.

254. A. V. Kulikov *et al.*, *Dokl. Akad. Nauk SSR* **262** (1981), 1177.

255. R. N. F. Thorneley and D. J. Lowe, in Reference 235, p. 221.

255a. F. A. Schultz, S. F. Gheller, and W. E. Newton, *Proc. Int. Symp. Redox Mech. Interfacial Prop. Mol. Biol. Importance* **3** (1988), 203.

256. B. K. Burgess and W. E. Newton, in A. Müller and W. E. Newton, eds., *Nitrogen Fixation: The Chemical-Biochemical-Genetic Interface*, Plenum, 1983, p. 83.

257. E. I. Stiefel and S. P. Cramer, in Reference 235, p. 88.

258. V. Shah and W. J. Brill, *Proc. Natl. Acad. Sci. USA* **74** (1977), 3249.

259. S. D. Conradson *et al.*, *J. Am. Chem. Soc.* **109** (1987), 7507.

259a. P. A. McLean *et al.*, *Biochemistry* **28** (1989), 9402.

259b. D. A. Wink *et al.*, *Biochemistry* **28** (1989), 9407.

260. M. A. Walters, S. K. Chapman, and W. H. Orme-Johnson, *Polyhedron* **5** (1986), 561.

261. P. A. McLean and R. A. Dixon, *Nature* **292** (1981), 655.

262. P. A. McLean and B. E. Smith, *Biochem. J.* **211** (1983), 589.

263. T. R. Hawkes, P. A. McLean, and B. E. Smith, *Biochem. J.* **217** (1984), 317.

264. T. R. Hoover *et al.*, *Biochemistry* **27** (1988), 3647.

265. T. R. Hoover *et al.*, *Biochemistry* **28** (1989), 2768.

265a. J. Liang *et al.*, *Biochemistry* **29** (1990), 8377.

265b. M. S. Madden *et al.*, *Proc. Natl. Acad. Sci. USA* **87** (1990), 6517.

266. A. C. Robinson, D. Dean, and B. K. Burgess, *J. Biol. Chem.* **262** (1989), 14327.

266a. A. C. Robinson *et al.*, *J. Biol. Chem.* **264** (1989), 10088.

266b. D. J. Scott *et al.*, *Nature* **343** (1990), 188.

266c. H. M. Kent *et al.*, *Biochem. J.* **264** (1989), 257.

267. W. R. Hagen *et al.*, *Eur. J. Biochem.* **169** (1987), 457.

268. P. A. McLean *et al.*, *J. Biol. Chem.* **262** (1987), 12900.

269. P. A. Lindahl *et al.*, *J. Biol. Chem.* **263** (1988), 19442.

270. G. D. Watt, A. Burns, and D. L. Tennent, *Biochemistry* **20** (1981), 7272; G. D. Watt and Z. C. Wang, *Biochemistry* **25** (1986), 5196.

271. R. Zimmermann *et al.*, *Biochim. Biophys. Acta.* **537** (1978), 185.

272. D. M. Kurtz *et al.*, *Proc. Natl. Acad. Sci. USA* **76** (1979), 4986.

273. R. A. Venters *et al.*, *J. Am. Chem. Soc.* **108** (1986), 3487.

273a. J. Bolin, in P. M. Gresshoff, L. E. Roth, G. Stacey, and W. E. Newton, eds., *Nitrogen Fixation: Achievements and Objectives,* Chapman and Hall, 1990, p. 111.

274. B. M. Hoffman, J. E. Roberts, and W. H. Orme-Johnson, *J. Am. Chem. Soc.* **104** (1982), 860.

275. A. E. True *et al.*, *J. Am. Chem. Soc.* **110** (1988), 1935.

275a. A. E. True *et al.*, *J. Am. Chem. Soc.* **112** (1990), 651.

276. G. N. George *et al.*, *Biochem. J.* **262** (1989), 349.

277. W. B. Mims and J. Peisach, in R. G. Shulman, ed., *Biological Applications of Magnetic Resonance,* Academic Press, 1980, p. 221.

278. W. H. Orme-Johnson *et al.*, in Reference 8, p. 79.

279. H. Thomann *et al.*, *J. Am. Chem. Soc.* **109** (1987), 7913.

280. H. Thomann *et al.*, *Proc. Natl. Acad. Sci. USA,* **88** (1991), 6620.

281. B. H. Huynh, E. Munck, and W. H. Orme-Johnson, *Biochim. Biophys. Acta* **527** (1979), 192.

282. B. H. Huynh *et al.*, *Biochim. Biophys. Acta* **623** (1980), 124.

283. E. Münck *et al.*, *Biochim. Biophys. Acta* **400** (1975), 32.

283a. W. E. Newton *et al.*, *Biochem. Biophys. Res. Commun.* **162** (1989), 882.

284. S. D. Conradson, B. K. Burgess, and R. H. Holm, *J. Biol. Chem.* **263** (1988), 13743.

285. W. R. Dunham *et al.*, *Eur. J. Biochem.* **146** (1985), 497.

286. S. P. Cramer *et al.*, *J. Am. Chem. Soc.* **100** (1978), 3398.

287. M. K. Eidsness *et al.*, *J. Am. Chem. Soc.* **108** (1986), 2746.

288. S. D. Conradson *et al.*, *J. Am. Chem. Soc.* **107** (1985), 7935; Reference 259.

289. B. Hedman *et al.*, *J. Am. Chem. Soc.* **110** (1988), 3798.

290. M. R. Antonio *et al.*, *J. Am. Chem. Soc.* **104** (1982), 4703.

291. J. M. Arber *et al.*, *Biochem. J.* **252** (1988), 421.

292. M. S. Weininger and L. E. Mortenson, *Proc. Natl. Acad. Sci. USA* **79** (1982), 378.

293. N. I. Sosfenov *et al.*, *Dokl. Akad. Nauk. SSSR* **291** (1986), 1123.

294. T. Yamane *et al.*, *J. Biol. Chem.* **257** (1982), 1221.

295. A. M. Flank *et al.*, *J. Am. Chem. Soc.* **108** (1986), 1049.

296. J. F. Rubinson *et al.*, *Biochemistry* **24** (1985), 273.

297. E. I. Stiefel, *Proc. Natl. Acad. Sci. USA* **70** (1973), 988.

298. K. L. Hadfield and W. A. Bulen, *Biochemistry* **8** (1969), 5103.

299. F. B. Simpson and R. Burris, *Science* **224** (1984), 1095.

300. B. K. Burgess *et al.*, *Biochemistry* **20** (1981), 5140.

301. S. Wherland *et al.*, *Biochemistry* **20** (1981), 5132.

302. J. H. Guth and R. H. Burris, *Biochemistry* **22** (1983), 5111.

303. Z.-C. Wang and G. D. Watt, *Proc. Natl. Acad. Sci. USA* **81** (1984), 376.

304. B. E. Smith *et al.*, *Phil. Trans. Roy. Soc. London* **B317** (1987), 131.

305. R. N. F. Thorneley, R. R. Eady, and D. J. Lowe, *Nature* **272** (1978), 557.

306. S. Wherland *et al.*, *Biochemistry* **20** (1981), 5132.

307. B. K. Burgess *et al.*, *Biochemistry* **20** (1981), 5140.

308. H. Bortels, *Arch. Mikrobiol.* **1** (1930), 333.

309. R. C. Bray, *Quart. Rev. Biophys.* **21** (1988), 299.

310. H. Bortels, *Zentbl. Bakt. Parasiten Abt. II* **95** (1935), 193.

311. C. E. McKenna, J. R. Benemann, and T. G. Traylor, *Biochem. Biophys. Res. Commun.* **41** (1970), 1501.

312. R. C. Burns, W. H. Fuchsman, and R. W. F. Hardy, *Biochem. Biophys. Res. Commun.* **42** (1971), 353.

313. J. R. Benemann *et al.*, *Biochim. Biophys. Acta* **264** (1972), 25.

314. P. E. Bishop, D. M. L. Jarlenski, and D. R. Hetherington, *Proc. Natl. Acad. Sci. USA* **77** (1980), 7342.

315. P. E. Bishop *et al.*, *Science* **232** (1986), 92.

316. B. J. Hales, D. J. Langosch, and E. E. Case, *J. Biol. Chem.* **261** (1986), 15301.

317. B. J. Hales *et al.*, *Biochemistry* **26** (1987), 1795.

318. J. Morningstar *et al.*, *Biochemistry* **26** (1987), 1795.

319. R. L. Robson *et al.*, *Nature* **322** (1986), 388.

320. R. R. Eady *et al.*, *Biochem. J.* **244** (1987), 197.

321. M. J. Dilworth *et al.*, *Nature* **327** (1987), 167.

322. R. N. Pau, *Trends Biochem. Res.* **14** (1989), 186; P. E. Bishop and R. D. Joerger, *Annu. Rev. Plant Physiol. Plant Mol. Biol.* **41** (1990), 109.

323. J. E. Morningstar and B. J. Hales, *J. Am. Chem. Soc.* **109** (1987), 6854.

324. J. M. Arber *et al.*, *Nature* **372** (1987), 325.

325. G. N. George *et al.*, *J. Am. Chem. Soc.* **110** (1988), 4057.

326. R. R. Eady *et al.*, *Recueil des Travaus Chim. des Pays-Bas* **106** (1987), 175.

327. M. J. Carney *et al.*, *J. Am. Chem. Soc.* **108** (1986), 3519.

328. R. H. Holm, *Chem. Soc. Rev.* (1981), 455.

329. G. Christou and C. D. Garner, *J. Chem. Soc. Dalton Trans.* (1980), 2354.

330. C. D. Garner *et al.*, *Phil. Trans. Roy. Soc. Lond.* **A308** (1982), 159.

330a. R. H. Holm and E. D. Simhon, in Reference 235a, p. 1.

331. W. H. Armstrong, P. K. Mascharak, and R. H. Holm, *Inorg. Chem.* **21** (1982), 1699.

332. R. E. Palermo and R. H. Holm, *J. Am. Chem. Soc.* **105** (1983), 4310.

332a. D. Coucouvanis, E. D. Simhon, and N. C. Baenziger, *J. Am. Chem. Soc.* **102** (1980), 6644.

332b. G. D. Friesen *et al.*, *Inorg. Chem.* **22** (1983), 2203.

332c. P. Stremple, N. C. Baenziger, and D. Coucouvanis, *J. Am. Chem. Soc.* **103** (1981), 4601.

332d. D. Coucouvanis *et al.*, *J. Am. Chem. Soc.* **102** (1980), 1732.

332e. A. Müller *et al.*, *Inorg. Chim. Acta* **148** (1988), 11.

332f. A. Müller *et al.*, *Angew. Chem. Int. Ed. Engl.* **21** (1982), 860.

332g. D. Coucouvanis *et al.*, *Inorg. Chem.* **27** (1988), 4066.

332h. P. A. Eldridge *et al.*, *J. Am. Chem. Soc.* **110** (1988), 5573.

332i. K. S. Bose *et al.*, *J. Am. Chem. Soc.* **111** (1989), 8953.

332j. J. A. Kovacs, J. K. Bashkin, and R. H. Holm, *Polyhedron* **6** (1987), 1445.

333. R. D. Sanner *et al.*, *J. Am. Chem. Soc.* **98** (1972), 8351.

333a. M. B. O'Regan *et al.*, *J. Am. Chem. Soc.* **112** (1990), 4331.

334. G. Pez, P. Apgar, and R. K. Crissey, *J. Am. Chem. Soc.* **104** (1982), 462.

335. K. Jones *et al.*, *J. Am. Chem. Soc.* **98** (1976), 74.

336. G. J. Leigh, *J. Mol. Catal.* **47** (1988), 363.

337. R. A. Henderson, G. J. Leigh, and C. J. Pickett, *Adv. Inorg. Chem. Radiochem.* **27** (1984), 198.

337a. M. Hidai and Y. Mizobe, in P. S. Braterman, ed., *Reactions of Coordinated Ligands,* Plenum, **2** (1989), 53.

337b. T. Yoshida *et al.*, *J. Am. Chem. Soc.* **110** (1988), 4872.

337c. T. Yoshida, T. Adachi, and T. Ueda, *Pure Appl. Chem.* **62** (1990), 1127.

338. D. Sellmann *et al.*, *Angew. Chem. Int. Ed. Engl.* **28** (1989), 1271.

339. A. E. Shilov, in M. Gratzel, ed., *Energy Resources through Chemistry and Catalysis,* Academic Press, 1983, p. 533.

340. W. E. Newton *et al.*, *Inorg. Chem.* **19** (1980), 1997.

341. T. R. Halbert, W.-H. Pan, and E. I. Stiefel, *J. Am. Chem. Soc.* **105** (1983), 5476.

342. M. Rakowski DuBois *et al.*, *J. Am. Chem. Soc.* **101** (1979), 5245.

343. R. H. Crabtree, *Inorg. Chim. Acta* **125** (1986), 27.

344. W. B. Mims and J. Peisach, in J. Berliner and J. Reuben, eds., *Biological Magnetic Resonance,* Plenum, **3** (1981), 213.

345. T. Yamane *et al.*, *J. Biol. Chem.* **257** (1982), 1221.

346. D. C. Rees and J. B. Howard, *J. Biol. Chem.* **2587** (1983), 12733.

347. R. B. Frankel *et al.*, *J. de Physique* **37** (1976), C6.

348. C. E. Johnson, *J. Appl. Phys.* **42** (1971), 1325.

349. P. Middleton *et al.*, *Eur. J. Biochem.* **88** (1978), 135.

350. K. Tagawa and D. I. Arnon, *Biochim. Biophys. Acta* **153** (1968), 602.

351. G. Palmer, R. H. Sands, and L. E. Mortenson, *Biochim. Biophys. Acta* **23** (1966), 357.

352. R. H. Sands and W. R. Dunham, *Quart. Rev. Biophys.* **4** (1975), 443.

353. R. Cammack, in M. J. Allen and P. N. R. Usherwood, eds., *Charge and Field Effects in Biosystems,* Abacus Press, 1984, p. 41.

354. J. Cardenas, L. E. Mortenson, and D. C. Yoch, *Biochim. Biophys. Acta* **434** (1976), 244.

355. R. Cammack, M. J. Barber, and R. C. Bray, *Biochem. J.* **157** (1976), 469.

356. R. C. Bray, in *The Enzymes,* 3d ed., **12** (1975), 299.

357. J. A. Fee *et al.*, *J. Biol. Chem.* **259** (1984), 124.

358. R. N. Mullinger *et al.*, *Biochem. J.* **151** (1975), 75.

359. R. Cammack *et al.*, *Biochim. Biophys. Acta* **490** (1977), 311.

360. F. A. Armstrong *et al.*, *FEBS Lett.* **234** (1988), 107.

361. W. R. Hagen *et al.*, *Biochim. Biophys. Acta* **828** (1985), 369.

362. E. deGryse, N. Glandsdorff, and A. Piérard, *Arch. Microbiol.* **117** (1978), 189.
363. V. M. Fernandez, E. C. Hatchikian, and R. Cammack, *Biochim. Biophys. Acta* **832** (1985), 69.
364. V. Niviére, *Biochem. Biophys. Res. Commun.* **139** (1986), 658.
365. M. Teixeira *et al.*, *Biochimie* **68** (1986), 75.
366. D. J. Lowe, B. E. Smith, and R. R. Eady, in N. S. Subba Rao, ed., *Recent Advances in Biological Nitrogen Fixation,* Arnold, 1980, p. 34.
367. R. C. Burns, R. D. Holsten, and R. W. F. Hardy, *Biochem. Biophys. Res. Commun.* **39** (1970), 90.
368. M. G. Yates and K. Planque, *Eur. J. Biochem.* **60** (1975), 467.
369. T. C. Huang, W. G. Zumft, and L. E. Mortenson, *J. Bact.* **113** (1973), 884.
370. P. C. Hallenback, P. J. Kostel, and J. R. Benemann, *Eur. J. Biochem.* **98** (1979), 275.
371. S. Norlund, U. Erikson, and H. Baltscheffsky, *Biochim. Biophys. Acta* **504** (1978), 248.
372. B. K. Burgess *et al.*, in Reference 242.
373. S. D. Conradson *et al.*, *J. Am. Chem. Soc.* **109** (1987), 7507.
374. S. A. Vaughn and B. K. Burgess, *Biochemistry* **28** (1989), 419.
375. R. W. Miller and R. R. Eady, *Biochim. Biophys. Acta* **952** (1988), 290.
376. B. K. Burgess, in Reference 235a, p. 161.
377. B. J. Hales *et al.*, *Biochemistry* **25** (1986), 7251.
378. M. M. Georgiadis *et al.*, *Science* **257** (1992), 1653.
379. J. T. Bolin *et al.*, in P. M. Greshoff *et al.*, eds., *Nitrogen Fixation: Achievements and Objectives,* Chapman and Hall, 1990, p. 117.
380. J. Kim and D. C. Rees, *Science* **257** (1992), 1677.
381. J. Kim and D. C. Rees, *Nature* **360** (1992), 553.
382. For allowing us to see and quote their work prior to publication, we are grateful to Prof. M. W. W. Adams, Prof. B. K. Burgess, Dr. R. Cammack, Prof. D. Coucouvanis, Prof. S. P. Cramer, Dr. S. J. George, Prof. J. N. Enemark, Prof. J. Lancaster, Dr. Michelle Millar, Prof. M. Maroney, Prof. W. E. Newton, Prof. D. C. Rees, Prof. Dieter Sellman, Prof. A. E. Shilov, Dr. Barry E. Smith, Dr. R. N. F. Thorneley, and Prof. G. D. Watt. We thank Pat Deuel for her superb efforts under difficult circumstances in the preparation of this manuscript.

8

Metal/Nucleic-Acid Interactions

JACQUELINE K. BARTON
Division of Chemistry and Chemical Engineering
California Institute of Technology

I. INTRODUCTION

The interest of the bioinorganic community in the field of metal/nucleic-acid interactions has burgeoned in the last decade. This interest and the resulting progress have come about primarily because of the tremendous advances that have occurred in nucleic-acid technology. We can now isolate, manipulate, and even synthesize nucleic acids of defined sequence and structure, as we would other molecules that chemists commonly explore. Furthermore, as may be evident already in other chapters of this book, bioinorganic chemistry has itself been evolving from a field focused on delineating metal centers in biology to one that includes also the application of inorganic chemistry *to probe* biological structures and function. In the past decades it has become clear that nucleic acids, structurally, functionally and even remarkably in terms of catalysis, play active and diverse roles in Nature. Transition-metal chemistry, both in the cell and in the chemist's test tube, provides a valuable tool both to accomplish and to explore these processes.

There are also many practical motivations behind the study of how metal ions and complexes interact with nucleic acids. Heavy-metal toxicity in our environment arises in part from the covalent interactions of heavy-metal ions with nucleic acids. In addition, these heavy metals interfere with metalloregulatory proteins and in so doing disrupt gene expression. We need to understand the functioning of the natural metalloregulators of gene expression and we need to design new metal-specific ligands, which, like the proteins themselves, capture heavy metals before their damage is done. Heavy-metal interactions with nucleic acids indeed have provided the basis also for the successful application of cisplatin and its derivatives as anticancer chemotherapeutic agents (see Chapter 9). The design of new pharmaceuticals like cisplatin requires a detailed understanding of how platinum and other metal ions interact with nucleic acids and nucleic-acid processing. Furthermore, we are finding that metal complexes can be uniquely useful in developing spectroscopic and reactive probes of nu-

455

cleic acids, and hence may become valuable in developing new diagnostic agents. Finally, Nature itself takes advantage of metal/nucleic-acid chemistry, from the biosynthesis of natural products such as bleomycin, which chelates redox-active metal ions to target and damage foreign DNA, to the development of basic structural motifs for eukaryotic regulatory proteins, the zinc-finger proteins, which bind to DNA and regulate transcription. In all these endeavors, we need first to develop an understanding of how transition-metal ions and complexes interact with nucleic acids and how this chemistry may best be exploited.

In this chapter we first summarize the "basics" needed to consider the interactions of metal ions and complexes with nucleic acids. What are the structures of nucleic acids? What is the basic repertoire of modes of association and chemical reactions that occur between coordination complexes and polynucleotides? We then consider in some detail the interaction of a simple family of coordination complexes, the tris(phenanthroline) metal complexes, with DNA and RNA to illustrate the techniques, questions, and applications of metal/nucleic-acid chemistry that are currently being explored. In this section, the focus on tris(phenanthroline) complexes serves as a springboard to compare and contrast studies of other, more intricately designed transition-metal complexes (in the next section) with nucleic acids. Last we consider how Nature uses metal ions and complexes in carrying out nucleic-acid chemistry. Here the principles, techniques, and fundamental coordination chemistry of metals with nucleic acids provide the foundation for our current understanding of how these fascinating and complex bioinorganic systems may function.

II. THE BASICS

A. Nucleic-Acid Structures[1]

Figure 8.1 displays a single deoxyribonucleotide and the four different nucleic-acid bases. As may be evident, each mononucleotide along a nucleic-acid polymer contains a variety of sites for interactions with metal ions, from electrostatic interactions with the anionic phosphate backbone to soft nucleophilic interactions with the purine heterocycles. The different nucleic-acid bases furthermore offer a range of steric and electronic factors to exploit. Coordination of a metal complex to the N7 nitrogen atom of a purine, for example, would position other coordinated ligands on the metal center for close hydrogen bonding to the O6 oxygen atom of guanine, but would lead to clashes with the amine hydrogen atoms of adenine.

The monomeric units strung together in a polynucleotide furthermore provide an array of polymeric conformers. Figure 8.2A *(See color plate section, pages C-14, C-15.)* shows three crystallographically characterized structures of double-helical DNA oligonucleotides,[2-4] Figure 8.2B a schematic illustration of other conformations of DNA, and Figure 8.2C the crystal structure [5] of yeast tRNA[Phe]. In double-helical DNA,[1] the two antiparallel polynucleotide strands

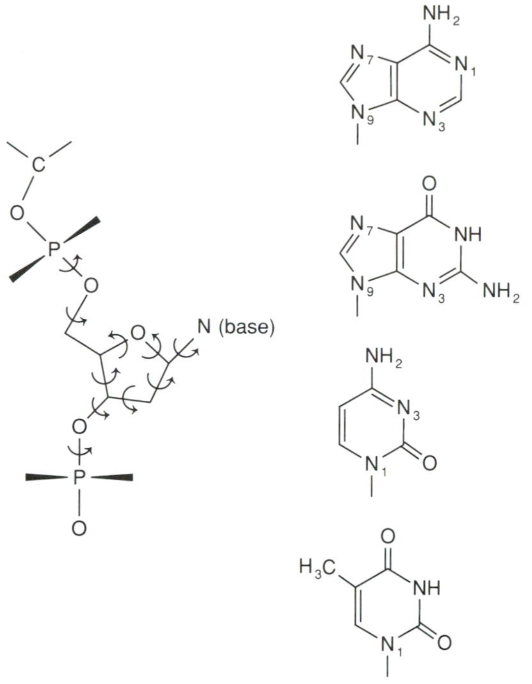

Figure 8.1
Illustration of a mononucleotide unit. Arrows indicate the various torsional angles within each unit that together generate the wide range of conformations available in the polymer. Also shown are the individual bases as well as the commonly employed numbering scheme.

are intertwined in a helix, stabilized through Watson-Crick hydrogen bonding between purines and pyrimidines, and through π-π stacking interactions among the bases arranged in the helical column. There are electrostatic repulsions between the anionic phosphate backbones of the polymer, causing a stiffening; each double-helical step has two formal negative charges. An atmosphere of metal ions condensed along the sugar-phosphate backbone serves partially to neutralize these electrostatic interactions. In the B-DNA conformation, the bases are stacked essentially perpendicular to the helical axis, and the sugars are puckered in general, with a C2'-endo geometry (the C2' carbon is to the same side as the C5' position relative to a plane in the sugar ring defined by the C1', C4', and O atoms). This conformer yields a right-handed helix with two distinct, well-defined grooves, termed the major and minor. The A-form helix, while still right-handed, is distinctly different in structure. The sugar rings are puckered generally in the C3'-endo conformation, causing the bases to be pushed out from the center of the helix toward the minor groove, and tilted relative to the helix perpendicular by almost 20°. What results is a shorter and fatter helix than the B-form; the helical pitch is 28.2 Å in A-DNA for an 11-residue helix and 33.8 Å for a 10-residue helix in B-DNA. The A-form helical shape is best characterized by the very shallow minor groove surface; what was the major groove in the B-form has been pulled deeply into the interior of the A-conformer and is really not accessible to binding by small molecules in solution. Transitions to the A-conformation are promoted by hydrophobic solvents or solutions of high ionic strength. The Z-conformation is perhaps most distinctive, owing to its left-handed helicity.[4] The conformer was dubbed Z-DNA because of the zig-zag in the helix. Alternations both in sugar puckering, between C2'-endo and C3'-endo, and in the rotation of the base about the glycosidic bond, anti or syn relative to the sugar, are evident, and lead to a dinucleoside repeating unit versus a mononucleoside repeat in the A- and B-helices. Alternating purine-pyrimidine sequences have the highest propensity to undergo transitions into the Z-form. It is actually this syn conformation of purines that leads to the left-handed helicity of the polymer. But it is not only its left-handedness that distinguishes the Z-conformation. The polymer is long and slender (the pitch is 45 Å for a 12-residue helix), and the major groove is a shallow and wide, almost convex, surface, whereas the minor groove is narrowed into a sharp and small crevice.

These crystal structures, shown in Figure 8.2A (see color plate section, page C-15), in fact each represent a family of conformations. The bases in a base pair often do not lie in the same plane, but are instead propeller-twisted with respect to one another. The local unwinding of the helix and tilting of the base pairs furthermore tend to vary with the local nucleic-acid sequence so as to maximize stacking or hydrogen-bonding interactions among the bases. Hence there is a variety of structures within each conformational family. Our understanding of these structural variations as a function of solution conditions and importantly of local sequence is still quite poor. But surely these structural variations affect and are affected by the binding of metal ions and complexes.

Even less defined structurally are other conformations of DNA, some of which are illustrated schematically in Figure 8.2B (see color plate section, page C-15). Double-helical DNA can bend,[6] form loops and cruciforms,[7] and fold back on itself into intramolecular triple helices, termed H-DNA.[8] At the ends of chromosomes, four strands may even come together in a unique conformation. These structures, characterized thus far by means of biochemical techniques, arise because of sequence and local torsional stress, or supercoiling. Many of these structures are stabilized by the binding of highly charged metal ions, probably because the highly charged metal center in a small volume can neutralize the electrostatic repulsions between polyanionic strands that are bundled together. Metal complexes can furthermore be extremely useful in targeting and characterizing these structures, as we will see. In chromosomes the DNA is packaged by histone proteins into even tighter bundles, with helical segments wrapped about the basic proteins to form superhelical nucleosomal units which are then arranged like beads on a string of more loosely packed DNA.[9]

This complexity in DNA structure is in fact small compared to that of RNA. Figure 8.2C (see color plate section, page C-15) shows the first crystallographically characterized structure[5] of an RNA polymer, yeast tRNA[Phe]. Ostensibly single-stranded RNAs do not exist as random coils, but instead fold up into well-defined three-dimensional structures, much like proteins. The structural variety, of course, bears some resemblance to that found in DNAs. Double-helical regions in the tRNA are A-like in conformation; helices fold together as one might imagine to occur in cruciforms, and even triple-helical segments are evident where three strands fold together in the polymer. But overall our ability to characterize structures of RNA thus far is lower than that with DNAs. RNAs are less stable in solution than is DNA, and fewer chemical as well as enzymatic tools are available for structural characterization. Yet the recent discovery of ribozymes,[10] the finding that RNAs can indeed catalyze nucleolytic reactions, makes our need to understand these structures even greater. Again transition-metal chemistry may participate in stabilizing, promoting, and probing these structures.

B. Fundamental Interactions with Nucleic Acids

Metal ions and complexes associate with DNA and RNA in a variety of ways, as illustrated in Figure 8.3. Both strong covalent interactions and weak noncovalent complexes are observed.[11] Each may yield a significant perturbation in the nucleic acid and/or may be exploited to obtain a site-specific response. Clearly there are some general guidelines, based on principles of coordination chemistry, that may be helpful in sorting out these interactions.

1. Coordination

Most prevalent among covalent complexes with DNA are those involving coordination between soft metal ions and nucleophilic positions on the bases. The structure[12] of cis-$(NH_3)_2$Pt-dGpG is an example: its platinum center coordinates to the N7 position of the guanine bases. In terms of interactions with the full polynucleotide, it is likely that the cis-diammineplatinum center, with two coordination sites available, would yield an intrastrand crosslink between neighboring guanine residues on a strand (see Chapter 9). Other nucleophilic sites targeted by soft metal ions on the bases include the N7 position of adenine, the N3 position on cytosine, and the deprotonated N3 position on thymine and uracil.[12,13] Some additional covalent binding to the N1 positions of the purines has also been observed. Indeed, coordination by the metal to one site on the heterocyclic base lowers the pK_a and increases the metal-binding affinity to secondary sites. It is noteworthy, however, that in base-paired double-helical DNA only the N7 positions on the purines are easily accessible in the major groove of the helix. Base binding at the purine N7 position is, of course, not limited to soft metal ions such as Pt(II), Pd(II), and Ru(II). Coordination at these sites has been evident also with first-row transition-metal ions such as Cu(II) and Zn(II).[13] For these, as is consistent with basic coordination chemistry, the lability of complexes formed is higher.

Transition-metal ions with decreasing softness are capable of coordinating also to the phosphate oxygen atoms. The ionic versus covalent character of these complexes clearly depends on the metal ions involved. In a classic study, examining the melting temperature of double-helical DNA in the presence of dif-

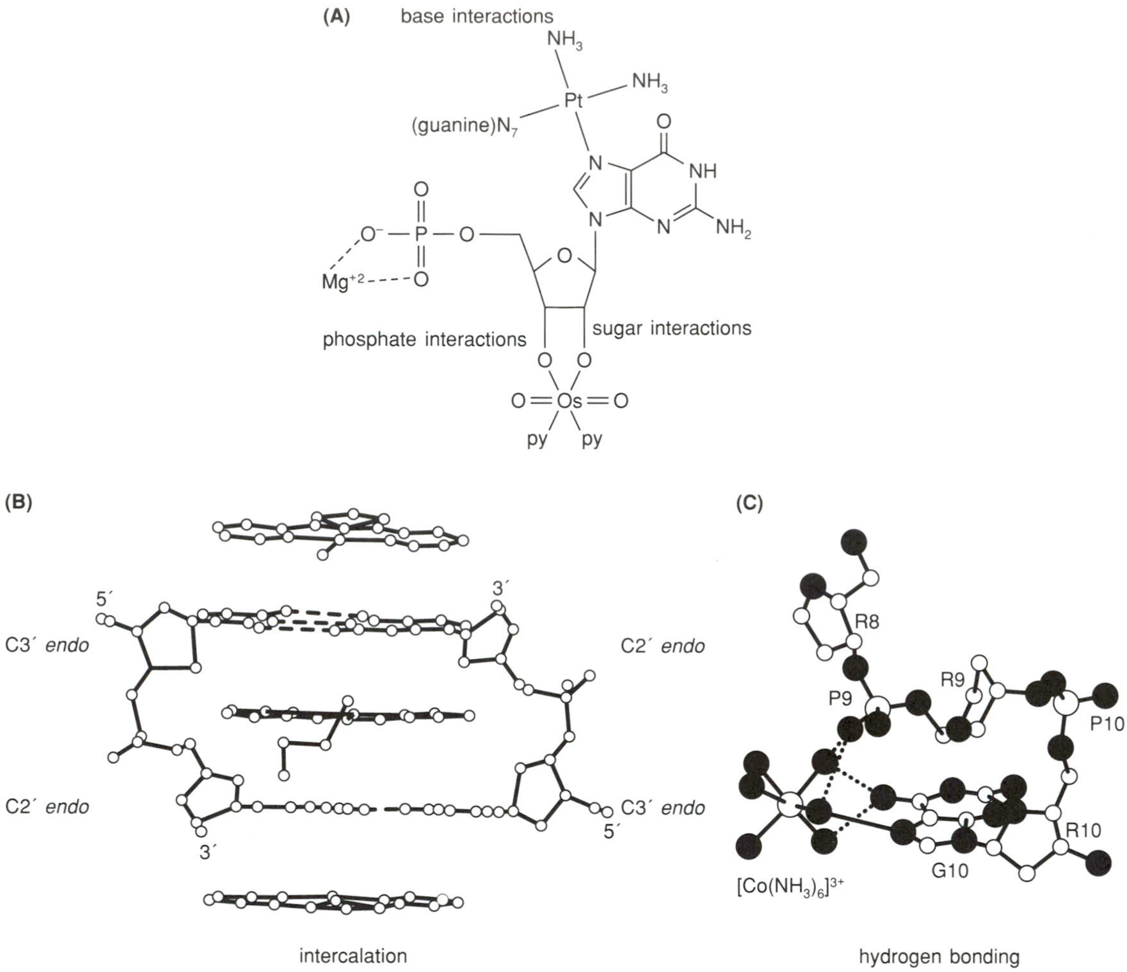

(A) base interactions

(guanine)N$_7$

phosphate interactions | sugar interactions

(B)

5′ | 3′

C3′ endo | C2′ endo

C2′ endo | C3′ endo

3′ | 5′

intercalation

(C)

R8

P9 | R9

P10

[Co(NH$_3$)$_6$]$^{3+}$

G10 | R10

hydrogen bonding

Figure 8.3

Covalent and noncovalent binding modes of metal complexes with DNA. (A) Representative covalent interactions. Shown schematically are examples of coordination to the DNA base, sugar, and phosphate moieties given by the covalent binding of *cis*-(diammine)platinum to the N7 nitrogen atom of neighboring guanine residues, the formation of an osmate ester with ribose hydroxyl groups, and the primarily electrostatic association between Mg(H$_2$O)$_6^{2+}$ and the guanosine phosphate, respectively. (B) Noncovalent intercalative stacking of a metal complex. Shown is the crystal structure[20b] of (terpyridyl)(2-hydroxyethanethiolate)platinum(II) intercalated and stacked above and below the base-paired dinucleotide d(CpG). (C) An illustration of hydrogen bonding of coordinated ligands. Shown is a partial view of the crystal structure[19] of Z-form d(CG)$_3$ with Co(NH$_3$)$_6^{3+}$ hydrogen-bonded both to the guanine base (G10) and phosphate backbone (P9).

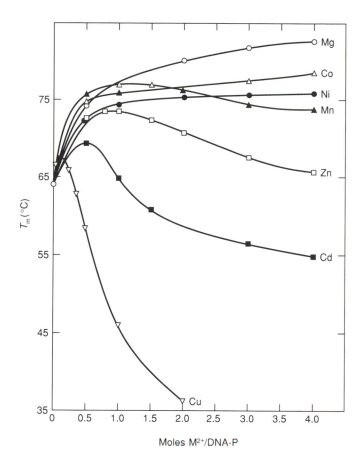

Figure 8.4
The effects of various metal ions on the melting temperature (T_m) of calf thymus DNA.[14]
Reproduced with permission from Reference 14.

ferent metal ions and as a function of their concentration, Eichhorn and coworkers established the preference of the metal ions for base versus phosphate binding
(Figure 8.4).[14] The preference for phosphate over base association was found
to decrease in the order Mg(II) > Co(II) >Ni(II) >Mn(II) >Zn(II) >Cd(II) >
Cu(II). This series arises from examination of DNA helix-melting temperatures,
since base interactions in general should destabilize the helical form [except
where interstrand crosslinking occurs, as may happen with Ag(I)], whereas
phosphate coordination and neutralization would increase the helix stability and
hence the melting temperature.

Also of interest, but less common, are covalent interactions with the sugar
moiety.[15,16] Although the pentose ring in general provides a poor ligand for
metal ions, osmate esters can form quite easily across the C2'-C3' positions in
ribose rings. This particular interaction has been suggested as a basis for heavy-
metal staining of RNA. In fact, OsO_4 is not restricted in its reactivity with the

sugar positions. Cisoid osmate esters form as well upon reaction of OsO_4 across the electron-rich C5-C6 double bonds of accessible pyrimidines on DNA.

2. Intercalation and hydrogen bonding

But important interactions of metal complexes with polynucleotides are not restricted to those involving direct coordination of the metal center to the polymer. Instead, an abundance of highly selective interactions arise from an ensemble of weaker noncovalent interactions between the ligands of coordinatively saturated metal complexes and the nucleic acid. Two primary examples of noncovalent association are given by metallointercalation and hydrogen-bonding interactions of coordinated ligands.[17,18] Planar aromatic heterocyclic ligands such as phenanthroline and terpyridine can stack in between the DNA base pairs, stabilized through dipole-dipole interactions. Here, depending on the complex and its extent of overlap with the base pairs, the free energy of stabilization can vary from ~2 to 10 kcal. Nonintercalative hydrophobic interactions of coordinated ligands in the DNA grooves also can occur, as we will see. Hydrogen-bonding interactions of coordinated ligands with the polynucleotide are quite common, and arise in particular with the phosphate oxygen atoms on the backbone. With cobalt hexaammine, for example, hydrogen bonding to an oligonucleotide occurs between the ammine hydrogens and both phosphate oxygen atoms and purine bases.[19]

A mix of covalent and noncovalent interactions is also possible. With *cis*-diammineplatinum(II) coordinated to the guanine N7 position, the ammine ligands are well-poised for hydrogen-bonding interactions with the phosphate backbone.[12] The steric constraints on the molecule must be considered, however. With Pt(terpy)Cl$^+$, both intercalation of the terpy ligand and direct coordination of the platinum center (after dissociation of the coordinated chloride) are available, but not simultaneously; coordination of the platinum to the base would likely position the terpyridyl ligand away from the base stack in the DNA major groove, precluding intercalation.[20] Sigel and coworkers[21] have studied the thermodynamics of noncovalent interactions coupled to direct coordination of simple first-row transition-metal complexes with mononucleotides, and these results illustrate well the interplay of weak noncovalent interactions and direct coordination in generating geometric specificity in complex formation.

C. Fundamental Reactions with Nucleic Acids

The reactions of transition-metal complexes with polynucleotides generally fall into two categories: (i) those involving a redox reaction of the metal complex that mediates oxidation of the nucleic acid; and (ii) those involving coordination of the metal center to the sugar-phosphate backbone so as to mediate hydrolysis of the polymer. Both redox and hydrolytic reactions of metal complexes with nucleic acids have been exploited with much success in the development of tools for molecular biology.

1. Redox chemistry

The simplest redox reaction with polynucleotides one might consider as an illustration is the Fenton reaction, which indirectly promotes DNA strand scission through radical reactions on the sugar ring. The reaction with $Fe(EDTA)^{2-}$ is shown in Figure 8.5A. As do other redox-active divalent metal ions, ferrous ion, in the presence of hydrogen peroxide, generates hydroxyl radicals, and in the presence of a reductant such as mercaptoethanol, the hydroxyl radical production can be made catalytic. Although ferrous ion itself does not appear to interact appreciably with a nucleic acid, especially when chelated in an anionic EDTA complex and repelled by the nucleic-acid polyanion, the hydroxyl radicals, produced in appreciable quantities catalytically, attack different sites on the sugar ring, indirectly yielding scission of the sugar-phosphate backbone. One such reaction that has been characterized in some detail is that involving

(A)

$$Fe(II) + H_2O_2 \longrightarrow Fe(III) + \boxed{OH\cdot} + OH^-$$

$$Fe(III) + e^- \longrightarrow Fe(II)$$

(B)

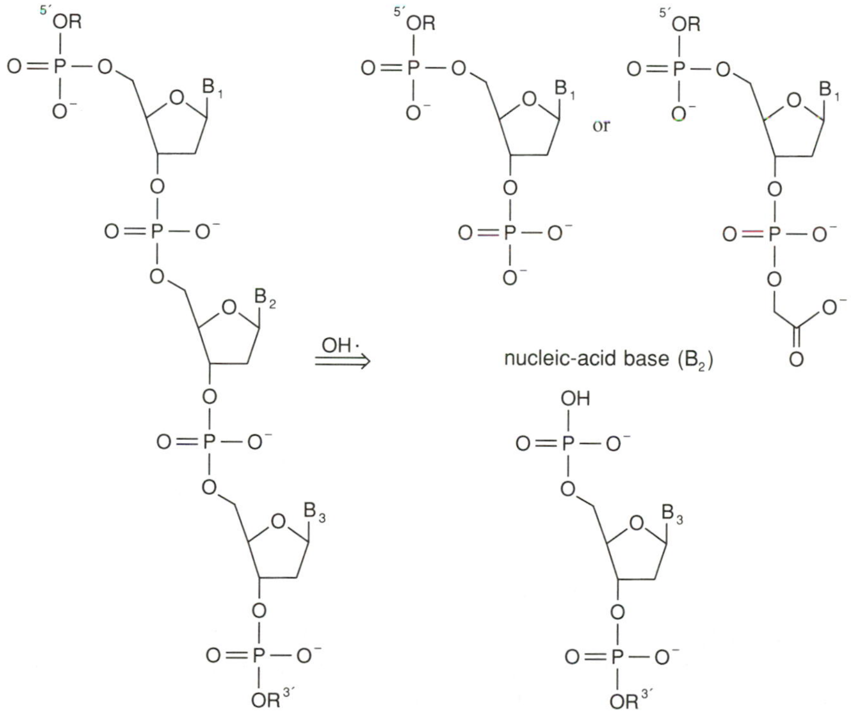

Figure 8.5
An illustration of DNA strand cleavage mediated by hydroxyl radicals produced by the Fenton reaction (A) of $Fe(EDTA)^{2-}$ with hydrogen peroxide. The cleavage scheme (B) shows the products obtained as a result of initial C4′-H abstraction by the hydroxyl radicals.

hydroxyl radical reaction at the C4′ position, the position most accessible to the diffusible radical in the minor groove of the helix.[22] As illustrated in Figure 8.5B, the products of this reaction include a 5′-phosphate, a mixture of 3′-phosphate and phosphoglycolates, and a mixture of free bases and base propenals. Reactions of the hydroxyl radical at other sites on the sugar ring are now being identified as well by isotope-labeling studies. Comparable reactions with RNA have also been described.[23]

The application of this Fenton chemistry to promote site-specific or sequence-neutral cleavage of DNA was first demonstrated[24] by Dervan and coworkers, and has provided the basis for the design of a tremendous range of new and valuable DNA cleavage agents. The development of this chemistry was originally based on modeling Fe-bleomycin, a natural product with antitumor and antibiotic activity, which binds and cleaves DNA.[25] The chemistry mediated by Fe-bleomycin, as we will discuss later, is likely to be far more complex, however, involving direct reaction of an intimately bound ferryl intermediate species with the nucleic acid, rather than net oxidation of the sugar mediated by a diffusing hydroxyl radical. Other metal ions such as Cu(II) can also promote redox-mediated cleavage of DNA[26,27] through reactions on the sugar ring; whether the oxidizing radical is still coordinated to the metal or is a dissociated and diffusing species is a topic of much debate.[26]

Metal ions can also be used to generate other oxidizing intermediates in aerated aqueous solution, such as superoxide ion and singlet oxygen. DNA strand-cleavage reactions mediated by superoxide have not thus far been demonstrated, however. Singlet oxygen may be produced by photosensitization of $Ru(phen)_3^{2+}$, and indeed photolysis of $Ru(phen)_3^{2+}$ bound to DNA yields oxygen-dependent, alkaline-sensitive strand cleavage.[28,29] For singlet oxygen, the oxidation occurs on the nucleic-acid base rather than on the sugar ring. As such, the reaction varies with base composition; guanine residues are most reactive. Furthermore, since the primary lesion is that of a base modification, piperidine treatment, or other weakly basic conditions, are needed to convert the base lesion into a strand-scission event.

Another scheme for oxidative cleavage of DNA mediated by metal complexes involves formation of a coordinated ligand radical bound to the helix that directly abstracts a hydrogen atom from the sugar ring. The photoreaction of $Rh(phen)_2phi^{3+}$ (phi = 9,10-phenanthrenequinone diimine) exemplifies this strategy.[30] Here photolysis promotes a ligand-to-metal charge transfer with formation of a phi-centered radical. Isotope-labeling studies and product analysis have shown that this phi radical bound intercalatively in the major groove of DNA directly abstracts the C3′-H (which sits in the major groove of the helix);[31] subsequent hydroxylation or dioxygen addition at this position promotes DNA strand scission without base treatment.

Some potent photooxidants can also produce outer-sphere electron transfer from the DNA. Here it is the guanine bases, likely those stacked with neighboring purines, that are most easily oxidized and hence most susceptible to attack. Again, this base modification requires alkaline treatment to convert the lesion to a strand breakage.[11b,17] The DNA double helix can furthermore also mediate electron-transfer reactions between bound metal complexes. The DNA polymer has, for example, been shown to catalyze photoinduced electron-transfer reactions between $Ru(phen)_3^{2+}$ and $Co(phen)_3^{3+}$ bound along the DNA strand.[32] Table 8.1 summarizes different redox reactions of metal complexes bound to DNA.

2. Hydrolytic chemistry

Hydrolysis reactions of nucleic acids mediated by metal ions are important elements in natural enzymatic reactions; chemists would like to exploit them in the design of artificial restriction endonucleases.[33] Hydrolysis reactions of the phosphodiester linkage of polynucleotides appear preferable to redox-mediated cleavage reactions, since in the hydrolytic reaction all information is preserved. In redox cleavage by sugar oxidation, for example, both a sugar fragment and free nucleic-acid base are released from the polymer, and, in contrast to hydrolytic chemistry, the direct religation of the fragments becomes practically impossible.

Table 8.1
Examples of metal complexes that cleave DNA through redox chemistry.

Complex	Target[a]	Chemistry[b]	Diffusibility[c]	DNA Binding[d]	Site Selectivity[e]
$Fe(EDTA)^{2-}$	sugar	OH·, Fenton	diffusible	none	none
MPE-Fe(II)	sugar (C4'-H)	OH·, Fenton	diffusible	sequence-neutral	none
$Co(NH_3)_6^{3+}$ *	base	photoelectron transfer	[f]	hydrogen-bonding	5'-G-pur-3'
$Cu(phen)_2^+$	sugar	Cu^{2+}-OH·	slight	AT-rich	AT-rich
Mn-Porphyrin	sugar	M=O	none	AT-rich	AT-rich
$U(O_2)(NO_3)_2$*	[f]	[f]	diffusible	[f]	none
$Ru(TMP)_3^{2+}$ *	base	$^1O_2$	diffusible	A-form	A-form, G
$Ru(phen)_3^{2+}$ *	base	$^1O_2$	diffusible	sequence-neutral	G
$Co(DIP)_3^{3+}$ *	sugar	ligand radical	none	Z-form (non-B)	Z-form (non-B)
$Rh(DIP)_3^{3+}$ *	sugar	ligand radical	none	Z, cruciforms	Z, cruciforms
$Rh(phen)_2phi^{3+}$ *	sugar (C3'-H)	ligand radical	none	open major groove	5'-pyr-pyr-pur-3'
$Rh(phi)_2bpy^{3+}$ *	sugar (C3'-H)	ligand radical	none	sequence-neutral	none

[a] DNA may be modified by attack either at the sugar or at the nucleotide base position.

[b] The reactive species involved in DNA cleavage, if known.

[c] Some reactive species are diffusible, producing broad patterns of DNA damage along the strand. Others are nondiffusible, resulting in cuts at single discrete sites.

[d] The site of metal complex binding to DNA, if known.

[e] The sites cleaved by the metal complex.

[f] Not known.

* Indicates an excited-state reaction requiring photoactivation.

Metal ions can be effective in promoting hydrolysis of the phosphodiester, since they can function as Lewis acids, polarizing the phosphorus-oxygen bond to facilitate bond breakage, and can also deliver the coordinated nucleophile to form the pentacoordinate phosphate intermediate. Figure 8.6 illustrates one crystallographically characterized model system developed by Sargeson and coworkers, where hydrolysis of a model phosphodiester was enhanced dramatically by taking advantage of both the acidic and the nucleophilic characteristics of the bound cobalt(III) species.[34] A whole series of model systems utilizing both cobalt and zinc ions has been designed to explore the hydrolytic reactions of simple phosphodiesters.[35] This strategy coupled to a DNA binding functionality has also been exploited, albeit inefficiently, in the hydrolytic cleavage of double-helical DNA by $Ru(DIP)_2Macro$ with Zn^{2+}, Cd^{2+}, or Pb^{2+} added *in situ*.[36] In this complex (see Figure 8.6), the central portion of the molecule, held together by the ruthenium(II), is responsible for DNA binding. Tethered onto the coordinatively saturated ruthenium complex are two diethylenetriamine functionalities (in the Macro ligand), however, and these serve to coordinate hydrolytically active metal ions such as Zn(II) and Co(II), which promote DNA hydrolysis once delivered to the sugar-phosphate backbone by the DNA-binding domain.

Perhaps simpler and certainly better understood are the hydrolytic reactions of RNAs mediated by metal ions. More than twenty years ago Eichhorn and coworkers showed that simple metal ions such as Zn(II) and Pb(II) promote the hydrolysis of RNA.[37] Figure 8.6 illustrates also the crystallographically characterized site-specific hydrolyis in tRNA by plumbous ion.[38] In tRNA, Pb(II) occupies three quite specific high-affinity binding sites, and at one of these sites, the metal ion becomes poised to promote strand cleavage. The crystal structure with bound Pb^{2+} suggests that the lead-coordinated hydroxide ion deprotonates the 2′-hydroxyl of one residue, so that the resulting 2′-oxygen nucleophile may attack the phosphate to give a pentavalent intermediate that decays to form the 2′,3′-cyclic phosphate and, after reprotonation, the 5′-hydroxide. This very specific cleavage reaction is already being used by biologists as a tool in probing structures of mutant tRNAs, since the reaction is exquisitely sensitive to the stereochemical alignment of the nucleic-acid residues, phosphate backbone, and associated metal ion. In hydrolytic reactions on RNA, it is commonly considered, though certainly not established, that the job of the metal ion may be simpler than with DNA, since the ribose provides a nearby nucleophile already in the 2′-hydroxide. The reaction of tRNA with Pb(II) nonetheless illustrates how a metal ion may be utilized in promoting highly specific chemistry on a nucleic-acid polymer.

Last, it must be mentioned that metal coordination to the purine N7 position can also indirectly promote strand cleavage, although not through direct hydrolytic reaction on the sugar-phosphate backbone. Metal ions such as Pd^{2+} and Cu^{2+}, through coordination at N7, promote depurination. The depurinated site then becomes easily susceptible to hydrolysis upon treatment with mild base.

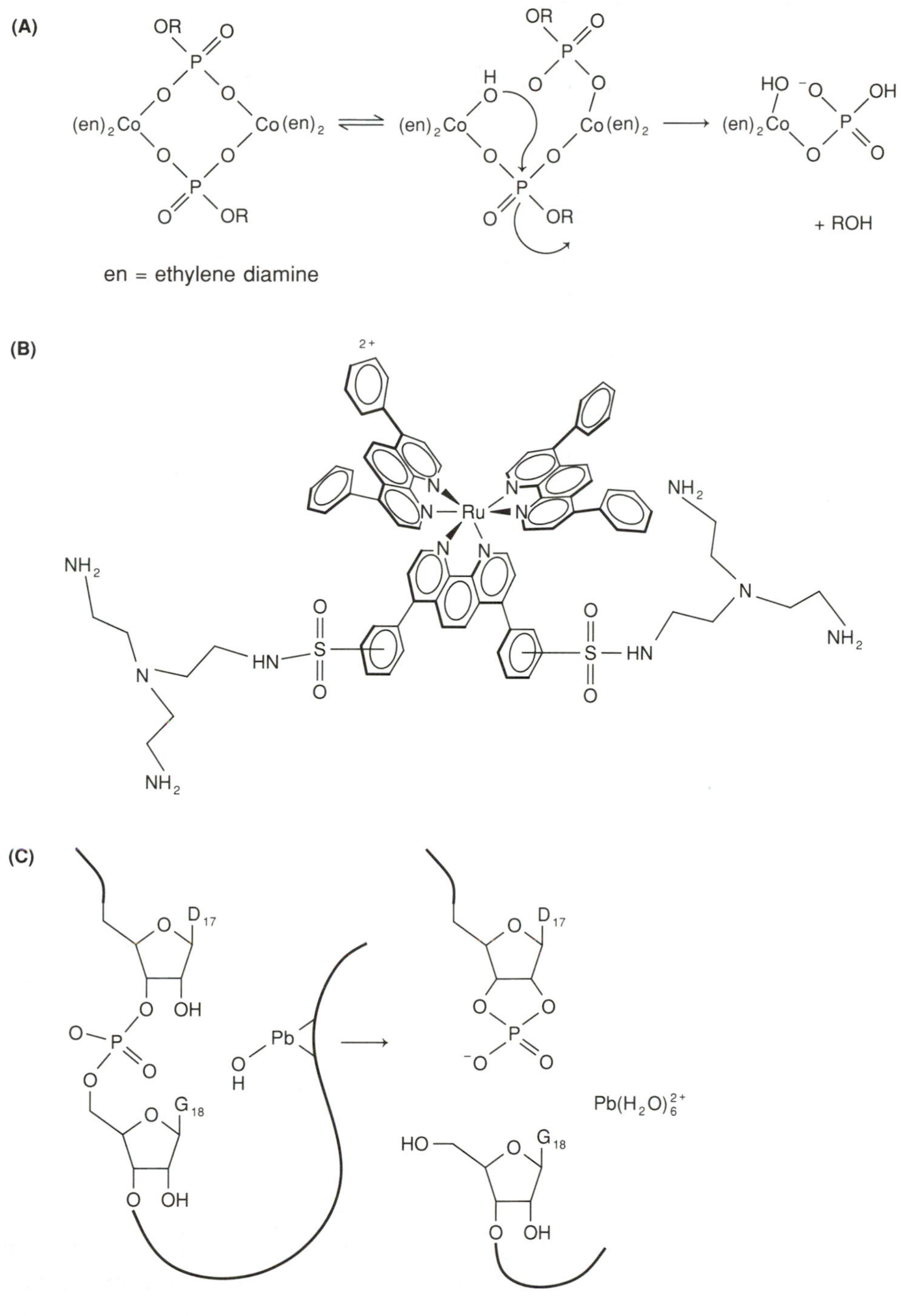

Figure 8.6
Hydrolysis reactions catalyzed by metal ions and complexes. (A) Illustration of a phosphate ester hydrolysis in a binuclear model complex catalyzed by coordinated cobaltic ions, with one metal ion functioning as a Lewis acid and the other functioning to deliver the coordinated hydroxide.[34] (B) Ru(DIP)$_2$Macro, a metal complex constructed to contain a central DNA-binding domain (Ru(DIP)$_3^{2+}$) with two tethered amine arms to chelate additional metal ions (Zn^{2+}) to deliver to the sugar-phosphate backbone and promote hydrolytic strand cleavage.[36] (C) RNA site-specifically hydrolyzed by lead ion. Diagram of the proposed mechanism of sugar-phosphate backbone cleavage between residue D$_{17}$ and G$_{18}$ in yeast RNA^{Phe}.[38]

III. A CASE STUDY: TRIS(PHENANTHROLINE) METAL COMPLEXES

Now we may examine in detail the interaction of one class of metal complexes with nucleic acids, how these complexes bind to polynucleotides, the techniques used to explore these binding interactions, and various applications of the complexes to probe biological structure and function. Tris(phenanthroline) metal complexes represent quite simple, well-defined examples of coordination complexes that associate with nucleic acids. Their examination should offer a useful illustration of the range of binding modes, reactivity, techniques for study, and applications that are currently being exploited and explored. In addition, we may contrast these interactions with those of other transition-metal complexes, both derivatives of the tris(phenanthroline) family and also some complexes that differ substantially in structure or reactivity.

A. Binding Interactions with DNA

Tris(phenanthroline) complexes of ruthenium(II), cobalt(III), and rhodium(III) are octahedral, substitutionally inert complexes, and as a result of this coordinative saturation the complexes bind to double-helical DNA through a mixture of noncovalent interactions. Tris(phenanthroline) metal complexes bind to the double helix both by intercalation in the major groove and through hydrophobic association in the minor groove.[11b,40] Intercalation and minor groove-binding are, in fact, the two most common modes of noncovalent association of small molecules with nucleic acids. In addition, as with other small molecules, a nonspecific electrostatic interaction between the cationic complexes and the DNA polyanion serves to stabilize association. Overall binding of the tris(phenanthroline) complexes to DNA is moderate (log $K = 4$).[41]

The extent of intercalative versus groove binding is seen to depend upon environmental conditions, such as temperature and ionic strength, the charge of the metal center, and the DNA base sequence; groove binding is favored at AT-rich sequences.[41] Second-generation mixed-ligand derivatives of the tris(phenanthroline) series have been prepared, and their interactions with DNA have provided useful insight into the factors important for promoting either intercalation or groove binding.[42] Aromatic heterocyclic ligands with increased surface areas that are planar bind DNA with increasing avidity through intercalation, irrespective of the charge on the metal center. Intercalative binding constants greater than $10^7 \, M^{-1}$ can be easily achieved with planar heterocyclic ligands that jut out from the metal center. Not surprisingly, complexes containing ligands of increasing hydrophobicity that are not planar favor minor-groove binding.[28]

Critically important as well in determining the binding mode is the chirality of the metal complex.[40] Intercalation into the right-handed helix favors the Δ-isomer, whereas groove binding favors the Λ-isomer. Figure 8.7 illustrates these symmetry-selective interactions. In intercalation, we consider that one phenan-

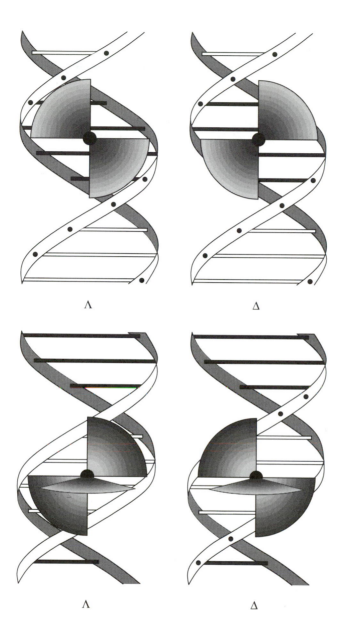

Λ   Δ

Λ   Δ

Figure 8.7
Enantiomeric discrimination in binding to DNA. Shown above is the basis for the preference for
Δ-Ru(phen)$_3$$^{2+}$ upon intercalation and Λ-Ru(phen)$_3$$^{2+}$ for surface binding against a right-handed
helix. With intercalation (top) the symmetry of the metal complex *matches* the symmetry of the
helix; steric interactions preclude a close association of the Λ-isomer. With groove binding (bottom), where the metal complex binds *against* the minor-groove helical surface, complementary
symmetries are required, and it is the Λ-isomer that is preferred.

throline inserts and stacks in between the base pairs, essentially perpendicular to the helix axis. For the Δ-isomer, once intercalated, the ancillary non-intercalated ligands are aligned along the right-handed groove of the helix. For the Λ-isomer, in contrast, with one ligand intercalated, the ancillary ligands are aligned in opposition to the right-handed groove, and steric interactions become evident between the phenanthroline hydrogen atoms and the phosphate oxygen atoms. Increasing the steric bulk on these phenanthrolines furthermore increases the enantioselective preference for intercalation of the Δ-isomer.[40,43] *For intercalation*, then, *the chiral discrimination depends on matching the symmetry of the metal complex to that of the DNA helix. For groove binding*, where the metal complex is thought to bind *against* the helix, instead *it is a complementary symmetry that is required*. In our model for groove binding of the tris(phenanthroline) metal complex, two phenanthroline ligands are likely bound against the right-handed helical groove, stabilized through hydrophobic association. For the Λ-isomer, bound in this fashion, the ligands lie against and complement the right-handed groove; with the Δ-isomer, the ligands oppose the groove, and no close surface contacts are made.

Intercalation of metal complexes in DNA is not uncommon. Lippard and coworkers first established metallointercalation by Pt(II) complexes in the 1970s.[18,20] Square-planar platinum(II) complexes containing the terpyridyl ligand were shown to intercalate into DNA. In an elegant series of x-ray diffraction experiments on DNA fibers, Lippard illustrated the requirement for planarity in the complex.[18,44] Although (phen)Pt(en)$^{2+}$ and (bpy)Pt(en)$^{2+}$ were shown to intercalate into the helix, (pyr)$_2$Pt(en)$^{2+}$, with pyridine ligands rotated out of the coordination plane, could not. Complex planarity is in itself insufficient to promote intercalation, however. *Cis*-(NH$_3$)$_2$PtCl$_2$ or even *cis*-(NH$_3$)$_2$Pt(en)$^{2+}$ does not appear to intercalate into a helix, despite full planarity. Instead, aromatic heterocyclic ligands must be included in order to promote dipole-dipole interactions with the heterocyclic bases stacked in the helix. Indeed, planarity of the full complex is not required. Intercalation is not restricted to coordination complexes that are square planar. The tris(phenanthroline) complexes represented the first examples of "three-dimensional intercalators" and illustrated that octahedral metal complexes could also intercalate into the helix.[40,45,46] Here one can consider the partial intercalation of one ligand into the helix, providing the remaining ligands on the complex an opportunity to enhance specificity or reactivity at a given site.

Curiously, one unique and apparently general characteristic of metallointercalators is their preference for intercalation from the *major groove* of the helix. Most small molecules associate with DNA from the minor groove, but metallointercalators, both those that are square planar, such as (terpyridyl)platinum(II) complexes, and those that are octahedral, such as the tris(phenanthroline) metal complexes, appear to intercalate into the major groove. This then mimics quite well the association of much larger DNA-binding proteins with the helix; DNA regulatory proteins generally appear to target the major groove. The reason why metallointercalators favor major groove association is still unclear.

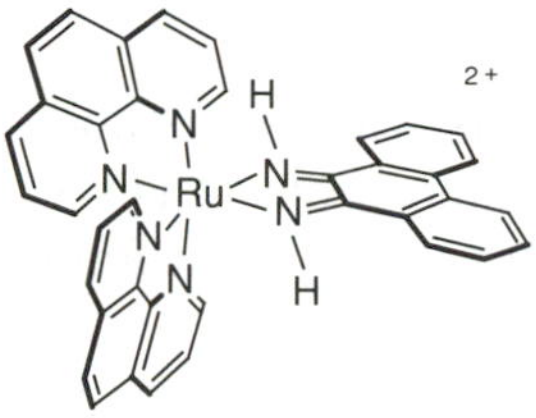

$[Ru(phen)_2\,phi]^{2+}$

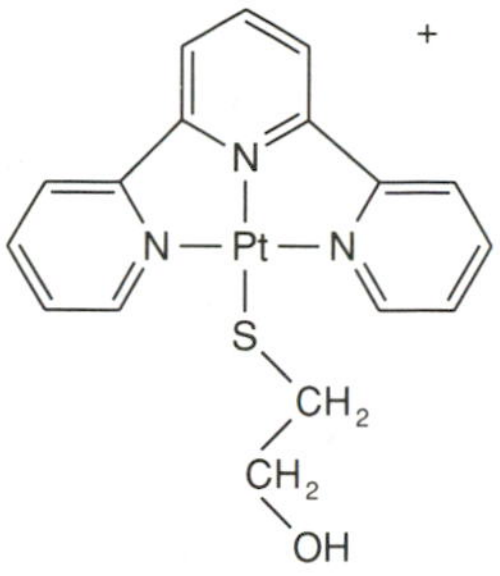

$[(terpy)Pt(HEt)]^{+}$

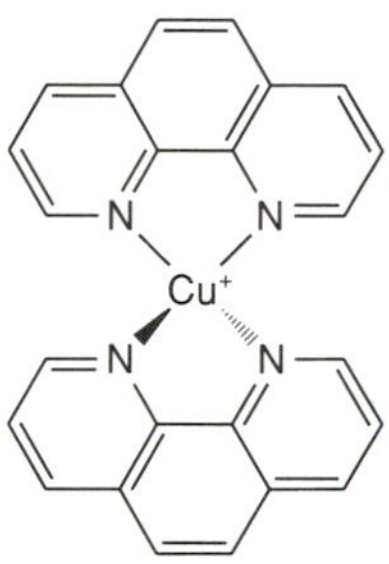

$Cu(phen)_2^{+}$

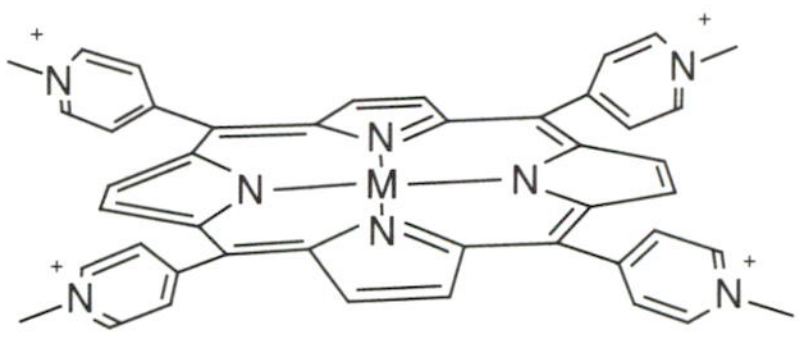

M-T4MPyP

Figure 8.8
Some metal complexes that bind DNA noncovalently primarily through intercalation (top) or binding in the minor groove (bottom). Some metalloporphyrins also primarily associate via intercalation.

Transition-metal complexes with aromatic ligands also generally associate by minor-groove binding or through the mix of intercalative and groove-bound interactions. $Cu(phen)_2{}^{+}$, a tetrahedral complex, appears to favor minor-groove binding over intercalation.[26] Perhaps the tetrahedral coordination does not permit appreciable overlap of the phenanthroline ring with the bases in an intercalative mode. Metalloporphyrins, despite their large expanse and the presence commonly of nonplanar substituents, appear to bind to double-helical DNA both by intercalation and by minor-groove binding at AT-rich sequences.[47] Occupation of the porphyrins by transition-metal ions, such as Cu(II), which bind axial ligands, leads to the favoring of groove binding over intercalation. Figure 8.8 illustrates some of the complexes that bind DNA noncovalently.

The tris(phenanthroline) metal complexes themselves do not offer an illustration of hydrogen-bonding interactions with the helix, since these ligands lack hydrogen-bonding donors and acceptors, but as mentioned already, hydrogen bonding of coordinated ligands to the helix can add some measure of stabilization, comparable to, but likely no greater in magnitude than, that provided by intercalative stacking, hydrophobic, or dispersive interactions. Indeed, mixed-ligand derivatives of the phenanthroline complexes have been prepared that in-clude hydrogen-bonding groups (amides, hydroxyls, and nitro substituents) on the ancillary phenanthroline ligands, and these have shown no greater avidity for double-helical DNA than their counterparts with hydrophobic substituents.[42] A large number of weak hydrogen-bonding interactions to DNA by one complex can be stabilizing, however, as with, for example, hexaamminecobalt(III) or hexaaquoterbium(III).

Tris(phenanthroline) metal complexes also do not offer an opportunity to explore covalent binding interactions with the helix in greater detail, but these interactions are, in fact, a major focus of Chapter 9, concerned with the mode of action of cisplatin. One derivative of the tris(phenanthroline) series, $Ru(phen)_2Cl_2$, has been shown to bind to DNA covalently.[48] In aqueous solution the dichlororuthenium(II) complex undergoes hydrolysis to form an equilibrium mixture of bis(phenanthroline) diaquo and chloroaquo species. These species bind covalently to DNA, with preferential reactivity at guanine sites. It is inter-esting that the same structural deformations in the DNA evident upon bind-ing *cis*-diammineplatinum units become apparent upon coordination of bis(phenanthroline)ruthenium(II). It is also noteworthy that the chiral preference in coordination is for the Λ-isomer. As with groove binding, direct coordination to base positions requires a complementary symmetry, with the the Λ-isomer binding *against* the right-handed groove. This preference for the Λ-isomer re-affirms that, rather than noncovalent intercalation (which would favor the Δ-isomer), covalent binding dominates the interaction. The energetic stabilization in direct coordination of the ruthenium(II) center is certainly more substantial than the weaker stabilization derived from intercalation. $Rh(phen)_2Cl_2{}^+$ and its derivatives have also been shown to bind covalently to DNA but only upon photoactivation, since light is needed to promote dissociation of the coordinated chloride and substitution of the nucleic acid base as a ligand.[49]

B. Techniques to Monitor Binding

Many of the same techniques employed in studying the basic chemistry of co-ordination complexes can be be used in following the binding of transition-metal complexes to nucleic acids, but biochemical methods, with their often exquisite sensitivity, become valuable aids as well in delineating specific binding inter-actions. Tris(phenanthroline) metal complexes are particularly useful to illustrate this point, since here the metal center in the complex is selected in terms of the technique used for examination.

Coordination complexes are often visibly colored, and these colorations pro-vide a useful and sensitive spectroscopic handle in following fundamental reac-tions. This notion holds as well with tris(phenanthroline) metal complexes in their interactions with nucleic acids. $Ru(phen)_3{}^{2+}$ and its derivatives are highly colored because of an intense metal-to-ligand charge-transfer band ($\lambda_{max} = 447$ nm, $\epsilon = 1.9 \times 10^4$ M^{-1}cm^{-1}). Furthermore, the complexes are highly photolumi-nescent ($\lambda_{em} = 610$ nm, $\tau = 0.6$ μs in aerated aqueous solution). On binding to nucleic acids these transitions are perturbed. Hypochromism is observed in the charge-transfer band, and intercalation leads to an increase in lifetime of the charge-transfer excited state.[43,46] Indeed, single-photon counting experiments show a biexponential decay in emission from $Ru(phen)_3^{2+}$ bound to double-helical DNA. The longer-lived component ($\tau = 2~\mu s$) has been assigned as the inter-calated component and the shorter-lived 0.6 μs component has been attributed to a mixture of free and groove-bound species. These spectroscopic perturbations permit one to define equilibrium-binding affinities for the different components of the interaction as a function of metal-center chirality and under different solution conditions.[41] One can also follow the polarization of emitted light from the complexes after excitation with polarized light, and these studies have been helpful in describing the dynamics of association of the complexes on the helix.[41,43] Mixed-ligand complexes of ruthenium(II) show similar spectroscopic perturbations, and these have been used to characterize binding affinities and chiral preferences, as well as the extent of intercalation versus groove binding as a function of ligand substitution on the metal center.[42] The spectroscopic handle of the metal center therefore affords a range of experiments to monitor and characterize the binding of the metal complexes to polynucleotides.

Binding interactions of metal complexes with oligonucleotides can also be followed by NMR, and here as well the metal center offers some useful characteristics to exploit. As with organic DNA-binding molecules, shifts in the ^{1}H-NMR resonances of both the DNA-binding molecule and the oligonucleotide become apparent as a function of increased association with the helix. These shift variations can be used empirically to watch the dynamics of association and to gain some structural insights into the binding modes of the complexes on the helix. These kinds of experiments have been performed with tris(phenanthroline) complexes of ruthenium(II) and rhodium(III), where it was observed that the double-helical oligonucleotide is an exceedingly good chiral-shift reagent to separate resonances in an enantiomeric mixture of the tris(phenanthroline) complexes.[50] For covalent binding molecules, such as cis-diammineplatinum(II), furthermore, the lowering of the pK_a of purine positions and therefore shifting of resonances as a function of coordination to an alternate site has been helpful as well in assigning the sites of covalent binding on the oligonucleotide.[51] But also in NMR experiments, special advantage can be taken of the metal center. For tris(phenanthroline) metal complexes, ^{1}H-NMR experiments[52] were performed on the paramagnetic analogues, $Ni(phen)_3^{2+}$ and $Cr(phen)_3^{3+}$. It was reasonable to assume the binding characteristics would be identical with their respective diamagnetic analogues, Ru(II) and Rh(III); yet paramagnetic broadening by the metal complexes of nearby resonances on the oligonucleotide would allow one to deduce where along the helix the complexes associate. Using this method the groove-binding interaction of the complexes was identified as occurring in the minor groove of the helix. Figure 8.9 illustrates the monitoring of DNA binding by tris(phenanthroline) metal complexes using both the luminescence characteristics of ruthenium(II) complexes and the paramagnetic characteristics of nickel(II).

474

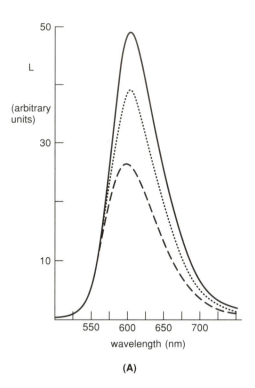

(A)

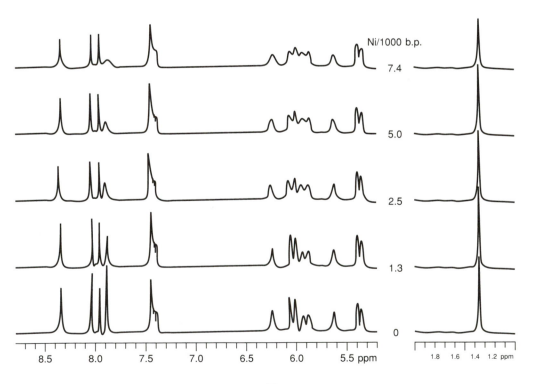

(B)

There are numerous other classic techniques of inorganic chemistry that have been or could be applied in studying the binding of metal complexes to nucleic acids. Coordination complexes have invariably been used in x-ray diffraction experiments because of the high electron density of the metal center. The tris(phenanthroline) metal complexes have not yet been applied in this context, but, as mentioned already, platinum metallointercalators were examined by fiber diffraction to delineate intercalation requirements. In fact, many nucleic-acid crystal structures have required specific metal ion additions for isomorphous heavy-metal derivatives to solve the structure. Such has certainly been true for the crystal structure of tRNA^{Phe}, where heavy-metal ions such as platinum, osmium, and mercury were targeted to specific base positions, and lanthanide ions were used to label phosphate positions around the periphery of the molecule.[53] Other techniques can also be exploited to monitor and characterize binding. A recent novel illustration is one from electrochemistry, which has been applied in monitoring the binding of $Co(phen)_3^{3+}$ to DNA.[54] Surely other techniques, from EXAFS to scanning tunneling microscopy, will be exploited in the future.

Biochemistry also provides very sensitive techniques that have been invaluable in characterizing interactions of metal complexes with nucleic acids. First are simply gel electrophoresis experiments, which permit an assessment of changes in the nucleic-acid conformation, through its changes in gel mobility, as a function of metal binding. A classic illustration is that of the unwinding of superhelical DNA as a function of intercalation. Closed circular DNA has much the same topological constraints on it as does a rope or a telephone cord; the DNA helices can wind up in coils. We define the duplex turning in a double helix as the secondary helical turns, and turns of the helices about one another as the supercoils or tertiary turns. As long as a DNA double helix is closed in a circle (form I), the total winding, that is, the total number of secondary and tertiary turns, is fixed. Molecules with differing extents of winding have different superhelical densities. In a circular molecule with one strand scission, what we call form II (nicked) DNA, the topological constraints are relaxed, and no supercoils are apparent. The same, by analogy, can be said of a telephone cord

Figure 8.9 *(facing page)*
An illustration of two spectroscopic techniques used to probe DNA. (A) The variation in luminescence characteristics of $Ru(phen)_3^{2+}$ with DNA binding. Shown is the emission spectrum of free $Ru(phen)_3^{2+}$ (————), Λ-$Ru(phen)_3^{2+}$ in the presence of DNA ($\cdots\cdots$), and Δ-$Ru(phen)_3^{2+}$ in the presence of DNA (————) illustrating the spectroscopic perturbation with DNA binding as well as the associated enantioselectivity in binding of the complexes to the helix. As is evident from the greater luminescence of the Δ isomer on binding, it is this Δ-isomer that intercalates preferentially into the right-handed helix. (B) An application of paramagnetic broadening by metal complexes in NMR experiments to obtain structural information on their association with nucleic acids. Shown is the ^{1}H-NMR spectrum of d(GTGCAC)$_2$ with increasing amounts of Λ-$Ni(phen)_3^{2+}$. Note the preferential broadening of the adenine AH2 resonance (7.8 ppm), indicating the association of this enantiomer in the minor groove of the helix.

off the phone receiver, which can turn about itself to relax its many supercoils. Now let us consider a DNA unwinding experiment, monitored by gel electrophoresis. Supercoiled form I DNA can be distinguished from nicked DNA (form II) in an agarose gel because of their differing mobilities; the wound-up supercoiled molecule moves easily through the gelatinous matrix to the positive pole, whereas the nicked species is more floppy and thus is inhibited in its travels down the gel. A closed circular molecule with no net supercoils (form I_0) comigrates with the nicked species. Consider now the supercoiled molecule in the presence of an intercalator. Since the intercalator unwinds the DNA base pairs, the number of secondary helical turns in the DNA is reduced. In a negatively supercoiled, closed circular DNA molecule, the number of supercoils must be increased in a compensatory fashion (the total winding is fixed); hence the total number of negative supercoils is reduced, and the molecule runs with slower mobility through the gel. As the intercalator concentration is increased still further, the mobility of the supercoiled species decreases until no supercoils are left, and the species comigrates with the nicked form II DNA. Increasing the bound intercalator concentration still further leads to the positive supercoiling of the DNA and an increase in mobility. Figure 8.10 illustrates the experiment with tris(phenanthroline)ruthenium(II) isomers.[46] This kind of unwinding experiment is an example of the sensitivity with which DNA structural changes can be monitored using biochemical methods; only low quantities ($<\mu$g) of materials are needed to observe these effects.

DNA strand scission can also be sensitively monitored, and even more importantly the specific nucleotide position cleaved can be pinpointed by biochemical methods. This methodology has been applied successfully in monitoring both the efficiency of DNA strand scission by metal complexes and the specific sites cleaved, and hence where the complexes are specifically bound on the helical strand.

Relative extents of cleavage of DNA by different metal complexes can be easily assayed in an experiment that is an extension of the unwinding experiment described above. One simply measures the conversion of supercoiled form I DNA to nicked form II species. One strand cleavage on the DNA circle releases the topological constraints on the circular molecule and relaxes the supercoils. Two cleavage events within 12 base pairs on opposite strands will convert the DNA to a linear form (III), which also has a distinguishable gel mobility. Photoactivated cleavage of DNA by tris(phenanthroline) complexes of cobalt(III) and rhodium(III) was first demonstrated using this assay.[55,56] Given the high sensitivity of this assay, redox-mediated cleavage of DNA by a wide range of metal complexes can be easily demonstrated. However, other techniques are required to analyze whether appreciable and significant cleavage results, and, if so, what products are obtained. Since the assay can monitor, in a short time using little sample, a single nick in a full 4,000-base-pair plasmid, reactions of very low, almost insignificant yield can be detected. The assay provides, however, a simple scheme to assess *relative* extents of cleavage by different metal

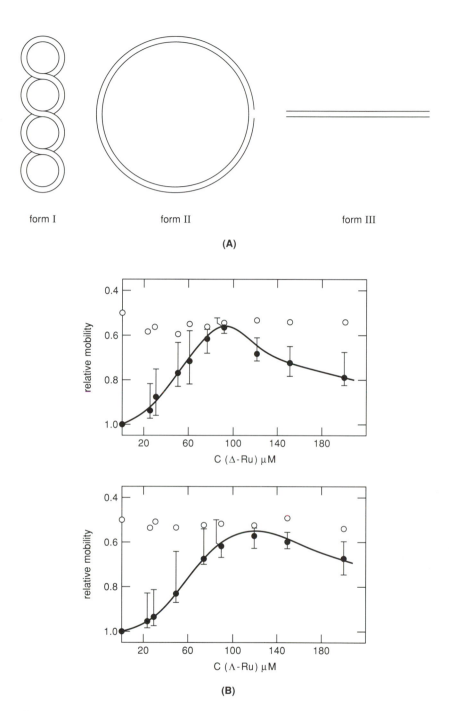

form I   form II   form III

(A)

(B)

Figure 8.10
The application of DNA supercoiling to probe metal-complex interactions with DNA. (A) A schematic representation of supercoiled DNA (form I), nicked DNA (form II) that, as a result of the single-strand scission, relaxes to a circular form lacking supercoils, and linear (form III) DNA. (B) Plots of the relative electrophoretic mobilities of form I (●) and form II (○) DNA as a function of increasing concentration of Δ-(top) and Λ-(bottom) Ru(phen)$_3^{2+}$ in the gel.[46] Increasing concentrations of bound intercalator unwind the negatively supercoiled DNA. Given the higher intercalative binding affinity of the Δ-isomer, slightly lower concentrations of this isomer are needed to unwind the plasmid to a totally relaxed state (where form I and II comigrate).

complexes, as well as a first indication that a cleavage reaction by a given metal complex occurs at all.

More informative are experiments on ^{32}P-labeled DNA fragments using high-resolution polyacrylamide gel electrophoresis, since these experiments allow one to find the exact nucleotide where the complexes break the sugar-phosphate backbone. Consider a cleavage reaction by a given metal complex on a DNA fragment of 100 base pairs in length that has been labeled enzymatically with ^{32}P on one end of one strand. If the metal complex cleaves the DNA at several different sites, then one can arrive at conditions where full cleavage is not obtained, but instead a population of molecules is generated where single cleavage events per strand are obtained, and cleavage at each of the sites is represented. After denaturation of the fragment, electrophoresis through a high-density polyacrylamide gel, and autoradiography, only fragments that are radioactively end-labeled are detected, and hence the population of sites cleaved is determined. The denatured cleaved fragments move through the gel according to their molecular weight. By measuring their length, using molecular-weight markers, one can find the specific position cleaved relative to the end of the fragment. By this route the specific sites cleaved by a molecule that binds and cleaves DNA, or end-labeled RNA, at unique positions may be identified. In a complementary experiment, using footprinting, where a molecule cleaves DNA nonspecifically at all sites along a fragment, one can find the binding positions of other molecules such as DNA-binding proteins. In this experiment, cleavage with the sequence-neutral cleaving reagent is carried out both in the presence and in the absence of the other binding molecule. In the absence of the protein, cleavage ideally occurs at all sites; hence a ladder of cleaved fragments is apparent on the autoradiograph. If cleavage is carried out in the presence of the protein, however, those sites that are bound by the protein are protected from cleavage by steric considerations, producing a shadow or footprint of the protein-binding site on the gel. Both the site-specific and footprinting experiments are illustrated schematically in Figure 8.11. This very powerful technology was first applied by Dervan and coworkers in demonstrating the application of methidium-propyl-FeEDTA (MPE-Fe) as a chemical footprinting reagent.[22,57] Tris(phenanthroline) metal complexes have been shown to cleave DNA nonspecifically, and their derivatives have been applied either as sensitive photofootprinting reagents, or as site-specific cleaving molecules, as we will see.

IV. APPLICATIONS OF DIFFERENT METAL COMPLEXES THAT BIND NUCLEIC ACIDS

Both the spectroscopy and the chemical reactivity of transition-metal complexes, coupled to biochemical assays, can therefore be exploited to obtain a wide range of useful reagents to probe nucleic acids. Here some specific applications are described.

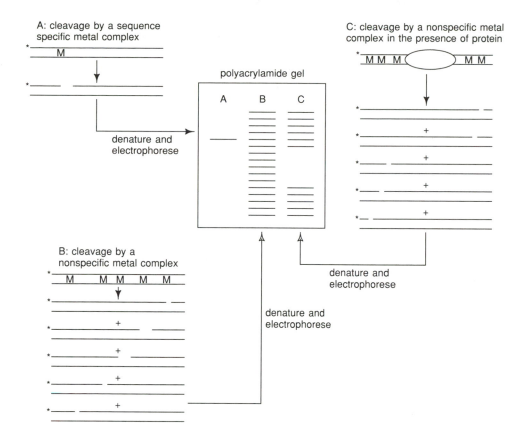

Figure 8.11

DNA cleavage by metal complexes. Shown schematically is the method used with single-base resolution to discover the sites where metal complexes are bound on double helical DNA. After the metal complex is bound to several sites on a radioactively end-labeled (*) DNA fragment and activated to permit strand cleavage at the binding sites, the nicked DNA is denatured and electrophoresed on a high-resolution polyacrylamide gel, and the gel submitted to autoradiography. From the molecular weights of the end-labeled denatured fragments, the positions of cleavage and therefore binding by the metal complex may be deduced. The results in lane A show the cleavage pattern observed for a metal complex that binds to a specific site. The results in lane B show cleavage observed for a nonspecifically bound metal complex that binds and therefore cleaves at every base site. Lane C illustrates a footprinting experiment. When protein is bound to DNA at a specific site, it protects the DNA from cleavage by the metal complex at its binding site, thus producing a "footprint" in the gel: the absence of end-labeled fragments of those lengths that are protected from cleavage as a result of protein binding.

A. Spectroscopic Probes

As discussed above, the tris(phenanthroline)ruthenium(II) complexes offer a novel spectroscopic probe of nucleic acids, since their luminescence is increased upon intercalation into the double helix. As a result the complexes provide a simple luminescent stain for DNA in fluorescent microscopy experiments. More interesting, perhaps, is the conformational selectivity of derivatives of tris(phenanthroline)ruthenium. $Ru(DIP)_3^{2+}$ (DIP = 4,7-diphenyl-1,10-phenanthroline) shows enantiospecificity in binding to B-form DNA.[40] Because of the steric bulk of the phenyl rings, detectable binding is seen only with the Δ-isomer in a right-handed helix; no binding is evident with the Λ-isomer. But with the left-handed Z-form helix, both isomers bind avidly.[40,58] The shallow left-handed major groove can accomodate the two enantiomers. A left-handed but more B-like helix shows selectivity instead for the Λ-isomer. Spectroscopic experiments that measure the chiral selectivity of $Ru(DIP)_3^{2+}$ isomers in binding to a given DNA then provide a novel probe for helical handedness. Indeed, Λ-$Ru(DIP)_3^{2+}$ was the first spectroscopic probe for Z-DNA (or other alternate conformations that are sufficiently unwound to permit binding by the bulky left-handed isomer).[58]

Another set of derivatives of the tris(phenanthroline) metal complexes that may become exceedingly useful as spectroscopic probes are $Ru(bpy)_2dppz^{2+}$ and $Ru(phen)_2dppz^{2+}$ (dppz = dipyridophenazine).[59] In these complexes the metal-to-ligand charge transfer is preferentially to the electron-accepting dppz ligand. In nonaqueous solutions, the complexes luminesce. However, in aqueous solution at pH 7, no luminescence is observed, likely because hydrogen bonding by water to the phenazine nitrogen atoms quenches the charge-transfer excited state. But the dppz ligand is also an expansive, aromatic heterocyclic ligand, and as a result both $Ru(bpy)_2dppz^{2+}$ and $Ru(phen)_2dppz^{2+}$ bind avidly to DNA by intercalation. Once intercalated, the phenazine ligand is protected from water. Therefore these complexes are luminescent when intercalated into DNA, whereas no luminescence is apparent from the complexes in the absence of DNA in aqueous solution. The enhancement factor is $>10^4$ with DNA. One might consider the ruthenium complexes as true ''molecular light switches'' for DNA.

Both simpler bipyridyl and phenanthroline derivatives as well as dppz complexes of ruthenium are currently being tethered onto other DNA binding moieties, in particular onto oligonucleotides, so as to develop new, nonradioactive luminescent probes for DNA sequences. These transition-metal complexes may provide the basis for the development of new families of DNA diagnostic agents, and many industrial laboratories are currently exploring routes to accomplish these goals. Figure 8.12 illustrates Λ-$Ru(DIP)_3^{2+}$ and $Ru(bpy)_2dppz^{2+}$, two complexes whose luminescence properties can be employed to probe nucleic acids.

Other transition-metal complexes besides those of ruthenium have shown some promise in spectroscopic applications with nucleic acids. Lanthanide ions have been applied both in NMR experiments and in luminescence experiments to probe tRNAs, and more recently with synthetic DNAs of differing sequence and structure.[60] Lanthanide ions have been exceedingly useful in probing Ca^{2+} binding sites in proteins, and one would hope that a parallel utility would be achieved with nucleic acids. Their poor absorptivity have made luminescent experiments difficult, however, requiring relatively high concentrations of material. Nonetheless, the sensitivity of luminescent lifetimes to coordination and indeed solvation is providing a novel spectroscopic handle to explore binding sites and structures of the macromolecules. Another quite novel luminescent

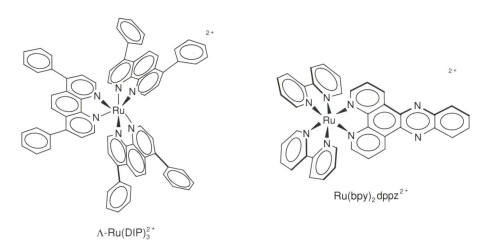

Figure 8.12
Two spectroscopic probes of nucleic acids: Λ-Ru(DIP)$_3{}^{2+}$ and Ru(bpy)$_2$dppz^{2+}.

handle has been phenanthroline and diphenylphenanthroline complexes of copper(I).[61] These complexes are extremely valuable cleavage probes, as we will see later; to characterize better their interactions with the helix, luminescence experiments are being explored. A problem here has been the nonphysiological conditions necessary to achieve detectable luminescence. Nonetheless, studies with the copper complexes demonstrate how the whole range of transition-metal chemistry and spectroscopy is beginning to be applied in sorting through nucleic-acid interactions.

B. Metallofootprinting Reagents

Probably the most widespread application of metal nucleic-acid chemistry in the biology community has been the utilization of metal complexes for chemical footprinting. The footprinting technique (Figure 8.11) was developed by biologists[62] as a means of locating protein-binding sites on DNA.[32]P-end-labeled double-stranded DNA fragments could be digested with a nuclease, such as DNAse, in the presence or absence of DNA-binding protein. After electrophoresis of the denatured digests and autoradiography, one would find a "footprint," that is, the inhibition of cleavage by DNAse, at the spot bound by protein, in comparison to a randomly cleaved pattern found on the DNA in the absence of binding protein. Although DNAse is still widely used, this footprinting reagent has some disadvantages: (i) the nuclease is not sequence-neutral in its cleavage, resulting in lots of noise in the footprinting background; and (ii) since the nuclease is itself a large protein, its ability to provide high-resolution footprinting patterns of smaller molecules is quite limited.

Several metal complexes now serve as high-resolution, sequence-neutral chemical footprinting reagents. Some of these reagents are shown in Figure 8.13. The first, as mentioned previously, was MPE-Fe(II).[57] The complex con-

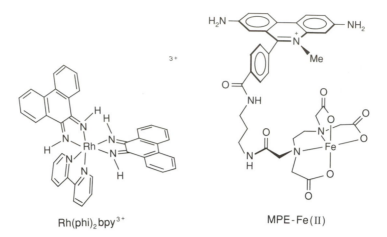

Figure 8.13
Examples of metallofootprinting reagents. Rh(phi)$_2$bpy^{3+}, a photofootprinting intercalator, and MPE-Fe(II), a sequence-neutral intercalating agent.

tains a sequence-neutral DNA binding moiety, the intercalator methidium, and a tethered DNA redox cleaving moiety, Fe(EDTA). The methidium, in binding nonspecifically to DNA, delivers the hydroxyl radicals, generated via Fenton chemistry at the Fe(II) center in the presence of peroxide and a reducing agent, to the DNA backbone in a random manner. Since the complex is small, high resolution can be achieved. Indeed, MPE-Fe(II) has been shown to footprint small natural products that bind to DNA, in addition to footprinting much larger DNA-binding peptides and proteins.

Perhaps simpler still and now very widely used as a footprinting reagent is Fe(EDTA)$^{2-}$ itself.[63] The concept here is that Fe(EDTA)$^{2-}$, as a dianion, is unlikely to associate at all with the DNA polyanion. Hence hydroxyl radicals, generated via Fenton chemistry at a distance from the helix, would likely diffuse to the helix with a uniform concentration along the helix and provide a completely sequence-neutral pattern of cleavage. Tullius and coworkers have demonstrated[63] this to be the case. The resolution is furthermore extremely high since the hydroxyl radical is sufficiently small that it can even diffuse within the DNA-binding protein to delineate binding domains. Some difficulties are found, however, with the high concentrations of activating reagents needed to activate a cleavage reagent that *does not* bind to the helix, and problems of course arise in trying to footprint metalloproteins. Nonetheless, Fe(EDTA)$^{2-}$, a reagent easily found on the biologist's shelf, is now finding great utility in labs as a chemical substitute for DNAse.

Other transition-metal complexes are also finding applications in chemical footprinting. Both Cu(phen)$_2$$^+$ and manganese porphyrins have been used to footprint DNA-binding proteins.[64,65] These complexes likely cleave DNA through either Fenton chemistry or direct reaction of a coordinated metal-oxo intermediate with the sugar-phosphate backbone. The complexes, however, appear to bind DNA predominantly along the surface of the DNA minor groove, and with some preference for AT-rich regions. The patterns obtained are actually quite similar to those found with DNAse, and thus the lack of high sequence neutrality is somewhat limiting. Furthermore, the complexes are most sensitive to binding moieties in the minor groove, rather than those in the major groove, where proteins bind. Intercalators such as MPE-Fe(II) can sense binding species in both grooves. $Cu(phen)_2{}^+$ has nonetheless proved to be quite effective in detecting hyperreactivities in the minor groove, owing to DNA structural perturbations that arise from protein binding in the major groove. Whether this sensitivity emanates from the intimate interaction of $Cu(phen)_2{}^+$ in the minor groove of the helix, or because of the characteristics of the reactive radical formed, is not known.

Inorganic photochemistry has also been applied in developing metal complexes as photofootprinting reagents. Uranyl acetate, for example, at high concentrations, upon photolysis, promotes DNA cleavage.[66] It is thought that the ions interact with the phosphates, generating some excited-state radical chemistry, although no detailed characterization of this chemistry has been undertaken. The cleavage reaction is nonetheless remarkably sequence-neutral and therefore shows some promise for photofootprinting applications. In fact, the applicability of uranyl acetate typifies how simple coordination chemistry and now even photochemistry may be helpful in the design of a variety of reagents that interact and cleave DNA, both nonspecifically and specifically. The biochemical techniques used to monitor such processes are sufficiently sensitive that even quite inefficient reactions in solution can be harnessed in developing useful reagents. The better our understanding of the chemistry of the coordination complex, the more effectively it may be utilized.

The best derivative of a tris(phenanthroline) metal complex currently being applied in footprinting experiments is $Rh(phi)_2bpy^{3+}$, a second-generation derivative of the tris(phenanthroline) series[67] that binds DNA avidly by intercalation and in the presence of light promotes direct strand cleavage by hydrogen-atom abstraction at the C3'-position on the sugar.[31] Since no diffusing intermediate is involved in this photocleavage reaction, the resolution of the footprinting pattern is to a single nucleotide. Here the excited-state transition-metal chemistry involves a ligand-to-metal charge transfer, producing a phi cation radical that directly abstracts the hydrogen from the sugar at the intercalated site. The high efficiency of this photoreaction and high sequence-neutral binding of the complex to double-stranded DNA add to the utility of this reagent in footprinting studies. Indeed, both DNA-binding proteins, bound in the major groove, and small natural products, associated with the minor groove, have been footprinted with $Rh(phi)_2bpy^{3+}$ to precisely that size expected based upon crystallographic results. One may hope that this and other photofootprinting reagents will soon find applications for footprinting experiments *in vivo*.

C. Conformational Probes

Metal complexes are also finding wide application in probing the local variations in conformation that arise along nucleic-acid polymers. X-ray crystallography has been critical in establishing the basic conformational families of double-helical DNA, and to some extent how conformations might vary as a function of nucleic-acid sequence. Yet many conformations have still not been described to high resolution, and only a few oligonucleotides have been crystallized. Other techniques are therefore required to bridge the small set of oligonucleotide crystal structures that point to plausible structures and the large array of structures that arise as a function of sequence on long helical polymers. Furthermore, only a very small number of RNA polymers has been characterized crystallographically; hence other chemical methods have been needed to describe the folding patterns in these important biopolymers. Metal complexes, mainly through specific noncovalent interactions, appear to be uniquely useful in probing the structural variations in nucleic acids.

1. Nonspecific reactions of transition-metal complexes

Hydroxyl radical cleavage with $Fe(EDTA)^{2-}$ illustrates again how simple metal complexes can be used in characterizing nucleic acids. One example involves efforts to describe the local structural variations in "bent" DNA. Biochemists had found that DNA fragments containing runs of adenines, such as in the tract dAAAAAA, possessed unusual gel-electrophoretic mobilities. Indeed, kinetoplast DNA isolated from mitochondria of trypanosomes showed a remarkable lacework pattern of structure, with loops and circles of DNA; these structures were found to be governed by the placement of these $d(A)_6$ tracts. By constructing a series of oligonucleotides with adenine runs positioned either in or out of phase relative to one another, researchers found that the adenine tracts caused a local bending of the DNA toward the minor groove.[6] But what were the detailed characteristics of these bent sites? Using hydroxyl radical cleavage of DNA, generated with $Fe(EDTA)^{2-}$, Tullius and coworkers found a distinctive pattern of cleavage across the adenine tracts, consistent with a locally perturbed structure.[68] Here the notion again was that $Fe(EDTA)^{2-}$ in the presence of peroxide would generate hydroxyl radicals at a distance from the helix, and thus careful densitometric analysis of the cleavage across ^{32}P-end-labeled DNA fragments would reveal any differential accessibility of sugar residues to cleavage mediated by the radicals caused by the bending. The cleavage patterns suggested a smooth bending of the DNA across the tract and indicated furthermore an asymmetry in structure from the 5'- to 3'-end of the adenine run.

The reactivities of other transition-metal reagents have also been used advantageously in probing nucleic-acid structures. As described in Section II, OsO_4 reacts across the 5,6 position of accessible pyrimidines to form a *cis*-osmate ester. Upon treatment with piperidine, this base modification can yield scission of the sugar-phosphate backbone. Hence DNAs containing unusual local con-

formations with prominent solvent-accessible pyrimidines can be probed with
OsO_4. The junction regions of Z-DNA, the single-stranded loops in cruciform
structures, and a segment of the dangling third strand in H-DNA, have all been
probed by means of the differential reactivity of osmium tetroxide with DNA
sites dependent upon their accessibility.[7,8,16,69] Surely other transition-metal
chemistry will become similarly applicable.

2. Transition-metal complexes as shape-selective probes

Transition-metal complexes have also been designed with three-dimensional
structures that target complementary structures along the helical polymer. This
recognition of DNA sites, based upon *shape selection*, has proved to be extremely useful both in demarcating and in characterizing structural variations
along the polymer and in developing an understanding of those factors important
to the recognition of specific polynucleotide sites. Complexes, basically derivatives of the tris(phenanthroline) metal series, have been designed that specifically target A- and Z-form helices, cruciforms, and even subtle variations such
as differential propeller twisting within B-form DNA.[11c] By appropriate substitution of the metal at the center of the coordinatively saturated complex, complexes that cleave the DNA at the binding site are obtained. Figure 8.14 shows
some of these shape-selective conformational probes.

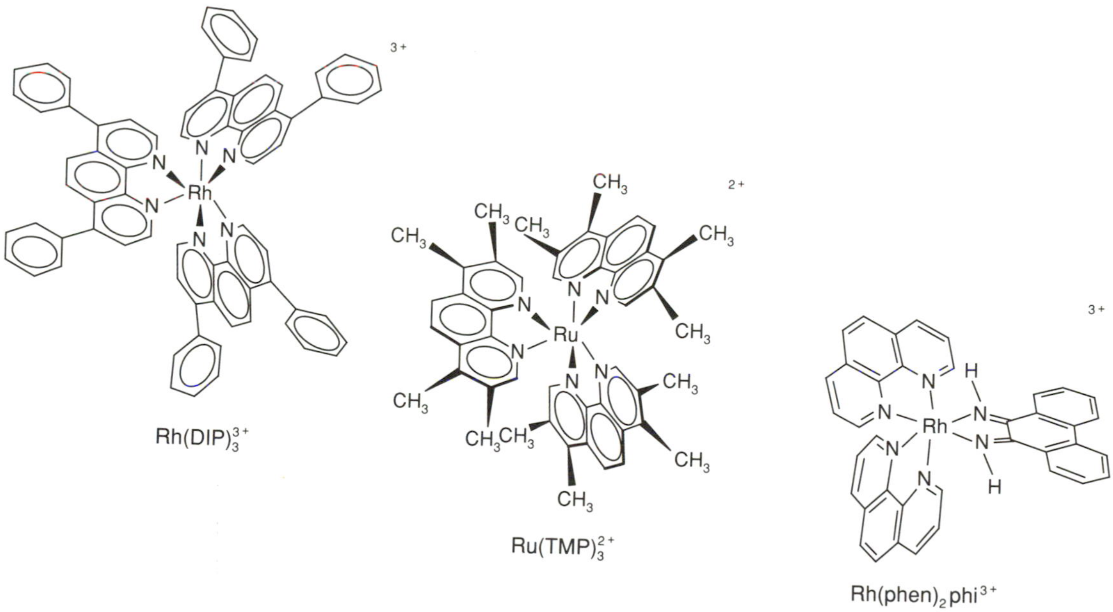

$Rh(DIP)_3^{3+}$ $Ru(TMP)_3^{2+}$ $Rh(phen)_2phi^{3+}$

Figure 8.14
Shape-selective probes that target local DNA conformations. $Rh(DIP)_3^{3+}$, which with photoactivation promotes double-stranded cleavage at cruciform sites; $Ru(TMP)_3^{2+}$, a photoactivated
probe for A-like conformations; and $Rh(phen)_2phi^{3+}$, which targets openings in the DNA major
groove.

One example of this shape-selective cleavage is apparent in reactions of Ru(TMP)$_3$$^{2+}$ (TMP = 3,4,7,8-tetramethylphenanthroline), a probe of the A-conformation.[28,29] The complex was designed by incorporating methyl groups about the periphery of each phenanthroline ligand to preclude intercalative binding of the complex to the helix, owing to the bulkiness of the methyl groups, and at the same time to promote hydrophobic groove binding. Importantly, however, this hydrophobic groove binding could not occur against the minor groove of B-DNA, given the width and depth of the groove versus the size of the complex. Instead, the shape of the complex was matched well to the shallow minor-groove surface of an A-form helix. Binding studies with synthetic polynucleotides of A, B, and Z-form were consistent with this scheme. Photolysis of the ruthenium complex, furthermore, as with Ru(phen)$_3$$^{2+}$, leads to the sensitization of singlet oxygen, and hence, after treatment with piperidine, to strand cleavage. Thus, photocleavage reactions with Ru(TMP)$_3$$^{2+}$ could be used to delineate A-like regions, with more shallow minor grooves, along a helical polymer. At such sites, Ru(TMP)$_3$$^{2+}$ would bind preferentially, and upon photolysis, generate *locally* higher concentrations of singlet oxygen to mediate cleavage of the sugar-phosphate backbone. This scheme revealed that homopyrimidine stretches along the helix adopt a more A-like conformation.[29]

The targeting of altered conformations such as Z-DNA has been described earlier[58] in the context of a spectroscopic probe, Λ-Ru(DIP)$_3$$^{2+}$. Substitution of a photoredox-active metal into the core of the tris(diphenylphenanthroline) unit leads also to a complex that both binds and, with photoactivation, cleaves at the altered conformation.[55] Both Co(III) polypyridyl and Rh(III) polypyridyl complexes have been shown to be potent photooxidants. Coupled to site-specific DNA binding, these metal complexes, with photoactivation, become conformationally selective DNA cleavage agents. Co(DIP)$_3$$^{3+}$, for example, has been shown to cleave specifically at Z-form segments inserted into DNA plasmids.[55,70] Perhaps even more interesting, on both natural plasmids and viral DNAs, the various sites cleaved by Co(DIP)$_3$$^{3+}$, corresponding both to Z-form sites and to other locally altered non-B-conformations, coincide with functionally important regions of the genome, e.g., regulatory sites, gene termination sites, and intron-exon joints.[70,71] The altered structures recognized by the metal complexes, therefore, appear to mark biologically important sites, those presumably recognized also by cellular proteins. Cleavage studies with these metal complexes, therefore, are providing some insight also into how Nature specifically targets and accesses the sequence information encoded along the DNA polymer, sequence information encoded indirectly through local structure.

The most striking example of the specificity to be derived from shape-selective targeting has been given by the double-stranded cleavage induced by Rh(DIP)$_3$$^{3+}$ at cruciforms.[72] Rh(DIP)$_3$$^{3+}$, like its Co(III) and Ru(II) congeners, binds to locally unwound, non-B-conformations such as Z-DNA, but interestingly this potent photooxidant yields the specific cleavage of *both* DNA strands at cruciform sites. Lacking any crystallographic information, our understanding of the local structure of a cruciform is poor. In these palindromic sites, a torsionally strained DNA extrudes two intrastrand hydrogen-bonded helices from the main helix (see Figure 8.2B). Clearly the structure is grossly altered and locally unwound. $Rh(DIP)_3^{3+}$ appears to bind into a pocket generated by the folding of the extruded helix onto the main helix. The recognition is of this intricately folded structure, not of the sequence used to generate the cruciform. Studies with the transition-metal complex on different cruciforms should be useful in helping to characterize this interesting tertiary DNA structure.

Shape-selective transition-metal probes have also been useful in delineating more subtle variations in structure, such as the propeller twisting and tilting evident in B-form DNA.[30,73] $Rh(phen)_2phi^{3+}$ was found to target preferentially sites in the major groove where the DNA base pairs are more open; this preferential recognition arises from the steric constraints at more-closed intercalation sites because of the bulkiness of the ancillary phenanthroline ligands above and below the intercalation plane. Two straightforward structural perturbations that lead to an opening of the major groove involve the propeller twisting of bases with respect to one another and the tilting of base pairs along the helix. Chiral discrimination in cleavage by $Rh(phen)_2phi^{3+}$ is now being used quantitatively to discriminate among these structural parameters. The goal in these studies is to use cleavage results with the shape-selective metal complexes to describe the three-dimensional structure of a long double-helical DNA sequence in solution. Probing structurally well-defined sequences with a whole family of shape-selective metal complexes may provide a route to this goal.

As mentioned above, describing the three-dimensional structures of RNAs is an even more complicated task than it is for double-stranded DNAs. Only a few tRNAs have been characterized crystallographically to high resolution, and for other larger RNA structures, such as 5S RNA, or any of the catalytic intervening-sequence RNAs, little is known about their folding characteristics. To understand the regulation and catalytic functions of these biopolymers, we need to develop chemical tools to explore these structures. Figure 8.15 shows the results of cleavage studies using the variety of transition-metal probes on tRNA[Phe]. Hydroxyl-radical cleavage mediated by $Fe(EDTA)^{2-}$ reveals the protection of solvent-inaccessible regions, the ''inside'' of the molecule.[23] MPE-Fe(II) appears to demarcate the double-helical regions,[74] $Cu(phen)_2^+$ shows the looped-out single-stranded segments,[75] and $Rh(phen)_2phi^{3+}$ seems to delineate those regions involved in triple-base interactions, the sites of tertiary folding.[76] Taken together, the full structure of the tRNA can be described based upon cleavage data with transition-metal complexes. It therefore seems as if this full family of coordination complexes might be generally useful in delineating RNA structures. Still more work is needed quantitatively to compare the patterns obtained with the few well-characterized structures. Nonetheless, an important role for these and possibly other transition-metal reagents is indicated.

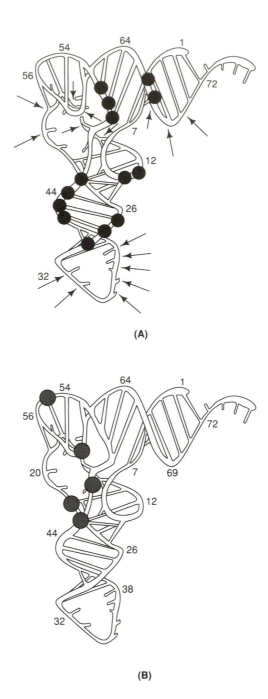

Figure 8.15
The diversity of cleavage sites for metal complexes on tRNA[Phe]. In (A) is shown cleavage by probes that primarily detect features of RNA secondary structure. Cu(phen)$_2$$^+$ (arrows), detecting single-stranded regions and MPE-Fe(II) (black dots), detecting double-helical segments. In (B) are shown probes that detect protected or more complex structures on tRNA. Inaccessible sites protected from OH· after treatment with Fe(EDTA)$^{2-}$ are shown as shaded portions of the molecule, and specific cleavage by Rh(phen)$_2$(phi)$^{3+}$ at tertiary folds is indicated by the circles.[11b]

D. Other Novel Techniques

Transition-metal ions can also be used advantageously tethered onto peptides, proteins, oligonucleotides, and other natural products, to provide a chemical probe for their binding interactions with nucleic acids. This strategy, termed *affinity cleavage*, was developed by Dervan and coworkers in preparing and characterizing distamycin-Fe(II)EDTA.[24] Distamycin is a known natural product that binds in the minor groove of DNA at AT-rich sequences. By tethering Fe(II)EDTA onto distamycin, the researchers converted the DNA-binding moiety into a DNA-cleaving moiety, since, as with MPE-Fe(II), in the presence of peroxide and a reductant, hydroxyl radical chemistry would be delivered to the distamycin binding site. Unlike MPE-Fe(II), however, the distamycin moiety shows preferential binding at some sites along the polymer, and hence only at those sites would the local hydroxyl-radical concentration be increased and cleavage be obtained. As a result the tethered $Fe(EDTA)^{2-}$ could be used as a cleavage probe, marking sites of specific binding.

Affinity cleaving has been generalized so that now $Fe(EDTA)^{2-}$ can be tethered onto both oligonucleotides and peptides to follow their interactions with nucleic acids. The sequence-specific binding of oligonucleotides to double-helical DNA through triple-helix formation is but one of many examples where the tethering of $Fe(EDTA)^{2-}$ has been applied advantageously.[77]

Other redox-active metals can be incorporated into DNA-binding moieties as well. Schemes have been developed to functionalize accessible lysine residues on DNA-binding repressor proteins with phenanthrolines, so that in the presence of copper ion, peroxide, and a reductant, the phenanthroline-bound copper on the protein would induce DNA strand cleavage. Through this scheme, again the conversion of a DNA-binding moiety into a cleaving moiety by incorporation of a redox-active metal, the specific binding sites of repressor proteins can be readily identified (far more quickly on large DNA than through footprinting).[78]

Another scheme, which perhaps takes advantage more directly of bioinorganic chemistry, involves engineering redox metal-binding sites into DNA-binding proteins and peptides. The DNA-binding domain of the protein Hin recombinase was synthesized chemically, and first, to examine the folding of the peptide on the DNA helix, EDTA was tethered onto the peptide for Fe(II) cleavage experiments.[79] But as is illustrated repeatedly in these chapters, Nature has already provided amino-acid residues for the chelation of metal ions into proteins. Thus the DNA-binding domain of Hin recombinase was synthesized again, now including at its terminus the residues Gly-Gly-His, a known chelating moiety for copper(II).[80] This chemically synthesized peptide, with now both DNA-binding and DNA-cleaving domains, as illustrated in Figure 8.16, specifically promotes cleavage at the Hin recombinase binding site in the presence of bound copper and ascorbate. Interestingly, the addition of nickel(II) also leads to specific strand cleavage, without diffusible intermediates. Using this approach, taking advan-

490

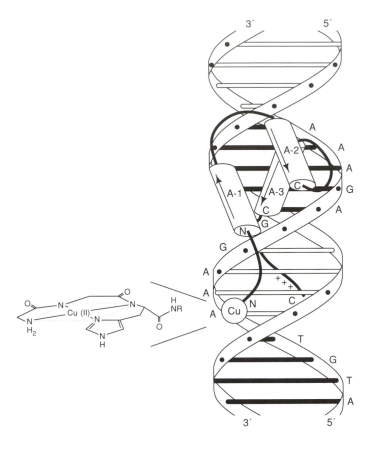

Figure 8.16
A schematic of a synthetic DNA-cleaving peptide bound to DNA that was constructed by synthesis of the DNA-binding domain of Hin Recombinase with Gly-Gly-His at the N-terminus to coordinate copper.[80] Reproduced with permission from Reference 80.

tage of the chelating abilities of amino acids and the cleaving abilities of different metal ions, one may prepare new synthetic, functional metalloproteins that bind and react with DNA.

V. NATURE'S USE OF METAL/NUCLEIC-ACID INTERACTIONS

In the context of what we understand about the fundamental interactions and reactions of metal ions and complexes with nucleic acids, and also in comparison to how chemists have been exploiting these interactions in probing nucleic acids, we can also consider how Nature has taken advantage of metal ions in the construction of metalloproteins, nucleic-acid assemblies, and smaller natural products containing metal ions that interact with DNA and RNA.

A. A Structural Role

One of the chief functions attributed to metal ions in biological systems is their ability to provide a structural center to direct the folding of a protein. Just as shape-selective recognition has been helpful in targeting metal complexes to specific sites on DNA, it appears that one element of the recognition of sites by DNA regulatory proteins may also involve the recognition of complementary shapes. Furthermore, metal ions appear to be used in these proteins to define the shape or folding pattern of the peptide domain that interacts specifically with the nucleic acid.

The DNA-binding metalloproteins that have received the greatest attention recently have been the ''zinc-finger'' regulatory proteins. It was discovered in 1983 that zinc ions played a role in the functioning of the nucleic acid-binding transcription factor IIIA (TFIIIA) from *Xenopus laevis*, which binds specifically both DNA, the internal control region of the 5S rRNA gene, and RNA, the 5S RNA itself.[81] The protein (actually the 7S storage particle) was found to contain two to three equivalents of zinc ion. Dialysis removed both the associated zinc ions and the nucleic-acid-binding ability of the protein. Importantly, treatment with zinc ion, or in later studies with higher concentrations of Co^{2+}, restored the specific binding ability. Hence, zinc ion was shown to be functionally important in these eukaryotic regulatory proteins.

The notion of a ''zinc-finger structural domain'' was first provided by Klug and coworkers, after examination of the amino-acid sequence in TFIIIA.[82] It was found that TFIIIA contained nine imperfect repeats of a sequence of approximately 30 amino acids, and furthermore that each repeat contained two cysteine residues, two histidine residues, and three hydrophobic residues, in conserved positions. In addition, subsequent metal analyses were revealing higher zinc contents (7 to 11 equivalents) associated with the protein, and protein-digestion experiments indicated that several repeated structural domains existed in the protein. The two cysteine thiolates and two histidine imidazoles in each repeated domain could certainly serve to coordinate a zinc ion. Thus it was proposed that each peptide repeat formed an independent nucleic-acid-binding domain, stabilized in its folded structure through coordination of a zinc ion. The peptide unit was termed a ''zinc finger,'' which is illustrated schematically in Figure 8.17. TFIIIA was therefore proposed to contain nine zinc fingers, which would cooperatively bind in the internal control region of the 5S RNA gene.

An enormous number of gene sequences from a variety of eukaryotic regulatory proteins was then found to encode strikingly similar amino-acid sequences,[83] and many were dubbed zinc-finger proteins. The bioinorganic chemist, however, should be aware that chemical analyses supporting such assignments are first required. Nonetheless, several legitimate examples of eukaryotic nucleic-acid-binding zinc-finger proteins containing multiple zinc-binding peptide domains have emerged since the first study of TFIIIA, including the proteins Xfin from *Xenopus*, the Kruppel protein from *Drosophila*, the Sp1 transcription factor, and human testes-determining factor. It has therefore become clear that

492

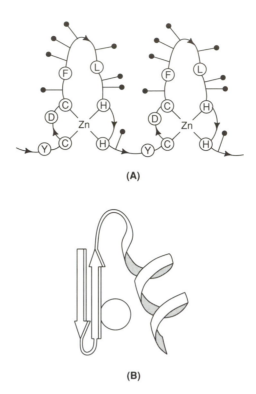

(A)

(B)

Figure 8.17
(A) A schematic of a zinc-finger peptide domain.[82] (B) The proposed schematic structure of a
zinc-finger domain based on comparisons to other structurally characterized metalloproteins.[85]

*the zinc-finger domain represents a ubiquitous structural motif for eukaryotic
DNA-binding proteins.*[84]

What is the structure of a zinc finger, and how is this structure important
for binding a specific nucleic-acid site? Based on a search of crystallographic
databases for metalloproteins and an examination of the consensus sequence
emerging for zinc fingers (that is, which residues were truly conserved and common to the different putative zinc fingers), Berg proposed a three-dimensional
structure for a zinc finger, shown schematically in Figure 8.17.[85] The proposed
structure included the tetrahedral coordination of zinc by the two cysteine and
histidine residues at the base of the finger and an α-helical region running almost the length of the domain. EXAFS studies also supported the tetrahedral
zinc site. Since this proposal, two detailed two-dimensional NMR studies have
been reported that are consistent with the tetrahedral zinc coordination and the
α-helical segment.[86] More recently, a crystal structure of a three-finger binding
domain associated with an oligonucleotide was determined.[87] The zinc fingers
lie in the major groove of DNA, the α-helical region being within the groove.
Not surprisingly, given basic coordination chemistry, the zinc does not interact

directly with the nucleic acid. Instead, the zinc ion must serve a structural role, defining the folding and three-dimensional structure of the protein scaffolding about it. This structure, defined by the metal at its center, like other coordination complexes, is able to recognize its complementary structure on the nucleic-acid polymer.

It should also be noted that this zinc-finger structural motif is not the only metal-containing or even zinc-containing structural motif important in nucleic-acid-binding proteins.[88] A clearly different domain is evident in the protein GAL4, a transcription factor required for galactose utilization in *S. cerevisiae*.[88a] A recent crystal structure of the protein bound to an oligonucleotide shows the protein to bind to DNA as a dimer; each monomer contains a binuclear zinc cluster with two zinc ions tetrahedrally coordinated by six cysteines (two cysteines are bridging), not dissimilar from proposed structures in metallothionein. Still another structural motif is found in the glucocorticoid receptor DNA-binding domain. Crystallography[89] has revealed that this domain also binds DNA as a dimer; here each monomer contains two zinc-nucleated substructures of distinct conformation. The zinc ions are each tetrahedrally coordinated to four cysteine residues. Likely this too represents another structural motif for proteins that bind nucleic acids, and one again in which the metal serves a structural role.

Lastly, one might consider why zinc ion has been used by Nature in these nucleic-acid binding proteins. Certainly, the natural abundance of zinc is an important criterion. But also important is the absence of any redox activity associated with the metal ion, activity that could promote DNA damage [as with Fe(II) or Cu(II), for example]. In addition, other softer, heavier metal ions might bind preferentially to the DNA bases, promoting sequence-specific covalent interactions. Zinc ion, therefore, is clearly well-chosen for the structural center of these various nucleic-acid-binding proteins.

B. A Regulatory Role

Metalloregulatory proteins, like the transcription factors described above, affect the expression of genetic information through structural interactions that depend upon the metal ions, but unlike the zinc-finger proteins, metalloregulatory proteins act as triggers, repressing or activating transcription given the presence or absence of metal ion. In some respects, even more than zinc fingers, these systems resemble the Ca^{2+}-activated proteins described in Chapter 3.

Consider the biological system that must respond to changing intracellular metal concentrations. At high concentrations many metal ions become toxic to the cell; hence, a full system of proteins must be synthesized that will chelate and detoxify the bound-metal-ion pool. In order to actively engage these proteins, the genes that encode them must be rapidly transcribed. But at the same time, the DNA itself must be protected from the high concentrations of metal ion. Hence the need for these metalloregulatory proteins, which bind DNA in the absence of metal ion, usually repressing transcription, but in the presence of metal ion bind the metal ion tightly and specifically, and as a consequence amplify transcription.

Perhaps the best-characterized metalloregulatory system thus far is the MerR system, regulating mercury resistance in bacteria.[90] An inducible set of genes arranged in a single operon is under the control of the metal-sensing MerR protein, and it is this system that mediates mercury resistance. Mercury resistance depends upon the expression of these genes to import toxic Hg(II), reduce Hg(II) to the volatile Hg(0) by NADPH, and often additionally to cleave organomercurials to their corresponding hydrocarbon and Hg(II) species. The MerR protein regulates mercury resistance both negatively and positively. As illustrated in Figure 8.18, MerR in the absence of Hg(II) binds tightly and sequence-specifically to the promoter. In so doing, MerR inhibits binding to the promoter by RNA polymerase. When Hg^{2+} is added at low concentrations, the metal ion binds specifically and with high affinity to the DNA-bound MerR, and causes a DNA conformational change detectable by using other metal reagents as conformational probes. This conformational change now facilitates the binding of RNA polymerase and hence activates expression of the gene family.

What are the structural requirements in the metal-binding site? Certainly one requirement is Hg(II) specificity, so that other metal ions will not also trigger transcriptional activation. Another is high metal-binding affinity to protect the DNA from direct coordination of the Hg(II). The MerR protein is dimeric, and contains four cysteine residues per monomer. Site-directed mutagenesis studies[91] have indicated that three of these four cysteine residues are needed for Hg(II) binding, and EXAFS studies[92] have been consistent with tricoordinate ligation, clearly a well-designed system for Hg(II) specificity. Perhaps even more interesting, the site-directed mutagenesis studies[91] on heterodimers (a mixture of mutant and wild-type monomers) have indicated that the coordinated Hg(II) bridges the dimer, ligating two cysteines of one monomer and one cysteine of the other. This scheme may provide the basis also for the kinetic lability needed in a rapidly responsive cellular system.

Model systems are also being constructed to explore metal modulation of DNA binding. One system involves the assembly of two dipeptides linked by a central acyclic metal-binding polyether ligand, with $Fe(EDTA)^{2-}$ tethered to one end to mark site-specific binding.[93] In the presence of alkaline earth cations, which induce a conformational change that generates a central macrocycle, the linked peptides become oriented to promote sequence-specific binding in the minor groove. In the absence of the alkaline earth ion, no site-specific binding, or cleavage of DNA, is evident. One might consider this system as a simple, first-order synthetic model for the metalloregulatory proteins.

MerR is clearly only one natural metalloregulatory system. Other metal ions bind regulatory factors to mediate the regulation of gene expression in a metal-specific manner. Two examples include the Fe(II)-binding Fur protein from enteric bacteria[94] and the copper-binding protein ACE1/CUP2 from *S. cerevisiae*.[95] Both copper and iron are essential trace elements for which high concentrations are toxic; for nucleic acids this toxicity is certainly the result of redox-mediated

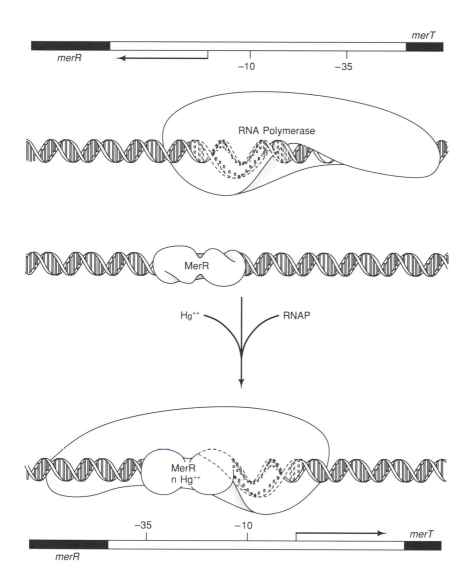

Figure 8.18
A model for MerR metalloregulation.[90a] In the absence of MerR, RNA polymerase binds and transcribes the MerR promoter. In the presence of MerR, the preferential binding of MerR to the promoter is observed that inhibits transcription by the polymerase. The addition of Hg^{2+} then leads to a conformational change that promotes binding of the polymerase, substantially increasing transcription. Reproduced with permission from Reference 90a.

strand damage. Other metal-specific regulatory systems are surely present as well. Both the MerR and the synthetic system may exemplify how these various systems function, how Nature might construct a ligand system to facilitate toxic-metal-specific binding in the presence of DNA that then alters or triggers how other moieties bind and access the nucleic acid.

C. A Pharmaceutical Role

With the exception of cisplatin (see Chapter 9), most pharmaceuticals currently being used as DNA-binding agents were first isolated as natural products from bacteria, fungi, plants, or other organisms. For the most part they represent complex organic moieties, including peptide and/or saccharide functionalities, and often a unique functionality, such as the ene-diyne in calichimycin. These natural products bind DNA quite avidly, through intercalation, groove binding, or a mixture thereof. Often the efficacy of these antitumor antibiotics stems from subsequent alkylation or DNA strand-cleavage reactions that damage the DNA.

Among the various natural products used clinically as antitumor antibiotics are bleomycins, a family of glycopeptide-derived species isolated from cultures of *Streptomyces*.[25,96] The structure of bleomycin A_2 is shown schematically in Figure 8.19. The molecular mode of action of these species clearly involves binding to DNA and the promotion of single-stranded cleavage at GT and GC sequences. Importantly, as demonstrated by Horwitz, Peisach, and coworkers, this DNA cleavage requires the presence of Fe(II) and oxygen.[97] Thus, one might consider Fe-bleomycin as a naturally occurring inorganic pharmaceutical.

What is the role of the metal ion in these reactions? As one might imagine, based upon our earlier discussions of metal-promoted DNA cleavage, the iron center is essential for the oxidative cleavage of the strand through reaction with the sugar moiety. The reaction of Fe(II)-bleomycin can, however, clearly be distinguished from the $Fe(EDTA)^{2-}$ reactions discussed earlier in that here no diffusible intermediate appears to be involved. Instead of generating hydroxyl radicals, the Fe center must be positioned near the sugar-phosphate backbone and activated in some fashion to promote strand scission *directly*.

Despite extensive studies, in fact little is known about how Fe(II)-bleomycin is oriented on the DNA. Indeed, the coordination about the metal is the subject of some debate. The structure of Cu(II)-P-3A,[98] a metallobleomycin derivative, is also shown in Figure 8.19. On the basis of this structure and substantive spectroscopic studies on Fe-bleomycin itself, it is likely that, as with Cu(II), in the Fe(II)-bleomycin complex the metal coordinates the β-hydroxyimidazole nitrogen, the secondary amine of β-aminoalanine, and the N1 nitrogen of the pyrimidine. Whether in addition the primary amines of the amino alanine and of the histidine coordinate the metal is still not settled. Possibly bithiazole coordination or some coordination of the sugar moieties is involved. Nonetheless, given five different coordination sites to the bleomycin, the sixth axial site is available for direct coordination of dioxygen. How is this $Fe-O_2$ complex oriented on the DNA? It is likely that at least in part the complex binds against the minor groove of the helix. There is some evidence that suggests that the bithiazole moiety intercalates in the helix. It is now becoming clear, however, that the structure of the metal complex itself, its three-dimensional shape, rather than simply the tethered bithiazole or saccharide, is needed for the sequence selectivity associated with its mode of action.

Although the coordination and orientation of the metal complex are still not understood, extensive studies have been conducted concerning the remarkable chemistry of this species.[96,99] The overall mechanism of action is described in Figure 8.19. In the presence of oxygen, the Fe(II) O_2 species is formed and is likely rapidly converted to a ferric superoxide species. The one-electron reduction of this species, using either an organic reductant or another equivalent of Fe(II)-bleomycin, leads formally to an Fe(III)-peroxide, which then undergoes O—O bond scission to form what has been termed "activated bleomycin." This species might be best described as Fe(V)=O (or [Fe O]$^{3+}$). This species is comparable in many respects to activated cytochrome P-450 or perhaps even more closely to the Fe center in chloroperoxidase (see Chapter 5). Like these systems, activated bleomycin can also epoxidize olefins and can generally function as an oxo transferase. In contrast to these systems, Fe-bleomycin clearly lacks a heme. How this species can easily shuttle electrons in and out, forming and reacting through a high-valent intermediate, without either the porphyrin sink or another metal linked in some fashion, is difficult to understand. In fact, understanding this process, even independently of our fascination with how the reaction is exploited on a DNA helix, forms the focus of a substantial effort of bioinorganic chemists today.

What has been elucidated in great detail is the reaction of activated bleomycin with DNA. It has been established that the activated species promotes hydrogen abstraction of the C4'-H atom, which is positioned in the minor groove of the helix (Figure 8.19). Addition of another equivalent of dioxgen to this C4'-radical leads to degradation of the sugar to form a 5'-phosphate, a 3'-phosphoglycolate, and free base propenal. Alternatively, oxidation of the C4'-radical followed by hydroxylation in the absence of oxygen yields, after treatment with base, a 5'-phosphate, an oxidized sugar phosphate, and free base.

Other metals such as copper and cobalt can also activate bleomycins, although their mechanistic pathways for strand scission are clearly different from that of Fe(II)-bleomycin. Whether other natural products that bind DNA also chelate metal ions and exploit them for oxidative strand cleavage is not known, but several systems provide hints that they do. Furthermore, such a fact would not be surprising given our understanding of the utility of metal ions in promoting this chemistry. An even more detailed understanding of this chemistry might lead to the development of second-generation synthetic transition-metal pharmaceuticals that specifically and efficiently target and cleave DNA sites.

D. A Catalytic Role

In addition to serving structural and modulating roles in proteins which bind nucleic acids, metal ions also appear to be essential to the functioning of various complex enzymes that act on nucleic acids. At this stage our understanding of the participation of the metal ion in the catalytic chemistry of these enzymes is somewhat sketchy, and we are relying more on our current understanding of the

(A)

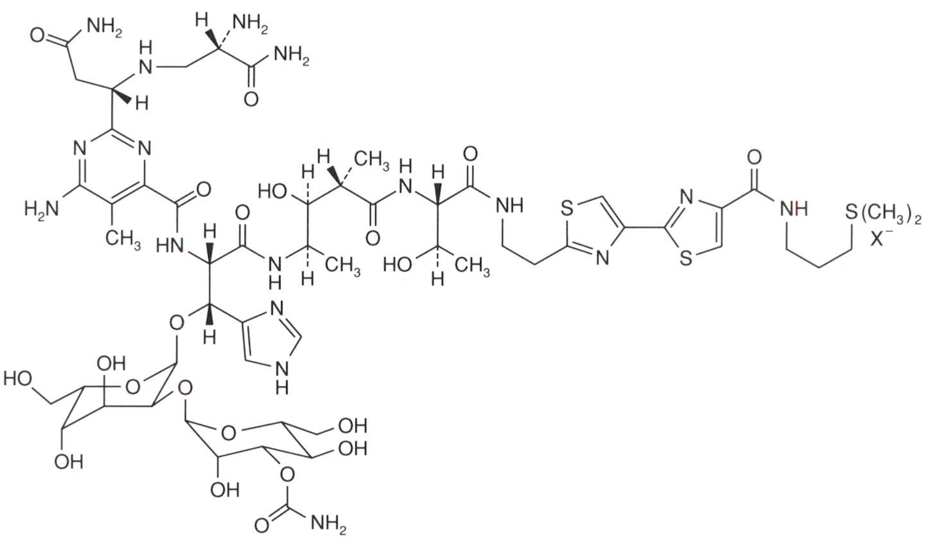

bleomycin

(B)

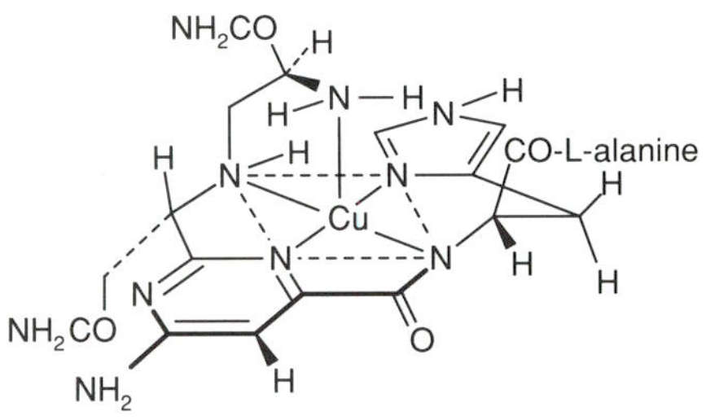
NH2CO
H
H
N
H
N
H
N
H
H
N
CO-L-alanine
H
Cu
N
N
H
H
N
NH2CO
O
NH2
H

(C)

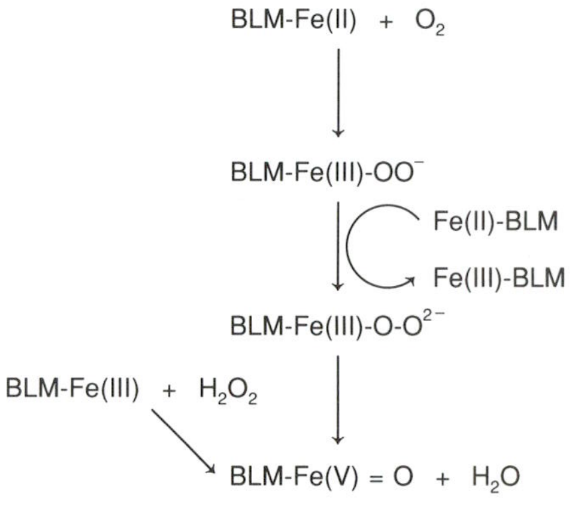
BLM-Fe(II) + O2
BLM-Fe(III)-OO⁻
Fe(II)-BLM
Fe(III)-BLM
BLM-Fe(III)-O-O2⁻
BLM-Fe(III) + H2O2
BLM-Fe(V) = O + H2O
"activated bleomycin"

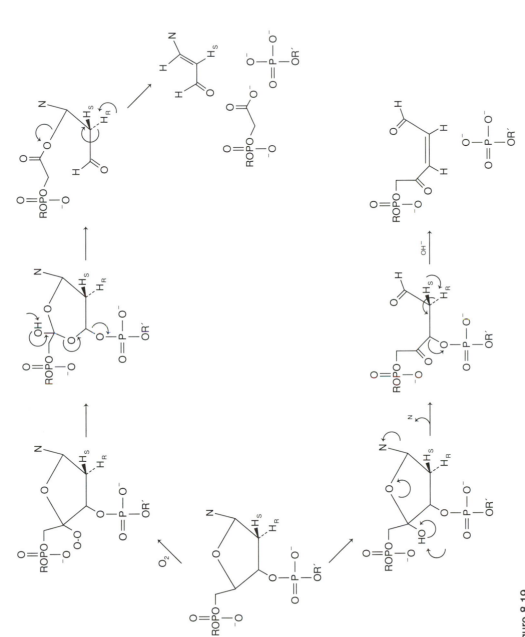

Figure 8.19

(A) The structure of bleomycin A₂. (B) The crystallographically determined structure[98] of a copper derivative of bleomycin, P3A. (C) A scheme to generate "activated bleomycin." (D) The proposed mechanism of action of activated Fe-bleomycin.

[499]

possible roles where metal ions may prove advantageous. These remain areas of biochemical focus where the inorganic chemist could make a major contribution.

For example, zinc ion appears to be essential to the functioning of both RNA polymerases and DNA topoisomerases.[100—102] These multisubunit enzymes perform quite complex tasks. RNA polymerase must bind site-specifically to its DNA template, bind its nucleotide and primer substrates, and form a new phosphodiester bond in elongating the growing RNA. Two zinc ions appear to be involved. One may be involved in orienting the nucleotide substrate, and the other structurally in template recognition. It would not be surprising, however—indeed, it might be advantageous—if one or both metal ions also participated in the polymerization step. Our mechanistic understanding of how topoisomerases function is even more cursory. These complex enzymes bind supercoiled DNA, sequentially break one strand through hydrolytic chemistry, move the strand around the other (releasing one tertiary turn), and religate the strand. Again, the zinc ion might participate in the hydrolytic chemistry, the ligation step, or both; alternatively, the metal might again serve a structural role in recognition of the site of reaction.

We do have some understanding of the role of metal ions in several endonucleases and exonucleases. As discussed in Section II.C, metal ions may effectively promote phosphodiester hydrolysis either by serving as a Lewis acid or by delivering a coordinated nucleophile. Staphylococcal nuclease[103] is an extracellular nuclease of *Staphylococcus aureus* that can hydrolyze both DNA and RNA in the presence of Ca^{2+}. The preference of the enzyme is for single-stranded DNA, in which it attacks the 5'-position of the phosphodiester linkage, cleaving the 5'-P-O bond to yield a 5'-hydroxyl and 3'-phosphate terminus. Ca^{2+} ions are added as cofactors and are strictly required for activity. The structure of staphylococcal nuclease, determined by x-ray crystallography and crystallized in the presence of Ca^{2+} and the enzyme competitive inhibitor pdTp, as well as subsequent NMR and EPR studies on mutant enzymes using Mn^{2+} as a substitute for the Ca^{2+} ion, have provided the basis for a detailed structural analysis of the mechanism of this enzyme. In this phosphodiester hydrolysis, the metal ion appears to function primarily as an electrophilic catalyst, polarizing the P-O bond, and stabilizing through its positive charge the evolving negative charge on the phosphorus in the transition state. The base is thought here not to be directly coordinated to the metal; instead, action of a general base is invoked.

Metal ions also participate in the functioning of other nucleases, although the structural details of their participation are not nearly as established as those for staphylococcal nuclease. DNAse I also requires Ca^{2+} for its catalytic activity.[104] S1 endonuclease, mung bean nuclease, and *Physarum polycephalem* nuclease require zinc ion either as cofactors or intrinsically for nuclease activity, and the restriction enzyme EcoRI may also require intrinsically bound zinc ion.[33] In terms of how the zinc ion might function in these enzymes, one can look both to staphylococcal nuclease and to bacterial alkaline phosphatase[105] for some illustrations. One would expect that this metal ion could serve both as an electrophilic catalyst and also in the delivery of a zinc-coordinated hydroxide, as it does in alkaline phosphatase, directly attacking the phosphate ester. More work needs to be done to establish the mechanisms by which zinc ion promotes phosphodiester hydrolysis in these enzymatic systems.

Probably most intriguing and mysterious at this stage is the metal participation in the very complex DNA-repair enzyme endonuclease III from *E. coli* (similar enzymes have also been isolated from eukaryotic sources). This enzyme is involved in the repair of DNA damaged by oxidizing agents and UV irradiation, and acts through an N-glycosylase activity to remove the damaged base, and through an apurinic/apyrimidinic endonuclease activity to cleave the phosphodiester bond adjacent to the damaged site. Although more complex in terms of recognition characteristics, this enzyme functions in hydrolyzing the DNA phosphodiester backbone. What is so intriguing about this enzyme is that it contains a 4Fe-4S cluster (see Chapter 7) that is essential for its activity![106] We think generally that Fe-S clusters best serve as electron-transfer agents. In the context of this repair enzyme, the cluster may be carrying out both an oxidation and a reduction, to effect hydrolysis, or alternatively perhaps a completely new function for this metal cluster will emerge. (Fe-S clusters may represent yet another structural motif for DNA-binding proteins and one which has the potential for regulation by iron concentration.) Currently the basic biochemical and spectroscopic characterization of the enzyme is being carried out. Understanding this very novel interaction of a metal center and nucleic acid will require some new ideas, and certainly represents one new challenge for the bioinorganic chemist.

VI. REFERENCES

1. W. Saenger, *Principles of Nucleic Acid Structure*, Springer-Verlag, 1984; J. K. Barton, *Chem. Eng. News* **66** (Sept. 26, 1988), 30.
2. M. McCall, T. Brown, and O. Kennard, *J. Mol. Biol.* **183** (1985), 385.
3. R. Wing *et al.*, *Nature* **287** (1980), 755.
4. A. H.-J. Wang *et al.*, *Nature* **282** (1979), 680.
5. S. H. Kim *et al.*, *Proc. Natl. Acad. Sci. USA* **71** (1974), 4970.
6. E. N. Trifonov and J. L. Sussman, *Proc. Natl. Acad. Sci USA* **77** (1980), 3816; J. C. Marini *et al.*, *Proc. Natl. Acad. Sci. USA* **79** (1982), 7664; H.-S. Koo, H.-M. Wu, and D. M. Crothers, *Nature* **320** (1986), 501.
7. M. Gellert *et al.*, *Cold Spring Harbor Symps. Quant. Biol.* **43** (1979), 35; D. M. J. Lilley, *Proc. Natl. Acad. Sci. USA* **77** (1980), 6468.
8. J. S. Lee *et al.*, *Nucleic Acids Res.* **12** (1984), 6603; V. I. Lyamichev, *J. Biomol. Struct. Dyn.* **3** (1986), 667; H. Htun and J. E. Dahlberg, *Science* **241** (1988), 1791.
9. R. D. Kornberg, *Annu. Rev. Biochem.* **46** (1977), 931; A. Klug *et al.*, *Nature* **287** (1980), 509.
10. T. R. Cech, *Science* **236** (1987), 1532.
11. (a) J. K. Barton and S. J. Lippard, *Metal Ions in Biol.* **1** (1980), 31; (b) A. M. Pyle and J. K. Barton, *Prog. Inorg. Chem.* **38** (1990), 413; (c) C. S. Chow and J. K. Barton, *Meth. Enzym.* **212** (1992), 219.
12. S. E. Sherman *et al.*, *Science* **230** (1985), 412.
13. D. Hodgson, *Prog. Inorg. Chem.* **23** (1977), 211.
14. G. L. Eichhorn and Y. A. Shin, *J. Am. Chem. Soc.* **90** (1968), 7323.
15. L. G. Marzilli, *Prog. Inorg. Chem.* **23** (1977), 255.

16. C. H. Chang, M. Beer, and L. G. Marzilli, *Biochemistry* **16** (1977), 33; G. C. Glikin *et al.*, *Nucleic Acids Res.* **12** (1984), 1725.

17. M. B. Fleisher, H. Y. Mei, and J. K. Barton, *Nucleic Acids and Mol. Biol.* **2** (1988), 65.

18. S. J. Lippard, *Acc. Chem. Res.* **11** (1978), 211.

19. R. V. Gessner *et al.*, *Biochemistry* **24** (1985), 237.

20. (a) K. W. Jennette *et al.*, *Proc. Natl. Acad. Sci. USA* **71** (1974), 3839; (b) A. H. Wang *et al.*, *Nature* **276** (1978), 471.

21. H. Sigel, in T. D. Tullius, ed., *Metal-DNA Chemistry*, ACS Symposium **402** (1989), 159.

22. R. P. Hertzberg and P. B. Dervan, *J. Am. Chem. Soc.* **104** (1982), 313; R. P. Hertzberg and P. B. Dervan, *Biochemistry* **23** (1984), 3934.

23. J. A. Latham and T. R. Cech, *Science* **245** (1989), 276.

24. P. B. Dervan, *Science* **232** (1986), 464.

25. S. M. Hecht, ed., *Bleomycin*, Springer-Verlag, 1979.

26. D. S. Sigman, *Acc. Chem. Res.* **19** (1986), 180; S. Goldstein and G. Czapski, *J. Am. Chem. Soc.* **108** (1986), 2244.

27. For other examples of metal-mediated redox cleavage of DNA, see also: N. Grover and H. H. Thorp, *J. Am. Chem. Soc.* **113** (1991), 7030; X. Chen, S. E. Rokita, and C. J. Burrows, *J. Am. Chem. Soc.* **113** (1991), 5884.

28. H. Y. Mei and J. K. Barton, *J. Am. Chem. Soc.* **108** (1986), 7414.

29. H. Y. Mei and J. K. Barton, *Proc. Natl. Acad. Sci. USA* **85** (1988), 1339.

30. A. M. Pyle, E. C. Long, and J. K. Barton, *J. Am. Chem. Soc.* **111** (1989), 4520.

31. A. Sitlani *et al.*, *J. Am. Chem. Soc.* **114** (1992), 2303.

32. M. D. Purugganan *et al.*, *Science* **241** (1988), 1645.

33. L. A. Basile and J. K. Barton, *Metal Ions Biol. Syst.*, **25** (1989), 31; J. K. Barton, in *Frontiers of Chemistry: Biotechnology*, Chem. Abstr. Serv., **5** (1989).

34. D. R. Jones, L. F. Lindoy, and A. M. Sargeson, *J. Am. Chem. Soc.* **106** (1984), 7807.

35. S. H. Gellman, R. Petter, and R. Breslow, *J. Am. Chem. Soc.* **108** (1986), 2388; J. Chin and X. Zhou, *J. Am. Chem. Soc.* **110** (1988), 223; J. R. Morrow and W. C. Trogler, *Inorg. Chem.* **27** (1988), 3387.

36. L. A. Basile, A. L. Raphael, and J. K. Barton, *J. Am. Chem. Soc.* **109** (1987), 7550.

37. G. L. Eichhorn and Y. A. Shin, *J. Am. Chem. Soc.* **90** (1968), 7322.

38. R. S. Brown, J. C. Dewan, and A. Klug, *Biochemistry* **24** (1985), 4785.

39. L. Behlen *et al.*, *Biochemistry* **29** (1990), 2515.

40. J. K. Barton, *Science* **233** (1986), 727.

41. J. K. Barton *et al.*, *J. Am. Chem. Soc.* **108** (1986), 2081.

42. A. M. Pyle *et al.*, *J. Am. Chem. Soc.* **111** (1989), 3051.

43. C. V. Kumar, J. K. Barton, and N. J. Turro, *J. Am. Chem. Soc.* **107** (1985), 5518.

44. S. J. Lippard *et al.*, *Science* **194** (1976), 726.

45. J. K. Barton, J. J. Dannenberg, and A. L. Raphael, *J. Am. Chem. Soc.* **104** (1982), 4967.

46. J. K. Barton, A. T. Danishefsky, and J. M. Goldberg, *J. Am. Chem. Soc.* **106** (1984), 2172.

47. R. F. Pasternack, E. J. Gibbs, and J. J. Villafranca, *Biochemistry* **22** (1983), 2406; R. F. Pasternack and E. J. Gibbs in T. D. Tullius, ed., *Metal-DNA Chemistry*, ACS Symposium **402** (1989), 59.

48. J. K. Barton and E. Lolis, *J. Am. Chem. Soc.* **107** (1985), 708.

49. R. E. Mahnken *et al.*, *Photochem. Photobiol.* **49** (1989), 519.

50. J. P. Rehmann and J. K. Barton, *Biochemistry* **29** (1990), 1701.

51. J. C. Caradonna *et al.*, *J. Am. Chem. Soc.* **104** (1982), 5793.

52. J. P. Rehmann and J. K. Barton, *Biochemisty* **29** (1990), 1710.

53. A. Jack *et al.*, *J. Mol. Biol.* **111** (1977), 315.

54. M. T. Carter and A. J. Bard, *J. Am. Chem. Soc.* **109** (1987), 7528.

55. J. K. Barton and A. L. Raphael, *J. Am. Chem. Soc.* **106** (1984), 2466.

56. M. B. Fleisher *et al.*, *Inorg. Chem.* **25** (1986), 3549.

57. M. W. van Dyke, R. P. Hertzberg, and P. B. Dervan, *Proc. Natl. Acad. Sci. USA* **79** (1982), 5470; M. W. van Dyke and P. B. Dervan, *Nucleic Acids Res.* **11** (1983), 5555.

58. J. K. Barton *et al.*, *Proc. Natl. Acad. Sci. USA* **81** (1984), 1961; A. E. Friedman *et al.*, *Nucleic Acids. Res.* **19** (1991), 2595.

59. A. E. Friedman *et al.*, *J. Am. Chem. Soc.* **112** (1990), 4960; R. Hartshorn and J. K. Barton, *J. Am. Chem. Soc.* **114** (1992), 5919.

60. W. deHorrocks and S. Klakamp, *Biopolymers* **30** (1990), 33.

61. R. Tamilarasan, S. Ropertz, and D. R. McMillin, *Inorg. Chem.* **27** (1988), 4082.

62. D. J. Galas and A. Schmitz, *Nucleic Acids Res.* **5** (1978), 3157.

63. T. D. Tullius *et al.*, *Methods in Enzym.* **155** (1987), 537.

64. R. Law *et al.*, *Proc. Natl. Acad. Sci. USA* **84** (1987), 9160; C. L. Peterson and K. L. Calane, *Mol. Cell Biol.* **7** (1987), 4194.

65. J. C. Dabrowiak, B. Ward, and J. Goodisman, *Biochemistry* **28** (1989), 3314.

66. P. E. Nielsen, C. Jeppesen, and O. Buchardt, *FEBS Lett.* **235** (1988), 122; C. Jeppesen and P. E. Nielsen, *Nucleic Acids Res.* **17** (1989), 4947.

67. K. Uchida *et al.*, *Nucleic Acids Res.* **17** (1989), 10259.

68. A. M. Burkhoff and T. D. Tullius, *Cell* **48** (1987), 935; A. M. Burkhoff and T. D. Tullius, *Nature* **331** (1988), 455.

69. B. H. Johnston and A. Rich, *Cell* **42** (1985), 713; E. Palacek, E. Rasovka, and P. Boublikova, *Biochem. Biophys. Res. Comm.* **150** (1988), 731.

70. J. K. Barton and A. L. Raphael, *Proc. Natl. Acad. Sci. USA* **82** (1985), 6460.

71. B. C. Muller, A. L. Raphael, and J. K. Barton, *Proc. Natl. Acad. Sci. USA* **84** (1987), 1764; I. Lee and J. K. Barton, *Biochemistry* **32** (1993), 6121.

72. M. R. Kirshenbaum, R. Tribolet, and J. K. Barton, *Nucleic Acids Res.* **16** (1988), 7948.

73. A. M. Pyle, T. Morii, and J. K. Barton, *J. Am. Chem. Soc.* **112** (1990), 9432.

74. J. M. Kean, S. A. White, and D. E. Draper, *Biochemistry* **24** (1985), 5062.

75. G. J. Murakawa *et al.*, *Nucleic Acids Res.* **17** (1989), 5361

76. C. S. Chow and J. K. Barton, *J. Am. Chem. Soc.* **112** (1990), 2839; C. S. Chow *et al.*, *Biochemistry* **31** (1992), 972.

77. H. E. Moser and P. B. Dervan, *Science* **238** (1987), 645.

78. C. B. Chen and D. S. Sigman, *Science* **237** (1987), 1197.

79. J. P. Sluka *et al.*, *Science* **238** (1987), 1129.

80. D. P. Mack, B. L. Iverson, and P. B. Dervan, *J. Am. Chem. Soc.* **110** (1988), 7572.

81. J. S. Heras *et al.*, *J. Biol. Chem.* **258** (1983), 14120.

82. J. Miller, A. D. McLachlan, and A. Klug, *EMBO* **4** (1985), 1609.

83. J. M. Berg, *Science* **232** (1986), 485.

84. A. Klug and D. Rhodes, *Trends Biochem. Sci.* **12** (1987), 464; R. M. Evans and S. M. Hollenberg, *Cell* **52** (1988), 1.

85. J. M. Berg, *Proc. Natl. Acad. Sci. USA* **85** (1988), 99.

86. M. S. Lee *et al.*, *Science* **245** (1989), 635; G. Parraga *et al.*, *Science* **241** (1988), 1489.

87. N. P. Pavletich and C. O. Pabo, *Science* **252** (1991), 809.

88. (a) T. Pan and J. E. Coleman, *Proc. Natl. Acad. Sci. USA* **86** (1989), 3145; (b) R. Marmorstein *et al.*, *Nature* **356** (1992), 408.

89. B. F. Luisi *et al.*, *Nature* **352** (1991), 497.

90. (a) T. V. O'Halloran, *Metal Ions Biol. Syst.* **25** (1989), 105; (b) C. T. Walsh *et al.*, *FASEB* **2** (1988), 124; T. V. O'Halloran and C. T. Walsh, *Science* **235** (1987), 211.

91. J. D. Helmann, B. T. Ballard, and C. T. Walsh, *Science* **248** (1990), 946.

92. J. E. Penner-Hahn *et al.*, *Physica B* **158** (1989), 117.

93. J. H. Griffin and P. B. Dervan, *J. Am. Chem. Soc.* **109** (1987), 6840.

94. A. Bagg and J. B. Neilands, *Microbiol. Rev.* **51** (1987), 509.

95. P. Furst *et al.*, *Cell* **55** (1988), 705; C. Buchman *et al.*, *Mol. Cell Biol.* **9** (1989).

96. J. Stubbe and J. W. Kozarich, *Chem. Rev.* **87** (1987), 1107.

97. E. A. Sausville, J. Peisach, and S. B. Horwitz, *Biochem. Biophys. Res. Comm.* **73** (1976), 814.

98. Y. Iitaka *et al.*, *J. Antibiot.* **31** (1978), 1070.

99. S. M. Hecht, *Acc. Chem. Res.* **19** (1986), 383.

100. C. W. Wu, F. Y. Wu, and D. C. Speckhard, *Biochemistry* **16** (1977), 5449.

101. D. P. Giedroc and J. E. Coleman, *Biochemistry* **25** (1986), 4946.

102. J. E. Coleman and D. P. Giedroc, *Metal Ions Biol. Syst.* **25** (1989), 171.

103. A. S. Mildvan and E. H. Serpersu, *Metal Ions Biol. Syst.* **25** (1989), 309.

104. D. Suck and C. Oefner, *Nature* **321** (1986), 620.

105. J. E. Coleman, *Metal Ions in Biol.* **5** (1983), 219.

106. H. Asahara *et al.*, *Biochemistry* **28** (1989), 4444.

107. I am grateful to my students and coworkers for their scientific contributions to some of the work described in this chapter and for their critical review of the manuscript. I also thank in particular Dr. Sheila David for preparation of the figures.

9

Metals in Medicine

STEPHEN J. LIPPARD

Department of Chemistry
Massachusetts Institute of Technology

I. INTRODUCTION AND OVERVIEW

Metal ions are required for many critical functions in humans. Scarcity of some metal ions can lead to disease. Well-known examples include pernicious anemia resulting from iron deficiency, growth retardation arising from insufficient dietary zinc, and heart disease in infants owing to copper deficiency. The ability to recognize, to understand at the molecular level, and to treat diseases caused by inadequate metal-ion function constitutes an important aspect of medicinal bioinorganic chemistry.

Metal ions can also induce toxicity in humans, classic examples being heavy-metal poisons such as mercury and lead. Even essential metal ions can be toxic when present in excess; iron is a common household poison in the United States as a result of accidental ingestion, usually by children, of the dietary supplement ferrous sulfate. Understanding the biochemistry and molecular biology of natural detoxification mechanisms, and designing and applying ion-specific chelating agents to treat metal overloads, are two components of a second major aspect of the new science that is evolving at the interface of bioinorganic chemistry and medicine.

Less well known than the fact that metal ions are required in biology is their role as pharmaceuticals. Two major drugs based on metals that have no known natural biological function, Pt (cisplatin) and Au (auranofin), are widely used

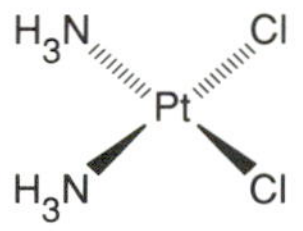

cis-diamminedichloroplatinum(II)
(cisplatin, or cis-DDP)

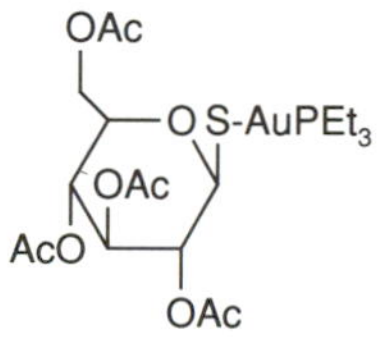

2,3,4,5-tetra-O-acetyl-1-1-β-D-
thioglucose(triethylphosphine)gold(I)
(auranofin)

505

for the treatment of genitourinary and head and neck tumors and of rheumatoid arthritis, respectively. In addition, compounds of radioactive metal ions such as ^{99m}Tc and complexes of paramagnetic metals such as Gd(III) are now in widespread use as imaging agents for the diagnosis of disease. Many patients admitted overnight to a hospital in the U.S. will receive an injection of a ^{99m}Tc compound for radiodiagnostic purposes. Yet, despite the obvious success of metal complexes as diagnostic and chemotherapeutic agents, few pharmaceutical or chemical companies have serious in-house research programs that address these important bioinorganic aspects of medicine.

This chapter introduces three broad aspects of metals in medicine: nutritional requirements and diseases related thereto; the toxic effects of metals; and the use of metals for diagnosis and chemotherapy. Each area is discussed in survey form, with attention drawn to those problems for which substantial chemical information exists. Since there is only a primitive understanding at the molecular level of the underlying biochemical mechanisms for most of the topics, this field is an important frontier area of bioinorganic chemistry. The major focus of this chapter is on the platinum anticancer drug cisplatin, which is presented as a case study exemplifying the scope of the problem, the array of methodologies employed, and the progress that can be made in understanding the molecular basis of a single, if spectacular, metal complex used in medicine today.

II. METAL DEFICIENCY AND DISEASE[1]

A. Essential Metals

Four main group (Na, K, Mg, and Ca) and ten transition (V, Cr, Mn, Fe, Co, Ni, Cu, Zn, Mo, and Cd) metals are currently known or thought to be required for normal biological functions in humans. Table 9.1 lists these elements, their relative abundances, and the medical consequences of insufficient quantities where known. The nutritional requirements for selected members of the essential metals are discussed in the following sections.

B. Anemia and Iron[2]

Anemia results from insufficient oxygen supply, often because of a decrease in hemoglobin (Hb) blood levels. Approximately 65 to 70 percent of total body iron resides in Hb. In the U.S., many foods, especially those derived from flour, are enriched in iron. In third-world countries, however, scarcity of dietary iron is a major contributor to anemia. This information illustrates one important fact about disease that results from metal deficiency, namely, the need for an adequate supply of essential metals in food. A related aspect, one of greater interest for bioinorganic chemistry, is the requirement that metals be adequately absorbed by cells, appropriately stored, and ultimately inserted into the proper environment to carry out the requisite biological function. For iron, these tasks,

Table 9.1
Essential metals and medical consequences resulting from their
deficiency.[a]

| Metal | Abundance | | Diseases resulting from metal deficiency |
	Sea Water mg/1 (ppm)	Earth's Crust mg/1 (ppm)	
Na	1.05×10^4	2.83×10^4	
K	380	2.59×10^4	
M-	1.35×10^3	2.09×10^4	
Ca	400	3.63×10^4	bone deterioration
V	2×10^{-3}	135	
Cr	5×10^{-5}	100	glucose tolerance (?)
Mn	2×10^{-3}	950	
Fe	1×10^{-2}	5.00×10^4	anemia
Co	1×10^{-4}	25	anemia
Ni	2×10^{-3}	75	
Cu	3×10^{-3}	55	brain disease, anemia, heart disease
Zn	1×10^{-2}	70	growth retardation, skin changes
Mo	1×10^{-2}	1.5	
Cd	1.1×10^{-4}	0.2	

[a] Data taken from E.-i. Ochiai, *Bioinorganic Chemistry*, Allyn & Bacon, 1977, p. 6.

among others, are performed by specific iron-chelating agents, the storage protein ferritin and the transport protein transferrin, the bioinorganic chemistry of which is extensively discussed in Chapter 1.

Another cause of anemia exists in individuals who have a mutant variety of hemoglobin, HbS, in which valine has been substituted for glutamic acid in the sixth position of the β subunits.[3] Interestingly, extensive studies have shown that this phenomenon, which leads to sickling of the red blood cells, does not result from failure of the protein to bind heme or from changes in the O_2 binding constant of the iron atom. Rather, deoxy HbS polymerizes into soluble, ordered fibrous structures that lower the ability of blood to carry oxygen effectively to the tissues. These results illustrate the importance of structural features remote from the metal-binding domain in determining the functional characteristics of a metalloprotein.

C. Causes and Consequences of Zinc Deficiency[4-6]

The average adult contains $\sim$ 2 g of zinc and requires a daily intake of 15 to 20 mg, only half of which is absorbed, to maintain this level. Although food in many technologically advanced societies contains sufficient zinc to afford this balance, zinc deficiencies occur in certain populations where there is either an unbalanced diet or food that inhibits zinc absorption. An especially interesting example of the latter phenomenon is found in certain villages in the Middle East where phytates, organic phosphates present in unleavened bread, chelate zinc ion and render it inaccessible. Zinc deficiency produces growth retardation, testicular atrophy, skin lesions, poor appetite, and loss of body hair. Little is known about the biochemical events that give rise to these varied consequences, although the three most affected enzymes are alkaline phosphatase, carboxypeptidase, and thymidine kinase. About 30 percent of zinc in adults occurs in skin and bones, which are also likely to be affected by an insufficient supply of the element. Zinc deficiency is readily reversed by dietary supplements such as $ZnSO_4$, but high doses (>200 mg) cannot be given without inducing secondary effects of copper, iron, and calcium deficiency.

D. Copper Deficiency[7]

More copper is found in the brain and heart than in any other tissue except for liver, where it is stored as copper thionein and released as ceruloplasmin or in the form of a complex with serum albumin. The high metabolic rate of the heart and brain requires relatively large amounts of copper metalloenzymes including tyrosinase, cytochrome c oxidase, dopamine-β-hydroxylase, pyridoxal-requiring monamine oxidases, and Cu-Zn superoxide dismutase. Copper deficiency, which can occur for reasons analogous to those discussed above for Fe and Zn, leads to brain disease in infants, anemia (since cytochrome oxidase is required for blood formation), and heart disease. Few details are known about the molecular basis for copper uptake from foods.

E. Summary

From the above anecdotal cases, for which similar examples may be found for the other metals in Table 9.1, the biological consequences of metal deficiency are seen to result from a breakdown in one or more of the following steps: adequate supply in ingestible form in foodstuffs; absorption and circulation in the body; uptake into cells; insertion into critical proteins and enzymes requiring the element; adequate storage to supply needed metal in case of stress; and an appropriate mechanism to trigger release of the needed element under such circumstances. Only for iron, and to a lesser extent copper and zinc, is there a reasonably satisfying picture of the molecular processes involved in this chain of events. The elucidation of the detailed mechanisms of these phenomena, for example, the insertion of iron into ferritin, remains an exciting challenge for the bioinorganic chemist (see Chapter 1).

III. TOXIC EFFECTS OF METALS

A. Two Classes of Toxic Metal Compounds

As intimated in the previous section, the presence of excess quantities of an essential metal can be as deleterious as insufficient amounts. This situation can arise from accidental ingestion of the element or from metabolic disorders leading to the incapacitation of normal biochemical mechanisms that control uptake and distribution phenomena. These possibilities constitute one major class of metal toxicity. The other broad class results from entry of nonessential metals into the cell through food, skin absorption, or respiration. The toxicities associated with this latter class have received much recent attention because of the public health risks of chemical and radioisotopic environmental pollutants.

In this section, we survey examples of both categories, and discuss ways in which bioinorganic chemistry can contribute to the removal of toxic metals and restoration of normal function. One way involves chelation therapy, in which metal-specific chelating agents are administered as drugs to complex and facilitate excretion of the unwanted excess element. The use of desferrioxamine to treat iron poisoning is one example of this approach. A second role of bioinorganic chemistry is to identify fundamental biological mechanisms that regulate metal detoxification, and to apply the principles that emerge to help control the toxic effects of metal ions in the environment. Recent studies of mercury resistance and detoxification in bacteria provide an elegant example of the way in which biochemistry and molecular biology can be used to elucidate events at the molecular level. This work, which has uncovered the existence of metalloregulatory proteins, is described in some detail in Section III.F below. It represents a benchmark by which other investigations into the mechanisms of metal-detoxification phenomena may be evaluated.

B. Copper Overload and Wilson's Disease[8]

Wilson's disease results from a genetically inherited metabolic defect in which copper can no longer be tolerated at normal levels. The clinical manifestations are liver disease, neurological damage, and brown or green (Kayser-Fleischer) rings in the cornea of the eyes. Patients suffering from Wilson's disease have low levels of the copper-storage protein ceruloplasmin; the gene and gene products responsible for the altered metabolism have not yet been identified. Chelation therapy, using $K_2Ca(EDTA)$, the Ca^{2+} ion being added to replenish body calcium stores depleted by EDTA coordination, 2,3-dimercaptopropan-1-ol (BAL, British Anti-Lewisite), or d-penicillamine to remove excess copper, causes the symptoms to disappear. The sulfhydryl groups of the latter two compounds presumably effect removal of copper as Cu(I) thiolate complexes. Wilson's disease offers an excellent opportunity for modern methodologies to isolate and clone the gene responsible for this altered Cu metabolism, ultimately providing a rational basis for treatment.

C. Iron Toxicity[9]

Chelation therapy is also used to treat iron overload. Acute iron poisoning, such as that resulting from accidental ingestion of $FeSO_4$ tablets, results in corrosion of the gastrointestinal tract. Chronic iron poisoning, or hemochromatosis, arises from digestion of excess iron usually supplied by vessels used for cooking. A classic case of the latter is siderosis induced in members of the Bantu tribe in South Africa, who consume large quantities of beer brewed in iron pots and who suffer from deposits of iron in liver, kidney, and heart, causing failure of these organs. The chelating agent of choice for iron toxicity is the siderophore desferrioxamine, a polypeptide having a very high affinity for Fe(III) but not for other metals. Ferrioxamine chelates occur naturally in bacteria as iron-transport agents. Attempts to mimic and improve upon the natural systems to provide better ligands for chelation therapy constitutes an active area of bioinorganic research (see Chapter 1).

D. Toxic Effects of Other Essential Metals[10,11]

When present in concentrations above their normal cellular levels, most of the other metals listed in Table 9.1 are toxic. Calcium levels in the body are controlled by vitamin D and parathyroid hormones. Failure to regulate Ca^{2+} leads to calcification of tissue, the formation of stones and cataracts, a complex process about which little is understood (see Chapter 3). Chronic manganese poisoning, which can occur following ingestion of metal-oxide dust, e.g., among miners in Chile, produces neurological symptoms similar to Parkinson's disease. Neuron damage has been demonstrated. Although Zn toxicity is rare, it can lead to deficiencies in other essential metals, notably calcium, iron, and copper. Cobalt poisoning leads to gastrointestinal distress and heart failure. Metal poisoning by those elements has been treated by chelating agents, most frequently $CaNa_2(EDTA)$, but the selectivity offered by the ferrioxamine class of ligands available for iron has not even been approached. Fortunately, there are few cases involving these metals.

E. Plutonium: A Consequence of the Nuclear Age[12]

Some of the chelating agents developed to treat iron toxicity have found application as therapeutics for plutonium poisoning. Diethylenetriaminepentaacetic acid (DTPA) salts and siderophores are especially effective. Some improvement over the naturally occurring chelates has been made by tailoring the ligand to encapsulate completely the eight-coordinate Pu(IV) center. Although few individuals have been affected, ingestion of ^{239}Pu, for example, as small particles of PuO_2, at nuclear-reactor sites can have dire consequences. ^{239}Pu emits high energy α particles, leading to malignancies of bone, liver, lung and lymph nodes, to which tissues it is transported by transferrin. With a maximum tolerated dose of only 1.5 μg, plutonium is among the most toxic metals known. We turn now to other, more classic examples of such industrial pollutants.

F. Mercury Toxicity[13] and Bacterial Resistance[14-17]

Mercury is released into the environment as Hg(II) ions through weathering of its most common ore, HgS, red cinnabar. Organomercurials of general formula

RHgX used in agriculture have also entered the environment as toxic waste. Both RHgX and HgX_2 compounds bind avidly to sulfhydryl groups in proteins, which can lead to neurological disease and kidney failure. Metallothionein is a favored protein target, which may help to limit mercury toxicity. A highly publicized case occurred in 1953 at Minimata, Japan, where 52 people died after eating mercury-contaminated fish and crustaceans near a factory waste outlet. The volatile, elemental form of mercury, Hg(0), is reportedly nontoxic, but its conversion to alkylmercury compounds by anaerobic microorganisms utilizing a vitamin B-12 biosynthetic pathway constitutes a serious health hazard.

Because of the high affinity of mercury for sulfur-donor ligands, mercury poisoning is treated by BAL; N-acetylpenicillamine has also been proposed. Recently, a very interesting natural detoxification system has been discovered in bacteria resistant to mercury; this system, when fully elucidated, might provide important strategies for treating heavy-metal poisoning in humans.

Presumably under environmental pressure, bacteria have developed mechanisms of resistance to HgX_2 and RHgX compounds in which mercury is recycled back to Hg(0). At least five gene products are involved in the bacterial mercury-resistance mechanism. MerT and MerP mediate the specific uptake of mercury compounds. MerB, organomercury lyase, and MerA, mercuric reductase, catalyze two of the reactions, given in Equations (9.1) and (9.2). Plasmids encoding the genes for these two proteins have been isolated. A typical arrangement of genes in the *mer* operon

$$\text{RHgX} + \text{H}^+ + \text{X}^- \xrightarrow[\text{lyase}]{\text{organomercury}} \text{HgX}_2 + \text{RH} \qquad (9.1)$$

$$\text{Hg(SR)}_2 + \text{NADPH} + \text{H}^+ \xrightarrow[\text{reductase}]{\text{mercuric}} \text{Hg(0)} + \text{NADP}^+ + 2\text{RSH} \qquad (9.2)$$

region of these plasmids is shown in Figure 9.1. The most thoroughly studied gene product is MerR, a metalloregulatory protein that controls transcription of the *mer* genes. In the absence of Hg(II) the MerR protein binds to DNA as a repressor, preventing transcription of the *merT, P, A*, and *B* genes (Figure 9.1) and negatively autoregulating its own synthesis. When Hg(II) is present, transcription of these genes is turned on. Interestingly, the MerR protein remains bound to the same site on DNA whether acting as an activator in the presence

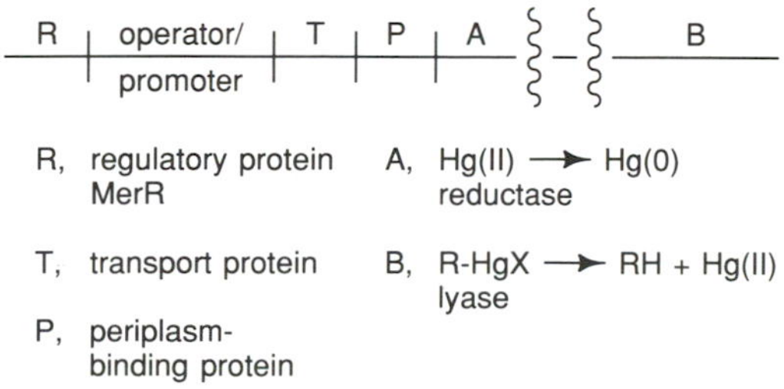

Figure 9.1
Arrangement of genes in *mer* operon of a gram negative bacterium (adapted from Figure 1, Reference 14).

of Hg(II) or as a repressor in its absence. Random and site-specific mutagenesis studies implicate several cysteine residues in the carboxyl terminal region of the protein as candidates for the mercury-binding site.

Organomercury lyase, encoded by the *merB* gene, achieves the remarkable enzymatic step of breaking Hg-C bonds (Equation 9.1). It is a 22-kDa protein with no metals or cofactors. Two cysteine-sulfhydryl groups on the protein have been postulated to effect this chemistry, as depicted in Equation (9.3). Stereochemical studies of the Hg-C bond cleavage revealed retention of configuration, indicating that cleavage of the Hg-C bond probably does not proceed by a radical pathway. A novel concerted S_E2 mechanism has been suggested. The enzyme turnover numbers, ranging from 1 min^{-1} for CH$_3$HgCl to 240 min^{-1} for butenylmercuric chloride, although slow, are $\sim 10^5$–10^8-fold faster than the nonenzymatic rate.

$$ \tag{9.3} $$

Mercuric ion reductase, the FAD-containing *merA* gene product, has several pairs of conserved cysteines. From site-specific mutagenesis studies, cysteine residues in the sequence 134-Thr-Cys-Val-Asn-Val-Gly-Cys-140 are known to comprise a redox-active disulfide group; in addition, a redox-inactive pair of cysteines near the carboxyl terminus is also required for the selective reduction of Hg(II). Exactly how the enzyme achieves the chemistry shown in Equation (9.2) is currently uncertain, but the redox activities of the flavin and disulfide/ thiol centers are undoubtedly involved. This enzyme serves both to detoxify mercury supplied directly from the environment as Hg(II) salts and to complete clearance of Hg^{2+} generated by the MerB protein from RHgX compounds. Clearly, Nature has invented a remarkable system to detoxify mercury in this fascinating class of Hg-resistant bacteria.

G. Cadmium and Lead Toxicity [18]

Gastrointestinal, neurological, and kidney toxicity are among the symptoms experienced by acute or chronic exposure to these heavy metals. The use of unleaded gasoline and the removal of lead-containing pigments from paint have substantially diminished the quantity of this element released to the environment each year. Cadmium sources include alkaline batteries, pigments, and plating.

Lead poisoning can be treated by chelation therapy using $CaNa_2(EDTA)$ (acute) or penicillamine (chronic). Although both Cd(II) and Pb(II) bind to sulfhydryl groups in thionein, we have little information at the molecular level on the mechanisms by which these elements induce toxicity.

H. Metals as Carcinogens[19,20]

Although most metal ions have been reported to be carcinogenic, the three most effective cancer-causing metals are Ni, Cr, and, to a lesser extent, Cd. Nickel subsulfide, Ni_2S_3, found in many nickel-containing ores, has been extensively studied and shown to be carcinogenic in humans and other animals. In short-term bioassays including mutagenesis, enhanced infidelity of gene replication *in vitro* and altered bacterial DNA repair were observed. Chromium is most carcinogenic as chromate ion ($CrO_4{}^{2-}$), which enters cells by the sulfate uptake pathway and is ultimately reduced to Cr(III) via a Cr(V)-glutathione intermediate species. The latter complex binds to DNA to produce a kinetically inert and potentially damaging lesion. Despite the fact that much information is available about metal-DNA interactions, molecular mechanisms of metal-induced carcinogenesis have not been elucidated. Two aspects of the problem are tumor initiation and tumor development, which are likely to involve different pathways. As new methods become available for studying the molecular events responsible for cancer (oncogenesis), it should be possible for bioinorganic chemists to unravel details of how metals act as carcinogens and as mutagens. Since cancer has genetic origins, metal/nucleic-acid chemistry is likely to be prominent in such mechanisms. As discussed later, metal-DNA interactions are an important aspect of the antitumor drug mechanism of *cis*-$[Pt(NH_3)_2Cl_2]$.

I. Summary

Toxicity can arise from excessive quantities of either an essential metal, possibly the result of a metabolic deficiency, or a nonessential metal. Both acute and chronic exposure can be treated by chelation therapy, in which hard-soft acid-base relationships are useful in the choice of chelating agent. Since chelates can also remove essential metals not present in toxic amounts, ligands with high specificity are greatly desired. The design and synthesis of such ligands for chelation therapy remains an important objective for the medicinal bioinorganic chemist. Until recently, studies of the toxic effects of metals and their removal, sometimes categorized under "environmental chemistry," have been empirical, with little insight at the molecular level. Application of the new tools of molecular biology to these problems has the potential to change this situation, as illustrated by rapid progress made in cloning the genes and studying the gene products of the mercury-resistance phenotype in bacteria. The discovery of such resistance phenomena in mammalian cells, and even the remote prospect of transferring Hg-resistant genes from bacteria to humans, are exciting possibilities for the future.

IV. SURVEY OF METALS USED FOR DIAGNOSIS AND CHEMOTHERAPY

A. Radiodiagnostic Agents[21,22]

Metal complexes having radioactive nuclei find many applications in medicine, such as in tumor, organ, and tissue imaging. Early detection of cancer, for example, by selective uptake and imaging of the tumor using a radioactive metal compound can facilitate surgical removal or chemotherapeutic treatment before the disease reaches an advanced stage. Ideally, radioisotopes used for diagnostic purposes should be short-lived, emit low-energy γ photons, and emit no α or β particles. Table 9.2 lists the radionuclides most commonly employed for this purpose in nuclear medicine. Among these, ^{99m}Tc is perhaps the most desirable,[23] for it gives off a 140-keV γ ray that is readily detected by scintillation cameras and produces clear images. This radionuclide is prepared from an alumina column loaded with $^{99}MoO_4^{2-}$, which decays to form $^{99m}TcO_4^{-}$, which in turn may be selectively eluted from the column with saline solution, owing to its lower charge. Subsequent treatment with a reducing agent in the presence of the appropriate ligands produces technetium radiopharmaceuticals with desired water solubility, stability, and tissue-distribution properties. Such complexes may be injected at concentrations of 10^{-6}–10^{-8} M. For example, isocyanide complexes such as $[Tc(CNR)_6]^{+}$ (R = t-Bu, $CH_2CO_2Bu^t$, etc.) have been found to be taken up selectively into heart tissue and thus have the potential to be used as heart-imaging agents. Figure 9.2 displays skeletal bone as imaged by a ^{99m}Tc bone agent. The dark portions correspond to surface areas of high metabolic activity, which can be used to diagnose injury or disease. One

Table 9.2
Radionuclides most commonly employed in diagnostic nuclear medicine.[a]

Radionuclide	Half-Life	Energy (keV)
^{57}Co	271 d	836
^{67}Ga	78 h	1,001
^{99m}Tc	6 h	140
^{111}In	67 h	172, 247
^{113m}In	104 m	392
^{123}I	13 h	1,230
^{169}Yb	32 d	207
^{197}Hg	64 h	159
^{201}Tl	72 h	135, 167

[a] Data are from Table of the Isotopes in D. R. Lide, ed., *CRC Handbook of Chemistry and Physics*, CRC Press, 71st ed., 1990–91, pp. 11–33 ff.

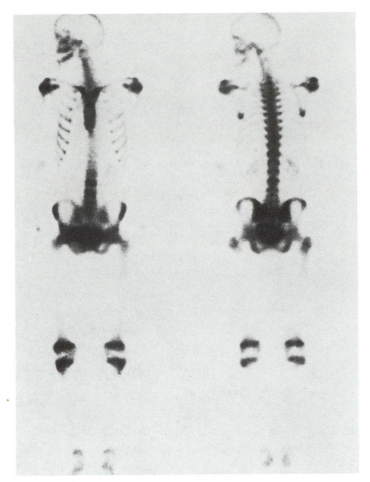

Figure 9.2
Human skeleton (bone) imaged with ^{99m}Tc. Both anterior (left)
and posterior (right) views are shown.

goal of research in this field is to provide real-time images of myocardial in-
farcts or clogged arteries for physicians who can watch the patient's heart on a
video monitor during surgery. Although chemical details responsible for the
selective tissue uptake of Tc isocyanide, phosphine, and other complexes are
largely unknown, synthetic modifications are possible and have provided many
new compounds for clinical evaluation.

Among the few molecules known to be absorbed selectively by tumor cells
is the antitumor antibiotic bleomycin (BLM),[24,25] the structure of which is por-
trayed in Figure 9.3. Bleomycin binds most radioactive metal ions, but the
^{57}Co(III) complex has the best tumor-to-blood distribution ratio. Unfortunately,
the long ^{57}Co half-life (Table 9.2) has limited its clinical utility. Attempts to
prepare stable ^{99m}Tc complexes of BLM with selective uptake properties approaching that of the cobalt complex have not yet been successful, although the

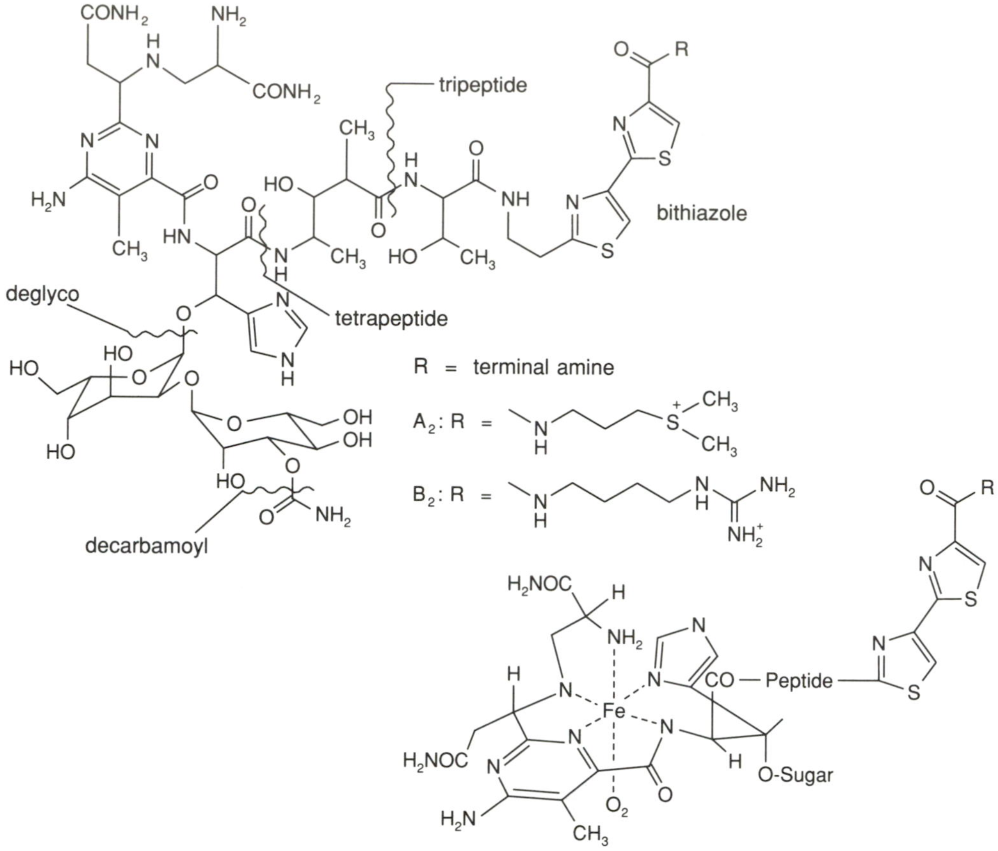

Figure 9.3
Structure of bleomycin and its proposed iron complex (reproduced by permission from Reference 25).

target molecule would be a most valuable radiodiagnostic agent. One imaginative solution[26] to this problem was achieved by covalent attachment of an EDTA moiety to the terminal thiazole ring of BLM (Figure 9.3). The resulting Co(III) BLM-EDTA molecule was radiolabeled with ^{111}In^{3+} and found to be useful for diagnosis of cancer in humans. Also used for tumor imaging are ^{99m}Tc and ^{67}Ga citrate complexes, the latter being the agent of choice for many applications. Again, there is little known at the molecular level about the mechanism of tumor-cell specificity.

An alternative approach to radionuclide-based tumor-imaging agents for diagnosis of disease is to modify, with metal chelating agents, antibodies raised against a biological substance, such as a tumor-cell antigen, hormone, or other target. Antibodies are proteins that are synthesized by specialized cells of the immune system in response to an external stimulant, or antigen. The high specificity and affinity of antibodies for the antigen can be used to target the antibody to a particular biological site, such as a site on the membrane of a particular cell type. Chelating agents are now routinely attached to antibodies and used to bind radioactive metal ions. The resulting radionuclide-labeled products are currently under extensive study in diagnostic medicine.[26]

B. Magnetic Resonance Imaging (MRI)[27]

Nuclear magnetic resonance (NMR) spectroscopy can be used to image specific tissues of biological specimens because of differences in the relaxation times of water proton resonances, usually brought about by paramagnetic metal ions. An early, pioneering example was the demonstration that Mn(II) salts localize in normal heart-muscle tissue in dogs rather than in regions affected by blocked coronary arteries. Since the paramagnetism of the d^5 Mn(II) ions alters the relaxation rate of nearby water protons, the normal and diseased tissue could be distinguished. Of the various metal ions surveyed in attempts to provide clinically useful NMR images in humans, Gd(III), Fe(III), and Mn(II) were found to give the best proton-relaxation enhancements. The gadolinium complex $[Gd(DTPA)(H_2O)]^{2-}$, an agent currently used in the clinic, has been successfully employed to image brain tumors. Ferric chloride improves gastrointestinal tract images in humans and, as already mentioned, manganous salts can be used for heart imaging. NMR imaging methodologies have advanced to the stage where increases as small as 10 to 20 percent in T_1^{-1}, the inverse nuclear-spin relaxation time, can be detected. As with radionuclide labeling, the complexes must be soluble and stable in biological fluids and relatively nontoxic, and are of greatest value when able to target a specific tissue. Even more important than targeting, however, is that proton relaxivity be maximally enhanced, an objective that depends not only upon the local binding constant but also upon large magnetic moments, long electron-spin relaxation (T_{1e}) values, access to and the residence lifetime in the inner and outer coordination spheres by water molecules, and the rotational correlation time of the complex at its binding site. An obvious advantage of paramagnetic NMR over radioisotopic imaging agents is that there is no possibility of radiation damage; on the other hand, the need for 10–100 μM concentrations at the site of imaging is a distinct drawback. Both methods are likely to continue to be used in the future, and both will benefit from the design of new stable chelates that are selectively absorbed by the tissue to be diagnosed.

C. Lithium and Mental Health[28–31]

One in every 1,000 people in the United States currently receives lithium, as Li_2CO_3, for the treatment and prophylaxis of manic-depressive behavior. Doses of 250 mg to 2 g per day are administered in order to maintain a 0.5 to 2.0 mM concentration window, outside of which the drug is either toxic or ineffective. The detailed molecular mechanism by which Li^+ ion brings about its remarkable chemotherapeutic effects is largely unknown, but there are various theories. One theory proposes that lithium binds to inositol phosphates, inhibiting their breakdown to inositol, and so reducing inositol-containing phospholipids. A consequence of this chain of events would be disruption of the neurotransmission pathway based on inositol 1,4,5-triphosphate and 1,2-diacylglycerol, reducing neuronal communication, which is most likely hyperactivated in the manic state. This theory does not account for the antidepressive action of the drug, however. An alternative explanation is that lithium inhibits cyclic adenosine monophosphate (AMP) formation, again interfering with neurotransmission by intercepting this key intracellular signaling molecule. Recent experiments indicate that lithium affects the activation of G-proteins, a class of guanosine triphosphate (GTP)-binding proteins involved in information transduction. Possibly these effects result from displacement by Li^+ of Mg^{2+} from GTP and/or from protein-binding sites normally required for activation. Use of ^7Li NMR spectroscopy to study lithium transport in human erythrocytes suggests that it might be possible to apply this method to unravel details of the bioinorganic chemistry of lithium associated with the management of manic depression.

D. Gold and Rheumatoid Arthritis[23,32,33]

Gold compounds have been used in medicine for centuries, an application known as chrysotherapy. Since 1940, however, complexes of gold have been used most successfully to treat arthritic disorders in humans and other animals. Au(I) compounds are currently the only class of pharmaceuticals known to halt the progression of rheumatoid arthritis.

Until recently, gold compounds used to treat arthritis were painfully administered as intramuscular injections. Included were colloidal gold metal, colloidal gold sulfides, $Na_3[Au(S_2O_3)_2]$ (Sanocrysin), gold thiomalate and its sodium and calcium salts (Myochrisin), and polymeric gold thioglucose (Solganol, approved by the FDA). It was discovered, however, that triethylphosphinegold(I) tetra-O-acetylthioglucose (auranofin, Figure 9.4, approved by the FDA) was equally effective against rheumatoid arthritis and could be orally administered. The availability of this compound has sparked many studies of its biodistribution, stability, and possible metabolism that lead to antiarthritic activity. The mode

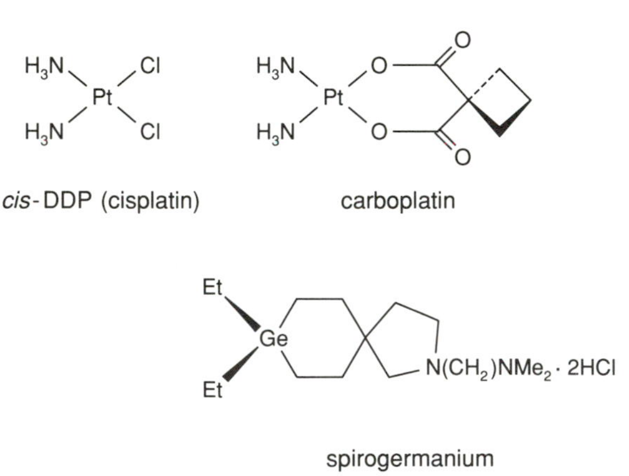

cis-DDP (cisplatin)   carboplatin

spirogermanium

Figure 9.4
Structures and trivial names of metal-based antitumor drugs.

of action of antiarthritic gold drugs is largely unknown, but it may involve binding of Au(I) to protein thiol groups, a process that inhibits the formation of disulfide bonds, and could lead to denaturation and subsequent formation of macroglobulins.

E. Anticancer Drugs

1. Platinum ammine halides[34,35]

The discovery that *cis*-diamminedichloroplatinum(II), *cis*-DDP or cisplatin (Figure 9.4), has anticancer activity in mice, and its subsequent clinical success in the treatment of genitourinary and head and neck tumors in humans, constitutes the most impressive contribution to the use of metals in medicine. Given in combination chemotherapy as an intravenous injection together with large amounts of saline solution to limit kidney toxicity, cisplatin treatment results in long-term (>5 yr) survival for more than 90 percent of testicular cancer patients. In a typical course, ~ 5 mg/kg body weight of the drug is administered once a week for four weeks. Extensive studies of platinum ammine halide analogues led to a series of empirical rules governing their chemotherapeutic potential. Specifically, it was concluded that active compounds should:

(1) be neutral, presumably to facilitate passive diffusion into cells;
(2) have two leaving groups in a *cis* configuration;
(3) contain nonleaving groups with poor trans-labilizing ability, similar to that of NH_3 or organic amines;
(4) have leaving groups with a "window of lability" centered on chloride.

These early structure-activity relationships have had to be modified somewhat, however, since chelating dicarboxylate ligands such as 1,1-dicarboxylatocyclobutane can replace the two chloride ions, and since cationic complexes with only one labile ligand, specifically, cis-$[Pt(NH_3)_2Cl(4-X-py)]^+$, where X = H, Br, CH_3, etc., showed activity in some tumor screens. The two compounds shown in Figure 9.4, cisplatin and carboplatin (Figure 9.4), were the first to be approved for clinical use. Of particular interest to the bioinorganic chemist is that complexes having a *trans* disposition of leaving groups are inactive *in vivo*. This difference suggests the presence of a specific cellular receptor that, when identified, should facilitate the design of new, metal-based anticancer drugs. Present evidence strongly points to DNA as being the relevant cellular target molecule. Section V of this chapter expands on this topic in considerable detail.

2. Metallocenes and their halides: Ti, V, Fe[36,37]

Several compounds in this category, including $[(C_5H_5)_2TiX_2]$ (X = Cl, Br, O_2CCl_3), $[(C_5H_5)_2VCl_2]$, $[(C_5H_5)_2NbCl_2]$, $[(C_5H_5)_2MoCl_2]$, and $[(C_5H_5)_2Fe]^+$ salts, exhibit significant activity against experimental animal tumors. Higher quantities

(200 mg/kg) of these compounds than of *cis*-DDP can be tolerated with fewer toxic side effects, but their failure in two mouse leukemia screens commonly used to predict the success of platinum anticancer agents appears to have delayed their introduction into human clinical trials. Studies of Ehrlich ascites tumor cells treated with [$(C_5H_5)_2VCl_2$] *in vitro* revealed selective inhibition of incorporation of radiolabeled thymidine, versus uridine or leucine, indicating that the complex blocks DNA replication. Unlike cisplatin, however, metallocene halides undergo rapid hydrolysis reactions in aqueous media, forming oxo-bridged and aqua complexes that may have a higher affinity for phosphate oxygen atoms than the heterocyclic nitrogen atoms of the bases in DNA.[38] Exactly how the ferrocenium ion might bind to DNA is even more obscure, although partial metallointercalation and groove binding are more likely than covalent attachment of the chemically unmodified cation. From the limited information available, metallocenes and their halides appear to behave fundamentally differently from platinum antitumor compounds. As a class, they provide a promising new opportunity to expand the scope of metal complexes used in cancer chemotherapy.

3. Gold and other metal phosphines[39]

Following the successful entry of the soluble gold-phosphine complex auranofin (Figure 9.4) into the metal-based pharmaceuticals industry, several gold-phosphine complexes were examined for possible anticancer activity. Although auranofin itself was active in only a small fraction of the mouse tumor models tested, biological activity approaching that of cisplatin was discovered for many analogues, most notably the diphosphine bridged complex [$ClAu(PPh_2CH_2CH_2$ $PPh_2)AuCl$]. Attempts to replace the phosphine with As or S donor ligands, to increase or decrease the length of the 2-carbon bridge, or to replace the phenyl with alkyl groups all led to diminished activity. Most noteworthy is that the diphosphine ligands themselves have activity very similar to that of their gold complexes, and that Ag(I) and Cu(I) analogues are also effective. These results strongly imply that the phosphine ligands are the chemical agents responsible for the anticancer properties of these compounds. Coordination to a metal presumably serves to protect phosphines against oxidation to the phosphine oxides, which independent investigations have proved to be ineffective. A possible role for the metal in the cytotoxicity of the compounds cannot be ruled out, however.

4. Other main group and transition-metal compounds[36,40,41]

Several main group metal complexes exhibit anticancer activity. Gallium(III) nitrate is active against human lymphomas, but with dose-limiting side effects on the kidneys and gastrointestinal tract. Tin complexes of general formula $R_2L_2SnX_2$, where R = alkyl or phenyl, $L_2 = py_2$, bpy, or phen, and X_2 = two *cis*-oriented halide or pseudohalide leaving groups, are active against the mouse P388 leukemia tumor. The *cis* disposition of the leaving groups suggests a possible mechanism analogous to that of cisplatin (see below). Organogermanium compounds are also active, notably the derivative spirogermanium shown in Figure 9.4. Nothing is known about the mechanism of action of any of these compounds.

Following the discovery of activity for cisplatin, several thousand platinum and nearly 100 other transition-metal complexes have been screened in various tumor model systems in the hope of achieving better activity against a broader range of tumors. Among the classes of nonplatinum compounds showing some activity are ruthenium complexes cis-[RuCl$_2$(DMSO)$_4$], [Ru(NH$_3$)$_5$(Asc)](CF$_3$SO$_3$), where Asc is ascorbate dianion, and fac-[Ru(NH$_3$)$_3$Cl$_3$], all of which are believed to bind to DNA; binuclear rhodium complexes [Rh$_2$(O$_2$CR)$_4$L$_2$]; octahedral Pd(IV) complexes such as cis-[Pd(NH$_3$)$_2$Cl$_4$]; and such miscellaneous molecules as the iron(II) complex of 2-formylpyridine thiosemicarbazone, the site of action of which is thought to be ribonucleotide reductase. These examples illustrate the broad scope encompassed by this field, which has a potential for developing fundamental information about metal-biomolecule interactions as well as novel anticancer drugs. Much remains to be explored.

F. Miscellaneous Metals in Medicine

Numerous other anecdotal and some fairly elaborate studies have been reported for metal complexes as medicinal agents. The use of zinc applied topically to promote the healing of wounds dates back to around 1500 B.C., and silver is now commonly applied to prevent infection in burn patients.[42,43] Osmium carbohydrate polymers have been reported to have antiarthritic activity.[44] Transition-metal complexes have a long history of use as antibacterial and antiviral agents; for example, Zn^{2+} is used to treat herpes, possibly by inhibiting the viral DNA polymerase.[45] Early transition-metal (e.g., tungsten) polyoxoanions have been employed to treat AIDS patients.[46] Numerous reports have appeared detailing the anti-inflammatory, antiulcer, and analgesic activities of copper carboxylate complexes.[7] As in the previous section, these reports and others like them require more serious attention from bioinorganic chemists to elucidate the molecular events responsible for such a fascinating menu of biologically active metal complexes.

G. Summary and Prospectus

The clinical successes of platinum anticancer and gold antiarthritic drugs have changed the attitudes of many who doubted that heavy-metal compounds, notorious for their deleterious effects on human health, would ever play a serious role in chemotherapy. Indeed, we have seen that Hg^{2+}, Pb^{2+}, and Cd^{2+} are toxic elements. Even essential metals can be highly toxic if present in excess, either because of chronic or acute poisoning or because of metabolic defects that deregulate their control in the cell. An important common theme running throughout this discussion is selectivity. For a drug to be effective, it must be selectively toxic to diseased tissue while leaving normal tissue alone; or it must selectively kill harmful microorganisms at levels where it fails to deplete helpful ones. For a chelating agent to be useful in limiting the toxic effects of metals, it must bind as selectively as possible to the deleterious ion while coordinating only weakly, if at all, to others. For a diagnostic metal complex to be useful, it must be taken up (or excluded) selectively from diseased cells relative to normal ones, or to one tissue type versus another. Rarely has such selectivity been designed in advance of the discovery of a useful metal-based pharmaceutical, although spectacular advances in biology, such as monoclonal antibodies, may be hastening the day when such an objective might be common. Interestingly, the successes of such unlikely compounds as cis-[Pt(NH$_3$)$_2$Cl$_2$] and [(Et$_3$P)Au(OAc$_4$-thioglucose)] in chemotherapy were driven by the personal involvement of individuals like B. Rosenberg for the former and B. Sutton for the latter. Like Hollywood producers, these men mustered every conceivable resource to promote the compounds for testing, introduction into human clinical trials, and eventually approval by the FDA. Such zeal requires years, usually more than a decade, of sustained personal effort, and may be the reason why other promising metal complexes, such as those mentioned above, have not had the impact of a cisplatin or an auranofin. On average, only one of 7,000 such compounds makes it from the laboratory bench to the patient, at an average cost of 250 million dollars and a time interval of 13 years.

Another significant component of the evolving field of metals in medicine, however, is that, once a compound has proved its utility in the clinic, how does it work? This question is deceptively simple, for coordination chemistry *in vivo*, and the ability of cells to respond to unnatural external stimuli such as metal complexes, are complex matters about which we are only just beginning to learn. As progress is made in this latter area, it should become possible to design drugs in a rational way to achieve the required selectivity.

The remainder of this chapter focuses on a single case study where some progress in unraveling the molecular mechanism of a metal-based drug, cisplatin, is being made. If nothing else, this discussion will elucidate strategic guidelines that may be employed to attack similar questions about other chemotherapeutic metal compounds discussed earlier in this section. Unfortunately, there is very little information available about the molecular mechanisms of these other complexes. At this major transition point in our discussion, we move from general considerations to a specific, in-depth analysis. The reader must here take time to become familiar with the biological aspects of the new material.

V. PLATINUM ANTICANCER DRUGS: A CASE STUDY

A. History of the Discovery[47]

cis-[Pt(NH$_3$)$_2$Cl$_2$], a molecule known since the mid-19th century, has been a subject of considerable importance in the development of inorganic stereochemistry and substitution reaction kinetics.[48] Its biological activity was discovered by accident. In the mid-1960s, biophysicist Barnett Rosenberg, at Michigan State University, was studying the effects of electric fields on the growth of

Escherichia coli cells in culture. They had hypothesized that, during cell division, the orientation of the mitotic spindle might be affected by local electric fields which they hoped to perturb. Instead, they observed growth without cell division, the result being elongated, spaghetti-like bacterial filaments approaching 1 cm in length. After much detective work, they realized that small amounts of platinum from the electrodes used to apply the electric fields had reacted with NH_4Cl in their buffer to produce various platinum ammine halide compounds. Two of these, *cis*-[Pt(NH$_3$)$_2$Cl$_2$] and *cis*-[Pt(NH$_3$)$_2$Cl$_4$], were capable of inducing filamentous growth in the absence of any electric field. Since chemicals that produce filamentation in bacteria had been known to exhibit antitumor activity, Rosenberg was eager to have his platinum compounds tested. Unable to convince existing agencies like the National Cancer Institute (who to their credit later spearheaded the development of cisplatin) that a heavy-metal complex could actually be beneficial to animals, the Michigan State group set up their own animal-tumor screens. The results were nothing short of spectacular. Injection of *cis*-DDP directly into the abdominal cavity of mice into which a solid Sarcoma-180 tumor had been implanted led within a few days to a blackening (necrosis), reduction in size, and eventual disappearance of the tumor (Figure 9.5). The cured mouse enjoyed a normal lifespan. From these and other animal

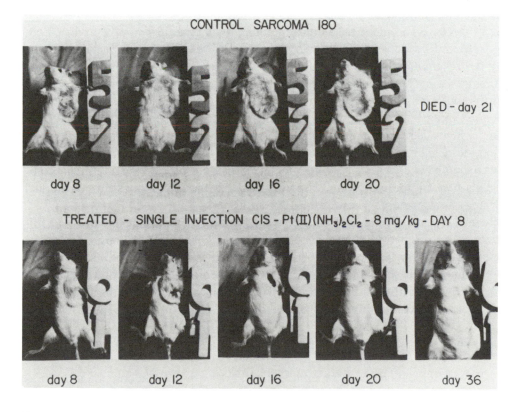

Figure 9.5
Photographic demonstration of the dramatic ability of cisplatin to cure a Sarcoma-180 murine tumor (reproduced by permission from Reference 47).

studies, physicians became convinced that administering platinum compounds to cancer patients might be worthwhile, and a new field involving bioinorganic chemistry and cancer chemotherapy was born. The drug, marketed as Platinol with the generic name cisplatin, received FDA approval in 1979 and is today one of the leading anticancer agents.

B. Principles that Underlie Drug Development

1. Strategic considerations

There are two general routes to the development of inorganic complexes, and indeed most chemical compounds, as drugs. One, illustrated by cisplatin, arises from an empirical observation of biological activity followed by attempts to optimize the clinical efficacy through investigations of structure-activity relationships (SAR). The goals are to minimize toxicity, to develop cell culture and animal screens for testing related compounds, and ultimately to elucidate the molecular mechanism. Knowledge of the molecular mechanism might eventually lead to a rational strategy for designing better drugs.

The second general approach to drug design begins with known biochemistry. For example, ribonucleotide reductase is required in the first committed step in the biosynthesis of DNA, the conversion of ribo- to deoxyribonucleoside diphosphates. The mammalian enzyme contains a binuclear non-heme iron center required for activity. Compounds that would selectively inhibit this enzyme by destroying this center are potentially useful as antiviral or antitumor agents. Another example is the enzyme reverse transcriptase, encoded by the HIV (AIDS) virus and required for its integration into the genome of the host cell. Compounds like 3'-azidothymidine (AZT) are accepted by the enzyme as substrates but, when added to the growing DNA chain, cannot be linked to the next nucleotide. Chain termination therefore occurs and the replication process becomes permanently interrupted. Attempts to find organic molecules or inorganic complexes that are more effective chain terminators than AZT constitute a rational strategy for developing new anti-AIDS drugs.

In the remainder of this chapter we describe research that has evolved following the discovery of biological activity for cisplatin. Although the initial breakthrough was serendipitous, subsequent studies have revealed many aspects of the molecular mechanism. From this known biochemistry we may one day be in a position to design more effective anticancer drugs and therapies based upon the fundamental bioinorganic chemistry of cisplatin.

2. Pre-clinical and clinical trials[49]

Predicting the chemotherapeutic potential of an inorganic compound such as *cis*-DDP prior to its introduction into human cancer patients is an important objective. Compounds are most easily tested for their cytotoxic effects on bacterial or mammalian cells in culture. Shown in Figure 9.6 are results for the

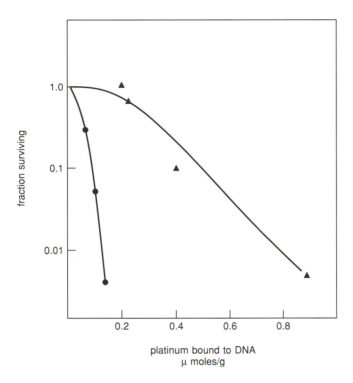

Figure 9.6
Differential toxicity of *cis*- (●) and *trans*-DDP (▲) on HeLa cells
growing in culture (reproduced by permission from Reference 51).

survival of cultured L1210 cells in the presence of increasing amounts of *cis*-
or *trans*-DDP.[50,51] The data reveal the markedly greater toxicity of the *cis* isomer, which is a much better anticancer agent than its stereoisomer. Unfortunately, no single assay has yet been found that can predict the chemotherapeutic
potential of platinum compounds in humans. The best that can be obtained are
results relative to those for *cis*-DDP, in which case toxicity at low dose is usually scored positive.

The next level of testing, often employed directly without first examining
cell-culture results, involves animal (usually mammals, excluding human)
screens.[49] Among the most popular measures of the chemotherapeutic activity
of platinum compounds has been their ability to prolong the survival of mice
bearing the L1210 or P388 leukemia. A suspension of cells is inoculated intra-
peritoneally (i.p., in the abdominal cavity), producing a leukemia that eventually progresses to the generalized disease. In one commonly used protocol, platinum compounds are dissolved in physiological saline (0.85 percent NaCl) or
sterile H_2O and injected i.p. 24 h, 5, 9, and 13 days after inoculation of the
leukemia cells. Several indices of antitumor activity and toxicity have been defined. The percent I.L.S., or increased lifespan, measures the mean survival of
treated versus control animals that were given no platinum drug. A related index

is the median survival, percent T/C (Test/Control), which is 100 + percent I.L.S. The LD_{50} value measures toxicity as mean lethal dose, the amount of drug (usually in mg/kg body weight) required to kill half the animals. Potency is defined by ID_{90}, the inhibiting dose at which 90 percent of the tumor cells are killed. From these values, a therapeutic index (TI) = LD_{50}/ID_{90} is sometimes defined, which should be substantially greater than one. Typical values for *cis*-DDP are 85 percent I.L.S. at 8 mg/kg for the L1210 tumor, 13.0 mg/kg LD_{50}, and 1.6 mg/kg ID_{90} resulting in a TI of 8.1.

In addition to being tested in mice, cisplatin and related compounds have been screened in other mammals, specifically dogs and monkeys, mainly to look for possible dose-limiting side effects. Severe vomiting, once thought to be an insurmountable obstacle, was monitored by using ferrets. None of the animal screens can substitute for the ultimate test, however, which is human clinical trials. In 1972, such trials were initiated using terminally ill cancer patients. It was determined that intravenous (i.v.) injection, rather than i.p. or oral administration, was the preferred method for giving the drug. Further details of the clinical development of cisplatin are discussed in a later section.

From the animal screens emerged the set of structure-activity relationships enumerated earlier (Section IV.E.1). Both cisplatin and carboplatin conform to these rules, and to date no compounds with demonstrably better antitumor activity have been tested in humans. The decision to move an experimental drug into the clinic is a difficult one, however, and it may be that molecules such as *cis*-[Pt(NH₃)₂(4-Br-py)Cl]Cl (see Section V.D.7.c) would be effective for tumors that are refractory to cisplatin chemotherapy. In any case, the foregoing chain of events, from studying the effects of a compound on cells in culture through animal screens and eventually to humans, constitutes the principal route for introducing a new anticancer drug. The process can take more than a decade.

3. Mechanism of action studies

Once a class of compounds has been identified as biologically active, studies to elucidate the molecular mechanism of action can be undertaken. A first step is to identify the major cellular target or targets responsible for the chemotherapeutic properties of the drug. These investigations must also focus on chemical transformations that might take place in the solutions being administered and in the biological fluids that transport the drug to its ultimate target site. The next major step is to characterize the adduct or family of adducts made with the biological target molecule. The structure and kinetic lifetime of these adducts need to be investigated. Once this information is in hand, the effect of the adducts on the structure, stability, and function of the biological target molecule must be studied. Here many powerful new methodologies of modern molecular biology, genetics, and immunology can be brought to bear on the problem. The ultimate goals are to translate the molecular events elucidated into a realistic mechanism for how the drug molecule brings about its toxic effects selectively at the sites responsible for the disease and to use this information to design even better drugs.

Having progressed this far, we next need to bridge the gap between fundamental knowledge gained in studies of the mechanism of action and the SAR gleaned through pre-clinical and clinical trials. Whether such a happy situation can be reached for cisplatin remains to be seen, but there are encouraging signs, as we hope to demonstrate in the following discussion.

C. Clinical Picture for Cisplatin and Carboplatin[49,52]

1. Responsive tumors and combination chemotherapy

It was an early observation that the best responses to cisplatin occurred in patients with genitourinary tumors. For testicular cancer, once a leading cause of death for males of age 20–40, cisplatin cures nearly all patients with stage A (testes alone) or B (metastasis or retroperitoneal lymph nodes) carcinomas. Platinum is usually given in combination with other drugs, commonly vinblastine and bleomycin for testicular cancer. This combination chemotherapy, as it is known, has several objectives. Some tumors have a natural or acquired resistance to one class of drugs and, by applying several, it is hoped that an effective reduction in tumor mass can be achieved. In addition, various drugs are known to affect different phases of the cell cycle, so several are applied simultaneously to allow for this possibility. Finally, synergism, where the response is greater than expected from simple additive effects, can occur, although it is rare. In addition to testicular cancer, platinum chemotherapy has produced responses in patients with ovarian carcinomas (>90 percent), head and neck cancers, non-small-cell lung cancer, and cervical cancer. Cisplatin is also effective when combined with radiation therapy.

2. Dose-limiting problems; toxicology

An early and quite worrisome adverse side effect of cisplatin was kidney toxicity. This problem, not commonly encountered with the older cancer drugs, nearly prevented its widespread use and eventual FDA approval. The major breakthrough here was made by E. Cvitkovic, who, while working at Sloan-Kettering Memorial Hospital in New York, administered large quantities of water by intravenous injection to patients, together with an osmotic diuretic agent such as D-mannitol. The rationale was that such hydration could ameliorate kidney toxicity by flushing out the heavy-metal complex. This simple idea worked, and the dose of the platinum compound could be increased threefold without accompanying nephrotoxicity. Hydration therapy is now commonly employed when cisplatin is administered. Among the other toxic effects encountered in cisplatin chemotherapy are nausea and vomiting, but this problem has also been controlled by use of antiemetic agents. Patients have also been known to experience bone-marrow suppression, a ringing in the ears, and occasional allergic reactions.

More recently, attempts have been made to extend cisplatin treatment to other broad classes of tumors by raising the dose above the ~ 5 mg/kg body weight levels given by i.v. injection every few weeks. Direct injection into the peritoneal cavity has been employed for refractory ovarian tumors. These more aggressive therapeutic protocols have been frustrated by drug resistance, a phenomenon by which cells learn to tolerate a toxic agent and for which many mechanisms exist, and by the return of the usual cisplatin side effects, most notably kidney toxicity and neurotoxicity. In order to combat toxic effects to the kidneys, chemoprotector drugs have been introduced. Based on the known affinity of platinum(II) complexes for sulfur-donor ligands, sodium diethyldithiocarbamate (DDTC) has been given both to experimental animals and to humans by i.v. infusion over about an hour following cisplatin administration.[53] DDTC inhibits many of the toxic side effects, particularly to the kidneys and bone marrow, without itself producing long-term side effects or apparently inhibiting the antitumor properties of cis-DDP. Similar efforts have been made to reduce the toxic effects of cisplatin with other sulfur-containing compounds including thiosulfate and the naturally occurring biomolecules glutathione, cysteine, and methionine. The relative amounts of the latter three molecules can be controlled by drugs that affect their normal cellular concentrations.

Another approach to reducing cisplatin toxicity is to develop new classes of platinum drugs or different routes of their administration. Carboplatin (Figure 9.4) is one result of these efforts. The bidentate chelating dicarboxylate leaving group in carboplatin presumably retards the rates of reactions leading to toxicity, but does not adversely interfere with the chemistry required for antitumor activity. Recently, promising platinum compounds for oral administration have been developed.[54] In Pt(IV) complexes of the kind cis, trans, cis-[Pt(NH$_3$)(C$_6$H$_{11}$NH$_2$) · (O$_2$CCH$_3$)$_2$Cl$_2$], where C$_6$H$_{11}$NH$_2$ is cyclohexylamine, have been found to be effective in preclinical screens. The greater kinetic inertness of these complexes apparently renders them sufficiently stable to the chemically harsh environment of the gastrointestinal tract. Once absorbed into the bloodstream, these compounds are metabolized to the Pt(II) analogues, cis-[Pt(NH$_3$)(C$_6$H$_{11}$NH$_2$)Cl$_2$], which are presumed to be the active form of the drug. The Pt(IV) compound has recently entered clinical trials.

Although impressive inroads have been made in the management of human tumors by platinum chemotherapy, the fact remains that, apart from testicular and to a lesser extent ovarian cancer, the median survival times are measured in months. Clearly, there is much room for improvement.

3. Pharmacology[49,52]

Solutions of cisplatin are usually given in physiological saline (NaCl), since hydrolysis reactions occur that can modify the nature of the compound and its reactions in vivo (see below). Cisplatin is rapidly cleared from the plasma after injection, 70–90 percent of the platinum being removed within the first 15 minutes. It has been found that more than half the platinum binds to serum proteins and is excreted. Most of the platinum exits the body via the urine within a few days. These results account for the use of multiple-dose chemotherapy at intervals of several weeks. Animal studies employing *cis*-DDP labeled with ^{195m}Pt, a 99 keV γ-emitter with a 4.1-day half-life, reveal retention half-times in various tissues of 8.4 (kidney), 6.0 (ileum), 4.1 (liver), 2.8 (tumor), and 1.9 (serum) days following a single dose. Platinum distributes widely to all tissue, with kidney, uterus, liver, and skin having the most, and muscles, testes, and brain the least amount of the compound. There is no evidence for selective uptake into normal versus tumor cells.

D. Bioinorganic Chemistry of Platinum Anticancer Drugs; How Might They Work?

The material in this section constitutes the major portion of this chapter. One important goal of the discussion is to illustrate, by means of an in-depth analysis of a single case history, the questions that must be addressed to elucidate the molecular mechanism of an inorganic pharmaceutical. Another is to introduce the techniques that are required to answer these questions, at least for the chosen case. The inorganic chemist reading this material with little or no biological background may find the experience challenging, although an attempt has been made to explain unfamiliar terms as much as possible. It is strongly advised that material in Chapter 8 be read before this section. Toward the end of this section, the results obtained are used to speculate about a molecular mechanism to account for the biological activity of the drug. Experiments directed toward evaluating the various hypotheses are delineated. Once the mechanism or mechanisms are known, it should be possible to design new and better antitumor drugs which, if successful, would be the ultimate proof of the validity of the hypotheses. This topic is discussed in the next and final section of the chapter. Such an analysis could, in principle, be applied to probe the molecular mechanisms of the other metals used in medicine described previously. In fact, it is hoped that the approach will prove valuable to students and researchers in these other areas, where much less information is currently available at the molecular level.

The material in this section has been organized in the following manner. First we discuss the relevant inorganic chemistry of platinum complexes in biological media. Next we summarize the evidence that DNA is a major target of cisplatin in the cancer cell, responsible for its antitumor activity. The chemical, physical, and biological consequences of damaging DNA by the drug are then described, followed by a presentation of the methodologies used to map its binding sites on DNA. The detailed structures of the DNA adducts of both active and inactive platinum complexes are then discussed, together with the way in which the tertiary structure of the double helix can modulate these structures. Finally, the response of cellular proteins to cisplatin-damaged DNA is presented, leading eventually to hypotheses about how tumor cells are selectively destroyed by the drug. Together these events constitute our knowledge of the ''molecular mechanism,'' at least as it is currently understood.

1. Reactions of cis-DDP and related compounds in aqueous, biological, and other media

cis-Diamminedichloroplatinum(II) is a square-planar d^8 complex. As such, it belongs to a class of compounds extensively investigated by coordination chemists.[55] Typically, such compounds are relatively inert kinetically, do not usually expand their coordination numbers, and undergo ligand substitution reactions by two independent pathways with the rate law as given by Equation (9.4). The rate constants k_1 and k_2 correspond to first-order (solvent-assisted) and second-order (bimolecular) pathways;

$$\text{Rate} = (k_1 + k_2[Y])\,[\text{complex}] \tag{9.4}$$

$[Y]$ is the concentration of the incoming ligand. Usually, $k_1 << k_2$ by several orders of magnitude. In biological fluids, however, the concentration of a potential target molecule could be $\sim 10^{-6}$ M, in which case $k_1 \gtrsim k_2[Y]$. Substitution of ligands in cis-DDP, required for binding to a cellular target molecule, is therefore likely to proceed by the solvent-assisted pathway. Such a pathway is assumed in the ensuing discussion.

For the hydrolysis of the first chloride ion from cis- or trans-DDP,

$$[\text{Pt(NH}_3)_2\text{Cl}_2] + \text{H}_2\text{O} \rightleftharpoons [\text{Pt(NH}_3)_2\text{Cl(OH}_2)]^+ + \text{Cl}^- \tag{9.5}$$

the k_1 values at 25°C are 2.5×10^{-5} and $9.8 \times 10^{-5}\,\text{s}^{-1}$, respectively.[55] These hydrolyzed complexes can undergo further equilibrium reactions, summarized by Equations (9.6) to (9.9).

$$[\text{Pt(NH}_3)_2\text{Cl(OH}_2)]^+ \rightleftharpoons [\text{Pt(NH}_3)_2\text{Cl(OH)}] + \text{H}^+ \tag{9.6}$$

$$[\text{Pt(NH}_3)_2\text{Cl(OH}_2)]^+ + \text{H}_2\text{O} \rightleftharpoons [\text{Pt(NH}_3)_2(\text{OH}_2)_2]^{2+} + \text{Cl}^- \tag{9.7}$$

$$[\text{Pt(NH}_3)_2(\text{OH}_2)_2]^{2+} \rightleftharpoons [\text{Pt(NH}_3)_2(\text{OH}_2)(\text{OH})]^+ + \text{H}^+ \tag{9.8}$$

$$[\text{Pt(NH}_3)_2(\text{OH}_2)(\text{OH})]^+ \rightleftharpoons [\text{Pt(NH}_3)_2(\text{OH})_2] + \text{H}^+ \tag{9.9}$$

The formation of dimers such as $[\text{Pt(NH}_3)_2(\text{OH})]_2^{2+}$ and higher oligomers can also occur,[56,57] but such reactions are unlikely to be important at the low platinum concentrations encountered in biological media. Reactions (9.5) to (9.9), which depend on pH and the chloride-ion concentrations, have been followed by ^{195}Pt ($I = \frac{1}{2}$, 34.4 percent abundance) and ^{15}N (using enriched compounds) NMR spectroscopy. The latter method has revealed for the cis-diammine complexes pK_a values of 6.70 ± 0.10 at 25°C for Reaction (9.6) and of 5.95 ± 0.1 and 7.85 ± 0.1 at 5°C for Reactions (9.8) and (9.9), respectively.[58]

The effects of pH and Cl$^-$ ion concentration on the species distribution of platinum compounds have been used to fashion the following plausible argument for the chemistry of cis-DDP in vivo.[59] With the use of thermodynamic data for the ethylenediamine (en) analogue $[\text{Pt(en)Cl}_2]$, the relative concentrations of hydrolyzed species at pH 7.4 were estimated (see Table 9.3) for blood plasma and cytoplasm (Figure 9.7). The higher chloride ion concentration in

Table 9.3

Distribution of various adducts formed between *cis*-DDP or [^{3}H][Pt(en)Cl$_2$][a] and DNA *in vitro* and *in vivo*.[118–122]

	Total incubation time	Adducts formed				Remaining platinum[c]
D/N ratio		*cis*-[PtA$_2${d(pGpG)}][b]	*cis*-[PtA$_2${d(pApG)}][b]	*cis*-[PtA$_2${d(GMP)}$_2$][b]	Mono-functional adducts	
In vitro						
0.055[c]	5 h (50°C)	47—50%	23—28%	8—10%	2—3%	10%
0.022[d]	5 h (50°C)	60—65%	20%	~4%	~2%	9—14%
0.01[e]	16 h (37°C)	62%	21%	7%	—	10%
ef	30 m (37°C)	36%	3%	8%	40%	13%
ef	2 h (37°C)	54%	9%	9%	14%	14%
ef	3 h (37°C)	57%	15%	9%	4%	15%
In vivo						
dg	1 h (37°C)	35.9 ± 4.7%[h]	<34%[i]	3.1 ± 1.6%[h]	38.5%[i]	~22%
dg	25 h (37°C)	46.6 ± 6.8%[h]	<48%[i]	3.0 ± 0.9%[h]	<14.5%[i]	~50%

[a] A radiolabeled analogue of *cis*-DDP, [^{3}H]dichloroethylenediamineplatinum(II).
[b] A$_2$ represents either (NH$_3$)$_2$ or ethylenediamine.
[c] By difference.
[d] Percentage of adducts based on total amount of platinum eluted from the separation column.
[e] Percentage of adducts based on total amount of radioactivity eluted from the separation column.
[f] Not given.
[g] Chinese hamster ovary cells treated with 83 μM *cis*-DDP.
[h] Results from ELISA.
[i] Results from AAS. Where the signal was too weak for reliable quantitation, the maximal amount possible is given. Adapted from Table I in Reference 81.

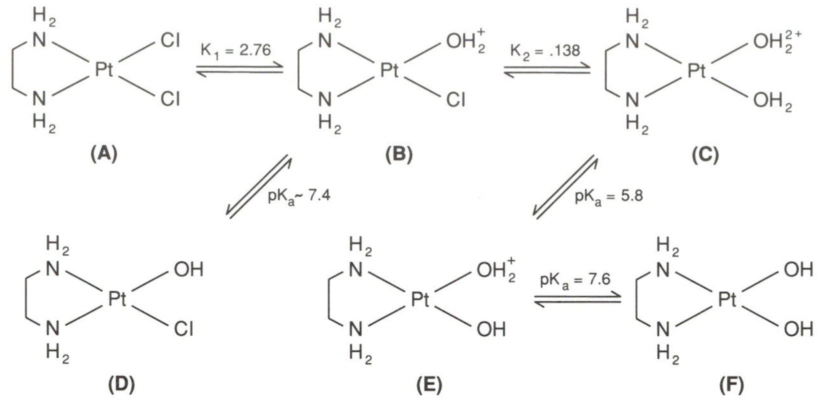

Figure 9.7
Hydrolysis reactions of antitumor platinum complexes and an estimate of the species present in plasma and cytoplasm (reproduced by permission from Reference 51).

plasma preserves the complex as the neutral molecule *cis*-DDP, which passively diffuses across cell membranes. The lower intracellular chloride ion concentration facilitates hydrolysis reactions such as Equations (9.5) to (9.9), thereby activating the drug for binding to its biological target molecules. There is, of course, a reasonable probability that *cis*-DDP and species derived from it will encounter small molecules and macromolecules *in vivo* that divert it from this route to the target. We have already seen such cases; cisplatin binds to serum proteins, and there is good evidence that intracellular thiols react with the drug.[60] Glutathione, for example, is present in millimolar concentrations in cells. How, one might ask, does cisplatin swim through such a sea of sulfur donors to find its target in the tumor cell? Is it possible that a modified form of the drug, in which a Pt-Cl bond has been displaced by thiolate to form a Pt-S bond, is the actual species responsible for its activity? Although these questions have not yet been satisfactorily answered, there is reason to believe that such reactions are not directly involved in the molecular mechanism of action. As evident from structure-activity relationship studies, the most active compounds have two labile ligands in *cis* positions. If Pt-S bonds were required, then compounds already having such linkages would be expected to exhibit activity and they do not. Rather, it seems most likely that the antitumor activity of cisplatin results from surviving species of the kind written in Equations (9.5) to (9.9) that find their way to the target molecule, and that the induced toxicity must arise from a significantly disruptive structural consequence of drug binding. Since only *cis* complexes are active, it is reasonable for the coordination chemist to infer that the stereochemistry of this interaction is of fundamental importance.

Reactions of platinum compounds with components in media used to dissolve them can give and undoubtedly have given rise to misleading results, both in fundamental mechanistic work and in screening studies. A particularly noteworthy example is dimethylsulfoxide (DMSO), which even recently has been used to dissolve platinum compounds, presumably owing to their greater solubility in DMSO compared to water. As demonstrated by ^{195}Pt NMR spectroscopy, both *cis*- and *trans*-DDP react rapidly ($t_{1/2}$ = 60 and 8 min at 37°C, respectively) to form $[Pt(NH_3)_2Cl(DMSO)]^+$ complexes with chemical and biological reactivity different from those of the parent ammine halides.[61]

2. Evidence that DNA is the target

Two early sets of experiments pointed to interactions of cisplatin with DNA, rather than the many other possible cellular receptors, as an essential target responsible for cytotoxicity and antitumor properties.[62,63] Monitoring the uptake of radiolabeled precursors for synthesizing DNA, RNA, and proteins, showed that [^{3}H]thymidine incorporation was most affected by therapeutic levels of cisplatin for both cells in culture and Ehrlich ascites cells in mice. Since independent studies showed that *cis*-DDP binding to DNA polymerase does not alter its ability to synthesize DNA, it was concluded that platination of the template and not the enzyme was responsible for the inhibition of replication.

In a second kind of experiment demonstrating that DNA is a target of cisplatin, hydrolyzed forms of the drug in low concentrations were added to a strain of *E. coli* K12 cells containing a sex-specific factor F.[64,65] After free platinum was removed, these F$^+$ cells were conjugated with a strain of *E. coli* K12 cells lacking this factor that had previously been infected with lambda bacteriophage. Addition of *cis*-DDP, but not *trans*-DDP, directly to the latter infected F$^-$ cells had been shown in a separate study to accelerate cell lysis. Conjugation with the platinum-treated F$^+$ cells produced the same effect, strongly suggesting that Pt had been transferred from the F$^+$ to the F$^-$ cells. Since only DNA is passed between the F$^+$ and F$^-$ strains, it was concluded that Pt was attached to the DNA and that this modification was essential for the observed lysis of the cell. Further studies showed a good correlation between cell lysis by platinum compounds and their antitumor properties.

Various other observations are consistent with the notion that platinum binding to DNA in the cell is an event of biological consequence.[66] The filamentous bacterial growth observed in the original Rosenberg experiment is one such piece of evidence, since other known DNA-damaging agents, for example, alkylating drugs and x-irradiation, also elicit this response. Another is the greater sensitivity toward *cis*-DDP of cells deficient in their ability to repair DNA. Finally, quantitation of the amount of platinum bound to DNA, RNA, and proteins revealed that, although more Pt was bound to RNA per gram biomolecule, much more Pt was on the DNA when expressed as a per-molecule basis. In the absence of any selective interaction of Pt with a specific molecule, only one out of every 1,500 protein molecules (average M.W. ~ 60 kDa) in a cell will contain a single bound platinum atom, whereas hundreds or thousands of Pt atoms are coordinated to DNA (M.W. ~ 10^{11}). If the replication apparatus cannot bypass these lesions, then cell division will not occur, and tumor growth is inhibited.

Although these and other results all point to DNA as an important cellular target of cisplatin, most likely responsible for its anticancer activity, this information does not explain why tumor cells are more affected by *cis*-DDP than non-tumor cells of the same tissue. Moreover, why is *trans*-DDP, which also enters cells, binds DNA, and inhibits replication, albeit at much higher doses (see discussion below), not an active anticancer drug? What causes cisplatin to kill cells and not merely to arrest tumor growth? The latter can be explained by DNA synthesis inhibition, but not necessarily the former. Very recent studies have begun to address these questions using powerful new methodologies of molecular and cell biology, as described in subsequent sections of this chapter. The results, although preliminary, continue to point to DNA as the most important cellular target of cisplatin.

3. Aspects of platinum binding to DNA

Given that DNA is a major target of platinum binding in cells, it is incumbent upon the bioinorganic chemist to investigate the nature of these interactions and their biological consequences. Of all the ligands studied in coordination chemistry, DNA is surely among the most complex. In the ensuing discussion, we first present experiments that delineate the chemical steps involved in *cis*- and *trans*-DDP binding to DNA as well as the chemical consequences of the adducts formed. We next describe the physical changes in the double helix that accompany platinum binding, and then we discuss the biological consequences that attend the platination of DNA. Subsequent sections describe the major adducts formed, in other words the regiospecificity of the drug, the three-dimensional structures of the adducts, and the way in which different structures within DNA can modulate platinum binding. Finally, we consider the response of the cell to Pt-DNA adducts, including studies with site-specifically modified DNA, and speculate about how this chemistry might relate to the antitumor drug mechanism.

a. Kinetics of Platinum Binding to DNA The binding of *cis*- and *trans*-DDP to DNA has been studied[67] by ^{195}Pt NMR spectroscopy with the use of isotopically enriched ^{195}Pt, which has a nuclear spin $I = \frac{1}{2}$. The DNA used in this experiment was obtained from chicken red blood cell chromosomes that had been enzymatically degraded to relatively small pieces ranging from 20 to 60 base pairs in length (molecular-weight range 13 to 30 kDa). Since the ^{195}Pt chemical shifts are very sensitive to chemical environment, this NMR study provided important details about the kinetics and mechanism of platinum binding to the biopolymer. The rate-determining step in platination of the DNA is loss of chloride ion (Equation 9.5) to form the monoaqua complex, which rapidly coordinates to a nitrogen donor on the nucleic acid. The identification of the coordinating atom as nitrogen was possible because the ^{195}Pt chemical shift is characteristic of species having one chloride and three nitrogen ligands bound to Pt(II).[67] The spectroscopic changes that accompany the formation of the family of monofunctional adducts are shown in Figure 9.8. Subsequent hydrolysis

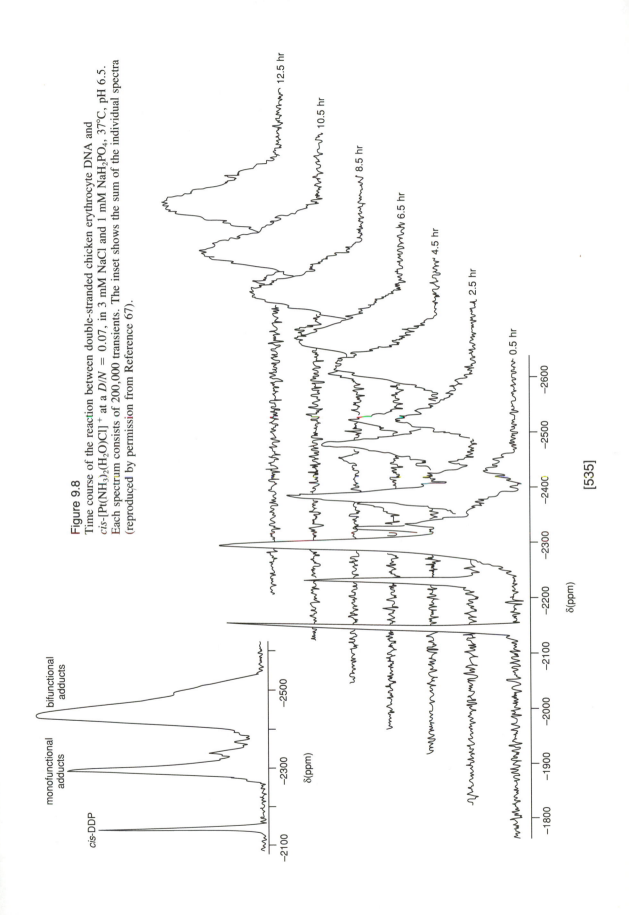

Figure 9.8

Time course of the reaction between double-stranded chicken erythrocyte DNA and cis-[Pt(NH$_3$)$_2$(H$_2$O)Cl]$^+$ at a D/N = 0.07, in 3 mM NaCl and 1 mM NaH$_2$PO$_4$, 37°C, pH 6.5. Each spectrum consists of 200,000 transients. The inset shows the sum of the individual spectra (reproduced by permission from Reference 67).

[535]

of the second chloride ion leads to the formation of a second bond with DNA. This sequence of events affords bifunctional adducts and is similarly accompanied by discrete ^{195}Pt spectral changes (Figure 9.8). From the ^{195}Pt chemical-shift range of the final products, it was apparent that the cis-$\{Pt(NH_3)_2\}^{2+}$ moiety is bound primarily to two nitrogen donors on the nucleic acid. This chemistry is summarized in Equation (9.10), together with the half-lives for the mono-

$$[Pt(NH_3)_2Cl_2] \xrightarrow{-Cl^-} [Pt(NH_3)_2Cl(H_2O)]^+ \xrightarrow{DNA}$$
cis- or trans-DDP

$$[Pt(NH_3)_2Cl]\text{-}DNA \xrightarrow{k_{obs}} [Pt(NH_3)_2]\text{=}DNA$$
monofunctional adducts   bifunctional adducts

$$\xrightarrow{-Cl}$$

$$[Pt(NH_3)_2(H_2O)]\text{-}DNA$$
$t_{1/2}$: cis-DDP, 1.9 h; trans-DDP, 2.0 h

(9.10)

functional adducts. The half-lives were calculated from a kinetic analysis of the time-dependence of the ^{195}Pt spectral changes. As can be seen, the rates of closure of mono- to bifunctional adducts for the two isomers are quite similar, suggesting that their different biological properties are not a consequence of the kinetics of binding to DNA.

The next logical question to address is what donor atoms on DNA are coordinating to platinum in the mono- and bifunctional adducts. This important issue is discussed in considerable detail in Sections V.D.4 and V.D.5. As will be shown, the N7 atoms of the purine bases adenine and guanine are the principal binding sites. Alkylation of DNA at these positions facilitates depurination. Platinum binding to N7 atoms of purines (Figure 9.9), however, stabilizes the glycosidic (N9-C1′) linkage.[68–70] Presumably the positive charge is better distributed over the platinum atom and its ligands in the adduct than over a purine alkylated at N7. On the other hand, platinum binding to N7 of guanine does perturb the charge distribution in the purine ring, as evidenced by the lowering of the pK_a of N1 by ≈ 2 units from its value in the unplatinated nucleotide (usually from $pK_a \approx 10$ to $pK_a \approx 8$).[71,72] This effect has been used to assign platinum binding sites in DNA fragments, as discussed below.

What are the chemical changes at the platinum center when cis-DDP binds to DNA? Both chloride ions are lost from the coordination sphere, as already indicated. Platinum EXAFS studies of calf-thymus DNA modified with cis-DDP revealed no chlorine backscattering features characteristic of Pt-Cl bonds.[73] The spectra were consistent with the presence of four Pt-N/O linkages, since the technique is unable to distinguish between the two low-Z elements oxygen and nitrogen. Various studies reveal that, under most circumstances, the NH_3 ligands are not lost from DNA upon the binding of platinum ammine halides. For example, when ^{14}C-labeled cis-$[Pt(NH_2CH_3)_2Cl_2]$ was allowed to bind to T7 (47 percent GC content) or M. luteus (73 percent GC content) DNA, no loss of radiolabel was found to accompany platinum binding.[74] In vivo, however, loss of amine ligands has been observed. Injection of ^{195m}Pt and ^{14}C doubly labeled

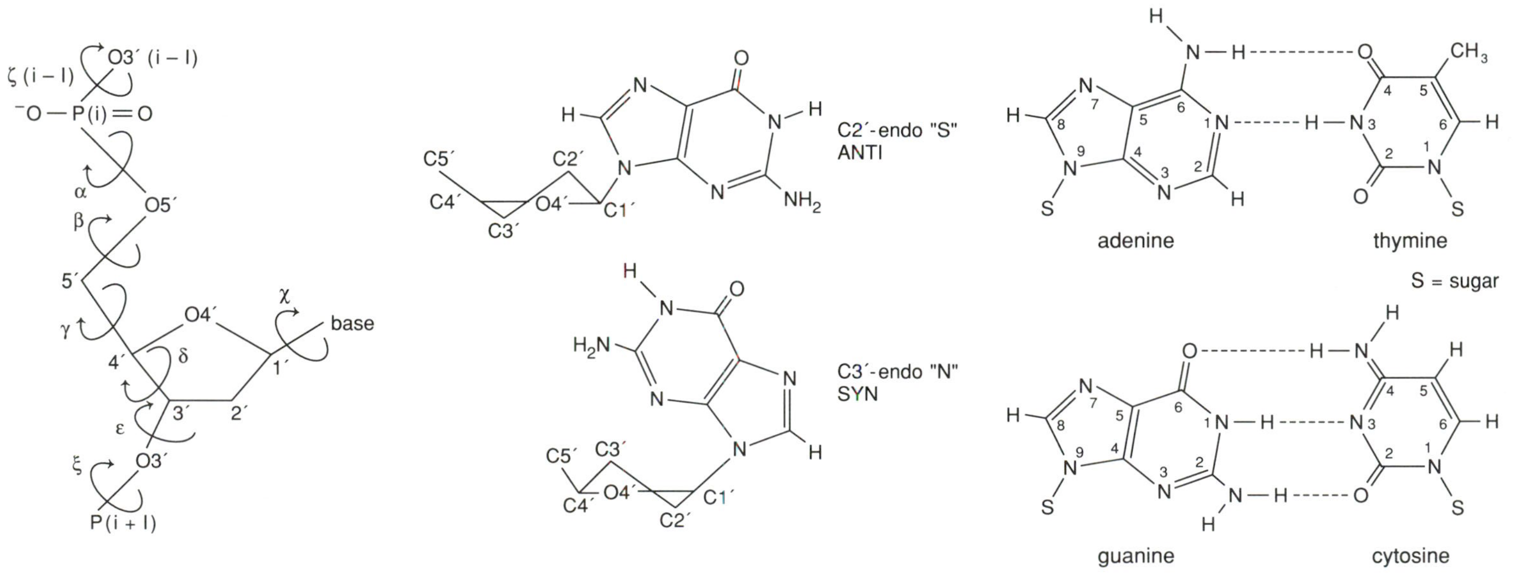

Figure 9.9
Structural components of DNA. The left panel shows a schematic of the DNA backbone, including the deoxyribose ring numbering scheme and torsion angles. The center panel gives the two main classes of sugar pucker and base orientations. The right panel gives the bases in their Watson-Crick, A-T, and G-C pairs and the base numbering scheme, viewed down the helix axis, with the major grooves pointing toward the top of the figure and the minor grooves toward the bottom.

[537]

[Pt(en)Cl$_2$] into tumor-bearing mice resulted in unequal distribution of the two labels in various biochemical fractions, but there is no reason to believe that this result is relevant to the antitumor mechanism.[75] Metabolic inactivation of the drug could occur in a variety of ways unrelated to anticancer activity. The best evidence that ammine loss does not occur at the critical biological target of cisplatin is the finding, by using antibodies specific for *cis*-{Pt(NH$_3$)$_2$}$^{2+}$ nucleotide complexes (see Section V.D.4.c), that DNA, extracted from cells in culture or from human cancer patients treated with *cis*-DDP and subsequently degraded, contains intact Pt-NH$_3$ linkages.[76]

Once bonds are made between Pt and its targets on DNA, they are relatively inert kinetically. Platinum-DNA complexes can be subjected to various physical methods of separation and purification, including gel electrophoresis, ethanol precipitation, centrifugation, and chromatography, as well as to enzymatic and even chemical degradation procedures that digest the DNA, without releasing the platinum. Platinum can be removed, however, either by use of cyanide ion, to form the very stable ($K \sim 10^{41}$) [Pt(CN)$_4$]$^{2-}$ complex, or by excess thiourea.[77,78] These properties have proved to be extremely valuable in facilitating localization and characterization of the major *cis*- and *trans*-DDP binding sites on DNA.

Although Pt-DNA linkages are, generally speaking, kinetically inert, sometimes a particular adduct will rearrange into a more stable linkage isomer. One interesting example is the product of the reaction of *trans*-DDP with the dodecanucleotide 5'-d(TCTACGCGTTCT).[79] Initially the platinum coordinates to the two guanosine residues, forming a *trans*-[Pt(NH$_3$)$_2${d(GCG)}] 1,3-intrastrand crosslink. This complex rearranges to a more stable *trans*-[Pt(NH$_3$)$_2${d(CGCG)}] 1,4-intrastrand crosslink with a half-life of 129 h at 30°C or 3.6 h at 62°C. In this rearrangement product the platinum is coordinated to a cytosine and a guanosine residue.

As just described, the binding of bifunctional platinum complexes to DNA proceeds in a stepwise fashion. The second step is sufficiently slow (a few hours), however, that various reagents such as NH$_3$, nucleobases, and low concentrations of thiourea can coordinate to the fourth site and trap the monoadducts. Generally speaking, however, given sufficient time both *cis*- and *trans*-DDP will bind DNA in a bifunctional manner. As such, they bear some resemblance to organic alkylating agents, such as the nitrogen mustards, which have been employed as anticancer agents.[80]

b. Crosslinking Reactions of Platinum Complexes   There are three broad classes of DNA adducts that can be made by bifunctional platinum complexes. As illustrated for *cis*-DDP in Figure 9.10, they are DNA-protein crosslinks, interstrand DNA-DNA crosslinks, and intrastrand crosslinks.[81] A fourth possibility for platinum complexes is bidentate chelate ring formation utilizing two donor atoms on a nucleotide. For many years, a favored such postulated mode of binding was chelation by the N7-O6 positions of the guanine base (Figure 9.9), since this structure could be formed only by *cis*- and not by *trans*-DDP.[82,83]

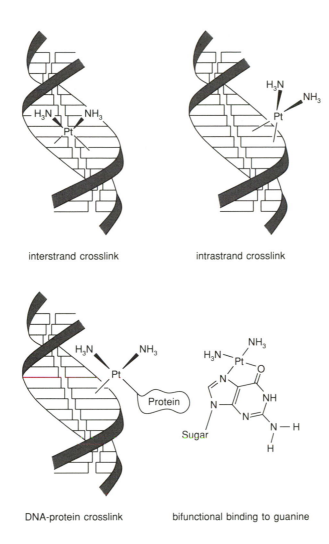

interstrand crosslink

intrastrand crosslink

DNA-protein crosslink

bifunctional binding to guanine

Figure 9.10
Possible bifunctional binding modes of *cis*-DDP with DNA
(reproduced by permission from Reference 81).

Such a structure has never been observed for *cis*-DDP binding to DNA, however. DNA-protein and interstrand crosslinks formed by platinum complexes have been the focus of many attempts to explain cytotoxicity and antitumor behavior.[84] The technique of alkaline elution, in which crosslinked DNA-DNA strands or DNA-protein molecules bind to filter paper following denaturation under basic conditions, sensitively and easily reveals such adducts. *trans*-DDP forms such adducts more rapidly than the *cis* isomer, perhaps because of its faster chloride-ion hydrolysis rates (see above) and a more favorable geometry, but they also seem to be repaired more rapidly in cells. As will be shown, interstrand and DNA-protein crosslinks are a small minority of adducts formed by cisplatin, and their contribution to the cytotoxic and anticancer properties of the drug remains to be established. In studies of SV40 replication *in vivo*, DNA-protein crosslinking by *cis*- and *trans*-DDP was shown not to be correlated with the inhibition of DNA replication.[85]

What proteins form crosslinks to DNA? One possibility is the histones that make up the spools around which DNA is wound when packaged into chromatin in the nucleus. Studies[86] of *cis*- and *trans*-DDP binding to nucleosome core particles (each particle made up of eight histone proteins; around each particle is wound a 146-bp piece of DNA in a shallow superhelix of 1.75 turns) revealed the DNA binding to be little affected by the protein core. Both DNA-protein and specific histone crosslinked species were observed; from the latter it was suggested that DDP complexes might be useful crosslinking probes of biological structures. Other proteins likely to form crosslinks to DNA in the presence of platinum complexes are DNA-processing enzymes, or enzymes requiring a DNA template for normal function. In the *in vivo* SV40 study, for example, T antigen was one of the proteins found to be crosslinked to SV40 DNA by cisplatin.[85] Other nuclear proteins such as the high-mobility group (HMG) class are also crosslinked to DNA in the presence of *cis*-DDP. In all cases so far, DNA-protein crosslinking has occurred when platinum was added to cells. There is as yet no evidence that transfection of platinated DNA into cells results in such crosslinking or that crosslinks form during *in vitro* enzymatic digestions of platinated DNAs.

c. Physical Effects of Platinum-DNA Binding

(i). Unwinding, shortening, and bending of the double helix   Early studies of *cis*- and *trans*-DDP binding to DNA employed closed and nicked circular plasmids.[87] As was described in more detail in Chapter 8, closed circular DNAs are topologically constrained such that any change in the number of helical turns must result in an equal and opposite number of superhelical turns. Any reagent that unwinds the double helix reduces the number of helical turns. Consider, for example, a stretch of DNA that is 360 base pairs (bp) long. Normal B-DNA has ≈ 10.5 bp per turn or a helical winding angle of $\approx 34.3°$ per bp. Suppose the DNA is unwound, so that there are now 12 bp per turn or a winding angle of $\approx 30°$. Instead of 34.3 helical turns ($360 \div 10.5$), the DNA now has only 30 ($360 \div 12$). If this DNA molecule were in the form of a covalently closed circle, the helical unwinding of -4.3 turns would be accompanied by a super-helical winding of $+4.3$ turns.

Planar organic dyes such as ethidium bromide (EtdBr) and inorganic complexes such as [Pt(terpy)(HET)]$^+$ (Figure 9.11) bind to DNA by intercalation, inserting between the base pairs and unwinding the double helix by $\sim 26°$ per molecule bound (Figure 9.12).[88] This unwinding can be measured by monitoring changes in the superhelicity of closed circular DNA. This kind of DNA is subjected to certain topological constraints that lead to the formation of supercoils and superhelical winding that dramatically alter the hydrodynamic properties of the DNA. Either gel electrophoresis or analytical ultracentrifugation can be used to measure this phenomenon. The platinum complexes *cis*- and *trans*-DDP also

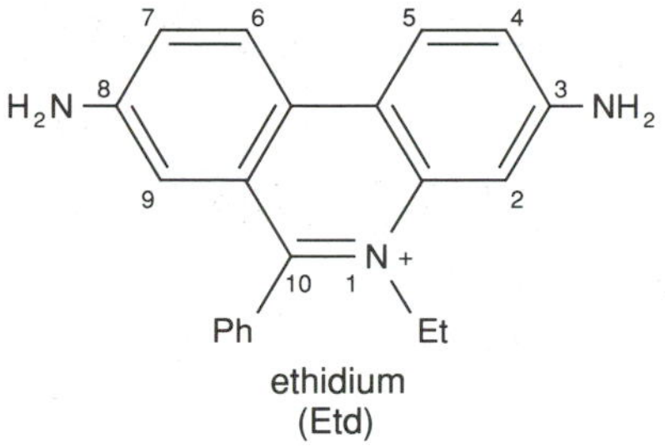

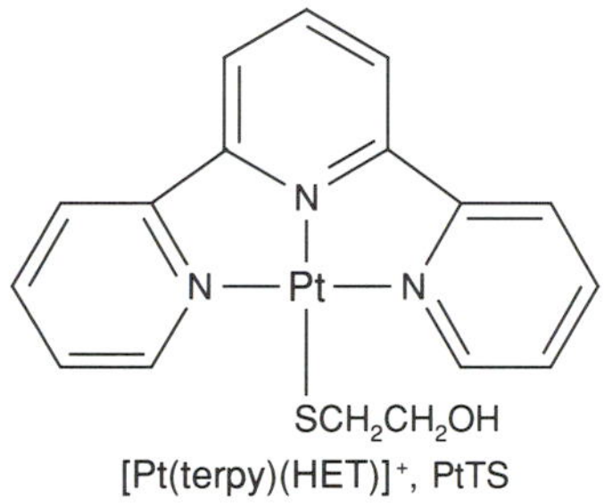

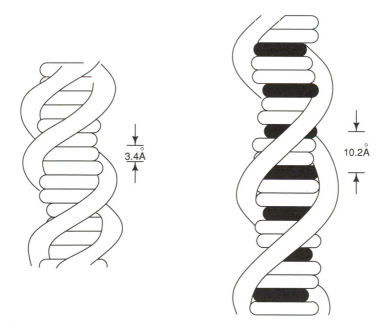

Figure 9.11
Organic (top) and inorganic (bottom) intercalators.

Figure 9.12
Schematic representation of double-stranded DNA without (left)
and with (shaded area, right) a bound intercalator (reproduced
with permission from Reference 51).

542

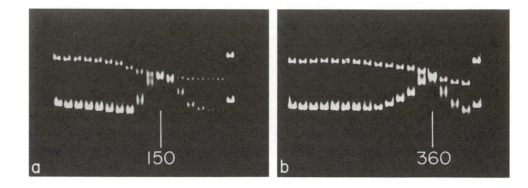

Figure 9.13
Electrophoresis in 1 percent agarose gels of nicked and closed circular PSM1 DNA incubated with (a) *cis*- and (b) *trans*-[Pt(NH$_3$)$_2$Cl$_2$] as function of time. After electrophoresis, gels were stained with ethidium bromide. Reproduced with permission from Reference 51.

produce changes in the superhelix density when bound to closed circular DNA.[87] As shown in Figure 9.13, increasing concentrations of platinum bound per nucleotide on the DNA first retard its mobility and then increase its mobility through the gel. These interesting alterations in gel mobility occur because the negatively coiled superhelix unwinds first into an open, or untwisted, form and then into a positively supercoiled form. The conformational changes, which are depicted in Figure 9.14, are directly proportional to the drug-per-nucleotide, or $(D/N)_b$, ratio. In addition to superhelical winding, both platinum complexes increase the mobility of nicked circular DNA in the gels (Figure 9.13). Nicked DNA has one or more breaks in the sugar-phosphate backbone, which relieve the topological constraint and prohibit the DNA from twisting into superhelical structures.

What could be the cause of these physical changes in the DNA structure upon *cis*- or *trans*-DDP binding? Intercalation can be excluded, not only because the compounds do not have the aromatic character normally associated with intercalators (Figure 9.11), but also through studies of the manner by which these and other platinum complexes inhibit the intercalative binding of EtdBr to DNA.[89,90] Platinum metallointercalators such as [Pt(terpy)(HET)]$^+$ are competitive inhibitors of EtdBr binding, as measured by fluorescence Scatchard plots, whereas the non-intercalators *cis*- and *trans*-DDP are not. Moreover, intercalation tends to lengthen and stiffen the double helix, whereas the mobility changes of nicked circular DNAs upon binding of *cis*- or *trans*-DDP were shown by electron microscopy experiments to arise from a pronounced shortening of the DNA with increased Pt binding.

One manner by which *cis*- or *trans*-DDP might produce these physical alterations in DNA structure is by kinking the double helix at or near the binding site. Such an effect could be produced by the bidentate attachment of platinum; the monofunctional [Pt(dien)Cl]$^+$ complex does not have these pronounced ef-

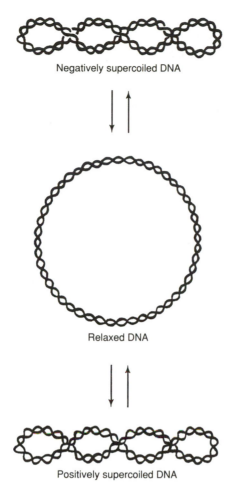

Negatively supercoiled DNA

Relaxed DNA

Positively supercoiled DNA

Figure 9.14
Topological forms of closed circular DNA.

fects on DNA secondary structure.[91] Recently, it has been demonstrated that *cis*-DDP binding to DNA does indeed produce a pronounced bend in the helix axis.[92,93] The proof employed a gel-electrophoretic method of analysis that had previously been used to study DNA bending at naturally occurring specific sequences called A-tracts, consisting of five or six adenosine nucleosides in a row followed by about the same number of thymidine residues.[94] When these $d(A_5T_5)_2$ sequences are positioned in the center of a DNA restriction fragment of, say, 150 bp, the mobility of the DNA through polyacrylamide electrophoresis gels is greatly retarded compared to that of a similar DNA fragment where the A-tract is at the end. For the former fragment, the bent molecules presumably cannot snake their way through the pores of the polyacrylamide as well as the molecules whose bends are at the ends and have little effect on the linear structure. It was further shown that A-tracts bend the duplex toward the minor groove of the DNA. Moreover, in a DNA containing multiple A-tracts, the bends must be separated by integral numbers of helical turns ($\sim$ 10.5 bp) or else the effect will cancel and the gel mobility will be that of normal DNA of similar length. This latter phenomenon has been referred to as phasing.

With this background information in mind, we can now discuss the experiments with *cis*-DDP that demonstrated bending.[92,93] By methods described in Section V.D.8, a 22-bp oligonucleotide (22-mer) containing self-complementary overhanging ends (''sticky ends'') was synthesized with a single *cis*-diammineplatinum(II) moiety linking adjacent guanosine residues (Figure 9.15A). A 22-mer was chosen since it has approximately two helical turns, accounting for some platinum-induced unwinding, and will thus have phased bends when polymerized. This platinated DNA was then labeled with ^{32}P and treated with the enzyme DNA ligase, which seals the ends, producing oligomers of the 22-mer having lengths 22, 44, 66, 88, 110, etc., bp. In these oligomers, the platinum atoms are spaced apart approximately by integral numbers of helical turns. As shown in Figure 9.15, studies of this family of oligomers by gel electrophoresis revealed a pronounced retardation compared to the mobility of unplatinated DNA oligomers of comparable size (line P22 in Figure 9.15B). The plots in this figure show the relative mobilities (R_L) of the different length multimers, compared to a control in which the top strand is not platinated, as a function of the length in base pairs. From the resulting curves may be extracted the extent of cooperative bending. When oligomers of a platinated DNA fragment in which the metal atoms were spaced apart by 27 bp were examined, their relative mobilities were found to be nearly the same as unplatinated control molecules (line P27 in Figure 9.15B). These experiments unequivocally established that platinum kinks the double helix. As with A-tract-induced bends, the platinum atoms must be phased in order to induce cooperative bending. Comparison of the magnitudes of the gel mobility changes made it evident that *cis*-{Pt(NH$_3$)$_2$}$^{2+}$ binding produces a bend comparable to that of two A-tracts, $\approx$ 34°.

In a related series of experiments,[93] the platinated 22-mer was copolymerized with various A-tract-containing 11-mers to produce ladders of oligomers in which the phasing of Pt with respect to the center of the A-tract was varied, but the Pt atoms were always in phase. The results of these studies showed that maximum gel-mobility retardation occurred when the Pt and A-tract center were spaced apart by half-integral numbers of helical turns (Figure 9.15C). Since A-tracts bend the DNA into the minor groove, this result implies that platinum bends the DNA into the major groove. Only when phased by $n/2$ (n = integer) turns will copolymers of species situated alternatively in the major and minor grooves of DNA exhibit such cooperative bending. It will be shown later that helix bending of *cis*-DDP-DNA adducts into the major groove is in accord with their known structures.

The ability to prepare site-specifically platinated oligonucleotides (see Section V.D.8) has provided a means for measuring the extent to which *cis*-DDP produces local unwinding of the double helix.[95] When the platinum atoms are positioned with respect to one another, or phased, by exactly integral numbers

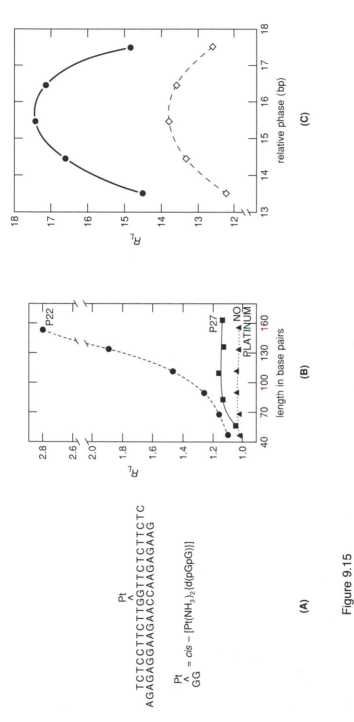

(A)

```
              Pt
              ∧
TCTCCTTCTTGGTTCTCTTCTC
AGAGAGGAAGAACCAAGAGAAG
```

$$\underset{GG}{\overset{Pt}{\wedge}} = cis - [Pt(NH_3)_2\{d(pGpG)\}]$$

(B)

(C)

Figure 9.15

Experiment to demonstrate that cis-[Pt(NH$_3$)$_2$\{d(pGpG)\}] intrastrand crosslinks bend duplex DNA by ≈34°. Panel A shows the platinated 22-mer sequence, panel B the effect of platination on the gel-electrophoresis mobility of the 22-mer (P22) and a control 27-mer (P27) oligomers, and panel C the mobility of copolymerized DNAs containing cis-DDP and A-tract induced bends (●, 128-bp Pt + A-tract DNA; ◇, 96-bp Pt + A-tract DNA) that maximize at approximately half-integral helical turns corresponding to their phasing. For more detail consult Reference 93.

[545]

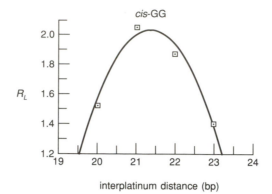

Figure 9.16
Plot showing the relative mobility (R_L) versus actual length curves
for the oligomers *cis*-GG-20, *cis*-GG-21, *cis*-GG-22, and *cis*-GG-23,
denoted as 20, 21, 22, and 23, respectively.

of helical turns, the retardation of the DNA multimers in the gel is maximized. This phenomenon is illustrated in Figure 9.16, where the R_L values are plotted as a function of the interplatinum spacing for oligonucleotides containing the *cis*-{Pt(NH$_3$)$_2$d(GpG)} intrastrand crosslink. When the resulting curve was analyzed, the maximum was found to occur at 21.38 bp. Since normal B-DNA has a helical repeat of 10.5 bp, one can compute the effect of platination from the expression [(21.38 − 2(10.5)] bp = 0.38 bp. From the fact that one helical turn of DNA comprises 360° and 10.5 bp, the unwinding of the DNA double helix due to the presence of a single *cis*-{Pt(NH$_3$)$_2$d(GpG)} intrastrand crosslink can be calculated as

$$(0.38/10.5) \times 360 = 13°.$$

Similar studies of DNA platinated with *trans*-DDP have been carried out. In these, oligonucleotides containing the 1,3-*trans*-{Pt(NH$_3$)$_2$d(GpNpG)} intrastrand crosslink were examined. The electrophoresis gels of polymerized 15-mers and 22-mers containing this adduct showed cooperative bending. This result indicates that bends at the sites of platination by *trans*-DDP are not phase sensitive, and has been interpreted to imply the formation of a "hinge joint" at these positions.[92,95] The directed bends and local unwinding of DNA produced by cisplatin could be an important structural element that triggers a response by cellular proteins. This subject is discussed in greater detail in Section V.D.7.d.

d. Biological Consequences of Platinum-DNA Binding

(i). Inhibition of replication   Binding of *cis*-DDP to DNA inhibits replication both *in vivo* and *in vitro*, as shown by a variety of assays. Inhibition of replication of SV40 viral DNA in African green monkey cells as a function of the concentration of added *cis*-DDP is shown in Figure 9.17. When SV40 virus infects monkey cells, it does not integrate its DNA into the genome of the host.

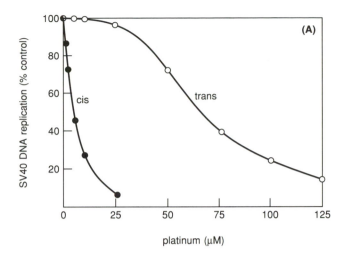

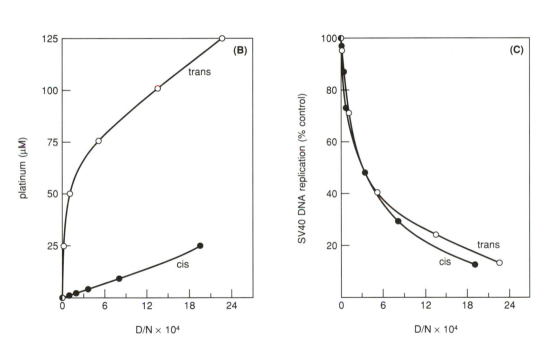

Figure 9.17

SV40 DNA replication in CV-1 cells as a function of platinum concentration in the medium (panel A) or D/N (panel C). In panel B, D/N is plotted as a function of platinum concentration in the medium. SV40-infected cells were treated with *cis*-DDP (●) or *trans*-DDP (o) at the indicated concentrations for 40 h. SV40 DNA replication relative to control (untreated) cells was measured by incorporation of [^{3}H]thymidine, added after the first 24 h of platinum treatment, and Pt in isolated SV40 chromosomes was measured by AAS. The data shown are from a representative experiment. Experiments were carried out in quadruplicate. Reproduced by permission from Reference 85.

Instead, it forms its own chromosomes in the cell nucleus. These so-called mini-chromosomes consist of $\approx$ 20 nucleosomes, fundamental chromosome building blocks. SV40 has its own life cycle, using virally encoded and cellular proteins to replicate and, ultimately, reassemble virus particles before lysing the cell and departing to infect neighboring cells.

In the experiment shown in Figure 9.17, the SV40-infected cells were treated with cisplatin. After 24 h, [^{3}H]thymidine was added and, after 24 more hours, the cells were harvested, and SV40 DNA was isolated; the amount of DNA synthesis was recorded by comparing incorporated radiolabel with results from control experiments where no platinum was present. The data show that, when 25 μM platinum was present, SV40 DNA replication was reduced to about 5 percent of control. Quantitation reveals that, at $\approx$2 platinum atoms bound per thousand nucleotides (drug-per-nucleotide, or $(D/N)_b$, = 0.002), synthesis is only 10 percent that of control.

Recently, a related series of experiments has been carried out that can monitor DNA synthesis from templates platinated *in vitro*.[96] In this work, DNA plasmids containing the SV40 origin of replication are added to cellular extracts prepared from human kidney cells previously infected with adenovirus. In the presence of large T antigen, a virally encoded protein required for replication, SV40 DNA is synthesized from the plasmid templates. Synthesis can be conveniently monitored by [^{32}P]dATP incorporation. At a $(D/N)_b$ ratio of only 1.7×10^{-3}, DNA synthesis is about 5 percent of control, in agreement with the results of the *in vivo* study.

The binding of *cis*-DDP to DNA has also been measured for normal and tumor cells implanted in nude mice and in cells obtained from the ascites fluid of patients with ovarian carcinoma 24 h after their last dose.[97] The data for mouse bone marrow and a human pancreatic tumor xenograft show that, at a dose of 10 mg/kg, $(D/N)_b$ platinum binding levels of 3.3×10^{-6} and 1.82×10^{-6} reduce survival to 20 and 10 percent of control, respectively. These ratios are in good accord with platinum levels required to inhibit DNA synthesis in mammalian cells, as revealed by various studies, but substantially less than that needed for replication inhibition in the SV40 experiments described above. The difference can be readily explained, illustrating an important point. The SV40 genome, like most other DNAs of viral or plasmid origin, consists of only 15,000 nucleotides whereas the nuclear DNA of mammals has about 10^9 nucleotides. Thus, $(D/N)_b$ levels of $\sim 10^{-6}$ would leave 99 out of 100 SV40 DNA molecules with no platinum at all, and replication would hardly be affected. For the mammalian genome, $(D/N)_b$ values of 10^{-6} place 10^3 platinum atoms on each DNA genome, sufficient to inhibit replication and reduce cell survival. Platinum-DNA binding levels of this magnitude are found for ovarian ascites cells taken from patients receiving cisplatin chemotherapy.[97]

(ii). Mutagenesis and repair Apart from inhibition of DNA synthesis, what are the other biological consequences of cisplatin binding to DNA? One such consequence is mutagenesis, in which a normal base in the sequence is replaced

by a different base. This phenomenon has been demonstrated for *cis*-DDP-treated cells in a variety of studies. What brings about such mutagenesis? There are several possibilities. One is that errors are introduced in DNA strands during attempts of the replication apparatus to synthesize past a platinum lesion. Another is that the platinum-damaged DNA is recognized by cellular repair systems that, in attempting to eliminate the platinated stretch of DNA, incorporate one or more incorrect nucleotides. Platinum-induced mutagenesis can lead to deleterious long-term health problems in patients treated with cisplatin. It is therefore important to understand the mechanism by which cellular DNA becomes mutated following platination, and to devise strategies for minimizing or eliminating this mutagenesis.

The foregoing considerations bring up another biological consequence of *cis*-DDP binding, namely, DNA repair. Removal of platinum from DNA by cellular repair mechanisms has been demonstrated by several groups. For example, in studying *cis*-DDP-treated human fibroblast cells in culture, it was found that the amount of bound platinum per nucleotide decreased according to first-order kinetics, from $(D/N)_b$ of 2.3×10^{-5} to 3.3×10^{-6} over a six-day period. Since Pt-DNA adducts are stable with respect to dissociation from DNA under physiological conditions (see above discussion), loss of platinum was attributed to DNA repair.[98]

How does the cell remove platinum from DNA? One mechanism is by a process known as excision repair, whereby the sugar-phosphate backbone on the platinated strand is hydrolyzed (''nicked'') on either side of the damage and the remaining, unplatinated strand is used as a template for new DNA synthesis. The platinated oligonucleotide is displaced and the resulting gap filled in. In support of this picture is the fact that, in xeroderma pigmentosum (XP) human fibroblast cells, known to be deficient in excision repair, there is very little removal of platinum during post-treatment incubation.[99] Recent studies of *in vitro* repair of cisplatin-DNA adducts by a defined enzyme system, the ABC excision nuclease of *E. coli*, have provided some details at the molecular level about the process.[100,101] [^{32}P]-Labeled double-stranded DNA fragments containing $\{Pt(NH_3)_2\}^{2+}$ or $\{Pt(en)\}^{2+}$ adducts at random or defined sites were incubated with the enzyme. Cleavage of the platinated strand occurred at the 8th phosphodiester bond 5′, and the 4th phosphodiester bond 3′, to the GG or AG intrastrand crosslink. Further details about the identification and construction of such specific crosslinks will be given later in this chapter.

(iii). Drug resistance   Another biological consequence of DNA-platinum interactions, probably related to the repair phenomenon, is resistance. Resistance of a cell to a chemotherapeutic agent, which can be inherent or acquired, is a phenotypical ability of the cell to tolerate doses of a drug that would be toxic to normal, or parent, cells.[102] Resistance is often acquired by prolonged exposure of cells in culture to the drug or, in patients, to repeated doses of drug therapy. There is not yet any direct proof that platinum-DNA interactions are responsible for acquired resistance to cisplatin. Studies of sensitive and resistant

tumors in rats have shown, however, that after intravenous injection of 10 mg/kg of the drug, the platinum levels were the same after an hour, but after 24 hours a larger proportion of adducts had been removed in the resistant cells.[103] Similar results have been found for studies of Pt-DNA adducts in cultured L1210 cells of varying levels of resistance to cisplatin where, in the 18 h following a 6-hour incubation with the drug, the resistant cells had up to fourfold more platinum removed than the sensitive cells.[104]

Experiments have also been carried out showing that cis-DDP binding to DNA inhibits transcription, the formation of RNA from a gene, and that this phenomenon is less efficiently reversed for parent versus resistant L1210 cells in culture.[105] The assay involves transfection (the process whereby free or viral DNA or RNA is taken into a cell) of pRSVcat plasmid DNA into L1210 cells. The plasmid contains the bacterial cat gene in a position that permits its expression in mammalian cells. The cat gene encodes the enzyme chloroamphenicol acyltransferase (CAT), an activity readily measured following lysis of the cells. Transfection of the cis-DDP-damaged plasmid into resistant L1210 cells showed that up to eight times the amount of platinum was required in the resistant versus sensitive cells to produce a mean lethal hit (63 percent reduction in activity). This result is consistent with greater repair of platinum-DNA adducts in the resistant cells.

These results should not be construed to mean that DNA repair is the only mechanism of cisplatin resistance. There is evidence that relative amounts of glutathione are increased in cisplatin-resistant cells.[106] Glutathione presumably uses its thiol moiety to coordinate platinum and diminish the amount that can bind to DNA. Reduced influx or increased efflux of a drug constitutes additional mechanisms by which cells become resistant. Further studies are required to ascertain which of these possibilities is most important for the cisplatin resistance phenomenon.

The discovery that cells can become resistant to cisplatin by repairing DNA lesions suggests a way to explain the selectivity of the drug for certain tumor tissue, and even the selective cytotoxicity of the drug for tumor versus normal cells of the same tissue. Tumor cells that cannot repair platinum-DNA adducts would be most affected by cis-DDP. This idea forms one of the central hypotheses about the molecular mechanism of action of cis-DDP, details of which can be probed by bioinorganic chemists. Specifically, it is important to inquire what DNA adducts formed by cis-DDP are both cytotoxic and repairable, what enzymes are responsible for such repair in mammalian cells, by what mechanisms these enzymes operate, and how this knowledge can be used to design better metal-based antitumor drugs and chemotherapeutic protocols.

(iv). DNA-protein interactions   Most of the phenomena discussed in this section, inhibition of replication, DNA repair, drug resistance, and mutagenesis, probably involve interaction of a protein or group of proteins with platinated DNA. These interactions are clearly important in determining the biological consequences of DNA templates containing bound platinum. Very recent experiments have uncovered the existence of proteins from a variety of mammalian sources that bind specifically to DNA platinated with *cis*- but not *trans*-DDP.[107] Identifying the nature and function of these factors may provide important clues about the mechanisms of antitumor activity, drug resistance, or repair. Study of protein-DNA-drug interactions is an essential feature of the bioinorganic chemistry of platinum chemotherapeutic agents.

4. Mapping the major adducts of *cis*- and *trans*-DDP on DNA; sequence specificity

As we have seen, the antitumor activity of cisplatin is most likely the result of its DNA-binding properties. But what are the adducts? The human genome has more than a billion nucleotides. Does platinum recognize any special regions of the DNA or any particular sequences? In other words, is binding simply random or is there at least a regioselectivity? In this section, we discuss the best strategies for answering these questions, strategies that evolved in pursuit of learning how *cis*-DDP binds to DNA. We also illustrate their power in elucidating the DNA-binding properties of other metal complexes of interest to bioinorganic chemists.

a. Early Strategic Approaches   The first experiments to imply the sequence preferences of *cis*-DDP binding to DNA employed synthetic polymers.[108,109] Specifically, the buoyant density of poly(dG)·poly(dC), poly(dG·dC), and their *cis*-DDP adducts was studied in the analytical ultracentrifuge. The greatest shift in buoyant density was seen for the platinum adducts of poly(dG)·poly(dC), from which it was concluded that platinum forms an intrastrand crosslink between two neighboring guanosine nucleosides on the same strand. This interpretation was suggested by the known preference of metal ions, and especially platinum, for binding at the N7 position on the guanine base (Figure 9.9), information available from model studies of metal-nucleobase chemistry. Although other interpretations of the buoyant-density shift were possible, especially since the amount of platinum bound was not quantitated, the conclusion proved to be correct, as confirmed by later investigations. Interestingly, *trans*-DDP did not selectively increase the buoyant density of poly(dG)·poly(dC).

Following these initial experiments, the regioselectivity of *cis*-DDP binding was investigated by studying the inhibition of enzymatic digestion of platinated DNA. For example, the platinum complex inhibits the cleavage of DNA by restriction enzymes that recognize specific sequences and cut both strands of the double helix.[110] The resulting fragments are readily identified on electrophoresis gels. One such restriction enzyme is Bam HI. As shown by the arrows in Scheme (9.11), Bam HI cleaves a six-bp palindromic sequence at the phosphodiester bonds between two guanosine nucleosides. Formation of an intrastrand crosslink between the two adjacent guanosine nucleosides inhibits digestion by the enzyme. Another method, termed exonuclease mapping, involves digestion of the strands of duplex DNA from its 3'-ends.[111,112] When the enzyme encounters a bound platinum atom, it is unable to proceed further. Analysis of the digestion products by gel electrophoresis reveals the presence of discrete bands caused by the inhibition of digestion by bound platinum at specific sequences. Results from experiments of this kind were the most definitive at this time in demonstrating the profound regioselectivity of cisplatin for adjacent guanosines, and strongly supported the earlier conclusion that the drug was making an intrastrand d(GpG) crosslink.

$$
\begin{array}{cc}
5' & \overset{\downarrow}{G}\text{ G A T C C} - \quad 3' \\
& -\text{C C T A G }\underset{\uparrow}{G} - \\
3' & \qquad\qquad\quad 5'
\end{array}
\tag{9.11}
$$

A third enzymatic strategy for exploring the regioselectivity of *cis*- and *trans*-DDP binding to DNA is outlined in Figure 9.18. Platinum is first bound to a single-stranded DNA template, in this example from bacteriophage M13mp18, to which is next annealed a short, complementary oligonucleotide termed a "primer" for DNA synthesis. Addition of the large (Klenow) fragment of *E. coli* DNA polymerase I and deoxynucleoside triphosphates, one of which bears a ^{32}P label, [α-^{32}P]dATP, initiates replication. When the enzyme encounters a platinum adduct, the chain is terminated. By running out the newly synthesized DNA strands on a sequencing gel, the sites of platinum binding can be detected by comparing the positions of the radiolabeled fragments with those obtained from sequencing ladders. The results of this procedure, which has been termed "replication mapping," confirmed that *cis*-DDP binds selectively to $(dG)_n$ ($n \geq 2$) sequences. In addition, they showed that *trans*-DDP blocks replication, in a much less regioselective manner, in the vicinity of sequences of the kind d(GpNpG), where N is an intervening nucleotide. These data afforded the first clear insight into the sequence preferences for *trans*-DDP on DNA. A control experiment run with DNA platinated by the monofunctional complex [Pt(dien)Cl]$^+$ gave the interesting result that DNA synthesis was virtually unaffected.

In yet another approach to the problem, DNA containing *cis*- or *trans*-DDP adducts was electrostatically coupled to bovine serum albumin, to enhance its

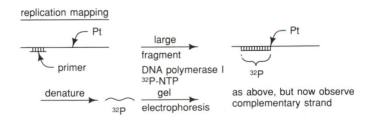

Figure 9.18
Diagram illustrating the replication mapping experiment. To a single-stranded, platinated template is annealed a short primer for DNA synthesis using DNA polymerase I (Klenow fragment) and radiolabeled nucleotides. Sites of platinum binding are revealed as bands on gel electrophoresis where chain termination occurs (see text for details).

antigenicity, and injected into rabbits.[113,114] The resulting antisera and antibodies were then studied for their ability to recognize and bind specifically to platinated DNAs having defined sequences, such as poly(dG)·poly(dC) and poly[d(GC)]·poly[d(GC)].

From experiments of this kind, the major cis-DDP adduct recognized by the antibody was found to be cis-[Pt(NH$_3$)$_2$\{d(GpG)\}], in accord with the findings of the enzymatic mapping experiments. Unplatinated DNA was not recognized, nor was DNA platinated with trans-DDP. On the other hand, the antibody recognized DNA platinated with antitumor-active compounds [Pt(en)Cl$_2$] and [Pt(DACH)(CP)], where DACH = 1,2-diaminocyclohexane and CP = 4-carboxyphthalate. This result revealed that the antibody recognized the structural change in DNA that accompanies formation of d(GpG) intrastrand crosslinks, irrespective of the diamine ligand in the coordination sphere of the platinum atom. The antibody is also capable of distinguishing adducts formed by active versus inactive platinum complexes. Most importantly, DNA isolated from the cells of mice bearing the L1210 tumor five hours after cisplatin injection, was recognized.[113,115] Subsequent studies[116] revealed that these antibodies could detect cisplatin-DNA adducts formed in the white blood cells of patients receiving platinum chemotherapy. Thus, the antibody work linked the regiospecificity of platination chemistry in vitro with that occurring in vivo and in a clinically relevant manner.

Additional studies with monoclonal antibodies generated using DNA platinated with cis- or trans-DDP further confirmed and extended these results.[117] This later work indicated that intrastrand crosslinked d(ApG) and d(GpG) sequences possess a common structural determinant produced by cis-DDP platination, and that carboplatin is also capable of inducing the same DNA structure. For trans-DDP-platinated DNA, a monoclonal antibody was obtained that appeared to have the intrastrand d(GpTpG) adduct as its major recognition site. In all these studies, the primary structural determinant appears to be DNA duplex opposite the site of platination, since fairly major stereochemical changes could be made in the amine ligands with no appreciable effect on antibody binding.

b. Degradation, Chromatographic Separation, and Quantitation of DNA Adducts Experiments in which DNA platinated with cis-DDP is degraded to chromatographically separable, well-defined adducts have been invaluable in revealing the spectrum of products formed. In a typical experiment, platinated DNA is digested with DNAse I, nuclease P1, and alkaline phosphatase. These enzymatic digestions degrade DNA into nucleosides that can be readily separated by high-performance liquid chromatography (HPLC). Detection of the adducts can be accomplished by the UV absorption of the nucleoside bases at 260 nm or, for platinum complexes containing a radioactively labeled ligand such as [^{14}C]ethylenediamine,[118] by monitoring counts. In addition to peaks corresponding to dA, dC, dG, and dT, the chromatographic trace contains additional peaks corresponding to specific platinum nucleobase adducts such as cis-[Pt(NH$_3$)$_2$(dG)$_2$]. The precise nature of these adducts was established by comparison with chemically synthesized compounds structurally characterized by NMR spectroscopy.[118-121] An alternative method for identifying the adducts employed antibodies raised against specific platinum-nucleobase complexes.[122]

This approach has revealed the relative amounts of various adducts formed by a variety of platinum complexes; selected results are summarized in Table 9.4. Usually, for cisplatin, the relative amounts of the various adducts formed varies according to the series cis-[Pt(NH$_3$)$_2${d(pGpG)}] > cis-[Pt(NH$_3$)$_2${d(pApG)}] > cis-[Pt(NH$_3$)$_2${d(GMP)}$_2$] > monofunctional adducts. Only when the total incubation time was short, less than an hour, were the monofunctional adducts more prevalent, as expected from the kinetic studies of cis-DDP binding to DNA discussed previously. It is noteworthy that no d(pGpA) adducts were detected. This result, which is consistent with information obtained by enzymatic mapping, can be understood on stereochemical grounds.[123] If the guanosine nucleoside N7 position is the most-preferred binding site on DNA, closure to make an N7,N7 intrastrand crosslink between two adjacent purine nucleotides is more feasible in the 5' direction along the helix backbone (N7···N7 distance of ≈ 3 Å) than in the 3' direction (N7···N7 distance ≈ 5 Å). In addition, molecular-me-

Table 9.4
Geometric features of the platinum coordination spheres of cis-[Pt(NH$_3$)$_2${d(pGpG)}].

Bond distances and angles[a]	Molecule 1	Molecule 2	Molecule 3	Molecule 4
Pt-N1	2.03(2)	2.01(2)	2.08(2)	2.08(2)
Pt-N2	2.03(3)	2.09(2)	2.04(3)	2.06(3)
Pt-N7A	2.01(2)	2.02(2)	1.91(3)	1.93(3)
Pt-N7B	2.05(2)	1.95(3)	2.00(3)	2.06(3)
N7A-Pt-N1	88.6(9)	90.3(9)	91.0(1)	88.4(9)
N7A-Pt-N2	179(1)	173.3(8)	178(1)	177(1)
N7A-Pt-N7B	89.1(9)	90.0(1)	85(1)	89(1)
N1-Pt-N2	92(9)	90.8(9)	91(1)	93(1)
qN1-Pt-N7B	176.5(9)	179.0(1)	173(1)	175(1)
N2-Pt-N7B	90.3(9)	89.0(1)	93(1)	89(1)

Dihedral angles[b] Molecule	3'-Gua/5'-Gua	5'-Gua/PtN$_4$	3'-Gua/PtN$_4$
1	76.2(5)	110.6(5) [3.30(3)]	86.1(5)
2	81.0(5)	110.8(5) [3.49(3)]	95.5(5)
3	86.8(6)	81.0(6)	58.0(6) [3.11(4)]
4	80.6(5)	76.6(6)	59.6(6) [3.18(4)]

[a] Bond distances are in Ångstroms and angles are in degrees.

[b] Conventions used for assigning Base/Base and Base/PtN$_4$ dihedral angles can be found in J. D. Orbell, L. G. Marzilli, and T. J. Kistenmacher, J. Am. Chem. Soc. **103** (1981), 5126. The numbers in square brackets refer to the corresponding N(ammine)···O6 distance, in Å (see text).

chanics modeling studies[124] indicate that a highly unfavorable steric clash occurs between the 6-amino group of the 3'-adenosine residue in a d(pGpA) crosslink and the platinum ammine ligand, whereas in the platinated d(pApG) sequence, the 6-oxo group forms a stabilizing hydrogen bond to this ligand. A 28 kJ mol^{-1} preference of *cis*-DDP for binding d(pApG) over d(pGpA) was calculated.

There are two likely sources of *cis*-[Pt(NH$_3$)$_2${d(GMP)}$_2$] in the spectrum of adducts. This species could arise from long-range intrastrand crosslinks, where the two coordinated guanosines are separated by one or more nucleotides. In support of this possibility is the fact that digestion of chemically synthesized *cis*-[Pt(NH$_3$)$_2${d(GpNpG)}], where N = C or A, led to *cis*-[Pt(NH$_3$)$_2${d(Gua)}-{d(GMP)}] and mononucleotides.[118,119,121] The other source of this product is interstrand crosslinked DNA, known to occur from the alkaline elution studies.

As indicated in Table 9.4, in all the experiments there was platinum that was unaccounted for in the quantitation procedures, which employed either antibodies, platinum atomic absorption spectroscopy, or a radiolabeled ethylenediamine ligand. Some of this material was assigned to oligonucleotides having high platinum content, resistant to enzymatic degradation.

Two important points emerge from the quantitation of adducts by this method. One is that intrastrand d(GpG) and d(ApG) crosslinks constitute the major adducts (>90 percent of total platination) made by cisplatin on DNA *in vivo*. Because they were identified by an antibody specific for their structures, no chemical change brought about by cellular metabolism has occurred. Secondly, the preponderance of these adducts far exceeds the frequency of adjacent guanosine or guanosine/adenosine nucleosides in DNA. This latter result implies a kinetic preference for, or recognition of, d(pGpG)- and d(pApG)-containing sequences by cisplatin.

c. Postscript: A Comment on Methodologies   With few exceptions, none of the experimental studies described in this section could have been carried out in 1969, when Rosenberg first demonstrated the anticancer activity of *cis*-DDP. The techniques of DNA sequencing, monoclonal antibody formation, oligonucleotide synthesis, HPLC, FPLC, and many of the higher resolution gel electrophoresis methodologies employed were the result of later developments driven by rapid advances in the fields of molecular biology and immunology. Future progress in elucidating the molecular mechanisms of action of cisplatin and other inorganic pharmaceuticals will no doubt benefit from new technological discoveries and inventions of this kind yet to come.

5. Structure of platinum-DNA complexes

a. NMR Studies of Platinated Oligonucleotides   Once the major spectrum of adducts formed by *cis*- and *trans*-DDP with DNA began to emerge, it was of immediate interest to learn to what positions on the nucleobases the platinum atom was coordinated. Proton NMR spectroscopy soon proved to be an invaluable tool for obtaining this information.[71,125,126] Several ribo- and deoxyribooli-gonucleotides containing GG, AG, or GNG sequences were synthesized, and allowed to react with *cis*-DDP or its hydrolysis products, and the resulting complexes were purified by chromatography. All GG-containing oligomers formed intrastrand crosslinks with the $\{Pt(NH_3)_2\}^{2+}$ moiety coordinated to the N7 atoms. This structure was deduced from several criteria. Most frequently studied were the nonexchangeable base protons H8 and H2 of adenine, H6 of thymine, H8 of guanine, and H5 and H6 of cytosine (Figure 9.9). Coordination of platinum to N7 of guanine causes a downfield shift of the H8 proton resonance. More importantly, however, it also lowers the pK_a of the N1 proton by ~ 2 units, because platination adds positive charge to the base. Thus, titration of the platinated oligonucleotide over a pH range, and comparison of the results to those obtained for the unplatinated oligomer, reveals a difference in the midpoint of the transition in chemical shift of the H8 proton by ~ 2 pH units if coordination occurs at N7. This effect is illustrated in Figure 9.19 for the adduct *cis*-[Pt(NH₃)₂{d(ApGpGpCpCpT)}N7-G(2),N7-G(3)], where the pK_a of N1 is seen to shift from ~ 10 to ~ 8 upon platination.[71] The pH titration in this example also reveals the pH-dependent chemical shift of the cytosine ¹H resonances at a pH of ~ 4.5, corresponding to protonation of the N3 atoms. The protonation of adenine N7 ($pK_a \sim 4$) is also frequently observed in these studies. These results conclusively demonstrate platinum coordination at N7 of the two guanosine nucleosides.

Although several of the oligonucleotides studied have self-complementary sequences, such that they can form a double helix when unplatinated, in no such case was a duplex observed for their platinated forms. The presence of the platinum-induced crosslink presumably decreases the stability of the double-stranded form of the oligonucleotide. Another interesting result is that all intra-strand $\{Pt(NH_3)_2\}^{2+}$ adducts of d(GpG) or d(ApG) have an altered deoxyribose-sugar ring conformation. In normal, unplatinated form, these single-stranded or duplex oligonucleotides have a C2'-endo sugar pucker (Figure 9.9). Upon platination, the 5'-nucleotide switches to C3'-endo. This change is readily monitored by the ring proton-coupling constants $J_{H1'-H2'}$ and $J_{H1'-H2''}$. These protons constitute an ABX spin system such that the sum, $\Sigma^3J = {}^3J_{1'2'} + {}^3J_{1'2''}$, is most easily measured as the separation between the outermost peaks in the multiplet. For the C2'-endo conformation, a pseudotriplet occurs with $\Sigma^3J = 13.6$ Hz, and for C3'-endo, $\Sigma^3J = 7.5$ Hz. The 3'-guanosines in the adducts show greater conformational flexibility, having ~ 70 to 80 percent C2'-endo sugar puckers, depending upon the temperature.

Another conformational feature that could be deduced from ¹H NMR studies of all *cis*-DDP-platinated oligonucleotides containing an embedded d(GpG) sequence is that both guanosine nucleosides retain the anti orientation of the base around the C1'-N9 glycosidic linkage (Figure 9.9). This result was deduced from the lack of a pronounced nuclear Overhauser effect (NOE) between H8 and the H1' protons, such as would occur in the syn conformation. An NOE between H8 resonances on the two coordinated nucleosides was observed for

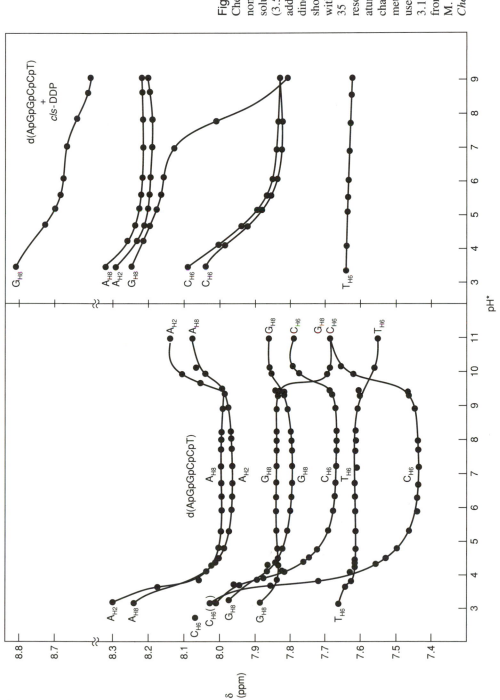

Figure 9.19

Chemical shift (δ) vs. pH* of the nonexchangeable base protons of D_2O solutions of $[d(ApGpGpCpCpT)]_2$ (3.5 mM, 35°C) and its cis-DDP adduct (2.5 mM, 70°C). The pyrimidine resonances of the latter sample show no chemical shift changes with temperature over the range $35 < T < 70$°C while the purine resonances show a slight temperature-dependent chemical shift change of up to 0.1 ppm. Tetramethylammonium chloride was used as the internal standard (δ 3.180). Reproduced by permission from J. C. Caradonna, S. J. Lippard, M. J. Gait, and M. Singh, *J. Am. Chem. Soc.* **104** (1982), 5793.

[557]

adducts of d(ApTpGpG) and d(CpGpG), indicating that the two bases are in a head-to-head orientation with respect to the platinum coordination plane. In other words, both O6 atoms lie on the same side of that plane. Two oligonucleotides containing cis-[Pt(NH₃)₂(ApG)] adducts have been examined; their structural properties closely resemble those of the (GpG) adducts, with platinum coordinated to N7 of both purine bases.

In order to study double-stranded DNAs platinated on one strand, it was necessary to adopt a special strategy. First, the desired oligonucleotide is synthesized. It is preferable that the DNA strands not be self-complementary, since the affinity of such an oligomer for itself is so much greater than that for its platinated form that the desired, singly platinated duplex will not form. After the platinated single strand is synthesized and purified, the complementary strand is added. Several duplex oligonucleotide-containing cis-[Pt(NH₃)₂{d(pGpG)}]-embedded adducts prepared in this manner have been studied by ^{1}H NMR spectroscopy. With the use of two-dimensional and temperature-dependent techniques, both the nonexchangeable base and sugar protons as well as the exchangeable (guanine N1 and thymine N3) N-H (imino) proton resonances were examined. The last are useful, since they give some measure of the extent to which the double helix remains intact. When not base-paired to their complements in the other strand, these protons exchange more rapidly with solvent (water) protons, leading at moderate exchange rates to broadening of the resonances and, at high exchange rates ($>10^7$ s^{-1}), disappearance of the signals. Several interesting results were obtained in these studies. In all of them, platination of the d(GpG) sequence brought about the same C2'-endo → C3'-endo sugar-ring pucker switch for the 5'-guanosine as seen in the single-stranded adducts. Head-to-head, anti conformations were also observed. At low temperatures, below the melting transition temperature, above which the duplex becomes single-stranded, the imino proton resonances were observed. This result was interpreted to mean that normal, Watson-Crick base pairs can still exist between the cis-DDP d(GpG) adduct and the d(CpC) sequence on the complementary strand. In the case of [d(TpCpTpCpG*pG*pTpCpTpC)]·[d(GpApGpApCpCpGpApGpA)], where the asterisks refer to the sites of platination, the imino proton resonances were assigned with the assistance of NOE experiments.[125] Temperature-dependent studies showed that, in the range $-4° < T < 46°$C, the imino resonances of the coordinated guanosine nucleosides broadened first with increasing temperature. Apparently the base pairs of the intrastrand crosslinked, platinated duplex DNA are disrupted, or "melted," outward from the point of platination as well as from the ends. Since the amino hydrogen atoms involved in base pairing were not observed in this study, a completely definitive structural analysis was not possible. Nevertheless, the authors proposed that the duplex would be kinked by an angle of $\sim 60°$ at the cis-DDP binding site in order to preserve full duplex character.

Another useful NMR nucleus for monitoring cis-DDP-DNA interactions is ^{195}Pt, which is 34 percent abundant with $I = \frac{1}{2}$. When used in conjunction with ^{15}N ($I = \frac{1}{2}$) enriched NH₃ ligands, ^{195}Pt NMR resonances provide a powerful means for characterizing complexes in solution. The ^{195}Pt and ^{15}N chemical shifts are both sensitive to the ligand trans to the NH_3 group, as is the ^{195}Pt-^{15}N coupling constant.[127] ^{195}Pt NMR studies of cis-DDP binding have been carried out using nucleobases, small oligonucleotides, and even double-stranded fragments of 20 to 40 bp in length, as previously described (Section V.D.1). The major contribution of this method is to show whether platinum coordinates to a nitrogen or an oxygen donor atom on the DNA, since the ^{195}Pt chemical shift is sensitive to this difference in ligation.

b. X-ray Structural Studies In recent years several oligonucleotide duplexes have been crystallized and characterized by x-ray diffraction methods. The probability of forming suitable single crystals of DNA fragments is disappointingly low, however, with only 1 in 10 such attempts being successful. Correspondingly, it has been difficult to crystallize platinated oligonucleotides. An alternative approach has been to soak nucleic-acid crystals of known structure with the platinum reagent in the hope of forming an isomorphous derivative, the structure of which could be obtained by using the changes in phases from the native material. In attempts to characterize a cis-DDP nucleic acid adduct, crystals of tRNA^{Phe} and the self-complementary dodecamer d(CpGpCpGpApApTpTpCpGpCpG) were soaked with cisplatin solutions in the hope of obtaining useful metric information.[123,128,129] These efforts have thus far failed to produce a high-resolution structure, although they confirm the predilection for platinum to coordinate to the N7 position of purine rings. Addition of cis-DDP tends to disorder the crystal, with platinum going to several sites of partial occupancy.

A more fruitful approach has been to crystallize a purified oligonucleotide containing the coordinated cis-${Pt(NH_3)_2}^{2+}$ moiety. The first x-ray structure to be deciphered through such a strategy was that of cis-[Pt(NH_3)_2{d(pGpG)}].[130] This compound crystallizes with water solvent and glycine buffer molecules in the lattice. The crystals were grown at pH 3.8, where the terminal phosphate is monoprotonated in order to provide a neutral complex of diminished solubility. Two crystalline forms have been obtained, and both structures solved, one to 0.94 Å resolution. The latter contains four crystallographically independent molecules, which, although complicating the structure solution, afforded four independent views of the major adduct formed by cis-DDP with DNA. The four molecules form an aggregate, held together by hydrogen bonding and intermolecular base-base stacking interactions (Figure 9.20). There are two conformationally distinct classes that comprise molecules 1 and 2, and molecules 3 and 4; within each class, the molecules are related by an approximate C_2 symmetry axis.

The molecular structure of molecule 1 is displayed in Figure 9.21; geometric information about all four molecules is contained in Table 9.4. As expected from the NMR studies, platinum coordinates to N7 atoms of the guanine bases, which are completely destacked (dihedral angles range from 76.2 to 86.7°), to form a square-planar geometry. The bases have a head-to-head configuration

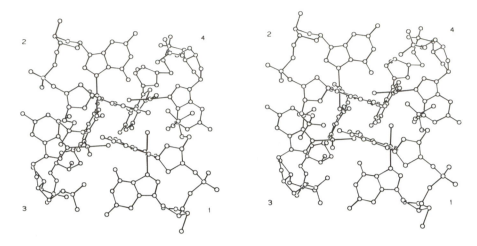

Figure 9.20
Stereoview of aggregate of four *cis*-[Pt(NH$_3$)$_2${d(pGpG)}]
molecules (reproduced by permission from Reference 130).

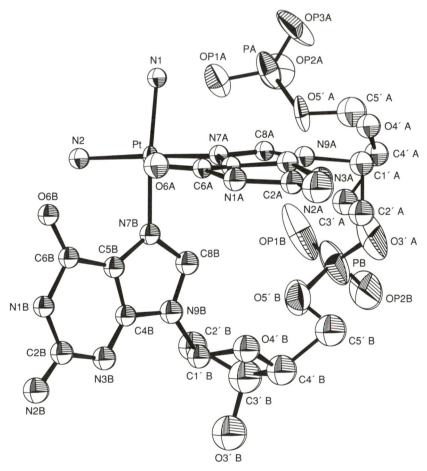

Figure 9.21
Molecular structure of *cis*-[Pt(NH$_3$)$_2${d(pGpG)}].

and conformational angles χ (Table 9.4 and Figure 9.9) that fall in the anti range. The sugar puckers of the 5′-deoxyribose rings for all four molecules have a C3′-endo conformation, and some of the 3′-sugar carbon atoms exhibit large thermal parameters suggestive of a less well-ordered structure. These results further demonstrate the similarity of the structure as detected in the solid state by x-ray diffraction and in the solution state by NMR spectroscopy.

An interesting additional feature of the cis-[Pt(NH$_3$)$_2${d(pGpG)}] crystal structure is a hydrogen bonding interaction between an ammine ligand and the oxygen atom of the terminal phosphate group (OP1A···N1, Figure 9.21). This intramolecular hydrogen bond is prominent in three of the four molecules in the asymmetric unit. Although the relevance of this hydrogen bonding interaction to the solution structure and molecular mechanism of cisplatin is presently unknown, it is interesting to note that the antitumor activity of platinum amine halide complexes is reduced when protons on coordinated NH$_3$ are replaced by alkyl groups.[34]

A second cis-DDP-oligonucleotide adduct characterized by x-ray crystallography is the neutral molecule cis-[Pt(NH$_3$)$_2${d(CpGpG)}].[131] Here again, there are several (three) molecules in the asymmetric unit. Although determined at lower resolution, the structure is similar in most respects to that of cis-[Pt(NH$_3$)$_2${d(pGpG)}] except for the presence of some weak NH$_3$···O6(guanosine) intramolecular hydrogen bonding interactions and a few unusual sugar-phosphate backbone torsional angles. Also, no NH$_3$(H)···phosphate(O) hydrogen bonds were observed.

From the foregoing discussion, it is apparent that adequate x-ray structure information is available for the cis-{Pt(NH$_3$)$_2$}$^{2+}$/d(pGpG) intrastrand crosslink. What is needed now are structures of the minor adducts and, most importantly, of adducts in double-stranded DNA. Very recently, dodecanucleotide duplexes containing cis-{Pt(NH$_3$)$_2$}$^{2+}$/d(pGpG) adducts have been crystallized, the structures of which are currently being investigated.[132]

c. Molecular Mechanics Calculations on Platinated Duplexes   As a supplement to x-ray structural information on double-stranded oligonucleotides containing an embedded cis-[Pt(NH$_3$)$_2${d(pGpG)}] adduct, several models have been constructed by using a molecular mechanics approach.[133] In this work, a set of coordinates was first obtained by amalgamation of structural information about standard double-helical DNAs and the platinated d(pGpG) fragment. Various starting structures were assumed, both linear and bent. The models were then refined according to various charge and stereochemical constraints built into the calculation. The results, which can reveal only what is feasible and not necessarily what actually happens, for both linear and bent structures are depicted in Figure 9.22 for two of the duplex sequences studied. In the linear model, the 5′-coordinated guanosine is rotated out of the stack, and its hydrogen bonding to the cytosine on the complementary strand is seriously disrupted. The imino N-H group is still involved in H-bonding, however; so this structure is not inconsistent with the NMR results. Two classes of kinked platinated duplex structures were encountered, with bending angles of 61 and 50°. In one of these, all

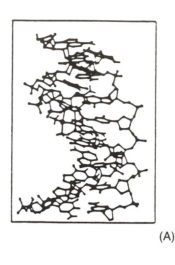
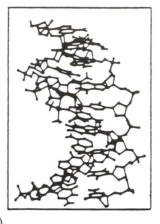

(A)

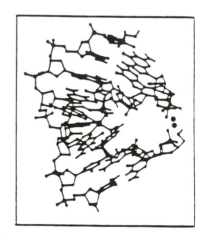

(B)

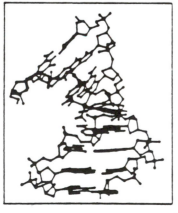
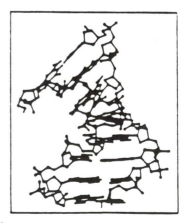

(C)

Watson-Crick hydrogen bonds remain intact. These kinked structures are supported most strongly by the gel-electrophoresis experiments discussed in Section V.D.3.b.v.

Molecular mechanics and the related molecular dynamics calculations are a potentially valuable tool for the bioinorganic chemist interested in how metal complexes might perturb the structures of biopolymers. Analysis of the results for cisplatin-DNA binding reveals that, compared with the sum of all contributions from the biopolymer, the Pt-DNA interactions constitute a small part of the overall energy. For the most accurate results, it is important to know the charge distributions on the metal and its ligands as well as the effects of solvent interactions. Much work needs to be done in these areas before the results of molecular mechanics and dynamics calculations can be used reliably to predict or analyze structures. At present, however, they are far superior to examination of space-filling molecular models, for example, and produce quantitatively revealing structural diagrams.

d. Platinum-Nucleobase Model Complexes   Several studies have been carried out of the *cis*-diammineplatinum(II) moiety coordinated to nucleobases in which the N9 (purine) or N3 (pyrimidine) positions either have been alkylated, to simulate the glycosidic linkages, or in which the actual nucleotide (AMP, dGMP, etc.) is employed.[134] These investigations are in many respects analogous to the synthesis and characterization by bioinorganic chemists of model complexes for the active site of a metalloenzyme. Their purpose is to simplify the problem, revealing kinetic, thermodynamic, and structural preferences of the primary building blocks involved in the metallodrug-biopolymer interaction, without the profound stereochemical constraints of the latter. Early studies of *cis*- and *trans*-DDP adducts with nucleobases (i) revealed the kinetic preferences for platinum binding to GMP and AMP, (ii) mapped out the preferred sites of platination (N7 of A and G; N1 of A; N3 of C; no N7-O6 chelate; no ribose or deoxyribose binding; only rare binding to phosphate oxygen atoms), (iii) demonstrated that Pt-N7 binding to G lowered the pK_a of N1-H by ~ 2 units, and (iv) led to the discovery of interesting new classes of coordination complexes such as the *cis*-diammineplatinum pyrimidine blues and metal-metal bonded diplatinum(III) complexes.

Figure 9.22 *(facing page)*
(A) Stereoscopic view of the unkinked, platinated model of duplex
d(TpCpTpCpG*pG*pTpCpTpC) from molecular-mechanics calculations. Counter ions
used to stabilize the negative charge of the phosphates are not shown. (B) Stereoscopic view
of the "high-salt" kinked, platinated model of duplex d(GpGpCpCpG*pG*pCpC) from
molecular-mechanics calculations. Counter-ions are not depicted, with the exception of the
bridging ion. (C) Stereoscopic view of the "low-salt" kinked, platinated model of duplex
d(GpGpCpCpG*pG*pCpC). Counter-ions are not depicted. Reproduced with
permission from Reference 133.

Attempts to model the intrastrand d(GpG) crosslink with nucleobases have met with only moderate success. Usually the O6 atoms of the two guanosine rings are on opposite sides of the platinum coordination plane ("head-to-tail" isomer). Only for cis-[Pt(NH$_3$)$_2$(9-EtG)$_2$]$^{2+}$ was the correct isomer obtained. Nucleobase complexes of the cis-diammineplatinum(II) moiety have been valuable for testing the controversial proposal of N7,O6 chelate formation, which to date has not been observed. Several interesting discoveries of metal-nucleobase chemistry are that metal binding can stabilize rare tautomers, for example, the 4-imino, 2-oxo form of cytosine, through N4 binding, that coordination of platinum often produces unusual base pairing, and that metal migration from one donor site to another on an isolated nucleobase can occur. These model studies will continue to provide valuable insights into the possible chemistry of platinum antitumor drugs with DNA.

e. trans-DDP-DNA Adducts   Because *trans*-DDP is biologically inactive, it has received less attention than the *cis* isomer. Nevertheless, knowledge of its binding to DNA is important to have as a reference point for mechanistic comparison with the active compounds. Shortly after replication mapping experiments established that *trans*-DDP binds preferentially to d(GpNpG) and d(ApNpG) sequences,[135] several synthetic oligonucleotides containing such sequences were prepared and used to investigate reactions with the *trans* isomer.[136–138] Kinetic studies of *trans*-DDP with d(GpCpG) and d(ApGpGpCpCpT) revealed the presence of, presumably monofunctional, intermediates that closed to form both intra- and interstrand products. In the reaction with d(GpCpG), the 1,3-intrastrand G-G chelate accounted for 70 percent of the product, and 21 percent of the remaining material was unreacted oligonucleotide. Proton NMR studies of purified *trans*-[Pt(NH$_3$)$_2${d(GpCpG)}] as well as the d(GpTpG) analog established platinum binding to N7 positions of the two *trans* guanosine nucleosides. As with the cis-[Pt(NH$_3$)$_2${d(pGpG)}] adducts, the 5'-guanosine residue no longer retained the normal B-DNA type conformation; instead, the sugar ring pucker switched to C3'-endo. A fairly detailed ^{1}H NMR characterization of *trans*-[Pt(NH$_3$)$_2${d(ApGpGpCpCpT)-N7-A(1),N7-G(3)}] revealed very similar features. This example nicely illustrates the different stereoselectivity of *cis*- and *trans*-DDP binding to DNA. The *cis* isomer forms exclusively an intrastrand d(GpG) crosslink, whereas the *trans* isomer makes a 1,3-d(A*pGpG*) adduct. A schematic depiction of the *trans*-{Pt(NH$_3$)$_2$}$^{2+}$ adduct is shown in Figure 9.23. As can be seen, the two purine rings enclose a large, 23-membered ring, the central guanosine residue is "bulged out," and the 5'-residue has a C3'-endo sugar pucker. This structure may be compared with that of cis-[Pt(NH$_3$)$_2${d(pGpG)}] (Figure 9.21), where the platinum is part of a smaller, 17-membered ring. Both space-filling model building studies and molecular mechanics calculations reveal that it would be stereochemically very unfavorable for the *trans*-{Pt(NH$_3$)$_2$}$^{2+}$ fragment to replace the *cis* analogue in an intrastrand crosslinked d(GpG) structure of the kind shown in Figure 9.21. Thus, for bidentate adducts, it seems clear that the important difference between *cis*- and *trans*-DDP binding to sin-

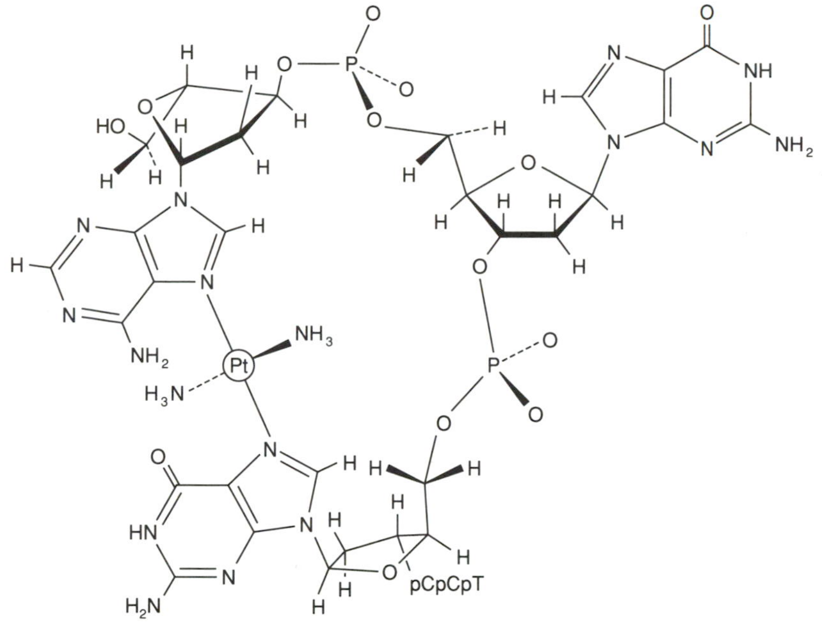

Figure 9.23
Structure of the intrastrand 1,3-d(A*pGpG*) crosslink formed in the reaction of *trans*-DDP with d(ApGpGpCpCpT). Reproduced by permission from Reference 136.

gle-stranded DNA is revealed by the structures shown in Figures 9.21 and 9.23, respectively.

Information about *trans*-DDP binding to double-stranded DNA is scanty, but very recent studies indicate that the *trans*-[Pt(NH$_3$)$_2${d(GpApG)-N7-G(1),N7-G(3)}] intrastrand crosslinked fragment can be embedded in duplex dodecamers.[139] Interestingly, for one sequence the melting temperature (T_M) of this duplex is not reduced over that of the unplatinated DNA fragment, in contrast to results for *cis*-DDP intrastrand d(GpG) adducts. This intriguing result, which agrees with earlier T_M studies of DNA platinated by *trans*-DDP, does not yet have a structural rationale. It is possibly relevant to the processing of bifunctional *trans*-DDP-DNA adducts *in vivo*.

f. Effects of Platination on DNA Structure   It is valuable to summarize at this stage all that has been learned concerning the changes in DNA structure that occur upon *cis*- or *trans*-DDP binding. *cis*-DDP intrastrand crosslinks result in unstacking of neighboring bases and a switch in the sugar pucker of the 5'-nucleoside from C2'-endo, the standard B-DNA conformation, to C3'-endo, a conformation encountered in A-DNA. These various forms of DNA have already been introduced in the previous chapter. Watson-Crick base pairing, although weakened, is probably maintained. Evidence that base pairing is altered comes from studies with antinucleoside antibodies that bind appreciably better to DNA platinated with *cis*-DDP than to unmodified DNA. These antibodies recognize the nucleobases much better in platinated than in unplatinated DNA, presumably because platination disrupts the double helix. Additional support for base-pair disruption comes from gradient gel-denaturation experiments using site-specifically platinated DNA (see Section V.D.8.b). Intrastrand crosslinking by *cis*-DDP also bends the helix by about 34° and unwinds the duplex by 13°. When *trans*-DDP forms 1,3-intrastrand crosslinks, the nucleotides situated between the platinated residues may be bulged out; consistent with this picture is the fact that they present an especially good target for antinucleoside antibodies. In 1,3-intrastrand d(GpNpG) or d(ApNpG) adducts, the 5'-nucleoside sugar pucker is altered to C3'-endo. Intrastrand crosslink formation by *trans*-DDP also leads to DNA bending, but the platinum serves as the locus for a hinge joint and not for cooperative bending. These different effects of platination on DNA structure brought about by the two isomers are likely to be related to their different biological activities.

6. Effects of DNA structure on platinum binding

a. A-, B-, and Z-DNA [140]   As discussed in more detail in Chapter 8, double-helical DNA can adopt different polymorphic forms depending on the conditions in solution or polycrystalline fiber. Even within a given DNA molecule, there can be sequence-dependent local secondary and tertiary structural differences that constitute important signals for cellular DNA binding and processing molecules. An example already discussed is the recognition of palindromic sequences by type II restriction endonucleases. As shown in Figure 9.24, three such DNA polymorphs are the right-handed A- and B- and the left-handed Z-forms. Most commonly encountered in solution is B-DNA, characterized by well-classified major and minor grooves designated by arrows in Figure 9.24. The targets of platinum binding, guanine N7 atoms, are situated in the major groove.

To what extent do sequence-dependent local structural modulations affect platinum binding? Although no general answer to this question can be given, there are several interesting anecdotal pieces of information worth mentioning. Z-DNA, a form favored by alternating purine-pyrimidine sequences such as in poly d(GC), does not constitute a particularly good target for *cis*-DDP binding. For one thing, it lacks the preferred d(GpG) or d(ApG) sequences. The monofunctional [Pt(dien)Cl]$^+$ complex, however, facilitates the B-DNA $\rightarrow$ Z-DNA conformational transition, as demonstrated by circular dichroism and ^{31}P NMR spectroscopic data. [141] In Z-DNA, the guanosine nucleoside adopts the syn conformation (Figure 9.9), which is presumably favored by placing a bulky {Pt(dien)}$^{2+}$ moiety on N7. Moreover, the local charge density on DNA is greater in Z- than B-DNA, owing to the closer proximity of the phosphate groups, and the former is presumably stabilized by the $+2$ charge on the platinum complex.

b. Effects of Local Sequence and of Free and Linked Intercalators on Platinum Binding   Of more interest perhaps to anticancer drug-DNA interactions is the fact that some d(GpG), d(ApG), and even d(GpA) targets for *cis*-DDP bind-

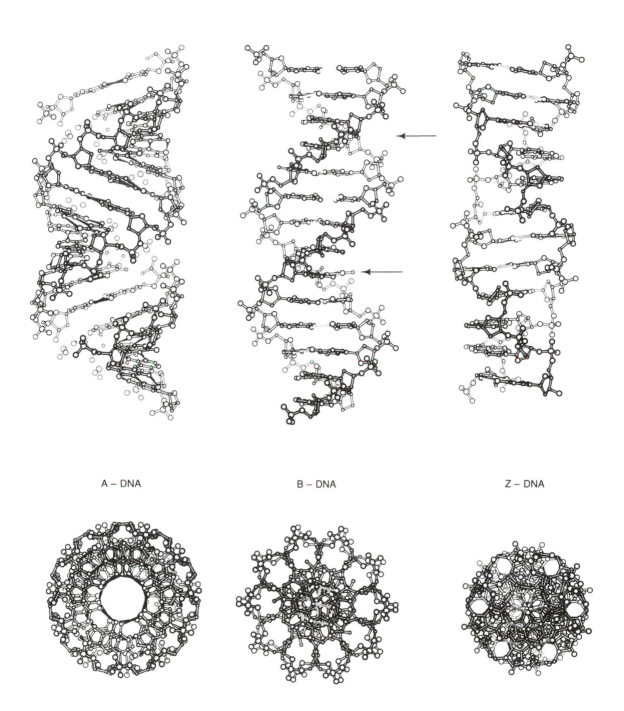

A – DNA

B – DNA

Z – DNA

Figure 9.24
Representations of side (above) and top (below) views of three major classes of double-stranded DNA. For B-DNA, arrows near the top and bottom of the helix designate the minor and major grooves, respectively. Reproduced with permission from Reference 140.

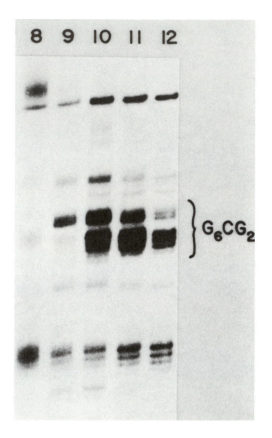

Figure 9.25
Autoradiograph of Exo III mapping results for *cis*-DDP
binding to a 165-bp DNA restriction fragment. The
Pt/nucleotide ratio is 0.05. Lanes 8—12 contain DNA
platinated in the presence of 0, 0.012, 0.057, 0.12,
and 0.23 Etd/nucleotide, respectively. For more details,
see Reference 142.

ing are very sensitive to the sequences in which they are embedded. This phenomenon was first discovered during exonuclease III mapping studies of cisplatin binding to a 165-bp restriction fragment from pBR322 DNA.[142] Although
cis-DDP binding stops the enzyme at G_3, G_5, and GAGGGAG sequences, at a
$(D/N)_b$ ratio of 0.05 there is little evidence for coordination to an apparently
favored G_6CG_2 sequence (Figure. 9.25, lane 8). When platination was carried
out in the presence of the DNA intercalator EtdBr (Figure 9.11), however, the
G_6CG_2 sequence became a Pt binding site (Figure 9.25, lanes 9-12). A more
extensive exonuclease III mapping study of this phenomenon suggested that
d(CGG)-containing sequences in general are less well platinated by *cis*-DDP.[143,144]
Moreover, only EtdBr, and not other acridine or phenanthridinium type intercalators, was able to promote an enzyme-detectable *cis*-DDP binding to these
sequences. A suggested explanation for these results is that local d(purCGG)
sequences might have an A-DNA-type structure (Figure 9.24) in which the major groove is narrow, inhibiting access of platinum to N7 of guanosine nucleosides. In the presence of the intercalator EtdBr, the local DNA structure might
be altered in such a manner as to permit platinum binding.[143,144]

In accord with this interpretation, and further to delineate a possible reason
why acridines and deaminated ethidium cations do not promote cisplatin binding
to d(purCGG) sequences, NMR studies were performed that revealed the mean

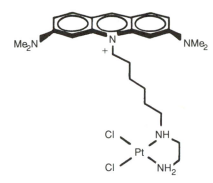

Figure 9.26
Structure of AO-Pt in which the {Pt(en)Cl$_2$} moiety is linked by a hexamethylene chain to acridine orange.

residence time of EtdBr on DNA to be 6 to 21 times longer than that of any of the other intercalators examined.[144] Thus, for these latter intercalators, the local DNA structure presumably can relax back to one unfavored for cisplatin binding before it can diffuse to the site. Moreover, when acridine orange (AO), one of the five intercalators studied that does not promote cisplatin binding to excluded sites, was covalently attached to dichloroethylenediamineplatinum(II) (Figure 9.26) via a hexamethylene linker chain, the resulting AO-Pt molecule was able to bind to all d(CpGpG) sites, as determined by exonuclease III mapping. In the tethered molecule, the high local platinum concentration near the intercalator binding site facilitates attachment of the {Pt(en)}$^{2+}$ moiety to DNA before the acridine orange fragment can diffuse away and the structure can relax to reform the excluded site.

Subsequently, the excluded site phenomenon was found for cis-DDP binding, as assayed by the 3'-5'-exonuclease activity of T4 DNA polymerase.[145] Enzyme stopping sites were observed at all platinated d(GpG) sequences, but only weakly when at a d(GTGGTC) site. Similarly, d(ApG) was not modified when embedded in pyGAGCpy and pyGAGCA sequences. Although most d(GpA) sequences were not platinated, as detected by T4 mapping, a few were. These results further underscore the importance of local sequence modulation of cisplatin binding to DNA.

c. DNA-Promoted Reaction Chemistry In the EtdBr-enhanced binding of cis-DDP to DNA, a small fraction (<5 percent) of the intercalator is strongly bound and can be dialyzed out only very slowly.[145] The detailed structure of this DNA-cisplatin-EtdBr ternary complex has been established, and involves cis-{Pt(NH$_3$)$_2$}$^{2+}$ binding to the exocyclic amino groups of ethidium as well as to donor sites on DNA.[146] This assignment was proved by synthesizing cis-[Pt(NH$_3$)$_2$(Etd)Cl]$^{2+}$ complexes in dimethylformamide solution and then allowing them to react with DNA. The optical spectra of the resulting adducts were identical to that of the ternary complex. The reaction of cis-DDP, EtdBr, and DNA to form the ternary complex is promoted by the favorable orientation of the exocyclic amino group of intercalated Etd with respect to the coordination plane of platinum bound to the double helix. The N-8 exocyclic amino group of ethidium, bound intercalatively at a site adjacent to a purine N-7 coordinated cis-{Pt(NH$_3$)$_2$Cl}$^+$ moiety, is positioned above the platinum atom in a structure resembling the transition state for a square-planar substitution reaction. The structure of this transition state has been modeled in a molecular mechanics calculation (Figure 9.27 *See color plate section, page C-16.*),[147] and evidence has been obtained that indicates selective binding of platinum to the Etd N-8 amino position.[148]

Since cisplatin is usually administered in combination chemotherapy with other drugs, many of which contain intercalating functionalities, strong covalent, DNA-promoted interactions between drug molecules at a target site must be considered as possibly relevant to the molecular mechanism of action. In such a situation, there must be a strong binding preference for both drug molecules for the same target sequences, since on probability grounds alone it is unlikely that both would migrate to the same site by random diffusion at the low concentrations found *in vivo*.

d. Effects of DNA Function on Platinum Binding   Although there is yet little known about this topic for cisplatin, it is worth pointing out that other DNA-targeted drugs, such as bifunctional alkylating agents, bind preferentially to actively transcribed genes. It is therefore possible that platinum exhibits such preferences, for example, to single-stranded DNA at or beyond the transcription fork, compared to duplex DNA in chromatin structures. Or, perhaps, it too binds selectively to actively transcribed DNA. Investigation of these possibilities seems worthwhile.

7. Speculations about the molecular mechanism

a. Is There a Single Mechanism?   Most investigators now agree that DNA is the cytotoxic target of cisplatin. We have seen that the drug inhibits DNA replication by binding to the template and halting the processive action of DNA polymerase. Less well-studied is the inhibition of transcription by platinum-DNA adducts, but recent evidence clearly indicates that they can do so. Studies of the effects of platinum on cells growing in culture reveal that DNA replication and cell growth can continue without cell division in the presence of low levels (1 μg/mL) of *cis*-DDP; cells are arrested at the G2 phase, the stage of cell growth just preceding division.[105] The G2 arrest was reversible, but at higher cisplatin levels (8 μg/mL), cell death occurred. These observations led to the speculation that, perhaps, post-replication DNA repair can handle the toxicity associated with a platinum-damaged template, at least for DNA synthesis, but that there is no known pathway by which transcription can circumvent Pt-DNA lesions. Possibly, inhibition of transcription is ultimately a more lethal event than inhibition of replication. This idea is inconsistent with the well-established fact that thymidine incorporation into DNA is more affected by low levels of cisplatin than is uridine incorporation into RNA. Might there be more than one biochemical pathway by which cisplatin manifests its anticancer activity? Further work is necessary to address this intriguing question.

b. Is There a "Critical Lesion"?   We now have an excellent understanding of the major DNA adducts made by *cis*-DDP, their structures, and the corresponding DNA distortions. Information about adducts made by the inactive *trans* isomer, though not as complete, is also substantial. During the period when this knowledge was being accumulated, it was of interest to learn whether a "critical lesion," a specific DNA adduct with a unique molecular structure, might be responsible for the antitumor activity of the drug. At present, it appears that all bidentate adducts made by *cis*- and *trans*-DDP can inhibit replication, although they may not be equally efficient at doing so.[149] Even monofunctional adducts of the kind formed by *cis*-[Pt(NH$_3$)$_2$(4-Br-py)Cl]$^+$ can block replication.[96] Thus, it might be better to think about the concept of "critical lesion" in a functional sense, where the rates of adduct formation, removal, and enzyme inhibition together determine which family of adducts will exhibit antitumor activity and which will not. Here the biochemistry of the host cell will also be an important determinant. Clearly, more studies are required to delineate these possibilities.

c. Replication and Repair in the Tumor Cell[150]   If the anticancer activity of cisplatin arises from damaged DNA templates, then the drug could be selectively toxic to cancer versus normal cells of the same tissue if repair of DNA damage occurred more efficiently in the latter. The best way to study this phenomenon would be to measure the platinum-DNA levels and list the spectrum of adducts formed in tumor versus normal biopsy tissue obtained from patients undergoing cisplatin chemotherapy. As described previously, methodologies are now reaching the point where such experiments can be carried out in order to test the key hypotheses about the mechanism of action of cisplatin. In addition, powerful new methods have recently been developed to screen for DNA binding proteins. If one could identify proteins that bind selectively to *cis*-DDP-platinated DNA and determine their function, further insights into cellular replication and repair phenomena would be forthcoming. Such cellular factors that bind selectively to DNA containing cisplatin adducts have, in fact, recently been discovered.[107] The experiments that led to this finding and their possible implications for the molecular mechanism of cisplatin are described in the next section.

d. Structure-Specific (or Damage) Recognition Proteins[150]   If selective repair of platinum-DNA adducts in cells of different origin is an integral part of the anticancer mechanism of *cis*-DDP, then it is important to identify the cellular factors associated with this phenomenon. In bacteria, *cis*-DDP adducts on DNA are removed by excision repair, a process in which the lesion is first identified and then excised by the uvrABC excinuclease system.[151] In this process, the uvrA protein first binds to the adducted DNA. Subsequently, the uvrB and C proteins excise the damaged strand, which additional cellular proteins rebuild by copying the genetic information from the remaining strand.

The repair of *cis*-DDP intrastrand crosslinks in mammalian cells is much less well understood. Under the assumption that an analogue of the uvrA protein might exist in such cells, experiments were carried out to try to isolate and clone the gene for such a protein. In particular, the mobility of platinated DNA restriction fragments of defined length was found to be substantially retarded in electrophoresis gels following incubation in extracts from human HeLa cells.[107] This gel-mobility shift was attributed to the binding of factors termed ''damage recognition proteins'' (or DRPs). Subsequent studies with site-specifically platinated oligonucleotides (see V.D.8) revealed that the cisplatin DRP binds specifically to DNA containing the intrastrand *cis*-[Pt(NH$_3$)$_2${d(pGpG)}] or *cis*-[Pt(NH$_3$)$_2${d(pApG)}] crosslink. In parallel work, the gene encoding for a DRP was cloned[152] and used to demonstrate the occurrence of such a protein in nearly all eukaryotic cells. Since binding of the DRP to platinated DNA is not specific for the ammine ligands opposite the crosslinked nucleobases, the interaction is thought to involve recognition of local changes in the twist and bending of the double helix. Figure 9.28 depicts one possible structure for the complex formed between *cis*-DDP platinated DNA and a DRP. More recently, the cloned proteins were found to contain a high mobility group (HMG) protein box, and even HMG1 itself binds to cisplatin-modified DNA.[152] The class of proteins was renamed ''structure-specific recognition proteins'' (SSRPs).

The discovery of SSRPs that bind specifically to cisplatin-modified DNA raises several questions that are the subjects of current study. The first is to determine whether the proteins are an integral component in the mechanism of action of the drug. Although it has not yet been possible to induce the proteins by treating cells with cisplatin, nor have elevated or suppressed levels been found in platinum-resistant cells, deletion of an SSRP gene in yeast has recently afforded a mutant strain less sensitive to cisplatin than wild-type cells.[153] This result links a yeast SSRP with cellular sensitization to the drug. Such a protein could contribute to the molecular mechanism in one of several ways (Figure 9.28). It might be the analogue of uvrA, which, as mentioned above, recognizes damage and signals the cell to perform excision repair. If so, then one would

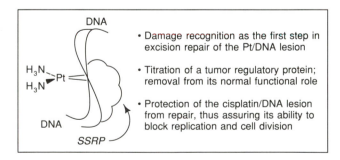

Figure 9.28
Model depicting the binding of an SSRP to cisplatin-damaged DNA and several hypotheses for its role in the molecular mechanism.

like to depress the levels of the protein in cancer cells to make them more sensitive to the drug. A second possibility is that the true role of the SSRP is to serve as a tumor-cell activator, and that cisplatin lesions titrate it away from its functionally active sites on the DNA. Alternatively, binding of the protein could protect cisplatin adducts from repair, preserving their lethality at the time of cell division and leading to the arrest of tumor growth. This last hypothesis would require more of the SSRP in cells sensitive to the drug. Studies are currently in progress to delineate these three and other hypotheses, and to learn whether the discovery of the SSRPs has heralded the final chapter in the quest for the molecular mechanism of cisplatin or merely been an entertaining sidelight.

e. Drug Resistance: What Do We Know?[150]   Perhaps the most serious problem for successful chemotherapy of tumors is drug resistance.[102] In most tumors there exists a subpopulation of cells that are naturally resistant to a given drug; as the sensitive cells are killed, these refractory clones take over. In addition, resistance can be acquired by tumor cells following repeated application of the drug. Attempts to identify mechanisms responsible for cisplatin resistance have therefore been the subjects of considerable research activity. Other DNA-damaging agents sometimes amplify genes as a mechanism of drug resistance. An example is the multidrug resistance phenomenon, in which a gene encoded for a P-glycoprotein is amplified in cells resistant to agents such as daunomycin. This protein is believed to increase efflux of the drug through the cell membrane by an ATP-dependent, energy-driven pump. There is currently an intensive search underway to see whether the cisplatin resistance phenomenon has a genetic origin. If a cisplatin resistance gene could be cloned and its phenotype identified, a powerful new avenue would be opened to overcome drug resistance.

8. Site-specifically platinated DNA[154]

a. The Problem   Much of the information obtained about the mechanism of action of cisplatin has been derived from experiments where Pt-DNA binding has occurred *in vivo* or *in vitro*, with the use of random-sequence DNA having all available targets for the drug. In these studies, platination is controlled by the inorganic chemistry of *cis*-DDP in the medium and the accessibility of target sites on the DNA, as already discussed in considerable detail. As such, this situation best represents drug action as it actually occurs in the tumor cell. On the other hand, the resultant spectrum of DNA adducts makes it difficult, if not impossible, to understand the structural and functional consequences of any specific adduct. In order to address this problem, a methodology has been developed in which a single platinum adduct is built into a unique position in the genome. This approach is powerful and has the potential to be extended to the study of many other metal-based drugs. In this section, we discuss the strategy used to construct such site-specifically platinated DNA molecules and the information obtained thus far from their study. Some uses have already been discussed.

574

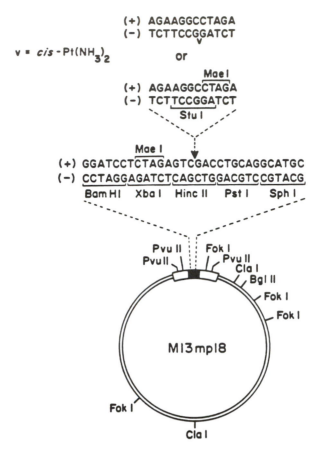

Figure 9.29
Map of the genome created by insertion of *cis*-DDP platinated or unplatinated
d(TpCpTpApGpGpCpCpTpTpCpT)-d(ApGpApApGpGpCpCpTpApGpA) into
the Hinc II restriction site of bacteriophage M13mp18 DNA.

b. Synthesis and Characterization   Figure 9.29 displays the map of a genome constructed by insertion of platinated or unplatinated dodecanucleotide duplexes d(pTpCpTpApGpGpCpCpTpTpCpT)·d(pApGpApApGpGpCpCpTpApGpA)
into DNA from bacteriophage M13mp18. This genome was constructed in the
following manner.[154] Double-stranded DNA from M13mp18 was first digested
with Hinc II, a restriction enzyme that recognizes a unique six-base-pair sequence in the DNA and cleaves the double helix there, leaving a blunt-ended
(no overhanging bases) cleavage site. The unplatinated dodecamer duplex was
next ligated into the Hinc II site, and the DNA amplified *in vivo*. The dodecamer can insert into the genome in two different orientations, the desired one of
which, termed M13-12A-Stu I, was identified by DNA sequencing. The presence of the insert in the new DNA was checked by its sensitivity to the restriction enzyme Stu I, which cleaves at the d(AGGCCT) sequence uniquely situated
in the dodecamer insert, and the absence of cleavage by Hinc II, the site for
which was destroyed. Next, Hinc II-linearized M13mp18 replicative form (RF)

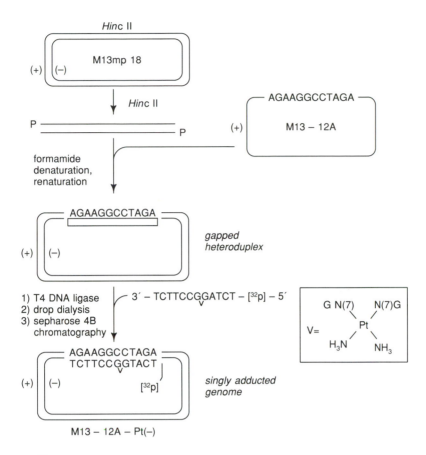

Figure 9.30
Scheme for constructing site-specifically platinated genomes via gapped
heteroduplex synthesis (reproduced by permission from Reference 154).

DNA was allowed to form a heteroduplex with excess viral DNA (which has
only the + strand) in the presence of the denaturant formamide, which was
dialyzed away during the experiment. The resulting circular DNA has a gap in
the minus strand into which the platinated d(TCTAG*G*CCTTCT) was ligated
(Figure 9.30). The latter material was prepared by the methods described in
Section V.D.5.a and characterized by ^{1}H NMR spectroscopy. The resulting site-
specifically platinated DNA contains a single cis-[Pt(NH$_3$)$_2${d(pGpG)}] intrastrand crosslink built into a unique position. The methodology is general, and
has been used to create other known platinum-DNA adducts site-specifically in
M13mp18.

The chemical properties of the platinated DNA, termed M13-12A-Pt(-)-Stu
I, were investigated by enzymatic, digestion and gel electrophoresis experiments. Platinum completely inhibits cleavage of the DNA by Stu I, as expected
from the earlier restriction enzyme mapping studies. In addition, the cis-
[Pt(NH$_3$)$_2${d(pGpG)}] and cis-[Pt(NH$_3$)$_2${d(pApG)}] intrastrand crosslinks were

found to inhibit a variety of DNA polymerases, with only a small amount of bypass of the platinum lesion.[149] These results indicate that the most abundant adducts of cisplatin on DNA are able to block replication efficiently.

c. Biological Properties   When M13-12A-Pt(-)-Stu I DNA was introduced into *E. coli* cells by transformation, DNA synthesis was uninterrupted, because the cell can both repair the damage and use the unmodified (+) strand for synthesis. Consequently, a slightly different strategy was used to construct single-stranded M13-12A-Pt(+)-Stu I DNA, the details of which are available elsewhere.[154] This platinated template, in which the damage can neither be repaired nor bypassed by known mechanisms *in vivo*, was then transformed into *E. coli* cells co-plated with GW5100 cells. Under these conditions, viral DNA replication is detected by the expression of the β-galactosidase gene, which, in the presence of appropriate reagents in the medium, leads to formation of blue plaques on a clear background. The results clearly indicate that many fewer plaques appear when M13-12A-Pt(+) is introduced into the cells than when M13-12A-u(+) was employed, where u stands for unmodified DNA. In three repeats of this experiment, survival of DNA containing only a single *cis*-[Pt(NH$_3$)$_2${d(pGpG)}] crosslink was only 11 ± 1 percent.

These data provide unambiguous proof that the most frequent DNA adduct formed by cisplatin is toxic, capable of inhibiting replication when only a single such lesion is present on a natural DNA template of 7,167 nucleotides. The fact that as many as 10 percent of the transformed cells can bypass or repair the lesion is also of interest, and parallels the results found *in vitro*. In related work, it was found that the *cis*-[Pt(NH$_3$)$_2${d(pGpG)}] intrastrand crosslink is not very mutagenic, but that *cis*-[Pt(NH$_3$)$_2${d(pApG)}] intrastrand adducts are considerably more so. This finding is important, since mutations could lead to long-term secondary tumor production in patients treated with cisplatin. The methodology affords a way to screen new compounds that one would like to be equally effective at inhibiting replication but less mutagenic. In addition, by using repair-deficient mutant cell lines, as well as cisplatin resistant cells, one can study the effects of varying the properties of the host cells. Incorporation of site-specifically platinated DNA sequences into appropriate shuttle vectors will also facilitate investigation of toxicity, repair, and resistance in mammalian cells.

d. Prospectus   The foregoing discussion illustrates the power of site-specifically platinated DNAs as a probe of the molecular mechanism of the drug. We recall that similar strategies were employed to obtain uniquely modified DNA in the bending[92,93] and unwinding[95] experiments discussed previously. In principle, this technique can be applied to examine other aspects of the molecular mechanism of other metallochemotherapeutic agents. The requirements are a synthetic route to the uniquely modified genome, for which both the inorganic coordination chemistry and molecular biology must be amenable, an adduct stable to the biological conditions for DNA synthesis, and a method (usually genetic) for scoring the biological effects being investigated. Site-specifically platinated

DNAs allow the bioinorganic chemist to have maximal control over the genetics and should continue to provide valuable information about the molecular mechanism of action of cisplatin.

E. Design of New Inorganic Anticancer Drugs

1. Objectives

Although chemotherapy has made significant contributions to cancer treatment, the effect of cisplatin on testicular cancer being a showcase example, early detection and surgical removal of all neoplastic tissue still remain the preferred means of combating most forms of the disease. What steps need to be taken to devise better chemotherapeutic agents? One answer is to understand the biochemical mechanisms that underlie the transformation of normal into neoplastic cells and to attack the disease on the basis of that knowledge. The value of this approach is indisputable, but it need not be the only one. We have seen that cis-[Pt(NH$_3$)$_2$Cl$_2$], a simple third-row transition-metal complex containing no carbon atoms, can contribute significantly to cancer chemotherapy. This example alone should lead us to search for improved inorganic drugs based on the evolving knowledge of the mechanism of action of cis-DDP. What then should our objectives be? Three answers are immediately apparent. First, we need to find compounds that are active against resistant cells. Such compounds are termed "second-generation" platinum drugs, and are the focus of much activity in the pharmaceutical industry. Their development will be facilitated by understanding the fundamental biochemistry of cisplatin drug resistance, designing complexes to circumvent the cellular resistance mechanisms. Second, there needs to be an improved spectrum of activity, to be provided by the so-called "third-generation" compounds. The major cancers of the colon, breast, and lung are not effectively diminished by cisplatin chemotherapy. Finally, cisplatin toxicity is often dose-limiting, and there is a need for agents with a greater chemotherapeutic index-to-toxicity ratio. Some of these objectives may ultimately be met by modifying the mode of delivery of cisplatin, for example, by encapsulating the drug in a tumor-seeking liposome or attaching it to a tissue-specific monoclonal antibody. A major step in alternative delivery has recently been taken with the development of a class of oral platinum complexes that have just entered clinical trials.[54] These complexes are platinum(IV) cycloalkylamine species of the kind cis, $trans$, cis-[Pt(NH$_3$)(C$_6$H$_5$NH$_2$)(O$_2$CCH$_3$)$_2$Cl$_2$]. The prospects are reasonably good that new platinum and other metal anticancer drugs can be designed in a bioinorganic chemical approach to the problem.

2. Strategies for drug development

a. Can We Build on Our Knowledge About Cisplatin?   If we consider what is known about the molecular mechanism of cisplatin, what properties are desirable in the design of new metal complexes for testing? The molecules should be reasonably stable kinetically and soluble in biological fluids, cross the cell membrane, bind covalently to DNA, and inhibit gene function. As described previously in this section of the chapter, powerful methods are now available to screen compounds for these properties in a relatively short time. But there are additional factors required for metallodrug anticancer activity, above and beyond these criteria; *trans*-DDP, after all, has all five of the above properties and is not active. Probably one should add to the list the requirement that the complex have two substitutionally labile *cis* sites for intrastrand crosslinking of adjacent DNA nucleotides; such a criterion would, of course, rule out molecules like *trans*-DDP. Recall, however, that cis-$[Pt(NH_3)_2(4\text{-}X\text{-}py)Cl]^+$ complexes (X = Br, Me) are active. These cations have the five properties listed above, but, as far as is currently known, bind only monofunctionally to DNA. The pyridine ring moiety of a covalently attached platinum atom could possibly intercalate into a neighboring interbase pair site on the DNA, making a pseudo-intrastrand crosslinked adduct structurally similar to the *cis*-DDP-d(pGpG) structure. Further information is required about these active, monofunctional cations before any firm conclusions can be drawn. Nevertheless, it is useful to remember that, if the requirement of two substitutionally labile *cis* ligands had been rigorously followed, this new class of monofunctional platinum complexes would not have been discovered.

Another rationale for designing new platinum or other metal antitumor drugs could emerge with a better understanding of the SSRPs in the mechanism of action of cisplatin. For example, if they serve to protect cisplatin lesions on DNA from repair, one would want to design complexes that form adducts that bind even more strongly to the purified protein. The strength of this binding interaction, having been a serendipitous discovery, surely cannot have been maximized. A tighter SSRP-platinated DNA complex would require the use of less platinum, and thus afford lower toxicities.

b. Is Platinum Uniquely Suited?   Given the above criteria, is platinum the only metal to be chosen for further drug development? The answer to this question is "probably not," but a few points need to be kept in mind. Given the assumption that the geometry of the cis-$[Pt(NH_3)_2\{d(GpG)\}]$ intrastrand crosslink was important for the antitumor activity of cisplatin, computer graphics methods were employed to probe the stereochemical consequences of modifying this structure.[155] Addition of axial chloride or water ligands in fifth and sixth coordination positions to form pseudo-octahedral adducts, for example, introduces several steric clashes with the guanosine O6 atoms. An octahedral complex, for example cis,cis,cis-$[Pt(NH_3)_2Cl_2(OH)_2]$, bifunctionally coordinated to DNA either would not form an intrastrand d(GpG) crosslink or would form an adduct structurally different from that made by *cis*-DDP. This octahedral Pt(IV) complex, known as "tetraplatin" in the pharmaceutical industry, is active, but is believed to be reduced *in vivo* to platinum(II) before coordinating to DNA.[156,157] These considerations might imply that the best strategy for inorganic drug development would be to employ square-planar d^8 complexes. Clearly there are as yet no definitive answers. Nevertheless, the criteria derived from the mechanism of action studies of cisplatin represent an excellent starting point for designing new antitumor metallodrugs.

   c. How Important Are Amine Ligands?   Here again, the answer is not un-equivocal, but amines (including NH_3) are probably ideally suited ligands for covalent DNA-binding metal complexes. Even completely inert complexes such as $[Co(NH_3)_6]^{3+}$ show sequence and DNA polymorph binding preferences,[158] suggesting that the N-H bonds orient toward the phosphate and heterocyclic nitrogen atoms in the major groove, forming hydrogen-bonding interactions. This chemistry is analogous to the binding and recognition of organic amines and polyamines, such as spermine and spermidine, by nucleic acids. Apart from amines, hydrophobic groove-binding and/or intercalating ligands such as o-phenanthroline and its derivatives should be considered. Molecules such as $[Rh(DIP)_3]^{3+}$, where DIP = 4,7-diphenyl-1,10-phenanthroline, bind to DNA and have proved to be useful structural probes (Chapter 8). Recent work has shown that $[Rh(DIP)_2Cl_2]^+$ binds preferentially to d(GpG) sequences, like cis-platin, although its antitumor properties have not yet been investigated.[159]

3. Second- and third-generation platinum anticancer drugs

   Improvements over cisplatin have been made, most notably the molecule carboplatin (Figure 9.4), which is less nephrotoxic and has been reported to be effective in some patients where cisplatin chemotherapy has failed. These prop-erties come solely from the dicarboxylate leaving group, which is kinetically more inert to substitution. Studies with monoclonal antibodies have shown the DNA adducts of carboplatin to be identical with those formed by *cis*-DDP.[117] Other platinum compounds that have undergone clinical trials are close ana-logues of cisplatin, *cis*-[PtA_2X_2], or tetraplatin, *cis,cis,cis*-[PtA_2X_2Y_2], that obey the classic structure-activity relationships. The activity of cationic triamines, *cis*-[Pt(NH_3)_2LCl]Cl, where L is pyridine, a substituted pyridine, pyrimidine, or purine, against S180 ascites and L1210 tumors in mice opens a new vista of possible structures to be tried. The intercalator-linked complex AO-Pt (Figure 9.26) has also been found to show activity in the S180 ascites system, suggest-ing a further class of complexes that could be studied. The oral compounds, *cis,trans,cis*-[Pt(NH_3)(C_6H_5NH_2)(O_2CCH_3)_2Cl_2], survive the digestive tract, are taken across the gastrointestinal mucosa, and metabolize to *cis*-[Pt(NH_3)(C_6H_5NH_2)Cl_2], a cisplatin analogue.[54] As such they are effective pro-drugs that could become the major platinum agent in clinical use. Until these recent advances, there was a general impression that, by chance, the best com-pound discovered was the first one, cisplatin. There is now sufficient reason to expect that innovative experimentation will lead to improved drugs, bearing in mind the comment made earlier (Section IV.G.) that sustained individual effort for up to a decade can be required to move a compound from the laboratory bench into the clinic.

4. Nonplatinum antitumor metal complexes

a. Soft Metals As mentioned in Section IV.E., some compounds of Pd(II), Au(I), Rh(II), and Ru(II or III) have been screened for antitumor activity, but much more work needs to be done in this arena. The higher metal-ligand exchange rates of Pd(II), $\sim 10^5$ faster than those of Pt(II), make these complexes potentially more toxic, as some preliminary animal studies have shown. By use of chelating or organometallic complexes, however, this problem might be avoided. The properties of Ru, Rh, and to a lesser extent Au amine and polypyridine complexes would seem to make them attractive candidates, and indeed there appears to be renewed interest in these molecules.[160] Inorganic chemists interested in pursuing drug development with these metals need to forge alliances with biological colleagues equipped to do the necessary animal screening and to develop in-house expertise for cell culture and related biochemical work. The techniques are not all that difficult, and it is actually fun to undertake studies of the biological consequences of metallodrug chemistry.

b. Metallocenes and Metallocene Dihalides[36,37] Although complexes such as $[(C_5H_5)_2TiCl_2]$ are superficially analogous to *cis*-DDP, in being potentially bifunctional DNA crosslinking agents, their hydrolytic reactions are sufficiently different to cast doubt on the value of this comparison. The fact that antitumor activity has been found for this very different class of inorganic compound, however, suggests that perhaps bioinorganic chemists have explored only a very small sample of possible metallodrugs.

VI. RETROSPECTIVE

The topics discussed in this chapter are helping to expand bioinorganic chemistry from a subject that arose chiefly from spectroscopic analysis of metal centers in proteins, because they were uniquely convenient functional groups, to a discipline where fundamental knowledge about metal functions and the application of metals as diagnostic and chemotherapeutic agents are making important contributions to medicine. As the case study of cisplatin is intended to demonstrate, progress in understanding how metals function in chemotherapy can be made only by the combined efforts of many disciplines, including synthetic and physical inorganic and organic chemistry, molecular and cell biology, immunology, pharmacology, toxicology, and clinical medicine. Although we have not yet reached the day where chemotherapeutic agents can be rationally designed from knowledge of a molecular mechanism, such a concept does not seem that farfetched. If nothing else, knowledge of fundamental bioinorganic processes related to metal-macromolecule interactions will continue to grow enormously through efforts to achieve this ultimate goal.

VII. REFERENCES

1. H. Sigel, ed., *Metal Ions in Biological Systems*, Dekker, vol. 14, 1982.
2. D. A. Brown, *Metal Ions Biol. Syst.* **14** (1982), 125.
3. A. D. Young and R. W. Noble, *Methods Enzymol.* **76** (1981), 792.
4. D. A. Phipps, *Metals and Metabolism*, Oxford University Press, 1976, p. 63.
5. G. J. Brewer, *Metal Ions Biol. Syst.* **14** (1982), 57.
6. A. S. Prasad, *Metal Ions Biol. Syst.* **14** (1982), 37.
7. J. R. J. Sorenson, *Metal Ions Biol. Syst.* **14** (1982), 77.
8. D. R. Williams, *An Introduction to Bioinorganic Chemistry*, C. C. Thomas, 1976, p. 327.
9. Reference 8, p. 371.
10. Reference 4, p. 56.
11. Reference 8, p. 372.
12. K. N. Raymond and W. L. Smith, *Struct. Bonding* **43** (1981), 159.
13. Reference 8, p. 366.
14. C. T. Walsh *et al.*, *FASEB* **2** (1988), 124.
15. J. G. Wright *et al.*, *Prog. Inorg. Chem.* **38** (1990), 323.
16. J. D. Helmann, L. M. Shewshuck, and C. T. Walsh, *Adv. Inorg. Biochem.* **8** (1990), 331.
17. M. J. Moore *et al.*, *Acc. Chem. Res.* **23** (1990), 301.
18. Reference 8, p. 363.
19. Reference 1, vol. 10, 1980.
20. K. E. Wetterhahn, *J. Am. Coll. Toxicol.* **8** (1989), 1275.
21. C. J. Mathias *et al.*, *Nucl. Med. Biol.* **15** (1988), 69.
22. A. Yokoyama and H. Saji, *Metal Ions Biol. Syst.* **10** (1980), 313.
23. R. C. Elder and M. K. Eidness, *Chem. Rev.* **87** (1987), 1027.
24. J. C. Dabrowiak, *Adv. Inorg. Chem.* **4** (1982), 70.
25. J. Stubbe and J. W. Kozarich, *Chem. Rev.* **87** (1987), 1107.
26. C. F. Meares and T. G. Wensel, *Acc. Chem. Res.* **17** (1984), 202.
27. R. B. Laufer, *Chem. Rev.* **87** (1987), 901.
28. N. J. Birch, *Metal Ions Biol. Syst.* **14** (1982), 257.
29. S. Avissar *et al.*, *Nature* **331** (1988), 440.
30. M. C. Espanol and D. Mota de Freitas, *Inorg. Chem.* **26** (1987), 4356.
31. P. F. Worley *et al.*, *Science* **239** (1988), 1428.
32. B. M. Sutton, in Reference 159, p. 355.
33. K. C. Dash and H. Schmidbauer, *Metal Ions Biol. Syst.* **14** (1982), 179.
34. M. J. Cleare and J. D. Hoeschele, *Bioinorg. Chem.* **2** (1973), 187.
35. B. Rosenberg *et al.*, *Nature* **222** (1969), 385.
36. P. Köpf-Maier and H. Köpf, *Chem. Rev.* **87** (1987), 1137.
37. P. Köpf-Maier and H. Köpf, *Struct. Bonding* **70** (1988), 105.
38. J. H. Toney, C. P. Brock, and T. J. Marks, *J. Am. Chem. Soc.* **108** (1986), 7263.
39. S. J. Berners-Price and P. J. Sadler, in Reference 41, p. 527.
40. Reference 1, vol. 11, 1980.
41. M. Nicolini, ed., *Platinum and Other Metal Coordination Compounds in Cancer Chemotherapy*, Nijhoff, 1988.
42. Reference 8, p. 316.
43. Reference 4, p. 60.
44. C. C. Hinckley *et al.*, in Reference 159, p. 421.
45. D. D. Perrin and H. Stünzi, *Metal Ions Biol. Syst.* **14** (1982), 207.
46. C. Hill, M. Weeks, and R. F. Schinazi, *J. Med. Chem.* **33** (1990), 2767.
47. B. Rosenberg, *Metal Ions in Biol.* **1** (1980), 1.
48. S. J. Lippard, *Science* **218** (1982), 1075.
49. B. Rosenberg, *Metal Ions Biol. Syst.* **11** (1980), 127.
50. J. M. Pascoe and J. J. Roberts, *Biochem. Pharmacol.* **23** (1974), 1345.
51. J. K. Barton and S. J. Lippard, *Metal Ions in Biol.* **1** (1980), 31.
52. P. J. Loehrer and L. H. Einhorn, *Ann. Intern. Med.* **100** (1984), 704.
53. R. F. Borch, in Reference 41, p. 216.

54. L. R. Kelland *et al., Cancer Res.* **52** (1992), 822.

55. M. E. Howe-Grant and S. J. Lippard, *Metal Ions Biol. Syst.* **11** (1980), 63.

56. R. Faggiani *et al., J. Am. Chem. Soc.* **99** (1977), 777.

57. R. Faggiani *et al., Inorg. Chem.* **16** (1977), 1192.

58. T. G. Appleton, J. R. Hall, and S. F. Ralph, in Reference 41, p. 634.

59. M. C. Lim and R. B. Martin, *J. Inorg. Nucl. Chem.* **38** (1976), 1911.

60. S. K. Mauldin *et al., Cancer Res.* **48** (1988), 5136.

61. W. I. Sundquist *et al., Inorg. Chem.* **26** (1987), 1524.

62. H. C. Harder and B. Rosenberg, *Int. J. Cancer* **6** (1970), 207.

63. J. H. Howle and G. R. Gale, *Biochem. Pharmacol.* **19** (1970), 2757.

64. S. Reslova, *Chem.-Biol. Interact.* **4** (1971), 66.

65. S. Reslova-Vasilukova, in T. A. Connors and J. J. Roberts, eds., *Recent Results in Cancer Research,* Springer, 1974, p. 105.

66. J. J. Roberts and A. J. Thomson, *Prog. Nucl. Acids Res. Mol. Biol.* **22** (1979), 71.

67. D. P. Bancroft, C. A. Lepre, and S. J. Lippard, *J. Am. Chem. Soc.* **112** (1990), 6860.

68. J.-P. Macquet, J.-L. Butour, and N. P. Johnson, in Reference 159, p. 75.

69. N. P. Johnson *et al., J. Am. Chem. Soc.* **107** (1985), 6376.

70. R. B. Ciccarelli, unpublished results.

71. J. C. Caradonna and S. J. Lippard, *Inorg. Chem.* **27** (1988), 1454.

72. K. Inagaki and Y. Kidani, *J. Inorg. Biochem.* **11** (1979), 39.

73. B. K. Teo *et al., J. Am. Chem. Soc.* **100** (1978), 3225.

74. H. M. Ushay, Ph.D. Dissertation, Columbia University, 1984.

75. A. B. Robins and M. O. Leach, *Cancer Treat. Rep.* **67** (1983), 245.

76. P. Bedford *et al., Cancer Res.* **48** (1988), 3019.

77. W. Bauer *et al., Biochemistry* **17** (1978), 1060.

78. J. Filipski *et al., Science* **204** (1979), 181.

79. K. M. Comess, C. E. Costello, and S. J. Lippard, *Biochemistry* **29** (1990), 2102.

80. W. A. Rembers, in *Antineoplastic Agents*, Wiley, 1984; p. 83.

81. S. E. Sherman and S. J. Lippard, *Chem. Rev.* **87** (1987), 1153.

82. J.-P. Macquet and T. Theophanides, *Bioinorg. Chem.* **5** (1975), 59.

83. D. M. L. Goodgame *et al., Biochim. Biophys. Acta* **378** (1975), 153.

84. L. A. Zwelling *et al., Cancer Res.* **38** (1978), 1762.

85. R. B. Ciccarelli *et al., Biochemistry* **24** (1985), 7533.

86. S. J. Lippard and J. D. Hoeschele, *Proc. Natl. Acad. Sci. USA* **76** (1979), 6091.

87. G. L. Cohen *et al., Science* **203** (1979), 1014.

88. S. J. Lippard, *Acc. Chem. Res.* **11** (1978), 211.

89. M. Howe-Grant *et al., Biochemistry* **15** (1976), 4339.

90. J.-L. Butour and J.-P. Macquet, *Eur. J. Biochemistry* **78** (1977), 455.

91. J.-P Macquet and J.-L. Butour, *Biochimie* **60** (1978), 901.

92. S. F. Bellon and S. J. Lippard, *Biophys. Chem.* **35** (1990), 179.

93. J. A. Rice *et al., Proc. Natl. Acad. Sci. USA* **85** (1988), 4158.

94. H.-M. Wu and D. M. Crothers, *Nature* **308** (1984), 509.

95. S. F. Bellon, J. H. Coleman, and S. J. Lippard, *Biochemistry* **30** (1991), 8026.

96. W. Heiger-Bernays, J. M. Essigmann, and S. J. Lippard, *Biochemistry* **29** (1990), 8461.

97. J. J. Roberts *et al.*, in Reference 41, p. 16.

98. M. F. Pera, C. J. Rawlings, and J. J. Roberts, *Chem.-Biol. Interact.* **37** (1981), 245.

99. A. M. J. Fichtinger-Schepman *et al.*, in Reference 41, p. 32.

100. D. J. Beck *et al., Nucleic Acids Res.* **13** (1985), 7395.

101. J. D. Page *et al., Biochemistry* **29** (1990), 1016.

102. G. A. Curt, N. J. Clendeninn, and B. A. Chabner, *Cancer Treat. Rep.* **68** (1984), 87.

103. W. H. DeJong *et al., Cancer Res.* **43** (1983), 4927.

104. A. Eastman and N. Schulte, *Biochemistry* **27** (1988), 4730.

105. A. Eastman *et al.*, in Reference 41, p. 178.

106. V. M. Richon, N. Schulte, and A. Eastman, *Cancer Res.* **47** (1987), 2056.

107. B. A. Donahue *et al., Biochemistry* **29** (1990), 5872.

108. P. J. Stone, A. D. Kelman, and F. M. Sinex, *J. Mol. Biol.* **104** (1976), 793.

109. P. J. Stone, A. D. Kelman, and F. M. Sinex, *Nature* **251** (1974), 736.

110. H. M. Ushay, T. D. Tullius, and S. J. Lippard, *Biochemistry* **20** (1981), 3744.

111. T. D. Tullius and S. J. Lippard, *J. Am. Chem. Soc.* **103** (1981), 4620.

112. B. Royer-Pokora, L. K. Gordon, and W. A. Haseltine, *Nucleic Acids Res.* **9** (1981), 4595.
113. M. C. Poirier *et al.*, *Proc. Natl. Acad. Sci USA* **79** (1982), 6443.
114. S. J. Lippard *et al.*, *Biochemistry* **22** (1983), 5165.
115. E. Reed *et al.*, *J. Clin. Invest.* **77** (1986), 545.
116. M. C. Poirier *et al.*, *Environ. Health Persp.* **62** (1985), 49.
117. W. I. Sundquist, S. J. Lippard, and B. D. Stollar, *Proc. Natl. Acad. Sci. USA* **84** (1987), 8225.
118. A. Eastman, *Biochemistry* **25** (1986), 3912.
119. A. M. J. Fichtinger-Schepman *et al.*, *Biochemistry* **24** (1985), 707.
120. A. M. J. Fichtinger-Schepman *et al.*, *Chem.-Biol. Interact.* **55** (1975), 275.
121. A. C. M. Plooy *et al.*, *Carcinogenesis* **6** (1985), 561.
122. A. M. J. Fichtinger-Schepman *et al.*, *Cancer Res.* **47** (1987), 3000.
123. J. C. Dewan, *J. Am. Chem. Soc.* **106** (1984), 7239.
124. T. W. Hambly, *J. Chem. Soc. Chem. Comm.* (1988), 221.
125. J. H. J. den Hartog *et al.*, *J. Biomol. Struct. Dynam.* **2** (1985), 1137.
126. B. van Hemelryck *et al.*, *Biochem. Biophys. Res. Comm.* **138** (1986), 758.
127. L. S. Hollis *et al.*, in Reference 41, p. 538.
128. R. M. Wing *et al.*, *EMBO* **3** (1984), 1201.
129. J. R. Rubin, M. Sabat, and M. Sundaralingam, *Nucleic Acids Res.* **11** (1983), 6571.
130. S. E. Sherman *et al.*, *Science* **230** (1985), 412.
131. G. Admiraal *et al.*, *J. Am. Chem. Soc.* **109** (1987), 592.
132. S. F. Bellon, T. Takahara, and S. J. Lippard, unpublished results.
133. J. Kozelka *et al.*, *Biopolymers* **26** (1987), 1245.
134. B. Lippert, *Prog, Inorg. Chem.* **37** (1989), 1.
135. A. L. Pinto and S. J. Lippard, *Proc. Natl. Acad. Sci. USA* **82** (1985), 4616.
136. C. A. Lepre, K. G. Strothkamp, and S. J. Lippard, *Biochemistry* **26** (1987), 5651.
137. D. Gibson and S. J. Lippard, *Inorg. Chem.* **26** (1987), 2275.
138. J. L. van der Veer *et al.*, *J. Am. Chem. Soc.* **108** (1986), 3860.
139. C. A. Lepre *et al.*, unpublished results.
140. W. Saenger, *Principles of Nucleic Acid Structure*, Springer, 1984.
141. B. Malfoy, B. Hartman, and M. Leng, *Nucleic Acids Res.* **9** (1981), 5659.
142. T. D. Tullius and S. J. Lippard, *Proc. Natl. Acad. Sci. USA* **79** (1982), 3489.
143. B. E. Bowler and S. J. Lippard, *Biochemistry* **25** (1986), 3031.
144. B. E. Bowler, Ph.D. Dissertation, Massachusetts Institute of Technology, 1987.
145. J.-M Malinge, A. Schwartz, and M. Leng, *Nucleic Acids Res.* **15** (1987), 1779.
146. W. I. Sundquist *et al.*, *J. Am. Chem. Soc.* **112** (1990), 1590.
147. W. I. Sundquist *et al.*, *J. Am. Chem. Soc.* **110** (1988), 8559.
148. D. P. Bancroft *et al.*, *J. Am. Chem. Soc.,* in press.
149. K. M. Comess, *Biochemistry* **25** (1992), 3975.
150. S. L. Bruhn, J. H. Toney, and S. J. Lippard, *Prog. Inorg. Chem.* **38** (1990), 477.
151. A. Sancar and G. B. Sancar, *Annu. Rev. Biochem.* **57** (1988), 29.
152. J. H. Toney *et al.*, *Proc. Natl. Acad. Sci. USA* **86** (1990), 8328; S. L. Bruhn *et al., Proc. Natl. Acad. Sci. USA* **89** (1992), 2307; P. M. Pil and S. J. Lippard, *Science* **256** (1992), 234.
153. S. J. Brown, P. J. Kellett, and S. J. Lippard, *Science* **261** (1993), 603.
154. L. J. Naser *et al.*, *Biochemistry* **27** (1988), 4357.
155. S. E. Sherman *et al.*, *J. Am. Chem. Soc.* **107** (1988), 7368.
156. E. Rotondo *et al.*, *Tumori* **69** (1983), 31.
157. V. Brabec, O. Vrana, and V. Kleinwächter, *Studia Biophys.* **114** (1986), 199.
158. G. E. Plum and V. A. Bloomfield, *Biopolymers* **27** (1988), 1045.
159. J. K. Barton, *Chem. Eng. News* **66** (Sept. 26, 1988), 30.
160. S. J. Lippard, *Platinum, Gold, and Other Metal Chemotherapeutic Agents*, American Chemical Society, 1983.
161. W. I. Sundquist and S. J. Lippard, *Coord. Chem. Rev.* **100** (1990), 293.
162. I am grateful to the Alexander von Humboldt Foundation for a U.S. Senior Scientist Award, Massachusetts Institute of Technology for sabbatical leave time, and Prof. Drs. W. Herrmann and K. Wieghardt for their kind hospitality, all of which were essential for the preparation of the first draft of this chapter during the spring of 1988. I very much appreciate help from the following individuals: A. Davison, for providing Figure 9.2, and D. L. Bancroft, S. F. Bellon, S. L. Bruhn, J. N. Burstyn, K. M. Comess, G. B. Jameson, C. A. Lepre, and J. T. Toney for commenting critically on the manuscript. I also thank M. Mason for typing the first draft.

Suggested Readings

I. General

Beveridge, T. J., and R. J. Doyle, eds. *Metal Ions and Bacteria*. New York: Wiley, 1989.

Ehrlich, H. L. *Geomicrobiology*. 2d ed. New York: Dekker, 1990.

Eichhorn, G., and L. Marzilli, series eds. *Advances in Inorganic Biochemistry*, Vol. 1. New York: Elsevier, 1979.

Frieden, E., series ed. *Biochemistry of the Elements*, Vol. 1. New York: Plenum, 1984.

Glusker, J., *et al.* Metalloproteins: Structural aspects. *Adv. Protein Chem.* **42** (1991).

Hausinger, R. P. Mechanisms of metal ion incorporation into metalloproteins. *BioFactors* **2** (1990), 179–184.

Hay, R. W., series ed. *Perspectives in Bioinorganic Chemistry*, Vol. 1. Greenwich, CT: JAI Press, 1991.

Hughes, M. N., and R. K. Poole. *Metals and Microorganisms*. New York: Chapman and Hall, 1989.

Ibers, J. A., and R. H. Holm. Modeling coordination sites in metallobiomolecules. *Science* **290** (1980), 223–235.

Irgolic, K. J., and A. E. Martell, eds. *Environmental Inorganic Chemistry*. Deerfield Beach, FL: VCH, 1985.

Legg, J. I. Substitution-inert metal ions as probes of biological function. *Coord. Chem. Rev.* **25** (1978), 103–132.

Leigh, G. J., ed. *The Evolution of Metalloenzymes, Metalloproteins, and Related Materials*. London: Symposium Press, 1977.

Lippard, S. J., ed. *Progress in Inorganic Chemistry, Vol. 38: Bioinorganic Chemistry*. New York: Wiley, 1990.

Loehr, T. M., ed. *Iron Carriers and Iron Proteins*. New York: VCH, 1989.

Lontie, R., ed. *Copper Proteins and Copper Enzymes*, Vols. 1–3. Boca Raton, FL: CRC Press, 1984.

Meares, C. F., and T. G. Wensel. Metal chelates as probes of biological systems. *Acc. Chem. Res.* **17** (1984), 202–209.

Que, L., Jr. *Metal Clusters in Proteins*. American Chemical Society Symposium Series no. 372. Washington, D.C.: American Chemical Society, 1988.

Schneider, W. Iron hydrolysis and the biochemistry of iron: The interplay of hydroxide and biogenic ligands. *Chimia* **42** (1988), 9–20.

Sigel, H., and R. B. Martin. Coordinating properties of the amide bond: Stability and structure of metal ion complexes of peptides and related ligands. *Chem. Rev.* **82** (1982), 385–426.

Sigel, H., and A. Sigel, series eds. *Metal Ions in Biological Systems*, Vol. 1. New York: Dekker, 1974.

Spiro, T., ed. *Copper Proteins*. New York: Wiley, 1981.

Thayer, J. S. *Organometallic Compounds and Living Organisms*. New York: Academic Press, 1984.

Williams, R. J. P. Missing information in bio-inorganic chemistry. *Coord. Chem. Rev.* **79** (1987), 175–193.

———. Structural aspects of metal toxicity. In J. O. Nriagu, ed., *Changing Metal Cycles and Human Health*. Dahlem Konferenzen, 1984; Berlin: Springer-Verlag, 251–263.

Wood, J. M. Biological cycles for elements in the environment. *Naturwissenschaften* **52** (1975), 357–364.

II. Techniques

Armstrong, F. A. Voltammetry of metal centres in proteins. *Persp. Bioinorg. Chem.* **1** (1991), 141–182.

Bertini, I., and C. Luchinat. *NMR of Paramagnetic Molecules in Biological Systems*. Menlo Park, CA: Benjamin/Cummings, 1986.

Cheesman, M. R., C. Greenwood, and A. J. Thomson. Magnetic circular dichroism of hemoproteins. *Adv. Inorg. Chem.* **35** (1991), 201–255.

Darnall, D. W., and R. G. Wilkins, eds. *Methods for Determining Metal Ion Environments in Proteins: Structure and Function of Metalloproteins*. New York: Elsevier, 1980.

Day, E. P., *et al.* Squid measurement of metalloprotein magnetization. *Biophys. J.* **52** (1987), 837–853.

Dooley, D. M., and J. H. Dawson. Bioinorganic applications of magnetic circular dichroism spectroscopy: Copper, rare-earth ions, cobalt, and non-heme iron systems. *Coord. Chem. Rev.* **60** (1984), 1–66.

Fairhurst, S. A., and L. H. Sutcliffe. The application of spectroscopy to the study of iron-containing biological molecules. *Prog. Biophys. Mol. Biol.* **34** (1978), 1–79.

Lever, A. B. P. *Inorganic Electronic Spectroscopy*. 2d ed. New York: Elsevier, 1984.

Palmer, G. The electron paramagnetic resonance of metalloproteins. *Biochem. Soc. Trans.* **13** (1985), 548–560.

Scott, R. A. Measurement of metal-ligand distance by EXAFS. *Methods Enzymol.* **177** (1985), 414–459.

Scott, R. A., and M. K. Eidsness. The use of x-ray absorption spectroscopy for detection of metal-metal interactions: Application to copper-containing enzymes. *Comments Inorg. Chem.* **7** (1988), 235–267.

Spiro, T. G., ed. *Biological Applications of Raman Spectroscopy*, Vols. 1–3. New York: Wiley, 1988.

Swartz, H. M., and S. M. Swartz. Biochemical and biophysical applications of electron-spin resonance. *Methods Biochem. Anal.* **29** (1983), 207–323.

Wilkins, R. G. *Kinetics and Mechanism of Reactions of Transition-Metal Complexes*, Second edition. New York: VCH, 1991.

———. Rapid-reaction techniques and bioinorganic reaction mechanisms. *Adv. Inorg. Bioinorg. Mech.* **2** (1983), 139–185.

Wüthrich, K. *NMR of Proteins and Nucleic Acids*. New York: Wiley, 1986.

Sigel, H., and A. Sigel, eds. Applications of nuclear magnetic resonance to paramagnetic species. *Metal Ions Biol. Syst.* **21** (1986).

Sigel, H., and A. Sigel, eds. ENDOR, EPR, and electron spin echo for probing coordination spheres. *Metal Ions Biol. Syst.* **22** (1987).

III. For Chapter 1

Baker, E. N., S. V. Rumball, and B. F. Anderson. Transferrins: Insights into structure and function from studies on lactoferrin. *Trends Biochem. Sci.* **12** (1987), 350–353.

Cousins, R. J. Absorption, transport, and hepatic metabolism of copper and zinc: Special reference to metallothionein and ceruloplasmin. *Physiological Rev.* **65** (1985), 238–309.

Crichton, R. R. *Inorganic Biochemistry of Iron Metabolism*. New York, E. Horwood, 1991.

Hamer, D. H. Metallothionein. *Annu. Rev. Biochem.* **55** (1986), 913–951.

Harrison, P. M., *et al.* Probing structure-function relations in ferritin and bacterioferritin. *Adv. Inorg. Chem.* **36** (1991), 449–487.

Kägi, J. H. R., and A. Schaffer. Biochemistry of metallothionein. *Biochemistry* **27** (1988), 8509–8515.

Lindenbaum, S., J. H. Rhytting, and L. A. Sternson. Ionophores. *Prog. Macrocyclic Chem.* **1** (1979), 219–254.

Lowenstam, H. A., and S. Weiner. *On Biomineralization*. New York: Oxford University Press, 1989.

Otvos, J. D., D. H. Petering, and C. F. Shaw. Structure-reactivity relationships of metallothionein, a unique metal-binding protein. *Comments Inorg. Chem.* **9** (1989), 1–35.

Ponka, P., H. M. Schulman, and R. C. Woodworth, eds. *Iron Transport and Storage*. Boca Raton, FL: CRC Press, 1990.

Theil, E. C. The ferritin family of iron-storage proteins. *Adv. Enzymol.* **63** (1990), 421–449.

Winkelmann, G., D. van der Helm, and J. B. Neilands, eds. *Iron Transport in Microbes, Plants, and Animals*. Weinheim, FRG: VCH, 1987.

IV. For Chapter 2

Bertini, I., and C. Luchinat. An insight on the active site of zinc enzymes through metal substitution. *Metal Ions Biol. Syst.* **15** (1982), 101–156.

Christianson, D. W., and W. N. Lipscomb. Carboxypeptidase A. *Acc. Chem. Res.* **22** (1989), 62–69.

Dolphin, D., ed. B_{12}, Vols. 1 and 2. New York: Wiley, 1982.

Fife, T. H. Metal-ion-catalyzed ester and amide hydrolysis. *Persp. Bioinorg. Chem.* **1** (1991), 43–93.

Matthews, B. W. Structural basis of the action of thermolysin and related zinc peptidases. *Acc. Chem. Res.* **21**, 333–340 (1988).

Spiro, T. G., ed. *Zinc Enzymes*. New York: Wiley, 1983.

Vallee, B. L. Zinc coordination, function, and structure of zinc enzymes and other proteins. *Biochemistry* **29** (1990), 5649–5659.

V. For Chapter 3

Cavaggoni, A. Calcium regulation in cell biology. *Bioscience Reports* **9** (1989), 421–436.

Christakos, S., C. Gabrielides, and W. B. Rhoten. Functional considerations of vitamin-D dependent calcium binding proteins. *Endocrine Rev.* **10** (1989), 3–26.

Haizmann, C. W., and W. Hunziker. Intracellular calcium-binding proteins. In F. Bonner, ed. *Intracellular Calcium Regulation*. New York: Wiley-Liss (1990), 211–248.

Mann, S., J. Webb, and R. J. P. Williams, eds. *Biomineralization*. Weinheim, FRG: VCH, 1990.

Strynadka, N. C. J., and M. N. G. James. Crystal structures of calcium-binding proteins. *Annu. Rev. Biochem.* **58** (1989), 951–998.

Tsien, R. Y., and M. Poenie. Fluorescence ratio imaging: a new window into intracellular ionic signaling. *Trends Biochem. Sci.* **11** (1986), 450–455.

VI. For Chapter 4

Buchler, J. W. Hemoglobin—An inspiration for research in coordination chemistry. *Angew. Chem. Intl. Ed. Eng.* **17** (1978), 407–423.

Dolphin, D., ed. *The Porphyrins.* New York: Academic Press, 1978.

Ellerton, H. D., N. F. Ellerton, and H. A. Robinson. Hemocyanin—A current perspective. *Prog. Biophys. Mol. Biol.* **41** (1983), 143–248.

Jameson, G. B., and J. A. Ibers. On carbon monoxide and dioxygen binding by iron(II) porphyrinato systems. *Comments Inorg. Chem.* **2** (1983), 97–126.

Jones, R. D., D. A. Summerville, and F. Basolo. Synthetic oxygen carriers related to biological systems. *Chem. Rev.* **79** (1979), 139–179.

Karlin, K. D. Binding and activation of molecular oxygen by copper complexes. *Prog. Inorg. Chem.* **35** (1987), 219–327.

Lamy, J., and J. Lamy, eds. *Invertebrate Oxygen-Binding Proteins.* New York: Dekker, 1981.

Lavallee, D. K. Kinetics and mechanisms of metalloporphyrin reactions. *Coord. Chem. Rev.* **61** (1985), 55–96.

Morgan, B., and D. Dolphin. Synthesis and structure of biomimetic porphyrins. *Struct. Bond.* **64**. Berlin and Heidelberg: Springer-Verlag, 1987.

Niederhoffer, E. C., J. H. Timmons, and A. E. Martell. Thermodynamics of oxygen binding in natural and synthetic dioxygen complexes. *Chem. Rev.* **84** (1984), 137–203.

Perutz, M. F., *et al.* Stereochemistry of cooperative mechanisms in hemoglobin. *Acc. Chem. Res.* **20** (1987), 309–321.

Scheidt, W. R., and C. A. Reed. Spin-state/stereochemical relationships in iron porphyrins: Implications for the hemoproteins. *Chem. Rev.* **81** (1981), 543–555.

Suslick, K., and T. J. Reinert. The synthetic analogs of O_2-binding heme proteins. *J. Chem. Educ.* **62** (1985), 974–983.

Woods, E. J. The oxygen transport and storage proteins of invertebrates. *Essays Biochem.* **16** (1980), 1–47.

VII. For Chapter 5

Babcock, G. T. and M. Wikstöm. Oxygen activation and the conservation of energy in cell respiration. *Nature* **356** (1992), 301–309.

Bruice, T. C. Reactions of hydroperoxides with metallotetraphenylporphyrins in aqueous solutions. *Acc. Chem. Res.* **24** (1991), 243–249.

Cadens, E. Biochemistry of oxygen toxicity. *Annu. Rev. Biochem.* **58** (1989), 79–110.

Chan, S. I., S. N. Witt, and D. F. Blair. The dioxygen chemistry of cytochrome *c* oxidase. *Chemica Scripta* **28A** (1988), 51–56.

Dix, T. A., and S. J. Benkovic. Mechanism of oxygen activation by pteridine-dependent monooxygenases. *Acc. Chem. Res.* **21** (1988), 101–107.

Everse, J., K. E. Everse, and M. B. Grisham, eds. *Peroxidases in chemistry and biology.* 2 Vols. Boca Raton, FL: CRC Press, 1991.

Fridovich, I. Superoxide dismutases: An adaptation to a paramagnetic gas. *J. Biol. Chem.* **264** (1989), 7761–7764.

Jefford, C. W., and P. A. Cadby. Molecular mechanisms of enzyme-catalyzed dioxygenation. *Prog. Chem. Nat. Prods.* **40** (1981), 191–265.

Kaufman, S. Aromatic amino-acid hydroxylases. *The Enzymes* **18** (1987), 217–282.

Malmström, B. G. Enzymology of oxygen. *Annu. Rev. Biochem.* **51** (1982), 21–59.

Malmström, B. G. Cytochrome and oxidase as a redox-linked proton pump. *Chem. Rev.* **90** (1990), 1247–1260.

Mansuy, D., P. Battioni, and J.-P. Battioni. Chemical model systems for drug-metabolizing cytochrome-P-450-dependent monooxygenases. *Eur. J. Biochem.* **184** (1989), 267–285.

Miller, D. M., G. R. Buettner, and S. D. Aust. Transition metals as catalysts of autoxidation reactions. *Free Radical Biol. Med.* **8** (1990), 95–108.

Ortiz de Montellano, P. R., ed. *Cytochrome P-450: Structure, Mechanism, and Biochemistry*. New York: Plenum, 1986.

Sawyer, D. T. *Oxygen Chemistry*. Oxford: Oxford Univ. Press, 1991.

Stadtman, E. R. Metal ion-catalyzed oxidation of proteins: Biochemical mechanism and biological consequences. *Free Radical Biol. Med.* **9** (1990), 315–325.

Stewart, L. C., and J. P. Klinman. Dopamine beta-hydroxylase of adrenal chromaffin granules: Structure and function. *Annu. Rev. Biochem.* **57** (1988), 551–592.

Vliegenthart, J. F. G., and G. A. Veldink. Lipoxygenases. *Free Radicals in Biol.* **5** (1982), 29–64.

VIII. For Chapter 6

Amesz, J., ed. *Photosynthesis*. Amsterdam: Elsevier, 1987.

Bertrand, P. ed. Long-range electron transfer in biology. *Struct. Bond.* **75** (1991), 1–47.

Bowler, B. E., A. L. Raphael, and H. B. Gray. Long-range electron transfer in donor (spacer) acceptor molecules and proteins. In S. J. Lippard, ed. *Progress in Inorganic Chemistry*, vol. *38: Bioinorganic Chemistry* (New York: Wiley, 1990), pp. 258–322.

DeVault, D. *Quantum-mechanical tunnelling in biological systems*. 2d ed. Cambridge: Cambridge Univ. Press, 1984.

Gray, H. B., and B. G. Malmström. Long-range electron transfer in multisite metalloproteins. *Biochemistry* **28** (1989), 7499–7505.

Gust, D., and T. A. Moore. Photosynthetic Model Systems. *Topics Curr. Chem.* **159** (1991), 103–151.

Marcus, R. A., and N. Sutin. Electron transfers in chemistry and biology. *Biochim. Biophys. Acta* **811** (1985), 265–322.

Moser, C. C., *et al.* Nature of biological electron transfer. *Nature* **355** (1992), 796.

Onuchic, J. N., *et al.* Pathway analysis of protein electron-transfer reactions. *Annu. Rev. Biophys. Biomol. Struct.* **21** (1992), 349–377.

Robinson, J. N., and D. J. Cole-Hamilton. Electron transfer across vesicle bilayers. *Chem. Soc. Rev.* **20** (1991), 49–94.

Scott, R. A., A. G. Mauk, and H. B. Gray. Experimental approaches to studying biological electron transfer. *J. Chem. Educ.* **52** (1985), 932–938.

Sigel, H., and A. Sigel, eds. Electron transfer reactions in metalloproteins. *Metal Ions Biol. Syst.* **27** (1991).

Sutin, N., and B. S. Brunschwig. Some aspects of electron transfer in biological systems. In M. K. Johnson *et al.*, eds., *Electron Transfer in Biology and the Solid State* (Washington, D.C.: American Chemical Society, 1990), pp. 65–88.

Wherland, S., and H. B. Gray. Electron-transfer mechanisms employed by metalloproteins. In A. W. Addison *et al.*, eds., *Biological Aspects of Inorganic Chemistry* (New York: Wiley, 1977), 289–368.

Winkler, J. R., and H. B. Gray. Electron transfer in ruthenium-modified proteins. *Chem. Rev.* **92** (1992), 369–379.

Wuttke, D. S., *et al.* Electron-tunneling pathways in cytochrome *c*. *Science* **256** (1992), 1007–1009.

IX. For Chapter 7

Bruschi, M., and F. Guerlesquin. Structure, function, and evolution of bacterial ferredoxins. FEMS *Microbiol. Rev.* **54** (1988), 155–176.

Burgess, B. K. The iron-molybdenum cofactor of nitrogenase. *Chem. Rev.* **90** (1990), 1377–1406.

Coucouvanis, D. Use of preassembled Fe/S and Fe/Mo/S clusters in the stepwise synthesis of potential analogues for the Fe/Mo/S site in nitrogenase. *Acc. Chem. Res.* **24** (1991), 1–8.

Holm, R. H. Identification of active sites in iron-sulfur proteins. In A. W. Addison *et al.*, *Biological Aspects of Inorganic Chemistry* (New York: Wiley-Interscience, 1977), 71–111.

Holm, R. H. Synthetic approaches to the active sites of iron-sulfur proteins. *Acc. Chem. Res.* **10** (1977), 427–434.

Holm, R. H., S. Ciurli, and J. A. Weigel. Subsite-specific structures and reactions in native and synthetic [4Fe-4S] cubane-type clusters. *Prog. Inorg. Chem.* **38** (1990), 1–74.

Odom, J. M., and H. D. Peck, Jr. Hydrogenase, electron-transfer proteins, and energy coupling in the sulfate-reducing bacteria desulfovibrio. *Annu. Rev. Microbiol.* **38** (1984), 551–592.

Postgate, J. R. *The Fundamentals of Nitrogen Fixation.* Cambridge: Cambridge Univ. Press, 1982.

Spiro, T., ed. *Iron-Sulfur Proteins.* New York: Wiley, 1982.

Sweeney, W. V., and J. C. Rabinowitz. Proteins containing 4Fe-4S clusters: An overview. *Annu. Rev. Biochem.* **49** (1980), 139–161.

X. For Chapter 8

Barton, J. K. Recognizing DNA. *Chem. Eng. News*, Sept. 26, 1988, pp. 30–41.

McGall, G. H., and J. Stubbe. Mechanistic studies of bleomycin-mediated DNA cleavage using isotope labeling. *Nucl. Acids Mol. Biol.* **2** (1989), 85–104.

Sigel, H., and A. Sigel, eds. Interrelations among metal ions, enzymes, and gene expression. *Metal Ions Biol. Syst.* **25** (1989).

Sigman, D. S., and A. Spassky. DNAse activity of 1,10-phenanthroline-copper ion. *Nucl. Acids Mol. Biol.* **3** (1989), 13–27.

Silver, S., R. A. Laddaga, and T. K. Misra. Plasmid-determined resistance to metal ions. In R. K. Poole and G. N. Gadd, eds., *Metal-microbe interactions* (Oxford: Oxford Univ. Press, 1989), pp. 49–63.

Tullius, T. D., ed. *Metal-DNA Chemistry.* American Chemical Society Symposium Series no. 402. Washington, D.C.: American Chemical Society, 1989.

XI. For Chapter 9

Blackburn, G. N., and M. J. Gait, ed. *Nucleic Acids in Chemistry and Biology.* Oxford: IRL Press, 1990.

Farrell, N. *Transition-Metal Complexes as Drugs and Chemotherapeutic Agents.* Dordrecht, The Netherlands: Kluwer, 1989.

Hacker, M. P., E. B. Douple, and I. H. Krakoff. *Platinum Coordination Complexes in Cancer Chemotherapy.* Boston: Martinus Nijhoff, 1984.

Lippard, S. J., ed. *Platinum, Gold, and Other Metal Chemotherapeutic Agents.* American Chemical Society Symposium Series no. 209. Washington, D.C.: American Chemical Society, 1983.

Nicolini, M., ed. *Platinum and Other Metal Coordination Compounds in Cancer Chemotherapy.* Boston: Martinus Nijhoff, 1988.

Nicolini, M., G. Bandoli, and U. Mazzi, eds. *Technetium Chemistry and Nuclear Medicine.* New York: Raven Press, 1986.

Saenger, W. *Principles of Nucleic-Acid Structure.* Heidelberg: Springer-Verlag, 1984.

Saenger, W., and U. Hinneman, eds. *Protein-Nucleic Acid Interaction.* Boca Raton, FL: CRC Press, 1989.

Tullius, T. D., ed. *Metal-DNA Chemistry.* American Chemical Society Symposium Series no. 402. Washington, D.C.: American Chemical Society, 1989.

XII. Related Topics

Berg, J. M. Metal-binding domains in nucleic acid-binding and gene-regulatory proteins. *Prog. Inorg. Chem.* **37** (1989), 143–185.

Bouwman, E., W. L. Driessen, and J. Reedijk. Model systems for type-1 copper proteins: Structures of copper coordination compounds with thioether and azole-containing ligands. *Coord. Chem. Rev.* **104** (1990), 143–172.

Brudvig, G. W., and R. H. Crabtree. Bioinorganic chemistry of manganese related to photosynthetic oxygen evolution. *Prog. Inorg. Chem.* **37** (1989), 99–142.

Chapman, S. K. Blue copper proteins. *Persp. Bioinorg. Chem.* **1** (1991), 95–140.

Chasteen, N. D., ed. *Vanadium in Biological Systems: Physiology and Biochemistry.* Boston: Kluwer Academic Publishers, 1990.

Christou, G. Manganese carboxylate chemistry and its biological relevance. *Acc. Chem. Res.* **22** (1989), 328–335.

Coughlan, M., ed. Molybdenum and molybdenum-containing enzymes. Oxford: Pergamon Press, 1980.

Eichhorn, G. L., and L. G. Marzilli, eds. Metal-ion induced regulation of gene expression. *Adv. Inorg. Biochem.* **9** (1990).

Evans, C. H. *Biochemistry of the Lanthanides.* New York: Plenum Press, 1990.

Frieden, E., ed. *Biochemistry of the Essential Ultratrace Elements.* New York: Plenum Press, 1984.

Hinton, S. M., and D. Dean. Biogenesis of molybdenum cofactors. *Crit. Rev. Microbiol.* **17** (1990), 169–188.

Jameson, R. F. Coordination chemistry of copper with regard to biological systems. *Metal Ions Biol. Syst.* **12** (1981), 1–30.

Keppler, B. K. Metal complexes as anticancer agents: The future role of inorganic chemistry in cancer therapy. *New J. Chem.* **14** (1990), 389–403.

Kurtz, D. M., Jr. Oxo- and hydroxo-bridged diiron complexes: A chemical perspective on a biological unit. *Chem. Rev.* **90** (1990), 585–606.

Lancaster, J. R., Jr., ed. *The Bioinorganic Chemistry of Nickel.* New York: VCH, 1988.

Lippard, S. J. Oxo-bridged polyiron centers in biology and chemistry. *Angew. Chem. Intl. Ed. Eng.* **27** (1988), 344–361.

Pecoraro, V. L., ed. *Manganese Redox Enzymes.* New York: VCH, 1992.

Que, L., Jr., and A. E. True. Dinuclear iron- and managanese-oxo sites in biology. *Prog. Inorg. Chem.* **38** (1990), 97–199.

Rajagopalan, K. V. Molybdenum: An essential trace element in human nutrition. *Annu. Res. Nutr.* **8** (1988), 401–427.

Rehder, D. The bioinorganic chemistry of vanadium. *Angew. Chem. Intl. Ed. Eng.* **30** (1991), 148–167.

Sigel, H., and A. Sigel, eds. Aluminum and its role in biology. *Metal Ions Biol. Syst.* **24** (1988).

Sigel, H., and A. Sigel, eds. Antibiotics and their complexes. *Metal Ions Biol. Syst.* **19** (1985).

Sigel, H., and A. Sigel, eds. Compendium on magnesium and its role in biology, nutrition, and physiology. *Metal Ions Biol. Syst.* **25** (1989).

Sigel, H., and A. Sigel, eds. Nickel and its role in biology. *Metal Ions Biol. Syst.* **23** (1988).

Spiro, T. G., ed. *Molybdenum Enzymes.* New York: Wiley, 1985.

Stadtman, T. C. Some selenium-dependent biochemical processes. *Adv. Enzymol.* **48** (1979), 1–28.

Stiefel, E. I. The coordination and bioinorganic chemistry of molybdenum. *Prog. Inorg. Chem.* **22** (1977), 1–223.

Sykes, A. G. Plastocyanin and the blue copper proteins. *Struct. Bond.* **75** (1991), 175–224.

Thorp, H. H., and G. W. Brudvig. The physical inorganic chemistry of manganese relevant to photosynthetic oxygen evolution. *New J. Chem.* **15** (1991), 479–490.

Vallee, B. L. Zinc: Biochemistry, physiology, toxicology, and clinical pathology. *BioFactors* **1** (1988), 31–36.

Vallee, B. L., J. E. Coleman, and D. S. Auld. Zinc fingers, zinc clusters, and zinc twists in DNA-binding protein domains. *Proc. Natl. Acad. Sci. USA* **88** (1991), 999–1003.

Vincent, J. B., and G. Christou. Higher oxidation state manganese biomolecules. *Adv. Inorg. Chem.* **33** (1989), 197–257.

Vincent, J. B., G. L. Olivier-Lilley, and B. A. Averill. Proteins containing oxo-bridged dinuclear iron centers: A bioinorganic perspective. *Chem. Rev.* **90** (1990), 1447–1467.

Wever, R., and K. Kustin. Vanadium: A biologically relevant element. *Adv. Inorg. Chem.* **35** (1990), 81–115.

Wieghardt, K. The active sites in manganese-containing metalloproteins and inorganic model complexes. *Angew. Chem. Intl. Ed. Eng.* **28** (1989), 1153–1172.

Index

Note: Page numbers preceded by a C indicate material in the color plate section.

593

Acknowledgment: HBG thanks Deborah Wuttke, Kara Bren, Gary Mines, and Paola Turano for assistance in preparing and checking the index.